Continuous and Discrete Time Signals and Systems

Signals and systems is a core topic for electrical and computer engineers. This textbook presents an introduction to the fundamental concepts of continuous-time (CT) and discrete-time (DT) signals and systems, treating them separately in a pedagogical and self-contained manner. Emphasis is on the basic signal processing principles, with underlying concepts illustrated using practical examples from signal processing and multimedia communications. The text is divided into three parts. Part I presents two introductory chapters on signals and systems. Part II covers the theories, techniques, and applications of CT signals and systems and Part III discusses these topics for DT signals and systems, so that the two can be taught independently or together. The focus throughout is principally on linear time invariant systems. Accompanying the book is a CD-ROM containing MATLAB code for running illustrative simulations included in the text; data files containing audio clips, images and interactive programs used in the text, and two animations explaining the convolution operation. With over 300 illustrations, 287 worked examples and 409 homework problems, this textbook is an ideal introduction to the subject for undergraduates in electrical and computer engineering. Further resources, including solutions for instructors, are available online at www.cambridge.org/9780521854559.

Mrinal Mandal is an Associate Professor at the Department of Electrical and Computer Engineering, University of Alberta, Edmonton, Canada. His main research interests include multimedia signal processing, medical image and video analysis, image and video compression, and VLSI architectures for real-time signal and image processing.

Amir Asif is an Associate Professor at the Department of Computer Science and Engineering, York University, Toronto, Canada. His principal research areas lie in statistical signal processing with applications in image and video processing, multimedia communications, and bioinformatics, with particular focus on video compression, array imaging detection, genomic signal processing, and block-banded matrix technologies.

Continuous and Discrete Time Signals and Systems

Mrinal Mandal
University of Alberta, Edmonton, Canada

and

Amir Asif
York University, Toronto, Canada

CAMBRIDGE
UNIVERSITY PRESS

CAMBRIDGE UNIVERSITY PRESS
Cambridge, New York, Melbourne, Madrid, Cape Town, Singapore, São Paulo

Cambridge University Press
The Edinburgh Building, Cambridge CB2 8RU, UK

Published in the United States of America by Cambridge University Press, New York

www.cambridge.org
Information on this title: www.cambridge.org/9780521854559

First published 2007

Printed in the United Kingdom at the University Press, Cambridge

A catalog record for this publication is available from the British Library

ISBN-13   978-0-521-85455-9 hardback

Contents

Preface

The book is primarily intended for instruction in an upper-level undergraduate or a first-year graduate course in the field of signal processing in electrical and computer engineering. Practising engineers would find the book useful for reference or for self study. Our main motivation in writing the book is to deal with continuous-time (CT) and discrete-time (DT) signals and systems separately. Many instructors have realized that covering CT and DT systems in parallel with each other often confuses students to the extent where they are not clear if a particular concept applies to a CT system, to a DT system, or to both. In this book, we treat DT and CT signals and systems separately. Following Part I, which provides an introduction to signals and systems, Part II focuses on CT signals and systems. Since many students are familiar with the theory of CT signals and systems from earlier courses, Part II can be taught to such students with relative ease. For students who are new to this area, we have supplemented the material covered in Part II with appendices, which are included at the end of the book. Appendices A–F cover background material on complex numbers, partial fraction expansion, differential equations, difference equations, and a review of the basic signal processing instructions available in MATLAB. Part III, which covers DT signals and systems, can either be covered independently or in conjunction with Part II.

The book focuses on linear time-invariant (LTI) systems and is organized as follows. Chapters 1 and 2 introduce signals and systems, including their mathematical and graphical interpretations. In Chapter 1, we cover the classification between CT and DT signals and we provide several practical examples in which CT and DT signals are observed. Chapter 2 defines systems as transformations that process the input signals and produce outputs in response to the applied inputs. Practical examples of CT and DT systems are included in Chapter 2. The remaining fifteen chapters of the book are divided into two parts. Part II constitutes Chapters 3–8 of the book and focuses primarily on the theories and applications of CT signals and systems. Part III comprises Chapters 9–17 and deals with the theories and applications of DT signals and systems. The organization of Parts II and III is described below.

Chapter 3 introduces the time-domain analysis of the linear time-invariant continuous-time (LTIC) systems, including the convolution integral used to evaluate the output in response to a given input signal. Chapter 4 defines the continuous-time Fourier series (CTFS) as a frequency domain representation for the CT periodic signals, and Chapter 5 generalizes the CTFS to aperiodic signals and develops an alternative representation, referred to as the continuous-time Fourier transform (CTFT). Not only do the CTFT and CTFS representations provide an alternative to the convolution integral for the evaluation of the output response, but also these frequency representations allow additional insights into the behavior of the LTIC systems that are exploited later in the book to design such systems. While the CTFT is useful for steady state analysis of the LTIC systems, the Laplace transform, introduced in Chapter 6, is used in control applications where transient and stability analyses are required. An important subset of LTIC systems are frequency-selective filters, whose characteristics are specified in the frequency domain. Chapter 7 presents design techniques for several CT frequency-selective filters including the Butterworth, Chebyshev, and elliptic filters. Finally, Chapter 8 concludes our treatment of LTIC signals and systems by reviewing important applications of CT signal processing.

The coverage of CT signals and systems concludes with Chapter 8 and a course emphasizing the CT domain can be completed at this stage. In Part III, Chapter 9 starts our consideration of DT signals and systems by providing several practical examples in which such signals are observed directly. Most DT sequences are, however, obtained by sampling CT signals. Chapter 9 shows how a band-limited CT signal can be accurately represented by a DT sequence such that no information is lost in the conversion from the CT to the DT domain. Chapter 10 provides the time-domain analysis of linear time-invariant discrete-time (LTID) systems, including the convolution sum used to calculate the output of a DT system. Chapter 11 introduces the frequency domain representations for DT sequences, namely the discrete-time Fourier series (DTFS) and the discrete-time Fourier transform (DTFT). The discrete Fourier transform (DFT) samples the DTFT representation in the frequency domain and is convenient for digital signal processing of finite-length sequences. Chapter 12 introduces the DFT, while Chapter 13 is devoted to a discussion of the z-transform. As for CT systems, DT systems are generally specified in the frequency domain. A particular class of DT systems, referred to as frequency-selective digital filters, is introduced in Chapter 14. Based on the length of the impulse response, digital filters can be further classified into finite impulse response (FIR) and infinite impulse response (IIR) filters. Chapter 15 covers the design techniques for the FIR filters, and Chapter 16 presents the design techniques for the IIR filters. Chapter 17 concludes the book by motivating the students with several applications of digital signal processing in audio and music, spectral analysis, and image and video processing.

Although the book has been designed to be as self-contained as possible, some basic prerequisites have been assumed. For example, an introductory background in mathematics which includes trigonometry, differential calculus, integral calculus, and complex number theory, would be helpful. A course in electrical circuits, although not essential, would be highly useful as several examples of electrical circuits have been used as systems to motivate the students. For students who lack some of the required background information, a review of the core background materials such as complex numbers, partial fraction expansion, differential equations, and difference equations is provided in the appendices.

The normal use of this book should be as follows. For a first course in signal processing, at say sophomore or junior level, a reasonable goal is to teach Part II, covering continuous-time (CT) signals and sysems. Part III provides the material for a more advanced course in discrete-time (DT) signal processing. We have also spent a great deal of time experimening with different presentations for a single-semester signals and systems course. Typically, such a course should include Chapters 1, 2, 3, 10, 4, 5, 11, 6, and 13 in that order. Below, we provide course outlines for a few traditional signal processing courses. These course outlines should be useful to an instructor teaching this type of material or using the book for the first time.

(1) Continuous-time signals and systems: Chapters 1–8.
(2) Discrete-time signals and systems: Chapters 1, 2, 9–17.
(3) Traditional signals and systems: Chapters 1, 2, (3, 10), (4, 5, 11), 6, 13.
(4) Digital signal processing: Chapters 10–17.
(5) Transform theory: Chapters (4, 5, 11), 6, 13.

Another useful feature of the book is that the chapters are self-contained so that they may be taught independent of each other. There is a significant difference between reading a book and being able to apply the material to solve actual problems of interest. An effective use of the book must include a fair coverage of the solved examples and problem solving by motivating the students to solve the problems included at the end of each chapter. As such, a major focus of the book is to illustrate the basic signal processing concepts with examples. We have included 287 worked examples, 409 supplementary problems at the ends of the chapters, and more than 300 figures to explain the important concepts. Wherever relevant, we have extensively used MATLAB to validate our analytical results and also to illustrate the design procedures for a variety of problems. In most cases, the MATLAB code is provided in the accompanying CD, so the students can readily run the code to satisfy their curiosity. To further enhance their understanding of the main signal processing concepts, students are encouraged to program extensively in MATLAB. Consequently, several MATLAB exercises have been included in the Problems sections.

Any suggestions or concerns regarding the book may be communicated to the authors; email addresses are listed at http://www.cambridge.org/9780521854559. Future updates on the book will also be available at the same website.

A number of people have contributed in different ways, and it is a pleasure to acknowledge them. Anna Littlewood, Irene Pizzie, and Emily Yossarian of Cambridge University Press contributed significantly during the production stage of the book. Professor Tyseer Aboulnasr reviewed the complete book and provided valuable feedback to enhance its quality. In addition, Mrinal Mandal would like to thank Gencheng Guo, Meghna Singh, Wen Chen, Saeed S. Tehrani, Sanjukta Mukhopadhayaya, and Professor Thomas Sikora for their help in the overall preparation of the book. On behalf of Amir Asif, special thanks are due to Professor José Moura, who introduced the fascinating field of signal processing to him for the first time and has served as his mentor for several years. Lastly, Mrinal Mandal thanks his parents, Iswar Chandra Mandal (late) and Mrs Kiran Bala Mandal, and his wife Rupa, and Amir Asif thanks his parents, Asif Mahmood (late) and Khalida Asif, his wife Sadia, and children Maaz and Sannah for their continuous support and love over the years.

Introduction to signals and systems

1 Introduction to signals

Signals are detectable quantities used to convey information about time-varying physical phenomena. Common examples of signals are human speech, temperature, pressure, and stock prices. Electrical signals, normally expressed in the form of voltage or current waveforms, are some of the easiest signals to generate and process.

Mathematically, signals are modeled as functions of one or more independent variables. Examples of independent variables used to represent signals are time, frequency, or spatial coordinates. Before introducing the mathematical notation used to represent signals, let us consider a few physical systems associated with the generation of signals. Figure 1.1 illustrates some common signals and systems encountered in different fields of engineering, with the physical systems represented in the left-hand column and the associated signals included in the right-hand column. Figure 1.1(a) is a simple electrical circuit consisting of three passive components: a capacitor C, an inductor L, and a resistor R. A voltage $v(t)$ is applied at the input of the RLC circuit, which produces an output voltage $y(t)$ across the capacitor. A possible waveform for $y(t)$ is the sinusoidal signal shown in Fig. 1.1(b). The notations $v(t)$ and $y(t)$ includes both the dependent variable, v and y, respectively, in the two expressions, and the independent variable t. The notation $v(t)$ implies that the voltage v is a function of time t. Figure 1.1(c) shows an audio recording system where the input signal is an audio or a speech waveform. The function of the audio recording system is to convert the audio signal into an electrical waveform, which is recorded on a magnetic tape or a compact disc. A possible resulting waveform for the recorded electrical signal is shown in Fig 1.1(d). Figure 1.1(e) shows a charge coupled device (CCD) based digital camera where the input signal is the light emitted from a scene. The incident light charges a CCD panel located inside the camera, thereby storing the external scene in terms of the spatial variations of the charges on the CCD panel. Figure 1.1(g) illustrates a thermometer that measures the ambient temperature of its environment. Electronic thermometers typically use a *therm*al resi*stor*, known as a *thermistor*, whose resistance varies with temperature. The fluctuations in the resistance are used to measure the temperature. Figure 1.1(h)

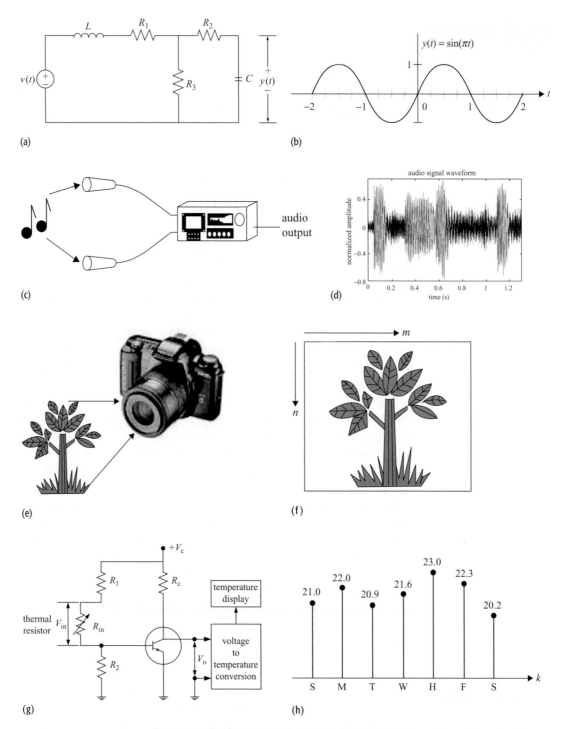

Fig. 1.1. Examples of signals and systems. (a) An electrical circuit; (c) an audio recording system; (e) a digital camera; and (g) a digital thermometer. Plots (b), (d), (f), and (h) are output signals generated, respectively, by the systems shown in (a), (c), (e), and (g).

Fig. 1.2. Processing of a signal by a system.

plots the readings of the thermometer as a function of discrete time. In the aforementioned examples of Fig. 1.1, the RLC circuit, audio recorder, CCD camera, and thermometer represent different systems, while the information-bearing waveforms, such as the voltage, audio, charges, and fluctuations in resistance, represent signals. The output waveforms, for example the voltage in the case of the electrical circuit, current for the microphone, and the fluctuations in the resistance for the thermometer, vary with respect to only one variable (time) and are classified as one-dimensional (1D) signals. On the other hand, the charge distribution in the CCD panel of the camera varies spatially in two dimensions. The independent variables are the two spatial coordinates (m, n). The charge distribution signal is therefore classified as a two-dimensional (2D) signal.

The examples shown in Fig. 1.1 illustrate that typically every system has one or more signals associated with it. A *system* is therefore defined as an entity that processes a set of signals (called the *input* signals) and produces another set of signals (called the *output* signals). The voltage source in Fig. 1.1(a), the audio sound in Fig. 1.1(c), the light entering the camera in Fig. 1.1(e), and the ambient heat in Fig. 1.1(g) provide examples of the input signals. The voltage across capacitor C in Fig. 1.1(b), the voltage generated by the microphone in Fig. 1.1(d), the charge stored on the CCD panel of the digital camera, displayed as an image in Fig. 1.1(f), and the voltage generated by the thermistor, used to measure the room temperature, in Fig. 1.1(h) are examples of output signals.

Figure 1.2 shows a simplified schematic representation of a signal processing system. The system shown processes an input signal $x(t)$ producing an output $y(t)$. This model may be used to represent a range of physical processes including electrical circuits, mechanical devices, hydraulic systems, and computer algorithms with a single input and a single output. More complex systems have multiple inputs and multiple outputs (MIMO).

Despite the wide scope of signals and systems, there is a set of fundamental principles that control the operation of these systems. Understanding these basic principles is important in order to analyze, design, and develop new systems. The main focus of the text is to present the theories and principles used in signals and systems. To keep the presentations simple, we focus primarily on signals with one independent variable (usually the time variable denoted by t or k), and systems with a single input and a single output. The theories that we develop for single-input, single-output systems are, however, generalizable to multidimensional signals and systems with multiple inputs and outputs.

1.1 Classification of signals

A signal is classified into several categories depending upon the criteria used for its classification. In this section, we cover the following categories for signals:

   (i)   continuous-time and discrete-time signals;
  (ii)   analog and digital signals;
 (iii)   periodic and aperiodic (or nonperiodic) signals;
  (iv)   energy and power signals;
   (v)   deterministic and probabilistic signals;
  (vi)   even and odd signals.

1.1.1 Continuous-time and discrete-time signals

If a signal is defined for all values of the independent variable t, it is called a *continuous-time* (CT) signal. Consider the signals shown in Figs. 1.1(b) and (d). Since these signals vary continuously with time t and have known magnitudes for all time instants, they are classified as CT signals. On the other hand, if a signal is defined only at discrete values of time, it is called a *discrete-time* (DT) signal. Figure 1.1(h) shows the output temperature of a room measured at the same hour every day for one week. No information is available for the temperature in between the daily readings. Figure 1.1(h) is therefore an example of a DT signal. In our notation, a CT signal is denoted by $x(t)$ with regular parenthesis, and a DT signal is denoted with square parenthesis as follows:

$$x[kT], \quad k = 0, \pm1, \pm2, \pm3, \ldots,$$

where T denotes the time interval between two consecutive samples. In the example of Fig. 1.1(h), the value of T is one day. To keep the notation simple, we denote a one-dimensional (1D) DT signal x by $x[k]$. Though the sampling interval is not explicitly included in $x[k]$, it will be incorporated if and when required.

Note that all DT signals are not functions of time. Figure 1.1(f), for example, shows the output of a CCD camera, where the discrete output varies spatially in two dimensions. Here, the independent variables are denoted by (m, n), where m and n are the discretized horizontal and vertical coordinates of the picture element. In this case, the two-dimensional (2D) DT signal representing the spatial charge is denoted by $x[m, n]$.

Fig. 1.3. (a) CT sinusoidal signal $x(t)$ specified in Example 1.1; (b) DT sinusoidal signal $x[k]$ obtained by discretizing $x(t)$ with a sampling interval $T = 0.25$ s.

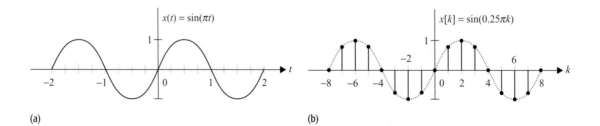

(a)   (b)

Example 1.1

Consider the CT signal $x(t) = \sin(\pi t)$ plotted in Fig. 1.3(a) as a function of time t. Discretize the signal using a sampling interval of $T = 0.25$ s, and sketch the waveform of the resulting DT sequence for the range $-8 \leq k \leq 8$.

Solution

By substituting $t = kT$, the DT representation of the CT signal $x(t)$ is given by

$$x[kT] = \sin(\pi k \times T) = \sin(0.25\pi k).$$

For $k = 0, \pm1, \pm2, \ldots$, the DT signal $x[k]$ has the following values:

$$x[-8] = x(-8T) = \sin(-2\pi) = 0, \qquad\qquad x[1] = x(T) = \sin(0.25\pi) = \frac{1}{\sqrt{2}},$$

$$x[-7] = x(-7T) = \sin(-1.75\pi) = \frac{1}{\sqrt{2}}, \qquad x[2] = x(2T) = \sin(0.5\pi) = 1,$$

$$x[-6] = x(-6T) = \sin(-1.5\pi) = 1, \qquad\qquad x[3] = x(3T) = \sin(0.75\pi) = \frac{1}{\sqrt{2}},$$

$$x[-5] = x(-5T) = \sin(-1.25\pi) = \frac{1}{\sqrt{2}}, \qquad x[4] = x(4T) = \sin(\pi) = 0,$$

$$x[-4] = x(-4T) = \sin(-\pi) = 0, \qquad\qquad x[5] = x(5T) = \sin(1.25\pi) = -\frac{1}{\sqrt{2}},$$

$$x[-3] = x(-3T) = \sin(-0.75\pi) = -\frac{1}{\sqrt{2}}, \quad x[6] = x(6T) = \sin(1.5\pi) = -1,$$

$$x[-2] = x(-2T) = \sin(-0.5\pi) = -1, \qquad\quad x[7] = x(7T) = \sin(1.75\pi) = -\frac{1}{\sqrt{2}},$$

$$x[-1] = x(-T) = \sin(-0.25\pi) = -\frac{1}{\sqrt{2}}, \quad x[8] = x(8T) = \sin(2\pi) = 0,$$

$$x[0] = x(0) = \sin(0) = 0.$$

Plotted as a function of k, the waveform for the DT signal $x[k]$ is shown in Fig. 1.3(b), where for reference the original CT waveform is plotted with a dotted line. We will refer to a DT plot illustrated in Fig. 1.3(b) as a *bar* or a *stem* plot to distinguish it from the CT plot of $x(t)$, which will be referred to as a *line* plot.

Example 1.2

Consider the rectangular pulse plotted in Fig. 1.4. Mathematically, the rectangular pulse is denoted by

$$x(t) = \text{rect}\left(\frac{t}{\tau}\right) = \begin{cases} 1 & |t| \leq \tau/2 \\ 0 & |t| > \tau/2. \end{cases}$$

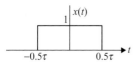

Fig. 1.4. Waveform for CT rectangular function. It may be noted that the rectangular function is discontinuous at $t = \pm\tau/2$.

From the waveform in Fig. 1.4, it is clear that $x(t)$ is continuous in time but has discontinuities in magnitude at time instants $t = \pm0.5\tau$. At $t = 0.5\tau$, for example, the rectangular pulse has two values: 0 and 1. A possible way to avoid this ambiguity in specifying the magnitude is to state the values of the signal $x(t)$ at $t = 0.5\tau^-$ and $t = 0.5\tau^+$, i.e. immediately before and after the discontinuity. Mathematically, the time instant $t = 0.5\tau^-$ is defined as $t = 0.5\tau - \varepsilon$, where ε is an infinitely small positive number that is close to zero. Similarly, the time instant $t = 0.5\tau^+$ is defined as $t = 0.5\tau + \varepsilon$. The value of the rectangular pulse at the discontinuity $t = 0.5\tau$ is, therefore, specified by $x(0.5\tau^-) = 1$ and $x(0.5\tau^+) = 0$. Likewise, the value of the rectangular pulse at its other discontinuity $t = -0.5\tau$ is specified by $x(-0.5\tau^-) = 0$ and $x(-0.5\tau^+) = 1$.

A CT signal that is continuous for all t except for a finite number of instants is referred to as a piecewise CT signal. The value of a piecewise CT signal at the point of discontinuity t_1 can either be specified by our earlier notation, described in the previous paragraph, or, alternatively, using the following relationship:

$$x(t_1) = 0.5 \left[x(t_1^+) + x(t_1^-) \right]. \tag{1.1}$$

Equation (1.1) shows that $x(\pm 0.5\tau) = 0.5$ at the points of discontinuity $t = \pm 0.5\tau$. The second approach is useful in certain applications. For instance, when a piecewise CT signal is reconstructed from an infinite series (such as the Fourier series defined later in the text), the reconstructed value at the point of discontinuity satisfies Eq. (1.1). Discussion of piecewise CT signals is continued in Chapter 4, where we define the CT Fourier series.

1.1.2 Analog and digital signals

A second classification of signals is based on their amplitudes. The amplitudes of many real-world signals, such as voltage, current, temperature, and pressure, change continuously, and these signals are called *analog* signals. For example, the ambient temperature of a house is an analog number that requires an infinite number of digits (e.g., 24.763 578...) to record the readings precisely. Digital signals, on the other hand, can only have a finite number of amplitude values. For example, if a digital thermometer, with a resolution of $1\,°C$ and a range of $[10\,°C, 30\,°C]$, is used to measure the room temperature at discrete time instants, $t = kT$, then the recordings constitute a digital signal. An example of a digital signal was shown in Fig. 1.1(h), which plots the temperature readings taken once a day for one week. This digital signal has an amplitude resolution of $0.1\,°C$, and a sampling interval of one day.

Figure 1.5 shows an analog signal with its digital approximation. The analog signal has a limited dynamic range between $[-1, 1]$ but can assume any real value (rational or irrational) within this dynamic range. If the analog signal is sampled at time instants $t = kT$ and the magnitude of the resulting samples are quantized to a set of finite number of known values within the range $[-1, 1]$, the resulting signal becomes a digital signal. Using the following set of eight uniformly distributed values,

$$[-0.875, -0.625, -0.375, -0.125, 0.125, 0.375, 0.625, 0.875],$$

within the range $[-1, 1]$, the best approximation of the analog signal is the digital signal shown with the stem plot in Fig. 1.5.

Another example of a digital signal is the music recorded on an audio compact disc (CD). On a CD, the music signal is first sampled at a rate of 44 100

Fig. 1.5. Analog signal with its digital approximation. The waveform for the analog signal is shown with a line plot; the quantized digital approximation is shown with a stem plot.

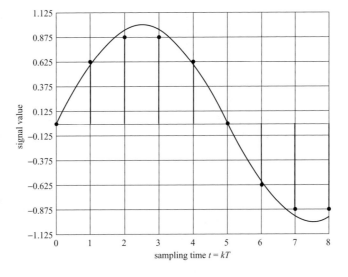

samples per second. The sampling interval T is given by $1/44\,100$, or 22.68 microseconds (µs). Each sample is then quantized with a 16-bit uniform quantizer. In other words, a sample of the recorded music signal is approximated from a set of uniformly distributed values that can be represented by a 16-bit binary number. The total number of values in the discretized set is therefore limited to 2^{16} entries.

Digital signals may also occur naturally. For example, the price of a commodity is a multiple of the lowest denomination of a currency. The grades of students on a course are also discrete, e.g. 8 out of 10, or 3.6 out of 4 on a 4-point grade point average (GPA). The number of employees in an organization is a non-negative integer and is also digital by nature.

1.1.3 Periodic and aperiodic signals

A CT signal $x(t)$ is said to be *periodic* if it satisfies the following property:

$$x(t) = x(t + T_0), \tag{1.2}$$

at all time t and for some positive constant T_0. The smallest positive value of T_0 that satisfies the periodicity condition, Eq. (1.2), is referred to as the *fundamental period* of $x(t)$.

Likewise, a DT signal $x[k]$ is said to be *periodic* if it satisfies

$$x[k] = x[k + K_0] \tag{1.3}$$

at all time k and for some positive constant K_0. The smallest positive value of K_0 that satisfies the periodicity condition, Eq. (1.3), is referred to as the fundamental period of $x[k]$. A signal that is not periodic is called an *aperiodic* or *non-periodic* signal. Figure 1.6 shows examples of both periodic and aperiodic

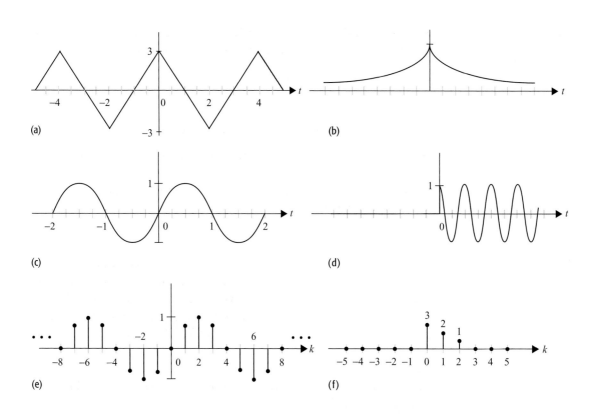

(a)

(b)

(c)

(d)

(e)

(f)

Fig. 1.6. Examples of periodic ((a), (c), and (e)) and aperiodic ((b), (d), and (f)) signals. The line plots (a) and (c) represent CT periodic signals with fundamental periods T_0 of 4 and 2, while the stem plot (e) represents a DT periodic signal with fundamental period $K_0 = 8$.

signals. The reciprocal of the fundamental period of a signal is called the *fundamental frequency*. Mathematically, the fundamental frequency is expressed as follows

$$f_0 = \frac{1}{T_0}, \text{ for CT signals, } \quad \text{or } \quad f_0 = \frac{1}{K_0}, \text{ for DT signals,} \qquad (1.4)$$

where T_0 and K_0 are, respectively, the fundamental periods of the CT and DT signals. The frequency of a signal provides useful information regarding how fast the signal changes its amplitude. The unit of frequency is *cycles per second* (c/s) or *hertz* (Hz). Sometimes, we also use *radians per second* as a unit of frequency. Since there are 2π radians (or 360°) in one cycle, a frequency of f_0 hertz is equivalent to $2\pi f_0$ radians per second. If radians per second is used as a unit of frequency, the frequency is referred to as the *angular frequency* and is given by

$$\omega_0 = \frac{2\pi}{T_0}, \text{ for CT signals, } \quad \text{or } \quad \Omega_0 = \frac{2\pi}{K_0}, \text{ for DT signals.} \qquad (1.5)$$

A familiar example of a periodic signal is a sinusoidal function represented mathematically by the following expression:

$$x(t) = A\sin(\omega_0 t + \theta).$$

The sinusoidal signal $x(t)$ has a fundamental period $T_0 = 2\pi/\omega_0$ as we prove next. Substituting t by $t + T_0$ in the sinusoidal function, yields

$$x(t + T_0) = A\sin(\omega_0 t + \omega_0 T_0 + \theta).$$

Since

$$x(t) = A\sin(\omega_0 t + \theta) = A\sin(\omega_0 t + 2m\pi + \theta), \text{ for } m = 0, \pm 1, \pm 2, \ldots,$$

the above two expressions are equal iff $\omega_0 T_0 = 2m\pi$. Selecting $m = 1$, the fundamental period is given by $T_0 = 2\pi/\omega_0$.

The sinusoidal signal $x(t)$ can also be expressed as a function of a complex exponential. Using the Euler identity,

$$e^{j(\omega_0 t + \theta)} = \cos(\omega_0 t + \theta) + j\sin(\omega_0 t + \theta), \tag{1.6}$$

we observe that the sinusoidal signal $x(t)$ is the imaginary component of a complex exponential. By noting that both the imaginary and real components of an exponential function are periodic with fundamental period $T_0 = 2\pi/\omega_0$, it can be shown that the complex exponential $x(t) = \exp[j(\omega_0 t + \theta)]$ is also a periodic signal with the same fundamental period of $T_0 = 2\pi/\omega_0$.

Example 1.3

(i) CT sine wave: $x_1(t) = \sin(4\pi t)$ is a periodic signal with period $T_1 = 2\pi/4\pi = 1/2$;

(ii) CT cosine wave: $x_2(t) = \cos(3\pi t)$ is a periodic signal with period $T_2 = 2\pi/3\pi = 2/3$;

(iii) CT tangent wave: $x_3(t) = \tan(10t)$ is a periodic signal with period $T_3 = \pi/10$;

(iv) CT complex exponential: $x_4(t) = e^{j(2t+7)}$ is a periodic signal with period $T_4 = 2\pi/2 = \pi$;

(v) CT sine wave of limited duration: $x_6(t) = \begin{cases} \sin 4\pi t & -2 \le t \le 2 \\ 0 & \text{otherwise} \end{cases}$ is an aperiodic signal;

(vi) CT linear relationship: $x_7(t) = 2t + 5$ is an aperiodic signal;

(vii) CT real exponential: $x_4(t) = e^{-2t}$ is an aperiodic signal.

Although all CT sinusoidals are periodic, their DT counterparts $x[k] = A\sin(\Omega_0 k + \theta)$ may not always be periodic. In the following discussion, we derive a condition for the DT sinusoidal $x[k]$ to be periodic.

Assuming $x[k] = A\sin(\Omega_0 k + \theta)$ is periodic with period K_0 yields

$$x[k + K_0] = \sin(\Omega_0(k + K_0) + \theta) = \sin(\Omega_0 k + \Omega_0 K_0 + \theta).$$

Since $x[k]$ can be expressed as $x[k] = \sin(\Omega_0 k + 2m\pi + \theta)$, the value of the fundamental period is given by $K_0 = 2\pi m/\Omega_0$ for $m = 0, \pm 1, \pm 2, \ldots$ Since we are dealing with DT sequences, the value of the fundamental period K_0 must be an integer. In other words, $x[k]$ is periodic if we can find a set of values for

m, $K_0 \in Z^+$, where we use the notation Z^+ to denote a set of positive integer values. Based on the above discussion, we make the following proposition.

Proposition 1.1 *An arbitrary DT sinusoidal sequence* $x[k] = A \sin(\Omega_0 k + \theta)$ *is periodic iff* $\Omega_0/2\pi$ *is a rational number.*

The term *rational number* used in Proposition 1.1 is defined as a fraction of two integers. Given that the DT sinusoidal sequence $x[k] = A \sin(\Omega_0 k + \theta)$ is periodic, its fundamental period is evaluated from the relationship

$$\frac{\Omega_0}{2\pi} = \frac{m}{K_0} \tag{1.7}$$

as

$$K_0 = \frac{2\pi}{\Omega_0}m. \tag{1.8}$$

Proposition 1.1 can be extended to include DT complex exponential signals. Collectively, we state the following.

(1) The fundamental period of a sinusoidal signal that satisfies Proposition 1.1 is calculated from Eq. (1.8) with m set to the smallest integer that results in an integer value for K_0.
(2) A complex exponential $x[k] = A \exp[j(\Omega_0 k + \theta)]$ must also satisfy Proposition 1.1 to be periodic. The fundamental period of a complex exponential is also given by Eq. (1.8).

Example 1.4

Determine if the sinusoidal DT sequences (i)–(iv) are periodic:

(i) $f[k] = \sin(\pi k/12 + \pi/4)$;
(ii) $g[k] = \cos(3\pi k/10 + \theta)$;
(iii) $h[k] = \cos(0.5k + \phi)$;
(iv) $p[k] = e^{j(7\pi k/8 + \theta)}$.

Solution

(i) The value of Ω_0 in $f[k]$ is $\pi/12$. Since $\Omega_0/2\pi = 1/24$ is a rational number, the DT sequence $f[k]$ is periodic. Using Eq. (1.8), the fundamental period of $f[k]$ is given by

$$K_0 = \frac{2\pi}{\Omega_0}m = 24m.$$

Setting $m = 1$ yields the fundamental period $K_0 = 24$.

To demonstrate that $f[k]$ is indeed a periodic signal, consider the following:

$$f[k + K_0] = \sin(\pi[k + K_0]/12 + \pi/4).$$

Substituting $K_0 = 24$ in the above equation, we obtain

$$f[k + K_0] = \sin(\pi[k + K_0]/12 + \pi/4) = \sin(\pi k/12 + 2\pi + \pi/4)$$
$$= \sin(\pi k/12 + \pi/4) = f[k].$$

(ii) The value of Ω_0 in $g[k]$ is $3\pi/10$. Since $\Omega_0/2\pi = 3/20$ is a rational number, the DT sequence $g[k]$ is periodic. Using Eq. (1.8), the fundamental period of $g[k]$ is given by

$$K_0 = \frac{2\pi}{\Omega_0}m = \frac{20m}{3}.$$

Setting $m = 3$ yields the fundamental period $K_0 = 20$.

(iii) The value of Ω_0 in $h[k]$ is 0.5. Since $\Omega_0/2\pi = 1/4\pi$ is not a rational number, the DT sequence $h[k]$ is not periodic.

(iv) The value of Ω_0 in $p[k]$ is $7\pi/8$. Since $\Omega_0/2\pi = 7/16$ is a rational number, the DT sequence $p[k]$ is periodic. Using Eq. (1.8), the fundamental period of $p[k]$ is given by

$$K_0 = \frac{2\pi}{\Omega_0}m = \frac{16m}{7}.$$

Setting $m = 7$ yields the fundamental period $K_0 = 16$.

Example 1.3 shows that CT sinusoidal signals of the form $x(t) = \sin(\omega_0 t + \theta)$ are always periodic with fundamental period $2\pi/\omega_0$ irrespective of the value of ω_0. However, Example 1.4 shows that the DT sinusoidal sequences are not always periodic. The DT sequences are periodic only when $\Omega_0/2\pi$ is a rational number. This leads to the following interesting observation.

Consider the periodic signal $x(t) = \sin(\omega_0 t + \theta)$. Sample the signal with a sampling interval T. The DT sequence is represented as $x[k] = \sin(\omega_0 kT + \theta)$. The DT signal will be periodic if $\Omega_0/2\pi = \omega_0 T/2\pi$ is a rational number. In other words, if you sample a CT periodic signal, the DT signal need not always be periodic. The signal will be periodic only if you choose a sampling interval T such that the term $\omega_0 T/2\pi$ is a rational number.

1.1.3.1 Harmonics

Consider two sinusoidal functions $x(t) = \sin(\omega_0 t + \theta)$ and $x_m(t) = \sin(m\omega_0 t + \theta)$. The fundamental angular frequencies of these two CT signals are given by ω_0 and $m\omega_0$ radians/s, respectively. In other words, the angular frequency of the signal $x_m(t)$ is m times the angular frequency of the signal $x(t)$. In such cases, the CT signal $x_m(t)$ is referred to as the mth harmonic of $x(t)$. Using Eq. (1.5), it is straightforward to verify that the fundamental period of $x(t)$ is m times that of $x_m(t)$.

Figure 1.7 plots the waveform of a signal $x(t) = \sin(2\pi t)$ and its second harmonic. The fundamental period of $x(t)$ is 1 s with a fundamental frequency of 2π radians/s. The second harmonic of $x(t)$ is given by $x_2(t) = \sin(4\pi t)$. Likewise, the third harmonic of $x(t)$ is given by $x_3(t) = \sin(6\pi t)$. The fundamental

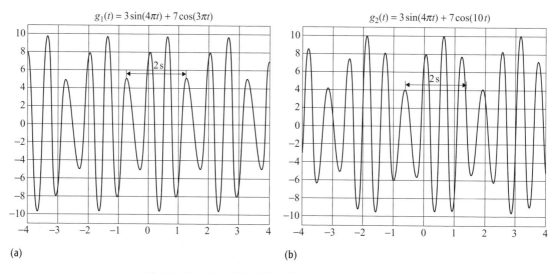

Fig. 1.8. Signals (a) $g_1(t)$ and (b) $g_2(t)$ considered in Example 1.5. Signal $g_1(t)$ is periodic with a fundamental period of 2 s, while $g_2(t)$ is not periodic.

1.1.4 Energy and power signals

Before presenting the conditions for classifying a signal as an energy or a power signal, we present the formulas for calculating the energy and power in a signal.

The *instantaneous power* at time $t = t_0$ of a real-valued CT signal $x(t)$ is given by $x^2(t_0)$. Similarly, the instantaneous power of a real-valued DT signal $x[k]$ at time instant $k = k_0$ is given by $x^2[k]$. If the signal is complex-valued, the expressions for the instantaneous power are modified to $|x(t_0)|^2$ or $|x[k_0]|^2$, where the symbol $|\cdot|$ represents the absolute value of a complex number.

The *energy* present in a CT or DT signal within a given time interval is given by the following:

CT signals $\qquad E_{(T_1, T_2)} = \displaystyle\int_{T_1}^{T_2} |x(t)|^2 dt$ in interval $t = (T_1, T_2)$ with $T_2 > T_1$;

$$(1.10a)$$

DT sequences $\quad E_{[N_1, N_2]} = \displaystyle\sum_{k=N_1}^{N_2} |x[k]|^2$ in interval $k = [N_1, N_2]$ with $N_2 > N_1$.

$$(1.10b)$$

The *total energy* of a CT signal is its energy calculated over the interval $t = [-\infty, \infty]$. Likewise, the total energy of a DT signal is its energy calculated over the range $k = [-\infty, \infty]$. The expressions for the total energy are therefore given by the following:

CT signals $\qquad\qquad\qquad\qquad E_x = \displaystyle\int_{-\infty}^{\infty} |x(t)|^2 dt;$ $\qquad\qquad$ (1.11a)

DT sequences
$$E_x = \sum_{k=-\infty}^{\infty} |x[k]|^2.$$
(1.11b)

Since power is defined as energy per unit time, the *average power* of a CT signal $x(t)$ over the interval $t = (-\infty, \infty)$ and of a DT signal $x[k]$ over the range $k = [-\infty, \infty]$ are expressed as follows:

CT signals
$$P_x = \lim_{T \to \infty} \frac{1}{T} \int_{-T/2}^{T/2} |x(t)|^2 dt.$$
(1.12)

DT sequences
$$P_x = \lim_{K \to \infty} \frac{1}{2K+1} \sum_{k=-K}^{K} |x[k]|^2.$$
(1.13)

Equations (1.12) and (1.13) are simplified considerably for periodic signals. Since a periodic signal repeats itself, the average power is calculated from one period of the signal as follows:

CT signals
$$P_x = \frac{1}{T_0} \int_{\langle T_0 \rangle} |x(t)|^2 dt = \frac{1}{T_0} \int_{t_1}^{t_1+T_0} |x(t)|^2 dt,$$
(1.14)

DT sequences
$$P_x = \frac{1}{K_0} \sum_{k=\langle K_0 \rangle} |x[k]|^2 = \frac{1}{K_0} \sum_{k=k_1}^{k_1+K_0-1} |x[k]|^2,$$
(1.15)

where t_1 is an arbitrary real number and k_1 is an arbitrary integer. The symbols T_0 and K_0 are, respectively, the fundamental periods of the CT signal $x(t)$ and the DT signal $x[k]$. In Eq. (1.14), the duration of integration is one complete period over the range $[t_1, t_1 + T_0]$, where t_1 can take any arbitrary value. In other words, the lower limit of integration can have any value provided that the upper limit is one fundamental period apart from the lower limit. To illustrate this mathematically, we introduce the notation $\int_{\langle T_0 \rangle}$ to imply that the integration is performed over a complete period T_0 and is independent of the lower limit. Likewise, while computing the average power of a DT signal $x[k]$, the upper and lower limits of the summation in Eq. (1.15) can take any values as long as the duration of summation equals one fundamental period K_0.

A signal $x(t)$, or $x[k]$, is called an *energy signal* if the total energy E_x has a non-zero finite value, i.e. $0 < E_x < \infty$. On the other hand, a signal is called a *power signal* if it has non-zero finite power, i.e. $0 < P_x < \infty$. Note that a signal cannot be both an energy and a power signal simultaneously. The energy signals have zero average power whereas the power signals have infinite total energy. Some signals, however, can be classified as neither power signals nor as energy signals. For example, the signal $e^{2t} u(t)$ is a growing exponential with infinite energy whose average power cannot be calculated. Such signals are generally of little interest to us.

Most periodic signals are typically power signals. For example, the average power of the CT sinusoidal signal, or $A \sin(\omega_0 t + \theta)$, is given by $A^2/2$ (see

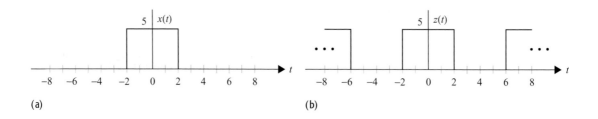

Fig. 1.9. CT signals for Example 1.6.

Problem 1.6). Similarly, the average power of the complex exponential signal $A \exp(j\omega_0 t)$ is given by A^2 (see Problem 1.8).

Example 1.6

Consider the CT signals shown in Figs. 1.9(a) and (b). Calculate the instantaneous power, average power, and energy present in the two signals. Classify these signals as power or energy signals.

Solution

(a) The signal $x(t)$ can be expressed as follows:

$$x(t) = \begin{cases} 5 & -2 \leq t \leq 2 \\ 0 & \text{otherwise.} \end{cases}$$

The instantaneous power, average power, and energy of the signal are calculated as follows:

instantaneous power $\quad P_x(t) = \begin{cases} 25 & -2 \leq t \leq 2 \\ 0 & \text{otherwise;} \end{cases}$

energy $\qquad\qquad E_x = \int_{-\infty}^{\infty} |x(t)|^2 \mathrm{d}t = \int_{-2}^{2} 25\, \mathrm{d}t = 100;$

average power $\qquad P_x = \lim_{T \to \infty} \frac{1}{T} E_x = 0.$

Because $x(t)$ has finite energy $(0 < E_x = 100 < \infty)$ it is an energy signal.

(b) The signal $z(t)$ is a periodic signal with fundamental period 8 and over one period is expressed as follows:

$$z(t) = \begin{cases} 5 & -2 \leq t \leq 2 \\ 0 & 2 < |t| \leq 4, \end{cases}$$

with $z(t + 8) = z(t)$. The instantaneous power, average power, and energy of the signal are calculated as follows:

instantaneous power $\quad P_z(t) = \begin{cases} 25 & -2 \leq t \leq 2 \\ 0 & 2 < |t| \leq 4 \end{cases}$

and $P_z(t + 8) = P_z(t)$;

average power $P_z = \dfrac{1}{8} \displaystyle\int_{-4}^{4} |z(t)|^2 \, dt = \dfrac{1}{8} \displaystyle\int_{-2}^{2} 25 \, dt = \dfrac{100}{8} = 12.5$;

energy $E_z = \displaystyle\int_{-\infty}^{\infty} |z(t)|^2 \, dt = \infty.$

Because the signal has finite power $(0 < P_z = 12.5 < \infty)$, $z(t)$ is a power signal.

Example 1.7
Consider the following DT sequence:

$$f[k] = \begin{cases} e^{-0.5k} & k \geq 0 \\ 0 & k < 0. \end{cases}$$

Determine if the signal is a power or an energy signal.

Solution
The total energy of the DT sequence is calculated as follows:

$$E_f = \sum_{k=-\infty}^{\infty} |f[k]|^2 = \sum_{k=0}^{\infty} |e^{-0.5k}|^2 = \sum_{k=0}^{\infty} (e^{-1})^k = \frac{1}{1 - e^{-1}} \approx 1.582.$$

Because E_f is finite, the DT sequence $f[k]$ is an energy signal.

In computing E_f, we make use of the geometric progression (GP) series to calculate the summation. The formulas for the GP series are considered in Appendix A.3.

Example 1.8
Determine if the DT sequence $g[k] = 3\cos(\pi k/10)$ is a power or an energy signal.

Solution
The DT sequence $g[k] = 3\cos(\pi k/10)$ is a periodic signal with a fundamental period of 20. All periodic signals are power signals. Hence, the DT sequence $g[k]$ is a power signal.

Using Eq. (1.15), the average power of $g[k]$ is given by

$$P_g = \frac{1}{20} \sum_{k=0}^{19} 9\cos^2\left(\frac{\pi k}{10}\right) = \frac{9}{20} \sum_{k=0}^{19} \frac{1}{2}\left[1 + \cos\left(\frac{2\pi k}{10}\right)\right]$$

$$= \underbrace{\frac{9}{40} \sum_{k=0}^{19} 1}_{\text{term I}} + \underbrace{\frac{9}{40} \sum_{k=0}^{19} \cos\left(\frac{2\pi k}{10}\right)}_{\text{term II}}.$$

Clearly, the summation represented by term I equals $9(20)/40 = 4.5$. To compute the summation in term II, we express the cosine as follows:

$$\text{term II} = \frac{9}{40} \sum_{k=0}^{19} \frac{1}{2}[e^{j\pi k/5} + e^{-j\pi k/5}] = \frac{9}{80} \sum_{k=0}^{19} (e^{j\pi/5})^k + \frac{9}{80} \sum_{k=0}^{19} (e^{-j\pi/5})^k.$$

Using the formulas for the GP series yields

$$\sum_{k=0}^{19} (e^{j\pi/5})^k = \frac{1 - (e^{j\pi/5})^{20}}{1 - (e^{j\pi/5})} = \frac{1 - e^{j\pi 4}}{1 - (e^{j\pi/5})} = \frac{1 - 1}{1 - (e^{j\pi/5})} = 0$$

and

$$\sum_{k=0}^{19} (e^{-j\pi/5})^k = \frac{1 - (e^{-j\pi/5})^{20}}{1 - (e^{j\pi/5})} = \frac{1 - e^{-j\pi 4}}{1 - (e^{j\pi/5})} = \frac{1 - 1}{1 - (e^{j\pi/5})} = 0.$$

Term II, therefore, equals zero. The average power of $g[k]$ is therefore given by

$$P_g = 4.5 + 0 = 4.5.$$

In general, a periodic DT sinusoidal signal of the form $x[k] = A \cos(\omega_0 k + \theta)$ has an average power $P_x = A^2/2$.

1.1.5 Deterministic and random signals

If the value of a signal can be predicted for all time (t or k) in advance without any error, it is referred to as a *deterministic signal*. Conversely, signals whose values cannot be predicted with complete accuracy for all time are known as *random signals*.

Deterministic signals can generally be expressed in a mathematical, or graphical, form. Some examples of deterministic signals are as follows.

(1) CT sinusoidal signal: $x_1(t) = 5 \sin(20\pi t + 6)$;
(2) CT exponentially decaying sinusoidal signal: $x_2(t) = 2e^{-t} \sin(7t)$;
(3) CT finite duration complex exponential signal: $x_3(t) = \begin{cases} e^{j4\pi t} & |t| < 5 \\ 0 & \text{elsewhere}; \end{cases}$
(4) DT real-valued exponential sequence: $x_4[k] = 4e^{-2k}$;
(5) DT exponentially decaying sinusoidal sequence: $x_5[k] = 3e^{-2k} \times \sin\left(\frac{16\pi k}{5}\right)$.

Unlike deterministic signals, random signals cannot be modeled precisely. Random signals are generally characterized by statistical measures such as means, standard deviations, and mean squared values. In electrical engineering, most meaningful information-bearing signals are random signals. In a digital communication system, for example, data are generally transmitted using a sequence of zeros and ones. The binary signal is corrupted with interference from other channels and additive noise from the transmission media, resulting in a received signal that is random in nature. Another example of a random signal in electrical engineering is the thermal noise generated by a resistor. The intensity of the thermal noise depends on the movement of billions of electrons and cannot be predicted accurately.

The study of random signals is beyond the scope of this book. We therefore restrict our discussion to deterministic signals. However, most principles and techniques that we develop are generalizable to random signals. The readers are advised to consult more advanced books for analysis of random signals.

1.1.6 Odd and even signals

A CT signal $x_e(t)$ is said to be an even signal if

$$x_e(t) = x_e(-t). \tag{1.16}$$

Conversely, a CT signal $x_o(t)$ is said to be an odd signal if

$$x_o(t) = -x_o(-t). \tag{1.17}$$

A DT signal $x_e[k]$ is said to be an even signal if

$$x_e[k] = x_e[-k]. \tag{1.18}$$

Conversely, a DT signal $x_o[k]$ is said to be an odd signal if

$$x_o[k] = -x_o[-k]. \tag{1.19}$$

The even signal property, Eq. (1.16) for CT signals or Eq. (1.18) for DT signals, implies that an even signal is symmetric about the *vertical* axis ($t = 0$). Likewise, the odd signal property, Eq. (1.17) for CT signals or Eq. (1.19) for DT signals, implies that an odd signal is antisymmetric about the *vertical* axis ($t = 0$). The symmetry characteristics of even and odd signals are illustrated in Fig. 1.10. The waveform in Fig 1.10(a) is an even signal as it is symmetric about the y-axis and the waveform in Fig. 1.10(b) is an odd signal as it is anti-symmetric about the y-axis. The waveforms shown in Figs. 1.6(a) and (b) are additional examples of even signals, while the waveforms shown in Figs. 1.6(c) and (e) are examples of odd signals.

Most practical signals are neither odd nor even. For example, the signals shown in Figs. 1.6(d) and (f), and 1.8(a) do not exhibit any symmetry about the y-axis. Such signals are classified in the "neither odd nor even" category.

Fig. 1.10. Example of (a) an even signal and (b) an odd signal.

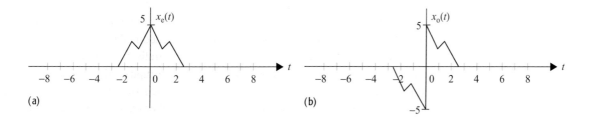

(a)

(b)

Neither odd nor even signals can be expressed as a sum of even and odd signals as follows:

$$x(t) = x_e(t) + x_o(t),$$

where the even component $x_e(t)$ is given by

$$x_e(t) = \frac{1}{2}[x(t) + x(-t)], \tag{1.20}$$

while the odd component $x_o(t)$ is given by

$$x_o(t) = \frac{1}{2}[x(t) - x(-t)]. \tag{1.21}$$

Example 1.9
Express the CT signal

$$x(t) = \begin{cases} t & 0 \le t < 1 \\ 0 & \text{elsewhere} \end{cases}$$

as a combination of an even signal and an odd signal.

Solution
In order to calculate $x_e(t)$ and $x_o(t)$, we need to calculate the function $x(-t)$, which is expressed as follows:

$$x(-t) = \begin{cases} -t & 0 \le -t < 1 \\ 0 & \text{elsewhere} \end{cases} = \begin{cases} -t & -1 < t \le 0 \\ 0 & \text{elsewhere}. \end{cases}$$

Using Eq. (1.20), the even component $x_e(t)$ of $x(t)$ is given by

$$x_e(t) = \frac{1}{2}[x(t) + x(-t)] = \begin{cases} \frac{1}{2}t & 0 \le t < 1 \\ -\frac{1}{2}t & -1 \le t < 0 \\ 0 & \text{elsewhere}, \end{cases}$$

while the odd component $x_o(t)$ is evaluated from Eq. (1.21) as follows:

$$x_o(t) = \frac{1}{2}[x(t) - x(-t)] = \begin{cases} \frac{1}{2}t & 0 \le t < 1 \\ \frac{1}{2}t & -1 \le t < 0 \\ 0 & \text{elsewhere}. \end{cases}$$

The waveforms for the CT signal $x(t)$ and its even and odd components are plotted in Fig. 1.11.

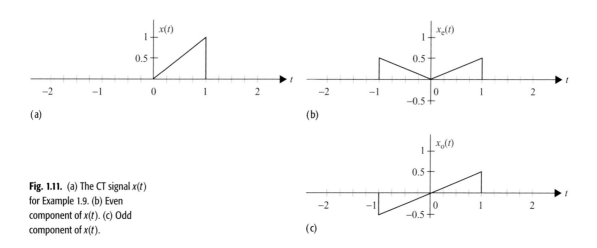

Fig. 1.11. (a) The CT signal $x(t)$ for Example 1.9. (b) Even component of $x(t)$. (c) Odd component of $x(t)$.

1.1.6.1 Combinations of even and odd CT signals

Consider $g_e(t)$ and $h_e(t)$ as two CT even signals and $g_o(t)$ and $h_o(t)$ as two CT odd signals. The following properties may be used to classify different combinations of these four signals into the even and odd categories.

(i) Multiplication of a CT even signal with a CT odd signal results in a CT odd signal. The CT signal $x(t) = g_e(t) \times g_o(t)$ is therefore an odd signal.

(ii) Multiplication of a CT odd signal with another CT odd signal results in a CT even signal. The CT signal $h(t) = g_o(t) \times h_o(t)$ is therefore an even signal.

(iii) Multiplication of two CT even signals results in another CT even signal. The CT signal $z(t) = g_e(t) \times h_e(t)$ is therefore an even signal.

(iv) Due to its antisymmetry property, a CT odd signal is always zero at $t = 0$. Therefore, $g_o(0) = h_o(0) = 0$.

(v) Integration of a CT odd signal within the limits $[-T, T]$ results in a zero value, i.e.

$$\int_{-T}^{T} g_o(t)\mathrm{d}t = \int_{-T}^{T} h_o(t)\mathrm{d}t = 0. \tag{1.22}$$

(vi) The integral of a CT even signal within the limits $[-T, T]$ can be simplified as follows:

$$\int_{-T}^{T} g_e(t)\mathrm{d}t = 2\int_{0}^{T} g_e(t)\mathrm{d}t. \tag{1.23}$$

It is straightforward to prove properties (i)–(vi). Below we prove property (vi).

Proof of property (vi)

By expanding the left-hand side of Eq. (1.23), we obtain

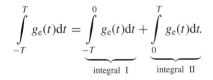

$$\int_{-T}^{T} g_e(t)dt = \underbrace{\int_{-T}^{0} g_e(t)dt}_{\text{integral I}} + \underbrace{\int_{0}^{T} g_e(t)dt}_{\text{integral II}}.$$

Substituting $\alpha = -t$ in integral I yields

$$\text{integral I} = \int_{T}^{0} g_e(-\alpha)(-d\alpha) = \int_{0}^{T} g_e(\alpha)d\alpha = \int_{0}^{T} g_e(t)dt = \text{integral II},$$

which proves Eq. (1.23).

1.1.6.2 Combinations of even and odd DT signals

Properties (i)–(vi) for CT signals can be extended to DT sequences. Consider $g_e[k]$ and $h_e[k]$ as even sequences and $g_o[k]$ and $h_o[k]$ are as odd sequences. For the four DT signals, the following properties hold true.

(i) Multiplication of an even sequence with an odd sequence results in an odd sequence. The DT sequence $x[k] = g_e[k] \times g_o[k]$, for example, is an odd sequence.

(ii) Multiplication of two odd sequences results in an even sequence. The DT sequence $h[k] = g_o[k] \times h_o[k]$, for example, is an even sequence.

(iii) Multiplication of two even sequences results in an even sequence. The DT sequence $z[k] = g_e[k] \times h_e[k]$, for example, is an even sequence.

(iv) Due to its antisymmetry property, a DT odd sequence is always zero at $k = 0$. Therefore, $g_o[0] = h_o[0] = 0$.

(v) Adding the samples of a DT odd sequence $g_o[k]$ within the range $[-M, M]$ is 0, i.e.

$$\sum_{k=-M}^{M} g_o[k] = 0 = \sum_{k=-M}^{M} h_o[k]. \tag{1.24}$$

(vi) Adding the samples of a DT even sequence $g_e[k]$ within the range $[-M, M]$ simplifies to

$$\sum_{k=-M}^{M} g_e[k] = g_e[0] + 2 \sum_{k=1}^{M} g_e[k]. \tag{1.25}$$

1.2 Elementary signals

In this section, we define some elementary functions that will be used frequently to represent more complicated signals. Representing signals in terms of the elementary functions simplifies the analysis and design of linear systems.

1.2.1 Unit step function

The CT unit step function $u(t)$ is defined as follows:

$$u(t) = \begin{cases} 1 & t \geq 0 \\ 0 & t < 0. \end{cases} \tag{1.26}$$

The DT unit step function $u[k]$ is defined as follows:

$$u[k] = \begin{cases} 1 & k \geq 0 \\ 0 & k < 0. \end{cases} \tag{1.27}$$

The waveforms for the unit step functions $u(t)$ and $u[k]$ are shown, respectively, in Figs. 1.12(a) and (b). It is observed from Fig. 1.12 that the CT unit step function $u(t)$ is piecewise continuous with a discontinuity at $t = 0$. In other words, the rate of change in $u(t)$ is infinite at $t = 0$. However, the DT function $u[k]$ has no such discontinuity.

1.2.2 Rectangular pulse function

The CT rectangular pulse $\text{rect}(t/\tau)$ is defined as follows:

$$\text{rect}\left(\frac{t}{\tau}\right) = \begin{cases} 1 & |t| \leq \tau/2 \\ 0 & |t| > \tau/2 \end{cases} \tag{1.28}$$

and it is plotted in Fig. 1.12(c). The DT rectangular pulse $\text{rect}(k/(2N + 1))$ is defined as follows:

$$\text{rect}\left(\frac{k}{2N+1}\right) = \begin{cases} 1 & |k| \leq N \\ 0 & |k| > N \end{cases} \tag{1.29}$$

and it is plotted in Fig. 1.12(d).

1.2.3 Signum function

The *signum* (or *sign*) function, denoted by $\text{sgn}(t)$, is defined as follows:

$$\text{sgn}(t) = \begin{cases} 1 & t > 0 \\ 0 & t = 0 \\ -1 & t < 0. \end{cases} \tag{1.30}$$

The CT sign function $\text{sgn}(t)$ is plotted in Fig. 1.12(e). Note that the operation $\text{sgn}(\cdot)$ can be used to output the sign of the input argument. The DT signum

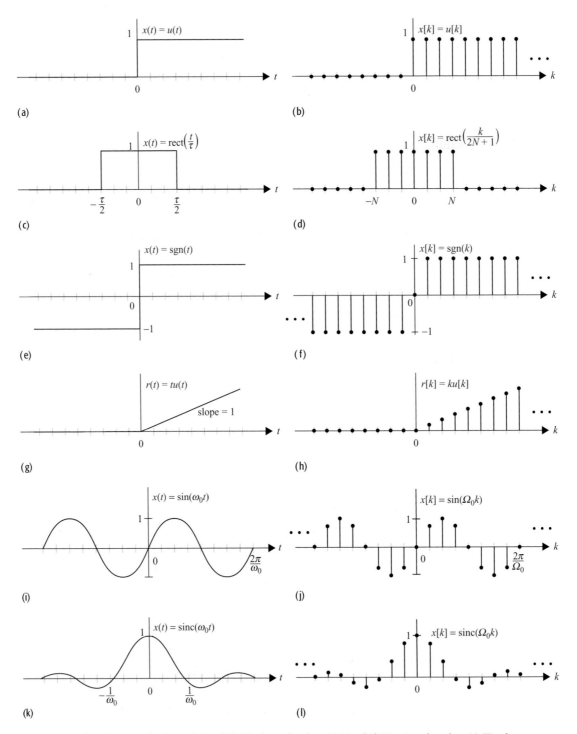

Fig. 1.12. CT and DT elementary functions. (a) CT and (b) DT unit step functions. (c) CT and (d) DT rectangular pulses. (e) CT and (f) DT signum functions. (g) CT and (h) DT ramp functions. (i) CT and (j) DT sinusoidal functions. (k) CT and (l) DT sinc functions.

function, denoted by sgn(k), is defined as follows:

$$sgn[k] = \begin{cases} 1 & k > 0 \\ 0 & k = 0 \\ -1 & k < 0 \end{cases} \tag{1.31}$$

and it is plotted in Fig. 1.12(f).

1.2.4 Ramp function

The CT ramp function $r(t)$ is defined as follows:

$$r(t) = tu(t) = \begin{cases} t & t \geq 0 \\ 0 & t < 0, \end{cases} \tag{1.32}$$

which is plotted in Fig. 1.12(g). Similarly, the DT ramp function $r[k]$ is defined as follows:

$$r[k] = ku[k] = \begin{cases} k & k \geq 0 \\ 0 & k < 0, \end{cases} \tag{1.33}$$

which is plotted in Fig. 1.12(h).

1.2.5 Sinusoidal function

The CT sinusoid of frequency f_0 (or, equivalently, an angular frequency $\omega_0 = 2\pi f_0$) is defined as follows:

$$x(t) = \sin(\omega_0 t + \theta) = \sin(2\pi f_0 t + \theta), \tag{1.34}$$

which is plotted in Fig. 1.12(i). The DT sinusoid is defined as follows:

$$x[k] = \sin(\Omega_0 k + \theta) = \sin(2\pi f_0 k + \theta), \tag{1.35}$$

where Ω_0 is the DT angular frequency. The DT sinusoid is plotted in Fig. 1.12(j). As discussed in Section 1.1.3, a CT sinusoidal signal $x(t) = \sin(\omega_0 t + \theta)$ is always periodic, whereas its DT counterpart $x[k] = \sin(\Omega_0 k + \theta)$ is not necessarily periodic. The DT sinusoidal signal is periodic only if the fraction $\Omega_0/2\pi$ is a rational number.

1.2.6 Sinc function

The CT sinc function is defined as follows:

$$sinc(\omega_0 t) = \frac{\sin(\pi \omega_0 t)}{\pi \omega_0 t}, \tag{1.36}$$

which is plotted in Fig. 1.12(k). In some text books, the sinc function is alternatively defined as follows:

$$sinc(\omega_0 t) = \frac{\sin(\omega_0 t)}{\omega_0 t}.$$

In this text, we will use the definition in Eq. (1.36) for the sinc function. The DT sinc function is defined as follows:

$$\text{sinc}(\Omega_0 k) = \frac{\sin(\pi \Omega_0 k)}{\pi \Omega_0 k}, \qquad (1.37)$$

which is plotted in Fig. 1.12(l).

1.2.7 CT exponential function

A CT exponential function, with complex frequency $s = \sigma + j\omega_0$, is represented by

$$x(t) = e^{st} = e^{(\sigma + j\omega_0)t} = e^{\sigma t}(\cos \omega_0 t + j \sin \omega_0 t). \qquad (1.38)$$

The CT exponential function is, therefore, a complex-valued function with the following real and imaginary components:

real component $\qquad \text{Re}\{e^{st}\} = e^{\sigma t} \cos \omega_0 t;$

imaginary component $\qquad \text{Im}\{e^{st}\} = e^{\sigma t} \sin \omega_0 t.$

Depending upon the presence or absence of the real and imaginary components, there are two special cases of the complex exponential function.

Case 1 Imaginary component is zero ($\omega_0 = 0$)
Assuming that the imaginary component ω of the complex frequency s is zero, the exponential function takes the following form:

$$x(t) = e^{\sigma t},$$

which is referred to as a real-valued exponential function. Figure 1.13 shows the real-valued exponential functions for different values of σ. When the value of σ is negative ($\sigma < 0$) then the exponential function decays with increasing time t.

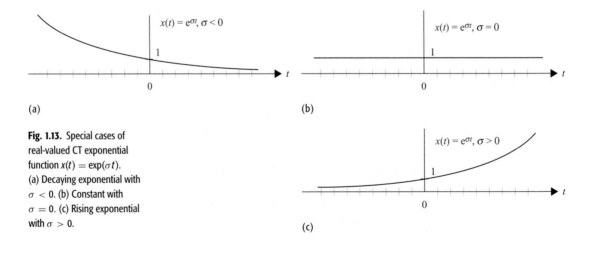

(a)

(b)

(c)

Fig. 1.13. Special cases of real-valued CT exponential function $x(t) = \exp(\sigma t)$. (a) Decaying exponential with $\sigma < 0$. (b) Constant with $\sigma = 0$. (c) Rising exponential with $\sigma > 0$.

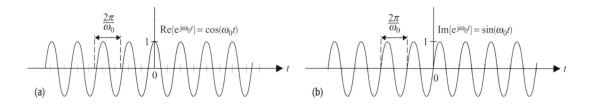

Fig. 1.14. CT complex-valued exponential function $x(t) = \exp(j\omega_0 t)$. (a) Real component; (b) imaginary component.

The exponential function for $\sigma < 0$ is referred to as a decaying exponential function and is shown in Fig. 1.13(a). For $\sigma = 0$, the exponential function has a constant value, as shown in Fig. 1.13(b). For positive values of σ ($\sigma > 0$), the exponential function increases with time t and is referred to as a rising exponential function. The rising exponential function is shown in Fig. 1.13(c).

Case 2 Real component is zero ($\sigma = 0$)
When the real component σ of the complex frequency s is zero, the exponential function is represented by

$$x(t) = e^{j\omega_0 t} = \cos \omega_0 t + j \sin \omega_0 t.$$

In other words, the real and imaginary parts of the complex exponential are pure sinusoids. Figure 1.14 shows the real and imaginary parts of the complex exponential function.

Example 1.10
Plot the real and imaginary components of the exponential function $x(t) = \exp[(j4\pi - 0.5)t]$ for $-4 \leq t \leq 4$.

Solution
The CT exponential function is expressed as follows:

$$x(t) = e^{(j4\pi - 0.5)t} = e^{-0.5t} \times e^{j4\pi t}.$$

The real and imaginary components of $x(t)$ are expressed as follows:

real component $\text{Re}\{(t)\} = e^{-0.5t} \cos(4\pi t)$;
imaginary component $\text{Im}\{(t)\} = e^{-0.5t} \sin(4\pi t)$.

To plot the real component, we multiply the waveform of a cosine function with $\omega_0 = 4\pi$, as shown in Fig. 1.14(a), by a decaying exponential $\exp(-0.5t)$. The resulting plot is shown in Fig. 1.15(a). Similarly, the imaginary component is plotted by multiplying the waveform of a sine function with $\omega_0 = 4\pi$, as shown in Fig. 1.14(b), by a decaying exponential $\exp(-0.5t)$. The resulting plot is shown in Fig. 1.15(b).

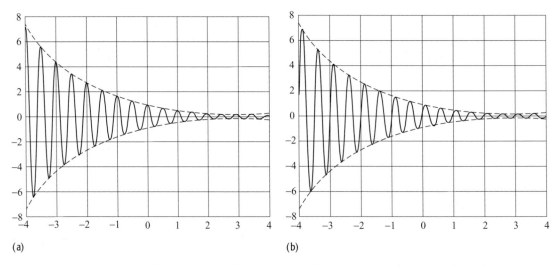

Fig. 1.15. Exponential function $x(t) = \exp[(j4\pi - 0.5)t]$. (a) Real component; (b) imaginary component.

1.2.8 DT exponential function

The DT complex exponential function with radian frequency Ω_0 is defined as follows:

$$x[k] = e^{(\sigma + j\Omega_0)k} = e^{\sigma t}(\cos \Omega_0 k + j \sin \Omega_0 k.) \tag{1.39}$$

As an example of the DT complex exponential function, we consider $x[k] = \exp(j0.2\pi - 0.05k)$, which is plotted in Fig. 1.16, where plot (a) shows the real component and plot (b) shows the imaginary part of the complex signal.

Case 1 Imaginary component is zero ($\Omega_0 = 0$). The signal takes the following form:

$$x[k] = e^{\sigma k}$$

when the imaginary component Ω_0 of the DT complex frequency is zero. Similar to CT exponential functions, the DT exponential functions can be classified as rising, decaying, and constant-valued exponentials depending upon the value of σ.

Case 2 Real component is zero ($\sigma = 0$). The DT exponential function takes the following form:

$$x[k] = e^{j\Omega_0 k} = \cos \Omega_0 k + j \sin \Omega_0 k.$$

Recall that a complex-valued exponential is periodic iff $\Omega_0/2\pi$ is a rational number. An alternative representation of the DT complex exponential function

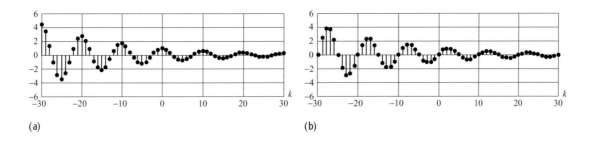

(a)   (b)

Fig. 1.16. DT complex exponential function $x[k] =$ exp($j0.2\pi k - 0.05k$). (a) Real component; (b) imaginary component.

is obtained by expanding

$$x[k] = \left(e^{(\sigma + j\Omega_0)}\right)^k = \gamma^k, \tag{1.40}$$

where $\gamma = (\sigma + j\Omega_0)$ is a complex number. Equation (1.40) is more compact than Eq. (1.39).

1.2.9 Causal exponential function

In practical signal processing applications, input signals start at time $t = 0$. Signals that start at $t = 0$ are referred to as causal signals. The causal exponential function is given by

$$x(t) = e^{st}u(t) = \begin{cases} e^{st} & t \geq 0 \\ 0 & t < 0, \end{cases} \tag{1.41}$$

where we have used the unit step function to incorporate causality in the complex exponential functions. Similarly, the causal implementation of the DT exponential function is defined as follows:

$$x[k] = e^{sk}u[k] = \begin{cases} e^{sk} & k \geq 0 \\ 0 & k < 0. \end{cases} \tag{1.42}$$

The same concept can be extended to derive causal implementations of sinusoidal and other non-causal signals.

Example 1.11
Plot the DT causal exponential function $x[k] = e^{(j0.2\pi - 0.05)k}u[k]$.

Solution
The real and imaginary components of the non-causal signal $e^{(j0.2\pi - 0.05)k}$ are plotted in Fig. 1.16. To plot its causal implementation, we multiply $e^{(j0.2\pi - 0.05)k}$ by the unit step function $u[k]$. This implies that the causal implementation will be zero for $k < 0$. The real and imaginary components of the resulting function are plotted in Fig. 1.17.

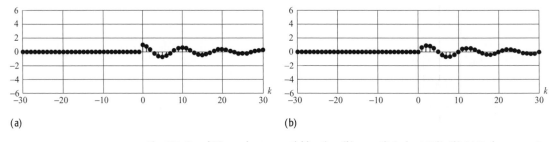

Fig. 1.17. Causal DT complex exponential function $x[k] = \exp(j0.2\pi k - 0.05k)u[k]$. (a) Real component; (b) imaginary component.

1.2.10 CT unit impulse function

The *unit impulse* function $\delta(t)$, also known as the *Dirac delta* function[†] or simply the *delta* function, is defined in terms of two properties as follows:

(1) amplitude $\qquad\qquad\qquad\qquad \delta(t) = 0, \quad t \neq 0;$ $\qquad\qquad\qquad$ (1.43a)

(2) area enclosed $\qquad\qquad\qquad \int_{-\infty}^{\infty} \delta(t)dt = 1.$ $\qquad\qquad\qquad$ (1.43b)

Direct visualization of a unit impulse function in the CT domain is difficult. One way to visualize a CT impulse function is to let it evolve from a rectangular function. Consider a tall narrow rectangle with width ε and height $1/\varepsilon$, as shown in Fig. 1.18(a), such that the area enclosed by the rectangular function equals one. Next, we decrease the width and increase the height at the same rate such that the resulting rectangular functions have areas $= 1$. As the width $\varepsilon \to 0$, the rectangular function converges to the CT impulse function $\delta(t)$ with an infinite amplitude at $t = 0$. However, the area enclosed by CT impulse function is finite and equals one. The impulse function is illustrated in our plots by an arrow pointing vertically upwards; see Fig. 1.18(b). The height of the arrow corresponds to the area enclosed by the CT impulse function.

Properties of impulse function

(i) The impulse function is an even function, i.e. $\delta(t) = \delta(-t)$.

(ii) Integrating a unit impulse function results in one, provided that the limits of integration enclose the origin of the impulse. Mathematically,

$$\int_{-T}^{T} A\delta(t - t_0)dt = \begin{cases} A & \text{for } -T < t_0 < T \\ 0 & \text{elsewhere.} \end{cases} \qquad (1.44)$$

[†] The unit impulse function was introduced by Paul Adrien Maurice Dirac (1902–1984), a British electrical engineer turned theoretical physicist.

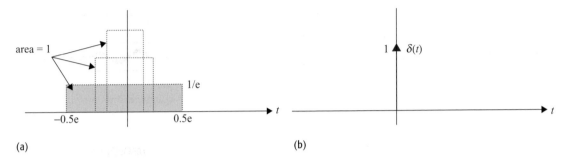

(a)

(b)

Fig. 1.18. Impulse function $\delta(t)$. (a) Generating the impulse function $\delta(t)$ from a rectangular pulse. (b) Notation used to represent an impulse function.

(iii) The scaled and time-shifted version $\delta(at + b)$ of the unit impulse function is given by

$$\delta(at + b) = \frac{1}{a}\delta\left(t + \frac{b}{a}\right) \tag{1.45}$$

(iv) When an arbitrary function $\phi(t)$ is multiplied by a shifted impulse function, the product is given by

$$\phi(t)\delta(t - t_0) = \phi(t_0)\delta(t - t_0). \tag{1.46}$$

In other words, multiplication of a CT function and an impulse function produces an impulse function, which has an area equal to the value of the CT function at the location of the impulse. Combining properties (ii) and (iv), it is straightforward to show that

$$\int_{-\infty}^{\infty} \phi(t)\delta(t - t_0)\mathrm{d}t = \phi(t_0). \tag{1.47}$$

(v) The unit impulse function can be obtained by taking the derivative of the unit step function as follows:

$$\delta(t) = \frac{\mathrm{d}u}{\mathrm{d}t} \tag{1.48}$$

(vi) Conversely, the unit step function is obtained by integrating the unit impulse function as follows:

$$u(t) = \int_{-\infty}^{t} \delta(\tau)\mathrm{d}\tau. \tag{1.49}$$

Example 1.12
Simplify the following expressions:

(i) $\dfrac{5 - jt}{7 + t^2}\delta(t)$

(ii) $\displaystyle\int_{-\infty}^{\infty} (t + 5)\delta(t - 2)\mathrm{d}t$;

(iii) $\displaystyle\int_{-\infty}^{\infty} e^{j0.5\pi\omega + 2}\delta(\omega - 5)\mathrm{d}\omega.$

Solution

(i) Using Eq. (1.46) yields $\dfrac{5-jt}{7+t^2}\delta(t) = \left[\dfrac{5-jt}{7+t^2}\right]_{t=0}\delta(t) = \dfrac{5}{7}\delta(t).$

(ii) Using Eq. (1.46) yields

$$\int_{-\infty}^{\infty} (t+5)\delta(t-2)\mathrm{d}t = \int_{-\infty}^{\infty} [(t+5)]_{t=2}\delta(t-2)\mathrm{d}t = 7\int_{-\infty}^{\infty} \delta(t-2)\mathrm{d}t.$$

Since the integral computes the area enclosed by the unit step function, which is one, we obtain

$$\int_{-\infty}^{\infty} (t+5)\delta(t-2)\mathrm{d}t = 7\int_{-\infty}^{\infty} \delta(t-2)\mathrm{d}t = 7.$$

(iii) Using Eq. (1.46) yields

$$\int_{-\infty}^{\infty} e^{j0.5\pi\omega+2}\delta(\omega-5)\mathrm{d}\omega = \int_{-\infty}^{\infty} [e^{j0.5\pi\omega+2}]_{\omega=5}\delta(\omega-5)\mathrm{d}\omega$$

$$= e^{j2.5\pi+2}\int_{-\infty}^{\infty} \delta(\omega-5)\mathrm{d}\omega.$$

Since $\exp(j2.5\pi+2) = j\exp(2)$, we obtain

$$\int_{-\infty}^{\infty} e^{j0.5\pi\omega+2}\delta(\omega-5)\mathrm{d}\omega = je^2.$$

1.2.11 DT unit impulse function

The DT impulse function, also referred to as the Kronecker delta function or the DT unit sample function, is defined as follows:

$$\delta[k] = u[k] - u[k-1] = \begin{cases} 1 & k = 0 \\ 0 & k \neq 0. \end{cases} \tag{1.50}$$

Unlike the CT unit impulse function, the DT impulse function has no ambiguity in its definition; it is well defined for all values of k. The waveform for a DT unit impulse function is shown in Fig. 1.19.

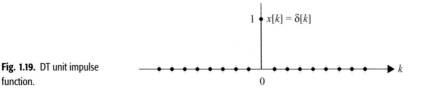

Fig. 1.19. DT unit impulse function.

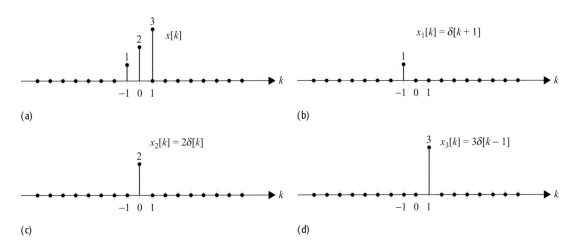

Fig. 1.20. The DT functions in Example 1.13: (a) x[k], (b), x[k], (c) $x_2[k]$, and (d) $x_3[k]$. The DT function in (a) is the sum of the shifted DT impulse functions shown in (b), (c), and (d).

Example 1.13

Represent the DT sequence shown in Fig. 1.20(a) as a function of time-shifted DT unit impulse functions.

Solution

The DT signal $x[k]$ can be represented as the summation of three functions, $x_1[k]$, $x_2[k]$, and $x_3[k]$, as follows:

$$x[k] = x_1[k] + x_2[k] + x_3[k],$$

where $x_1[k]$, $x_2[k]$, and $x_3[k]$ are time-shifted impulse functions,

$$x_1[k] = \delta[k + 1], \quad x_2[k] = 2\delta[k], \quad \text{and} \quad x_3[k] = 3\delta[k - 1],$$

and are plotted in Figs. 1.20(b), (c), and (d), respectively. The DT sequence $x[k]$ can therefore be represented as follows:

$$x[k] = \delta[k + 1] + 2\delta[k] + 3\delta[k - 1].$$

1.3 Signal operations

An important concept in signal and system analysis is the transformation of a signal. In this section, we consider three elementary transformations that are performed on a signal in the time domain. The transformations that we consider are time shifting, time scaling, and time inversion.

1.3.1 Time shifting

The *time-shifting* operation delays or advances forward the input signal in time. Consider a CT signal $\phi(t)$ obtained by shifting another signal $x(t)$ by T time units. The time-shifted signal $\phi(t)$ is expressed as follows:

$$\phi(t) = x(t + T).$$

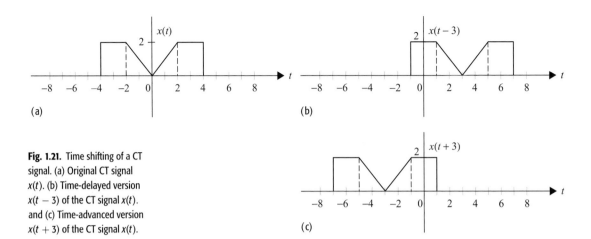

Fig. 1.21. Time shifting of a CT signal. (a) Original CT signal $x(t)$. (b) Time-delayed version $x(t - 3)$ of the CT signal $x(t)$. and (c) Time-advanced version $x(t + 3)$ of the CT signal $x(t)$.

In other words, a signal time-shifted by T is obtained by substituting t in $x(t)$ by $(t + T)$. If $T < 0$, then the signal $x(t)$ is delayed in the time domain. Graphically this is equivalent to shifting the origin of the signal $x(t)$ towards the right-hand side by duration T along the t-axis. On the other hand, if $T > 0$, then the signal $x(t)$ is advanced forward in time. The plot of the time-advanced signal is obtained by shifting $x(t)$ towards the left-hand side by duration T along the t-axis.

Figure 1.21(a) shows a CT signal $x(t)$ and the corresponding two time-shifted signals $x(t - 3)$ and $x(t + 3)$. Since $x(t - 3)$ is a delayed version of $x(t)$, the waveform of $x(t - 3)$ is identical to that of $x(t)$, except for a shift of three time units towards the right-hand side. Similarly, $x(t + 3)$ is a time-advanced version of $x(t)$. The waveform of $x(t + 3)$ is identical to that of $x(t)$ except for a shift of three time units towards the left-hand side.

The theory of the CT time-shifting operation can also be extended to DT sequences. When a DT signal $x[k]$ is shifted by m time units, the delayed signal $\phi[k]$ is expressed as follows:

$$\phi[k] = x[k + m].$$

If $m < 0$, the signal is said to be delayed in time. To obtain the time-delayed signal $\phi[k]$, the origin of the signal $x[k]$ is shifted towards the right-hand side along the k-axis by m time units. On the other hand, if $m > 0$, the signal is advanced forward in time. The time-advanced signal $\phi[k]$ is obtained by shifting $x[k]$ towards the left-hand side along the k-axis by m time units.

Figure 1.22 shows a DT signal $x[k]$ and the corresponding two time-shifted signals $x[k - 4]$ and $x[k + 4]$. The waveforms of $x[k - 4]$ and $x[k + 4]$ are identical to that of $x[k]$. The time-delayed signal $x[k - 4]$ is obtained by shifting $x[k]$ towards the right-hand side by four time units. The time-advanced signal $x[k + 4]$ is obtained by shifting $x[k]$ towards the left-hand side by four time units.

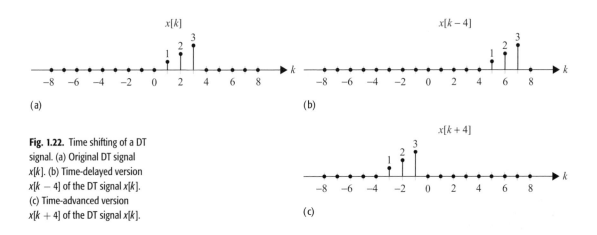

Fig. 1.22. Time shifting of a DT signal. (a) Original DT signal $x[k]$. (b) Time-delayed version $x[k-4]$ of the DT signal $x[k]$. (c) Time-advanced version $x[k+4]$ of the DT signal $x[k]$.

Example 1.14

Consider the signal $x(t) = e^{-t}u(t)$. Determine and plot the time-shifted versions $x(t-4)$ and $x(t+2)$.

Solution

The signal $x(t)$ can be expressed as follows:

$$x(t) = e^{-t}u(t) = \begin{cases} e^{-t} & t \geq 0 \\ 0 & \text{elsewhere,} \end{cases} \qquad (1.51)$$

and is shown in Fig. 1.23(a). To determine the expression for $x(t-4)$, we substitute t by $(t-4)$ in Eq. (1.51). The resulting expression is given by

$$x(t-4) = \begin{cases} e^{-(t-4)} & (t-4) \geq 0 \\ 0 & \text{elsewhere} \end{cases}$$

$$= \begin{cases} e^{-(t-4)} & t \geq 4 \\ 0 & \text{elsewhere.} \end{cases}$$

The function $x(t-4)$ is plotted in Fig. 1.23(b).

Similarly, we can calculate the expression for $x(t+2)$ by substituting t by $(t+2)$ in Eq. (1.51). The resulting expression is given by

$$x(t+2) = \begin{cases} e^{-(t+2)} & (t+2) \geq 0 \\ 0 & \text{elsewhere} \end{cases}$$

$$= \begin{cases} e^{-(t+2)} & t \geq -2 \\ 0 & \text{elsewhere.} \end{cases}$$

The function $x(t+2)$ is plotted in Fig. 1.23(c). From Fig. 1.23, we observe that the waveform for $x(t-4)$ can be obtained directly from $x(t)$ by shifting the waveform of $x(t)$ by four time units towards the right-hand side. Similarly, the waveform for $x(t+2)$ can be obtained from $x(t)$ by shifting the waveform of $x(t)$ by two time units towards the left-hand side. This is the result expected from our previous discussion.

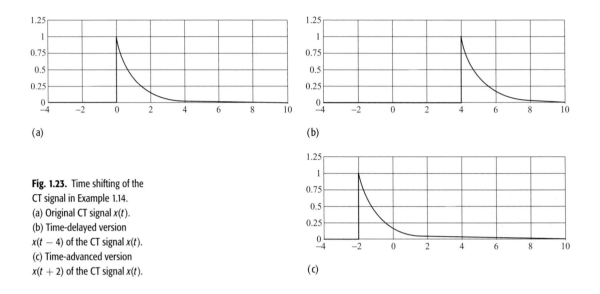

(a)

(b)

Fig. 1.23. Time shifting of the
CT signal in Example 1.14.
(a) Original CT signal $x(t)$.
(b) Time-delayed version
$x(t - 4)$ of the CT signal $x(t)$.
(c) Time-advanced version
$x(t + 2)$ of the CT signal $x(t)$.

(c)

Example 1.15

Consider the signal $x[k]$ defined as follows:

$$x[k] = \begin{cases} 0.2k & 0 \le k \le 5 \\ 0 & \text{elsewhere.} \end{cases} \qquad (1.52)$$

Determine and plot signals $p[k] = x[k - 2]$ and $q[k] = x[k + 2]$.

Solution

The signal $x[k]$ is plotted in Fig. 1.24(a). To calculate the expression for $p[k]$, substitute $k = m - 2$ in Eq. (1.52). The resulting equation is given by

$$x[m - 2] = \begin{cases} 0.2(m - 2) & 0 \le (m - 2) \le 5 \\ 0 & \text{elsewhere.} \end{cases}$$

By changing the independent variable from m to k and simplifying, we obtain

$$p[k] = x[k - 2] = \begin{cases} 0.2(k - 2) & 2 \le k \le 7 \\ 0 & \text{elsewhere.} \end{cases}$$

The non-zero values of $p[k]$ for $-2 \le k \le 7$, are shown in Table 1.1, and the stem plot $p[k]$ is plotted in Fig. 1.24(b). To calculate the expression for $q[k]$, substitute $k = m + 2$ in Eq. (1.52). The resulting equation is as follows:

$$x[m + 2] = \begin{cases} 0.2(m + 2) & 0 \le (m + 2) \le 5 \\ 0 & \text{elsewhere.} \end{cases}$$

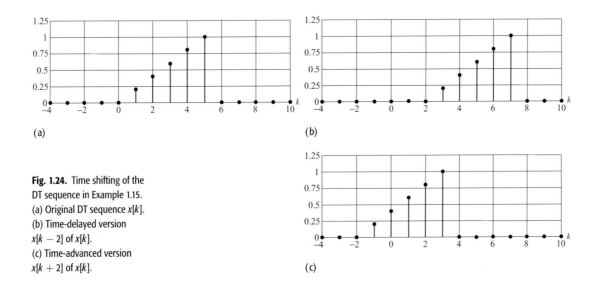

Fig. 1.24. Time shifting of the DT sequence in Example 1.15. (a) Original DT sequence $x[k]$. (b) Time-delayed version $x[k-2]$ of $x[k]$. (c) Time-advanced version $x[k+2]$ of $x[k]$.

Table 1.1. Values of the signals $p[k]$ and $q[k]$

k	-2	-1	0	1	2	3	4	5	6	7
$p[k]$	0	0	0	0	0	0.2	0.4	0.6	0.8	1
$q[k]$	0	0.2	0.4	0.6	0.8	1	0	0	0	0

By changing the independent variable from m to k and simplifying, we obtain

$$q[k] = x[k+2] = \begin{cases} 0.2(k+2) & -2 \leq k \leq 3 \\ 0 & \text{elsewhere.} \end{cases}$$

Values of $q[k]$, for $-2 \leq k \leq 7$, are shown in Table 1.1, and the stem plot for $q[k]$ is plotted in Fig. 1.24(c).

As in Example 1.14, we observe that the waveform for $p[k] = x[k-2]$ can be obtained directly by shifting the waveform of $x[k]$ towards the right-hand side by two time units. Similarly, the waveform for $q[k] = x[k+2]$ can be obtained directly by shifting the waveform of $x[k]$ towards the left-hand side by two time units.

1.3.2 Time scaling

The *time-scaling* operation compresses or expands the input signal in the time domain. A CT signal $x(t)$ scaled by a factor c in the time domain is denoted by $x(ct)$. If $c > 1$, the signal is compressed by a factor of c. On the other hand, if $0 < c < 1$ the signal is expanded. We illustrate the concept of time scaling of CT signals with the help of a few examples.

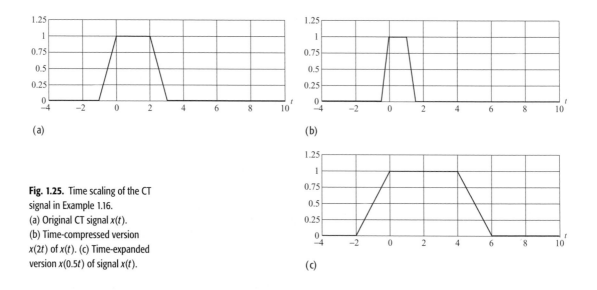

(a)

(b)

Fig. 1.25. Time scaling of the CT
signal in Example 1.16.
(a) Original CT signal $x(t)$.
(b) Time-compressed version
$x(2t)$ of $x(t)$. (c) Time-expanded
version $x(0.5t)$ of signal $x(t)$.

(c)

Example 1.16

Consider a CT signal $x(t)$ defined as follows:

$$x(t) = \begin{cases} t+1 & -1 \leq t \leq 0 \\ 1 & 0 \leq t \leq 2 \\ -t+3 & 2 \leq t \leq 3 \\ 0 & \text{elsewhere,} \end{cases} \tag{1.53}$$

as plotted in Fig. 1.25(a). Determine the expressions for the time-scaled signals
$x(2t)$ and $x(t/2)$. Sketch the two signals.

Solution

Substituting t by 2α in Eq. (1.53), we obtain

$$x(2\alpha) = \begin{cases} 2\alpha+1 & -1 \leq 2\alpha \leq 0 \\ 1 & 0 \leq 2\alpha \leq 2 \\ -2\alpha+3 & 2 \leq 2\alpha \leq 3 \\ 0 & \text{elsewhere.} \end{cases}$$

By changing the independent variable from α to t and simplifying, we obtain

$$x(2t) = \begin{cases} 2t+1 & -0.5 \leq t \leq 0 \\ 1 & 0 \leq t \leq 1 \\ -2t+3 & 1 \leq t \leq 1.5 \\ 0 & \text{elsewhere,} \end{cases}$$

which is plotted in Fig. 1.25(b). The waveform for $x(2t)$ can also be obtained
directly by *compressing* the waveform for $x(t)$ by a factor of 2. It is important
to note that compression is performed with respect to the y-axis such that the
values $x(t)$ and $x(2t)$ at $t = 0$ are the same for both waveforms.

Substituting t by $\alpha/2$ in Eq. (1.53), we obtain

$$x(\alpha/2) = \begin{cases} \alpha/2 + 1 & -1 \le \alpha/2 \le 0 \\ 1 & 0 \le \alpha/2 \le 2 \\ -\alpha/2 + 3 & 2 \le \alpha/2 \le 3 \\ 0 & \text{elsewhere.} \end{cases}$$

By changing the independent variable from α to t and simplifying, we obtain

$$x(t/2) = \begin{cases} t/2 + 1 & -2 \le t \le 0 \\ 1 & 0 \le t \le 4 \\ -t/2 + 3 & 4 \le t \le 6 \\ 0 & \text{elsewhere,} \end{cases}$$

which is plotted in Fig. 1.25(c). The waveform for $x(0.5t)$ can also be obtained directly by *expanding* the waveform for $x(t)$ by a factor of 2. As for compression, expansion is performed with respect to the y-axis such that the values $x(t)$ and $x(t/2)$ at $t = 0$ are the same for both waveforms.

A CT signal $x(t)$ can be scaled to $x(ct)$ for any value of c. For the DTFT, however, the time-scaling factor c is limited to integer values. We discuss the time scaling of the DT sequence in the following.

1.3.2.1 Decimation

If a sequence $x[k]$ is compressed by a factor c, some data samples of $x[k]$ are lost. For example, if we decimate $x[k]$ by 2, the decimated function $y[k] = x[2k]$ retains only the alternate samples given by $x[0]$, $x[2]$, $x[4]$, and so on. Compression (referred to as decimation for DT sequences) is, therefore, an irreversible process in the DT domain as the original sequence $x[k]$ cannot be recovered precisely from the decimated sequence $y[k]$.

1.3.2.2 Interpolation

In the DT domain, expansion (also referred to as interpolation) is defined as follows:

$$x^{(m)}[k] = \begin{cases} x\left[\dfrac{k}{m}\right] & \text{if } k \text{ is a multiple of integer } m \\ 0 & \text{otherwise.} \end{cases} \tag{1.54}$$

The interpolated sequence $x^{(m)}[k]$ inserts $(m - 1)$ zeros in between adjacent samples of the DT sequence $x[k]$. Interpolation of the DT sequence $x[k]$ is a reversible process as the original sequence $x[k]$ can be recovered from $x^{(m)}[k]$.

Example 1.17
Consider the DT sequence $x[k]$ plotted in Fig. 1.26(a). Calculate and sketch $p[k] = x[2k]$ and $q[k] = x[k/2]$.

Part I Introduction to signals and systems

Table 1.2. Values of the signal $p[k]$ for $-3 \leq k \leq 3$

k	-3	-2	-1	0	1	2	3
$p[k]$	$x[-6]=0$	$x[-4]=0.2$	$x[-2]=0.6$	$x[0]=1$	$x[2]=0.6$	$x[4]=0.2$	$x[6]=0$

Table 1.3. Values of the signal $q[k]$ for $-10 \leq k \leq 10$

k	-10	-9	-8	-7	-6	-5	-4
$q[k]$	$x[-5]=0$	0	$x[-4]=0.2$	0	$x[-3]=0.4$	0	$x[-2]=0.6$

k	-3	-2	-1	0	1	2	3
$q[k]$	0	$x[-1]=0.8$	0	$x[0]=1$	0	$x[1]=0.8$	0

k	4	5	6	7	8	9	10
$q[k]$	$x[2]=0.6$	0	$x[3]=0.4$	0	$x[4]=0.2$	0	$x[5]=0$

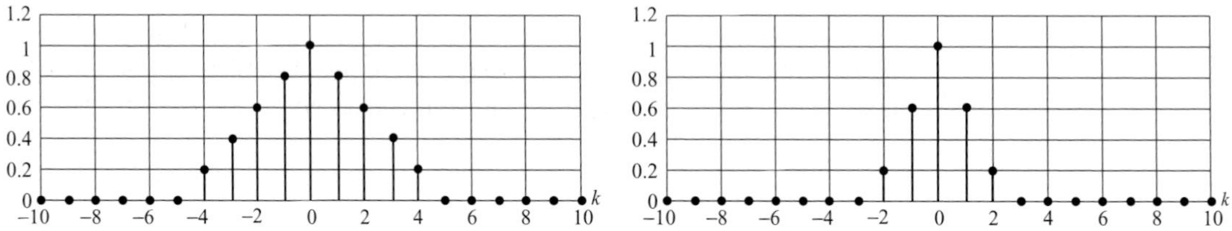

(a)

(b)

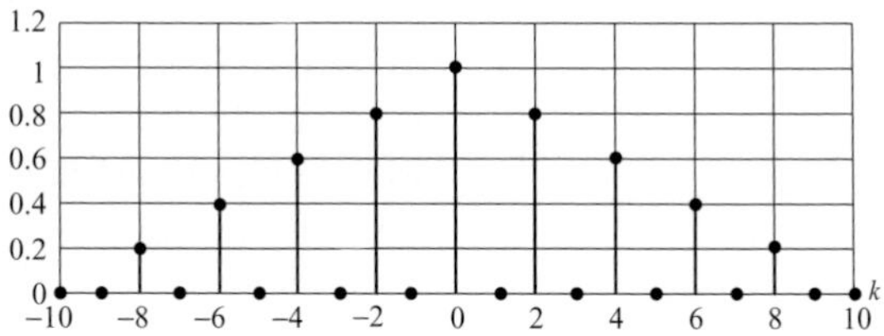

Fig. 1.26. Time scaling of the DT
signal in Example 1.17.
(a) Original DT sequence $x[k]$.
(b) Decimated version $x[2k]$, of
$x[k]$. (c) Interpolated version
$x[0.5k]$ of signal $x[k]$.

(c)

Solution

Since $x[k]$ is non-zero for $-5 \leq k \leq 5$, the non-zero values of the decimated
sequence $p[k] = x[2k]$ lie in the range $-3 \leq k \leq 3$. The non-zero values of
$p[k]$ are shown in Table 1.2. The waveform for $p[k]$ is plotted in Fig. 1.26(b).

The waveform for the interpolated sequence $p[k]$ can be obtained by directly
compressing the waveform for $x[k]$ by a factor of 2 about the y-axis. While
performing the compression, the value of $x[k]$ at $k = 0$ is retained in $p[k]$. On
both sides of the $k = 0$ sample, every second sample of $x[k]$ is retained in $p[k]$.

To determine $q[k] = x[k/2]$, we first determine the range over which $x[k/2]$
is non-zero. The non-zero values of $q[k] = x[k/2]$ lie in the range $-10 \leq k \leq$
10 and are shown in Table 1.3. The waveform for $q[k]$ is plotted in Fig. 1.26(c).

The waveform for the interpolated sequence $q[k]$ can be obtained by directly
expanding the waveform for $x[k]$ by a factor of 2 about the y-axis. During

Table 1.4. Values of the signal $q_2[k]$ for $-10 \leq k \leq k$

k	-10	-9	-8	-7	-6	-5	-4
$q_2[k]$	$x[-5]=0$	0.1	$x[-4]=0.2$	0.3	$x[-3]=0.4$	0.5	$x[-2]=0.6$
k	-3	-2	-1	0	1	2	3
$q_2[k]$	0.7	$x[-1]=0.8$	0.9	$x[0]=1$	0.9	$x[1]=0.8$	0.7
k	4	5	6	7	8	9	10
$q_2[k]$	$x[2]=0.6$	0.5	$x[3]=0.4$	0.3	$x[4]=0.2$	0.1	$x[5]=0$

expansion, the value of $x[k]$ at $k = 0$ is retained in $q[k]$. The even-numbered samples, where k is a multiple of 2, of $q[k]$ equal $x[k/2]$. The odd-numbered samples in $q[k]$ are set to zero.

While determining the interpolated sequence $x[mk]$, Eq. (1.54) inserts $(m - 1)$ zeros in between adjacent samples of the DT sequence $x[k]$, where $x[k]$ is not defined. Instead of inserting zeros, we can possibly interpolate the undefined values from the neighboring samples where $x[k]$ is defined. Using linear interpolation, an interpolated sequence can be obtained using the following equation:

$$
x^{(m)}[k] =
\begin{cases}
x\left[\dfrac{k}{m}\right] & \text{if } k \text{ is a multiple of integer } m \\
(1 - \alpha)x\left[\left\lfloor\dfrac{k}{m}\right\rfloor\right] + \alpha\, x\left[\left\lceil\dfrac{k}{m}\right\rceil\right] & \text{otherwise,}
\end{cases}
\tag{1.55}
$$

where $\left\lfloor\frac{k}{m}\right\rfloor$ denotes the nearest integer less than or equal to (k/m), $\left\lceil\frac{k}{m}\right\rceil$ denotes the nearest integer greater than or equal to (k/m), and $\alpha = (k \bmod m)/m$. Note that mod is the modulo operator that calculates the remainder of the division k/m. For $m = 2$, Eq. (1.55) simplifies to the following:

$$
x^{(2)}[k] =
\begin{cases}
x\left[\dfrac{k}{2}\right] & \text{if } k \text{ is even} \\
0.5\left(x\left[\dfrac{k-1}{2}\right] + x\left[\dfrac{k+1}{2}\right]\right) & \text{if } k \text{ is odd.}
\end{cases}
$$

Although, Eq. (1.55) is useful in many applications, we will use Eq. (1.54) to denote an interpolated sequence throughout the book unless explicitly stated otherwise.

Example 1.18
Repeat Example 1.17 to obtain the interpolated sequence $q_2[k] = x[k/2]$ using the alternative definition given by Eq. (1.55).

Solution
The non-zero values of $q_2[k] = x[k/2]$ are shown in Table 1.4, where the values of the odd-numbered samples of $q_2[k]$, highlighted with the gray background, are obtained by taking the average of the values of the two neighboring

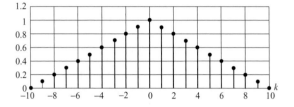

samples at k and $k - 1$ obtained from $x[k]$. The waveform for $q_2[k]$ is plotted in Fig. 1.27.

1.3.3 Time inversion

The *time inversion* (also known as *time reversal* or *reflection*) operation reflects the input signal about the vertical axis ($t = 0$). When a CT signal $x(t)$ is time-reversed, the inverted signal is denoted by $x(-t)$. Likewise, when a DT signal $x[k]$ is time-reversed, the inverted signal is denoted by $x[-k]$. In the following we provide examples of time inversion in both CT and DT domains.

Example 1.19

Sketch the time-inverted version of the causal decaying exponential signal

$$x(t) = e^{-t}u(t) = \begin{cases} e^{-t} & t \geq 0 \\ 0 & \text{elsewhere,} \end{cases} \tag{1.56}$$

which is plotted in Fig. 1.28(a).

Solution

To derive the expression for the time-inverted signal $x(-t)$, substitute $t = -\alpha$ in Eq. (1.56). The resulting expression is given by

$$x(-\alpha) = e^{\alpha}u(-\alpha) = \begin{cases} e^{\alpha} & -\alpha \geq 0 \\ 0 & \text{elsewhere.} \end{cases}$$

Simplifying the above expression and expressing it in terms of the independent variable t yields

Fig. 1.28. Time inversion of the CT signal in Example 1.19.
(a) Original CT signal $x(t)$.
(b) Time-inverted version $x(-t)$.

$$x(-t) = \begin{cases} e^{t} & t \leq 0 \\ 0 & \text{elsewhere.} \end{cases}$$

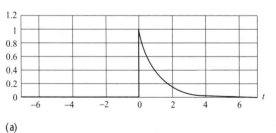

(a)

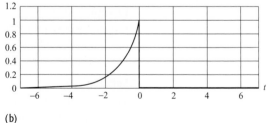

(b)

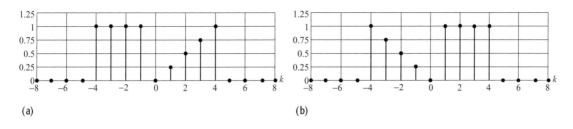

(a)   (b)

Fig. 1.29. Time inversion of the
DT signal in Example 1.20.
(a) Original CT sequence $x[k]$.
(b) Time-inverted version $x[-k]$.

The time-reversed signal $x(-t)$ is plotted in Fig. 1.28(b). Signal inversion can also be performed graphically by simply flipping the signal $x(t)$ about the y-axis.

Example 1.20
Sketch the time-inverted version of the following DT sequence:

$$x[k] = \begin{cases} 1 & -4 \leq k \leq -1 \\ 0.25k & 0 \leq k \leq 4 \\ 0 & \text{elsewhere,} \end{cases} \quad (1.57)$$

which is plotted in Fig. 1.29(a).

Solution
To derive the expression for the time-inverted signal $x[-k]$, substitute $k = -m$ in Eq. (1.57). The resulting expression is given by

$$x[-m] = \begin{cases} 1 & -4 \leq -m \leq -1 \\ -0.25m & 0 \leq -m \leq 4 \\ 0 & \text{elsewhere.} \end{cases}$$

Simplifying the above expression and expressing it in terms of the independent variable k yields

$$x[-m] = \begin{cases} 1 & 1 \leq m \leq 4 \\ -0.25m & -4 \leq m \leq 0 \\ 0 & \text{elsewhere.} \end{cases}$$

The time-reversed signal $x[-k]$ is plotted in Fig. 1.29(b).

1.3.4 Combined operations

In Sections 1.3.1–1.3.3, we presented three basic time-domain transformations. In many signal processing applications, these operations are combined. An arbitrary linear operation that combines the three transformations is expressed as $x(\alpha t + \beta)$, where α is the time-scaling factor and β is the time-shifting factor. If α is negative, the signal is inverted along with the time-scaling and time-shifting operations. By expressing the transformed signal as

$$x(\alpha t + \beta) = x\left(\alpha\left[t + \frac{\beta}{\alpha}\right]\right), \quad (1.58)$$

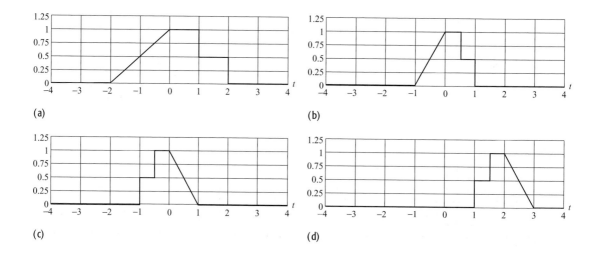

(a)   (b)   (c)   (d)

Fig. 1.30. Combined CT operations defined in Example 1.21. (a) Original CT signal $x(t)$. (b) Time-scaled version $x(2t)$. (c) Time-inverted version $x(-2t)$ of (b). (d) Time-shifted version $x(4 - 2t)$ of (c).

we can plot the waveform graphically for $x(\alpha t + \beta)$ by following steps (i)–(iii) outlined below.

(i)   Scale the signal $x(t)$ by $|\alpha|$. The resulting waveform represents $x(|\alpha|t)$.
(ii) If α is negative, invert the scaled signal $x(|\alpha|t)$ with respect to the $t = 0$ axis. This step produces the waveform for $x(\alpha t)$.
(iii) Shift the waveform for $x(\alpha t)$ obtained in step (ii) by $|\beta/\alpha|$ time units. Shift towards the right-hand side if (β/α) is negative. Otherwise, shift towards the left-hand side if (β/α) is positive. The waveform resulting from this step represents $x(\alpha t + \beta)$, which is the required transformation.

Example 1.21
Determine $x(4 - 2t)$, where the waveform for the CT signal $x(t)$ is plotted in Fig. 1.30(a).

Solution
Express $x(4 - 2t) = x(-2[t - 2])$ and follow steps (i)–(iii) as outlined below.

(i)   Compress $x(t)$ by a factor of 2 to obtain $x(2t)$. The resulting waveform is shown in Fig. 1.30(b).
(ii) Time-reverse $x(2t)$ to obtain $x(-2t)$. The waveform for $x(-2t)$ is shown in Fig. 1.30(c).
(iii) Shift $x(-2t)$ towards the right-hand side by two time units to obtain $x(-2[t - 2]) = x(4 - 2t)$. The waveform for $x(4 - 2t)$ is plotted in Fig. 1.30(d).

Example 1.22
Sketch the waveform for $x[-15 - 3k]$ for the DT sequence $x[k]$ plotted in Fig. 1.31(a).

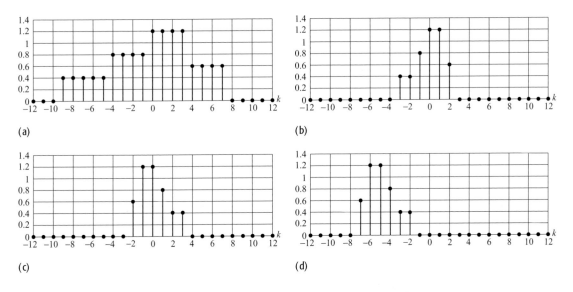

Fig. 1.31. Combined DT operations defined in Example 1.22. (a) Original DT signal x[k]. (b) Time-scaled version x[3k]. (c) Time-inverted version x[−3k] of (b). (d) Time-shifted version x[−15 − 3k] of (c).

Solution

Express $x[-15 - 3k] = x[-3(k + 5)]$ and follow steps (i)–(iii) as outlined below.

(i) Compress $x[k]$ by a factor of 3 to obtain $x[3k]$. The resulting waveform is shown in Fig. 1.31(b).

(ii) Time-reverse $x[3k]$ to obtain $x[-3k]$. The waveform for $x[-3k]$ is shown in Fig. 1.31(c).

(iii) Shift $x[-3k]$ towards the left-hand side by five time units to obtain $x[-3(k + 5)] = x[-15 - 3k]$. The waveform for $x[-15 - 3k]$ is plotted in Fig. 1.31(d).

1.4 Signal implementation with MATLAB

MATLAB is used frequently to simulate signals and systems. In this section, we present a few examples to illustrate the generation of different CT and DT signals in MATLAB. We also show how the CT and DT signals are plotted in MATLAB. A brief introduction to MATLAB is included in Appendix E.

Example 1.23

Generate and sketch in the same figure each of the following CT signals using MATLAB. Do not use the "for" loops in your code. In each case, the horizontal axis t used to sketch the CT should extend only for the range over which the three signals are defined.

(a) $x_1(t) = 5 \sin(2\pi t) \cos(\pi t - 8)$ for $-5 \leq t \leq 5$;

(b) $x_2(t) = 5e^{-0.2t} \sin(2\pi t)$ for $-10 \leq t \leq 10$;

(c) $x_3(t) = e^{(j4\pi - 0.5)t} u(t)$ for $-5 \leq t \leq 15$.

Solution

The MATLAB code for the generation of signals (a)–(c) is as follows:

```
>> %%%%%%%%%%%%
>> % Part(a)    %          % Clear any existing figure
>> %%%%%%%%%%%%
>> clf                     % Clear any existing figure
>> t1 = [-5:0.001:5];      % Set the time from -5 to 5
                               % with a sampling
                           % rate of 0.001s

>> x1 = 5*sin(2*pi*t1).    % compute function x1
     *cos(pi*t1-8);
>> % plot x1(t)
>> subplot(2,2,1);         % select the 1st out of 4
                               % subplots

>> plot(t1,x1);            % plot a CT signal
>> grid on;                % turn on the grid
>> xlabel('time (t)');     % Label the x-axis as time
>> ylabel('5sin(2\pi t)    % Label the y-axis
     cos(\pi t - 8)');
>> title('Part (a)');      % Insert the title
>> %%%%%%%%%%%%
>> %  Part(b)    %
>> %%%%%%%%%%%%
>> t2 = [-10:0.002:10];    % Set the time from -10 to
                               % 10 with a sampling
                           % rate of 0.002s

>> x2 = 5*exp(-0.2*t2).    % compute function x2
     *sin(2*pi*t2);
>> % plot x2(t)
>> subplot(2,2,2);         % select the 2nd out of 4
                               % subplots

>> plot(t2,x2);            % plot a CT signal
>> grid on;                % turn on the grid
>> xlabel('time (t)');     % Label the x-axis as time
>> ylabel('5exp(-0.2t)     % Label the y-axis
     sin(2\pi t)');
>> title('Part (b)');      % Insert the title
>> %%%%%%%%%%%%
>> %Part(c)%
>> %%%%%%%%%%%%
>> t3 = [-5:0.001:15];     % Set the time from -5 to
                               % 15 with a sampling
                           % rate of 0.001s
```

```
>> x3 = exp((j*4*pi-0.5)*t3).    % compute function x3
      *(t3>=0);
>> % plot the real component
      of x3(t)
>> subplot(2,2,3);                       % select the 3rd out of 4
                                         % subplots
>> plot(t3,real(x3));            % plot a CT signal
>> grid on;                      % turn on the grid
>> xlabel('time (t)')            % Label the x-axis as time
>> ylabel('e^{(j*4\pi-0.5)       % Label the y-axis
      t}u(t)');
>> title('Part (c): Real         % Insert the title
      Component');
>> subplot(2,2,4);                       % select the 4th out of 4
                                         % subplots
>> plot(t3,imag(x3));            % plot the imaginary
                                         % component of a CT
                                         % signal
>> grid on;                      % turn on the grid
>> xlabel('time (t)');           % Label the x-axis as time
>> ylabel('e^{(j*4\pi-0.5)       % Label the y-axis
      t}u(t)');
>> title('Part (d): Imaginary    % Insert the title
      Component');
```

The resulting MATLAB plot is shown in Fig. 1.32.

Example 1.24

Repeat Example 1.23 for the following DT sequences:

(a) $f_1[k] = -0.92 \sin(0.1\pi k - 3\pi/4)$ for $-10 \le k \le 20$;
(b) $f_2[k] = 2.0(1.1)^{1.8k} - 2.1(0.9)^{0.7k}$ for $-5 \le k \le 25$;
(c) $f_3[k] = (-0.93)^k e^{j\pi k/\sqrt{350}}$ for $0 \le k \le 50$.

Solution

The MATLAB code for the generation of signals (a)–(c) is as follows:

```
>> %%%%%%%%%%%%
>> % Part(a) %
>> %%%%%%%%%%%%
>> clf                           % clear any existing figure
>> k = [-10:20];                 % set the time index from
                                         % -10 to 20
>> f1 = -0.92 * sin(0.1*pi*k - 3*pi/4);
                                 % compute function f1
```

Fig. 1.32. MATLAB plot for
Example 1.23. (a) $x_1(t)$;
(b) $x_2(t)$; (c) Re$\{x_3(t)\}$;
(d) Im$\{x_3(t)\}$.

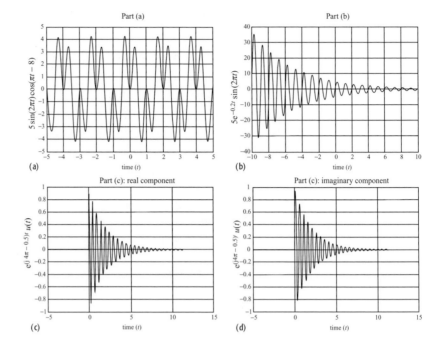

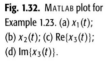

Fig. 1.32. MATLAB plot for
Example 1.23. (a) $x_1(t)$;
(b) $x_2(t)$; (c) Re$\{x_3(t)\}$;
(d) Im$\{x_3(t)\}$.

```
>> % plot function 1
>> subplot(2,2,1), stem(k, f1, 'filled'), grid
>> xlabel('k')
>> ylabel('-9.2sin(0.1\pi k-0.75\pi')
>> title('Part (a)')
>> %%%%%%%%%%%%
>> % Part(b) %
>> %%%%%%%%%%%%
>> k  = [-5:25];
>> f2 = 2 * 1.1.^(-1.8*k) - 2.1 * 0.9.^(0.7*k);
>> subplot(2,2,2), stem(k, f2, 'filled'), grid
>> xlabel('k')
>> ylabel('2(1.1)^{-1.8k} - 2.1(0.9)^0.7k')
>> title('Part (b)')
>> %%%%%%%%%%%%
>> % Part(c) %
>> %%%%%%%%%%%%
>> k  = [0:50];
>> f3 = (-0.93).^k .* exp(j*pi*k/sqrt(350));
>> subplot(2,2,3), stem(k, real(f3), 'filled'), grid
```

Fig. 1.33. MATLAB plot for
Example 1.24.

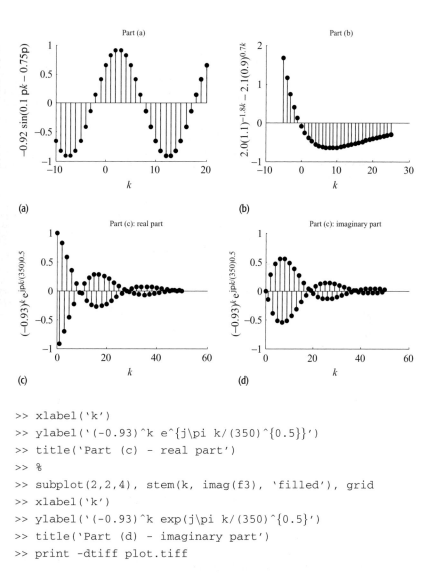

Fig. 1.33. MATLAB plot for
Example 1.24.

```
>> xlabel('k')
>> ylabel('(-0.93)^k e^{j\pi k/(350)^{0.5}}')
>> title('Part (c) - real part')
>> %
>> subplot(2,2,4), stem(k, imag(f3), 'filled'), grid
>> xlabel('k')
>> ylabel('(-0.93)^k exp(j\pi k/(350)^{0.5}')
>> title('Part (d) - imaginary part')
>> print -dtiff plot.tiff
```

The resulting MATLAB plots are shown in Fig. 1.33.

1.5 Summary

In this chapter, we have introduced many useful concepts related to signals
and systems, including the mathematical and graphical interpretations of signal
representation. In Section 1.1, we classified signals in six different categories:
CT versus DT signals; analog versus digital signals; periodic versus aperiodic
signals; energy versus power signals; deterministic versus probabilistic signals;
and even versus odd signals. We classified the signals based on the following
definitions.

(1) A time-varying signal is classified as a continuous time (CT) signal if it is defined for all values of time t. A time-varying discrete time (DT) signal is defined for certain discrete values of time, $t = kT_s$, where T_s is the sampling interval. In our notation, a CT signal is represented by $x(t)$ and a DT signal is denoted by $x[k]$.

(2) An analog signal is a CT signal whose amplitude can take any value. A digital signal is a DT signal that can only have a discrete set of values. The process of converting a DT signal into a digital signal is referred to as quantization.

(3) A periodic signal repeats itself after a known fundamental period, i.e. $x(t) = x(t + T_0)$ for CT signals and $x[k] = x[k + K_0]$ for DT signals. Note that CT complex exponentials and sinusoidal signals are always periodic, whereas DT complex exponentials and sinusoidal signals are periodic only if the ratio of their DT fundamental frequency Ω_0, to 2π is a rational number.

(4) A signal is classified as an energy signal if its total energy has a non-zero finite value. A signal is classified as a power signal if it has non-zero finite power. An energy signal has zero average power whereas a power signal has an infinite energy. Periodic signals are generally power signals.

(5) A deterministic signal is known precisely and can be predicted in advance without any error. A random signal cannot be predicted with 100% accuracy.

(6) A signal that is symmetric about the vertical axis $(t = 0)$ is referred to as an even signal. An odd signal is antisymmetric about the vertical axis $(t = 0)$. Mathematically, this implies $x(t) = x(-t)$ for the CT even signals and $x(t) = -x(-t)$ for the CT odd signals. Likewise for the DT signals.

In Section 1.2, we introduced a set of 1D elementary signals, including rectangular, sinusoidal, exponential, unit step, and impulse functions, defined both in the DT and CT domains. We illustrated through examples how the elementary signals can be used as building blocks for implementing more complicated signals. In Section 1.3, we presented three fundamental signal operations, namely time shifting, scaling, and inversion that operate on the independent variable. The time-shifting operation $x(t - T)$ shifts signal $x(t)$ with respect to time. If the value of T in $x(t - T)$ is positive, the signal is delayed by T time units. For negative values of T, the signal is time-advanced by T time units. The time-scaling, $x(ct)$, operation compresses $(c > 0)$ or expands $(c < 0)$ signal $x(t)$. The time-inversion operation is a special case of the time-scaling operation with $c = -1$. The waveform for the time-scaled signal $x(-t)$ is the reflection of the waveform of the original signal $x(t)$ about the vertical axis $(t = 0)$. The three transformations play an important role in the analysis of linear time-invariant (LTI) systems, which will be covered in Chapter 2. Finally, in Section 1.4, we used MATLAB to generate and analyze several CT and DT signals.

Problems

1.1 For each of the following representations:

(i) $z[m, n, k]$,

(ii) $I(x, y, z, t)$,

establish if the signal is a CT or a DT signal. Specify the independent and dependent variables. Think of an information signal from a physical process that follows the mathematical representation given in (i). Repeat for the representation in (ii).

1.2 Sketch each of the following CT signals as a function of the independent variable t over the specified range:

(i) $x1(t) = \cos(3\pi t/4 + \pi/8)$ for $-1 \leq t \leq 2$;

(ii) $x2(t) = \sin(-3\pi t/8 + \pi/2)$ for $-1 \leq t \leq 2$;

(iii) $x3(t) = 5t + 3\exp(-t)$ for $-2 \leq t \leq 2$;

(iv) $x4(t) = (\sin(3\pi t/4 + \pi/8))^2$ for $-1 \leq t \leq 2$;

(v) $x5(t) = \cos(3\pi t/4) + \sin(\pi t/2)$ for $-2 \leq t \leq 3$;

(vi) $x6(t) = t\exp(-2t)$ for $-2 \leq t \leq 3$.

1.3 Sketch the following DT signals as a function of the independent variable k over the specified range:

(i) $x1[k] = \cos(3\pi k/4 + \pi/8)$ for $-5 \leq k \leq 5$;

(ii) $x2[k] = \sin(-3\pi k/8 + \pi/2)$ for $-10 \leq k \leq 10$;

(iii) $x3[k] = 5k + 3^{-k}$ for $-5 \leq k \leq 5$;

(iv) $x4[k] = |\sin(3\pi k/4 + \pi/8)|$ for $-6 \leq k \leq 10$;

(v) $x5[k] = \cos(3\pi k/4) + \sin(\pi k/2)$ for $-10 \leq k \leq 10$;

(vi) $x6[k] = k4^{-|k|}$ for $-10 \leq k \leq 10$.

1.4 Prove Proposition 1.2.

1.5 Determine if the following CT signals are periodic. If yes, calculate the fundamental period T_0 for the CT signals:

(i) $x1(t) = \sin(-5\pi t/8 + \pi/2)$;

(ii) $x2(t) = |\sin(-5\pi t/8 + \pi/2)|$;

(iii) $x3(t) = \sin(6\pi t/7) + 2\cos(3t/5)$;

(iv) $x4(t) = \exp(j(5t + \pi/4))$;

(v) $x5(t) = \exp(j3\pi t/8) + \exp(\pi t/86)$;

(vi) $x6(t) = 2\cos(4\pi t/5)^* \sin^2(16t/3)$;

(vii) $x7(t) = 1 + \sin 20t + \cos(30t + \pi/3)$.

1.6 Determine if the following DT signals are periodic. If yes, calculate the fundamental period N_0 for the DT signals:

(i) $x1[k] = 5 \times (-1)^k$;

(ii) $x2[k] = \exp(j(7\pi k/4)) + \exp(j(3k/4))$;

(iii) $x3[k] = \exp(j(7\pi k/4)) + \exp(j(3\pi k/4))$;

(iv) $x4[k] = \sin(3\pi k/8) + \cos(63\pi k/64)$;

(v) $x5[k] = \exp(j(7\pi k/4)) + \cos(4\pi k/7 + \pi)$;

(vi) $x6[k] = \sin(3\pi k/8)\cos(63\pi k/64)$.

1.7 Determine if the following CT signals are energy or power signals or neither. Calculate the energy and power of the signals in each case:

(i) $x1(t) = \cos(\pi t)\sin(3\pi t)$;

(ii) $x2(t) = \exp(-2t)$;

(iii) $x3(t) = \exp(-j2t)$;

(iv) $x4(t) = \exp(-2t)u(t)$;

(v) $x5(t) = \begin{cases} \cos(3\pi t) & -3 \leq t \leq 3; \\ 0 & \text{elsewhere}; \end{cases}$

(vi) $x6(t) = \begin{cases} t & 0 \leq t \leq 2 \\ 4 - t & 2 \leq t \leq 4 \\ 0 & \text{elsewhere}. \end{cases}$

1.8 Repeat Problem 1.7 for the following DT sequences:

(i) $x1[k] = \cos\left(\dfrac{\pi k}{4}\right)\sin\left(\dfrac{3\pi k}{8}\right)$;

(ii) $x2[k] = \begin{cases} \cos\left(\dfrac{3\pi k}{16}\right) & -10 \leq k \leq 0 \\ 0 & \text{elsewhere}; \end{cases}$

(iii) $x3[k] = (-1)^k$;

(iv) $x4[k] = \exp(j(\pi k/2 + \pi/8))$;

(v) $x5[k] = \begin{cases} 2^k & 0 \leq k \leq 10 \\ 1 & 11 \leq k \leq 15 \\ 0 & \text{elsewhere}. \end{cases}$

1.9 Show that the average power of the CT periodic signal $x(t) = A\sin(\omega_0 t + \theta)$, with real-valued coefficient A, is given by $A^2/2$.

1.10 Show that the average power of the CT signal $y(t) = A_1\sin(\omega_1 t + \phi_1) + A_2\sin(\omega_2 t + \phi_2)$, with real-valued coefficients A_1 and A_2, is given by

$$P_y = \begin{cases} \dfrac{A_1^2}{2} + \dfrac{A_2^2}{2} & \omega_1 \neq \omega_2 \\ \dfrac{A_1^2}{2} + \dfrac{A_2^2}{2} + A_1 A_2 \cos(\phi_1 - \phi_2) & \omega_1 = \omega_2. \end{cases}$$

1.11 Show that the average power of the CT periodic signal $x(t) = D\exp[j(\omega_0 t + \theta)]$ is given by $|D|^2$.

1.12 Show that the average power of the following CT signal:

$$x(t) = \sum_{n=1}^{N} D_n e^{j\omega_n t}, \quad \omega_p \neq \omega_r \text{ if } p \neq r,$$

for $1 \leq p, r \leq N$, is given by

$$P_x = \sum_{n=1}^{N} |D_n|^2.$$

1.13 Calculate the average power of the periodic function shown in Fig. P1.13 and defined as

$$x(t)|_{t=(0,1]} = \begin{cases} 1 & 2^{-2m-1} < t \leq 2^{-2m} \\ 0 & 2^{-2m-2} < t \leq 2^{-2m-1} \end{cases}$$

$m \in Z^+$ and $x(t) = x(t+1)$.

Fig. P1.13. The CT function $x(t)$ in Problem 1.13.

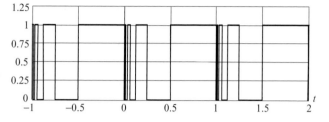

1.14 Determine if the following CT signals are even, odd, or neither even nor odd. In the latter case, evaluate and sketch the even and odd components of the CT signals:

(i) $x1(t) = 2\sin(2\pi t)[2 + \cos(4\pi t)]$;

(ii) $x2(t) = t^2 + \cos(3t)$;

(iii) $x3(t) = \exp(-3t)\sin(3\pi t)$;

(iv) $x4(t) = t\sin(5t)$;

(v) $x5(t) = tu(t)$;

(vi) $x6(t) = \begin{cases} 3t & 0 \le t < 2 \\ 6 & 2 \le t < 4 \\ 3(-t+6) & 4 \le t \le 6 \\ 0 & \text{elsewhere.} \end{cases}$

1.15 Determine if the following DT signals are even, odd, or neither even nor odd. In the latter case, evaluate and sketch the even and odd components of the DT signals:

(i) $x1[k] = \sin(4k) + \cos(2\pi k/3)$;

(ii) $x2[k] = \sin(\pi k/3000) + \cos(2\pi k/3)$;

(iii) $x3[k] = \exp(j(7\pi k/4)) + \cos(4\pi k/7 + \pi)$;

(iv) $x4[k] = \sin(3\pi k/8)\cos(63\pi k/64)$;

(v) $x5[k] = \begin{cases} (-1)^k & k \ge 0 \\ 0 & k < 0. \end{cases}$

1.16 Consider the following signal:

$$x(t) = 3\sin\left(\frac{2\pi(t-T)}{5}\right).$$

Determine the values of T for which the resulting signal is (a) an even function, and (b) an odd function of the independent variable t.

1.17 By inspecting plots (a), (b), (c), and (d) in Fig. P1.17, classify the CT waveforms as even versus odd, periodic versus aperiodic, and energy versus power signals. If the waveform is neither even nor odd, then determine the even and odd components of the signal. For periodic signals, determine the fundamental period. Also, compute the energy and power present in each case.

1.18 Sketch the following CT signals:

(i) $x1(t) = u(t) + 2u(t-3) - 2u(t-6) - u(t-9)$;

(ii) $x2(t) = u(\sin(\pi t))$;

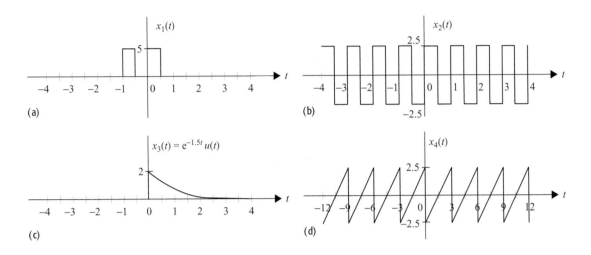

(a)

(b)

(c)

(d)

Fig. P1.17. Waveforms for
Problem 1.17.

(iii) $x3(t) = \mathrm{rect}(t/6) + \mathrm{rect}(t/4) + \mathrm{rect}(t/2)$;

(iv) $x4(t) = r(t) - r(t - 2) - 2u(t - 4)$;

(v) $x5(t) = (\exp(-t) - \exp(-3t))u(t)$;

(vi) $x6(t) = 3\,\mathrm{sgn}(t) \cdot \mathrm{rect}(t/4) + 2\delta(t + 1) - 3\delta(t - 3)$.

1.19 (a) Sketch the following functions with respect to the time variable (if a function is complex, sketch the real and imaginary components separately). (b) Locate the frequencies of the functions in the 2D complex plane.

(i) $x1(t) = e^{j2\pi t + 3}$;

(ii) $x2(t) = e^{j2\pi t + 3t}$;

(iii) $x3(t) = e^{-j2\pi t + j3t}$;

(iv) $x4(t) = \cos(2\pi t + 3)$;

(v) $x5(t) = \cos(2\pi t + 3) + \sin(3\pi t + 2)$;

(vi) $x6(t) = 2 + 4\cos(2\pi t + 3) - 7\sin(5\pi t + 2)$.

1.20 Sketch the following DT signals:

(i) $x1[k] = u[k] + u[k - 3] - u[k - 5] - u[k - 7]$;

(ii) $x2[k] = \displaystyle\sum_{m=0}^{\infty} \delta[k - m]$;

(iii) $x3[k] = (3^k - 2^k)u[k]$;

(iv) $x4[k] = u[\cos(\pi k/8)]$;

(v) $x5[k] = ku[k]$;

(vi) $x6[k] = |k|\,(u[k + 4] - u[k - 4])$.

1.21 Evaluate the following expressions:

(i) $\dfrac{5 + 2t + t^2}{7 + t^2 + t^4}\delta(t - 1)$;

(ii) $\dfrac{\sin(t)}{2t}\delta(t)$;

(iii) $\dfrac{\omega^3 - 1}{\omega^2 + 2}\delta(\omega - 5)$.

1.22 Evaluate the following integrals:

(i) $\displaystyle\int_{-\infty}^{\infty} (t-1)\delta(t-5)dt;$

(ii) $\displaystyle\int_{-\infty}^{6} (t-1)\delta(t-5)dt;$

(iii) $\displaystyle\int_{6}^{\infty} (t-1)\delta(t-5)dt;$

(iv) $\displaystyle\int_{-\infty}^{\infty} (2t/3-5)\delta(3t/4-5/6)dt;$

(v) $\displaystyle\int_{-\infty}^{\infty} \exp(t-1)\sin(\pi(t+5)/4)\delta(1-t)dt;$

(vi) $\displaystyle\int_{-\infty}^{\infty} [\sin(3\pi t/4)+\exp(-2t+1)]\delta(-t-1)dt;$

(vii) $\displaystyle\int_{-\infty}^{\infty} [u(t-6)-u(t-10)]\sin(3\pi t/4)\delta(t-5)dt;$

(viii) $\displaystyle\int_{-21}^{21} \left(\sum_{m=-\infty}^{\infty} t\delta(t-5m)\right)dt.$

1.23 In Section 1.2.8, the Dirac delta function was obtained as a limiting case of the rectangular function, i.e. $\delta(t) = \lim_{\varepsilon\to 0} \frac{1}{\varepsilon}\text{rect}\left(\frac{t}{\varepsilon}\right)$. Show that the Dirac delta function can also be obtained from each of the following functions (i.e. that Eq. (1.43) is satisfied by each of the following functions):

(i) $\displaystyle\lim_{\varepsilon\to 0} \frac{\varepsilon}{\pi(t^2+\varepsilon^2)};$

(iii) $\displaystyle\lim_{\varepsilon\to 0} \frac{1}{\pi t}\sin \varepsilon t;$

(ii) $\displaystyle\lim_{\varepsilon\to 0} \frac{2\varepsilon}{4\pi^2 t^2+\varepsilon^2};$

(iv) $\displaystyle\lim_{\varepsilon\to 0} \frac{1}{\varepsilon\sqrt{2\pi}}\exp\left(-\frac{t^2}{2\varepsilon^2}\right).$

1.24 Consider the following signal:

$$x(t) = \begin{cases} t+2 & -2 \le t \le -1 \\ 1 & -1 \le t \le 1 \\ -t+2 & 1 < t \le 2 \\ 0 & \text{elsewhere.} \end{cases}$$

(a) Sketch the functions: (i) $x(t-3)$; (ii) $x(-2t-3)$; and (iii) $x(-0.75t-3)$.

(b) Determine the analytical expressions for each of the four functions.

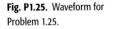

Fig. P1.25. Waveform for Problem 1.25.

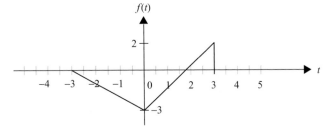

Fig. P1.26. Waveform for Problem 1.26.

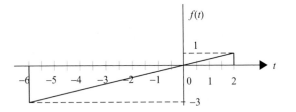

Fig. P1.27. Waveform for Problem 1.27.

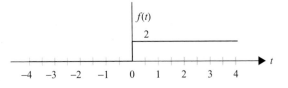

1.25 Consider the function $f(t)$ shown in Fig. P1.25.
  (i) Sketch the function $g(t) = f(-3t + 9)$.
  (ii) Calculate the energy and power of the signal $f(t)$. Is it a power signal or an energy signal?
  (iii) Repeat (ii) for $g(t)$.

1.26 Consider the function $f(t)$ shown in Fig. P1.26.
  (i) Sketch the function $g(t) = f(-2t + 6)$.
  (ii) Represent the function $f(t)$ as a summation of an even and an odd signal. Sketch the even and odd parts.

1.27 Consider the function $f(t)$ shown in Fig. P1.27.
  (i) Sketch the function $g(t) = tf(t + 2) - tf(t - 2)$.
  (ii) Sketch the function $g(2t)$.

1.28 Consider the two DT signals

$$x_1[k] = |k|(u[k + 4] - u[k - 4])$$

and

$$x_2[k] = k(u[k + 5] - u[k - 5]).$$

Fig. P1.29. ECG pattern for Problem 1.29.

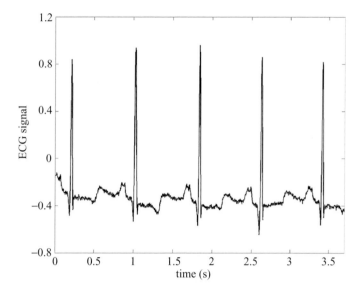

Sketch the following signals expressed as a function of $x_1[k]$ and $x_2[k]$:

(i) $x_1[k]$;   (vi) $x_2[3k]$;

(ii) $x_2[k]$;   (vii) $x_1[k/2]$;

(iii) $x_1[3-k]$;   (viii) $x_1[2k] + x_2[3k]$;

(iv) $x_1[6-2k]$;   (ix) $x_1[3-k]x_2[6-2k]$;

(v) $x_1[2k]$;   (x) $x_1[2k]x_2[-k]$.

1.29 In most parts of the human body, a small electrical current is often produced by movement of different ions. For example, in cardiac cells the electric current is produced by the movement of *sodium* (Na^+) and *potassium* (K^+) ions (during different phases of the heart beat, these ions enter or leave cells). The electric potential created by these ions is known as an ECG signal, and is used by doctors to analyze heart conditions. A typical ECG pattern is shown in Fig. P1.29.

Assume a hypothetical case in which the ECG signal corresponding to a normal human is available from birth to death (assume a longevity of 80 years). Classify such a signal with respect to the six criteria mentioned in Section 1.1. Justify your answer for each criterion.

1.30 It was explained in Section 1.2 that a complicated function could be represented as a sum of elementary functions. Consider the function $f(t)$ in Fig. P1.26. Represent $f(t)$ in terms of the *unit step* function $u(t)$ and the *ramp* function $r(t)$.

1.31 (MATLAB exercise) Write a set of MATLAB functions that compute and plot the following CT signals. In each case, use a sampling interval of 0.001 s.

(i) $x(t) = \exp(-2t)\sin(10\pi t)$ for $|t| \leq 1$.

(ii) A periodic signal $x(t)$ with fundamental period $T = 5$. The value over one period is given by

$$x(t) = 5t \quad 0 \leq t < 5.$$

Use the `sawtooth` function available in MATLAB to plot five periods of $x(t)$ over the range $-10 \leq t < 15$.

(iii) The unit step function $u(t)$ over $[-10, 10]$ using the `sign` function available in MATLAB.

(iv) The rectangular pulse function rect(t)

$$\text{rect}\left(\frac{t}{10}\right) = \begin{cases} 1 & -5 < t < 5 \\ 0 & \text{elsewhere} \end{cases}$$

using the unit step function implemented in (iii).

(v) A periodic signal $x(t)$ with fundamental period $T = 6$. The value over one period is given by

$$x(t) = \begin{cases} 3 & |t| \leq 1 \\ 0 & 1 < |t| \leq 3. \end{cases}$$

Use the `square` function available in MATLAB.

1.32 (MATLAB exercise) Write a MATLAB function `mydecimate` with the following format:

```
function [y] = mydecimate(x, M)
% MYDECIMATE: computes y[k] = x[kM]
% where
%   x is a column vector containing the DT input
%       % signal
%   M is the scaling factor greater than 1
%   y is a column vector containing the DT output time
%       % decimated by M
```

In other words, `mydecimate` accepts an input signal $x[k]$ and produces the signal $y[k] = x[kM]$.

1.33 (MATLAB exercise) Repeat Problem 1.30 for the transformation $y[k] = x[k/N]$. In other words, write a MATLAB function `myinterpolate` with the following format:

```
function [y] = myinterpolate(x, N)
% MYINTERPOLATE: computes y[k] = x[k/N]
% where
%   x is a column vector containing the DT input
%       % signal
%   N is the scaling factor greater than 1
```

```
%   y is a column vector containing the DT output
        % signal time expanded by N
```

Use linear interpolation based on the neighboring samples to predict any required unknown values in $x[k]$.

1.34 (MATLAB exercise) Construct a DT signal given by

$$x[k] = (1 - e^{-0.003k})\cos(\pi k/20) \quad \text{for } 0 \le k \le 120.$$

(i) Sketch the signal using the `stem` function.
(ii) Using the `mydecimate` (Problem P1.30) and `myinterpolate` (Problem P1.31) functions, transform the signal $x[k]$ based on the operation $y[k] = x[k/5]$ followed by the operation $z[k] = y[5k]$. What is the relationship between $x[k]$ and $z[k]$?
(iii) Repeat (ii) with the order of interpolation and decimation reversed.

In Eq. (2.7), the Boltzmann constant k equals 1.38×10^{-23} joules/kelvin, T is the absolute temperature measured in kelvin, and e is the negative charge contained in an electron. The value of e is 1.6×10^{-19} coulombs. At room temperature, 300 K, the value of the voltage equivalent V_T, computed using Eq. (2.7), is found to be 0.026 V. Substituting the values of the saturation current I_s and the voltage equivalent V_T, Eq. (2.6) simplifies to

$$i = 4.2 \times 10^{-15}[\exp(v/0.026) - 1]\,\text{A} = 0.0042[\exp(38.61v) - 1]\,\text{pA}, \quad (2.8)$$

which describes the relationship between the forward bias voltage v and the current i flowing through the semiconductor diode. Equation (2.8) is plotted in the first quadrant ($v > 0$ and $i > 0$) of Fig. 2.4(c).

In the reverse bias condition, the negative polarity of the voltage source is applied to the p region of the diode and the positive polarity is applied to the n region. When the diode is reverse biased, the current through the diode is negligibly small and is given by its saturation value, $I_s = 4.2 \times 10^{-15}$ A. The current–voltage relationship of a reverse biased diode is plotted in the third quadrant ($v < 0$ and $i < 0$) of Fig. 2.4(c), where we observe a relatively small value of current flowing through the diode.

As illustrated in Fig. 2.4(c), the input–output relationship of a semiconductor diode is highly non-linear. Compared to the linear electrical circuit discussed in Section 2.1.1, such non-linear systems are more difficult to analyze and are beyond the scope of this book.

2.1.3 Amplitude modulator

Modulation is the process used to shift the frequency content of an information-bearing signal such that the resulting modulated signal occupies a higher frequency range. Modulation is the key component in modern-day communication systems for two main reasons. One reason is that the frequency components of the human voice are limited to a range of around 4 kHz. If a human voice signal is transmitted directly by propagating electromagnetic radio waves, the communication antennas required to transmit and receive these radio signals would be impractically long. A second reason for modulation is to allow for simultaneous transmission of several voice signals within the same geographic region. If two signals within the same frequency range are transmitted together, they will interfere with each other. Modulation provides us with the means of separating the voice signals in the frequency domain by shifting each voice signal to a different frequency band. There are different techniques used to modulate a signal. Here we introduce the simplest form of modulation referred to as amplitude modulation (AM).

Consider an information-bearing signal $m(t)$ applied as an input to an AM system, referred to as an amplitude modulator. In communications, the input $m(t)$ to a modulator is called the modulating signal, while its output $s(t)$ is

Fig. 2.5. Amplitude modulation (AM) system.

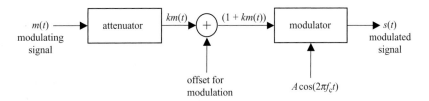

called the modulated signal. The steps involved in an amplitude modulator are illustrated in Fig. 2.5, where the modulating signal $m(t)$ is first processed by attenuating it by a factor k and adding a dc offset such that the resulting signal $(1 + km(t))$ is positive for all time t. The modulated signal is produced by multiplying the processed input signal $(1 + km(t))$ with a high-frequency carrier $c(t) = A \cos(2\pi f_c t)$. Multiplication by a sinusoidal wave of frequency f_c shifts the frequency content of the modulating signal $m(t)$ by an additive factor of f_c. Mathematically, the amplitude modulated signal $s(t)$ is expressed as follows:

$$s(t) = A[1 + km(t)] \cos(2\pi f_c t), \tag{2.9}$$

where A and f_c are, respectively, the amplitude and the fundamental frequency of the sinusoidal carrier.

It may be noted that the amplitude A and frequency f_c of the carrier signal, along with the attenuation factor k used in the modulator, are fixed; therefore, Eq. (2.9) provides a direct relationship between the input and the output signals of an amplitude modulator. For example, if we set the attenuation factor k to 0.2 and use the carrier signal $c(t) = \cos(2\pi \times 10^8 t)$, Eq. (2.9) simplifies to

$$s(t) = [1 + 0.2m(t)] \cos(2\pi \times 10^8 t). \tag{2.10}$$

Amplitude modulation is covered in more detail in Chapter 7.

2.1.4 Mechanical water pump

The mechanical pump shown in Fig. 2.6 is another example of a linear CT system. Water flows into the pump through a valve V1 controlled by an electrical circuit. A second valve V2 works mechanically as the outlet. The rate of the outlet flow depends on the height of the water in the mechanical pump. A higher level of water exerts more pressure on the mechanical valve V2, creating a wider opening in the valve, thus releasing water at a faster rate. As the level of water drops, the opening of the valve narrows, and the outlet flow of water is reduced.

A mathematical model for the mechanical pump is derived by assuming that the rate of flow F_{in} of water at the input of the pump is a function of the input

Fig. 2.6. Mechanical water
pumping system.

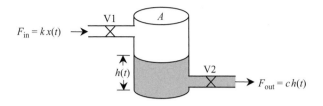

water flow $x(t)$:

$$F_{\text{in}} = kx(t), \qquad (2.11)$$

where k is the linearity constant. Valve V2 is designed such that the outlet flow
rate F_{out} is given by

$$F_{\text{out}} = ch(t), \qquad (2.12)$$

where c denotes the outlet flow constant and $h(t)$ is the height of the water
level. Denoting the total volume of the water inside the tank by $V(t)$, the rate
of change in the volume of the stored water is dV/dt, which must be equal to
the difference between the input flow rate, Eq. (2.11), and the outlet flow rate,
Eq. (2.12). The resulting equation is as follows:

$$\frac{dV}{dt} = F_{\text{in}} - F_{\text{out}} = kx(t) - ch(t). \qquad (2.13)$$

Expressing $V(t)$ as the product of the cross-sectional area A of the water tank
and the height $h(t)$ of the water yields

$$A\frac{dh}{dt} + ch(t) = kx(t), \qquad (2.14)$$

which is a first-order, constant-coefficient differential equation describing the
relationship between the water flow $x(t)$ and height $h(t)$ of water in the mechanical pump. It may be noted that the input–output relationship in the electrical
circuit, discussed in Section 2.1.1, was also a constant-coefficient differential equation. In fact, most CT linear systems are often modeled with linear,
constant-coefficient differential equations.

2.1.5 Mechanical spring damper system

The spring damping system shown in Fig. 2.7 is another classical example of a
linear mechanical system. An application of such a mechanical damping system
is in the shock absorber installed in an automobile. Figure 2.7 models a spring
damping system where mass M, which is attached to a rigid body through a
mechanical spring with a spring constant of k, is pulled downward with force
$x(t)$. Assuming that the vertical displacement from the initial location of mass

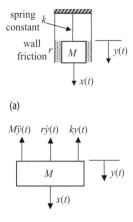

(a)

$M\ddot{y}(t)$ $r\dot{y}(t)$ $ky(t)$

(b)

Fig. 2.7. (a) Mechanical spring damper system. (b) Free-body diagram illustrating the opposing forces acting on mass M of the mechanical spring damping system.

M is given by $y(t)$, the three upward forces opposing the external downward force $x(t)$ are given by

inertial (or accelerating) force $\quad\quad F_i = M\dfrac{d^2 y}{dt^2};$ $\quad\quad$ (2.15a)

frictional (or damping) force $\quad\quad F_f = r\dfrac{dy}{dt};$ $\quad\quad$ (2.15b)

spring (or restoring) force $\quad\quad F_s = ky(t),$ $\quad\quad$ (2.15c)

where r is the damping constant for the medium surrounding the mass. Applying Newton's third law of motion, the input–output relationship of the spring damping system is given by

$$M\frac{d^2 y}{dt^2} + r\frac{dy}{dt} + ky(t) = x(t),\quad\quad (2.16)$$

which is a linear, constant-coefficient second-order differential equation. Equation (2.16) describes the relationship between the applied force $x(t)$ and the resulting vertical displacement $y(t)$. As in the case of the RLC circuit, a second-order differential equation is used to model the mechanical spring damper system.

2.1.6 Numerical differentiation and integration

Numerical methods are widely used in calculus for finding approximate values of derivatives and definite integrals. Here, we present examples of differentiation and integration of a CT function $x(t)$. The system representing integrator and differentiator are shown in Fig. 2.8. We show that the numerical approximations of a CT differentiator and integrator lead to finite difference equations that are frequently used to describe DT systems.

To discretize a derivative over a continuous interval $[0, T]$, the time interval T is divided into intervals of duration Δt, resulting in the sampled values $x(k\Delta t)$ for $k = 0, 1, 2, \ldots, K$, with K given by the ratio $T/\Delta t$. Using a single-step backward finite-difference scheme, the time derivative can be approximated as follows:

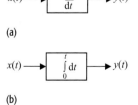

(a)

(b)

Fig. 2.8. Schematics of (a) a differentiator and (b) an integrator. Finite-difference schemes are often used to compute the values of derivatives and finite integrals numerically.

$$\left.\frac{dx}{dt}\right|_{t=k\Delta t} \approx \frac{x(k\Delta t) - x((k-1)\Delta t)}{\Delta t},\quad\quad (2.17)$$

which yields

$$y(k\Delta t) = \frac{x(k\Delta t) - x((k-1)\Delta t)}{\Delta t}\quad\quad (2.18)$$

or,

$$y(k\Delta t) = C_1(x(k\Delta t) - x((k-1)\Delta t)),\quad\quad (2.19)$$

where $x(k\Delta t)$ is the sampled value of $x(t)$ at $t = k\Delta t$ and C_1 is a constant, equal to $1/\Delta t$. The CT signal $y(t) = dx/dt$ and represents the result of differentiation.

Usually, the sampling interval Δt in Eq. (2.19) is omitted, resulting in the following expression:

$$y[k] = C_1 x[k] - C_1 x[k-1], \qquad (2.20)$$

which is a finite-difference representation of the differentiator shown in Fig. 2.8(a).

To integrate a function, we use Euler's formula, which approximates the integral by the following:

$$\int_{(k-1)\Delta t}^{k\Delta t} x(t)\mathrm{d}t \approx \Delta t x((k-1)\Delta t). \qquad (2.21)$$

In other words, the area under $x(t)$ within the range $[(k-1)\Delta t, k\Delta t]$ is approximated by a rectangle with width Δt and height $x((k-1)\Delta t)$. Expressing the integral as follows:

$$y(t)|_{t=k\Delta t} = \int_0^t x(t)\mathrm{d}t = \underbrace{\int_0^{(k-1)\Delta t} x(t)\mathrm{d}t}_{y((k-1)\Delta t)} + \underbrace{\int_{(k-1)\Delta t}^{k\Delta t} x(t)\mathrm{d}t}_{\Delta t x((k-1)\Delta t)} \qquad (2.22)$$

and simplifying, we obtain

$$y(k\Delta t) = y((k-1)\Delta t) + \Delta t x((k-1)\Delta t). \qquad (2.23)$$

Again, omitting the sampling interval Δt in Eq. (2.23) yields

$$y[k] - y[k-1] = C_2 x[k-1], \qquad (2.24)$$

where $C_2 = \Delta t$. Equation (2.24) is a first-order finite-difference equation modeling an integrator and can be solved iteratively to compute the integral at discrete time instants $k\Delta t$. Systems represented by finite-difference equations of the form of Eqs. (2.20) or (2.24) are referred to as DT systems and are the focus of our discussion in the second half of the book. In the case of DT systems, a difference equation, along with the ancillary conditions, provides a complete description of the DT systems.

2.1.7 Delta modulation

In a digital communication system, the information-bearing analog signal is first transformed into a binary sequence of zeros and ones, referred to as a digital signal, which is then transferred using a digital communication technique from a transmitter to a receiver. Compared to analog transmission, digital communications operate with a lower signal-to-noise ratio (SNR) and can therefore provide almost error-free performance over long distances. In addition, digital

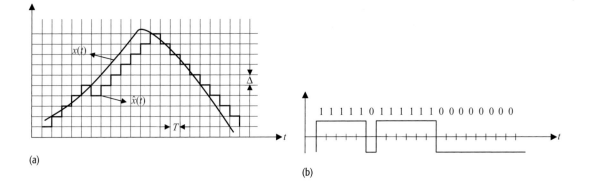

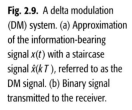

Fig. 2.9. A delta modulation (DM) system. (a) Approximation of the information-bearing signal $x(t)$ with a staircase signal $\hat{x}(kT)$, referred to as the DM signal. (b) Binary signal transmitted to the receiver.

communications allow for other data processing features such as error correction, data encryption, and jamming resistance, which can be exploited for secure data transmission. In this section, we study a basic waveform coding procedure, referred to as delta modulation (DM), which is widely used to transform an analog signal into a digital signal.

The process of DM is illustrated in Fig. 2.9, where an information-bearing analog signal $x(t)$ is approximated by a delta modulated signal $\hat{x}(t)$. The analog signal $x(t)$ is uniformly sampled at time instants $t = kT$. At each sampling instant, the sampled value $x(kT)$ of the analog signal is compared with the amplitude of the DM signal $\hat{x}(kT)$. If the magnitude of the sampled signal $x(kT)$ is greater than the corresponding magnitude of the DM signal $\hat{x}(kT)$, then the DM signal is increased by a fixed amplitude, say Δ, at $t = kT$. Bit 1 is transmitted to the receiver to indicate the increase in the amplitude of the DM signal. On the other hand, if the amplitude of the sampled signal $x(kT)$ is less than the magnitude of the DM signal $\hat{x}(kT)$, then the DM signal is decreased by Δ. Bit 0 is transmitted to the receiver to indicate the decrease in the amplitude of the DM signal. In other words, a single bit is used at each time instant $t = kT$ to indicate an increase or decrease in the amplitude of the information-bearing signal.

A major advantage of DM is the simple structure of the receiver. At the receiving end, the signal $\hat{x}(t)$ is reconstructed using the following simple relationship:

$$\hat{x}(kT) = \hat{x}((k-1)T) + b_k\Delta, \tag{2.25}$$

where $b_k = 1$ if bit 1 is received and $b_k = -1$ if bit 0 is received. Solving for $\hat{x}(kT)$, Eq. (2.25) is represented as follows:

$$\hat{x}(kT) = \sum_{k=0}^{n} b_k\Delta + \hat{x}(0), \tag{2.26}$$

where $\hat{x}(0)$ represents the initial value used at $t = 0$ in the DM signal. Equation (2.26) implies that the DM signal $\hat{x}(kT)$ is obtained by accumulating the values of the $b_k \Delta$'s. Such a DT system that accumulates the values of the input is referred to as an accumulator. It may be noted that the receiver of a DM system is a linear system as it can be modeled by a constant-coefficient difference equation.

2.1.8 Digital filter

Digital images are made up of tiny "dots" obtained by sampling a two-dimensional (2D) analog image. Each dot is referred to as a picture element, or a pixel. A digital image, therefore, can be modeled with a 2D array, $x[m, n]$, where the index (m, n) refers to the spatial coordinate of a pixel with m being the number of the row and n being the number of the column. In a monochrome image, the value $x[m, n]$ of a pixel indicates its intensity value. When the pixels are placed close to each other and illuminated according to their intensity values on the computer monitor, a continuous image is perceived by the human eye.

In digital image processing, spatial averaging is frequently used for smoothing noise, lowpass filtering, and subsampling of images. In spatial averaging, the intensity of each pixel is replaced by a weighted average of the intensities of the pixels in the neighborhood of the reference pixel. Using a unidirectional fourth-order neighborhood, the reference pixel $x[m, n]$ is replaced by the spatially averaged value:

$$y[m, n] = \frac{1}{4}(x[m, n] + x[m, n-1] + x[m-1, n] + x[m-1, n-1]),$$
(2.27)

where $y[m, n]$ represents the 2D output image of the spatial averaging system. Equation (2.27) is an example of a 2D finite-difference equation and it models a 2D DT system with input $x[m, n]$ and output $y[m, n]$.

In this section, we have considered some interesting applications of signal processing in CT and DT systems. Our goal has been to motivate the reader to learn about the techniques and basic concepts required to investigate one or more of these application areas. Each of the discussed areas is a subject of considerable study. Nevertheless, certain fundamentals are central to most applications, and many of these basic concepts will be discussed in the chapters that follow.

2.2 Classification of systems

In the analysis or design of a system, it is desirable to classify the system according to some generic properties that the system satisfies. In this segment we introduce a set of basic properties that may be used to categorize a system. For a system to possess a given property, the property must hold true for all possible input signals that can be applied to the system. If a property holds for some input signals but not for others, the system does not satisfy that property.

In this section, we classify systems into six basic categories:

(i) linear and non-linear systems;
(ii) time-invariant and time-varying systems;
(iii) systems with and without memory;
(iv) causal and non-causal systems;
(v) invertible and non-invertible systems;
(vi) stable and unstable systems.

In the following discussion, we make use of the notation given in Eqs. (2.1) and (2.2), which we repeat here:

CT system $\qquad\qquad\qquad\qquad x(t) \rightarrow y(t)$;

DT system $\qquad\qquad\qquad\qquad x[k] \rightarrow y[k]$;

to refer to output $y(t)$ resulting from input $x(t)$ for a CT system and to output $y[k]$ resulting from input $x[k]$ for a DT system.

2.2.1 Linear and non-linear systems

A CT system with the following sets of inputs and outputs:

$$x_1(t) \rightarrow y_1(t) \quad \text{and} \quad x_2(t) \rightarrow y_2(t)$$

is linear iff it satisfies the additive and the homogeneity properties described below:

additive property $\qquad x_1(t) + x_2(t) \rightarrow y_1(t) + y_2(t)$; (2.28)

homogeneity property $\qquad \alpha x_1(t) \rightarrow \alpha y_1(t)$; (2.29)

for any arbitrary value of α and all possible combinations of inputs and outputs. The additive and homogeneity properties are collectively referred to as the principle of superposition. Therefore, linear systems satisfy the principle of superposition. Based on the principle of superposition, the properties in Eqs. (2.28) and (2.29) can be combined into a single statement as follows. A CT system with the following sets of inputs and outputs:

$$x_1(t) \rightarrow y_1(t) \quad \text{and} \quad x_2(t) \rightarrow y_2(t)$$

is linear iff

$$\alpha x_1(t) + \beta x_2(t) \rightarrow \alpha y_1(t) + \beta y_2(t)$$ (2.30)

for any arbitrary set of values for α and β, and for all possible combinations of inputs and outputs.

Likewise, a DT system with

$$x_1[k] \rightarrow y_1[k] \quad \text{and} \quad x_2[k] \rightarrow y_2[k],$$

is linear iff

$$\alpha x_1[k] + \beta x_2[k] \rightarrow \alpha y_1[k] + \beta y_2[k] \qquad (2.31)$$

for any arbitrary set of values for α and β, and for all possible combinations of inputs and outputs.

A consequence of the linearity property is the special case when the input x to a linear CT or DT system is zero. Substituting $\alpha = 0$ in Eq. (2.29) yields

$$0 \cdot x_1(t) = 0 \rightarrow 0 \cdot y_1(t) = 0. \qquad (2.32)$$

In other words, if the input $x(t)$ to a linear system is zero, then the output $y(t)$ must also be zero for all time t. This property is referred to as the *zero-input, zero-output property*. Both CT and DT systems that are linear satisfy the zero-input, zero-output property for all time t. Note that Eq. (2.32) is a necessary condition and is not sufficient to prove linearity. Many non-linear systems satisfy this property as well.

Example 2.1
Consider the CT systems with the following input–output relationships:

(a) differentiator $\qquad\qquad\qquad y(t) = \dfrac{dx(t)}{dt}$; $\qquad\qquad\qquad$ (2.33)

(b) exponential amplifier $\qquad\qquad y(t) = e^{x(t)}$; $\qquad\qquad\qquad$ (2.34)

(c) amplifier $\qquad\qquad\qquad\qquad y(t) = 3x(t)$; $\qquad\qquad\qquad$ (2.35)

(d) amplifier with additive bias $\qquad y(t) = 3x(t) + 5$. $\qquad\qquad$ (2.36)

Determine whether the CT systems are linear.

Solution
(a) From Eq. (2.33), it follows that

$$x_1(t) \rightarrow \frac{dx_1(t)}{dt} = y_1(t)$$

and

$$x_2(t) \rightarrow \frac{dx_2(t)}{dt} = y_2(t),$$

which yields

$$\alpha x_1(t) + \beta_1 x_2(t) \rightarrow \frac{d}{dt}\{\alpha x_1(t) + \beta_1 x_2(t)\} = \alpha\frac{dx_1(t)}{dt} + \beta\frac{dx_2(t)}{dt}.$$

Since

$$\alpha\frac{dx_1(t)}{dt} + \beta\frac{dx_2(t)}{dt} = \alpha y_1(t) + \beta y_2(t),$$

the differentiator as represented by Eq. (2.33) is a linear system.

(b) From Eq. (2.34), it follows that

$$x_1(t) \rightarrow e^{x_1(t)} = y_1(t)$$

and

$$x_2(t) \rightarrow e^{x_2(t)} = y_2(t),$$

giving

$$\alpha x_1(t) + \beta x_2(t) \rightarrow e^{\alpha x_1(t) + \beta x_2(t)}.$$

Since

$$e^{\alpha x_1(t) + \beta x_2(t)} = e^{\alpha x_1(t)} \cdot e^{\beta x_2(t)} = [y_1(t)]^\alpha + [y_2(t)]^\beta \neq \alpha y_1(t) + \beta y_2(t),$$

the exponential amplifier represented by Eq. (2.34) is not a linear system.

(c) From (2.35), it follows that

$$x_1(t) \rightarrow 3x_1(t) = y_1(t)$$

and

$$x_2(t) \rightarrow 3x_2(t) = y_2(t),$$

giving

$$\alpha x_1(t) + \beta x_2(t) \rightarrow 3\{\alpha x_1(t) + \beta x_2(t)\} = 3\alpha x_1(t) + 3\beta x_2(t)$$
$$= \alpha y_1(t) + \beta y_2(t).$$

Therefore, the amplifier of Eq. (2.35) is a linear system.

(d) From Eq. (2.36), we can write

$$x_1(t) \rightarrow 3x_1(t) + 5 = y_1(t)$$

and

$$x_2(t) \rightarrow 3x_2(t) + 5 = y_2(t),$$

giving

$$\alpha x_1(t) + \beta x_2(t) \rightarrow 3[\alpha x_1(t) + \beta x_2(t)] + 5.$$

Since

$$3[\alpha x_1(t) + \beta x_2(t)] + 5 \neq \alpha y_1(t) + \beta y_2(t)$$

the amplifier with an additive bias as specified in Eq. (2.36) is not a linear system.

An alternative approach to check if a system is non-linear is to apply the zero-input, zero-output property. For system (b), if $x(t) = 0$, then $y(t) = 1$. System (b) does not satisfy the zero-input, zero-output property, hence system (b) is non-linear. Likewise, for system (d), if $x(t) = 0$ then $y(t) = 5$. Therefore, system (d) is not a linear system.

If a system does not satisfy the zero-input, zero-output property, we can safely classify the system as a non-linear system. On the other hand, if it satisfies the zero-input, zero-output property, it can be linear or non-linear. Satisfying the zero-input, zero-output property is not a sufficient condition to prove the

Fig. 2.10. Incrementally linear
system expressed as a linear
system with an additive offset.

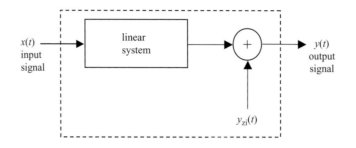

Fig. 2.10. Incrementally linear system expressed as a linear system with an additive offset.

linearity of a system. A CT system $y(t) = x^2(t)$ is clearly a non-linear system, yet it satisfies the zero-input, zero-output property. For the system to be linear, it must satisfy Eq. (2.30).

Incrementally linear system   In Example 2.1, we proved that the amplifier $y(t) = 3x(t)$ represents a linear system, while the amplifier with additive bias $y(t) = 3x(t) + 5$ represents a non-linear system. System $y(t) = 3x(t) + 5$ satisfies a different type of linearity. For two different inputs $x_1(t)$ and $x_2(t)$, the respective outputs of system $y(t) = 3x(t) + 5$ are given by

| input $x_1(t)$ | $y_1(t) = 3x_1(t) + 5$; |
| input $x_2(t)$ | $y_2(t) = 3x_2(t) + 5$. |

Calculating the difference on both sides of the above equations yields

$$y_2(t) - y_1(t) = 3[x_2(t) - x_1(t)]$$

or

$$\Delta y(t) = 3\Delta x(t).$$

In other words, the change in the output of system $y(t) = 3x(t) + 5$ is linearly related to the change in the input. Such systems are called incrementally linear systems.

An incrementally linear system can be expressed as a combination of a linear system and an adder that adds an offset $y_{zi}(t)$ to the output of the linear system. The value of offset $y_{zi}(t)$ is the zero-input response of the original system. System S_1, $y(t) = 3x(t) + 5$, for example, can be expressed as a combination of a linear system S_2, $y(t) = 3x(t)$, plus an offset given by the zero-input response of S_1, which equals $y_{zi}(t) = 5$. Figure 2.10 illustrates the block diagram representation of an incrementally linear system in terms of a linear system and an additive offset $y_{zi}(t)$.

Example 2.2
Consider two DT systems with the following input–output relationships:

(a) differencing system $y[k] = 3(x[k] - x[k-2])$;   (2.37)

(b) sinusoidal system $y[k] = \sin(x[k])$.   (2.38)

Determine if the DT systems are linear.

Solution

(a) From Eq. (2.37), it follows that:

$$x_1[k] \rightarrow 3x_1[k] - 3x_1[k-2] = y_1[k]$$

and

$$x_2[k] \rightarrow 3x_2[k] - 3x_2[k-2] = y_2[k],$$

giving

$$\alpha x_1[k] + \beta x_2[k] \rightarrow 3\alpha x_1[k] - 3\alpha x_1[k-2] + 3\beta x_2[k] - 3\beta x_2[k-2].$$

Since

$$3\alpha x_1[k] - 3\alpha x_1[k-2] + 3\beta x_2[k] - 3\beta x_2[k-2] = \alpha y_1[k] + \beta y_2[k],$$

the differencing system, Eq. (2.37), is linear.

To illustrate the linearity property graphically, we consider two DT input signals $x_1[k]$ and $x_2[k]$ shown in the two top-left subplots in Figs. 2.11(a) and (c). The resulting outputs $y_1[k]$ and $y_2[k]$ for the two inputs applied to the differencing system, Eq. (2.37), are shown in the two top-right stem subplots in Figs. 2.11(b) and (d), respectively. A linear combination, $x_3[k] = x_1[k] + 2x_2[k]$, of the two inputs is shown in the bottom-left subplot in Fig. 2.11(e). The resulting output $y_3[k]$ of the system for input signal $x_3[k]$ is shown in the bottom-right subplot in Fig. 2.11(f). By looking at the subplots, it is clear that the output $y_3[k] = y_1[k] + 2y_2[k]$. In other words, the output $y_3[k]$ can be determined by using the same linear combination of outputs $y_1[k]$ and $y_2[k]$ as the linear combination used to obtain $x_3[k]$ from $x_1[k]$ and $x_2[k]$.

(b) From Eq. (2.38), it follows that:

$$x_1[k] \rightarrow \sin(x_1[k]) = y_1[k], \quad x_2[k] \rightarrow \sin(x_2[k]) = y_2[k],$$

giving

$$\alpha x_1[k] + \beta x_2[k] \rightarrow \sin(\alpha x_1[k] + \beta x_2[k]) \neq \alpha y_1[k] + \beta y_2[k];$$

therefore, the sinusoidal system in Eq. (2.38) is not linear.

To illustrate graphically that system (b) indeed does not satisfy the linearity property, we consider two input signals $x_1[k]$ and $x_2[k]$ shown, respectively, in Figs. 2.12(a) and (c). Their corresponding outputs, $y_1[k]$ and $y_2[k]$, are shown in Figs. 2.12(b) and (d). The output $y_3[k]$ of the system for the input signal $x_3[k] = x_1[k] + 2x_2[k]$, obtained by combining $x_1[k]$ and $x_2[k]$, is shown in Fig. 2.12(f). Comparing Fig. 2.12(f) with Figs. 2.12(b) and (d), we note that output $y_3[k] \neq y_1[k] + 2y_2[k]$. To check, we select $k = 4$. From Fig. 2.12, inputs $x_1[4] = 0$ and $x_2[4] = 2$. Using Eq. (2.38), outputs $y_1[4] = \sin(0) = 0$ and $y_2[4] = \sin(2) = 0.91$. The linear combination $y_1[k] + 2y_2[k]$ of $y_1[k]$ and $y_2[k]$ at $k = 4$ gives a value of 1.82. If the system is linear, we should get $y_3[4] = 1.82$ from the combined input $x_3[k] = x_1[k] + 2x_2[k] = 4$ at $k = 4$. Substituting in Eq. (2.38), we obtain $y_3[4] = \sin(4) = -0.76$. Since the value of output $y_3[k]$ at $k = 4$ obtained from the linear combination of individual

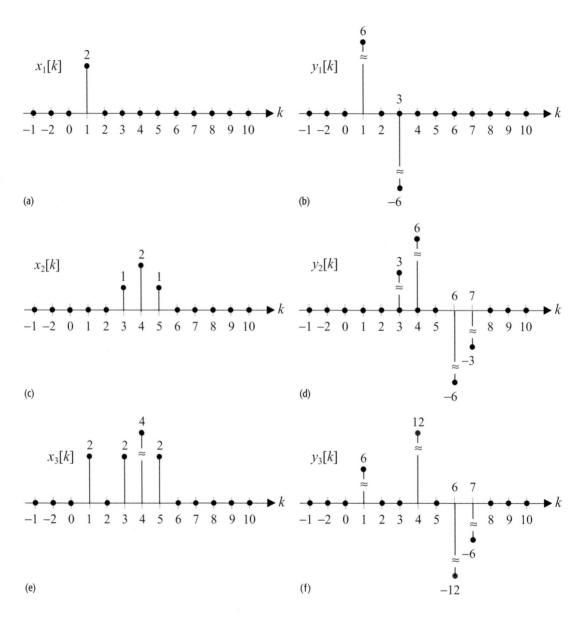

Fig. 2.11. Input–output pairs of the linear DT system specified in Example 2.2(a). Parts (a)–(f) are discussed in the text.

outputs $y_1[k]$ and $y_2[k]$ is different from the value obtained directly by applying the combined input, we may say that the system in Fig. 2.12(b) is not linear. The graphical result is in accordance with the mathematical proof.

Example 2.3

Consider the AM system with input–output relationship given by

$$s(t) = [1 + 0.2m(t)]\cos(2\pi \times 10^8 t). \tag{2.39}$$

Determine if the AM system is linear.

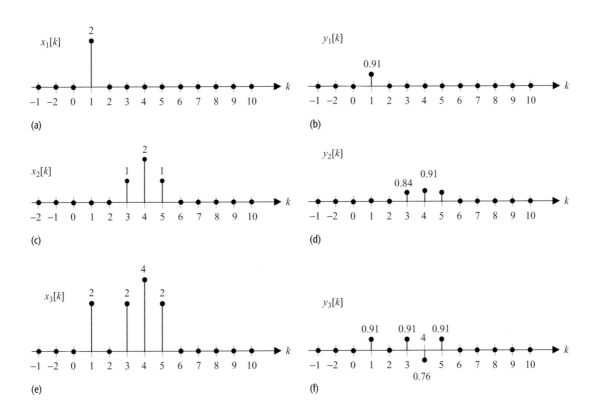

Fig. 2.12. Input–output pairs of the linear DT system specified in Example 2.2(b). Parts (a)–(f) are discussed in the text.

Solution

From Eq. (2.39), it follows that:

$$m_1(t) \rightarrow [1 + 0.2m_1(t)]\cos(2\pi \times 10^8 t) = s_1(t)$$

and

$$m_2(t) \rightarrow [1 + 0.2m_2(t)]\cos(2\pi \times 10^8 t) = s_2(t),$$

giving

$$\alpha m_1(t) + \beta m_2(t) \rightarrow [1 + 0.2\{\alpha m_1(t) + \beta m_2(t)\}]\cos(2\pi \times 10^8 t)$$
$$\neq \alpha s_1(t) + \beta s_2(t).$$

Therefore, the AM system is not linear.

2.2.2 Time-varying and time-invariant systems

A system is said to be time-invariant (TI) if a time delay or time advance of the input signal leads to an identical time-shift in the output signal. In other words, except for a time-shift in the output, a TI system responds exactly the same way no matter when the input signal is applied. We now define a TI system formally.

A CT system with $x(t) \to y(t)$ is time-invariant iff

$$x(t - t_0) \to y(t - t_0) \tag{2.40}$$

for any arbitrary time-shift t_0. Likewise, a DT system with $x[k] \to y[k]$ is time-invariant iff

$$x[k - k_0] \to y[k - k_0] \tag{2.41}$$

for any arbitrary discrete time shift k_0.

Example 2.4

Consider two CT systems represented mathematically by the following input–output relationship:

(i) system I $\qquad\qquad\qquad y(t) = \sin(x(t));$ $\qquad\qquad\qquad$ (2.42)

(ii) system II $\qquad\qquad\qquad y(t) = t \sin(x(t)).$ $\qquad\qquad\qquad$ (2.43)

Determine if systems (i) and (ii) are time-invariant.

Solution

(i) From Eq. (2.42), it follows that:

$$x(t) \to \sin(x(t)) = y(t)$$

and

$$x(t - t_0) \to \sin(x(t - t_0)) = y(t - t_0).$$

Since $\sin[x(t - t_0)] = y(t - t_0)$, system I is time-invariant. We demonstrate the time-invariance property of system I graphically in Fig. 2.13, where a time-shifted version $x(t - 1)$ of input $x(t)$ produces an equal shift of one time unit in the original output $y(t)$ obtained from $x(t)$.

(ii) From Eq. (2.43), it follows that:

$$x(t) \to t \sin(x(t)) = y(t).$$

If the time-shifted signal $x(t - t_0)$ is applied at the input of Eq. (2.43), the new output is given by

$$x(t - t_0) \to t \sin(x(t - t_0)).$$

The shifted output $y(t - t_0)$ is given by

$$y(t - t_0) = (t - t_0) \sin(x(t - t_0)).$$

Since $t \sin[x(t - t_0)] \neq y(t - t_0)$, system II is not time-invariant. The time-invariance property of system II is demonstrated in Fig. 2.14, where we observe that a right shift of one time unit in input $x(t)$ alters the shape of the output $y(t)$.

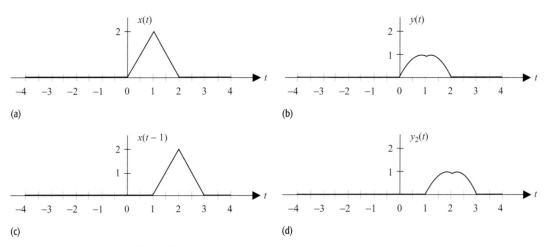

Fig. 2.13. Input–output pairs of the CT time-invariant system specified in Example 2.4(i). (a) Arbitrary signal $x(t)$. (b) Output of system for input signal $x(t)$. (c) Signal $x(t-1)$. (d) Output of system for input signal $x(t-1)$. Note that except for a time-shift, the two output signals are identical.

Example 2.5

Consider two DT systems with the following input–output relationships:

(i) system I $y[k] = 3(x[k] - x[k-2])$;   (2.44)

(ii) system II $y[k] = k\,x[k]$.   (2.45)

Determine if the systems are time-invariant.

Solution

(i) From Eq. (2.44), it follows that:

$$x[k] \rightarrow 3(x[k] - x[k-2]) = y[k]$$

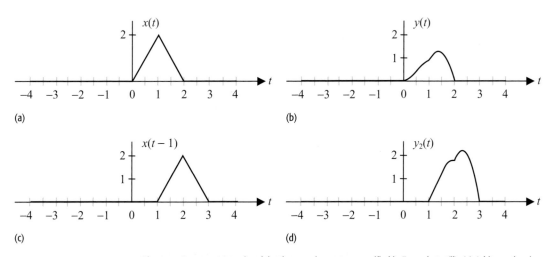

Fig. 2.14. Input–output pairs of the time-varying system specified in Example 2.4(ii). (a) Arbitrary signal $x(t)$. (b) Output of system for input signal $x(t)$. (c) Signal $x(t-1)$. (d) Output of system for input signal $x(t-1)$. Note that the output for time-shifted input $x(t-1)$ is different from the output $y(t)$ for the original input $x(t)$.

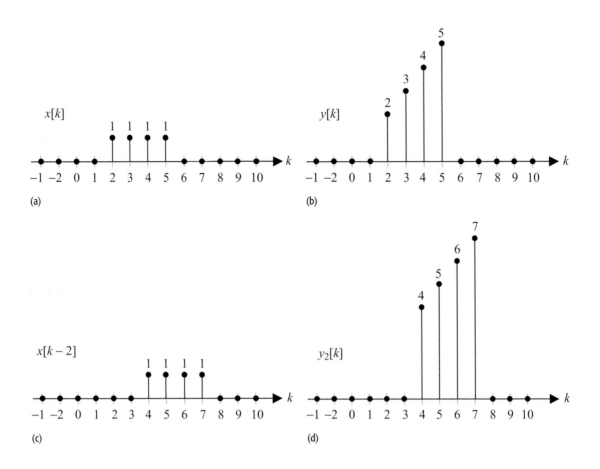

(a)

(b)

(c)

(d)

Fig. 2.15. Input–output pairs of the DT time-varying system specified in Example 2.5(ii). The output $y_2[k]$ for the time-shifted input $x_2[k] = x[k - 2]$ is different in shape from the output $y[k]$ obtained for input $x[k]$. Therefore the system is time-variant. Parts (a)–(d) are discussed in the text.

and

$$x[k - k_0] \rightarrow 3(x[k - k_0] - x[k - k_0 - 2]) = y[k - k_0].$$

Therefore, the system in Eq. (2.44) is a time-invariant system.

(ii) From Eq. (2.45), it follows that:

$$x[k] \rightarrow kx[k] = y[k]$$

and

$$x[k - k_0] \rightarrow kx[k - k_0] \neq y[k - k_0] = (k - k_0)x[k - k_0].$$

Therefore, system II is not time-invariant. In Fig. 2.15, we plot the outputs of the DT system in Eq. (2.45) for input $x[k]$, shown in Fig. 2.15(a) and a shifted version $x[k - 2]$ of the input, shown in Fig. 2.15(c). The resulting outputs are plotted, respectively, in Figs. 2.15(b) and (d). As expected, the Fig. 2.15(d) is not a delayed version of Fig. 2.15(b) since the system is time-variant.

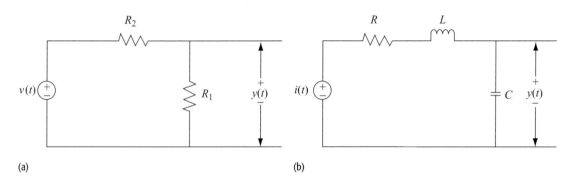

(a)   (b)

Fig. 2.16. (a) Passive electrical circuit comprising resistors R_1 and R_2. (b) Active electrical circuit comprising resistor R, inductor L, and capacitor C. Both inductor L and capacitor C are storage components, and hence lead to a system with memory.

2.2.3 Systems with and without memory

A CT system is said to be *without memory* (*memoryless* or *instantaneous*) if its output $y(t)$ at time $t = t_0$ depends only on the values of the applied input $x(t)$ at the same time $t = t_0$. On the other hand, if the response of a system at $t = t_0$ depends on the values of the input $x(t)$ in the past or in the future of time $t = t_0$, it is called a *dynamic* system, or a system *with memory*. Likewise, a DT system is said to be memoryless if its output $y[k]$ at instant $k = k_0$ depends only on the value of its input $x[k]$ at the same instant $k = k_0$. Otherwise, the DT system is said to have memory.

Example 2.6
Determine if the two electrical circuits shown in Figs. 2.16(a) and (b) are memoryless.

Solution
The relationship between the input voltage $v(t)$ and the output voltage $y(t)$ across resistor R_1 in the electrical circuit of Fig. 2.16(a) is given by

$$y(t) = \frac{R_1}{R_1 + R_2} v(t). \tag{2.46}$$

For time $t = t_0$, the output $y(t_0)$ depends only on the value $v(t_0)$ of the input $v(t)$ at $t = t_0$. The electrical circuit shown in Fig. 2.16(a) is, therefore, a memoryless system.

The relationship between the input current $i(t)$ and the output voltage $y(t)$ in Fig. 2.16(b) is given by

$$y(t) = \frac{1}{C} \int_{-\infty}^{t} i(\tau) d\tau. \tag{2.47}$$

Table 2.1. Examples of CT and DT systems with and without memory

Continuous-time		Discrete-time	
Memoryless systems	Systems with memory	Memoryless systems	Systems with memory
$y(t) = 3x(t) + 5$	$y(t) = x(t - 5)$	$y[k] = 3x[k] + 7$	$y[k] = x[k - 5]$
$y(t) = \sin\{x(t)\} + 5$	$y(t) = x(t + 2)$	$y[k] = \sin(x[k]) + 3$	$y[k] = x[k + 3]$
$y(t) = e^{x(t)}$	$y(t) = x(2t)$	$y[k] = e^{x[k]}$	$y[k] = x[2k]$
$y(t) = x^2(t)$	$y(t) = x(t/2)$	$y[k] = x^2[k]$	$y[k] = x[k/2]$

To compute the output voltage $y(t_0)$ at time t_0, we require the value of the current source for the time range $(-\infty, t_0]$, which includes the entire past. Therefore, the electrical circuit in Fig. 2.16(b) is not a memoryless system.

In Table 2.1, we consider several examples of memoryless and dynamic systems. The reader is encouraged to verify mathematically the classifications made in Table 2.1.

As a side note to our discussion on memoryless systems, we consider another class of systems with memory that require only a limited set of values of input $x(t)$ in $t_0 - T \leq t \leq t_0$ to compute the value of output $y(t)$. Such CT systems, whose response $y(t)$ is completely determined from the values of input $x(t)$ over the most recent past T time units, are referred to as *finite-memory* or *Markov systems* with memory of length T time units. Likewise, a DT system is called a finite-memory or a Markov system with memory of length M if output $y[k]$ at $k = k_0$ depends only on the values of input $x[k]$ for $k_0 - M \leq k \leq k_0$ in the most recent past.

2.2.4 Causal and non-causal systems

A CT system is *causal* if the output at time t_0 depends only on the input $x(t)$ for $t \leq t_0$. Likewise, a DT system is *causal* if the output at time instant k_0 depends only on the input $x[k]$ for $k \leq k_0$. A system that violates the causality condition is called a *non-causal* (or *anticipative*) system. Note that all memoryless systems are causal systems because the output at any time instant depends only on the input at that time instant. Systems with memory can either be causal or non-causal.

Example 2.7

  (i) CT time-delay system $y(t) = x(t - 2) \Rightarrow$ causal system;

 (ii) CT time-forward system $y(t) = x(t + 2) \Rightarrow$ non-causal system;

(iii) DT time-delay system $y[k] = x[k - 2] \Rightarrow$ causal system;

 (iv) DT time-advance system $y[k] = x[k + 2] \Rightarrow$ non-causal system;

  (v) DT linear system $y[k] = x[k - 2] + x[k + 10] \Rightarrow$ non-causal system.

Table 2.2. Examples of causal and non-causal systems

The CT and DT systems are represented using their input–output relationships. Note that all systems in the table have memory.

CT systems		DT systems	
Causal	Non-causal	Causal	Non-causal
$y(t) = x(t-5)$	$y(t) = x(t+2)$	$y[k] = 3x[k-1]+7$	$y[k] = x[k+3]$
$y(t) = \sin\{x(t-4)\}+3$	$y(t) = \sin\{x(t+4)\}+3$	$y[k] = \sin(x[k-4])+3$	$y[k] = \sin(x[k+4])+3$
$y(t) = e^{x(t-2)}$	$y(t) = x(2t)$	$y[k] = e^{x[k-2]}$	$y[k] = x[2k]$
$y(t) = x^2(t-2)$	$y(t) = x(t/2)$	$y[k] = x^2[k-5]$	$y[k] = x[k/2]$
$y(t) = x(t-2)+x(t-5)$	$y(t) = x(t-2)+x(t+2)$	$y[k] = x[k-2]+x[k-8]$	$y[k] = x[k+2]+x[k-8]$

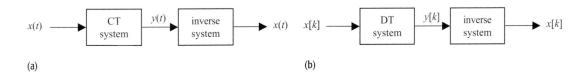

(a)

(b)

Fig. 2.17. Invertible systems. (a) Inverse of a CT system. (b) Inverse of a DT system.

Causality is a required condition for the system to be physically realizable. A non-causal system is a predictive system and cannot be implemented physically. Table 2.2 presents examples of causal and non-causal systems in CT and DT domains.

2.2.5 Invertible and non-invertible systems

A CT system is *invertible* if the input signal $x(t)$ can be uniquely determined from the output $y(t)$ produced in response to $x(t)$ for all time $t \in (-\infty, \infty)$. Similarly, a DT system is called *invertible* if, given an arbitrary output response $y[k]$ of the system for $k \in (-\infty, \infty)$, the corresponding input signal $x[k]$ can be uniquely determined for all time $k \in (-\infty, \infty)$. To be invertible, two different inputs cannot produce the same output since, in such cases, the input signal cannot be uniquely determined from the output signal.

A direct consequence of the invertibility property is the determination of a second system that restores the original input. A system is said to be invertible if the input to the system can be recovered by applying the output of the original system as input to a second system. The second system is called the inverse of the original system. The relationship between the original system and its inverse is shown in Fig. 2.17.

Example 2.8

Determine if the following CT systems are invertible.

   (i) Incrementally linear system:

$$y(t) = 3x(t) + 5.$$

(iii) Increasing ramped output:

$$y[k] = k \, x[k].$$

The input–output relationship is expressed as follows:

$$x[k] = \frac{1}{k} y[k].$$

The input signal can be uniquely determined for all time instant k, except at $k = 0$. Therefore, the system is not invertible.

(iv) Summer:

$$y[k] = x[k] + x[k-1].$$

Following the procedure used in Example 2.8(iv), the input signal is expressed as an infinite sum of the output $y[k]$ as follows:

$$x[k] = y[k] - y[k-1] + y[k-2] - y[k-3] + - \cdots$$

$$= \sum_{m=0}^{\infty} (-1)^m y[k-m].$$

The input signal $x[k]$ can be reconstructed if $y[m]$ is known for all $m \le k$. Therefore, the system is invertible.

(v) Accumulator:

$$y[k] = \sum_{m=-\infty}^{k} x[m].$$

We express the accumulator as follows:

$$y[k] = x[k] + \sum_{m=-\infty}^{k-1} x[m] = x[k] + y[k-1]$$

or

$$x[k] = y[k] - y[k-1].$$

Therefore, the system is invertible.

2.2.6 Stable and unstable systems

Before defining the stability criteria for a system, we define the bounded property for a signal. A CT signal $x(t)$ or a DT signal $x[k]$ is said to be *bounded in magnitude* if

CT signal	$\|x(t)\| \le B_x < \infty \quad$ for $t \in (-\infty, \infty)$;	(2.48)
DT signal	$\|x[k]\| \le B_x < \infty \quad$ for $k \in (-\infty, \infty)$,	(2.49)

where B_x is a finite number. Next, we define the stability criteria for CT and DT systems.

A system is referred to as bounded-input, bounded-output (BIBO) stable if an arbitrary bounded-input signal always produces a bounded-output signal. In other words, if an input signal $x(t)$ for CT systems, or $x[k]$ for DT systems, satisfying either Eq. (2.48) or Eq. (2.49), is applied to a stable CT or DT system, it is always possible to find an finite number $B_y < \infty$ such that

CT system $\qquad |y(t)| \le B_y < \infty \quad$ for $t \in (-\infty, \infty)$; $\qquad$ (2.50)

DT system $\qquad |y[k]| \le B_y < \infty \quad$ for $k \in (-\infty, \infty)$. $\qquad$ (2.51)

Example 2.10

Determine if the following CT systems are stable.

　(i) Incrementally linear system:

$$y(t) = 50x(t) + 10. \qquad (2.52)$$

Assume $|x(t)| \le B_x$ for all t. Based on Eq. (2.52), it follows that:

$$y(t) \le 50B_x + 10 = B_y \quad \text{for all } t.$$

As the magnitude of $y(t)$ does not exceed $50B_x + 10$, which is a finite number, the incrementally linear system given in Eq. (2.52) is a stable system.

　(ii) Integrator:

$$y(t) = \int_{-\infty}^{t} x(\tau)d\tau. \qquad (2.53)$$

This system integrates the input signal from $t = -\infty$ to t. Assume that a unit-step function $x(t) = u(t)$ is applied at the input of the integrator. The output of the system is given by

$$y(t) = tu(t) = \begin{cases} 0 & t < 0 \\ t & t \ge 0. \end{cases}$$

Signal $y(t)$ is plotted in Fig. 2.18(b). It is observed that $y(t)$ increases steadily for $t > 0$ and that there is no upper bound of $y(t)$. Hence, the integrator is not a BIBO stable system.

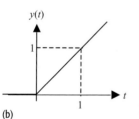

(a)

(b)

Fig. 2.18. Input and output of the unstable system in Example 2.10(ii). (a) Input $x(t)$ to the system. (b) Output $y(t)$ of the system. The input $x(t)$ is bounded for all t, but the output $y(t)$ is unbounded as $t \to \infty$.

Example 2.11

Determine if the following DT systems are stable.

　(i) $\qquad\qquad\qquad\qquad y[k] = 50 \sin(x[k]) + 10. \qquad (2.54)$

Note that $\sin(x[k])$ is bounded between $[-1, 1]$ for any arbitrary choice of $x[k]$. The output $y[k]$ is therefore bounded within the interval $[-40, 60]$. Therefore, system (i) is stable.

　(ii) $\qquad\qquad\qquad\qquad y[k] = e^{x[k]}. \qquad (2.55)$

Assume $|x[k]| \le B_x$ for all t. Based on Eq. (2.52), it follows that:

$$y[k] \le e^{B_x} = B_y \quad \text{for all } k.$$

Therefore, system (ii) is stable.

(iii)
$$y[k] = \sum_{m=-2}^{2} x[k - m]. \qquad (2.56)$$

The output is expressed as follows:

$$y[k] = x[k - 2] + x[k - 1] + x[k] + x[k + 1] + x[k + 2].$$

If $|x[k]| \leq B_x$ for all k, then $|y[k]| \leq 5B_x$ for all k. Therefore, the system is stable.

(iv)
$$y[k] = \sum_{m=-\infty}^{k} x[m]. \qquad (2.57)$$

The output is calculated by summing an infinite number of input signal values. Hence, there is no guarantee that the output will be bounded even if all the input values are bounded. System (iv) is, therefore, not a stable system.

2.3 Interconnection of systems

In signal processing, complex structures are formed by interconnecting simple linear and time-invariant systems. In this section, we describe three widely used configurations for developing complex systems.

2.3.1 Cascaded configuration

As shown in Fig. 2.19(a), a series or cascaded configuration between two systems is formed by interconnecting the output of the first system S_1 to the input of the second system S_2. If the interconnected systems S_1 and S_2 are linear, it is straightforward to show that the overall cascaded system is also linear. Likewise, if the two systems S_1 and S_2 are time-invariant, then the overall cascaded system is also time-invariant. Another feature of the cascaded configuration is that the order of the two systems S_1 and S_2 may be interchanged without changing the output response of the overall system.

Example 2.12
Determine the relationship between the overall output and input signals if the two cascaded systems in Fig. 2.19(a) are specified by the following relationships:

(i) $S_1 : \dfrac{dw}{dt} + 2w(t) = x(t)$ with $w(0) = 0$

and

$S_2 : \dfrac{dy}{dt} + 3y(t) = w(t)$ with $y(0) = 0;$

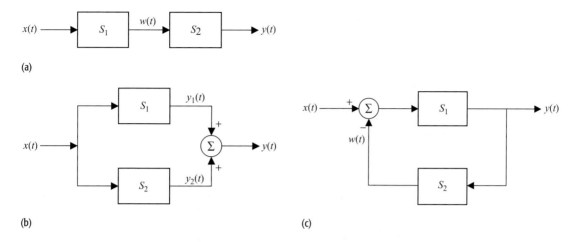

(a)

(b)   (c)

Fig. 2.19. Interconnection of systems: (a) cascaded configuration; (b) parallel configuration; (c) feedback configuration. Although these diagrams are for CT systems, the DT systems can be interconnected to form the three configurations in exactly the same manner.

(ii) $S_1 : w[k] - w[k-1] = x[k]$ with $w[0] = 0$

and

$S_2 : y[k] - 2y[k-1] = w[k]$ with $y[0] = 0$.

Solution

(i) Differentiating both sides of the differential equation modeling system S_2 with respect to t yields

$$S_2 : \frac{d^2 y}{dt^2} + 3\frac{dy}{dt} = \frac{dw}{dt}.$$

Multiplying the differential equation modeling system S_2 by 2 and adding the result to the above equation yields

$$\frac{d^2 y}{dt^2} + 5\frac{dy}{dt} + 6y(t) = \underbrace{\frac{dw}{dt} + 2w(t)}_{x(t)}.$$

Based on the differential equation modeling system S_1, the right-hand side of the equation equals $x(t)$. The overall relationship of the cascaded system is, therefore, given by

$$\frac{d^2 y}{dt^2} + 5\frac{dy}{dt} + 6y(t) = x(t).$$

(ii) Substituting $k = p - 1$ in the difference equation modeling system S_2 yields

$$S_2 : y[p-1] - 2y[p-2] = w[p-1],$$

or, in terms of time index k,

$$S_2 : y[k-1] - 2y[k-2] = w[k-1].$$

Subtracting the above equation from the original difference equation modeling system S_2 yields

$$S_2 : y[k] - 3y[k-1] + 2y[k-2] = \underbrace{w[k] - w[k-1]}_{x[k]}.$$

Based on the difference equation modeling system S_1, the right-hand side of the above equation equals $x[k]$. The overall relationship of the cascaded system is, therefore, given by

$$y[k] - 3y[k-1] + 2y[k-2] = x[k].$$

2.3.2 Parallel configuration

The parallel configuration is shown in Fig. 2.19(b), where a single input is applied simultaneously to two systems S_1 and S_2. The overall output response is obtained by adding the outputs of the individual systems. In other words, if

$$S_1 : x(t) \rightarrow y_1(t) \text{ and } S_2 : x(t) \rightarrow y_2(t), \text{ then } S_{\text{parallel}} : x(t) \rightarrow y_1(t) + y_2(t).$$

As for the series configuration, the system formed by a parallel combination of two linear systems is also linear. Similarly, if the two systems S_1 and S_2 are time-invariant, then the overall parallel system is also time-invariant.

Example 2.13

Determine the relationship between the overall output and input signals if the two parallel systems in Fig. 2.19(b) are specified by the following relationships:

(i) $S_1 : y_1(t) = x(t) + \dfrac{dx}{dt}$ and $S_2 : y_2(t) = x(t) + 3\dfrac{dx}{dt} + 5\dfrac{d^2x}{dt^2}$;

(ii) $S_1 : y_1[k] = x[k] - x[k-1]$ and $S_2 : y_2[k] = x[k] - 2x[k-1] - x[k-2]$.

Solution

(i) The response of the overall system is obtained by adding the two differential equations modeling the individual systems. The resulting expression is given by

$$y_1(t) + y_2(t) = 2x(t) + 4\frac{dx}{dt} + 5\frac{d^2x}{dt^2}.$$

Since $y(t) = y_1(t) + y_2(t)$, the response of the overall system is given by

$$y(t) = 2x(t) + 4\frac{dx}{dt} + 5\frac{d^2x}{dt^2}.$$

(ii) The response of the overall system is obtained by adding the two difference equations modeling the individual systems. The resulting expression is given by

$$y_1[k] + y_2[k] = 2x[k] - 3x[k-1] - x[k-2].$$

Since $y[k] = y_1[k] + y_2[k]$, the response of the overall system is given by

$$y[k] = 2x[k] - 3x[k-1] - x[k-2].$$

2.3.3 Feedback configuration

The feedback configuration is shown in Fig. 2.19(c), where the output of system S_1 is fed back, processed by system S_2, and then subtracted from the input signal. Such systems are difficult to analyze in the time domain and will be considered in Chapter 6 after the introduction of the Laplace transform.

2.4 Summary

In this chapter we presented an overview of CT and DT systems, classifying the systems into several categories. A CT system is defined as a transformation that operates on a CT input signal to produce a CT output signal. In contrast, a DT system transforms a DT input signal into a DT output signal. In Section 2.1, we presented several examples of systems used to abstract everyday physical processes. Section 2.2 classified the systems into different categories: linear versus non-linear systems; time-invariant versus variant systems; memoryless versus dynamic systems; causal versus non-causal systems; invertible versus non-invertible systems; and stable versus unstable systems. We classified the systems based on the following definitions.

(1) A system is linear if it satisfies the principle of superposition.
(2) A system is time-invariant if a time-shift in the input signal leads to an identical shift in the output signal without affecting the shape of the output.
(3) A system is memoryless if its output at $t = t_0$ depends only on the value of input at $t = t_0$ and no other value of the input signal.
(4) A system is causal if its output at $t = t_0$ depends on the values of the input signal in the past, $t \leq t_0$, and does not require any future value ($t > t_0$) of the input signal.
(5) A system is invertible if its input can be completely determined by observing its output.
(6) A system is BIBO stable if all bounded inputs lead to bounded outputs.

An important subset of systems is described by those that are both linear and time-invariant (LTI). By invoking the linearity and time-invariance properties, such systems can be analyzed mathematically with relative ease compared with non-linear systems. In Chapters 3–8, we will focus on linear time-invariant CT (LTIC) systems and study the time-domain and frequency-domain techniques used to analyze such systems. DT systems and the techniques used to analyze them will be presented in Part III, i.e. Chapters 9–17.

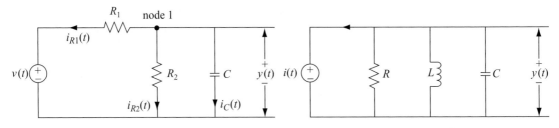

Fig. P2.1. RC circuit consisting of two resistors (R_1 and R_2) and a capacitor C.

Fig. P2.2. Resonator in an AM modulator.

Fig. P2.3. AM demodulator. The input signal is represented by $v_1(t) = A_c \cos(2\pi f_c t) + m(t)$, where $A_c \cos(2\pi f_c t)$ is the carrier and $m(t)$ is the modulating signal.

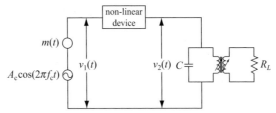

Problems

2.1 The electrical circuit shown in Fig. P2.1 consists of two resistors R_1 and R_2 and a capacitor C.

   (i) Determine the differential equation relating the input voltage $v(t)$ to the output voltage $y(t)$.

   (ii) Determine whether the system is (a) linear, (b) time-invariant; (c) memoryless; (d) causal, (e) invertible, and (f) stable.

2.2 The resonant circuit shown in Fig. P2.2 is generally used as a resonator in an amplitude modulation (AM) system.

   (i) Determine the relationship between the input $i(t)$ and the output $y(t)$ of the AM modulator.

   (ii) Determine whether the system is (a) linear, (b) time-invariant; (c) memoryless; (d) causal, (e) invertible, and (f) stable.

2.3 Figure P2.3 shows the schematic of a square-law demodulator used in the demodulation of an AM signal. Demodulation is the process of extracting the information-bearing signal from the modulated signal. The input–output relationship of the non-linear device is approximated by (assuming $v_1(t)$ is small)

$$v_2(t) = c_1 v_1(t) + c_2 v_1^2(t),$$

where c_1 and c_2 are constants, and $v_1(t)$ and $v_2(t)$ are, respectively, the input and output signals.

   (i) Show that the demodulator is a non-linear device.

   (ii) Determine whether the non-linear device is (a) time-invariant, (b) memoryless, (c) invertible, and (d) stable.

2.4 The amplitude modulation (AM) system covered in Section 2.1.3 is widely used in communications as in the AM band on radio tuner sets. Assume that the sinusoidal tone $m(t) = 2\sin(2\pi \times 100t)$ is modulated by the carrier $c(t) = 5\cos(2\pi \times 10^6 t)$.

  (i) Determine the value of the modulation index k that will ensure $(1 + km(t)) \geq 0$ for all t.

  (ii) Derive the expression for the AM signal $s(t)$ and express it in the form of Eq. (2.10).

  (iii) Using the following trigonometric relationship:

$$2\sin\theta_1\cos\theta_2 = \sin(\theta_1 + \theta_2) + \sin(\theta_1 - \theta_2),$$

  show that the frequency of the sinusoidal tone is shifted to a higher frequency range in the frequency domain.

2.5 Equation (2.16) describes a linear, second-order, constant-coefficient differential equation used to model a mechanical spring damper system.

  (i) By expressing Eq. (2.16) in the following form:

$$\frac{d^2 y}{dt^2} + \frac{\omega_n}{Q}\frac{dy}{dt} + \omega_n^2 y(t) = \frac{1}{M}x(t),$$

  determine the values of ω_n and Q in terms of mass M, damping factor r, and the spring constant k.

  (ii) The variable ω_n denotes the natural frequency of the spring damper system. Show that the natural frequency ω_n can be increased by increasing the value of the spring constant k or by decreasing the mass M.

  (iii) Determine whether the system is (a) linear, (b) time-invariant, (c) memoryless, (d) causal, (e) invertible, and (f) stable.

2.6 The solution to the following linear, second-order, constant-coefficient differential equation:

$$\frac{d^2 y}{dt^2} + 5\frac{dy}{dt} + 6y(t) = x(t) = 0,$$

with input signal $x(t) = 0$ and initial conditions $y(0) = 3$ and $\dot{y}(0) = -7$, is given by

$$y(t) = [e^{-3t} + 2e^{-2t}]u(t).$$

  (i) By using the backward finite-difference scheme

$$\left.\frac{dy}{dt}\right|_{t=k\Delta t} \approx \frac{y(k\Delta t) - y((k-1)\Delta t)}{\Delta t}$$

  and

$$\left.\frac{d^2 y}{dt^2}\right|_{t=k\Delta t} \approx \frac{y(k\Delta t) - 2y((k-1)\Delta t) + y((k-2)\Delta t)}{(\Delta t)^2}$$

show that the finite-difference representation of the differential equation is given by

$$(1 + 5\Delta t + 6(\Delta t)^2)y[k] + (-2 - 5\Delta t)y[k-1] + y[k-2] = 0.$$

(ii) Show that the ancillary conditions for the finite-difference scheme are given by

$$y[0] = 3 \quad \text{and} \quad y[-1] = 3 + 7\Delta t.$$

(iii) By iteratively computing the finite-difference scheme for $\Delta t = 0.02$ s, show that the computed result from the finite-difference equation is the same as the result of the differential equation.

2.7 Assume that the delta modulation scheme, presented in Section 2.1.7, uses the following design parameters:

sampling period $T = 0.1$ s   and   quantile interval $\Delta = 0.1$ V.

Sketch the output of the receiver for the following binary signal:

$$111110111111100000000.$$

Assume that the initial value $x(0)$ of the transmitted signal $x(t)$ at $t = 0$ is $x(0) = 0$ V.

2.8 Determine if the digital filter specified in Eq. (2.27) is an invertible system. If yes, derive the difference equation modeling the inverse system. If no, explain why.

2.9 The following CT systems are described using their input–output relationships between input $x(t)$ and output $y(t)$. Determine if the CT systems are (a) linear, (b) time-invariant, (c) stable, and (d) causal. For the non-linear systems, determine if they are incrementally linear systems.

(i) $y(t) = x(t-2)$;

(ii) $y(t) = x(2t - 5)$;

(iii) $y(t) = x(2t) - 5$;

(iv) $y(t) = tx(t+10)$;

(v) $y(t) = \begin{cases} 2 & x(t) \geq 0 \\ 0 & x(t) < 0; \end{cases}$

(vi) $y(t) = \begin{cases} 0 & t < 0 \\ x(t) - x(t-5) & t \geq 0; \end{cases}$

(vii) $y(t) = 7x^2(t) + 5x(t) + 3$;

(viii) $y(t) = \text{sgn}(x(t))$;

(ix) $y(t) = \displaystyle\int_{-t_0}^{t_0} x(\lambda)d\lambda + 2x(t)$;

(x) $y(t) = \displaystyle\int_{-\infty}^{t_0} x(\lambda)d\lambda + \frac{dx}{dt}$;

(xi) $\dfrac{d^4 y}{dt^4} + 3\dfrac{d^3 y}{dt^3} + 5\dfrac{d^2 y}{dt^2} + 3\dfrac{dy}{dt} + y(t) = \dfrac{d^2 x}{dt^2} + 2x(t) + 1.$

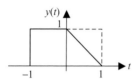

Fig. P2.11. CT output $y(t)$ for Problem 2.11.

2.10 The following DT systems are described using their input–output relationships between input $x[k]$ and output $y[k]$. Determine if the DT systems are (a) linear, (b) time-invariant, (c) stable, and (d) causal. For the non-linear systems, determine if they are incrementally linear systems.

(i) $y[k] = ax[k] + b$;

(ii) $y[k] = 5x[3k - 2]$;

(iii) $y[k] = 2^{x[k]}$;

(iv) $y[k] = \displaystyle\sum_{m=-\infty}^{k} x[m]$;

(v) $y[k] = \displaystyle\sum_{m=k-2}^{k+2} x[m] - 2|x[k]|$;

(vi) $y[k] + 5y[k-1] + 9y[k-2] + 5y[k-3] + y[k-4]$
$= 2x[k] + 4x[k-1] + 2x[k-2]$.

(vii) $y[k] = 0.5x[6k - 2] + 0.5x[6k + 2]$.

2.11 For an LTIC system, an input $x(t)$ produces an output $y(t)$ as shown in Fig. P2.11. Sketch the outputs for the following set of inputs:

(i) $5x(t)$;

(ii) $0.5x(t-1) + 0.5x(t+1)$;

(iii) $x(t+1) - x(t-1)$;

(iv) $\dfrac{dx(t)}{dt} + 3x(t)$.

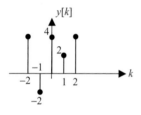

Fig. P2.12. DT output $y[k]$ for Problem 2.12.

2.12 For a DT linear, time-invariant system, an input $x[k]$ produces an output $y[k]$ as shown in Fig. P2.12. Sketch the outputs for the following set of inputs:

(i) $4x[k-1]$;

(ii) $0.5x[k-2] + 0.5x[k+2]$;

(iii) $x[k+1] - 2x[k] + x[k-1]$;

(iv) $x[-k]$.

2.13 Determine if the following CT systems are invertible. If yes, find the inverse systems.

(i) $y(t) = 3x(t+2)$;

(ii) $y(t) = \displaystyle\int_{-\infty}^{t} x(\tau - 10)d\tau$;

(iii) $y(t) = |x(t)|$;

(iv) $\dfrac{dy(t)}{dt} + y(t) = x(t)$;

(v) $y(t) = \cos(2\pi x(t))$.

2.14 Determine if the following DT systems are invertible. If yes, find the inverse systems.

(i) $y[k] = (k+1)x[k+2]$;

(ii) $y[k] = \displaystyle\sum_{m=0}^{|k|} x[m+2]$;

(iii) $y[k] = x[k] \displaystyle\sum_{m=-\infty}^{\infty} \delta[k - 2m]$;

(iv) $y[k] = x[k+2] + 2x[k+1] - 6x[k] + 2x[k-1] + x[k-2]$;

(v) $y[k] + 2y[k-1] + y[k-2] = x[k]$.

2.15 For an LTIC system, if $x(t) \rightarrow y(t)$, show that $\dfrac{dx(t)}{dt} \rightarrow \dfrac{dy(t)}{dt}$. Assume that both $x(t)$ and $y(t)$ are differentiable functions.

Fig. P2.16. (a) Input–output pair for an LTI CT system. (b) Periodic input to the LTI system.

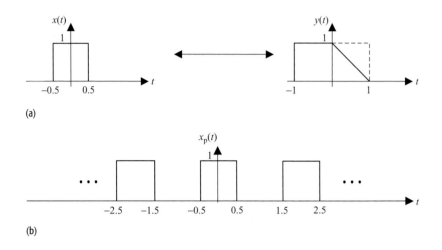

(a)

(b)

2.16 Figure P2.16(a) shows an input–output pair of an LTI CT system. Calculate the output $y_p(t)$ of the system for the periodic signal $x_p(t)$ shown in Fig. P2.16(b).

2.17 The output $h(t)$ of a CT LTI system in response to a unit impulse function $\delta(t)$ is referred to as the impulse response of the system. Calculate the impulse response of the CT LTI systems defined by the following input–output relationships:

(i) $y(t) = x(t + 2) - 2x(t) + 2x(t - 2)$;

(ii) $y(t) = \displaystyle\int_{t-t_0}^{t+t_0} x(\tau - 4)\,d\tau$;

(iii) $y(t) = \displaystyle\int_{-\infty}^{t} e^{-2(t-\tau)} x(\tau - 4)\,d\tau$;

(iv) $y(t) = \displaystyle\int_{-\infty}^{\infty} f(T - \tau) x(t - \tau)\,d\tau$ where $f(t)$ is a known signal and T is a constant.

2.18 The output $h[k]$ of a DT LTI system in response to a unit impulse function $\delta[k]$ is shown in Fig. P2.18. Find the output for the following set of inputs:

(i) $x[k] = \delta[k + 1] + \delta[k] + \delta[k - 1]$;

(ii) $x[k] = \displaystyle\sum_{m=-\infty}^{\infty} \delta[k - 4m]$;

(iii) $x[k] = u[k]$.

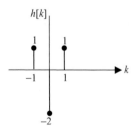

Fig. P2.18. Output $h[k]$ for input $x[k] = \delta[k]$ in Problem 2.18.

2.19 A DT LTI system is described by the following difference equation:

$$y[k] = x[k] - 2x[k - 1] + x[k - 2].$$

Determine the output $y[k]$ of the system if the input $x[k]$ is given by

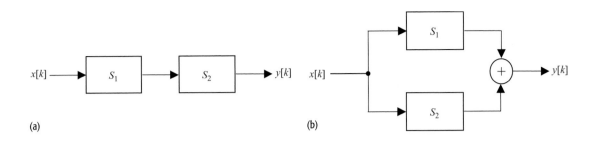

(a)   (b)

Fig. P2.21. (a) Series
configuration; (b) parallel
configuration.

(i) $x[k] = \delta[k]$;

(ii) $x[k] = \delta[k-1] + \delta[k+1]$;

(iii) $x[k] = \begin{cases} |k| & |k| \leq 3 \\ 0 & \text{elsewhere.} \end{cases}$

2.20 A five-point running average DT system is defined by the following input–output relationship:

$$y[k] = \frac{1}{5} \sum_{m=0}^{4} x[k-m].$$

(i) Show that the five-point running average DT system is an LTI system.

(ii) Calculate the impulse response $h[k]$ of the system when input $x[k] = \delta[k]$.

(iii) Compute the output $y[k]$ of the system for $-10 \leq k \leq 10$ if the input $x[k] = u[k]$, where $u[k]$ is a unit step function.

(iv) Based on your answer to (iii), calculate the impulse response $h[k]$ of the system using the property $\delta[k] = u[k] - u[k-1]$. Compare your answer to $h[k]$ obtained in (ii).

2.21 The series and parallel configurations of systems S_1 and S_2 are shown in Fig. P2.21. The two systems are specified by the following input–output relationships:

$$S_1 : y[k] = x[k] - 2x[k-1] + x[k-2];$$
$$S_2 : y[k] = x[k] + x[k-1] - 2x[k-2].$$

(i) Show that S_1 and S_2 are LTI systems.

(ii) Calculate the input–output relationship for the series configuration of systems S_1 and S_2 as shown in Fig. P2.21(a).

(iii) Calculate the input–output relationship for the parallel configuration of systems S_1 and S_2 as shown in Fig. P2.21(b).

(iv) Show that the series and parallel configurations of systems S_1 and S_2 are LTI systems.

Continuous-time signals and systems

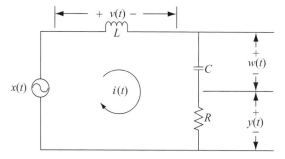

where coefficients a_k, for $0 \leq k \leq (n-1)$, and b_k, for $0 \leq k \leq m$, are parameters characterized by the linear system. If the linear system is also time-invariant, then the a_k and b_k coefficients are constants. We will use the compact notation $\dot{y}$ to denote the first derivative of $y(t)$ with respect to t. Thus $\dot{y} = \mathrm{d}y/\mathrm{d}t$, $\ddot{y} = \mathrm{d}^2y/\mathrm{d}t^2$, and so on for the higher derivatives. We now consider an electrical circuit that is modeled by a differential equation.

Example 3.1

Determine the input–output representations of the series RLC circuit shown in Fig. 3.1 for the three outputs $v(t)$, $w(t)$, and $y(t)$.

Solution

Figure 3.1 illustrates an electrical circuit consisting of three passive components: resistor R, inductor L, and capacitor C. Applying Kirchhoff's voltage law, the relationship between the input voltage $x(t)$ and the loop current $i(t)$ is given by

$$x(t) = L\frac{\mathrm{d}i}{\mathrm{d}t} + Ri(t) + \frac{1}{C}\int_{-\infty}^{t} i(t)\mathrm{d}t. \tag{3.2}$$

Differentiating Eq. (3.2) with respect to t yields

$$L\frac{\mathrm{d}^2i}{\mathrm{d}t^2} + R\frac{\mathrm{d}i}{\mathrm{d}t} + \frac{1}{C}i(t) = \frac{\mathrm{d}x}{\mathrm{d}t}. \tag{3.3}$$

We consider three different outputs of the RLC circuit in the following discussion, and for each output we derive the differential equation modeling the input–output relationship of the LTIC system.

Relationship between $x(t)$ and $v(t)$ The output voltage $v(t)$ is measured across inductor L. Expressed in terms of the loop current $i(t)$, the voltage $v(t)$ is given by

$$v(t) = L\frac{\mathrm{d}i}{\mathrm{d}t}.$$

Integrating the above equation with respect to t yields

$$i(t) = \frac{1}{L} \int v(t) dt.$$

By substituting the value of $i(t)$ into Eq. (3.3), we obtain

$$\frac{dv}{dt} + \frac{R}{L} v(t) + \frac{1}{LC} \int v(t) dt = \frac{dx}{dt}.$$

The above input–output relationship includes both differentiation and integration operations. The integral operator can be eliminated by calculating the derivative of both sides of the equation with respect to t. This results in the following equation:

$$\frac{d^2 v}{dt^2} + \frac{R}{L} \frac{dv}{dt} + \frac{1}{LC} v(t) = \frac{d^2 x}{dt^2}, \tag{3.4}$$

which models the input–output relationship between the input voltage $x(t)$ and the output voltage $v(t)$ measured across inductor L. Equation (3.4) is a linear, second-order differential equation with constant coefficients. In fact, it can be shown that *an LTIC system can always be modeled by a linear, constant-coefficient differential equation with the appropriate initial conditions.*

Relationship between $x(t)$ and $w(t)$ The output voltage $w(t)$, measured across capacitor C, is given by

$$w(t) = \frac{1}{C} \int_{-\infty}^{t} i(t) dt,$$

which is expressed as follows:

$$i(t) = C \frac{dw}{dt}.$$

Substituting the value of $i(t)$ into Eq. (3.3) yields

$$LC \frac{d^3 w}{dt^3} + RC \frac{d^2 w}{dt^2} + \frac{dw}{dt} = \frac{dx}{dt}, \tag{3.5}$$

which specifies the relationship between the input voltage $x(t)$ and the output voltage $w(t)$ measured across capacitor C. Equation (3.5) can be further simplified by integrating both sides with respect to t. The resulting equation is simplified to

$$LC \frac{d^2 w}{dt^2} + RC \frac{dw}{dt} + w(t) = x(t), \tag{3.6}$$

which is a linear, second-order, constant-coefficient differential equation.

Relationship between $x(t)$ and $y(t)$ Finally, we measure the output voltage $y(t)$ across resistor R. Using Ohm's law, the output voltage $y(t)$ is given by

$y(t) = i(t)R$. Substituting the value of $i(t) = y(t)/R$ into Eq. (3.3) yields

$$\frac{L}{R}\frac{d^2 y}{dt^2} + \frac{dy}{dt} + \frac{1}{RC}y(t) = \frac{dx}{dt}, \qquad (3.7)$$

which is a linear, second-order, constant-coefficient, differential equation modeling the relationship between the input voltage $x(t)$ and the output voltage $y(t)$ measured across resistor R.

A more compact representation for Eq. (3.1) is obtained by denoting the differentiation operator d/dt by D:

$$D^n y + a_{n-1}D^{n-1}y + \cdots + a_1 Dy + a_0 y(t)$$
$$= b_m D^m y + b_{m-1}D^{m-1}y + \cdots + b_1 Dy + b_0 x(t).$$

By treating D as a differential operator, we obtain

$$\underbrace{(D^n + a_{n-1}D^{n-1} + \cdots + a_1 D + a_0)}_{Q(D)} y(t)$$
$$= \underbrace{(b_m D^m + b_{m-1}D^{m-1} + \cdots + b_1 D + b_0)}_{P(D)} x(t), \qquad (3.8)$$

or

$$Q(D)y(t) = P(Q)x(t), \qquad (3.9)$$

where $Q(D)$ is the nth-order differential operator, $P(D)$ is the mth-order differential operator, and the a_i and b_i are constants. Equation (3.9) is used extensively to describe an LTIC system.

To compute the output of an LTIC system for a given input, we must solve the constant-coefficient differential equation, Eq. (3.9). If the reader has little or no background in differential equations, it will be helpful to read through Appendix C before continuing. Appendix C reviews the direct method for solving linear, constant-coefficient differential equations and can be used as a quick look-up of the theory of differential equations. In the material that follows, it is assumed that the reader has adequate background in solving linear, constant-coefficient differential equations.

From the theory of differential equations, we know that output $y(t)$ for Eq. (3.9) can be expressed as a sum of two components:

$$y(t) = \underbrace{y_{zi}(t)}_{\text{zero-input response}} + \underbrace{y_{zs}(t)}_{\text{zero-state response}}, \qquad (3.10)$$

where $y_{zi}(t)$ is the *zero-input response* of the system and $y_{zs}(t)$ is the *zero-state response* of the system. Note that the zero-input component $y_{zi}(t)$ is the response produced by the system because of the initial conditions (and not due to any external input), and hence $y_{zi}(t)$ is also known as the natural response of the system. For example, the initial conditions may include charges stored in a capacitor or energy stored in a mechanical spring. The zero-input response $y_{zi}(t)$ is evaluated by solving a homogeneous equation obtained by setting the input signal $x(t) = 0$ in Eq. (3.9). For Eq. (3.9), the homogeneous equation is given by

$$Q(D)y(t) = 0.$$

The zero-state response $y_{zs}(t)$ arises due to the input signal and does not depend on the initial conditions of the system. In calculating the zero-state response, the initial conditions of the system are assumed to be zero. The zero-state response is also referred to as the *forced response* of the system since the zero-state response is forced by the input signal. For most stable LTIC systems, the zero-input response decays to zero as $t \to \infty$ since the energy stored in the system decays over time and eventually becomes zero. The zero-state response, therefore, defines the steady state value of the output.

Example 3.2
Consider the RLC series circuit shown in Fig. 3.1. Assume that the inductance $L = 0$ H (i.e. the inductor does not exist in the circuit), resistance $R = 5\,\Omega$, and capacitance $C = 1/20$ F. Determine the output signal $y(t)$ when the input voltage is given by $x(t) = \sin(2t)$ and the initial voltage $y(0^-) = 2$ V across the resistor.

Solution
Substituting $L = 0$, $R = 5$, and $C = 1/20$ in Eq. (3.7) yields

$$\frac{dy}{dt} + 4y(t) = \frac{dx}{dt} = 2\cos(2t). \tag{3.11}$$

Zero-input response of the system Using the procedure outlined in Appendix C, we determine the characteristic equation for Eq. (3.11) as

$$(s + 4) = 0,$$

which has a root at $s = -4$. The zero-input response of Eq. (3.11) is given by

zero input response $\qquad\qquad y_{zi}(t) = Ae^{-4t}$,

where A is a constant. The value of A is obtained from the initial condition $y(0^-) = 2$ V. Substituting $y(0^-) = 2$ V in the above equation yields $A = 2$. The zero-input response is given by $y_{zi}(t) = 2e^{-4t}$.

Zero-state response of the system The zero-state response is calculated by solving Eq. (3.11) with a zero initial condition, $y(0^-) = 0$. The homogeneous component of the zero-state response of Eq. (3.11) is similar to the zero input response and is given by

$$y_{zs}^{(h)}(t) = Ce^{-4t},$$

where C is a constant. The particular component of the zero-state response of Eq. (3.11) for input $x(t) = \sin(2t)$ is of the following form:

$$y_{zs}^{(p)}(t) = K_1 \cos(2t) + K_2 \sin(2t).$$

Substituting the particular component in Eq. (3.11) gives $K_1 = 0.4$ and $K_2 = 0.2$. The overall zero-state response of the system is as follows:

zero state response $y_{zs}(t) = Ce^{-4t} + 0.2\sin(2t) + 0.4\cos(2t),$

with zero initial condition, i.e. $y_{zs}(t) = 0$. Substituting the initial condition in the zero-state response yields $C = -0.4$. The total response of the system is the sum of the zero-input and zero-state responses and is given by

$$y(t) = 1.6e^{-4t} + 0.2\sin(2t) + 0.4\cos(2t). \tag{3.12}$$

Theorem 3.1 states the total response of a LTIC system modeled with a first-order, constant-coefficient, linear differential equation.

Theorem 3.1 *The output of a first-order differential equation,*

$$\frac{dy}{dt} + f(t)y(t) = r(t), \tag{3.13}$$

resulting from input r(t) is given by

$$y(t) = e^{-p}\left[\int e^p r\, dt + c\right], \tag{3.14}$$

where function p is given by

$$p(t) = \int f(t)dt \tag{3.15}$$

and c is a constant.

Using Theorem 3.1 to solve Eq. (3.11), we obtain $p(t) = \int 4\, dt = 4t$. Substituting $p(t) = 4t$ into Eq. (3.14), we obtain

$$y(t) = e^{-4t}\left[\int e^{4t} 2\cos(2t)dt + c\right],$$

where the integral simplifies to (see Section A.5 of Appendix A)

$$2\int e^{4t}\cos(2t)dt = \frac{2}{2^2 + 4^2}[4e^{4t}\cos(2t) + 2e^{4t}\sin(2t)].$$

Based on Theorem 3.1, the output is therefore given by

$$y(t) = ce^{-4t} + 0.2\sin(2t) + 0.4\cos(2t).$$

The value of constant c in the above equation can be computed using the initial condition. Substituting $y(0^-) = 2$ V gives $c = 1.6$. The result is, therefore, the same as the solution in Eq. (3.12) obtained by following the formal procedure outlined in Appendix C.

Steady state value of the output The steady state value of $y(t)$ can be obtained by applying the limit $(t \to \infty)$ to $y(t)$. For the differential equation (3.11), the steady state solution is therefore obtained by applying the limit to Eq. (3.12), giving

$$y(t) = \lim_{t \to \infty} [1.6e^{-4t} + 0.2\sin(2t) + 0.4\cos(2t)] = 0.2\sin(2t) + 0.4\cos(2t),$$

or

$$y(t) = \sqrt{0.4^2 + 0.2^2}\, \sin\left(2t + \tan^{-1}\left(\frac{0.4}{0.2}\right)\right) = \sqrt{0.2}\,\sin(2t + 63.4°) \tag{3.16}$$

The steady state solution given by Eq. (3.16) can also be verified using results from the circuit theory. For sinusoidal inputs, the electrical circuit in Fig. 3.1 can be reduced to an equivalent impedance circuit by replacing capacitor C with a capacitive reactance of $1/(j\omega C)$ and inductor L with an inductive reactance of $j\omega L$, where ω is the fundamental frequency of the input sinusoidal signal $x(t) = \sin(2t)$. In our example, $\omega = 2$. Figure 3.1, therefore, becomes a voltage divider circuit with the steady state value of the output $y(t)$ given by

$$y(t) = \frac{R}{R + j\omega L + (1/j\omega C)} x(t). \tag{3.17}$$

In Example 3.2, the values of the components are set to $L = 0\,\text{H}$, $R = 5\,\Omega$, and $C = 1/20\,\text{F}$. Substituting these values into Eq. (3.17) yields

$$y(t) = \frac{5}{5 + (10/j)} x(t) = \frac{1}{1 - j2}\sin(2t) = \left|\frac{1}{1 - j2}\right| \sin(2t - \angle(1 - j2))$$

$$= \frac{1}{\sqrt{5}}\sin\left(2t + \tan^{-1}(2)\right) = \sqrt{0.2}\,\sin(2t + 63.4°),$$

which is the same solution as given in Eq. (3.16).

Example 3.3

Consider the electrical circuit shown in Fig. 3.1 with the values of inductance, resistance, and capacitance set to $L = 1/12\,\text{H}$, $R = 7/12\,\Omega$, and $C = 1\,\text{F}$. The circuit is assumed to be open before $t = 0$, i.e. no current is initially flowing through the circuit. However, the capacitor has an initial charge of 5 V. Determine

  (i) the zero-input response $w_{zi}(t)$ of the system;
 (ii) the zero-state response $w_{zs}(t)$ of the system; and
(iii) the overall output $w(t)$,

when the input signal is given by $x(t) = 2\exp(-t)u(t)$ and the output $w(t)$ is measured across capacitor C.

Solution

Substituting $L = 1/12$ H, $R = 7/12\,\Omega$, and $C = 1$ F into Eq. (3.6) and multiplying both sides of the equation by 12 yields

$$\frac{d^2w}{dt^2} + 7\frac{dw}{dt} + 12w(t) = 12x(t), \tag{3.18}$$

with initial conditions, $w(0^-) = 5$ and $\dot{w}(0^-) = 0$, and the input signal is given by $x(t) = 2e^{-t}u(t)$.

(i) Zero-input response of the system Based on Eq. (3.18), the characteristic equation of the LTIC system is given by

$$s^2 + 7s + 12 = 0,$$

which has roots at $s = -4, -3$. The zero-input response is therefore given by

$$w_{zi}(t) = (Ae^{-4t} + Be^{-3t})u(t),$$

where A and B are constants. To calculate the value of the constants, we substitute the initial conditions $w(0^-) = 5$ and $\dot{w}(0^-) = 0$ in the above equation. The resulting simultaneous equations are as follows:

$$A + B = 5,$$
$$4A + 3B = 0,$$

which have the solution $A = -15$ and $B = 20$. The zero-input response is therefore given by

$$w_{zi}(t) = (20e^{-3t} - 15e^{-4t})u(t).$$

(ii) Zero-state response of the system To calculate the zero-state response of the system, the initial conditions are assumed to be zero, i.e. the capacitor is assumed to be uncharged. Hence, the zero-state response $w_{zs}(t)$ can be calculated by solving the following differential equation:

$$\frac{d^2w}{dt^2} + 7\frac{dw}{dt} + 12w(t) = 12x(t), \tag{3.19}$$

with initial conditions, $w(0^-) = 0$ and $\dot{w}(0^-) = 0$, and input $x(t) = 2\exp(-t)u(t)$.

The homogeneous solution of Eq. (3.18) has the same form as the zero-input response and is given by

$$w_{zs}^{(h)}(t) = C_1 e^{-4t} + C_2 e^{-3t},$$

where C_1 and C_2 are constants. The particular solution for input $x(t) = 2e^{-t}u(t)$ is of the form $w_{zs}^{(p)}(t) = Ke^{-t}u(t)$. Substituting the particular solution into

Fig. 3.2. Output response of the
system considered in Example
3.3.

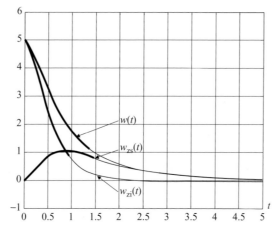

Eq. (3.19) and solving the resulting equation yields $K = 4$. The zero-state response of the system is, therefore, given by

$$w_{zs}(t) = (C_1 e^{-4t} + C_2 e^{-3t} + 4e^{-t})u(t).$$

To compute the values of constants C_1 and C_2, we use the initial conditions $w(0^-) = 0$ and $\dot{w}(0^-) = 0$. Substituting the initial conditions in $w_{zs}(t)$ leads to the following simultaneous equations:

$$C_1 + C_2 + 4 = 0,$$
$$-4C_1 - 3C_2 - 4 = 0,$$

with solutions $C_1 = 8$ and $C_2 = -12$. The zero-state solution of Eq. (3.18) is, therefore, given by

$$w_{zs}(t) = (8e^{-4t} - 12e^{-3t} + 4e^{-t})u(t).$$

(iii) Overall response of the system The overall response of the system can be obtained by summing up the zero-input and zero-state responses, and can be expressed as

$$w(t) = (-7e^{-4t} + 8e^{-3t} + 4e^{-t})u(t).$$

The zero-input, zero-state, and overall responses of the system are plotted in Fig. 3.2.

Section 3.1 presented the procedure for calculating the output response of a LTIC system by directly solving its input–output relationship expressed in the form of a differential equation. However, there is an alternative and more convenient approach to calculate the output based on the impulse response of a system. This approach is developed in the following sections.

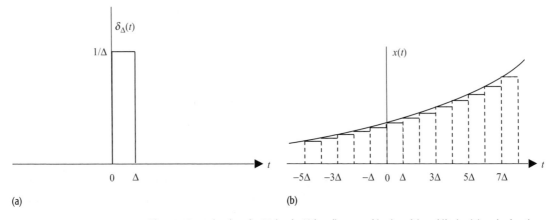

Fig. 3.3. Approximation of a CT signal $x(t)$ by a linear combination of time-shifted unit impulse functions. (a) Rectangular function $\delta_\Delta(t)$ used to approximate $x(t)$. (b) CT signal $x(t)$ and its approximation $\hat{x}(t)$ shown with the staircase function.

3.2 Representation of signals using Dirac delta functions

In this section we will show that any arbitrary signal $x(t)$ can be represented as a linear combination of time-shifted impulse functions. To illustrate our result, we define a new function $\delta_\Delta(t)$ as follows:

$$\delta_\Delta(t) = \begin{cases} 1/\Delta & 0 < t < \Delta \\ 0 & \text{otherwise.} \end{cases} \tag{3.20}$$

The waveform for $\delta_\Delta(t)$ is shown in Fig. 3.3(a); it resembles that of a rectangular pulse with width Δ and height $1/\Delta$. To approximate $x(t)$ as a linear combination of $\delta_\Delta(t)$, the time axis is divided into uniform intervals of duration Δ. Within a time interval of duration Δ, say $k\Delta < t < (k+1)\Delta$, $x(t)$ is approximated by a constant value $x(k\Delta)\delta_\Delta(t - k\Delta)\Delta$. Following the aforementioned procedure for the entire time axis, $x(t)$ can be approximated as follows:

$$\begin{aligned} \hat{x}(t) = \cdots &+ x(-k\Delta)\delta_\Delta(t + k\Delta) \cdot \Delta + \cdots + x(-\Delta)\delta_\Delta(t + \Delta) \cdot \Delta \\ &+ x(0)\delta_\Delta(t) \cdot \Delta + x(\Delta)\delta_\Delta(t - \Delta) \cdot \Delta + \cdots \\ &+ x(k\Delta)\delta_\Delta(t - k\Delta) \cdot \Delta + \cdots, \end{aligned} \tag{3.21}$$

which is shown as the staircase waveform in Fig. 3.3(b). For a given value of t, say $t = m\Delta$, only one term ($k = m$) on the right-hand side of Eq. (3.21) is non-zero. This is because only one of the shifted functions $\delta_\Delta(t - k\Delta)$ corresponding to $k = m$ is non-zero. Therefore, a more compact representation for Eq. (3.21) is obtained by using the following summation:

$$\hat{x}(t) = \sum_{k=-\infty}^{\infty} x(k\Delta)\delta_\Delta(t - k\Delta)\Delta. \tag{3.22}$$

Applying the limit $\Delta \to 0$, $\hat{x}(t)$ converges to $x(t)$, giving

$$x(t) = \lim_{\Delta \to 0} \sum_{K=-\infty}^{\infty} x(k\Delta)\delta_\Delta(t - k\Delta)[(k+1)\Delta - k\Delta], \qquad (3.23)$$

which is the same as

$$x(t) = \int_{-\infty}^{\infty} x(\tau)\delta(t - \tau)d\tau. \qquad (3.24)$$

Equation (3.24) is very important in the analysis of CT signals. It suggests that a CT function can be represented as a weighted superposition of time-shifted impulse functions. We will use Eq. (3.24) to calculate the output of an LTIC system.

The above procedure used to prove Eq. (3.24) illustrates the physical significance of the equation. A more compact proof of Eq. (3.24), based on the properties of the impulse function, is presented below.

Alternative proof for Eq. (3.24)

In the following discussion, we present a simpler proof of Eq. (3.24), which uses the properties of impulse functions. We start with the right-hand side of Eq. (3.24):

$$\text{RHS} = \int_{-\infty}^{\infty} x(\tau)\delta(t - \tau)d\tau.$$

Since $\delta(t - \tau) = \delta(\tau - t)$,

$$\text{RHS} = \int_{-\infty}^{\infty} x(\tau)\delta(\tau - t)d\tau.$$

Also, $x(\tau)\delta(\tau - t) = x(t)\delta(\tau - t)$; therefore

$$\text{RHS} = x(t) \int_{-\infty}^{\infty} \delta(\tau - t)d\tau,$$

which equals $x(t)$, as the area enclosed by the unit impulse function equals unity.

3.3 Impulse response of a system

In Section 3.1, a constant-coefficient differential equation is used to specify the input–output characteristics of an LTIC system. An alternative representation of an LTIC system can be obtained by specifying its impulse response. In this section, we will formally define the impulse response and illustrate how the

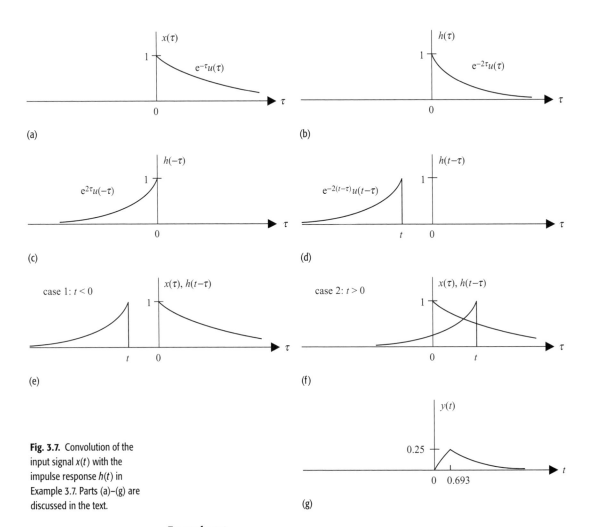

Fig. 3.7. Convolution of the input signal $x(t)$ with the impulse response $h(t)$ in Example 3.7. Parts (a)–(g) are discussed in the text.

Example 3.8

The input signal $x(t) = \exp(-t)u(t)$ is applied to an LTIC system whose impulse response is given by

$$h(t) = \begin{cases} 1 - t & 0 \leq t \leq 1 \\ 0 & \text{otherwise.} \end{cases}$$

Calculate the output of the system.

Solution

In order to calculate the output of the system, we need to calculate the convolution integral for the two functions $x(t)$ and $h(t)$. Functions $x(\tau)$, $h(\tau)$, and $h(-\tau)$ are plotted as a function of the variable τ in the top three subplots of Fig. 3.8(a)–(c). The function $h(t - \tau)$ is obtained by shifting the time-reflected function $h(-\tau)$ by t. Depending on the value of t, three special cases may arise.

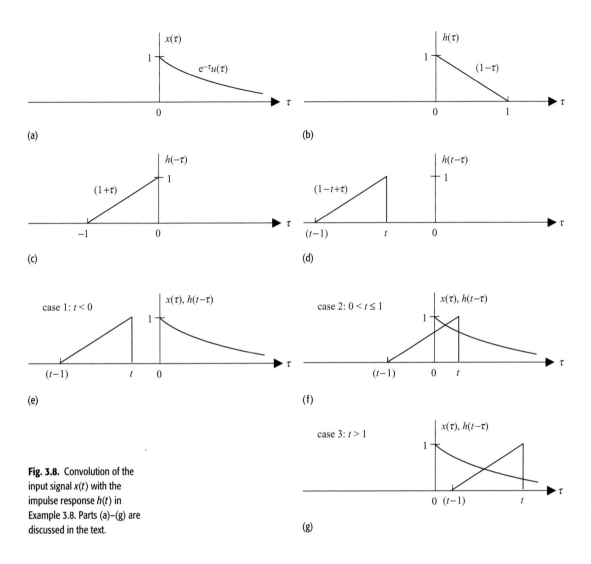

(a)

(b)

(c)

(d)

(e)

(f)

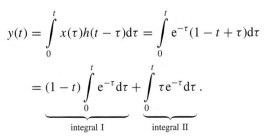

(g)

Fig. 3.8. Convolution of the input signal $x(t)$ with the impulse response $h(t)$ in Example 3.8. Parts (a)–(g) are discussed in the text.

Case 1 For $t < 0$, we see from Fig. 3.8(e) that the non-zero parts of $h(t - \tau)$ and $x(\tau)$ do not overlap. In other words, output $y(t) = 0$ for $t < 0$.

Case 2 For $0 \le t \le 1$, we see from Fig. 3.8(f) that the non-zero parts of $h(t - \tau)$ and $x(\tau)$ do overlap over the duration $\tau = [0, t]$. Therefore,

$$y(t) = \int_0^t x(\tau)h(t - \tau)\,d\tau = \int_0^t e^{-\tau}(1 - t + \tau)\,d\tau$$

$$= (1 - t)\underbrace{\int_0^t e^{-\tau}\,d\tau}_{\text{integral I}} + \underbrace{\int_0^t \tau e^{-\tau}\,d\tau}_{\text{integral II}}.$$

The two integrals simplify as follows:

$$\text{integral I} = (1-t)\left[-e^{-\tau}\right]_0^t = (1-t)(1-e^{-t});$$

$$\text{integral II} = \left[-\tau e^{-\tau} - e^{-\tau}\right]_0^t = 1 - e^{-t} - te^{-t}.$$

For $0 \le t \le 1$, the output $y(t)$ is given by

$$y(t) = (1 - t - e^{-t} + te^{-t}) + (1 - e^{-t} - te^{-t}) = (2 - t - 2e^{-t}).$$

Case 3 For $t > 1$, we see from Fig. 3.8(g) that the non-zero part of $h(t - \tau)$ completely overlaps $x(\tau)$ over the region $\tau = [t-1, t]$. The lower limit of the overlapping region in case 3 is different from the lower limit of the over-lapping region in case 2; therefore, case 3 results in a different convolution integral and is considered separately from case 2. The output $y(t)$ for case 3 is given by

$$y(t) = \int_{t-1}^{t} x(\tau)h(t-\tau)\mathrm{d}\tau = \int_{t-1}^{t} e^{-\tau}(1 - t + \tau)\mathrm{d}\tau$$

$$= (1-t)\underbrace{\int_{t-1}^{t} e^{-\tau}\mathrm{d}\tau}_{\text{integral I}} + \underbrace{\int_{t-1}^{t} \tau e^{-\tau}\mathrm{d}\tau}_{\text{integral II}}.$$

The two integrals simplify as follows:

$$\text{integral I} = (1-t)[-e^{-\tau}]_{t-1}^{t} = (1-t)(e^{-(t-1)} - e^{-t});$$

$$\text{integral II} = [-\tau e^{-\tau} - e^{-\tau}]_{t-1}^{t} = (t-1)e^{-(t-1)} + e^{-(t-1)} - te^{-t} - e^{-t}$$

$$= te^{-(t-1)} - te^{-t} - e^{-t}.$$

For $t > 1$, the output $y(t)$ is given by

$$y(t) = \left(e^{-(t-1)} - te^{-(t-1)} - e^{-t} + te^{-t}\right) + \left(te^{-(t-1)} - te^{-t} - e^{-t}\right)$$
$$= \left(e^{-(t-1)} - 2e^{-t}\right).$$

Combining the above three cases, we obtain

$$y(t) = \begin{cases} 0 & t < 0 \\ (2 - t - 2e^{-t}) & 0 \le t \le 1 \\ (e^{-(t-1)} - 2e^{-t}) & t > 1, \end{cases}$$

which is plotted in Fig. 3.9.

Example 3.9
Calculate the output for the following input signal and impulse response:

$$x(t) = \begin{cases} 1.5 & -2 \le t \le 3 \\ 0 & \text{otherwise} \end{cases} \quad \text{and} \quad h(t) = \begin{cases} 2 & -1 \le t \le 2 \\ 0 & \text{otherwise.} \end{cases}$$

Fig. 3.9. Output $y(t)$ computed in Example 3.8.

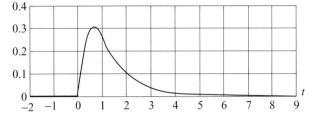

Solution

Functions $x(\tau)$, $h(\tau)$, $h(-\tau)$, and $h(t-\tau)$ are plotted in Figs. 3.10(a)–(d). Depending on the value of t, the convolution integral takes five different forms. We consider these five cases below.

Case 1 $(t < -3)$. As seen in Fig. 3.10(e), the non-zero parts of $h(t-\tau)$ and $x(\tau)$ do not overlap. Therefore, the output signal $y(t) = 0$.

Case 2 $(-3 \leq t \leq 0)$. As seen in Fig. 3.10(f), the non-zero part of $h(t-\tau)$ partially overlaps with $x(\tau)$ within the region $\tau = [-2, t+1]$. The product $x(\tau)h(t-\tau)$ becomes a rectangular function in the region with an amplitude of $1.5 \times 2 = 3$. Therefore, the output for $-3 \leq t \leq 0$ is given by

$$y(t) = \int_{-2}^{t+1} 3\,d\tau = 3(t+3).$$

Case 3 $(0 \leq t \leq 2)$. As seen in Fig. 3.10(g), the non-zero part of $h(t-\tau)$ overlaps completely with $x(\tau)$. The overlapping region is given by $\tau = [t-2, t+1]$. The product $x(\tau)h(t-\tau)$ is a rectangular function with an amplitude of 3 in the region $\tau = [t-2, t+1]$. The output for $0 \leq t \leq 2$ is given by

$$y(t) = \int_{t-2}^{t+1} 3\,d\tau = 9.$$

Case 4 $(2 \leq t \leq 5)$. The non-zero part of $h(t-\tau)$ overlaps partially with $x(\tau)$ within the region $\tau = [t-2, 3]$. Therefore, the output for $2 \leq t \leq 5$ is given by

$$y(t) = \int_{t-2}^{3} 3\,d\tau = 3(5-t).$$

Case 5 $(t \geq 5)$. We see from Fig. 3.10(i) that the non-zero parts of $h(t-\tau)$ and $x(\tau)$ do not overlap. Therefore, the output $y(t) = 0$.

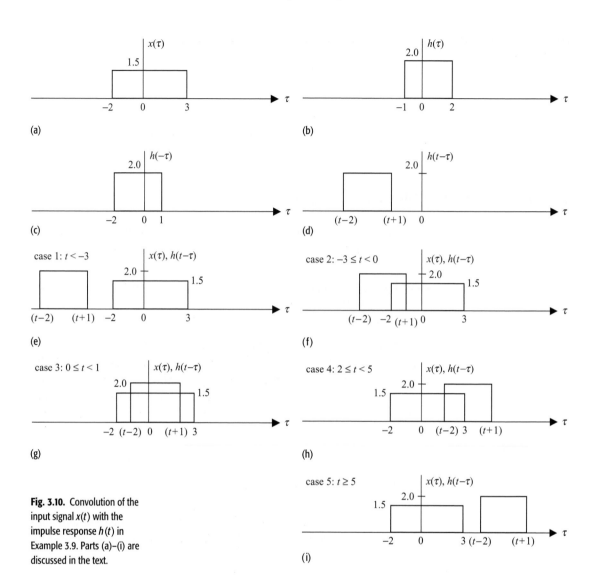

Fig. 3.10. Convolution of the input signal $x(t)$ with the impulse response $h(t)$ in Example 3.9. Parts (a)–(i) are discussed in the text.

Combining the five cases, we obtain

$$
y(t) = \begin{cases}
0 & t < -3 \\
3(t+3) & -3 \le t \le 0 \\
9 & 0 \le t \le 2 \\
3(5-t) & 2 \le t \le 5 \\
0 & t > 5.
\end{cases}
$$

The waveform for the output response is sketched in Fig. 3.11.

Fig. 3.11. Output $y(t)$ obtained in Example 3.9.

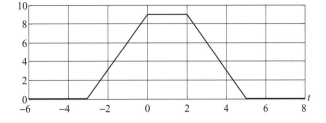

3.6 Properties of the convolution integral

The convolution integral has several interesting properties that can be used to simplify the analysis of LTIC systems. Some of these properties are presented in the following discussion.

Commutative property

$$x_1(t) * x_2(t) = x_2(t) * x_1(t). \tag{3.37}$$

The commutative property states that the order of the convolution operands does not affect the result of the convolution. In calculating the output of an LTIC system, the impulse response and input signal can be interchanged without affecting the output. The commutative property can be proved directly from the definition of the convolution integral by changing the dummy variable used for integration.

Proof

By definition,

$$x_1(t) * x_2(t) = \int\limits_{-\infty}^{\infty} x_1(\tau) x_2(t - \tau) d\tau.$$

Substituting $u = t - \tau$ gives

$$x_1(t) * x_2(t) = \int\limits_{\infty}^{-\infty} x_1(t - u) x_2(u)(-du).$$

By interchanging the order of the upper and lower limits, we obtain

$$x_1(t) * x_2(t) = \int\limits_{-\infty}^{\infty} x_1(t - u) x_2(u) du = x_2(t) * x_1(t).$$

Below, we list the remaining properties of convolution. Each of these properties can be proved by following the approach used in the proof for the commutative property. To avoid redundancy, the proofs for the remaining properties are not included.

Distributive property

$$x_1(t) * [x_2(t) + x_3(t)] = x_1(t) * x_2(t) + x_1(t) * x_3(t). \qquad (3.38)$$

The distributive property states that convolution is a linear operation.

Associative property

$$x_1(t) * [x_2(t) * x_3(t)] = [x_1(t) * x_2(t)] * x_3(t). \qquad (3.39)$$

This property states that changing the order of the convolution operands does not affect the result of the convolution integral.

Shift property If $x_1(t) * x_2(t) = g(t)$ then

$$x_1(t - T_1) * x_2(t - T_2) = g(t - T_1 - T_2), \qquad (3.40)$$

for any arbitrary real constants T_1 and T_2. In other words, if the two operands of the convolution integral are shifted, then the result of the convolution integral is shifted in time by a duration that is the sum of the individual time shifts introduced in the operands.

Duration of convolution Let the non-zero durations (or widths) of the convolution operands $x_1(t)$ and $x_2(t)$ be denoted by T_1 and T_2 time units, respectively. It can be shown that the non-zero duration (or width) of the convolution $x_1(t) * x_2(t)$ is $T_1 + T_2$ time units.

Convolution with impulse function

$$x(t) * \delta(t - t_0) = x(t - t_0). \qquad (3.41)$$

In other words, convolving a signal with a unit impulse function whose origin is at $t = t_0$ shifts the signal to the origin of the unit impulse function.

Convolution with unit step function

$$x(t) * u(t) = \int_{-\infty}^{\infty} x(\tau)u(t - \tau)d\tau = \int_{-\infty}^{t} x(\tau)d\tau. \qquad (3.42)$$

Equation (3.42) states that convolving a signal $x(t)$ with a unit step function produces the running integral of the original signal $x(t)$ as a function of time t.

Scaling property If $y(t) = x_1(t) * x_2(t)$, then $y(\alpha t) = |\alpha|x_1(\alpha t) * x_2(\alpha t)$. In other words, if we scale the two convolution operands $x_1(t)$ and $x_2(t)$ by a factor of α, then the result of convolution $x_1(t) * x_2(t)$ is (i) scaled by α and (ii) amplified by $|\alpha|$ to determine $y(\alpha t)$.

3.7 Impulse response of LTIC systems

In Section 2.2, we considered several properties of CT systems. Since an LTIC system is completely specified by its impulse response, it is therefore logical to assume that its properties are completely determined from its impulse response. In this section, we express some of the basic properties of LTIC systems defined in Section 2.2 in terms of the impulse response of the LTIC systems. We consider the memorylessness, causality, stability, and invertibility properties for such systems.

3.7.1 Memoryless LTIC systems

A CT system is said to be memoryless if its output $y(t)$ at time $t = t_0$ depends only on the value of the applied input signal $x(t)$ at the same time instant $t = t_0$. In other words, a memoryless LTIC system typically has an input–output relationship of the form

$$y(t) = kx(t),$$

where k is a constant. Substituting $x(t) = \delta(t)$, the impulse response $h(t)$ of a memoryless system can be obtained as follows:

$$h(t) = k\delta(t). \tag{3.43}$$

> An LTIC system will be memoryless if and only if its impulse response $h(t) = 0$ for $t \neq 0$.

3.7.2 Causal LTIC systems

A CT system is said to be causal if the output at time $t = t_0$ depends only on the value of the applied input signal $x(t)$ at and before the time instant $t = t_0$. The output of an LTIC system at time $t = t_0$ is given by

$$y(t_0) = \int_{-\infty}^{\infty} x(\tau)h(t_0 - \tau)\mathrm{d}\tau.$$

In a causal system, output $y(t_0)$ must not depend on $x(\tau)$ for $\tau > t_0$. This condition is only satisfied if the time-shifted and reflected impulse response $h(t_0 - \tau) = 0$ for $\tau > t_0$. Choosing $t_0 = 0$, the causality condition reduces to $h(-\tau) = 0$ for $\tau > 0$, which is equivalent to stating that $h(\tau) = 0$ for $\tau < 0$. Below we state the causality condition explicitly.

> An LTIC system will be causal if and only if its impulse response $h(t) = 0$ for $t < 0$.

3.7.3 Stable LTIC systems

A CT system is BIBO stable if an arbitrary bounded input signal produces a bounded output signal. Consider a bounded signal $x(t)$ with $|x(t)| < B_x$ for all t, applied as input to an LTIC system with impulse response $h(t)$. The magnitude of output $y(t)$ is given by

$$|y(t)| = \left| \int_{-\infty}^{\infty} h(\tau)x(t-\tau)d\tau \right|.$$

Using the Schwartz inequality, we can say that the output is bounded within the range

$$|y(t)| \leq \int_{-\infty}^{\infty} |h(\tau)||x(t-\tau)|d\tau.$$

Since $x(t)$ is bounded, $|x(t)| < B_x$, therefore the above inequality reduces to

$$|y(t)| \leq B_x \int_{-\infty}^{\infty} |h(\tau)|d\tau.$$

It is clear from the above expression that for the output $y(t)$ to be bounded, i.e. $|y(t)| < \infty$, the integral $\int h(\tau)d\tau$ within the limits $[-\infty, \infty]$ should also be bounded. The stability condition can, therefore, be stated as follows.

If the impulse response $h(t)$ of an LTIC system satisfies the following condition:

$$\int_{-\infty}^{\infty} |h(t)|dt < \infty, \tag{3.44}$$

then the LTIC system is BIBO stable.

Example 3.10
Determine if systems with the following impulse responses:

  (i) $h(t) = \delta(t) - \delta(t-2)$,
 (ii) $h(t) = 2\,\text{rect}(t/2)$,
(iii) $h(t) = 2\exp(-4t)u(t)$,
(iv) $h(t) = [1 - \exp(-4t)]u(t)$,

are memoryless, causal, and stable.

Solution

System (i)

Memoryless property. Since $h(t) \neq 0$ for $t \neq 0$, system (i) is not memoryless. The system has a limited memory as it only requires the values of the input signal within three time units of the time instant at which the output is being evaluated.

Causality property. Since $h(t) = 0$ for $t < 0$, system (i) is causal.

Stability property. To verify if system (i) is stable, we compute the following integral:

$$\int_{-\infty}^{\infty} |h(t)| dt = \int_{-\infty}^{\infty} |\delta(t) - \delta(t-2)| dt$$

$$\leq \int_{-\infty}^{\infty} |\delta(t)| dt + \int_{-\infty}^{\infty} |\delta(t-2)| dt = 2 < \infty,$$

which shows that system (i) is stable.

System (ii)

Memoryless property. Since $h(t) \neq 0$ for $t \neq 0$, system (ii) is not memoryless.

Causality property. Since $h(t) \neq 0$ for $t < 0$, system (ii) is not causal.

Stability property. To verify if system (ii) is stable, we compute the following integral:

$$\int_{-\infty}^{\infty} |h(t)| dt = \int_{-1}^{1} 2 \, dt = 4 < \infty,$$

which shows that system (ii) is stable.

System (iii)

Memoryless property. Since $h(t) \neq 0$ for $t \neq 0$, system (iii) is not memoryless. The memory of system (iii) is infinite, as the output at any time instant depends on the values of the input taken over the entire past.

Causality property. Since $h(t) = 0$ for $t < 0$, system (iii) is causal.

Stability property. To verify that system (iii) is stable, we solve the following integral:

$$\int_{-\infty}^{\infty} |h(t)| dt = \int_{0}^{\infty} 2e^{-4t} dt = -0.5 \times [e^{-4t}]_0^{\infty} = 0.5 < \infty,$$

which shows that system (iii) is stable.

System (iv)

Memoryless property. Since $h(t) \neq 0$ for $t \neq 0$, system (iv) is not memoryless.

Causality property. Since $h(t) = 0$ for $t < 0$, system (iv) is causal.

Stability property. To verify that system (iv) is stable, we solve the following integral:

$$\int_{-\infty}^{\infty} |h(t)|dt = \int_{0}^{\infty} (1 - e^{-4t})dt = [t + 0.25e^{-4t}]_{0}^{\infty} = \infty,$$

which shows that system (iv) is not stable.

3.7.4 Invertible LTIC systems

Consider an LTIC system with impulse response $h(t)$. The output $y_1(t)$ of the system for an input signal $x(t)$ is given by $y_1(t) = x(t) * h(t)$. For the system to be invertible, we cascade a second system with impulse response $h_i(t)$ in series with the original system. The output of the second system is given by

$$y_2(t) = y_1(t) * h_i(t).$$

For the second system to be an inverse of the original system, output $y_2(t)$ should be the same as $x(t)$. Substituting $y_1(t) = x(t) * h(t)$ in the above expression results in the following condition for invertibility:

$$x(t) = [x(t) * h(t)] * h_i(t) = x(t) * [h(t) * h_i(t)].$$

The above equation is true if and only if

$$h(t) * h_i(t) = \delta(t). \tag{3.45}$$

The existence of $h_i(t)$ proves that an LTIC system is invertible. At times, it is difficult to determine the inverse system $h_i(t)$ in the time domain. In Chapter 5, when we introduce the Fourier transform, we will revisit the topic and illustrate how the inverse system can be evaluated with relative ease in the Fourier-transform domain.

Example 3.11
Determine if systems with the following impulse responses:

(i) $h(t) = \delta(t - 2)$,
(ii) $h(t) = \delta(t) - \delta(t - 2)$,

are invertible.

Solution
(i) Since $\delta(t - 2) * \delta(t + 2) = \delta(t)$, system (i) is invertible. The impulse response of the inverse system is given by

$$h_i(t) = \delta(t + 2).$$

(ii) Assuming that the impulse response of the inverse system is $h_i(t)$, the stability condition is expressed as

$$h(t) * h_i(t) = [\delta(t) - \delta(t-2)] * h_i(t) = \delta(t).$$

By applying the convolution property, Eq. (3.41), the above expression simplifies to

$$h_i(t) - h_i(t-2) = \delta(t)$$

or

$$h_i(t) = \delta(t) + h_i(t-2).$$

The above expression can be solved iteratively. For example, $h_i(t-2)$ is given by

$$h_i(t-2) = \delta(t-2) + h_i(t-4).$$

Substituting the value of $h_i(t-2)$ in the earlier expression gives

$$h_i(t) = \delta(t) + \delta(t-2) + h_i(t-4),$$

leading to the iterative expression

$$h_i(t) = \sum_{m=0}^{\infty} \delta(t-2m).$$

To verify that $h_i(t)$ is indeed the impulse response of the inverse system, we convolve $h(t)$ with $h_i(t)$. The resulting expression is as follows:

$$h(t) * h_i(t) = [\delta(t) - \delta(t-2)] * \sum_{m=0}^{\infty} \delta(t-2m),$$

which simplifies to

$$h(t) * h_i(t) = \delta(t) * \sum_{m=0}^{\infty} \delta(t-2m) - \delta(t-2) * \sum_{m=0}^{\infty} \delta(t-2m)$$

or

$$h(t) * h_i(t) = \sum_{m=0}^{\infty} \delta(t-2m) - \sum_{m=0}^{\infty} \delta(t-2-2m) = \delta(t).$$

Therefore, $h_i(t)$ is indeed the impulse response of the inverse system.

3.8 Experiments with MATLAB

In this chapter, we have so far presented two approaches to calculate the output response of an LTIC system: the differential equation method and the convolution method. Both methods can be implemented using MATLAB. However, the convolution method is more convenient for MATLAB implementation in the discrete-time domain and this will be presented in Chapter 8. In this section, therefore, we present the method for constant-coefficient differential equations with initial conditions.

MATLAB provides several M-files for solving differential equations with known initial conditions. The list includes `ode23`, `ode45`, `ode113`, `ode15s`, `ode23s`, `ode23t`, and `ode23tb`. Each of these functions uses a finite-difference-based scheme for discretizing a CT differential equation and iterates the resulting DT finite-difference equation for the solution. A detailed analysis of the implementations of these MATLAB functions is beyond the scope of the text. Instead we will focus on the procedure for solving differential equations with MATLAB. Since the syntax used to name these M-files is similar, we illustrate the procedure for the function call using `ode23`. Any other M-file can be used instead of `ode23` by replacing `ode23` with the selected M-file. We will solve first- and second-order differential equations, and we compare the computed values with the analytical solution derived earlier.

Example 3.12

Compute the solution $y(t)$ for Eq. (3.11), reproduced below for convenience:

$$\frac{dy}{dt} + 4y(t) = 2\cos(2t)u(t),$$

with initial condition $y(0) = 2$ for $0 \leq t \leq 15$. Compare the computed solution with the analytical solution given by Eq. (3.12).

Solution

The first step towards solving Eq. (3.11) is to create an M-file containing the differential equation. We implement a reordered version of Eq. (3.11), given by

$$\frac{dy}{dt} = -4y(t) + 2\cos(2t)u(t),$$

where the derivative dy/dt is the output of the M-file based on the input y and time t. Calling the M-file `myfunc1`, the format for the M-file is as follows:

```
function [ydot] = myfunc1(t,y)
% MYFUNC1
% Computes first derivative in (3.11) given the value of
      % signal y and time t.
% Usage: ydot = myfunc1(t,y)
ydot = -4*y + 2*cos(2*t).*(t >= 0)
```

The above function is saved in a file named `myfunc1.m` and placed in a directory included within the defined paths of the MATLAB environment. To solve the differential equation defined in `myfunc1` over the interval $0 \leq t \leq 15$, we invoke `ode23` after initializing the input parameters in an M-file as shown:

Fig. 3.12. Solution $y(t)$ for Eq. (3.11) computed using MATLAB.

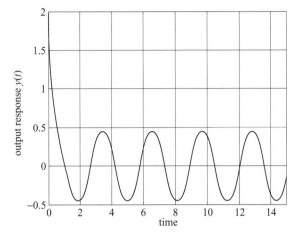

```
% MATLAB program to solve Equation (3.11) in Example 3.12
tspan = [0:0.01:15];              % duration with resolution
                                        % of 0.01s.
y0 = [2];                         % initial condition
[t,y] = ode23('myfunc1', tspan,y0);
                                  % solve ODE using ode23
plot(t,y)                         % plot the result
xlabel('time')                    % Label of X-axis
ylabel('Output Response y(t)')    % Label of Y-axis
```

The final plot is shown in Fig. 3.12 and is the same as the analytical solution given by Eq. (3.12).

Example 3.13
Compute the solution for the following second-order differential equation:

$$\ddot{y}(t) + 5\dot{y}(t) + 4y(t) = (3\cos t)u(t) \quad \text{with initial conditions}$$
$$y(0) = 2 \text{ and } \dot{y}(0) = -5,$$

for $0 \le t \le 20$ using MATLAB. Note that the analytical solution of this problem is presented in Appendix C (see Example C.6).

Solution
Higher-order differential equations are typically represented by a system of first-order differential equations before their solution can be computed in MATLAB. Assuming $y_2(t)$ to be the solution of the aforementioned differential equation, we obtain

$$\ddot{y}_2(t) + 5\dot{y}_2(t) + 4y_2(t) = (3\cos t)u(t).$$

To reduce the second-order differential equation into a system of two first-order differential equations, assume the following:

$$\dot{y}_2(t) = y_1(t). \tag{3.46}$$

Substituting $y_2(t)$ in the original equation and rearranging the terms yields

$$\dot{y}_1(t) = -5y_1(t) - 4y_2(t) + (3\cos t)u(t). \tag{3.47}$$

Equations (3.46) and (3.47) collectively define a system of first-order differential equations that simulate the original differential equation. The coupled system can be represented in the matrix-vector form as follows:

$$\begin{bmatrix} \dot{y}_1(t) \\ \dot{y}_2(t) \end{bmatrix} = \begin{bmatrix} -5y_1(t) - 4y_2(t) + (3\cos t)u(t) \\ y_1(t) \end{bmatrix}. \tag{3.48}$$

To simulate the above system, we write an M-file myfunc2 that computes the vector of derivatives on the left-hand side of Eq. (3.48) based on the input parameters t and vector y that contains the values of y_1 and y_2:

```
function [ydot] = myfunc2(t,y)
% The function computes first derivative of (3.48) from
     % vector y and time t.
% Usage: ydot = myfunc2(t,y)
ydot(1,1) = -5*y(1) - 4*y(2)+ 3*cos(t)*(t >= 0);
ydot(2,1) = y(1);
%---end of the function---------------------
```

Note that the output of the above M-file is the column vector ydot corresponding to Eq. (3.48). The M-file myfunc2.m should be placed in a directory included within the defined paths of the MATLAB environment. To solve the differential equation defined in myfunc2 over the interval $0 \le t \le 20$, we invoke ode23 after initializing the input parameters as given below:

```
% MATLAB program to solve Example 3.13
tspan = [0:0.02:20];          % duration with resolution of
                              % 0.02s.
y0 = [-5; 2];                 % initial conditions
[t,y] = ode23('myfunc2', tspan,y0);
                              % solve ODE using ode23
plot(t,y(:,2))                % plot the result
```

Note that the order of the initial conditions is reversed such that $\dot{y}_2(0) = -5$ is mentioned first and $y_2(0) = 2$ later in the initial condition vector y0. Looking at the structure of Eq. (3.48), it is clear that the top entry in the first row of ydot corresponds to $\dot{y}_1(t)$, which is equal to $\ddot{y}_2(t)$. Similarly, the entry in the second row of ydot contains the value of $\dot{y}_2(t)$. The function ode23 will integrate ydot returning the value in y. The vector y, therefore, contains the values of $\dot{y}_2(t)$ in the top row and the values of $y_2(t)$ in the bottom row. The order of the initial conditions is adjusted according to the returned values such that $\dot{y}_2(0) = -5$ is mentioned first and $y_2(0) = 2$ later in the initial condition vector y0. The solution of the differential equation is also contained in the second column of vector y.

Fig. 3.13. Solution $y(t)$ for Example 3.13 computed using MATLAB.

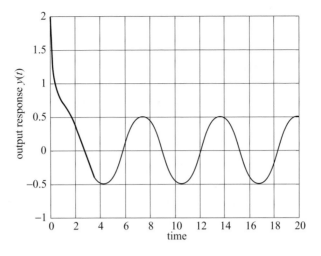

Fig. 3.13. Solution $y(t)$ for Example 3.13 computed using MATLAB.

The solution $y(t)$ is plotted in Fig. 3.13. It can be easily verified that the plot is same as the analytical solution given by Eq. (C.38), which is reproduced below

$$y(t) = \frac{1}{2}e^{-t} + \frac{21}{17}e^{-4t} + \frac{9}{34}\cos t + \frac{15}{34}\sin t \quad \text{for } t \geq 0.$$

3.9 Summary

In Chapter 3, we developed analytical techniques for LTIC systems. We saw that the output signal $y(t)$ of an LTIC system can be evaluated analytically in the time domain using two different methods. In Section 3.1, we determined the output of an LTIC by solving a linear, constant-coefficient differential equation. The solution of such a differential equation can be expressed as a sum of two components: zero-input response and zero-state response. The zero-input response is the output produced by the LTIC system because of the initial conditions. For stable LTIC systems, the zero-input response decays to zero with increasing time. The zero-state response is due to the input signal. The overall output of the LTIC system is the sum of the zero-input response and zero-state response.

An alternative representation for determining the output of an LTIC system is based on the impulse response of the system. In Section 3.3, we defined the impulse response $h(t)$ as the output of an LTIC system when a unit impulse $\delta(t)$ is applied at the input of the system. In Section 3.4, we proved that the output $y(t)$ of an LTIC system can be obtained by convolving the input signal $x(t)$ with its impulse response $h(t)$. The resulting convolution integral can either be solved analytically or by using a graphical approach. The graphical approach was illustrated through several examples in Section 3.5. The convolution integral satisfies the commutative, distributive, associative, time-shifting, and scaling properties.

(1) The commutative property states that the order of the convolution operands does not affect the result of the convolution.

(2) The distributive property states that convolution is a linear operation with respect to addition.

(3) The associative property is an extension of the commutative property to more than two convolution operands. It states that changing the order of the convolution operands does not affect the result of the convolution integral.

(4) The time-shifting property states that if the two operands of the convolution integral are shifted in time, then the result of the convolution integral is shifted by a duration that is the sum of the individual time shifts introduced in the convolution operands.

(5) The duration of the waveform produced by the convolution integral is the sum of the durations of the convolved signals.

(6) Convolving a signal with a unit impulse function with origin at $t = t_0$ shifts the signal to the origin of the unit impulse function.

(7) Convolving a signal with a unit step function produces the running integral of the original signal as a function of time t.

(8) If the two convolution operands are scaled by a factor α, then the result of the convolution of the two operands is scaled by α and amplified by $|\alpha|$.

In Section 3.7, we expressed the memoryless, causality, inverse, and stability properties of an LTIC system in terms of its impulse response.

(1) An LTIC system will be memoryless if and only if its impulse response $h(t) = 0$ for $t \neq 0$.

(2) An LTIC system will be causal if and only if its impulse response $h(t) = 0$ for $t < 0$.

(3) The impulse response of the inverse of an LTIC system satisfies the property $h_i(t) * h(t) = \delta(t)$.

(4) The impulse response $h(t)$ of a (BIBO) stable LTIC system is absolutely integrable, i.e.

$$\int_{-\infty}^{\infty} |h(t)| dt < \infty.$$

Finally, in Section 3.8 we presented a few MATLAB examples for solving constant-coefficient differential equations with initial conditions.

In Chapters 4 and 5, we will introduce the frequency representations for CT signals and systems. Such representations provide additional tools that simplify the analysis of LTIC systems.

Problems

3.1 Show that a system whose input $x(t)$ and output $y(t)$ are related by a linear differential equation of the form

$$\frac{d^n y}{dt^n} + a_{n-1}\frac{d^{n-1}y}{dt^{n-1}} + \cdots + a_1\frac{dy}{dt} + a_0 y(t)$$
$$= b_m\frac{d^m x}{dt^m} + b_{m-1}\frac{d^{m-1}x}{dt^{m-1}} + \cdots + b_1\frac{dx}{dt} + b_0 x(t)$$

is linear and time-invariant if the coefficients $\{a_r, 0 \le r \le n-1\}$ and $\{b_r, 0 \le r \le m\}$ are constants.

3.2 For each of the following differential equations modeling an LTIC system, determine (a) the zero-input response, (b) the zero-state response, (c) the overall response and (d) the steady state response of the system for the specified input $x(t)$ and initial conditions.

(i) $\ddot{y}(t) + 4\dot{y}(t) + 8y(t) = \dot{x}(t) + x(t)$ with $x(t) = e^{-4t}u(t)$,
$y(0) = 0$, and $\dot{y}(0) = 0$.

(ii) $\ddot{y}(t) + 6\dot{y}(t) + 4y(t) = \dot{x}(t) + x(t)$ with $x(t) = \cos(6t)u(t)$,
$y(0) = 2$, and $\dot{y}(0) = 0$.

(iii) $\ddot{y}(t) + 2\dot{y}(t) + y(t) = \ddot{x}(t)$ with $x(t) = [\cos(t) + \sin(2t)]u(t)$,
$y(0) = 3$, and $\dot{y}(0) = 1$.

(iv) $\ddot{y}(t) + 4y(t) = 5x(t)$ with $x(t) = 4te^{-t}u(t)$, $y(0) = -2$,
and $\dot{y}(0) = 0$.

(v) $\dddot{y}(t) + 2\ddot{y}(t) + y(t) = x(t)$ with $x(t) = 2u(t)$, $y(0) = \ddot{y}(0) = $
$\ddot{y}(0) = 0$, and $\dot{y}(0) = 1$.

3.3 Find the impulse responses for the following LTIC systems characterized by linear, constant-coefficient differential equations with zero initial conditions.

(i) $\dot{y}(t) = 2x(t)$;
(ii) $\dot{y}(t) + 6y(t) = x(t)$;
(iii) $2\dot{y}(t) + 5y(t) = \dot{x}(t)$;

(iv) $\dot{y}(t) + 3y(t) = 2\dot{x}(t) + 3x(t)$;
(v) $\ddot{y}(t) + 5\dot{y}(t) + 4y(t) = x(t)$;
(vi) $\ddot{y}(t) + 2\dot{y}(t) + y(t) = x(t)$.

3.4 The input signal $x(t) = e^{-\alpha t}u(t)$ is applied to an LTIC system with impulse response $h(t) = e^{-\beta t}u(t)$.

(i) Calculate the output $y(t)$ when $\alpha \ne \beta$.
(ii) Calculate the output $y(t)$ when $\alpha = \beta$.
(iii) Intuitively explain why the output signals are different in parts (i) and (ii).

3.5 Determine the output $y(t)$ for the following pairs of input signals $x(t)$ and impulse responses $h(t)$:

(i) $x(t) = u(t), h(t) = u(t)$;
(ii) $x(t) = u(-t), h(t) = u(-t)$;
(iii) $x(t) = u(t) - 2u(t-1) + u(t-2), h(t) = u(t+1) - u(t-1)$;

3.17 A sinusoidal signal $x(t) = A\sin(\omega_0 t + \theta)$ is applied at the input of an LTIC system with real-valued impulse response $h(t)$. By expressing the sinusoidal signal as the imaginary term of a complex exponential, i.e. as

$$jA\sin(\omega_0 t + \theta) = \text{Im}\left\{Ae^{j(\omega_0 + t)}\right\}, \quad A \in \Re,$$

show that the output of the LTIC system is given by

$$y(t) = A|H(\omega_0)|\sin(\omega_0 t + \theta + \arg(H(\omega_0))),$$

where $H(\omega)$ is the Fourier transform of the impulse response $h(t)$ as defined in Problem 3.16.

Hint: If $h(t)$ is real and $x(t) \rightarrow y(t)$, then $\text{Im}\{x(t)\} \rightarrow \text{Im}\{y(t)\}$.

3.18 Given that the LTIC system produces the output $y(t) = 5\cos(2\pi t)$ when the signal $x(t) = -3\sin(2\pi t + \pi/4)$ is applied at its input, derive the value of the tranfer function $H(\omega)$ at $\omega = 2\pi$. Hint: Use the result derived in Problem 3.17.

3.19 (a) Compute the solutions of the differential equations given in P3.2 for duration $0 \le t \le 20$ using MATLAB. (b) Compare the computed solution with the analytical solution obtained in P3.2.

4   Signal representation using Fourier series

In Chapter 3, we developed analysis techniques for LTIC systems using the convolution integral by representing the input signal $x(t)$ as a linear combination of time-shifted impulse functions $\delta(t)$. In Chapters 4 and 5, we will introduce alternative representations for CT signals and LTIC systems based on the weighted superpositions of complex exponential functions. The resulting representations are referred to as the continuous-time Fourier series (CTFS) and continuous-time Fourier transform (CTFT). Representing CT signals as superpositions of complex exponentials leads to frequency-domain characterizations, which provide a meaningful insight into the working of many natural systems. For example, a human ear is sensitive to audio signals within the frequency range 20 Hz to 20 kHz. Typically, a musical note occupies a much wider frequency range. Therefore, the human ear processes frequency components within the audible range and rejects other frequency components. In such applications, frequency-domain analysis of signals and systems provides a convenient means of solving for the response of LTIC systems to arbitrary input signals.

In this chapter, we focus on *periodic* CT signals and introduce the CTFS used to decompose such signals into their frequency components. Chapter 5 considers *aperiodic* CT signals and develops an equivalent Fourier representation, CTFT, for aperiodic signals. The organization of Chapter 4 is as follows. In Section 4.1, we define two- and three-dimensional orthogonal vector spaces and use them to motivate our introduction to orthogonal signal spaces in Section 4.2. We show that sinusoidal and complex exponential signals form complete sets of orthogonal functions. By selecting the sinusoidal signals as an orthogonal set of basis functions, Sections 4.3 and 4.4 present the trigonometric CTFS for a CT periodic signal. Section 4.5 defines the exponential representation for the CTFS based on using the complex exponentials as the basis functions. The properties of the exponential CTFS are presented in Section 4.6. The condition for the existence of CTFS is described in Section 4.7. Several interesting applications of the CTFS are presented in Section 4.8, which is followed by a summary of the chapter in Section 4.9.

4.1 Orthogonal vector space

From the theory of vector space, we know that an arbitrary M-dimensional vector can be represented in terms of its M orthogonal coordinates. For example, a two-dimensional (2D) vector $\vec{V}$ with coordinates (v_i, v_j) can be expressed as follows:

$$\vec{V} = v_i\vec{i} + v_j\vec{j}, \tag{4.1}$$

where $\vec{i}$ and $\vec{j}$ are the two basis vectors, respectively, along the x- and y-axis. A graphical representation for the 2D vector is illustrated in Fig. 4.1(a). The two basis vectors $\vec{i}$ and $\vec{j}$ have unit magnitudes and are perpendicular to each other, as described by the following two properties:

orthogonality property $\qquad \vec{i} \cdot \vec{j} = |\vec{i}||\vec{j}|\cos 90° = 0; \tag{4.2}$

unit magnitude property $\qquad \begin{cases} \vec{i} \cdot \vec{i} = |\vec{i}||\vec{i}|\cos 0° = 1 \\ \vec{j} \cdot \vec{j} = |\vec{j}||\vec{j}|\cos 0° = 1. \end{cases} \tag{4.3}$

In Eqs. (4.2) and (4.3), the operator $(\cdot)$ denotes the dot product between the two 2D vectors.

Similarly, an arbitrary three-dimensional (3D) vector $\vec{V}$, illustrated in Fig. 4.1(b), with Cartesian coordinates (v_i, v_j, v_k), is expressed as follows:

$$\vec{V} = v_i\vec{i} + v_j\vec{j} + v_k\vec{k}, \tag{4.4}$$

where $\vec{i}$, $\vec{j}$, and $\vec{k}$ represent the three basis vectors along the x-, y-, and z-axis, respectively. All possible dot product combinations of basis vectors satisfy the orthogonality and unit magnitude properties defined in Eqs. (4.2) and (4.3), i.e.

orthogonality property $\qquad \vec{i} \cdot \vec{j} = \vec{i} \cdot \vec{k} = \vec{k} \cdot \vec{j} = 0; \tag{4.5}$

unit magnitude property $\qquad \vec{i} \cdot \vec{i} = \vec{j} \cdot \vec{j} = \vec{k} \cdot \vec{k} = 1. \tag{4.6}$

(a)

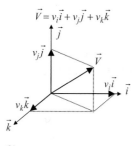

Collectively, the orthogonal and unit magnitude properties are referred to as the orthonormal property. Given vector $\vec{V}$, coordinates v_i, v_j, and v_k can be calculated directly from the dot product of vector $\vec{V}$ with the appropriate basis vectors. In other words,

$$v_u = \frac{\vec{V} \cdot \vec{u}}{\vec{u} \cdot \vec{u}} = \frac{|\vec{V}||\vec{u}|\cos\theta_{\vec{V}\vec{u}}}{|\vec{u}||\vec{u}|} \qquad \text{for } u \in \{i, j, k\}, \tag{4.7}$$

(b)

where $\theta_{\vec{V}\vec{u}}$ is the angle between $\vec{V}$ and $\vec{u}$. Just as an arbitrary vector can be represented as a linear combination of orthonormal basis functions, it is also possible to express an arbitrary signal as a weighted combination of orthornormal (or more generally, orthogonal) waveforms. In Section 4.2, we extend the principles of an orthogonal vector space to an orthogonal signal space.

Fig. 4.1. Representation of multidimensional vectors in Cartesian planes: (a) 2D vector; (b) 3D vector.

4.2 Orthogonal signal space

Definition 4.1 *Two non-zero signals $p(t)$ and $q(t)$ are said to be orthogonal over interval $t = [t_1, \ t_2]$ if*

$$\int_{t_1}^{t_2} p(t)q^*(t)\mathrm{d}t = \int_{t_1}^{t_2} p^*(t)q(t)\mathrm{d}t = 0, \qquad (4.8)$$

where the superscript $*$ denotes the complex conjugation operator. In addition to Eq. (4.8), if both signals $p(t)$ and $q(t)$ also satisfy the unit magnitude property:

$$\int_{t_1}^{t_2} p(t)p^*(t)\mathrm{d}t = \int_{t_1}^{t_2} q(t)q^*(t)\mathrm{d}t = 1, \qquad (4.9)$$

they are said to be orthonormal to each other over the interval $t = [t_1, t_2]$.

Example 4.1
Show that

  (i) functions $\cos(2\pi t)$ and $\cos(3\pi t)$ are orthogonal over interval $t = [0, \ 1]$;
 (ii) functions $\exp(\mathrm{j}2t)$ and $\exp(\mathrm{j}4t)$ are orthogonal over interval $t = [0, \pi]$;
(iii) functions $\cos(t)$ and t are orthogonal over interval $t = [-1, 1]$.

Solution
(i) Using Eq. (4.8), we obtain

$$\int_0^1 \cos(2\pi t)\cos(3\pi t)\mathrm{d}t = \frac{1}{2}\int_0^1 [\cos(\pi t) + \cos(5\pi t)]\mathrm{d}t$$

$$= \frac{1}{2}\left[\frac{1}{\pi}\sin(\pi t) + \frac{1}{5\pi}\sin(5\pi t)\right]_0^1 = 0.$$

Therefore, the functions $\cos(2\pi t)$ and $\cos(3\pi t)$ are orthogonal over interval $t = [0, 1]$.

   Figure 4.2 illustrates the graphical interpretation of the orthogonality condition for the functions $\cos(2\pi t)$ and $\cos(3\pi t)$ within interval $t = [0, 1]$. Equation (4.8) implies that the area enclosed by the waveform for $\cos(2\pi t) \times \cos(3\pi t)$ with respect to the t-axis within the interval $t = [0, 1]$, which is shaded in Fig. 4.2(c), is zero, which can be verified visually.

   (ii) Using Eq. (4.8), we obtain

$$\int_0^\pi \mathrm{e}^{\mathrm{j}2t}\mathrm{e}^{-\mathrm{j}4t}\mathrm{d}t = \int_0^\pi \mathrm{e}^{-\mathrm{j}2t}\mathrm{d}t = \frac{1}{-2\mathrm{j}}[\mathrm{e}^{-\mathrm{j}2t}]_0^\pi = -\frac{1}{2\mathrm{j}}[\mathrm{e}^{-\mathrm{j}2\pi} - 1] = 0,$$

implying that the functions $\exp(\mathrm{j}2t)$ and $\exp(\mathrm{j}4t)$ are orthogonal over interval $t = [0, \pi]$.

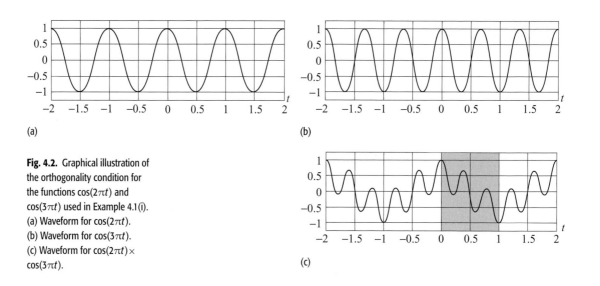

Fig. 4.2. Graphical illustration of the orthogonality condition for the functions $\cos(2\pi t)$ and $\cos(3\pi t)$ used in Example 4.1(i).
(a) Waveform for $\cos(2\pi t)$.
(b) Waveform for $\cos(3\pi t)$.
(c) Waveform for $\cos(2\pi t) \times \cos(3\pi t)$.

(iii) Using Eq. (4.8), we obtain

$$\int_{-1}^{1} t \cos(t) \mathrm{d}t = [t \sin(t) + \cos(t)]_{-1}^{1} = [1 \cdot \sin(1) + \cos(1)]$$

$$- [(-1) \cdot \sin(-1) + \cos(-1)] = 0,$$

implying that the functions $\cos(t)$ and t are orthogonal over interval $t = [-1, 1]$. Further, it is straightforward to verify that these functions are also orthogonal over any interval $t = [-L, \ L]$ for any real value of L.

We now extend the definition of orthogonality to a larger set of functions.

Definition 4.2 *A set of N functions $\{p_1(t), p_2(t), \ldots, p_N(t)\}$ is mutually orthogonal over the interval $t = [t_1, t_2]$ if*

$$\int_{t_1}^{t_2} p_m(t) p_n^*(t) \mathrm{d}t = \begin{cases} E_n \neq 0 & m = n \\ 0 & m \neq n \end{cases} \quad for\ 1 \leq m, n \leq N. \quad (4.10)$$

In addition, if $E_n = 1$ for all n, the set is referred to as an orthonormal *set.*

Definition 4.3 *An orthogonal set $\{p_1(t), p_2(t), \ldots, p_N(t)\}$ is referred to as a* complete orthogonal set *if no function $q(t)$ exists outside the set, which satisfies the orthogonality condition (4.6) with respect to the entries $p_n(t)$, $1 \leq n \leq N$, of the orthogonal set. Mathematically, function $q(t)$ exist if*

$$\int_{t_1}^{t_2} q(t) p_n^*(t) \mathrm{d}t \neq 0 \quad for\ at\ least\ one\ value\ of\ n \in \{1, \ldots, N\} \quad (4.11)$$

with

$$\int_{t_1}^{t_2} q(t)q^*(t)\mathrm{d}t \neq 0. \tag{4.12}$$

Definition 4.4 *If an orthogonal set is complete for a certain class of orthogonal functions within interval $t = [t_1, t_2]$, then any arbitrary function $x(t)$ can be expressed within interval $t = [t_1, t_2]$ as follows:*

$$x(t) = c_1 p_1(t) + c_2 p_2(t) + \cdots + c_n p_n(t) + \cdots + c_N p_N(t), \tag{4.13}$$

where coefficients $c_n, n \in [1, \ldots, N]$, are obtained using the following expression:

$$c_n = \frac{1}{E_n} \int_{t_1}^{t_2} x(t) p_n^*(t)\mathrm{d}t. \tag{4.14}$$

The constant E_n is calculated using Eq. (4.10). The integral Eq. (4.14) is the continuous time equivalent of the dot product in vector space, as represented in Eq. (4.7). The coefficient c_n is sometimes referred to as the nth Fourier coefficient of the function $x(t)$.

Definition 4.5 *A complete set of orthogonal functions $\{p_n(t)\}$, $1 \leq n \leq N$, that satisfies Eq. (4.10) is referred to as a set of basis functions.*

Example 4.2

For the three CT functions shown in Fig. 4.3

(a) show that the functions form an orthogonal set of functions;
(b) determine the value of T that makes the three functions orthonormal;
(c) express the signal

$$x(t) = \begin{cases} A & \text{for } 0 \leq t \leq T \\ 0 & \text{elsewhere} \end{cases}$$

in terms of the orthogonal set determined in (a).

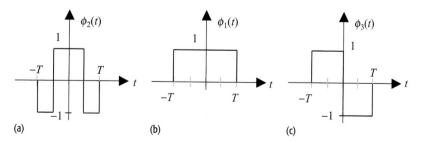

Fig. 4.3. Orthogonal functions for Example 4.2.   (a)   (b)   (c)

Solution

(a) We check for the unit magnitude and the orthogonality properties for all possible combinations of the basis vectors:

$$\text{unit magnitude property} \quad \int_{-T}^{T} |\phi_1(t)|^2 dt = \int_{-T}^{T} |\phi_2(t)|^2 dt = \int_{-T}^{T} |\phi_3(t)|^2 dt$$

$$= \int_{-T}^{T} 1 \, dt = 2T;$$

$$\text{orthogonality property} \quad \int_{-T}^{T} \phi_1(t)\phi_2^*(t) dt = \int_{-T}^{T} \phi_2^*(t) dt = 0,$$

$$\int_{-T}^{T} \phi_1(t)\phi_3^*(t) dt = \int_{-T}^{T} \phi_3^*(t) dt = 0,$$

and

$$\int_{-T}^{T} \phi_2(t)\phi_3^*(t) dt = \int_{0}^{T} \phi_2(t) dt - \int_{-T}^{0} \phi_2(t) dt = 0.$$

In other words,

$$\int_{-T}^{T} \phi_m(t)\phi_n^*(t) dt = \begin{cases} 2T \neq 0 & m = n \\ 0 & m \neq n, \end{cases}$$

for $1 \leq m, n \leq 3$. The three functions are orthogonal to each other over the interval $[-T, T]$.

(b) The three functions will be orthonormal to each other:

$$\int_{-T}^{T} \phi_m(t)\phi_n^*(t) dt = \begin{cases} 2T = 1 & m = n \\ 0 & m \neq n, \end{cases}$$

which implies that $T = 1/2$.

(c) Using Definition 4.4, the CT function $x(t)$ can be represented as $x(t) = c_1\phi_1(t) + c_2\phi_2(t) + c_3\phi_3(t)$ with the coefficients c_n, for $n = 1, 2$, and 3 given by

$$c_1 = \frac{1}{2T} \int_{-T}^{T} x(t)\phi_1(t) dt = \frac{1}{2T} \int_{0}^{T} A \, dt = \frac{A}{2},$$

$$c_2 = \frac{1}{2T} \int_{-T}^{T} x(t)\phi_2(t) dt = \frac{1}{2T} \int_{0}^{T} A\phi_2(t) dt$$

$$= \frac{1}{2T} \int_{0}^{T/2} A \, dt - \frac{1}{2T} \int_{T/2}^{T} A \, dt = 0,$$

and

$$c_3 = \frac{1}{2T} \int_{-T}^{T} x(t)\phi_3(t)dt = \frac{1}{2T} \int_{0}^{T} A(-1)dt = -\frac{A}{2}.$$

In other words, $x(t) = 0.5A[\phi_1(t) - \phi_3(t)]$.

Example 4.3

Show that the set $\{1, \cos(\omega_0 t), \cos(2\omega_0 t), \cos(3\omega_0 t), \dots, \sin(\omega_0 t), \sin(2\omega_0 t), \sin(3\omega_0 t), \dots\}$, consisting of all possible harmonics of sine and cosine waves with fundamental frequency of ω_0, is an orthogonal set over any interval $t = [t_0, t_0 + T_0]$, with duration $T_0 = 2\pi/\omega_0$.

Solution

It may be noted that the set $\{1, \cos(\omega_0 t), \cos(2\omega_0 t), \cos(3\omega_0 t), \dots, \sin(\omega_0 t), \sin(2\omega_0 t), \sin(3\omega_0 t), \dots\}$ contains three types of functions: 1, $\{\cos(m\omega_0 t)\}$, and $\{\sin(n\omega_0 t)\}$ for arbitrary integers $m, n \in Z^+$, where Z^+ is the set of positive integers. We will consider all possible combinations of these functions.

Case 1 The following proof shows that functions $\{\cos(m\omega_0 t), m \in Z^+\}$ are orthogonal to each other over interval $t = [t_0, t_0 + T_0]$ with $T_0 = 2\pi/\omega_0$. Equation (4.10) yields

$$\int_{\langle T_0 \rangle} \cos(m\omega_0 t)\cos(n\omega_0 t)dt = \int_{t_0}^{t_0+T_0} \cos(m\omega_0 t)\cos(n\omega_0 t)dt \quad \text{for any arbitrary } t_0.$$

Using the trigonometric identity $\cos(m\omega_0 t)\cos(n\omega_0 t) = (1/2)[\cos((m - n)\omega_0 t) + \cos((m + n)\omega_0 t)]$, the above integral reduces as follows:

$$\int_{\langle T_0 \rangle} \cos(m\omega_0 t)\cos(n\omega_0 t)dt = \begin{cases} \left[\dfrac{\sin(m - n)\omega_0 t}{2(m - n)\omega_0} + \dfrac{\sin(m + n)\omega_0 t}{2(m + n)\omega_0} \right]_{t_0}^{t_0+T_0} & m \neq n \\[3mm] \left[\dfrac{t}{2} + \dfrac{\sin 2m\omega_0 t}{4m\omega_0} \right]_{t_0}^{t_0+T_0} & m = n, \end{cases}$$

or

$$\int_{\langle T_0 \rangle} \cos(m\omega_0 t)\cos(n\omega_0 t)dt = \begin{cases} 0 & m \neq n \\ \dfrac{T_0}{2} & m = n, \end{cases} \tag{4.15}$$

for $m, n \in Z^+$. Equation (4.15) demonstrates that the functions in the set $\{\cos(m\omega_0 t), m \in Z^+\}$ are mutually orthogonal.

Case 2 By following the procedure outlined in case 1, it is straightforward to show that

$$\int_{\langle T_0 \rangle} \sin(m\omega_0 t) \sin(n\omega_0 t) dt = \begin{cases} 0 & m \neq n \\ \dfrac{T_0}{2} & m = n, \end{cases} \tag{4.16}$$

for $m, n \in Z^+$. Equation (4.16) proves that the set $\{\sin(n\omega_0 t), n \in Z^+\}$ contains mutually orthogonal functions over interval $t = [t_0, t_0 + T_0]$ with $T_0 = 2\pi/\omega_0$.

Case 3 To verify that functions $\{\cos(m\omega_0 t)\}$ and $\{\sin(n\omega_0 t)\}$ are mutually orthogonal, consider the following:

$$\int_{\langle T_0 \rangle} \cos(m\omega_0 t) \sin(n\omega_0 t) dt = \int_{t_0}^{t_0+T_0} \cos(m\omega_0 t) \sin(n\omega_0 t) dt$$

$$= \begin{cases} \dfrac{1}{2} \displaystyle\int_{t_0}^{t_0+T_0} [\sin((m+n)\omega_0 t) - \sin((m-n)\omega_0 t)] dt & m \neq n \\ \dfrac{1}{2} \displaystyle\int_{t_0}^{t_0+T_0} [\sin(2m\omega_0 t) dt & m = n \end{cases}$$

$$= \begin{cases} -\dfrac{1}{2}\left[\dfrac{\cos((m+n)\omega_0 t)}{(m+n)\omega_0}\right]_{t_0}^{t_0+T_0} + \dfrac{1}{2}\left[\dfrac{\cos((m-n)\omega_0 t)}{(m-n)\omega_0}\right]_{t_0}^{t_0+T_0} & m \neq n \\ -\dfrac{1}{2}\left[\dfrac{\cos(2m\omega_0 t)}{2m\omega_0}\right]_{t_0}^{t_0+T_0} & m = n \end{cases}$$

$$= \begin{cases} 0 & m \neq n \\ 0 & m = n, \end{cases} \tag{4.17}$$

for $m, n \in Z^+$, which proves that $\{\cos(m\omega_0 t)\}$ and $\{\sin(n\omega_0 t)\}$ are orthogonal over interval $t = [t_0, t_0 + T_0]$ with $T_0 = 2\pi/\omega_0$.

Case 4 The following proof demonstrates that the function "1" is orthogonal to $\cos(m\omega_0 t)\}$ and $\{\sin(n\omega_0 t)\}$:

$$\int_{\langle T_0 \rangle} 1 \cdot \cos(m\omega_0 t) dt = \left[\dfrac{\sin(m\omega_0 t)}{m\omega_0}\right]_{t_0}^{t_0+T_0}$$

$$= \left[\dfrac{\sin(m\omega_0 t_0 + 2m\pi) - \sin(m\omega_0 t_0)}{m\omega_0}\right] = 0 \tag{4.18}$$

and

$$\int_{\langle T_0 \rangle} 1 \cdot \sin(m\omega_0 t) dt = \left[-\dfrac{\cos(m\omega_0 t)}{m\omega_0}\right]_{t_0}^{t_0+T_0}$$

$$= -\left[\dfrac{\cos(m\omega_0 t_0 + 2m\pi) - \cos(m\omega_0 t_0)}{m\omega_0}\right] = 0 \tag{4.19}$$

for $m, n \in Z^+$. Combining Eqs. (4.15)–(4.19), it can be inferred that the set $\{1, \cos(\omega_0 t), \cos(2\omega_0 t), \cos(3\omega_0 t), \ldots, \sin(\omega_0 t), \sin(2\omega_0 t), \sin(3\omega_0 t), \ldots\}$ consists of mutually orthogonal functions. It can also be shown that this particular set is complete over $t = [t_0, t_0 + T_0]$ with $T_0 = 2\pi/\omega_0$. In other words, there exists no non-trivial function outside the set which is orthogonal to all functions in the set over the given interval.

Example 4.4

Show that the set of complex exponential functions $\{\exp(jn\omega_0 t), n \in Z\}$ is an orthogonal set over any interval $t = [t_0, t_0 + T_0]$ with duration $T_0 = 2\pi/\omega_0$. The parameter Z refers to the set of integer numbers.

Solution

Equation (4.10) yields

$$\int_{\langle T_0 \rangle} \exp(jm\omega_0 t)(\exp(jn\omega_0 t))^* dt$$

$$= \int_{t_0}^{t_0+T_0} \exp(j(m-n)\omega_0 t) dt = \begin{cases} [t]_{t_0}^{t_0+T_0} & m = n \\ \left[\dfrac{\exp(j(m-n)m\omega_0 t)}{j(m-n)m\omega_0}\right]_{t_0}^{t_0+T_0} & m \neq n \end{cases}$$

$$= \begin{cases} T_0 & m = n \\ 0 & m \neq n. \end{cases} \tag{4.20}$$

Equation (4.20) shows that the set of functions $\{\exp(jn\omega_0 t), n \in Z\}$ is indeed mutually orthogonal over interval $t = [t_0, t_0 + T_0]$ with duration $T_0 = 2\pi/\omega_0$. It can also be shown that this set is complete.

Examples 4.3 and 4.4 illustrate that the sinusoidal and complex exponential functions form two sets of complete orthogonal functions. There are several other orthogonal sets of functions, for example the Legendre polynomials (Problem 4.3), Chebyshev polynomials (Problem 4.4), and Haar functions (Problem 4.5). We are particularly interested in sinusoidal and complex exponential functions since these satisfy a special property with respect to the LTIC systems that is not observed for any other orthogonal set of functions. In Section 4.3, we discuss this special property.

4.3 Fourier basis functions

In Example 3.2, it was observed that the output response of an RLC circuit to a sinusoidal function was another sinusoidal function of the same frequency. The changes observed in the output sinusoidal function were only in its amplitude and phase. Below we illustrate that the property holds true for any LTIC system. Further, we extend the property to complex exponential signals proving that the output response of an LTIC system to a complex exponential function is another complex exponential with the same frequency, except for possible changes in its amplitude and phase.

Theorem 4.1 *If a complex exponential function is applied to an LTIC system with a real-valued impulse response function, the output response of the system is identical to the complex exponential function except for changes in amplitude and phase. In other words,*

$$k_1 e^{j\omega_1 t} \rightarrow A_1 k_1 e^{j(\omega_1 t + \phi_1)},$$

where A_1 and ϕ_1 are constants.

Proof
Assume that the complex exponential function $x(t) = k_1 \exp(j\omega_1 t)$ is applied to an LTIC system with impulse response $h(t)$. The output of the system, given by the convolution of the input signal $x(t)$ and the impulse response $h(t)$, can be calculated as

$$y(t) = \int_{-\infty}^{\infty} h(\tau) x(t - \tau) d\tau = k_1 e^{j\omega_1 t} \int_{-\infty}^{\infty} h(\tau) e^{-j\omega_1 \tau} d\tau. \qquad (4.21)$$

Defining

$$H(\omega) = \int_{-\infty}^{\infty} h(\tau) e^{-j\omega \tau} d\tau, \qquad (4.22)$$

Eq. (4.21) can be expressed as follows:

$$y(t) = k_1 e^{j\omega_1 t} H(\omega_1). \qquad (4.23)$$

From the definition in Eq. (4.22), we observe that $H(\omega_1)$ is a complex-valued constant, for a given value of ω_1, such that it can be expressed as $H(\omega_1) = A_1 \exp(j\phi_1)$. In other words, A_1 is the magnitude of the complex constant $H(\omega_1)$ and ϕ_1 is the phase of $H(\omega_1)$. Expressing $H(\omega_1) = A_1 \exp(j\phi_1)$ in Eq. (4.23), we obtain

$$y(t) = A_1 k_1 e^{j(\omega_1 t + \phi_1)},$$

which proves Theorem 4.1.

Corollary 4.1 The output response of an LTIC system, characterized by a real-valued impulse response $h(t)$, to a sinusoidal input is another sinusoidal function with the same frequency, except for possible changes in its amplitude and phase. In other words,

$$k_1 \sin(\omega_1 t) \rightarrow A_1 k_1 \sin(\omega_1 t + \phi_1) \qquad (4.24)$$

and

$$k_1 \cos(\omega_1 t) \rightarrow A_1 k_1 \cos(\omega_1 t + \phi_1), \qquad (4.25)$$

where constants A_1 and ϕ_1 are the magnitude and phase of $H(\omega_1)$ defined in Eq. (4.22) with ω set to ω_1.

Proof

The proof of Corollary 4.1 follows the same lines as the proof of Theorem 4.1. The sinusoidal signals can be expressed as real (Re) and imaginary (Im) components of a complex exponential function as follows:

$$\cos(\omega_1 t) = \text{Re}\{e^{j\omega_1 t}\} \quad \text{and} \quad \sin(\omega_1 t) = \text{Im}\{e^{j\omega_1 t}\}.$$

Because the impulse response function is real-valued, the output $y_1(t)$ to $k_1 \sin(\omega_1 t)$ is the imaginary component of $y(t)$ given in Eq. (4.23). In other words,

$$y_1(t) = \text{Im}\{A_1 k_1 e^{j(\omega_1 t + \phi_1)}\} = A_1 k_1 \sin(\omega_1 t + \phi_1).$$

Likewise, the output $y_2(t)$ to $k_1 \cos(\omega_1 t)$ is the real component of $y(t)$ given in Eq. (4.23). In other words,

$$y_2(t) = \text{Re}\{A_1 k_1 e^{j(\omega_1 t + \phi_1)}\} = A_1 k_1 \cos(\omega_1 t + \phi_1).$$

Example 4.5

Calculate the output response if signal $x(t) = 2 \sin(5t)$ is applied as an input to an LTIC system with impulse response $h(t) = 2e^{-4t} u(t)$.

Solution

Based on Corollary 4.1, we know that output $y(t)$ to the sinusoidal input $x(t) = 2 \sin(5t)$ is given by

$$y(t) = 2A_1 \sin(5t + \phi_1),$$

where A_1 and ϕ_1 are the magnitude and phase of the complex constant $H(\omega_1)$, given by

$$H(\omega) = \int_{-\infty}^{\infty} h(\tau) e^{-j\omega\tau} d\tau = 2 \int_{0}^{\infty} e^{-4\tau} e^{-j\omega\tau} d\tau = 2 \int_{0}^{\infty} e^{-(4+j\omega)\tau} d\tau = \frac{2}{4 + j\omega}.$$

The magnitude A_1 and phase ϕ_1 are given by

magnitude
$$A_1 A_1 = |H(\omega_1)| = \left| \frac{2}{4 + j\omega} \right|_{\omega=5} = \frac{2}{\sqrt{41}}.$$

phase ϕ_1
$$\phi_1 = <H(\omega_1) = <\frac{2}{4 + j\omega} \Big|_{\omega=5} = 0 - \tan^{-1}\left(\frac{5}{4}\right) = -51.34°.$$

The output response of the system is, therefore, given by

$$y(t) = \frac{4}{\sqrt{41}} \sin(5t - 51.34°).$$

As shown in Example 3.4, the LTIC system with impulse response $h(t) = 2e^{-4t} u(t)$ can alternatively be represented by the linear, constant-coefficient differential equation as follows:

$$\frac{dy}{dt} + 4y(t) = 2x(t).$$

Fig. 4.4. Output response of an LTIC system, with a real-valued impulse response, to sinusoidal inputs.

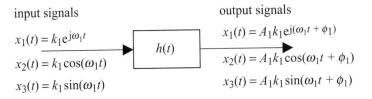

input signals

$x_1(t) = k_1 e^{j\omega_1 t}$

$x_2(t) = k_1 \cos(\omega_1 t)$

$x_3(t) = k_1 \sin(\omega_1 t)$

$h(t)$

output signals

$x_1(t) = A_1 k_1 e^{j(\omega_1 t + \phi_1)}$

$x_2(t) = A_1 k_1 \cos(\omega_1 t + \phi_1)$

$x_3(t) = A_1 k_1 \sin(\omega_1 t + \phi_1)$

Substituting $x(t) = 2\sin(5t)$ into this equation and solving the differential equation, we arrive at the same value of the output $y(t)$ obtained using the convolution approach.

Figure 4.4 illustrates Theorem 4.1 and Corollary 4.1 graphically. It may be noted that this property is not observed for any other input signal but only for the sinusoids and complex exponentials.

4.3.1 Generalization of Theorem 4.1

In the preceding discussion, we have restricted the input signal $x(t)$ to sinusoids or complex exponentials. In such cases, Theorem 4.1 or Corollary 4.1 simplifies the computation of the output response of a LTIC system. In cases where the input signal $x(t)$ is periodic but different from a sinusoidal or complex exponential function, we follow an indirect approach. We express the input signal $x(t)$ as a linear combination of complex exponentials:

$$x(t) = k_1 e^{j\omega_1 t} + k_2 e^{j\omega_2 t} + \cdots + k_N e^{j\omega_N t} = \sum_{n=1}^{N} k_n e^{j\omega_n t}. \qquad (4.26)$$

Applying Theorem 4.1 to each of the N complex exponential terms in Eq. (4.26), the output $y_m(t)$ to the complex exponential term $x_m(t) = k_m \exp(j\omega_m t)$ is given by $y_m(t) = A_m k_m \exp(j\omega_m t + \phi_m)$. Using the principle of superposition, the overall output $y(t)$ is the sum of the individual outputs and is expressed as follows:

$$y(t) = A_1 k_1 e^{j(\omega_1 t + \phi_1)} + A_2 k_2 e^{j(\omega_2 t + \phi_2)} + \cdots + A_N k_N e^{j(\omega_N t + \phi_N)}$$

$$= \sum_{n=1}^{N} A_n k_n e^{j(\omega_n t + \phi_n)}. \qquad (4.27)$$

In the above discussion, we have illustrated the advantage of expressing a periodic signal $x(t)$ as a linear combination of complex exponentials. Such a representation provides an alternative interpretation of the signal. This interpretation is referred to as the exponential CT Fourier series (CTFS).[†] Alternatively,

---

[†] The Fourier series is named after Jean Baptiste Joseph Fourier (1768–1830), a French mathematician and physicist who initiated its development and applied it to problems of heat flow for the first time.

an arbitrary periodic signal can also be expressed as a linear combination of sinusoidal signals:

$$x(t) = a_0 + \sum_{n=1}^{\infty} (a_n \cos(n\omega_0 t) + b_n \sin(n\omega_0 t)). \qquad (4.28)$$

Corollary 4.1 can then be applied to calculate the output $y(t)$. Expressing a periodic signal as a linear combination of sinusoidal signals leads to the trigonometric CTFS. The trigonometric and exponential CTFS representations of CT periodic signals are covered in Sections 4.4 and 4.5.

4.4 Trigonometric CTFS

Definition 4.6 *An arbitrary periodic function $x(t)$ with fundamental period T_0 can be expressed as follows:*

$$x(t) = a_0 + \sum_{n=1}^{\infty} (a_n \cos(n\omega_0 t) + b_n \sin(n\omega_0 t)), \qquad (4.29)$$

where $\omega_0 = 2\pi/T_0$ is the fundamental frequency of $x(t)$ and coefficients a_0, a_n, and b_n are referred to as the trigonometric CTFS coefficients. The coefficients are calculated as follows:

$$a_0 = \frac{1}{T_0} \int_{\langle T_0 \rangle} x(t) dt, \qquad (4.30)$$

$$a_n = \frac{2}{T_0} \int_{\langle T_0 \rangle} x(t) \cos(n\omega_0 t) dt, \qquad (4.31)$$

and

$$b_n = \frac{2}{T_0} \int_{\langle T_0 \rangle} x(t) \sin(n\omega_0 t) dt. \qquad (4.32)$$

From Eqs. (4.29)–(4.32), it is straightforward to verify that coefficient a_0 represents the average or mean value (also referred to as the dc component) of $x(t)$. Collectively, the cosine terms represent the even component of the zero mean signal $(x(t) - a_0)$. Likewise, the sine terms collectively represent the odd component of the zero mean signal $(x(t) - a_0)$.

Example 4.6
Calculate the trigonometric CTFS coefficients of the periodic signal $x(t)$ defined over one period $T_0 = 3$ as follows:

$$x(t) = \begin{cases} t+1 & -1 \leq t \leq 1 \\ 0 & 1 < t < 2. \end{cases} \qquad (4.33)$$

Fig. 4.5. Sawtooth periodic waveform $x(t)$ considered in Example 4.6.

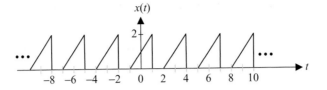

Solution

The periodic signal $x(t)$ is plotted in Fig. 4.5. Since $x(t)$ has a fundamental period $T_0 = 3$, the fundamental frequency $\omega_0 = 2\pi/3$. Using Eq. (4.30), the dc CTFS coefficient a_0 is given by

$$a_0 = \frac{1}{T_0} \int_{\langle T_0 \rangle} x(t) dt = \frac{1}{3} \int_{-1}^{1} (t+1) dt = \frac{1}{3} \left[\frac{1}{2}t^2 + t \right]_{-1}^{1} = \frac{2}{3}. \quad (4.34)$$

The CTFS coefficients a_n are given by

$$a_n = \frac{2}{T_0} \int_{\langle T_0 \rangle} x(t) \cos(n\omega_0 t) dt = \frac{2}{3} \int_{-1}^{1} (t+1) \cos(n\omega_0 t) dt$$

$$= \frac{2}{3} \underbrace{\int_{-1}^{1} t \cos(n\omega_0 t) \, dt}_{\text{odd function}} + \frac{2}{3} \underbrace{\int_{-1}^{1} \cos(n\omega_0 t) \, dt}_{\text{even function}}.$$

Since the integral of odd functions within the limit $[-t_0, t_0]$ is zero,

$$\int_{-1}^{1} t \cos(n\omega_0 t) dt = 0,$$

and the value of a_n is given by

$$a_n = \frac{2}{3} \int_{-1}^{1} \cos(n\omega_0 t) dt = \frac{4}{3} \int_{0}^{1} \cos(n\omega_0 t) dt = \frac{4}{3} \left[\frac{\sin(n\omega_0 t)}{n\omega_0} \right]_{0}^{1} = \frac{4 \sin(n\omega_0)}{3n\omega_0}.$$

Substituting $\omega_0 = 2\pi/3$, we obtain

$$a_n = \begin{cases} 0 & n = 3k \\ \dfrac{\sqrt{3}}{n\pi} & n = 3k+1 \\ -\dfrac{\sqrt{3}}{n\pi} & n = 3k+2, \end{cases} \quad (4.35)$$

for $k \in Z$. Similarly, the CTFS coefficients b_n are given by

$$b_n = \frac{2}{T_0} \int_{\langle T_0 \rangle} x(t) \sin(n\omega_0 t) dt = \frac{2}{3} \int_{-1}^{1} (t+1) \sin(n\omega_0 t) dt$$

$$= \frac{2}{3} \underbrace{\int_{-1}^{1} t \sin(n\omega_0 t) dt}_{\text{even function}} + \frac{2}{3} \underbrace{\int_{-1}^{1} \sin(n\omega_0 t) dt}_{\text{odd function}}.$$

Since the integral of odd functions within the limits $[-t_0, t_0]$ is zero,

$$\int_{-1}^{1} \sin(n\omega_0 t) dt = 0,$$

and the value of b_n is given by

$$b_n = \frac{2}{3} \int_{-1}^{1} t \sin(n\omega_0 t) dt = \frac{4}{3} \int_{0}^{1} t \sin(n\omega_0 t) dt$$

$$= \frac{4}{3} \left[-t\frac{\cos(n\omega_0 t)}{n\omega_0} + \frac{\sin(n\omega_0 t)}{(n\omega_0)^2} \right]_{0}^{1} = -\frac{4\cos(n\omega_0)}{3n\omega_0} + \frac{4\sin(n\omega_0)}{3(n\omega_0)^2}.$$

Substituting $\omega_0 = 2\pi/3$, we obtain

$$b_n = \begin{cases} -\dfrac{2}{n\pi} & n = 3k \\[2mm] \dfrac{1}{n\pi} + \dfrac{3\sqrt{3}}{2(n\pi)^2} & n = 3k+1 \\[2mm] \dfrac{1}{n\pi} - \dfrac{3\sqrt{3}}{2(n\pi)^2} & n = 3k+2, \end{cases} \tag{4.36}$$

for $k \in Z$. The periodic signal $x(t)$ is therefore expressed as follows:

$$x(t) = \underbrace{\frac{2}{3}}_{x_{av}(t)} + \underbrace{\sum_{n=1}^{\infty} a_n \cos\left(\frac{2n\pi}{3}t\right)}_{\text{Ev}\{x(t)-a_0\}} + \underbrace{\sum_{n=1}^{\infty} b_n \sin\left(\frac{2n\pi}{3}t\right)}_{\text{Odd}\{x(t)-a_0\}}, \tag{4.37}$$

where coefficients a_n and b_n are given in Eqs. (4.35) and (4.36). Coefficient a_0 represents the average value of signal $x(t)$, referred to as $x_{av}(t)$. The cosine terms collectively represent the zero-mean even component of signal $x(t)$, denoted by $\text{Ev}\{x(t) - a_0\}$, while the sine terms collectively represent the zero-mean odd component of $x(t)$, denoted by $\text{Odd}\{x(t) - a_0\}$. Based on the values of the coefficients, the three components of $x(t)$ are plotted in Fig. 4.6. It can be verified easily that the sum of these three components will indeed produce the original signal $x(t)$.

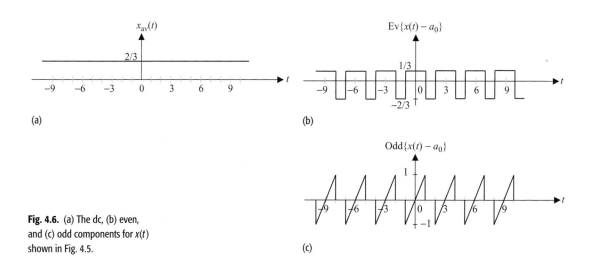

Fig. 4.6. (a) The dc, (b) even, and (c) odd components for $x(t)$ shown in Fig. 4.5.

4.4.1 CTFS coefficients for symmetrical signals

If the periodic signal $x(t)$ with angular frequency ω_0 exhibits some symmetry, then the computation of the CTFS coefficients is simplified considerably. Below, we list the properties of the trigonometric coefficients of the CTFS for symmetrical signals.

(1) If $x(t)$ is zero-mean, then $a_0 = 0$. In such cases, one does not need to calculate the dc coefficient a_0.

(2) If $x(t)$ is an even function, then $b_n = 0$ for all n. In other words, an even signal is represented by its dc component and a linear combination of a cosine function of frequency ω_0 and its higher-order harmonics.

(3) If $x(t)$ is an odd function, then $a_0 = a_n = 0$ for all n. In other words, an odd signal can be represented by a linear combination of a sine function of frequency ω_0 and its higher-order harmonics.

(4) If $x(t)$ is a real function, then the trigonometric CTFS coefficients a_0, a_n, and b_n are also real-valued for all n.

(5) If $g(t) = x(t) + c$ (where c is a constant) then the trigonometric DTFS coefficients $\{a_0^g, a_n^g, b_n^g\}$ of function $g(t)$ are related to the CTFS coefficients $\{a_0^x, a_n^x, b_n^x\}$ of $x(t)$ as follows:

$$\text{dc coefficient} \qquad\qquad a_0^g = a_0^x + c, \qquad\qquad (4.38)$$

$$\text{coefficients } a_n \qquad a_n^g = a_n^x \text{ for } n = 1, 2, 3, \ldots, \qquad (4.39)$$

$$\text{coefficients } b_n \qquad b_n^g = b_n^x \text{ for } n = 1, 2, 3, \ldots \qquad (4.40)$$

Application of the aforementioned properties is illustrated in the following examples.

Example 4.7

Consider the function $w(t) = \text{Ev}\{x(t) - a_0\}$ shown in Fig. 4.6(b). Express $w(t)$ as a trigonometric CTFS.

Solution

From inspection, we see $w(t)$ is even. Therefore, $b_n = 0$ for all n. Since $w(t)$ is periodic with a fundamental period $T_0 = 3$, $\omega_0 = 2\pi/3$. The area enclosed by one period of $w(t)$, say $t = [-1, \ 2]$, is given by $2(1/3) + 1(-2/3) = 0$. Function $w(t)$ is, therefore, zero-mean, which imples that $a_0 = 0$.

The value of a_n is calculated as follows:

$$a_n = \frac{2}{3} \int_{-1.5}^{1.5} w(t)\cos(n\omega_0 t)\mathrm{d}t = \frac{4}{3} \int_{0}^{1.5} w(t)\cos(n\omega_0 t)\mathrm{d}t,$$

which simplifies to

$$a_n = \frac{4}{3} \int_{0}^{1} \frac{1}{3}\cos(n\omega_0 t)\mathrm{d}t - \frac{4}{3} \int_{1}^{1.5} \frac{2}{3}\cos(n\omega_0 t)\mathrm{d}t$$

$$= \frac{4}{9} \left[\frac{\sin(n\omega_0 t)}{n\omega_0} \right]_{0}^{1} - \frac{8}{9} \left[\frac{\sin(n\omega_0 t)}{n\omega_0} \right]_{1}^{1.5},$$

or

$$a_n = \frac{4}{9} \frac{\sin(n\omega_0)}{n\omega_0} - \frac{8}{9} \frac{\sin(1.5 n\omega_0)}{n\omega_0} + \frac{8}{9} \frac{\sin(n\omega_0)}{n\omega_0}$$

$$= \frac{4}{3} \frac{\sin(n\omega_0)}{n\omega_0} - \frac{8}{9} \frac{\sin(1.5 n\omega_0)}{n\omega_0}.$$

Substituting $\omega_0 = 2\pi/3$, we obtain

$$a_n = \frac{2}{n\pi} \sin\left(\frac{2n\pi}{3} \right),$$

leading to the CTFS representation

$$w(t) = \sum_{n=1}^{\infty} \frac{2}{n\pi} \sin\left(\frac{2n\pi}{3} \right) \cos\left(\frac{2n\pi}{3} t \right), \tag{4.41}$$

which is same as the even component $\text{Ev}\{x(t) - a_0\}$ in Eq. (4.37) in Example 4.6. The CTFS coefficients a_n are plotted in Fig. 4.7.

From Example 4.7, we observe that a rectangular pulse train $w(t) = \text{Ev}\{x(t) - a_0\}$, as shown in Fig. 4.6(b), has a CTFS representation that includes a linear combination of an infinite number of cosine functions. A question that arises is why an infinite number of cosine functions are needed. The answer

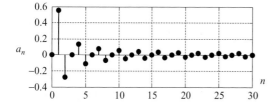

Fig. 4.7. DTFS coefficients a_n for the rectangular pulse in Example 4.7.

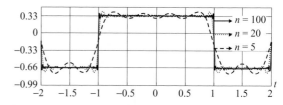

Fig. 4.8. Rectangular pulse reconstructed with a finite number n of DTFS coefficients a_n. Three different values $n = 5$, 20, and 100 are considered.

lies in the shape of the rectangular pulse that includes two constant values $(1/3, -2/3)$ separated by a discontinuity within one period. The discontinuity or the sharp transition in $w(t)$ is accounted for by a sinusoidal function with an infinite fundamental frequency. Generally, if a function has at least one discontinuity, the CTFS representation will contain an infinite number of sinusoidal functions.

Figure 4.7 shows the exponentially decaying value of the CTFS coefficients a_n. To obtain the precise waveform $w(t)$, an infinite number of the CTFS coefficients a_n are needed. Because of the decaying magnitude of the CTFS coefficients, however, a fairly reasonable approximation for $w(t)$ can be obtained by considering only a finite number of the CTFS coefficients a_n. Figure 4.8 shows the reconstruction of $w(t)$ obtained for three different values of n. We set $n = 5$, 20, and 100. It is observed that $w(t)$ provides a close approximation of $w(t)$ for $n = 20$. For $n = 100$, the approximated waveform is almost indistinguishable from the waveform of $w(t)$.

4.4.2 Jump discontinuity

Figure 4.8 shows that a CT function with a discontinuity can be approximated more accurately by including a larger number of CTFS coefficients. When approximating CT periodic functions with a finite number of CTFS coefficients, two errors arise because of the discontinuity. First, several ripples are observed in the approximated function. A careful observation of Fig. 4.8 reveals that, as more terms are added to the CTFS, the separation between the ripples becomes narrower and the approximated function is closer to the original function. The peak magnitude of the ripples, however, does not decrease with more CTFS terms. The presence of ripples near the discontinuity (i.e. around $t = \pm 1$ in Fig. 4.8) is a limitation of the CTFS representation of discontinuous signals, and is known as the *Gibbs phenomenon*.

Secondly, an approximation error is observed at the location of the discontinuity (i.e. at $t = \pm 1$ in Fig. 4.8). With a finite number of terms, it is impossible to reconstruct precisely the edge of a discontinuity. However, it is possible to calculate the value of the approximated function at the discontinuity. Suppose

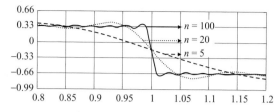

Fig. 4.9. Magnified sketch of Fig. 4.8 at $t = 1$.

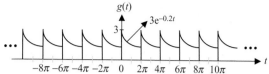

Fig. 4.10. CT periodic signal $g(t)$ with fundamental period $T_0 = 2\pi$ considered in Example 4.8.

$x(t)$ has a jump discontinuity at $t = t_j$. The reconstructed value for $x(t_j)$ is given by

$$\tilde{x}(t_j) = \frac{1}{2}[x(t_j+) + x(t_j-)]. \qquad (4.42)$$

For example, the reconstructed value of $w(t)$ in Fig. 4.8 at $t = 1$ is given by

$$\tilde{w}(1) = \frac{1}{2}[w(1-) + w(1+)] = \frac{1}{2}\left[\frac{1}{3} - \frac{2}{3}\right] = -\frac{1}{6}.$$

Figure 4.9 is an enlargement of part of Fig. 4.8 at $t = 1$, where it is observed that the reconstructed signals have a value of $-1/6$ at $t = 1$.

Example 4.8

Consider the periodic signal $g(t)$ shown in Fig. 4.10. Calculate the CTFS coefficients.

Solution

Because $T_0 = 2\pi$, the fundamental frequency $\omega_0 = 1$. The dc coefficient a_0 is given by

$$a_0 = \frac{1}{T_0}\int_{\langle T_0 \rangle} g(t)dt = \frac{1}{2\pi}\int_0^{2\pi} 3e^{-0.2t}dt = \frac{3}{2\pi}\left[\frac{e^{-0.2t}}{-0.2}\right]_0^{2\pi}$$

$$= \frac{15}{2\pi}[1 - e^{-0.4\pi}] \approx 1.7079.$$

The CTFS coefficients a_n are given by

$$a_n = \frac{2}{T_0}\int_{\langle T_0 \rangle} g(t)\cos(n\omega_0 t)dt = \frac{1}{\pi}\int_0^{2\pi} 3e^{-0.2t}\cos(nt)dt$$

$$= \frac{3}{\pi}\frac{1}{n^2 + 0.2^2}[e^{-0.2t}\{-0.2\cos(nt) + n\sin(nt)\}]_0^{2\pi}$$

or

$$a_n = \frac{3}{\pi}\frac{1}{n^2 + 0.04}[-0.2e^{-0.4\pi} + 0.2]$$

$$= \frac{0.6}{(n^2 + 0.04)\pi}[1 - e^{-0.4\pi}] \approx \frac{3.4157}{1 + 25n^2}.$$

Fig. 4.11. Periodic signal $f(t)$
considered in Example 4.9.

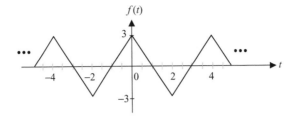

Similarly, the CTFS coefficients b_n are given by

$$b_n = \frac{2}{T_0} \int_{\langle T_0 \rangle} g(t) \sin(n\omega_0 t)\mathrm{d}t = \frac{1}{\pi} \int_0^{2\pi} 3e^{-0.2t} \sin(nt)\mathrm{d}t$$

$$= \frac{3}{\pi} \frac{1}{n^2 + 0.2^2} [e^{-0.2t}\{-0.2\sin(nt) - n\cos(nt)\}]_0^{2\pi}$$

or

$$b_n = \frac{3}{\pi} \frac{1}{n^2 + 0.04}[-ne^{-0.4\pi} + n] = \frac{3n}{(n^2 + 0.04)\pi}[1 - e^{-0.4\pi}] \approx \frac{17.0787n}{1 + 25n^2}.$$

The trigonometric CTFS representation of $g(t)$ is therefore given by

$$g(t) = 1.7079 + \sum_{n=1}^{\infty} \frac{3.4157}{1 + 25n^2} \cos(nt) + \sum_{n=1}^{\infty} \frac{17.0787}{1 + 25n^2} n \sin(nt).$$

Example 4.9
Consider the periodic signal $f(t)$ as shown in Fig. 4.11. Calculate the CTFS
coefficients.

Solution
Because $T_0 = 4$, the fundamental frequency $\omega_0 = \pi/2$. Since $f(t)$ is zero-mean,
the dc coefficient $a_0 = 0$. Also, since $f(t)$ is an even function, $b_n = 0$ for all n.
The CTFS coefficients a_n are given by

$$a_n = \frac{2}{4} \int_{-2}^{2} \underbrace{f(t)\cos(n\omega_0 t)}_{\text{even function}}\mathrm{d}t = \frac{4}{4} \int_0^2 (3 - 3t)\cos(n\omega_0 t)\mathrm{d}t$$

$$= \left[(3 - 3t)\frac{\sin(n\omega_0 t)}{n\omega_0} - 3\frac{\cos(n\omega_0 t)}{(n\omega_0)^2}\right]_0^2.$$

Substituting $\omega_0 = \pi/2$, we obtain

$$a_n = \left[(-3)\frac{\sin(n\pi)}{0.5n\pi} - 3\frac{\cos(n\pi)}{(0.5n\pi)^2} + 3\frac{1}{(0.5n\pi)^2} \right]$$

$$= 3\left[0 - \frac{(-1)^n}{(0.5n\pi)^2} + \frac{1}{(0.5n\pi)^2} \right] = \frac{12}{(n\pi)^2}[1 - (-1)^n]$$

or

$$a_n = \begin{cases} 0 & n \text{ is even} \\ \dfrac{24}{(n\pi)^2} & n \text{ is odd.} \end{cases}$$

The CTFS representation of $f(t)$ is given by

$$f(t) = \sum_{n=1,3,5,\ldots}^{\infty} \frac{24}{(n\pi)^2} \cos(0.5n\pi t)$$

$$= \frac{24}{\pi^2} \left[\cos(0.5\pi t) + \frac{1}{9}\cos(1.5\pi t) + \frac{1}{25}\cos(2.5\pi t) + \cdots \right].$$

Example 4.10
Calculate the CTFS coefficients for the following signal:

$$x(t) = 3 + \cos\left(4t + \frac{\pi}{4}\right) + \sin\left(10t + \frac{\pi}{3}\right).$$

Solution
The fundamental period of $\cos(4t + \pi/4)$ is given by $T_1 = \pi/2$, while the fundamental period of $\sin(10t + \pi/3)$ is given by $T_2 = \pi/5$. Since the ratio

$$\frac{T_1}{T_2} = \frac{5}{2}$$

is a rational number, Proposition 1.2 states that $x(t)$ is periodic with a fundamental period of π. The fundamental frequency ω_0 is therefore given by $\omega_0 = 2\pi/T_0 = 2$.

Since $x(t)$ is a linear combination of sinusoidal functions, the CTFS coefficients can be calculated directly by expanding the sine and cosine terms as follows:

$$x(t) = 3 + \cos(4t)\cos\left(\frac{\pi}{4}\right) - \sin(4t)\sin\left(\frac{\pi}{4}\right) + \sin(10t)\cos\left(\frac{\pi}{3}\right)$$
$$+ \cos(10t)\sin\left(\frac{\pi}{3}\right).$$

Substituting the values of $\sin(\pi/4)$, $\cos(\pi/4)$, $\sin(\pi/3)$, and $\cos(\pi/3)$, we obtain

$$x(t) = 3 + \frac{1}{\sqrt{2}}\cos(4t) - \frac{1}{\sqrt{2}}\sin(4t) + \frac{1}{2}\sin(10t) + \frac{\sqrt{3}}{2}\cos(10t).$$

Comparing the above equation with the CTFS expansion,

$$x(t) = a_0 + \sum_{n=1}^{\infty} (a_n \cos(n\omega_0 t) + b_n \sin(n\omega_0 t)),$$

with $\omega_0 = 2$, we obtain

$$a_0 = 3, a_2 = \frac{1}{\sqrt{2}}, a_5 = \frac{\sqrt{3}}{2}, b_2 = -\frac{1}{\sqrt{2}}, \text{ and } b_5 = \frac{1}{2}.$$

The CTFS coefficients a_n and b_n, for values of n other than $n = 0, 2$, and 5, are all zeros.

Example 4.11

A periodic signal is represented by the following CTFS:

$$x(t) = \frac{2}{\pi} \sum_{m=0}^{\infty} \frac{1}{2m + 1} \sin(4\pi(2m + 1)t).$$

- (i) From the CTFS representation, determine the fundamental period T_0 of $x(t)$.
- (ii) Comment on the symmetry properties of $x(t)$.
- (iii) Plot the function to verify if your answers to (i) and (ii) are correct.

Solution

(i) The CTFS representation is obtained by expanding the summation as follows:

$$\begin{aligned}
x(t) &= \frac{2}{\pi} \sum_{m=0}^{\infty} \frac{1}{2m + 1} \sin(4\pi(2m + 1)t) \\
&= \frac{2}{\pi} \left[\sin(4\pi t) + \frac{1}{3} \sin(12\pi t) + \frac{1}{5} \sin(20\pi t) + \frac{1}{7} \sin(28\pi t) + \cdots \right].
\end{aligned}$$

Note that the signal $x(t)$ contains the fundamental component $\sin(4\pi t)$ and its higher-order harmonics. Hence, the fundamental frequency is $\omega_0 = 4\pi$ with the fundamental period given by $T_0 = 2\pi/4\pi = 1/2$.

(ii) Because the CTFS contains only sine terms, $x(t)$ must be odd based on property (3) on page 156.

(iii) It is generally difficult to evaluate the function $x(t)$ manually. We use a MATLAB function ictfs.m (provided in the accompanying CD) to calculate $x(t)$. The function, reconstructed using the first 1000 CTFS coefficients, is plotted in Fig. 4.12 for $-1 \leq t \leq 1$. It is observed that the function is a rectangular pulse train with a fundamental period of 0.5. It is also observed that the function is odd.

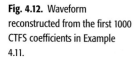

Fig. 4.12. Waveform reconstructed from the first 1000 CTFS coefficients in Example 4.11.

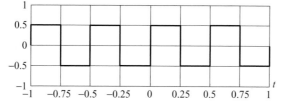

4.5 Exponential Fourier series

In Section 4.4, we considered the trigonometric CTFS expansion using a set of sinusoidal terms as the basis functions. An alternative expression for the CTFS is obtained if complex exponentials $\{\exp(jn\omega_0 t)\}$, for $n \in Z$, are used as the basis functions to expand a CT periodic signal. The resulting CTFS representation is referred to as the exponential CTFS, which is defined below.

Definition 4.7 *An arbitrary periodic function $x(t)$ with a fundamental period T_0 can be expressed as follows:*

$$x(t) = \sum_{n=0}^{\infty} D_n \mathrm{e}^{jn\omega_0 t}, \qquad (4.43)$$

where the exponential CTFS coefficients D_n are calculated as

$$D_n = \frac{1}{T_0} \int_{\langle T_0 \rangle} x(t) \mathrm{e}^{-jn\omega_0 t} \, \mathrm{d}t, \qquad (4.44)$$

ω_0 *being the fundamental frequency given by* $\omega_0 = 2\pi/T_0$.

Equation (4.43) is known as the exponential CTFS representation of $x(t)$. Since the basis functions corresponding to the trigonometric and exponential CTFS are related by Euler's identity,

$$\mathrm{e}^{-jn\omega_0 t} = \cos(n\omega_0 t) - \mathrm{j}\sin(n\omega_0 t),$$

it is intuitively pleasing to believe that the exponential and trigonometric CTFS coefficients are also related to each other. The exact relationship is derived by expanding the trigonometric CTFS series as follows:

$$x(t) = a_0 + \sum_{n=1}^{\infty} (a_n \cos(n\omega_0 t) + b_n \sin(n\omega_0 t))$$

$$= a_0 + \sum_{n=1}^{\infty} \frac{a_n}{2} (\mathrm{e}^{jn\omega_0 t} + \mathrm{e}^{-jn\omega_0 t}) + \sum_{n=1}^{\infty} \frac{b_n}{2\mathrm{j}} (\mathrm{e}^{jn\omega_0 t} - \mathrm{e}^{-jn\omega_0 t}).$$

Combining terms with the same exponential functions, we obtain

$$x(t) = a_0 + \frac{1}{2} \sum_{n=1}^{\infty} (a_n - \mathrm{j}b_n) \mathrm{e}^{jn\omega_0 t} + \frac{1}{2} \sum_{n=1}^{\infty} (a_n + \mathrm{j}b_n) \mathrm{e}^{-jn\omega_0 t}.$$

The second summation can be expressed as follows:

$$\sum_{n=1}^{\infty} (a_n + jb_n)e^{-jn\omega_0 t} = \sum_{n=-\infty}^{-1} (a_{-n} + jb_{-n})e^{jn\omega_0 t},$$

which leads to the following expression:

$$x(t) = a_0 + \frac{1}{2}\sum_{n=1}^{\infty} (a_n - jb_n)e^{jn\omega_0 t} + \frac{1}{2}\sum_{n=-\infty}^{-1} (a_{-n} + jb_{-n})e^{jn\omega_0 t}.$$

Comparing the above expansion with the definition of exponential CTFS, Eq. (4.31), yields

$$D_n = \begin{cases} a_0 & n = 0 \\ \frac{1}{2}(a_n - jb_n) & n > 0 \\ \frac{1}{2}(a_{-n} + jb_{-n}) & n < 0. \end{cases} \tag{4.45}$$

Example 4.12

Calculate the exponential CTFS coefficients for the periodic function $g(t)$ shown in Fig. 4.10.

Solution

By inspection, the fundamental period $T_0 = 2\pi$, which gives the fundamental frequency $\omega_0 = 2\pi/2\pi = 1$. The exponential CTFS coefficients D_n are given by

$$D_n = \frac{1}{T_0}\int_{\langle T_0 \rangle} g(t)e^{-jn\omega_0 t}\, dt = \frac{1}{2\pi}\int_0^{2\pi} 3e^{-0.2t}e^{-jn\omega_0 t}\, dt = \frac{3}{2\pi}\int_0^{2\pi} e^{-(0.2+jn\omega_0)\,t}\, dt$$

or

$$D_n = -\frac{3}{2\pi}\left[\frac{e^{-(0.2+jn\omega_0)t}}{(0.2+jn\omega_0)}\right]_0^{2\pi} = \frac{3}{2\pi}\frac{1}{(0.2+jn\omega_0)}\left[1 - e^{-(0.2+jn\omega_0)2\pi}\right].$$

Substituting $\omega_0 = 1$, we obtain the following expression for the exponential CTFS coefficients:

$$D_n = \frac{3}{2\pi(0.2+jn)}\left[1 - e^{-(0.2+jn)2\pi}\right]$$

$$= \frac{3}{2\pi(0.2+jn)}\left[1 - e^{-0.4\pi}\right] \approx \frac{0.3416}{(0.2+jn)}. \tag{4.46}$$

Example 4.13

Calculate the exponential CTFS coefficients for $f(t)$ as shown in Fig. 4.11.

Solution

Since the fundamental period $T_0 = 4$, the angular frequency $\omega_0 = 2\pi/4 = \pi/2$. The exponential CTFS coefficients D_n are calculated directly from the definition as follows:

$$D_n = \frac{1}{T_0} \int_{\langle T_0 \rangle} f(t)e^{-jn\omega_0 t}\,dt = \frac{1}{4} \int_{-2}^{2} f(t)e^{-jn\omega_0 t}\,dt$$

$$= \frac{1}{4} \int_{-2}^{2} \underbrace{f(t)\cos(n\omega_0 t)dt}_{\text{even function}} - j\frac{1}{4} \int_{-2}^{2} \underbrace{f(t)\sin(n\omega_0 t)dt}_{\text{odd function}}.$$

Since the integration of an odd function within the limits $[t_0, -t_0]$ is zero,

$$D_n = \frac{1}{4} \int_{-2}^{2} f(t)\cos(n\omega_0 t)dt = \frac{1}{2} \int_{0}^{2} (3 - 3t)\cos(n\omega_0 t)dt,$$

which simplifies to

$$D_n = \frac{1}{2} \left[(3 - 3t)\frac{\sin(n\omega_0 t)}{n\omega_0} - 3\frac{\cos(n\omega_0 t)}{(n\omega_0)^2} \right]_0^2$$

$$= \frac{3}{2} \left[-\frac{\sin(2n\omega_0)}{n\omega_0} - \frac{\cos(2n\omega_0)}{(n\omega_0)^2} + \frac{1}{(n\omega_0)^2} \right].$$

Substituting $\omega_0 = \pi/2$, we obtain

$$D_n = \frac{3}{2} \left[-\frac{\sin(n\pi_0)}{0.5n\pi} - \frac{\cos(n\pi)}{(0.5n\pi)^2} + \frac{1}{(0.5n\pi)^2} \right] = \frac{6}{(n\pi)^2}[1 - (-1)^n]$$

or

$$D_n = \begin{cases} 0 & n \text{ is even} \\ \dfrac{12}{(n\pi)^2} & n \text{ is odd}. \end{cases} \tag{4.47}$$

In Examples 4.12 and 4.13, the exponential CTFS coefficients can also be derived from the trigonometric CTFS coefficients calculated in Examples 4.8 and 4.9 using Eq. (4.45).

Example 4.14

Calculate the exponential Fourier series of the signal $x(t)$ shown in Fig. 4.13.

Solution

The fundamental period $T_0 = T$, and therefore the angular frequency $\omega_0 = 2\pi/T$. The exponential CTFS coefficients are given by

$$D_n = \frac{1}{T} \int_{-T/2}^{T/2} x(t)e^{-jn\omega_0 t}\,dt = \frac{1}{T} \int_{-\tau/2}^{\tau/2} 1 \cdot e^{-jn\omega_0 t}\,dt.$$

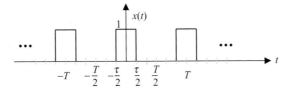

From integral calculus, we know that

$$\int e^{-jn\omega_0 t}\, dt = \begin{cases} -\dfrac{1}{jn\omega_0} e^{-jn\omega_0 t} + c & n \neq 0 \\ t + c & n = 0. \end{cases} \tag{4.48}$$

We consider the two cases separately.

Case I For $n = 0$, the exponential CTFS coefficients are given by

$$D_n = \frac{1}{T}[t]_{-\tau/2}^{\tau/2} = \frac{\tau}{T}.$$

Case II For $n \neq 0$, the exponential CTFS coefficients are given by

$$D_n = -\frac{1}{jn\omega_0 T}[e^{-jn\omega_0 t}]_{-\tau/2}^{\tau/2} = \frac{1}{n\pi}\sin\left(\frac{n\pi\tau}{T}\right)$$

or

$$D_n = \frac{\tau}{T}\frac{\sin\left(\pi\dfrac{n\tau}{T}\right)}{\left(\pi\dfrac{n\tau}{T}\right)} = \frac{\tau}{T}\operatorname{sinc}\left(\frac{n\tau}{T}\right).$$

In the above derivation, the CTFS coefficients are computed separately for $n = 0$ and $n \neq 0$. However, on applying the limit $n \to 0$ to the D_n in case II, we obtain

$$\lim_{n\to 0} D_n = \lim_{n\to 0}\frac{\tau}{T}\operatorname{sinc}\left(\frac{n\tau}{T}\right) = \frac{\tau}{T}\lim_{n\to 0}\operatorname{sinc}\left(\frac{n\tau}{T}\right) = \frac{\tau}{T}$$

$$\left[\because \lim_{x\to 0}\operatorname{sinc}(mx) = 1\right].$$

In other words, the value of D_n for $n = 0$ is covered by the value of D_n for $n \neq 0$. Therefore, combining the two cases, the CTFS coefficient for the function $x(t)$ is expressed as follows:

$$D_n = \frac{\tau}{T}\frac{\sin\left(\pi\dfrac{n\tau}{T}\right)}{\left(\pi\dfrac{n\tau}{T}\right)} = \frac{\tau}{T}\operatorname{sinc}\left(\frac{n\tau}{T}\right), \tag{4.49}$$

for $-\infty < n < \infty$. As a special case, we set $\tau = \pi/2$ and $T = 2\pi$. The resulting waveform for $x(t)$ is shown in Fig. 4.14(a). The CTFS coefficients for the

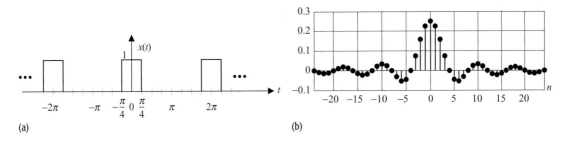

Fig. 4.14. Exponential CTFS coefficients for the signal $x(t)$ shown in Fig. 4.13 with $\tau = \pi/2$ and $T = 2\pi$. (a) Waveform for $x(t)$. (b) Exponential CTFS coefficients.

special case are given by

$$D_n = \frac{1}{4} \mathrm{sinc}\left(\frac{n}{4}\right),$$

for $-\infty < n < \infty$. The CTFS coefficients are plotted in Fig. 4.14(b).

As a side note to our discussion on exponential CTFS, we make the following observations.

(i) The exponential CTFS provide a more compact representation compared with the trigonometric CTFS. However, the exponential CTFS coefficients are generally complex-valued.

(ii) For real-valued functions, the coefficients D_n and D_{-n} are complex conjugates of each other. This is easily verified from Eq. (4.45) and the symmetry property (4) described in Section 4.4.

4.5.1 Fourier spectrum

The exponential CTFS coefficients provide frequency information about the content of a signal. However, it is difficult to understand the nature of the signal by looking at the values of the coefficients, which are generally complex-valued. Instead, the exponential CTFS coefficients are generally plotted in terms of their magnitude and phase. The plot of the magnitude of the exponential CTFS coefficients $|D_n|$ versus n (or $n\omega_0$) is known as the magnitude (or amplitude) spectrum, while the plot of the phase of the exponential CTFS $< D_n$ versus n (or $n\omega_0$) is referred to as the phase spectrum.

Example 4.15
Plot the magnitude and phase spectra of the signal $g(t)$ considered in Example 4.12.

Solution
From Example 4.12, we know that the exponential CTFS coefficients are given by

$$D_n = \frac{0.3416}{0.2 + jn}.$$

Table 4.1. Magnitude and phase of D_n for a few values of n given in Example 4.15

n	0	± 1	± 2	± 3	± 4	...	$\pm \infty$
$\|D_n\|$	1.7080	0.3350	0.1700	0.1136	0.0853	...	0
$<D_n$	0	$\mp 0.4372\pi$	$\mp 0.4683\pi$	$\mp 0.4788\pi$	$\mp 0.4841\pi$	...	$\mp 0.5\pi$

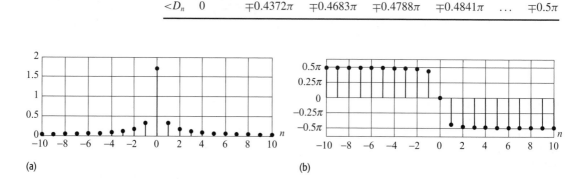

(a)   (b)

Fig. 4.15. CTFS coefficients of signal $g(t)$ shown in Fig. 4.10. (a) Magnitude spectrum. (b) Phase spectrum.

The magnitude and phase of the exponential CTFS coefficients are as follows:

magnitude
$$|D_n| = \frac{0.3416}{|(0.2 + jn)|} = \frac{0.3416}{\sqrt{0.04 + n^2}};$$

phase
$$<D_n = 0.3416 - \; <(0.2 + jn) = -\tan^{-1}(5n).$$

Table 4.1 shows the magnitude and phase of D_n for a few selected values of n. The phase values are expressed in radians. The magnitude and phase spectra are plotted in Fig. 4.15.

The magnitude of the exponential CTFS coefficients D_n indicates the strength of the frequency component $n\omega_0$ (i.e. the nth harmonic) in the signal $x(t)$. The phase of D_n provides additional information on how different harmonics should be shifted and added to reconstruct $x(t)$.

Example 4.16

Calculate and plot the amplitude and phase spectra of signal $x(t)$ considered in Example 4.14 for $\tau = \pi/2$ and $T = 2\pi$.

Solution

The exponential DTFS coefficients are given by

$$D_n = \frac{\tau}{T} \text{sinc}\left(\frac{n\tau}{T}\right).$$

Substituting $\tau = \pi/2$ and $T = 2\pi$, we obtain

$$D_n = \frac{1}{4} \text{sinc}\left(\frac{n}{4}\right),$$

which are plotted in Fig. 4.14. Note that the coefficients are all real-valued but periodically vary between positive and negative values. Because the CTFS coefficients D_n do not have imaginary components, the phase corresponding to

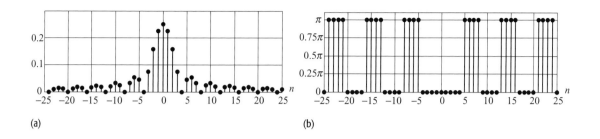

Fig. 4.16. (a) The amplitude and (b) the phase spectra of the function shown in Fig. 4.14 (see Example 4.14). The phase spectra are given in radians/s.

the CTFS coefficients is calculated from its sign as follows:

if $D_n \geq 0$, then the associated phase $<D_n = 0$;

if $D_n < 0$, then the associated phase $<D_n = \pi$ or $-\pi$.

The magnitude and phase spectra are plotted in Fig. 4.16. In Fig. 4.16(a), we observe that the magnitude spectrum is always positive, while the phase spectrum toggles between the values of 0 and π radians/s. Note that the phase plot is not unique since the phase of π is equivalent to the value of $-\pi$.

4.6 Properties of exponential CTFS

The exponential CTFS has several interesting properties that are useful in the analysis of CT signals. We list the important properties in the following discussion.

Symmetry property For real-valued periodic signals, the exponential CTFS coefficients D_n and D_{-n} are complex conjugates of each other.

Proof
Recall that the exponential CTFS coefficients are related to the trigonometric CTFS coefficients by Eq. (4.45), given below

$$D_n = \frac{1}{2}(a_n - jb_n) \qquad \text{for } n > 0$$

and

$$D_{-n} = \frac{1}{2}(a_n + jb_n) \qquad \text{for } n > 0.$$

For real-valued functions, property (4) of the symmetric functions in Section 4.4.1 states that the trigonometric Fourier coefficients a_n and b_n are always real. Based on the aforementioned equations, the exponential CTFS coefficients D_n and D_{-n} are therefore complex conjugates of each other.

As a corollary to this property, consider the magnitude and phase of the exponential CTFS coefficients:

$$|D_{-n}| = |D_n| = \frac{1}{4}\sqrt{a_n^2 + b_n^2} \qquad (4.50)$$

Fig. 4.17. Periodic signal $s(t)$ for Example 4.18.

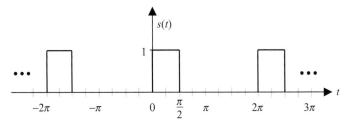

The two waveforms $s(t)$ and $x(t)$ have the same time period $T_0 = 2\pi$, which gives $\omega_0 = 1$. Based on the time-shifting property, we obtain

$$s(t) = x\left(t - \frac{\pi}{4}\right) \xleftrightarrow{\text{CTFS}} D_n e^{-jn\pi/4},$$

where D_n denotes the exponential CTFS coefficients of $x(t)$. Using the value of D_n from Example 4.14, the CTFS coefficients S_n for $s(t)$ are given by

$$S_n = \frac{1}{4}\,\text{sinc}\left(\frac{n}{4}\right) e^{-jn\pi/4},$$

for $-\infty < n < \infty$. From the above expression, it is clear that the magnitude $|S_n| = |D_n|$, but that the phase of S_n changes by an additive factor of $-n\pi/4$.

Time reversal If a periodic signal $x(t)$ is time-reversed, the amplitude spectrum remains unchanged. The phase spectrum changes by an exponential phase shift. Mathematically,

$$\text{if } x(t) \xleftrightarrow{\text{CTFS}} D_n \quad \text{then} \quad x(-t) \xleftrightarrow{\text{CTFS}} D_{-n}, \tag{4.59}$$

which implies that if a signal is time-reversed, the CTFS coefficients of a time-reversed signal are the time-reversed CTFS coefficients of the original signal.

Example 4.19
Calculate the exponential CTFS coefficients of the periodic signal $p(t)$ shown in Fig. 4.18. Represent the function as a CTFS.

Solution
From Fig. 4.18, it is observed that $p(t)$ is a time-reversed version of $s(t)$ plotted in Fig. 4.17. Therefore, the exponential CTFS coefficients can be obtained by

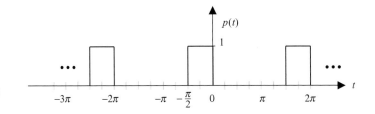

Fig. 4.18. The periodic signal $p(t)$ in Example 4.19.

applying the time-reversal property to the answer in Example 4.18. Using the latter approach, the CTFS coefficients P_n for $p(t)$ are given by

$$P_n = S_{-n} = \frac{1}{4}\,\text{sinc}\left(\frac{-n}{4}\right)\,e^{-j(-n)\pi/4} = \frac{1}{4}\,\text{sinc}\left(\frac{n}{4}\right)\,e^{jn\pi/4}. \qquad (4.60)$$

Equation (4.60) can also be obtained directly by applying the time-shifting property ($t_0 = -\pi/4$) to the waveform in Fig. 4.14(a) in Example 4.14.

The function $p(t)$ can now be represented as an exponential CTFS as follows:

$$p(t) = \sum_{n=-\infty}^{\infty} P_n e^{jn\omega_0 t} = \frac{1}{4}\sum_{n=-\infty}^{\infty}\text{sinc}\left(\frac{n}{4}\right)e^{jn\pi/4}e^{jnt}$$

$$= \frac{1}{4}\sum_{n=-\infty}^{\infty}\text{sinc}\left(\frac{n}{4}\right)e^{jn(t+\pi/4)},$$

where the fundamental frequency ω_0 is set to 1.

Time scaling   If a periodic signal $x(t)$ with period T_0 is time-scaled, the CTFS spectra are inversely time-scaled. Mathematically,

$$\text{if } x(t) \xleftrightarrow{\text{CTFS}} D_n \quad \text{then} \quad x\left(\frac{t}{a}\right) \xleftrightarrow{\text{CTFS}} D_{an}, \qquad (4.61)$$

where the time period of the time-scaled signal $x(t/a)$ is given by (T_0/a).

Example 4.20

Calculate the exponential CTFS coefficients of the periodic function $r(t)$ shown in Fig. 4.19. Represent the function as a CTFS.

Solution

From Fig. 4.19, it is observed that $r(t)$ (with $T_0 = \pi$) is a time-scaled version of $x(t)$ (with $T_0 = 2\pi$) plotted in Fig. 4.14. The relationship between $r(t)$ and $x(t)$ is given by

$$r(t) = 2x(2t).$$

Using the time-scaling and linearity properties,

$$\text{if } x(t) \xleftrightarrow{\text{CTFS}} D_n \quad \text{then} \quad 2x(2t) \xleftrightarrow{\text{CTFS}} 2D_{n/2}. \qquad (4.62)$$

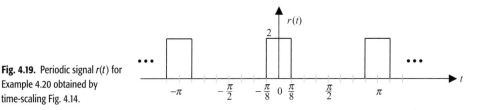

Fig. 4.19. Periodic signal $r(t)$ for Example 4.20 obtained by time-scaling Fig. 4.14.

Using the results obtained in Example 4.14, the CTFS coefficients R_n of $r(t)$ are given by

$$R_n = 2\frac{1}{4}\operatorname{sinc}\left(\frac{n/2}{4}\right) = \frac{1}{2}\operatorname{sinc}\left(\frac{n}{8}\right), \tag{4.63}$$

for $-\infty < n < \infty$. The function $r(t)$ can now be represented as an exponential CTFS as follows:

$$x(t) = \sum_{n=-\infty}^{\infty} R_n e^{jn\omega_0 t} = \frac{1}{2}\sum_{n=-\infty}^{\infty}\operatorname{sinc}\left(\frac{n}{8}\right)e^{j2nt},$$

where the fundamental frequency ω_0 is set to 2.

Differentiation and integration The exponential CTFS coefficients of the time-differentiated and time-integrated signal are expressed in terms of the exponential CTFS coefficients of the original signal as follows:

$$\text{if } x(t) \xleftrightarrow{\text{CTFS}} D_n \quad \text{then} \quad \frac{dx}{dt} \xleftrightarrow{\text{CTFS}} jn\omega_0 D_n \quad \text{and} \quad \int_{T_0} x(t)dt \xleftrightarrow{\text{CTFS}} \frac{D_n}{jn\omega_0}. \tag{4.64}$$

It may be noted that the signal obtained by differentiating or integrating a periodic signal $x(t)$ over one period T_0 has the same period T_0 as that of the original signal.

Example 4.21
Calculate the exponential CTFS coefficients of the periodic signal $g(t)$ shown in Fig. 4.20.

Solution
The function $g(t)$ can be obtained by differentiating $x(t)$ shown in Fig. 4.14. Therefore, the CTFS coefficients G_n can be expressed in terms of the CTFS coefficients D_n as follows:

$$G_n = jn\omega_0 D_n \quad \text{with} \quad \omega_0 = 1.$$

Substituting the value of

$$D_n = \frac{1}{4}\operatorname{sinc}\left(\frac{n}{4}\right)$$

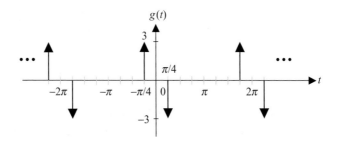

Fig. 4.20. Periodic signal $g(t)$ for Example 4.21.

yields

$$G_n = (jn)\frac{1}{4}\text{sinc}\left(\frac{n}{4}\right) = \frac{jn}{4}\text{sinc}\left(\frac{n}{4}\right).$$

The function $g(t)$ can now be represented as an exponential CTFS as follows:

$$g(t) = \sum_{n=-\infty}^{\infty} G_n e^{jn\omega_0 t} = \frac{1}{4}\sum_{n=-\infty}^{\infty} (jn)\,\text{sinc}\left(\frac{n}{4}\right) e^{jnt},$$

where the fundamental frequency ω_0 is set to 1.

4.6.1 CTFS with different periods

In this section, we consider the variation of the CTFS when the period of a function is changed. We use the rectangular pulse train for simplicity as its CTFS coefficients are real-valued.

Example 4.22
Consider the periodic function $x(t)$ in Fig. 4.13 (in Example 4.14) for the following three cases:

(a) $\tau = 1$ ms and $T = 5$ ms;
(b) $\tau = 1$ ms and $T = 10$ ms;
(c) $\tau = 1$ ms and $T = 20$ ms.

In each of the above cases, (i) determine the fundamental frequency, (ii) plot the CTFS coefficients, and (iii) determine the higher-order harmonics absent in the function.

Solution
It was shown in Example 4.14 that the exponential DTFS coefficients are given by

$$D_n = \frac{\tau}{T}\,\text{sinc}\left(\frac{n\tau}{T}\right).$$

(a) With $T = 5$ ms, the fundamental frequency is $f_0 = 1/T = 1/5$ ms $= 200$ Hz, while the fundamental angular frequency is $\omega_0 = 2\pi f_0 = 400\pi$ radians/s. The corresponding exponential CTFS coefficients are given by

$$D_n = \frac{1}{5}\,\text{sinc}\left(\frac{n}{5}\right),$$

which are plotted in Fig. 4.21(a) using two scales on the horizontal axis. The first scale represents the number n of the CTFS coefficients and the second scale

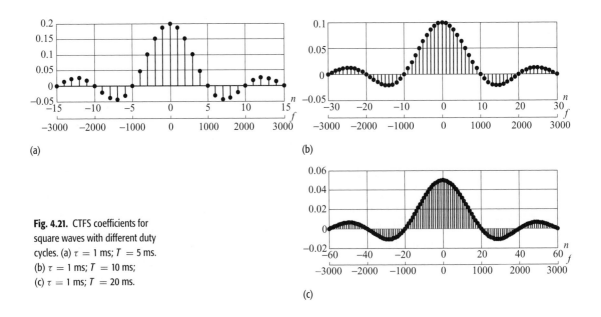

Fig. 4.21. CTFS coefficients for square waves with different duty cycles. (a) $\tau = 1$ ms; $T = 5$ ms. (b) $\tau = 1$ ms; $T = 10$ ms; (c) $\tau = 1$ ms; $T = 20$ ms.

represents the corresponding frequency $f = nf_0$ in hertz. The CTFS coefficient for $n = 0$ (or $f = 0$ Hz) has a value of 0.2, which is the strength of the dc component in the function. The spectrum at $n = 1$ (or $f = 200$ Hz) has a value of 0.19, which is the strength of the fundamental frequency (corresponding to 200 Hz, or 400π radians/s) in the function. The spectrum at $n = 2$ has a value of 0.15, which is the strength of the first harmonic corresponding to a frequency f of 400 Hz, or angular frequency ω_0 of 800π radians/s in the function.

From Fig. 4.21(a), we observe that the CTFS coefficients D_n are zero at $n = \pm 5, \pm 10, \pm 15, \ldots$, which correspond to frequencies ± 1000 Hz, ± 2000 Hz, ± 3000 Hz, $\ldots$ (i.e. nf_0), respectively. In other words, the missing harmonics will correspond to frequencies ± 1000 Hz, ± 2000 Hz, ± 3000 Hz, $\ldots$ or $m \times 10^3$ Hz, where m is a non-zero integer.

(b) With T set to 10 ms, the fundamental frequency $f_0 = 1/T = 1/10$ ms $= 100$ Hz, while the fundamental angular frequency is given by $\omega_0 = 2\pi f_0 = 200\pi$ radians/s. The exponential CTFS coefficients are now given by

$$D_n = \frac{1}{10} \operatorname{sinc}\left(\frac{n}{10}\right),$$

which are plotted in Fig. 4.21(b). The CTFS coefficient for $n = 0$ has a value of 0.1. With $T = 10$ ms, the harmonics corresponding to $n = \pm 10, \pm 20, \pm 30, \ldots$ are all equal to zero. Interestingly, the missing harmonics correspond to frequencies $f = nf_0$, which are given by $\pm 1000, \pm 2000, \pm 3000, \ldots$ Hz, have the same values as the frequency components missing in part (a).

(c) With T set to 20 ms, the new fundamental frequency $f_0 = 1/T = 1/20$ ms $= 50$ Hz, while the fundamental angular frequency is given by

$\omega_0 = 2\pi f_0 = 100\pi$ radians/s. The exponential CTFS coefficients are now given by

$$D_n = \frac{1}{20} \operatorname{sinc}\left(\frac{n}{20}\right),$$

which are plotted in Fig. 4.21(c). The CTFS coefficient for $n = 0$ has a value of 0.05. With $T = 20$ ms, the harmonics corresponding to $n = \pm20, \pm40, \pm60, \ldots$ are all equal to zero. As was the case in parts (a) and (b), the missing harmonics correspond to frequencies f of $\pm1000, \pm2000, \pm3000, \ldots$ Hz.

For a square wave, the ratio τ/T is referred to as the duty cycle, which is defined as the ratio between the time τ that the waveform has a high value and the fundamental period T. Cases (a)–(c) are illustrated in Figs. 4.21(a)–(c), where the duty cycle was reduced by keeping τ constant and increasing the value of the fundamental period T. Alternatively, the duty cycle may be decreased by reducing the value of τ, while maintaining the fundamental period T at a constant value. By changing the duty cycle, we observe the following variations in the exponential DTFS representation.

DC coefficient Since the dc coefficient represents the average value of the waveform, the value of the dc coefficient D_0 decreases as the duty cycle (τ/T) of the square wave is reduced.

Zero crossings As the duty cycle (τ/T) is decreased, the energy within one period of the waveform in the time domain is concentrated over a relatively narrower fraction of the time period. Based on the time-scaling property, the energy in the corresponding CTFS representations is distributed over a larger number of the CTFS coefficients. In other words, the width of the main lobe and side lobes of the discrete sinc function increases with a reduction in the duty cycle.

4.7 Existence of Fourier series

In Sections 4.4 and 4.5, the trigonometric and exponential CTFS representations of a periodic signal were covered. Because the CTFS coefficients are calculated by integration, there is a possibility that the integral may result in an infinite value. In this case, we state that the CTFS representation does not exist. Below we list the conditions for the existence of the CTFS representation.

Definition 4.8 *The CTFS representation (trigonometric or exponential) of a periodic function $x(t)$ exists if all CTFS coefficients are finite and the series converges for all n. In other words, there is no infinite value in the magnitude spectrum of the CTFS representation.*

For the CTFS representation to exist, the periodic signal $x(t)$ must satisfy the following three conditions.

(1) *Absolutely integrable.* The area under one period of $|x(t)|$ is finite, i.e.

$$\int_{T_0} |x(t)|\mathrm{d}t < \infty. \tag{4.65}$$

(2) *Bounded variation.* The periodic signal $x(t)$ has a finite number of maxima or minima in one period.

(3) *Finite discontinuities.* The period $x(t)$ has a finite number of discontinuities in one period. In addition, each of the discontinuity has a finite value.

The above conditions are known as the Dirichlet conditions.[†] If these conditions are satisfied, it is guaranteed that perfect reconstruction is obtained from the CTFS coefficients except at a few isolated points where the function $x(t)$ is discontinuous. The first condition is also known as the *weak* Dirichlet condition, whereas the second and third conditions are known as *strong* Dirichlet conditions. Most practical signals satisfy these three conditions. Examples of the CT functions that violate these conditions are included in the following discussion.

Example 4.23

Determine whether the following functions satisfy the Dirichlet conditions:

(i) $h(t) = \tan(\pi t)$; $\qquad\qquad\qquad\qquad\qquad\qquad\qquad\qquad$ (4.66)

(ii) $g(t) = \sin(0.5\pi/t)$ for $0 \leq t < 1$ and $g(t) = g(t+1)$; $\qquad$ (4.67)

(iii) $x(t) = \begin{cases} 1 & 2^{-2m-1} < t \leq 2^{-2m} \\ 0 & 2^{-2m-2} < t \leq 2^{-2m-1} \end{cases}$ $\qquad\qquad$ (4.68)

for $m \in Z^+$, $0 \leq t < 1$, and $x(t) = x(t+1)$.

Solution

(i) The CT function $h(t)$ is plotted in Fig. 4.22(a). We now proceed to determine if $h(t)$ satisfies the Dirichlet conditions. Condition (1) is violated because

$$\int_{T_0} |h(t)|\mathrm{d}t = \int_{-0.5}^{0.5} \tan(\pi t)\mathrm{d}t = \infty.$$

This is also apparent from the waveform of $\tan(\pi t)$, plotted in Fig. 4.22(a), where the waveform approaches $\pm\infty$ at each discontinuity. Condition (2) is satisfied as there are only one maximum and one minimum within a single period of $h(t)$. Condition (3) is violated. Although there is only one discontinuity within a single period of $h(t)$, the magnitude of the discontinuity is infinite.

(ii) The CT function $g(t)$ is plotted in Fig. 4.22(b). Condition (1) is satisfied as the area enclosed by $|g(t)|$ is finite. Condition (2) is violated as an

---

[†] These conditions were derived by Johann Peter Gustav Lejeune Dirichlet (1805–1859), a German mathematician.

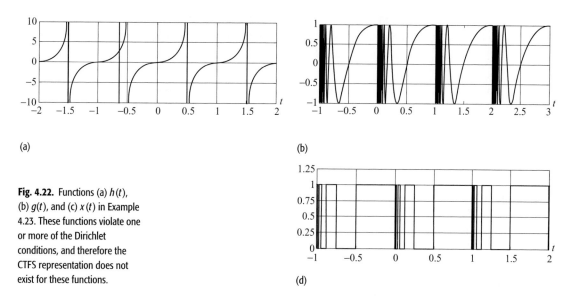

Fig. 4.22. Functions (a) $h(t)$, (b) $g(t)$, and (c) $x(t)$ in Example 4.23. These functions violate one or more of the Dirichlet conditions, and therefore the CTFS representation does not exist for these functions.

infinite number of maxima and minima exist within a single period of $g(t)$. Condition (3) is satisfied as there are no discontinuities within a single period of $g(t)$.

(iii) The CT function $x(t)$ is plotted in Fig. 4.22(c). Condition (1) is satisfied as the area enclosed by $|x(t)|$ is finite. Condition (2) is violated as there are an infinite number of maxima and minima within a single period of $g(t)$. Condition (3) is violated as an infinite number of discontinuities exist within a single period of $g(t)$.

4.8 Application of Fourier series

The exponential CTFS has several interesting applications. In Section 4.8.1, we highlight an application of the CTFS representation in calculating the sum of an infinite series. Section 4.8.2 considers the use of the CTFS representation in calculating the response of an LTIC system to a periodic signal. By using the CTFS representation, we avoid the convolution integral.

4.8.1 Computing the sum of an infinite series

The following example illustrates an application of the CTFS in calculating the sum of a series:

Example 4.24

Calculate the sum S of the following infinite series:

$$S = \sum_{n=0}^{\infty} \frac{1}{(2n+1)^4} = 1 + \frac{1}{3^4} + \frac{1}{5^4} + \frac{1}{7^4} + \frac{1}{9^4} + \frac{1}{11^4} + \cdots$$

Solution

To compute the sum S, we consider the periodic signal $f(t)$ shown in Fig. 4.11. As shown in Example 4.13, the exponential CTFS coefficients of $f(t)$ are given by

$$D_n = \begin{cases} 0 & n \text{ is even} \\ \dfrac{12}{(n\pi)^2} & n \text{ is odd.} \end{cases}$$

Using Parseval's theorem, the average power of $f(t)$ is given by

$$P_x = \sum_{n=-\infty}^{\infty} |D_n|^2 = |D_0|^2 + 2\sum_{n=1}^{\infty} |D_n|^2 = 2\sum_{\substack{n=1 \\ n=odd}}^{\infty} \frac{144}{\pi^4} \cdot \frac{1}{n^4} = \frac{288}{\pi^4} S. \tag{4.69}$$

Using the time-domain approach, it was shown in Example 4.17 that the average power of $f(t)$ is given by

$$P_f = \frac{1}{T_0} \int_{-2}^{2} |x(t)|^2 dt = 3. \tag{4.70}$$

Combining Eqs. (4.69) and (4.70) gives $(288/\pi^4)\, S = 3$ or

$$S = \sum_{n=0}^{\infty} \frac{1}{(2n+1)^4} = \frac{3\pi^4}{288} = \frac{\pi^4}{96} \approx 1.0147.$$

4.8.2 Response of an LTIC system to periodic signals

As a second application of the exponential CTFS representation, we consider the response $y(t)$ of an LTIC system with the impulse response $h(t)$ to an periodic input $x(t)$. The system is illustrated in Fig. 4.23. Assuming that the input signal $x(t)$ has the fundamental period T_0, the exponential CTFS representation of $x(t)$ is given by

$$x(t) = \sum_{m=0}^{\infty} D_n e^{jn\omega_0 t}, \tag{4.71}$$

where the fundamental frequency $\omega_0 = 2\pi/T_0$. The steps involved in calculating the output $y(t)$ are as follows.

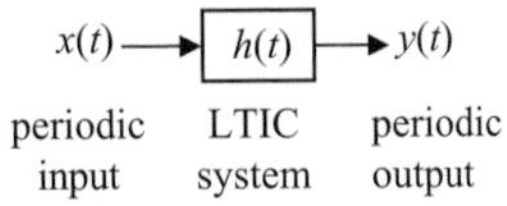

$x(t) \longrightarrow \boxed{h(t)} \longrightarrow y(t)$

periodic   LTIC   periodic
input   system   output

Fig. 4.23. Response of an LTIC system to a periodic input.

Step 1 Based on Theorem 4.1, the output of an LTIC system $y_n(t)$ to a complex exponential $x_n(t) = D_n \exp(jn\omega_0 t)$ is given by

$$y_n(t) = D_n H(n\omega_0) e^{jn\omega_0 t}, \tag{4.72}$$

where $H(n\omega_0) = H(\omega)$, evaluated at $\omega = n\omega_0$. The new term $H(\omega)$ is referred to as the transfer function of the LTIC system and is given by

$$H(\omega) = \int_{-\infty}^{\infty} h(t)e^{-j\omega t}\,dt. \qquad (4.73)$$

Step 2 Using the principle of superposition, the overall output $y(t)$ by adding individual outputs $y_n(t)$ is given by

$$y(t) = \sum_{n=-\infty}^{\infty} y_n(t) \qquad (4.74)$$

or

$$y(t) = \sum_{n=-\infty}^{\infty} D_n e^{jn\omega_0 t} H(\omega)|_{\omega=n\omega_0}. \qquad (4.75)$$

Step 3 Based on Eq. (4.75), it is clear that the response $y(t)$ of an LTIC system to a periodic input $x(t)$ is also periodic with the same fundamental period as $x(t)$. In addition, the exponential CTFS coefficients E_n of the output $y(t)$ are related to the CTFS coefficients D_n of the periodic input signal $x(t)$ by the following relationship

$$E_n = D_n H(\omega)|_{\omega=n\omega_0}. \qquad (4.76)$$

Example 4.25
Calculate the exponential CTFS coefficients of the output $y(t)$ if the square wave $x(t)$ illustrated in Fig. 4.14 is applied as the input to an LTIC system with impulse response $h(t) = \exp(-2t)u(t)$.

Solution
The exponential CTFS coefficients of the square wave $x(t)$ shown in Fig. 4.14(a) are given by (see Example 4.14)

$$D_n = \frac{1}{4}\,\mathrm{sinc}\left(\frac{n}{4}\right), \quad \text{for } -\infty < n < \infty.$$

The transfer function $H(\omega)$ of the LTIC is given by

$$H(\omega) = \int_{-\infty}^{\infty} h(t)e^{-j\omega t}\,dt = \int_{0}^{\infty} e^{-(2+j\omega)t}\,dt = \frac{1}{(2+j\omega)}. \qquad (4.77)$$

For $\omega_0 = 1$ radian/s, the exponential CTFS coefficients of the output $y(t)$ are given by

$$E_n = D_n H(\omega)|_{\omega=n} = \frac{1}{4}\,\mathrm{sinc}\left(\frac{n}{4}\right) \times \frac{1}{(2+jn)} = \frac{\mathrm{sinc}(n/4)}{8+j4n}, \qquad (4.78)$$

and the output $y(t)$ is given by

$$y(t) = \sum_{n=-\infty}^{\infty} E_n e^{jn\omega_0 t} = \sum_{n=-\infty}^{\infty} \frac{\mathrm{sinc}(n/4)}{8+j4n} e^{jnt}. \qquad (4.79)$$

Fig. 4.24. Response of the LTIC
system in Example 4.25.

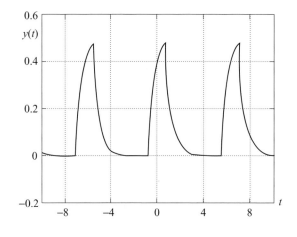

Using the MATLAB function ictfs.m (provided in the accompanying CD),
$y(t)$ is calculated and shown in Fig. 4.24. It is observed that $y(t)$ does not have
any sharp (rising or falling) edges. This is primarily because, at high frequencies,
the gain of the system ($|H(\omega)|$) is small. As the high-frequency components
of the inputs are suppressed by the system, the sharp edges are absent at the
output.

Example 4.25 used the CTFS to calculate the output $y(t)$ of a periodic signal
$x(t)$. Such a method is limited to periodic input signals. In Chapter 5, we show
how the continuous-time Fourier transform (CTFT) can be used to compute the
output of the LTIC systems for both periodic and aperiodic inputs. Since the
CTFT is more inclusive than the CTFS representation, our analysis of the LTIC
systems will be based primarily on the frequency decompositions using the
CTFT. The CTFS is, however, used indirectly to compute the CTFT of periodic
signals. We shall explore the relationship between the CTFS and CTFT more
fully in Chapter 5.

4.9 Summary

In Chapter 4, we introduced frequency-domain analysis of periodic signals based on the trigonometric and exponential CTFS representations. In
Sections 4.1 and 4.2, the basis functions are defined as a complete set $\{p_n(t)\}$,
for $1 \leq n \leq N$, of orthogonal functions satisfying the following orthogonality
properties over interval $[t_1, t_2]$:

orthogonality property
$$\int_{t_1}^{t_2} p_m(t)p_n^*(t)dt = \begin{cases} E_n \neq 0 & m = n \\ 0 & m \neq n \end{cases}$$

$$\text{for } 1 \leq m, n \leq N,$$

for any pair of functions taken from the set $\{p_n(t)\}$. Section 4.3 proves that the complex exponentials $\{\exp(jn\omega_0 t)\}$, for $-\infty < n < \infty$, and sinusoidal functions $\{\sin(n\omega_0 t), 1, \cos(m\omega_0 t)\}$, for $0 < n, m < \infty$, form two complete orthogonal sets over any interval $[t_1, t_1 + 2\pi/\omega_0]$ of duration $T_0 = 2\pi/\omega_0$. We refer to ω_0 as the angular frequency and to its inverse $T_0 = 2\pi/\omega_0$ as the fundamental period. Expressing a periodic signal $x(t)$ as a linear combination of the sinusoidal set of functions $\{\sin(n\omega_0 t), 1, \cos(m\omega_0 t)\}$ leads to the trigonometric representation of the CTFS. The trigonometric CTFS is defined as follows:

$$x(t) = a_0 + \sum_{n=1}^{\infty} (a_n \cos(n\omega_0 t) + b_n \sin(n\omega_0 t)),$$

where $\omega_0 = 2\pi/T_0$ is the fundamental frequency of $x(t)$ and coefficients a_0, a_n, and b_n are referred to as the trigonometric CTFS coefficients. The coefficients are calculated using the following formulas:

$$a_0 = \frac{1}{T_0} \int_{\langle T_0 \rangle} x(t) dt,$$

$$a_n = \frac{2}{T_0} \int_{\langle T_0 \rangle} x(t) \cos(n\omega_0 t) dt,$$

and

$$b_n = \frac{2}{T_0} \int_{\langle T_0 \rangle} x(t) \sin(n\omega_0 t) dt.$$

The trigonometric CTFS is presented in Section 4.4, while its counterpart, the exponential CTFS, is covered in Section 4.5. The exponential CTFS is obtained by expressing the periodic signal $x(t)$ as a linear combination of complex exponentials $\{\exp(jn\omega_0 t)\}$ and is given by

$$x(t) = \sum_{n=-\infty}^{\infty} D_n e^{jn\omega_0 t},$$

where the exponential CTFS coefficients D_n are calculated using the following expression:

$$D_n = \frac{1}{T_0} \int_{\langle T_0 \rangle} x(t) e^{-jn\omega_0 t} dt.$$

The exponential CTFS has several interesting properties that are useful in the analysis of CT signals.

(1) The *linearity* property states that the exponential CTFS coefficients of a linear combination of periodic signals are given by the same linear combination of the exponential CTFS coefficients of each of the periodic signals.
(2) A time shift of t_0 in the periodic signal does not affect the magnitude of the exponential CTFS coefficients. However, the phase changes by an additive

factor of $\pm n\omega_0 t_0$, the sign of the phase change depending on the direction of the shift. This property is referred to as the *time-shifting* property.

(3) The exponential CTFS coefficients of a *time-reversed* periodic signal are the time-reversed CTFS coefficients of the original signal.

(4) If a periodic signal is *time-scaled*, the exponential CTFS coefficients are inversely time-scaled.

(5) The exponential CTFS coefficients of a *time-differentiated* periodic signal are obtained by *multiplying* the CTFS coefficients of the original signal by a factor of $jn\omega_0$.

(6) The exponential CTFS coefficients of a *time-integrated* periodic signal are obtained by *dividing* the CTFS coefficients of the original signal by a factor of $jn\omega_0$.

(7) For *real-valued* periodic signals, the exponential CTFS coefficients D_n and D_{-n} are complex conjugates of each other.

(8) Based on *Parseval's property*, the power of a periodic signal $x(t)$ with the fundamental period of T_0 is computed directly from the exponential CTFS coefficients as follows:

$$P_x = \frac{1}{T_0} \int_{\langle T_0 \rangle} |x(t)|^2 \mathrm{d}t = \sum_{n=-\infty}^{\infty} |D_n|^2.$$

The plot of the magnitude $|D_n|$ of the exponential CTFS coefficients versus the coefficient number n is referred to as the magnitude spectrum, while the plot of the phase $<D_n$ of the exponential CTFS coefficients versus the coefficient number n is referred to as the phase spectrum of the periodic signal $x(t)$. Section 4.6 covers the conditions for the existence of the CTFS representations, and Section 4.7 concludes the chapter by calculating the output response $y(t)$ of an LTIC system to a periodic input $x(t)$. In such cases, the output $y(t)$ is given by

$$y(t) = \sum_{n=-\infty}^{\infty} D_n \mathrm{e}^{jn\omega_0 t} H(\omega)|_{\omega=n\omega_0},$$

where the transfer function $H(\omega)$ is obtained from the impulse response $h(t)$ of the LTIC system as follows:

$$H(\omega) = \int_{-\infty}^{\infty} h(t) \mathrm{e}^{-j\omega t} \mathrm{d}t.$$

The above expression also defines the continuous-time Fourier transform (CTFT) for *aperiodic* signals, which is covered in depth in Chapter 5.

Problems

4.1 Express the following functions in terms of the orthogonal basis functions specified in Example 4.2 and illustrated in Fig. 4.3.

(a) $x_1(t) = \begin{cases} A & 0 \le t \le T \\ -A & -T \le t \le 0; \end{cases}$

(b) $x_2(t) = \begin{cases} A & \dfrac{T}{2} \le |t| \le T \\ -A & 0 \le |t| \le \dfrac{T}{2}; \end{cases}$

(c) $x_3(t) = \begin{cases} A & \dfrac{T}{2} \le |t| \le T \\ 0 & 0 \le |t| \le \dfrac{T}{2}. \end{cases}$

4.2 For the functions

$$\phi_1(t) = e^{-2|t|} \quad \text{and} \quad \phi_2(t) = 1 - Ke^{-4|t|}$$

determine the value of K such that the functions are orthogonal over the interval $[-\infty, \infty]$.

4.3 The Legendre polynomials are widely used to approximate functions. An nth-order Legendre polynomial $P_n(x)$ is defined as

$$P_n(x) = \frac{1}{n!2^n}\frac{d^n}{dx^n}(x^2 - 1)^n = \sum_{m=0}^{M} a_{nm}x^m,$$

where the values of a_{nm} can be expressed as follows:

$$a_{nm} = \sum_{\substack{m=0 \\ n,m \text{ odd} \\ n,m \text{ even}}}^{n} (-1)^{(n-m)/2}\frac{(n+m)!}{2^n m!(n-m/2)!(n+m/2)!}$$

Note that a_{nm} is non-zero only when both n and m are either odd or even. For all other values of n and m, a_{nm} is zero. The first few orders of Legendre polynomials are given by

$$P_0(x) = 1; \quad P_2(x) = \frac{1}{2}(3x^2 - 1);$$

$$P_1(x) = x; \quad P_3(x) = \frac{1}{2}(5x^3 - 3x);$$

and are shown in Fig. P4.3.

The Legendre polynomials $\{P_n(x), n = 0, 1, 2, \ldots\}$ form a set of orthogonal functions over the interval $[-1, 1]$ by satisfying the following property:

$$\int_{-1}^{1} P_m(x)P_n(x)dx = \begin{cases} \dfrac{2}{2m+1} & m = n \\ 0 & m \ne n. \end{cases}$$

Verify the above orthogonality condition for $m, n = 0, 1, 2, 3$.

4.4 The Chebyshev polynomials of the first kind are used as the approximation to a least-squares fit. The nth-order polynomial $T_n(x)$ can be expressed as

Fig. P4.3. Legendre polynomials with order 0–3.

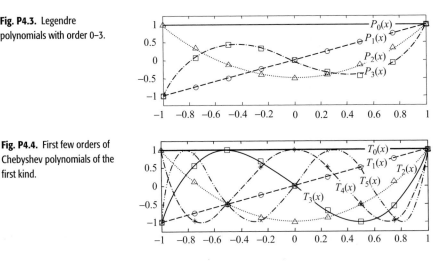

Fig. P4.4. First few orders of Chebyshev polynomials of the first kind.

follows:

$$T_n(x) = \frac{n}{2} \sum_{k=0}^{\lfloor n/2 \rfloor} (-1)^k \frac{(n-k-1)!}{k!(n-2k)!} (2x)^{n-2k}, \quad n = 0, 1, 2, 3, \ldots$$

The first few Chebyshev polynomials are given by

$$
\begin{aligned}
T_0(x) &= 1; & T_3(x) &= 4x^3 - 3x; \\
T_1(x) &= x; & T_4(x) &= 8x^4 - 8x^2 + 1; \\
T_2(x) &= 2x^2 - 1; & T_5(x) &= 16x^5 - 20x^3 + 5x;
\end{aligned}
$$

which satisfy the following relationship:

$$T_{n+1}(x) = 2x T_n(x) - T_{n-1}(x)$$

and are shown in Fig. P4.4.

The Chebyshev polynomials $\{T_n(x), n = 0, 1, 2, \ldots\}$ form an orthogonal set on the interval $[-1, 1]$ with respect to the weighting function by satisfying the following:

$$\int_{-1}^{1} \frac{1}{\sqrt{1-x^2}} T_m(x) T_n(x) \mathrm{d}x = \begin{cases} \pi & m = n = 0 \\ \pi/2 & m = n = 1, 2, 3 \\ 0 & m \neq n. \end{cases}$$

Verify the above orthogonality condition for $m, n = 0, 1, 2, 3, 4$.

4.5 The Haar functions are very popular in signal processing and wavelet applications. These functions are generated using a scale parameter (m) and a translation parameter (n). Let the mother Haar function ($m = n = 0$) be defined as follows:

$$H_{0,0}(t) = \begin{pmatrix} 1 & 0 \leq t < 0.5 \\ -1 & 0.5 \leq t \leq 1 \\ 0 & \text{otherwise.} \end{pmatrix}$$

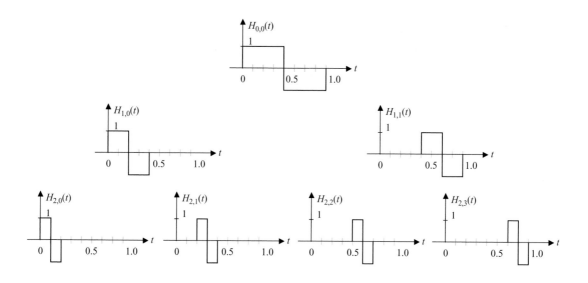

Fig. P4.5. Haar functions for
$m = 0$, 1, and 2.

The other Haar functions, at scale m and with translation n, are defined
using the mother Haar function as follows:

$$H_{m,n}(t) = H_{0,0}(2^m t - n), \quad n = 0, 1, \ldots, (2^m - 1).$$

The Haar functions for $m = 0, 1, 2$ are shown in Fig. P4.5.

Show that the Haar wavelet functions $\{H_{m,n}(t), \; m = 0, 1, 2, \ldots, \; n = 0, 1, 2, \ldots (2^m - 1)\}$ form a set of orthogonal functions over the interval
$[0, 1]$ by proving the following:

$$\int_0^1 H_{m,n}(t) H_{p,q}(t) dt = \begin{cases} 2^{-m} & m = p, n = q \\ 0 & \text{otherwise.} \end{cases}$$

4.6 Calculate the trigonometric CTFS coefficients for the periodic functions
shown in Figs. P4.6(a)–(e).

(a) Rectangular pulse train with period 2π:

$$x1(t) = \begin{cases} 3 & \text{for} \quad 0 \le t < \pi \\ 0 & \text{for} \quad \pi \le t < 2\pi. \end{cases}$$

(b) Raised square wave with period $2T$:

$$x2(t) = \begin{cases} 0.5 & \text{for} \quad \dfrac{-T}{2} \le t < \dfrac{T}{2} \\ 1 & \text{for} \quad \dfrac{T}{2} \le t < \dfrac{3T}{2}. \end{cases}$$

(c) Half sawtooth wave with period T:

$$x3(t) = 1 - \frac{t}{T} \quad \text{for} \quad 0 \le t < T.$$

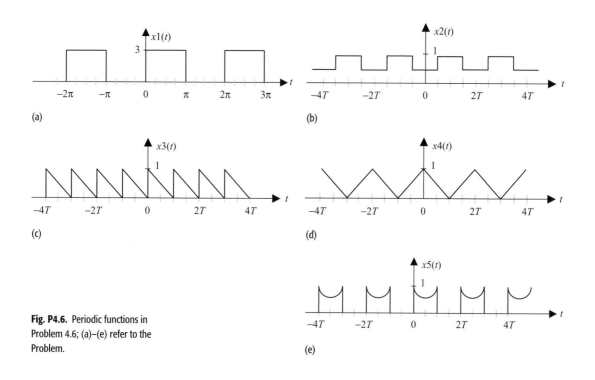

Fig. P4.6. Periodic functions in
Problem 4.6; (a)–(e) refer to the
Problem.

(d) Sawtooth wave with period $2T$:

$$x4(t) = 1 - \left| \frac{t}{T} \right| \quad \text{for} \quad -T \leq t < T.$$

(e) Periodic wave with period $2T$.

$$x5(t) = \begin{cases} 0 & \text{for} \quad -T \leq t < 0 \\ 1 - 0.5 \sin\left(\frac{\pi t}{T} \right) & \text{for} \quad 0 \leq t < T. \end{cases}$$

4.7 Calculate the trigonometric CTFS coefficients for the periodic function
shown in Fig. P4.7. Note that the function

$$s(t) = \sum_{k=-\infty}^{k=\infty} \delta(t - kT)$$

is known as the sampling function and that it is used to obtain a discrete-
time signal by sampling a continuous-time signal (see Chapter 9).

4.8 Calculate the trigonometric CTFS coefficients for the following functions:
   (i) $x_t(t) = \cos 7t + \sin(15t + \pi/2)$;
   (ii) $x_2(t) = 3 + \sin 2t + \cos(4t + \pi/4)$;
   (iii) $x_3(t) = 1.2 + e^{j2t+1} + e^{j(5t+2)} + e^{-j(3t+1)}$;
   (iv) $x_4(t) = e^{t+1} + e^{j(2t+3)}$.

4.9 Show that if $x(t)$ is an even periodic function with period T_0, the exponential CTFS coefficients can be calculated by evaluating the following

Fig. P4.7. Periodic function (an impulse train with period T) in Problem 4.7.

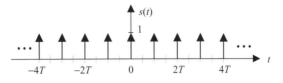

Fig. P4.7. Periodic function (an impulse train with period T) in Problem 4.7.

integral:

$$D_n = \frac{2}{T_0} \int_0^{T_0/2} x(t) \cos(n\omega_0 t)\mathrm{d}t,$$

where $\omega_0 = 2\pi/T_0$.

4.10 Show that if $x(t)$ is an odd periodic function with period T_0, the exponential CTFS coefficients can be calculated by evaluating the following integral:

$$D_n = \frac{-2\mathrm{j}}{T_0} \int_0^{T_0/2} x(t) \sin(n\omega_0 t)\mathrm{d}t,$$

where $\omega_0 = 2\pi/T_0$.

4.11 For the periodic functions shown in Fig. P4.6:
(i) calculate the exponential CTFS coefficients directly using Eq. (4.44);
(ii) plot the magnitude and phase spectra.

4.12 Repeat Problem 4.11 for the function shown in Fig. P4.7.

4.13 For the periodic functions shown in Fig. P4.6, calculate the exponential CTFS coefficients by applying Eq. (4.45) to the trigonometric CTFS coefficients calculated in Problem 4.6. Compare your answers with the CTFS coefficients obtained in Problem 4.11.

4.14 Consider the raised square wave shown in Fig. P4.6(b). Using the time-differentiation property and the exponential CTFS coefficients calculated in Problem 4.11, calculate the exponential CTFS coefficients of an impulse train with period $T_0 = 2T$, with impulses located at $T/2 + 2kT$ with $k \in Z$.

4.15 Calculate the exponential CTFS coefficients for the functions given in Problem 4.8.

4.16 The derivative of the square wave $x(t)$ shown in Fig. 4.14 can be expressed in terms of two shifted impulse trains as

$$\frac{\mathrm{d}x(t)}{\mathrm{d}t} = \sum_{k=-\infty}^{\infty} \delta\left(t + \frac{\pi}{4} - 2k\pi\right) - \delta\left(t - \frac{\pi}{4} - 2k\pi\right).$$

Using the time-shifting and time-scaling properties, express the exponential CTFS coefficients D_n for the square wave in terms of the exponential

CTFS coefficients E_n of the impulse train. Calculate the CTFS coefficients of the square wave and compare with the values evaluated in Example 4.14.

4.17 Repeat Example 4.22 with the following values of τ and T such that the duty cycle (τ/T) is fixed at 0.2:

(i) $\tau = 1$ ms, $\quad T = 5$ ms;

(ii) $\tau = 2$ ms, $\quad T = 10$ ms;

(iii) $\tau = 4$ ms, $\quad T = 20$ ms.

Discuss the changes in the CTFS representations for the above selections of τ and T.

4.18 For the periodic functions shown in Fig. P4.6:

(i) calculate the average power in the time domain, and

(ii) calculate the average power using Parseval's theorem. Verify your result with that obtained in step (i).

[Hint: If you find it difficult to calculate the summation $\sum\limits_{n=-\infty}^{n=\infty} |D_n|^2$ analytically, write a MATLAB program to calculate an approximate value of $\sum\limits_{n=-\infty}^{n=\infty} |D_n|^2$ for $-1000 \leq n \leq 1000$.]

4.19 Determine whether the periodic functions shown in Fig. P4.6 satisfy the Dirichlet conditions and have CTFS representation.

4.20 Determine if the following functions satisfy the Dirichlet conditions and have CTFS representation:

(i) $x(t) = 1/t, \quad t = (0, 2]$ and $x(t) = x(t + 2)$;

(ii) $g(t) = \cos(\pi/2t), \quad t = (0, 1]$ and $g(t) = g(t + 1)$;

(iii) $h(t) = \sin(\ln(t)), \quad t = (0, 1]$ and $h(t) = h(t + 1)$.

4.21 Consider the periodic signal $f(t)$ considered in Example 4.9 and shown in Fig. 4.11. From the CTFS representation, prove the following identity:

$$\frac{\pi^2}{8} = 1 + \frac{1}{3^2} + \frac{1}{5^2} + \frac{1}{7^2} + \cdots .$$

4.22 From the half sawtooth wave shown in Fig. P4.6(c) and its trigonometric CTFS coefficients (calculated in Problem 4.6(c)), prove the following identity:

$$\frac{\pi}{4} = 1 - \frac{1}{3} + \frac{1}{5} - \frac{1}{7} + \frac{1}{9} - \frac{1}{11} + \cdots .$$

[Hint: Evaluate the function at $t = T/4$.]

4.23 Using the exponential CTFS coefficients of the function shown in Fig. P4.6(c) (calculated in Problem 4.11) and Parseval's power theorem, prove the following identity:

$$\frac{\pi^2}{6} = 1 + \frac{1}{2^2} + \frac{1}{3^2} + \frac{1}{4^2} + \frac{1}{5^2} + \cdots .$$

4.24 The impulse response of an LTIC system is given by

$$h(t) = e^{-2|t|}.$$

(a) Based on Eq. (4.44), calculate the transfer function $H(\omega)$ of the LTIC system.
(b) The plot of magnitude $|H(\omega)|$ with respect to ω is referred to as the magnitude spectrum of the LTIC system. Plot the magnitude spectrum of the LTIC system for the range $(-\infty < \omega < \infty)$.

Calculate the output response $y(t)$ of the LTIC system if the impulse train shown in Fig. P4.7 is applied as an input to the LTIC system.

4.25 Repeat P4.24 for the following LTIC system:

$$h(t) = [e^{-2t} - e^{-4t}]u(t),$$

with the raised square wave function shown in Fig. P4.6(b) applied at the input of the LTIC system.

4.26 Repeat P4.24 for the following LTIC system:

$$h(t) = te^{-4t}u(t),$$

with the sawtooth wave function shown in Fig. P4.6(d) applied at the input of the LTIC system.

4.27 Consider the following periodic functions represented as CTFS:

(i) $x_1(t) = \dfrac{7}{\pi} \displaystyle\sum_{m=0}^{\infty} \dfrac{1}{2m+1} \sin[8\pi(2m+1)t];$

(ii) $x_2(t) = 1.5 + \displaystyle\sum_{m=0}^{\infty} \dfrac{1}{4m+1} \cos[2\pi(4m+1)t].$

(a) Determine the fundamental period of $x(t)$.
(b) Determine if $x(t)$ is an even signal or an odd signal.
(c) Using the `ictfs.m` function provided in the CD, calculate and plot the functions in the time interval $-1 \le t \le 1$. [Hint: You may calculate $x(t)$ for $t = [-1{:}0.01{:}1]$. The MATLAB "`plot`" function will give a smooth interpolated plot.]
(d) From the plot in step (c), determine the period of $x(t)$. Does it match your answer to part (a)?

4.28 Using the MATLAB function `ictfs.m` (provided in the CD), show that the periodic function $g(t)$ (shown in Fig. 4.10) considered in Example 4.8, can be reconstructed from its trigonometric Fourier series coefficients.

4.29 Using the MATLAB function `ictfs.m` (provided in the CD), show that the periodic function $f(t)$ (shown in Fig. 4.11) considered in Example 4.9, can be reconstructed from its trigonometric Fourier series coefficients.

4.30 Using the MATLAB function ictfs.m (provided in the CD), show that the periodic function $g(t)$ (shown in Fig. 4.10) considered in Example 4.12, can be reconstructed from its exponential Fourier series coefficients.

4.31 Using the MATLAB function ictfs.m (provided in the CD), show that the periodic function $f(t)$ (shown in Fig. 4.11) considered in Example 4.13, can be reconstructed from its exponential Fourier series coefficients.

4.32 Using the MATLAB function ictfs.m (provided in the CD), plot the output response $y(t)$ obtained in Problem 4.24 for $T = 1$ s.

4.33 Using the MATLAB function ictfs.m (provided in the CD), plot the output response $y(t)$ obtained in Problem 4.25 for $T = 1$ s.

5 Continuous-time Fourier transform

In Chapter 4, we introduced the frequency representations for periodic signals based on the trigonometric and exponential continuous-time Fourier series (CTFS). The exponential CTFS is useful in calculating the output response of a linear time-invariant (LTI) system to a periodic input signal. In this chapter, we extend the Fourier framework to continuous-time (CT) aperiodic signals. The resulting frequency decompositions are referred to as the continuous-time Fourier transform (CTFT) and are used to express both aperiodic and periodic CT signals in terms of linear combinations of complex exponential functions. We show that the convolution in the time domain is equivalent to multiplication in the frequency domain. The CTFT, therefore, provides an alternative analysis technique for LTIC systems in the frequency domain.

Chapter 5 is organized as follows. Section 5.1 considers the CTFT as a limiting case of the CTFS and formally defines the CTFT and its inverse. In Section 5.2, we provide several examples to illustrate the steps involved in the calculation of the CTFT for a number of elementary signals. Section 5.3 presents the look-up table and partial fraction methods for calculating the inverse CTFT. Section 5.4 lists the symmetry properties of the CTFT for real-valued, even, and odd signals, while Section 5.5 lists the CTFT properties arising due to linear transformations in the time domain. The condition for the existence of the CTFT is derived in Section 5.6, while the relationship between the CTFT and the CTFS for periodic signals is discussed in Sections 5.7 and 5.8. Section 5.9 applies the convolution property of the CTFT to evaluate the output response of an LTIC system to an arbitrary CT input signal. The gain and phase responses of LTIC systems are also defined in this section. Section 5.10 demonstrates how MATLAB is used to compute the CTFT, and Section 5.11 concludes the chapter.

5.1 CTFT for aperiodic signals

Consider the aperiodic signal $x(t)$ shown in Fig. 5.1(a). In order to extend the Fourier framework of the CTFS to aperiodic signals, we consider several

Fig. 5.1. Periodic extension of a time-limited aperiodic signal. (a) Aperiodic signal and (b) its periodic extension.

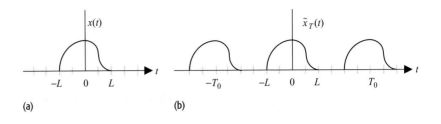

(a)   (b)

repetitions of $x(t)$ uniformly spaced from each other by duration T_0 such that there is no overlap between two adjacent replicas of $x(t)$. The resulting signal is denoted by $\tilde{x}_T(t)$ and is shown in Fig. 5.1(b). Clearly, the new signal $\tilde{x}_T(t)$ is periodic with the fundamental period of T_0 and in the limit

$$\lim_{T_0 \to \infty} \tilde{x}_T(t) = x(t). \tag{5.1}$$

Since $\tilde{x}_T(t)$ is a periodic signal with a fundamental frequency of $\omega_0 = 2\pi/T_0$ radians/s, its exponential CTFS representation is expressed as follows:

$$\tilde{x}_T(t) = \sum_{n=-\infty}^{\infty} \tilde{D}_n e^{jn\omega_0 t}, \tag{5.2}$$

where the exponential CTFS coefficients are given by

$$\tilde{D}_n = \frac{1}{T_0} \int_{\langle T_0 \rangle} \tilde{x}_T(t) e^{-jn\omega_0 t} \, dt. \tag{5.3}$$

The spectra of $\tilde{x}_T(t)$ are the magnitude and phase plots of the CTFS coefficients $\tilde{D}_n$ as a function of $n\omega_0$. Because n takes on integer values, the magnitude and phase spectra of $\tilde{x}_T(t)$ consist of vertical lines separated uniformly by ω_0. Applying the limit $T_0 \to \infty$ to $\tilde{x}_T(t)$ causes the spacing $\omega_0 = 2\pi/T_0$ in the spectral lines of the magnitude and phase spectra to decrease to zero. The resulting spectra represent the Fourier representation of the aperiodic signal $x(t)$ and are continuous along the frequency (ω) axis. The CTFT for aperiodic signals is, therefore, a continuous function of frequency ω. To derive the mathematical definition of the CTFT, we apply the limit $T_0 \to \infty$ to Eq. (5.3). The resulting expression is as follows:

$$\lim_{T_0 \to \infty} \tilde{D}_n = \lim_{T_0 \to \infty} \frac{1}{T_0} \int_{\langle T_0 \rangle} x(t) e^{-jn\omega_0 t} \, dt$$

or

$$D_n = \lim_{T_0 \to \infty} \frac{1}{T_0} \int_{-\infty}^{\infty} x(t) e^{-jn\omega_0 t} \, dt \quad \text{since } \lim_{T_0 \to \infty} \tilde{x}_T(t) = x(t). \tag{5.4}$$

In Eq. (5.4), the term D_n denotes the exponential CTFT coefficients of $x(t)$. Let us define a continuous function $X(\omega)$ (with the independent variable ω) as follows:

$$X(\omega) = \int_{-\infty}^{\infty} x(t)e^{-j\omega t}\,dt.$$

(5.5)

In terms of $X(\omega)$, Eq. (5.4) can, therefore, be expressed as follows:

$$D_n = \lim_{T_0 \to \infty} \frac{1}{T_0} X(n\omega_0).$$

(5.6)

Using the exponential CTFS definition, $x(t)$ can be evaluated from the CTFS coefficients D_n as follows:

$$x(t) = \sum_{n=-\infty}^{\infty} D_n e^{jn\omega_0 t} = \lim_{T_0 \to \infty} \sum_{n=-\infty}^{\infty} \frac{1}{T_0} X(n\omega_0)e^{jn\omega_0 t}.$$

(5.7)

As $T_0 \to \infty$, the fundamental frequency ω_0 approaches a small value denoted by $\Delta\omega$. The fundamental period T_0 is therefore given by $T_0 = 2\pi/\Delta\omega$. Substituting $T_0 = 2\pi/\Delta\omega$ as $\omega_0 \to \Delta\omega$ in Eq. (5.7) yields

$$x(t) = \frac{1}{2\pi} \lim_{\Delta\omega \to 0} \underbrace{\sum_{n=-\infty}^{\infty} X(n\Delta\omega)\,e^{jn\Delta\omega t}\,\Delta\omega}_{A}.$$

(5.8)

In Eq. (5.8), consider the term A as illustrated in Fig. 5.2. In the limit $\Delta\omega \to 0$, term A represents the area under the function $X(\omega)\exp(j\omega t)$. Therefore Eq. (5.8) can be rewritten as follows:

CTFT synthesis equation $x(t) = \dfrac{1}{2\pi} \int_{-\infty}^{\infty} X(\omega)e^{j\omega t}\,d\omega,$

(5.9)

which is referred to as the *synthesis* equation for the CTFT used to express any aperiodic signal in terms of complex exponentials, $\exp(j\omega t)$. The *analysis* equation of the CTFT is given by Eq. (5.5), which, for convenience of reference, is repeated below.

CTFT analysis equation $X(\omega) = \int_{-\infty}^{\infty} x(t)e^{-j\omega t}\,dt.$

(5.10)

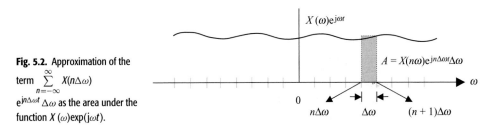

Fig. 5.2. Approximation of the term $\sum_{n=-\infty}^{\infty} X(n\Delta\omega)$ $e^{jn\Delta\omega t}\Delta\omega$ as the area under the function $X(\omega)\exp(j\omega t)$.

Collectively, Eqs. (5.9) and (5.10) form the CTFT pair, which is denoted by

$$x(t) \xleftrightarrow{\text{DTFT}} X(\omega). \tag{5.11}$$

Alternatively, the CTFT pair may also be represented as follows:

$$X(\omega) = \Im\{x(t)\} \tag{5.12}$$

or

$$x(t) = \Im^{-1}\{X(\omega)\}, \tag{5.13}$$

where $\Im$ denotes for the CTFT and $\Im^{-1}$ denotes the inverse of the CTFT. Based on Eqs. (5.9) and (5.10), we make the following observations about the CTFT.

(1) The frequency representation of a periodic signal $\tilde{x}(t)$ is obtained by expressing $\tilde{x}(t)$ in terms of the CTFS. The basis function of the CTFS consists of complex exponentials $\{\exp(jn\omega_0 t)\}$, which are defined at the fundamental frequency ω_0 and its harmonics $n\omega_0$. The frequency representation of an aperiodic signal $x(t)$ is obtained through the CTFT, where the complex exponential $\exp(jn\omega t)$ is the basis function. The variable ω in the basis function of the CTFT is a continuous variable and may have any value within the range $-\infty < \omega < \infty$. Unlike the CTFS, the CTFT is therefore defined for all frequencies ω.

(2) In general, the CTFT $X(\omega)$ is a complex function of the angular frequency ω. A great deal of information is obtained by plotting the magnitude and phase of $X(\omega)$ with respect to ω. The plots of magnitude $|X(\omega)|$ and phase $<X(\omega)$ with respect to ω are, respectively, referred to as the magnitude and phase spectra of the aperiodic function $x(t)$.

(3) In deriving the definition of the CTFT, we assumed that the aperiodic function $x(t)$ is time-limited such that $x(t) = 0$ for $|t| > L$. This is not a required condition for the existence of the CTFT. In other words, the function $x(t)$ may be infinitely long but its CTFT can exist.

5.2 Examples of CTFT

In Section 5.2, we calculate the forward and inverse CTFT of several well known functions. We assume that the CTFT exists in all cases. A general condition for the existence of the CTFT is derived in Section 5.6.

Example 5.1

Determine the CTFT of the following functions and plot the corresponding magnitude and phase spectra:

(i) $x_1(t) = \exp(-at)u(t)$, $a \in R^+$;

(ii) $x_2(t) = \exp(-a|t|)$, $a \in R^+$.

Table 5.3. Magnitude and phase spectra for selected elementary CT functions
Magnitude spectra are shown as lines and phase spectra are shown as dashed lines

Function	Time-domain waveform	Magnitude and phase spectra
(1) Constant $x(t) = 1$		
(2) Unit impulse function $x(t) = \delta(t)$		
(3) Unit step function $x(t) = u(t)$		
(4) Decaying exponential $x(t) = e^{-at}u(t)$		
(5) Two-sided decaying exponential $x(t) = e^{-a|t|}$		

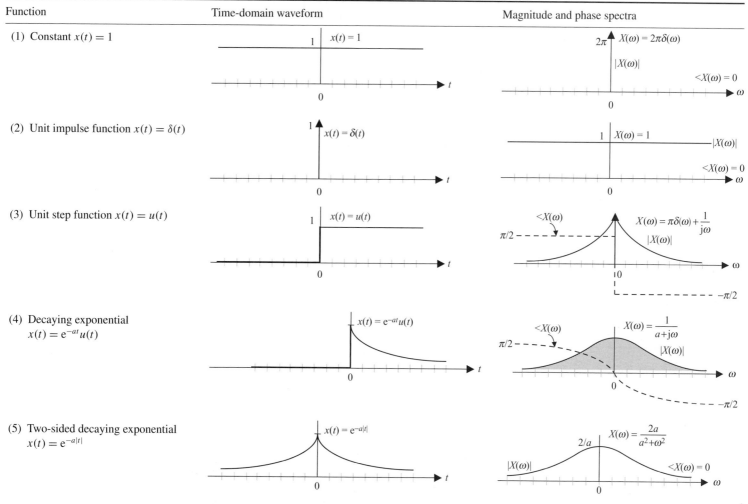

(cont.)

Table 5.3. (*cont.*)

Function	Time-domain waveform	Magnitude and phase spectra
(6) First-order time-rising decaying exponential function $x(t) = t e^{-at} u(t)$		
(7) Nth-order time-rising decaying exponential function $x(t) = t^n e^{-at} u(t)$		
(8) Sign function $\operatorname{sgn}(t) = \begin{cases} 1 & t > 0 \\ -1 & t < 0 \end{cases}$		
(9) Complex exponential function $x(t) = e^{j\omega_0 t}$		
(10) Cosine function $x(t) = \cos(\omega_0 t)$		

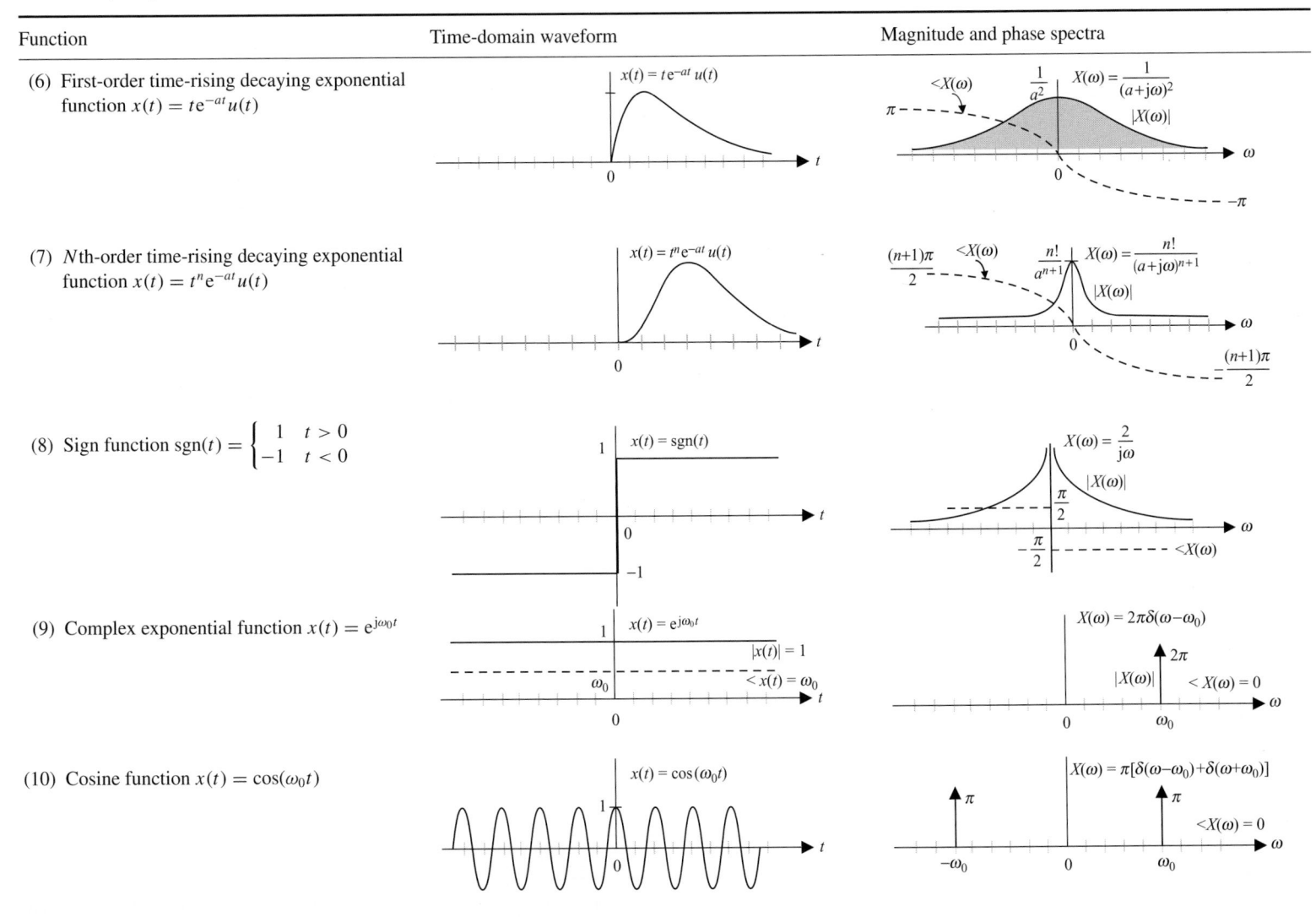

(11) Sine function $x(t) = \sin(\omega_0 t)$

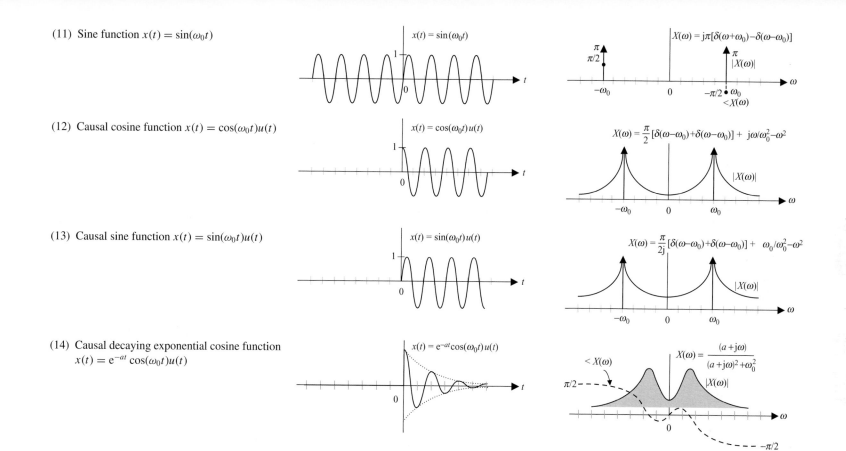

$x(t) = \sin(\omega_0 t)$

$X(\omega) = j\pi[\delta(\omega+\omega_0)-\delta(\omega-\omega_0)]$

$|X(\omega)|$

$<X(\omega)$

(12) Causal cosine function $x(t) = \cos(\omega_0 t)u(t)$

$x(t) = \cos(\omega_0 t)u(t)$

$X(\omega) = \dfrac{\pi}{2}[\delta(\omega-\omega_0)+\delta(\omega-\omega_0)] + j\omega/\omega_0^2-\omega^2$

$|X(\omega)|$

(13) Causal sine function $x(t) = \sin(\omega_0 t)u(t)$

$x(t) = \sin(\omega_0 t)u(t)$

$X(\omega) = \dfrac{\pi}{2j}[\delta(\omega-\omega_0)+\delta(\omega-\omega_0)] + \omega_0/\omega_0^2-\omega^2$

$|X(\omega)|$

(14) Causal decaying exponential cosine function
$x(t) = e^{-at}\cos(\omega_0 t)u(t)$

$x(t) = e^{-at}\cos(\omega_0 t)u(t)$

$<X(\omega)$

$X(\omega) = \dfrac{(a+j\omega)}{(a+j\omega)^2+\omega_0^2}$

$|X(\omega)|$

(15) Causal decaying exponential sine function
$x(t) = e^{-at}\sin(\omega_0 t)u(t)$

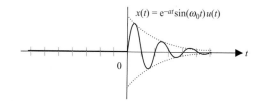

$x(t) = e^{-at}\sin(\omega_0 t)u(t)$

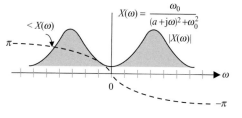

$<X(\omega)$

$X(\omega) = \dfrac{\omega_0}{(a+j\omega)^2+\omega_0^2}$

$|X(\omega)|$

(cont.)

Table 5.3. (*cont.*)

Function	Time-domain waveform	Magnitude and phase spectra

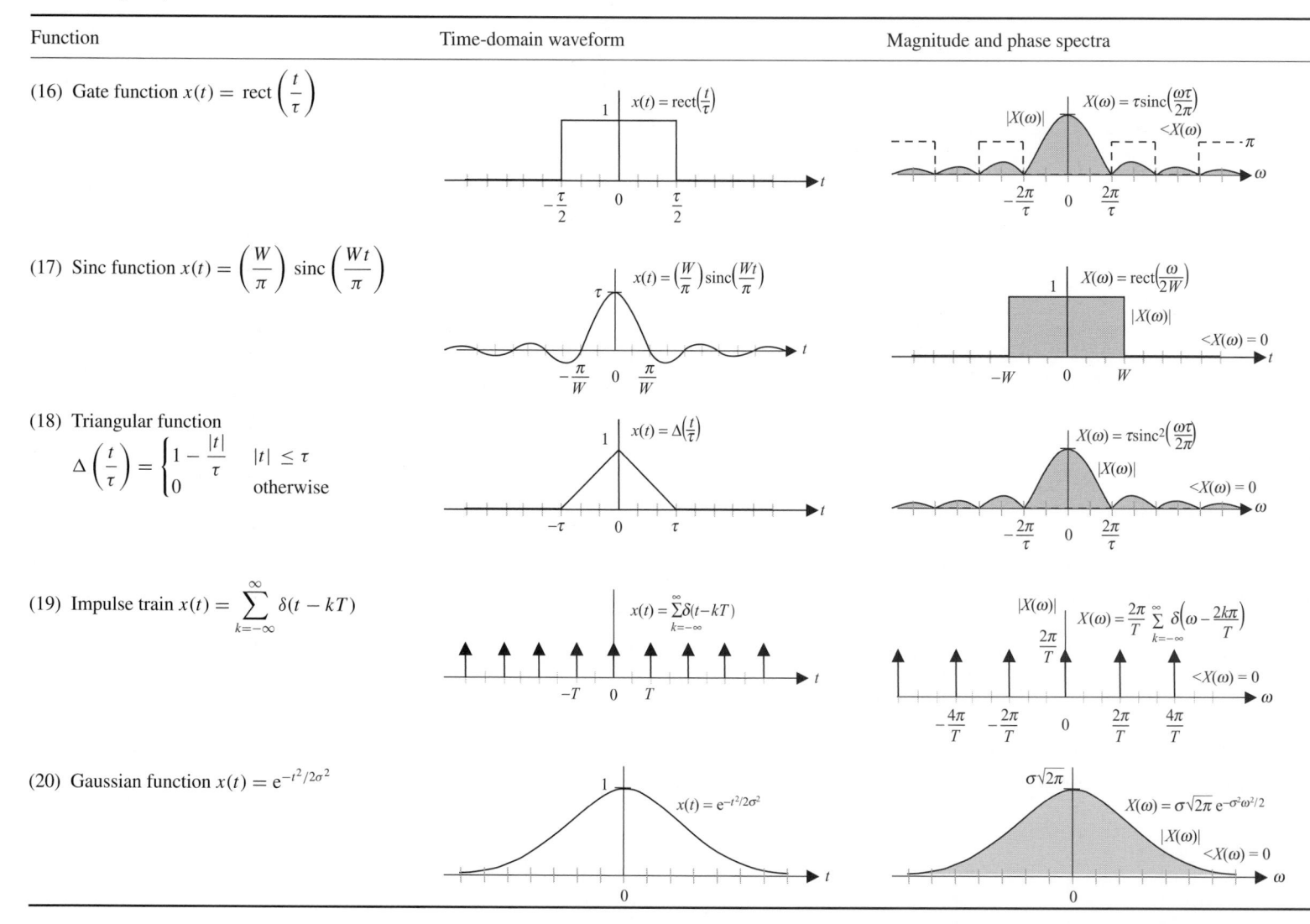

(16) Gate function $x(t) = \text{rect}\left(\dfrac{t}{\tau}\right)$

$x(t) = \text{rect}\left(\dfrac{t}{\tau}\right)$

$X(\omega) = \tau\,\text{sinc}\left(\dfrac{\omega\tau}{2\pi}\right)$

(17) Sinc function $x(t) = \left(\dfrac{W}{\pi}\right)\text{sinc}\left(\dfrac{Wt}{\pi}\right)$

$x(t) = \left(\dfrac{W}{\pi}\right)\text{sinc}\left(\dfrac{Wt}{\pi}\right)$

$X(\omega) = \text{rect}\left(\dfrac{\omega}{2W}\right)$

$\angle X(\omega) = 0$

(18) Triangular function

$\Delta\left(\dfrac{t}{\tau}\right) = \begin{cases} 1 - \dfrac{|t|}{\tau} & |t| \le \tau \\ 0 & \text{otherwise} \end{cases}$

$x(t) = \Delta\left(\dfrac{t}{\tau}\right)$

$X(\omega) = \tau\,\text{sinc}^2\left(\dfrac{\omega\tau}{2\pi}\right)$

$\angle X(\omega) = 0$

(19) Impulse train $x(t) = \displaystyle\sum_{k=-\infty}^{\infty} \delta(t - kT)$

$x(t) = \displaystyle\sum_{k=-\infty}^{\infty}\delta(t-kT)$

$X(\omega) = \dfrac{2\pi}{T}\displaystyle\sum_{k=-\infty}^{\infty}\delta\left(\omega - \dfrac{2k\pi}{T}\right)$

$\angle X(\omega) = 0$

(20) Gaussian function $x(t) = e^{-t^2/2\sigma^2}$

$x(t) = e^{-t^2/2\sigma^2}$

$X(\omega) = \sigma\sqrt{2\pi}\,e^{-\sigma^2\omega^2/2}$

$\angle X(\omega) = 0$

5.3 Inverse Fourier transform

Evaluation of the inverse CTFT is an important step in analysis of LTIC systems. There are three main approaches that may be taken to calculate the inverse CTFT:

(i)   using the synthesis equation;
(ii) using a look-up table;
(iii) using partial fraction expansion.

In the first approach, the inverse CTFT is calculated by solving the synthesis equation, Eq. (5.9). This method was used in Examples 5.3, 5.5, 5.7, and 5.8. However, this approach is difficult. We now present the second and third approaches. Approach (ii) is straightforward as it determines the inverse CTFT by comparing the entries with Table 5.2. We illustrate this with an example.

Example 5.9

Using the look-up table method, calculate the inverse CTFT of the following function:

$$X(\omega) = \frac{2(j\omega) + 24}{(j\omega)^2 + 4(j\omega) + 29}.$$

(5.27)

Solution

The function $X(\omega)$ is decomposed into simpler terms, whose inverse CTFT can be determined directly from Table 5.2. One possible decomposition is as follows:

$$X(\omega) = 2\frac{2 + (j\omega)}{(2 + j\omega)^2 + 5^2} + 4\frac{5}{(2 + j\omega)^2 + 5^2}.$$

(5.28)

From Entries (14) and (15) of Table 5.2, we know that

$$e^{-2t}\cos(5t)u(t) \xleftarrow{\quad\text{CTFT}\quad} \frac{2 + j\omega}{(2 + j\omega)^2 + 5^2}$$

and

$$e^{-2t}\sin(5t)u(t) \xleftarrow{\quad\text{CTFT}\quad} \frac{5}{(2 + j\omega)^2 + 5^2}.$$

Therefore, the inverse CTFT is calculated as follows:

$$x(t) = 2e^{-2t}\cos(5t)u(t) + 4e^{-2t}\sin(5t)u(t).$$

(5.29)

5.3.1 Partial fraction expansion

The look-up table approach is simple to use once a suitable decomposition is obtained. A major problem, however, is faced in the decomposition of the CTFT $X(\omega)$ in terms of simpler functions whose inverse CTFTs are listed in Table 5.2.

We now present approach (iii), which uses the partial fraction expansion to decompose systematically a rational function in simpler terms. Consider the CTFT

$$X(\omega) = \frac{N(\omega)}{D(\omega)} = \frac{b_m(j\omega)^m + b_{m-1}(j\omega)^{m-1} + \cdots + b_1(j\omega) + b_0}{(j\omega)^n + a_{n-1}(j\omega)^{n-1} + \cdots + a_1(j\omega) + a_0}, \quad (5.30)$$

where the numerator is an mth-order polynomial and the denominator is an nth-order polynomial. The partial fraction method is explained in more detail in Appendix D (see Section D.2). The main steps are summarized as follows.

(1) Factorize $D(\omega)$ into n first-order factors and express $X(\omega)$ as follows:

$$X(\omega) = \frac{N(\omega)}{(j\omega - p_1)(j\omega - p_2)\cdots(j\omega - p_n)}. \quad (5.31)$$

(2) If there are no repeated or complex roots in $D(\omega)$, $X(\omega)$ is expressed in terms of n partial fractions:

$$X(\omega) = \frac{k_1}{(j\omega - p_1)} + \frac{k_2}{(j\omega - p_2)} + \cdots + \frac{k_n}{(j\omega - p_n)}, \quad (5.32)$$

where the partial fraction coefficients are calculated using the Heaviside formula as follows:

$$k_r = [(j\omega - p_r)X(\omega)]_{j\omega = p_r}, \quad (5.33)$$

for $1 \le r \le n$. For repeated or complex roots, the partial fraction expansion is more complicated and is discussed in Appendix D.
(3) The inverse CTFT can then be calculated as follows:

$$x(t) = [k_1 e^{p_1 t} + k_2 e^{p_2 t} + \cdots + k_n e^{p_n t}]u(t). \quad (5.34)$$

Example 5.10

Using the partial fraction method, calculate the inverse CTFT of the following function:

$$X(\omega) = \frac{5(j\omega) + 30}{(j\omega)^3 + 17(j\omega)^2 + 80(j\omega) + 100}.$$

Solution

In terms of $j\omega$, the roots of $D(\omega) = (j\omega)^3 + 17(j\omega)^2 + 80(j\omega) + 100$ are given by $j\omega = -2, -5,$ and -10. The partial fraction expansion of $X(\omega)$ is given by

$$X(\omega) = \frac{5(j\omega) + 30}{(j\omega + 2)(j\omega + 5)(j\omega + 10)} \equiv \frac{k_1}{(j\omega + 2)} + \frac{k_2}{(j\omega + 5)} + \frac{k_3}{(j\omega + 10)},$$

where the partial fraction coefficients are given by

$$k_1 = (j\omega + 2)\frac{5(j\omega) + 30}{(j\omega + 2)(j\omega + 5)(j\omega + 10)}\bigg|_{j\omega=-2}= \frac{5(j\omega) + 30}{(j\omega + 5)(j\omega + 10)}\bigg|_{j\omega=-2}$$

$$= \frac{20}{(3)(8)} = \frac{5}{6},$$

$$k_2 = (j\omega + 5)\frac{5(j\omega) + 30}{(j\omega + 2)(j\omega + 5)(j\omega + 10)}\bigg|_{j\omega=-5} = \frac{5(j\omega) + 30}{(j\omega + 2)(j\omega + 10)}\bigg|_{j\omega=-5}$$

$$= \frac{5}{(-3)(5)} = -\frac{1}{3},$$

and

$$k_3 = (j\omega + 10)\frac{5(j\omega) + 30}{(j\omega + 2)(j\omega + 5)(j\omega + 10)}\bigg|_{j\omega=-10} = \frac{5(j\omega) + 30}{(j\omega + 2)(j\omega + 5)}\bigg|_{j\omega=-10}$$

$$= \frac{-20}{(-8)(-5)} = -\frac{1}{2}.$$

Therefore, the partial fraction expansion of $X(\omega)$ is given by

$$X(\omega) = \frac{5}{6(j\omega + 2)} - \frac{1}{3(j\omega + 5)} - \frac{1}{2(j\omega + 10)}. \qquad (5.35)$$

Using the CTFT pairs in Table 5.2 to calculate the inverse CTFT, the function $x(t)$ is calculated as

$$x(t) = \left[\frac{5}{6}e^{-2t} - \frac{1}{3}e^{-5t} - \frac{1}{2}e^{-10t}\right]u(t). \qquad (5.36)$$

5.4 Fourier transform of real, even, and odd functions

In Example 5.1, it was observed that the CTFT of a causal decaying exponential,

$$e^{-at}u(t) \xleftrightarrow{\text{CTFT}} \frac{1}{(a + j\omega)},$$

has an even magnitude spectrum, while the phase spectrum is odd. This is known as *Hermitian symmetry* and holds true for the CTFT of any real-valued function. In this section, we consider various properties of the CTFT for real-valued functions.

5.4.1 CTFT of real-valued functions

5.4.1.1 Hermitian symmetry property

The CTFT $X(\omega)$ of a real-valued signal $x(t)$ satisfies the following:

$$X(-\omega) = X^*(\omega), \qquad (5.37)$$

where $X^*(\omega)$ denotes the complex conjugate of $X(\omega)$.

Proof

By definition,

$$X^*(\omega) = [\Im\{x(t)\}]^* = \left[\int_{-\infty}^{\infty} x(t) e^{-j\omega t} \mathrm{d}t\right]^* = \int_{-\infty}^{\infty} [x(t) e^{-j\omega t}]^* \mathrm{d}t,$$

which simplifies to

$$X^*(\omega) = \int_{-\infty}^{\infty} x^*(t) e^{j\omega t} \mathrm{d}t.$$

Since $x(t)$ is a real-valued signal, $x^*(t) = x(t)$ and we obtain

$$X^*(\omega) = \int_{-\infty}^{\infty} x(t) e^{-j(-\omega)t} \mathrm{d}t = X(-\omega),$$

which completes the proof.

The Hermitian property can also be expressed in terms of: (i) the real and imaginary components of the CTFT $X(\omega)$, and (ii) the magnitude and phase of $X(\omega)$. These lead to alternative representations for the Hermitian property, which are listed below.

5.4.1.2   Alternative form I for Hermitian symmetry property

The real component of the CTFT $X(\omega)$ of a real-valued signal $x(t)$ is even, while its imaginary component is odd. Mathematically,

$$\mathrm{Re}\{X(-\omega)\} = \mathrm{Re}\{X(\omega)\} \quad \text{and} \quad \mathrm{Im}\{X(-\omega)\} = -\mathrm{Im}\{X(\omega)\}. \qquad (5.38)$$

Proof

Substituting $X(\omega) = \mathrm{Re}\{X(\omega)\} + j\,\mathrm{Im}\{X(\omega)\}$ in the Hermitian symmetry property, Eq. (5.37), yields

$$\mathrm{Re}\{X(-\omega)\} + j\,\mathrm{Im}\{X(-\omega)\} = \mathrm{Re}\{X(\omega)\} - j\,\mathrm{Im}\{X(\omega)\}.$$

Separating the real and imaginary components in the above expression proves the alternative form I of the Hermitian symmetry property.

5.4.1.3   Alternative form II for Hermitian symmetry property

The magnitude spectrum $|X(\omega)|$ of the CTFT $X(\omega)$ of a real-valued signal $x(t)$ is even, while its phase spectrum $<X(\omega)$ is odd. Mathematically,

$$|X(-\omega)| = |X(\omega)| \quad \text{and} \quad <X(-\omega) = -<X(-\omega). \qquad (5.39)$$

Proof

The magnitude of the complex function $X(-\omega) = \text{Re}\{X(-\omega)\} + j\,\text{Im}\{X(-\omega)\}$ is given by

$$|X(-\omega)| = \sqrt{(\text{Re}\{X(-\omega)\})^2 + (\text{Im}\{X(-\omega)\})^2}.$$

Substituting $\text{Re}\{X(-\omega)\} = \text{Re}\{X(\omega)\}$ and $\text{Im}\{X(-\omega)\} = -\text{Im}\{X(\omega)\}$, obtained from the alternative form I of the Hermitian symmetry property in the above expression, yields

$$|X(-\omega)| = \sqrt{(\text{Re}\{X(\omega)\})^2 + (-\text{Im}\{X(\omega)\})^2} = |X(\omega)|,$$

which proves that the magnitude spectrum $|X(\omega)|$ of a real-valued signal is even. Alternatively, consider the phase of the complex function $X(-\omega) = \text{Re}\{X(-\omega)\} + j\,\text{Im}\{X(-\omega)\}$ as given by

$$<X(-\omega) = \tan^{-1}\left(\frac{\text{Re}\{X(-\omega)\}}{\text{Im}\{X(-\omega)\}}\right).$$

Substituting $\text{Re}\{X(-\omega)\} = \text{Re}\{X(\omega)\}$ and $\text{Im}\{X(-\omega)\} = -\text{Im}\{X(\omega)\}$ yields

$$<X(-\omega) = \tan^{-1}\left(\frac{\text{Re}\{X(-\omega)\}}{-\text{Im}\{X(-\omega)\}}\right) = -<X(\omega),$$

which proves that the phase spectrum $<X(\omega)$ of a real-valued signal is odd.

Example 5.11

Consider a function $g(t)$ whose CTFT is given by $G(\omega) = 1 + 2\pi\delta(\omega - \omega_0)$. Determine if $g(t)$ is a real-valued function.

Solution

Substituting ω by $-\omega$ in the CTFT $G(\omega)$ yields

$$G(-\omega) = 1 + 2\pi\delta(-\omega - \omega_0) = 1 + 2\pi\delta(\omega + \omega_0).$$

The complex conjugate of $G(\omega)$ is given by

$$G^*(\omega) = [1 + 2\pi\delta(\omega - \omega_0)]^* = 1 + 2\pi\delta(\omega - \omega_0).$$

Comparing the two expressions, it is clear that $G^*(\omega) \neq G(-\omega)$, and therefore that $g(t)$ is not a real-valued function. In order to verify the result, we calculate the inverse CTFT of $G(\omega)$ as follows:

$$g(t) = \Im^{-1}\{G(\omega)\} = \Im^{-1}\{1 + 2\pi\delta(\omega - \omega_0)\} = \Im^{-1}\{1\} + 2\pi\Im^{-1}\{\delta(\omega - \omega_0)\},$$

which results in

$$g(t) = \delta(t) + e^{j\omega_0 t},$$

or

$$g(t) = \delta(t) + \cos(\omega_0 t) + j\sin(\omega_0 t),$$

verifying that $g(t)$ is indeed not real-valued. In deriving the inverse CTFT of $G(\omega)$, we have assumed that the CTFT satisfies the linearity property, which is formally proved in Section 5.5.

5.4.2 CTFT of real-valued even and odd functions

A second set of symmetry properties is obtained if we assume that, in addition to being real-valued, $x(t)$ is an even or odd function. Before expressing these properties, we show that the expression of the CTFT is simplified considerably if we assume that $x(t)$ is an even or odd function.

Using the Euler identity, the CTFT is expressed as follows:

$$X(\omega) = \int_{-\infty}^{\infty} x(t) e^{-j\omega t} dt = \int_{-\infty}^{\infty} x(t) \cos(\omega t) dt - j \int_{-\infty}^{\infty} x(t) \sin(\omega t) dt.$$

Case I If $x(t)$ is even, then $x(t) \cos(\omega t)$ is also an even function, while $x(t) \sin(\omega t)$ is an odd function. Therefore, the CTFT for the even-valued function can alternatively be calculated from

$$X(\omega) = 2 \int_{0}^{\infty} x(t) \cos(\omega t) dt. \qquad (5.40)$$

Case II If $x(t)$ is odd, then $x(t) \sin(\omega t)$ is an even function, while $x(t) \cos(\omega t)$ is an odd function. An alternative expression for the CTFT for the odd-valued function is given by

$$X(\omega) = -j2 \int_{0}^{\infty} x(t) \sin(\omega t) dt. \qquad (5.41)$$

By combining the Hermitian property with Eqs. (5.40) and (5.41), the following two properties are obtained.

Property 5.1 CTFT of real-valued, even functions The CTFT $X(\omega)$ of a real-valued, even function $x(t)$ is also real and even. In other words, $\text{Re}\{X(\omega)\} = \text{Re}\{X(-\omega)\}$ and $\text{Im}\{X(\omega)\} = 0$.

Property 5.2 CTFT of real-valued, odd functions The CTFT $X(\omega)$ of a real-valued, odd function $x(t)$ is imaginary and odd. In other words, $\text{Re}\{X(\omega)\} = 0$ and $\text{Im}\{X(-\omega)\} = -\text{Im}\{X(-\omega)\}$.

The proofs of Properties 5.1 and 5.2 are left as exercises for the readers. See Problems 5.6 and 5.7. The symmetry properties of the CTFT are summarized in Table 5.4.

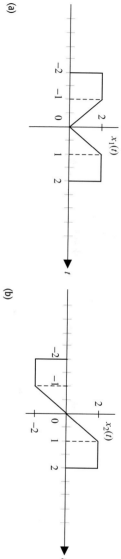

(a)

(b)

Fig. 5.9. CT signals used in Example 5.12. (a) $x_1(t)$; (b) $x_2(t)$.

Example 5.12

Calculate the Fourier transform of the functions $x_1(t)$ and $x_2(t)$ shown in Fig. 5.9.

Solution

(a) The mathematical expression for the CT function $x_1(t)$, illustrated in Fig. 5.9(a), is given by

$$x_1(t) = \begin{cases} 2|t| & -1 \le t \le 1 \\ 2 & 1 < |t| \le 2 \\ 0 & \text{elsewhere.} \end{cases}$$

Since $x_1(t)$ is an even function, its CTFT is calculated using Eq. (5.40) as follows:

$$X_1(\omega) = 2 \int_0^\infty x(t)\cos(\omega t)\,dt = 2 \int_0^1 (2t)\cos(\omega t)\,dt + 2 \int_1^2 2\cos(\omega t)\,dt,$$

which simplifies to

$$X_1(\omega) = 4 \left[t\,\frac{\sin(\omega t)}{\omega} + 1\,\frac{\cos(\omega t)}{\omega^2} - \frac{1}{\omega^2} \right]_0^1 + 4 \left[\frac{\sin(\omega t)}{\omega} \right]_1^2$$

or

$$X_1(\omega) = 4 \left[\frac{\sin(\omega)}{\omega} + \frac{\cos(\omega)}{\omega^2} - \frac{1}{\omega^2} \right] + 4 \left[\frac{\sin(2\omega)}{\omega} - \frac{\sin(\omega)}{\omega} \right]$$

$$= \frac{4}{\omega^2} \left[\omega \sin(2\omega) + \cos(\omega) - 1 \right]. \tag{5.42}$$

The above result validates the symmetry property for real-valued, even functions. Property 5.1 states that the CTFT of a real-valued, even function is real and even. This is indeed the case for $X_1(\omega)$ in Eq. (5.42).

(b) The function $x_2(t)$, shown in Fig. 5.9(b), is expressed as follows:

$$x_2(t) = \begin{cases} -2 & -2 \le t \le 1 \\ 2t & -1 \le t \le 1 \\ 2 & 1 < t \le 2 \\ 0 & \text{elsewhere.} \end{cases}$$

Since $x_2(t)$ is an odd function, its CTFT, based on Eq. (5.41), is given by

$$X_2(\omega) = -j2 \int_0^\infty x(t) \sin(\omega t) dt = -j2 \int_0^1 (2t) \sin(\omega t) dt - j2 \int_1^2 2 \sin(\omega t) dt,$$

which simplifies to

$$X_2(\omega) = -j4 \left[-t \frac{\cos(\omega t)}{\omega} + 1 \frac{\sin(\omega t)}{\omega^2} \right]_0^1 - j4 \left[-\frac{\cos(\omega t)}{\omega} \right]_1^2$$

or

$$X_2(\omega) = j4 \left[\frac{\cos(\omega)}{\omega} - \frac{\sin(\omega)}{\omega^2} \right] + j4 \left[\frac{\cos(2\omega)}{\omega} - \frac{\cos(\omega)}{\omega} \right]$$

$$= j\frac{4}{\omega^2} \left[\omega \cos(2\omega) - \sin(\omega) \right]. \tag{5.43}$$

The above result validates the symmetry property for real-valued odd functions. Property 5.2 states that the CTFT of a real-valued odd function is imaginary and odd. This is indeed the case for $X_2(\omega)$ in Eq. (5.43).

5.5 Properties of the CTFT

In Section 5.4, we covered the symmetry properties of the CTFT. In this section, we present the properties of the CTFT based on the transformations of the signals. Given the CTFT of a CT function $x(t)$, we are interested in calculating the CTFT of a function produced by a linear operation on $x(t)$ in the time domain. The linear operations being considered include superposition, time shifting, scaling, differentiation and integration. We also consider some basic non-linear operations like multiplication of two CT signals, convolution in the time and frequency domain, and Parseval's relationship. A list of the CTFT properties is included in Table 5.4.

5.5.1 Linearity

Often we are interested in calculating the CTFT of a signal that is a linear combination of several elementary functions whose CTFTs are known. In such a scenario, we use the linearity property to show that the overall CTFT is given by the same linear combination of the individual CTFTs used in the time domain. The linearity property is defined below.

If $x_1(t)$ and $x_2(t)$ are two CT signals with the following CTFT pairs:

$$x_1(t) \overset{\text{CTFT}}{\longleftrightarrow} X_1(\omega)$$

and

$$x_2(t) \overset{\text{CTFT}}{\longleftrightarrow} X_2(\omega)$$

Table 5.4. Symmetry and transformation properties of the CTFT

Transformation properties	Time domain $x(t) = \frac{1}{2\pi}\int_{-\infty}^{\infty} X(\omega)e^{j\omega t}\,d\omega$	Frequency domain $X(\omega) = \int_{-\infty}^{\infty} x(t)e^{-j\omega t}\,dt$	Comments
Linearity	$a_1 x_1(t) + a_2 x_2(t)$	$a_1 X_1(\omega) + a_2 X_2(\omega)$	$a_1, a_2 \in C$
Scaling	$x(at)$	$\frac{1}{|a|} X\left(\frac{\omega}{a}\right)$	$a \in \Re$, real-valued
Time shifting	$x(t - t_0)$	$e^{-j\omega t_0} X(\omega)$	$t_0 \in \Re$, real-valued
Frequency shifting	$e^{j\omega_0 t} x(t)$	$X(\omega - \omega_0)$	$\omega_0 \in \Re$, real-valued
Time differentiation	$\dfrac{d^n x}{dt^n}$	$(j\omega)^n X(\omega)$	provided dx/dt exists
Time integration	$\displaystyle\int_{-\infty}^{t} x(\tau)\,d\tau$	$\dfrac{X(\omega)}{j\omega} + \pi X(0)\delta(\omega)$	provided $\displaystyle\int_{-\infty}^{t} x(\tau)\,d\tau$ exists
Frequency differentiation	$t^n x(t)$	$(j)^n \dfrac{d^n X}{d\omega^n}$	provided $dX/d\omega$ exists
Duality	$X(t)$	$2\pi x(-\omega)$	
Time convolution	$x_1(t) * x_2(t)$	$X_1(\omega) X_2(\omega)$	convolution in time domain
Frequency convolution	$x_1(t) \times x_2(t)$	$\frac{1}{2\pi}[X_1(\omega) * X_2(\omega)]$	multiplication in time domain
Parseval's relationship	$E_x = \displaystyle\int_{-\infty}^{\infty} |x(t)|^2 dt = \frac{1}{2\pi}\int_{-\infty}^{\infty} |X(\omega)|^2 d\omega$		energy in a signal

Symmetry properties

CTFT: $X(-\omega) = X^*(\omega)$

Hermitian property	$x(t)$ is a real-valued function	real and imaginary components $\begin{cases} Re\{X(\omega)\} = Re\{X(-\omega)\} \\ Im\{X(\omega)\} = -Im\{X(-\omega)\} \end{cases}$ magnitude and phase spectra $\begin{cases} |X(-\omega)| = |X(\omega)| \\ <X(-\omega) = -<X(\omega) \end{cases}$	real component is even; imaginary component is odd; magnitude spectrum is even; phase spectrum is odd
Even function	$x(t)$ is even	$X(\omega) = 2\displaystyle\int_0^\infty x(t)\cos(\omega t)\,dt$	simplified CTFT expression for even signals
Odd function	$x(t)$ is odd	$X(\omega) = -j2\displaystyle\int_0^\infty x(t)\sin(\omega t)\,dt$	simplified CTFT expression for odd signals
Real-valued and even function	$x(t)$ is even and real-valued	$Re\{X(\omega)\} = Re\{X(-\omega)\}$ $Im\{X(\omega)\} = 0$	CTFT is real-valued and even
Real-valued and odd function	$x(t)$ is odd and real-valued	$Re\{X(\omega)\} = 0$ $Im\{X(\omega)\} = -Im\{X(-\omega)\}$	CTFT is imaginary and odd

then, for any arbitrary constants a_1 and a_2, the linearity property states that

$$a_1 x_1(t) + a_2 x_2(t) \xrightarrow{\text{CTFT}} a_1 X_1(\omega) + a_2 X_2(\omega), \quad \text{for } a_1, a_2 \in C, \quad (5.44)$$

where C denotes the set of complex numbers.

Proof

By Eq. (5.10), the CTFT of the linear combination $a_1 x_1(t)$ and $a_2 x_2(t)$ is given by

$$\Im\{a_1 x_1(t) + a_2 x_2(t)\} = \int_{-\infty}^{\infty} [a_1 x_1(t) + a_2 x_2(t)] e^{-j\omega t} dt$$

$$= a_1 \underbrace{\int_{-\infty}^{\infty} x_1(t) e^{-j\omega t} dt}_{X_1(\omega)} + a_2 \underbrace{\int_{-\infty}^{\infty} x_2(t) e^{-j\omega t} dt}_{X_2(\omega)}$$

or

$$\Im\{a_1 x_1(t) + a_2 x_2(t)\} = a_1 X_1(\omega) + a_2 X_2(\omega),$$

which completes the proof.

The application of the linearity property is demonstrated through the following example.

Example 5.13

Using the CTFT pairs given in Eqs. (5.25) and (5.26),

$$e^{j\omega_0 t} \xrightarrow{\text{CTFT}} 2\pi \delta(\omega - \omega_0)$$

and

$$e^{-j\omega_0 t} \xrightarrow{\text{CTFT}} 2\pi \delta(\omega + \omega_0),$$

calculate the CTFT of the cosine function $\cos(\omega_0 t)$.

Solution

Using Euler's formula,

$$\Im\{\cos(\omega_0 t)\} = \left\{ \frac{1}{2} [e^{j\omega_0 t} + e^{-j\omega_0 t}] \right\} = \frac{1}{2}\Im\{e^{j\omega_0 t}\} + \frac{1}{2}\Im\{e^{-j\omega_0 t}\}.$$

Using the aforementioned CTFT pairs for $\exp(j\omega_0 t)$ and $\exp(-j\omega_0 t)$, we obtain

$$\Im\{\cos(\omega_0 t)\} = \pi[\delta(\omega - \omega_0) + \delta(\omega + \omega_0)],$$

which is the same as the CTFT for the periodic cosine function in Table 5.2.

Fig. 5.10. Waveform $g(t)$ used in Example 5.14.

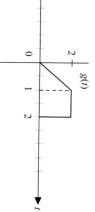

Example 5.14

Calculate the CTFT of the waveform $g(t)$ plotted in Fig. 5.10.

Solution

By inspection, the waveform $g(t)$ can be expressed as a linear combination of $x_1(t)$ and $x_2(t)$ from Fig. 5.9, as follows:

$$g(t) = \frac{1}{2}[x_1(t) + x_2(t)].$$

Using the linearity property, the CTFT of $g(t)$ is given by

$$G(\omega) = \frac{1}{2}X_1(\omega) + \frac{1}{2}X_2(\omega).$$

Based on Eqs. (5.42) and (5.43), the CTFT pairs for $x_1(t)$ and $x_2(t)$ are given by

$$X_1(\omega) = \frac{4}{\omega^2}[\omega \sin(2\omega) + \cos(\omega) - 1]$$

and

$$X_2(\omega) = j\frac{4}{\omega^2}[\omega \cos(2\omega) - \sin(\omega)].$$

The CTFT of $g(t)$ is therefore given by

$$
\begin{aligned}
G(\omega) &= \frac{2}{\omega^2}[\omega \sin(2\omega) + \cos(\omega) - 1] + j\frac{2}{\omega^2}[\omega \cos(2\omega) - \sin(\omega)] \\
&= \frac{2}{\omega^2}[j\omega e^{-j2\omega} + e^{-j\omega} - 1].
\end{aligned}
$$

5.5.2 Time scaling

In Section 1.4.1, we showed that the time-scaled version of a signal $x(t)$ is given by $x(at)$. If $a > 1$, the signal compresses in time. If $a < 1$, the signal expands in time. The time-scaling property expresses the CTFT of the time-scaled signal $x(at)$ in terms of the CTFT of the original signal $x(t)$.

If $x(t) \xleftarrow{\text{CTFT}} X(\omega)$ then

$$x(at) \xleftarrow{\text{CTFT}} \frac{1}{|a|}X\left(\frac{\omega}{a}\right), \quad \text{for } a \in \Re \quad \text{and} \quad a \neq 0, \tag{5.45}$$

where $\Re$ denotes the set of real values.

Fig. 5.11. Waveform $h(t)$ used in Example 5.15.

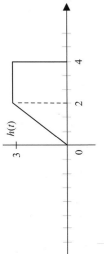

Proof

Equation (5.45) can be proved separately for the two cases $a > 0$ and $a < 0$. Case I ($a > 0$). By Eq. (5.10), the CTFT of the time-scaled signal $x(at)$ is given by

$$\Im\{x(at)\} = \int_{-\infty}^{\infty} x(at)e^{-j\omega t}\,dt.$$

Substituting $\tau = at$, the above integral reduces to

$$\Im\{x(at)\} = \int_{-\infty}^{\infty} x(\tau)e^{-j\omega\tau/a}\frac{d\tau}{a} = \frac{1}{a}X\left(\frac{\omega}{a}\right),$$

which proves Eq. (5.45) for $a > 0$. The proof for $a < 0$ follows the above procedure and is left as an exercise for the reader (see Problem 5.13).

Example 5.15

To illustrate the usefulness of the time-scaling property, let us calculate the CTFT of the function $h(t)$ shown in Fig. 5.11.

Solution

By inspection, the waveform $h(t)$ can be expressed as a scaled version of $g(t)$ illustrated in Fig. 5.10 as follows:

$$h(t) = \frac{3}{2}g\left(\frac{t}{2}\right) = \frac{3}{2}g(0.5t).$$

Applying the linearity and time-scaling properties with $a = 0.5$, the CTFT of $g(t)$ is given by

$$H(\omega) = \frac{3}{2}\left[\frac{1}{0.5}G\left(\frac{\omega}{0.5}\right)\right] = 3G(2\omega).$$

Based on the result of Example 5.14, $G(\omega) = (2/\omega^2)[j\omega e^{-j2\omega} + e^{-j\omega} - 1]$, which yields

$$H(\omega) = 3\frac{2}{(2\omega)^2}[j(2\omega)e^{-j(2\omega)} + e^{-j(2\omega)} - 1] = \frac{3}{2\omega^2}[j2\omega e^{-j4\omega} + e^{-j2\omega} - 1].$$

5.5.3 Time shifting

The time-shifting operation delays or advances the reference signal in time. Given a signal $x(t)$, the time-shifted signal is given by $x(t - t_0)$. If the value of the shift t_0 is positive, the reference signal $x(t)$ is delayed and shifted towards the right-hand side of the t-axis. On the other hand, if the value of the shift t_0 is negative, signal $x(t)$ advances forward and is shifted towards the left-hand side of the t-axis.

If $x(t) \xleftrightarrow{\text{CTFT}} X(\omega)$ then

$$g(t) = x(t - t_0) \xleftrightarrow{\text{CTFT}} e^{-j\omega t_0} X(\omega) \quad for\ t_0 \in \Re, \qquad (5.46)$$

where $\Re$ denotes the set of real values.

Proof

By Eq. (5.10), the CTFT of the time-shifted signal $x(t - t_0)$ is given by

$$\Im\{x(t - t_0)\} = \int_{-\infty}^{\infty} x(t - t_0)e^{-j\omega t}\,dt$$

$$= e^{-j\omega t_0} \int_{-\infty}^{\infty} x(\tau)e^{-j\omega \tau}\,d\tau \quad \text{by substituting } \tau = (t - t_0)$$

$$= e^{-j\omega t_0} X(\omega),$$

which proves the time-shifting property, Eq. (5.46).

The CTFT time-shifting property states that if a signal is shifted by t_0 time units in the time domain, the CTFT of the original signal is modified by a multiplicative factor of $\exp(-j\omega_0 t)$. The magnitude and phase of the CTFT of the time-shifted signal $g(t) = x(t - t_0)$ are given by

magnitude $\quad |G(\omega)| = |e^{-j\omega t_0} X(\omega)| = |e^{-j\omega t_0}||X(\omega)| = |X(\omega)|;$ (5.47)

phase $\quad <G(\omega) = <\{e^{-j\omega t_0} X(\omega)\} = <e^{-j\omega t_0} + <X(\omega) = -\omega t_0 + <X(\omega).$

(5.48)

Based on Eqs. (5.47) and (5.48), we can conclude that the time shifting does not change the magnitude spectrum of the original signal, while the phase spectrum is modified by an additive factor of $-\omega t_0$.

In Example 5.16, we illustrate the application of the time-shifting property by calculating the CTFT of the waveform illustrated in Fig. 5.12.

Example 5.16

Express the CTFT of the function $f(t)$ shown in Fig. 5.12 in terms of the CTFT of $g(t)$ shown in Fig. 5.10.

Fig. 5.12. Waveform $f(t)$ used in Example 5.16.

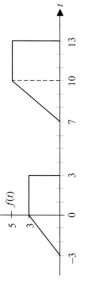

Solution

By inspection, $f(t)$ can be expressed in terms of $g(t)$ as

$$f(t) = \frac{3}{2} g\left(\frac{t+3}{3}\right) + \frac{5}{2} g\left(\frac{t-7}{3}\right).$$

We calculate the CTFT of each term in $f(t)$ separately. By considering the CTFT pair $g(t) \xrightarrow{\text{CTFT}} G(\omega)$ and applying the time-shifting property with $a = 3$, we obtain

$$g\left(\frac{t}{3}\right) \xrightarrow{\text{CTFT}} 3G(3\omega).$$

Using the time-shifting property,

$$g\left(\frac{t+3}{3}\right) \xrightarrow{\text{CTFT}} 3\mathrm{e}^{\mathrm{j}3\omega} G(3\omega) \quad \text{and} \quad g\left(\frac{t-7}{3}\right) \xrightarrow{\text{CTFT}} 3\mathrm{e}^{-\mathrm{j}7\omega} G(3\omega).$$

Finally, by applying the linearity property, we obtain

$$\frac{3}{2} g\left(\frac{t+3}{3}\right) + \frac{5}{2} g\left(\frac{t-7}{3}\right) \xrightarrow{\text{CTFT}} \frac{3}{2} \cdot 3\mathrm{e}^{\mathrm{j}3\omega} G(3\omega) + \frac{5}{2} \cdot 3\mathrm{e}^{-\mathrm{j}7\omega} G(3\omega).$$

Expressed in terms of the CTFT of $g(t)$, the CTFT $F(\omega)$ of the function $f(t)$ is therefore given by

$$F(w) = \frac{9}{2} \mathrm{e}^{\mathrm{j}3\omega} G(3\omega) + \frac{15}{2} \mathrm{e}^{-\mathrm{j}7\omega} G(3\omega).$$

5.5.4 Frequency shifting

In the time-shifting property, we observed the change in the CTFT when a signal $x(t)$ is shifted in the time domain. The frequency-shifting property addresses the converse problem of how a signal $x(t)$ is modified in the time domain if its CTFT is shifted in the frequency domain.

If $x(t) \xrightarrow{\text{CTFT}} X(\omega)$ then

$$h(t) = \mathrm{e}^{\mathrm{j}\omega_0 t} x(t) \xrightarrow{\text{CTFT}} X(\omega - \omega_0), \quad \text{for } \omega_0 \in \Re, \qquad (5.49)$$

where $\Re$ denotes the set of real values.

The frequency-shifting property can be proved directly from Eq. (5.10) by considering the CTFT of the signal $\exp(\mathrm{j}\omega_0 t)x(t)$. The proof is left as an exercise for the reader (see Problem 5.15).

By calculating the magnitude and phase of the term $\exp(j\omega_0 t)x(t)$ on the left-hand side of the CTFT pair shown in Eq. (5.49), we obtain

magnitude $\qquad |h(t)| = |e^{j\omega_0 t} x(t)| = |e^{j\omega_0 t}||x(t)| = |x(t)|;$ $\qquad$ (5.50)

phase $\qquad <h(t) = <e^{j\omega_0 t} x(t) = <e^{j\omega_0 t} + <x(t) = \omega_0 t + <x(t).$ $\qquad$ (5.51)

In other words, frequency shifting the CTFT of a signal does not change the amplitude $|x(t)|$ of the signal $x(t)$ in the time domain. The only change is in the phase $<x(t)$ of the signal $x(t)$, which is modified by an additive factor of $\omega_0 t$.

Example 5.17

In Section 2.1.3, we considered an amplitude modulator used in the AM band of the radio transmission to transmit an information signal $m(t)$ to the receiver. In Example 5.13, we calculated the CTFT of the $A\cos(\omega_0 t)$ as

$$s(t) = A[1 + km(t)]\cos(\omega_0 t).$$

Express the CTFT of the amplitude-modulated signal $s(t)$ in terms of the CTFT $M(\omega)$ of the information signal $m(t)$.

Solution

The amplitude-modulated signal is a sum of two terms: $A\cos(\omega_0 t)$ and $Akm(t)\cos(\omega_0 t)$. In Example 5.13, we calculated the CTFT of the $A\cos(\omega_0 t)$ as

$$A\cos(\omega_0 t) \xleftrightarrow{\text{CTFT}} A\pi[\delta(\omega - \omega_0) + \delta(\omega + \omega_0)].$$

By expanding $\cos(\omega_0 t)$, the second term $Akm(t)\cos(\omega_0 t)$ is expressed as follows:

$$Akm(t)\cos(\omega_0 t) = \frac{1}{2}Akm(t)[e^{j\omega_0 t} + e^{-j\omega_0 t}].$$

By using the frequency-shifting property, the CTFT of the terms $m(t)\exp(j\omega_0 t)$ and $m(t)\exp(-j\omega_0 t)$ are given by

$$m(t)e^{j\omega_0 t} \xleftrightarrow{\text{CTFT}} M(\omega - \omega_0) \quad \text{and} \quad m(t)e^{-j\omega_0 t} \xleftrightarrow{\text{CTFT}} M(\omega + \omega_0).$$

By using the linearity property, the CTFT of $Akm(t)\cos(\omega_0 t)$ is then given by

$$Akm(t)\cos(\omega_0 t) \xleftrightarrow{\text{CTFT}} \frac{1}{2}Ak[M(\omega - \omega_0) + M(\omega + \omega_0)].$$

By adding the CTFTs of the two terms, the CTFT of the amplitude-modulated signal is given by

$$s(t) \xleftrightarrow{\text{CTFT}} A\left[\pi\delta(\omega - \omega_0) + \pi\delta(\omega + \omega_0) + \frac{k}{2}M(\omega - \omega_0) + \frac{k}{2}M(\omega + \omega_0)\right].$$

5.5.5 Time differentiation

The time-differentiation property expresses the CTFT of a time-differentiated signal dx/dt in terms of the CTFT of the original signal $x(t)$. We state the time-differentiation property next.

If $x(t) \xrightarrow{\text{CTFT}} X(\omega)$ then

$$\frac{dx}{dt} \xrightarrow{\text{CTFT}} j\omega X(\omega) \qquad (5.52)$$

provided the derivative dx/dt exists at all time t.

Proof

From the CTFT synthesis equation, Eq. (5.9), we have

$$x(t) = \frac{1}{2\pi} \int_{-\infty}^{\infty} X(\omega) e^{j\omega t} \, d\omega.$$

Taking the derivative with respect to t on both sides of the equation yields

$$\frac{dx}{dt} = \frac{d}{dt} \left\{ \frac{1}{2\pi} \int_{-\infty}^{\infty} X(\omega) e^{j\omega t} \, d\omega \right\}.$$

Interchanging the order of differentiation and integration, we obtain

$$\frac{dx}{dt} = \frac{1}{2\pi} \int_{-\infty}^{\infty} X(\omega) \frac{d}{dt} \{e^{j\omega t}\} d\omega = \frac{1}{2\pi} \int_{-\infty}^{\infty} [j\omega X(\omega)] e^{j\omega t} \, d\omega.$$

Comparing this with Eq. (5.9), we obtain

$$\frac{dx}{dt} \xrightarrow{\text{CTFT}} j\omega X(\omega).$$

Corollary By repeatedly applying the time differentiation property, it is straightforward to verify that

$$\frac{d^n x}{dt^n} \xrightarrow{\text{CTFT}} (j\omega)^n X(\omega).$$

Example 5.18

In Example 5.11, we showed that the CTFT for the periodic cosine function is given by

$$\cos(\omega_0 t) \xrightarrow{\text{CTFT}} \pi [\delta(\omega - \omega_0) + \delta(\omega + \omega_0)].$$

Using the above CTFT pair, derive the CTFT for the periodic sine function $\sin(\omega_0 t)$.

Solution

Taking the derivative of the CTFT pair for the cosine function yields

$$\frac{d}{dt}\{\cos(\omega_0 t)\} \xleftrightarrow{\text{CTFT}} (j\omega)\pi[\delta(\omega - \omega_0) + \delta(\omega + \omega_0)].$$

By using the multiplicative property of the impulse function, $x(\omega)\delta(\omega + \omega_0) = x(-\omega_0)\delta(\omega + \omega_0)$, and by rearranging terms, we obtain

$$-\omega_0 \sin(\omega_0 t) \xleftrightarrow{\text{CTFT}} j\pi[\omega_0\delta(\omega - \omega_0) - \omega_0\delta(\omega + \omega_0)],$$

which can be expressed as

$$\omega_0 \sin(\omega_0 t) \xleftrightarrow{\text{CTFT}} \frac{\pi}{j}[\omega_0\delta(\omega - \omega_0) - \omega_0\delta(\omega + \omega_0)].$$

The CTFT of the periodic sine function is, therefore, given by

$$\sin(\omega_0 t) \xleftrightarrow{\text{CTFT}} \frac{\pi}{j}[\delta(\omega - \omega_0) - \delta(\omega + \omega_0)].$$

5.5.6 Time integration

The time-integration property expresses the CTFT of a time-integrated signal $\int x(t)dt$ in terms of the CTFT of the original signal $x(t)$.

If $x(t) \xleftrightarrow{\text{CTFT}} X(\omega)$, then

$$\int_{-\infty}^{t} x(\tau)d\tau \xleftrightarrow{\text{CTFT}} \frac{X(\omega)}{j\omega} + \pi X(0)\delta(\omega). \tag{5.53}$$

The proof of the time-integration property is left as an exercise for the reader (see Problem 5.14).

Example 5.19

Given $\delta(t) \xleftrightarrow{\text{CTFT}} 1$, calculate the CTFT of the unit step function $u(t)$ using the time-integration property.

Solution

Integrating the CTFT pair for the unit impulse function yields

$$\int_{-\infty}^{t} \delta(t)dt \xleftrightarrow{\text{CTFT}} \frac{1}{j\omega} + \pi\delta(\omega).$$

By noting that the left-hand side of the aforementioned CTFT pair represents the unit step function, we obtain

$$u(t) \xleftrightarrow{\text{CTFT}} \frac{1}{j\omega} + \pi\delta(\omega).$$

The above CTFT pair can be verified from Table 5.2.

5.5.7 Duality

The CTFTs of a constant signal $x(t) = 1$ and of an impulse function $x(t) = \delta(t)$ are given by the following CTFT pairs (see Table 5.2):

$$1 \xleftrightarrow{\text{CTFT}} 2\pi\delta(\omega) \quad \text{and} \quad \delta(t) \xleftrightarrow{\text{CTFT}} 1.$$

and

For the above examples, the CTFT exhibits symmetry across the time and frequency domains in the sense that the CTFT of a constant $x(t) = 1$ is an impulse function, while the CTFT of an impulse function $x(t) = \delta(t)$ is a constant. This symmetry extends to the CTFT of any arbitrary signal and is referred to as the duality property. We formally define the duality property below.

If $x(t) \xleftrightarrow{\text{CTFT}} X(\omega)$, then

$$X(t) \xleftrightarrow{\text{CTFT}} 2\pi x(-\omega) \tag{5.54}$$

is also a CTFT pair.

Proof

By the definition of the inverse CTFT, Eq. (5.9), we know that

$$x(t) = \frac{1}{2\pi} \int_{-\infty}^{\infty} X(r) e^{jrt} dr,$$

where the dummy variable r is used instead of ω. Substituting $t = -\omega$ in the above equation yields

$$2\pi x(-\omega) = \int_{-\infty}^{\infty} X(r) e^{-j\omega r} dr = \Im\{X(t)\}.$$

To illustrate the application of the duality property, consider the CTFT pair

$$\delta(t) \xleftrightarrow{\text{CTFT}} 1,$$

with $x(t) = \delta(t)$ and $X(\omega) = 1$. By interchanging the role of the independent variables t and ω, we obtain $X(t) = 1$ and $x(\omega) = \delta(\omega)$. Using the duality property, the converse CTFT pair is given by

$$1 \xleftrightarrow{\text{CTFT}} 2\pi\delta(-\omega) = 2\pi\delta(\omega),$$

which is indeed the CTFT of the constant signal $x(t) = 1$.

Example 5.20

As stated in Eq. (5.22), the following is a CTFT pair (see Example 5.6):

$$\text{rect}\left(\frac{t}{\tau}\right) \xleftrightarrow{\text{CTFT}} \tau \, \text{sinc}\left(\frac{\omega\tau}{2\pi}\right).$$

Calculate the CTFT of $x(t) = (W/\pi) \, \text{sinc}(Wt/\pi)$ using the duality property.

Solution

By interchanging the role of variables t and ω in the following CTFT pair:

$$x(t) = \text{rect}\left(\frac{t}{\tau}\right) \xleftrightarrow{\text{CTFT}} \tau \, \text{sinc}\left(\frac{\omega\tau}{2\pi}\right) = X(\omega),$$

we obtain $X(t) = \tau \, \text{sinc}(t\tau/2\pi)$ and $x(-\omega) = \text{rect}(-\omega/\tau)$. Using the duality property, we obtain

$$\tau \, \text{sinc}\left(\frac{t\tau}{2\pi}\right) \xleftrightarrow{\text{CTFT}} 2\pi \, \text{rect}\left(\frac{-\omega}{\tau}\right).$$

Substituting $\tau = 2W$ and dividing both sides of the above equation by 2π yields

$$\frac{W}{\pi} \, \text{sinc}\left(\frac{Wt}{\pi}\right) \xleftrightarrow{\text{CTFT}} \text{rect}\left(\frac{\omega}{2W}\right).$$

The above result was proved in Example 5.7 by deriving it directly from the definition of the CTFT.

5.5.8 Convolution

In Section 3.4, we showed that the output response of an LTIC system is obtained by convolving the input signal with the impulse response of the system. At times, the resulting convolution integral is difficult to solve analytically in the time domain. The convolution property provides us with an alternative approach, based on the CTFT, of calculating the output response. Below we define the convolution property and explain its application in calculating the output response of an LTIC system.

If $x_1(t) \xleftrightarrow{\text{CTFT}} X_1(\omega)$ and $x_2(t) \xleftrightarrow{\text{CTFT}} X_2(\omega)$, then

$$x_1(t) * x_2(t) \xleftrightarrow{\text{CTFT}} X_1(\omega)X_2(\omega) \qquad (5.55)$$

and

$$x_1(t)x_2(t) \xleftrightarrow{\text{CTFT}} \frac{1}{2\pi}[X_1(\omega) * X_2(\omega)]. \qquad (5.56)$$

In other words, convolution between two signals in the time domain is equivalent to the multiplication of the CTFTs of the two signals in the frequency domain. Conversely, convolution in frequency domain is equivalent to multiplication of the inverse CTFTs in the time domain. In the case of the frequency-domain convolution, one has to be careful in including a normalizing factor of $1/2\pi$. We prove the convolution property next.

Proof

To prove Eq. (5.55), consider the CTFT of the convolved signal $[x_1(t) * x_2(t)]$. By the definition in Eq. (5.9),

$$\Im\{x_1(t) * x_2(t)\} = \int_{-\infty}^{\infty} \{x_1(t) * x_2(t)\} e^{-j\omega t} dt.$$

Substituting the convolution $[x_1(t) * x_2(t)]$ by its integral, we obtain

$$\Im\{x_1(t) * x_2(t)\} = \int_{-\infty}^{\infty} \left[\int_{-\infty}^{\infty} x_1(\tau) x_2(t - \tau) d\tau \right] e^{-j\omega t} dt.$$

By changing the order of the two integrations, we obtain

$$\Im\{x_1(t) * x_2(t)\} = \int_{-\infty}^{\infty} x_1(\tau) \left[\int_{-\infty}^{\infty} x_2(t - \tau) e^{-j\omega t} dt \right] d\tau,$$

where the inner integral is given by

$$\int_{-\infty}^{\infty} x_2(t - \tau) e^{-j\omega t} dt = \Im\{x_2(t - \tau)\} = X_2(\omega) e^{-j\omega \tau}.$$

Therefore,

$$\Im\{x_1(t) * x_2(t)\} = X_2(\omega) \int_{-\infty}^{\infty} x_1(\tau) e^{-j\omega \tau} d\tau = X_2(\omega) X_1(\omega).$$

The convolution property, Eq. (5.56), in the frequency domain can be proved similarly by taking the inverse CTFT of $[X_1(\omega) * X_2(\omega)]$ and following the aforementioned procedure.

Equation (5.55) provides us with an alternative method to calculate the convolution integral using the CTFT. Expressed in terms of the CTFT pairs

$$x(t) \xleftrightarrow{\text{CTFT}} X(\omega), \quad h(t) \xleftrightarrow{\text{CTFT}} H(\omega), \quad \text{and} \quad y(t) \xleftrightarrow{\text{CTFT}} Y(\omega),$$

the output signal $y(t)$ is expressed in terms of the impulse response $h(t)$ and the input signal $x(t)$ as follows:

$$y(t) = x(t) * h(t) \xleftrightarrow{\text{CTFT}} Y(\omega) = X(\omega) H(\omega),$$

obtained by applying the convolution property in the time domain. In other words, the CTFT of the output signal is obtained by multiplying the CTFTs of the input signal and the impulse response. The procedure for evaluating the output $y(t)$ of an LTIC system in the frequency domain, therefore, consists of the following four steps.

(1) Calculate the CTFT $X(\omega)$ of the input signal $x(t)$.
(2) Calculate the CTFT $H(\omega)$ of the impulse response $h(t)$ of the LTIC system. The CTFT $H(\omega)$ is referred to as the transfer function of the LTIC system.
(3) Based on the convolution property, the CTFT $Y(\omega)$ of the output $y(t)$ is given by $Y(\omega) = X(\omega)H(\omega)$.
(4) Calculate the output $y(t)$ by taking the inverse CTFT of $Y(\omega)$ obtained in step (3).

The CTFT-based approach is convenient for three reasons. First, in most cases we can use Table 5.2 to look up the expression of the CTFTs and their inverses. In such cases, the CTFT-based approach is simpler to use than the time-domain approach based on the convolution integral. In cases where the CTFTs are difficult to evaluate analytically, they are obtained by using fast computational techniques for calculating the Fourier transform. The CTFT-based approach, therefore, allows the use of digital computers to calculate the output. Finally, the CTFT-based approach provides us with a meaningful insight into the behavior of many systems. An LTIC system is typically designed in the frequency domain.

Example 5.21

In Example 3.6, we showed that in response to the input signal $x(t) = e^{-t}u(t)$, the LTIC system with the impulse response $h(t) = e^{-2t}u(t)$ produces the following output:

$$y(t) = (e^{-t} - e^{-2t})u(t).$$

We will verify the above result using the CTFT-based approach.

Solution

Based on Table 5.2, the CTFTs for the input signal and the impulse response are as follows:

$$e^{-t}u(t) \xleftrightarrow{\text{CTFT}} \frac{1}{1+j\omega} \quad \text{and} \quad e^{-2t}u(t) \xleftrightarrow{\text{CTFT}} \frac{1}{2+j\omega}.$$

The CTFT of the output signal is therefore calculated as follows:

$$Y(\omega) = \Im\{e^{-t}u(t) * e[^{-2t}u(t)]\} = \Im\{e^{-t}u(t)\} \times \Im\{e^{-2t}u(t)\}.$$

Using the CTFT pair

$$e^{-at}u(t) \xleftrightarrow{\text{CTFT}} \frac{1}{a+j\omega},$$

we obtain

$$Y(\omega) = \frac{1}{1+j\omega} \times \frac{1}{2+j\omega},$$

Applying the triangle inequality in the CT domain, we obtain

$$|X(\omega)| \leq \int_{-\infty}^{\infty} |x(t)e^{-j\omega t}|\mathrm{d}t = \int_{-\infty}^{\infty} |x(t)||e^{-j\omega t}|\mathrm{d}t = \int_{-\infty}^{\infty} |x(t)|\mathrm{d}t,$$

which leads to the following condition for the existence of the CTFT.

Condition for existence of CTFT   The Fourier CTFT $X(\omega)$ of a function $x(t)$ *exists* if

$$\int_{-\infty}^{\infty} |x(t)|\mathrm{d}t < \infty. \qquad (5.59)$$

Equation (5.59) is a sufficient condition to verify the existence of the CTFT.

Example 5.23

Determine if the CTFTs exist for the following functions:

(i) causal decaying exponential function $f(t) = \exp(-at)u(t)$;
(ii) exponential function $g(t) = \exp(-at)$;
(iii) periodic cosine waveform $h(t) = \cos(\omega_0 t)$,

where $a, \omega_0 \in \Re^+$.

Solution

(i) Equation (5.59) yields

$$\int_{-\infty}^{\infty} |f(t)|\mathrm{d}t = \int_{-\infty}^{\infty} |e^{-at}u(t)|\mathrm{d}t = \int_{-\infty}^{\infty} e^{-at}u(t)\mathrm{d}t = \int_{0}^{\infty} e^{-at}\mathrm{d}t$$

$$= \frac{1}{-a}[e^{-at}]_0^{\infty} = \frac{1}{a} < \infty.$$

Therefore, the CTFT exists for the causal decaying exponential function.

(ii) Equation (5.59) yields

$$\int_{-\infty}^{\infty} |g(t)|\mathrm{d}t = \int_{-\infty}^{\infty} |e^{-at}|\mathrm{d}t = \int_{-\infty}^{\infty} e^{-at}\mathrm{d}t = \underbrace{\int_{-\infty}^{0} e^{-at}\mathrm{d}t}_{=\infty} + \underbrace{\int_{0}^{\infty} e^{-at}\mathrm{d}t}_{=1/a} = \infty.$$

Therefore, the CTFT does not exist for the exponential function.

(iii) Equation (5.59) reduces to

$$\int_{-\infty}^{\infty} |h(t)|\mathrm{d}t = \int_{-\infty}^{\infty} |\cos(\omega_0 t)|\mathrm{d}t = \infty.$$

Therefore, the CTFT does not exist for the periodic cosine function.

In part (iii), we proved that the CTFT does not exist for a periodic cosine function. This appears to be in violation of Table 5.2, which lists the following CTFT pair for the periodic cosine function:

$$\cos(\omega_0 t) \xleftrightarrow{\text{CTFT}} \pi[\delta(\omega - \omega_0) + \delta(\omega + \omega_0)].$$

Actually, the two statements do not contradict each other. The condition for the existence of the CTFT assumes that the CTFT must be finite for all values of ω. The above CTFT pairs indicate that the CTFT of the periodic cosine function consists of two impulses at $\omega = \pm\omega_0$. From the definition of the impulses, we know that the magnitudes of the two impulse functions in the aforementioned CTFT pair are infinite at $\omega = \pm\omega_0$, and therefore that the periodic cosine function violates the condition for the existence of the CTFT.

In Section 5.7, we show that the CTFTs of most periodic signals are derived from the CTFS representation of such signals, not directly from the CTFT definition. Therefore, we make an exception for periodic signals and ignore the condition of CTFT existence for periodic signals.

5.7 CTFT of periodic functions

Consider a periodic function $x(t)$ with a fundamental period of T_0. Using the exponential CTFS, the frequency representation of $x(t)$ is obtained from the following expression:

$$x(t) = \sum_{n=-\infty}^{\infty} D_n e^{jn\omega_0 t}, \qquad (5.60)$$

where $\omega_0 = 2\pi/T_0$ is the fundamental frequency of the periodic signal and D_n denotes the exponential CTFS coefficients D_n, given by

$$D_n = \frac{1}{T_0} \int_{(T_0)} x(t) e^{-jn\omega_0 t} \, dt.$$

Calculating the CTFT of both sides of Eq. (5.60), we obtain

$$X(\omega) = \Im\{x(t)\} = \Im\left\{ \sum_{n=-\infty}^{\infty} D_n e^{jn\omega_0 t} \right\}.$$

Using the linearity property, the above expression is simplified to

$$X(\omega) = \sum_{n=-\infty}^{\infty} D_n \Im\{e^{jn\omega_0 t}\} = 2\pi \sum_{n=-\infty}^{\infty} D_n \delta(\omega - n\omega_0). \qquad (5.61)$$

In other words, the CTFT of a periodic function $x(t)$ is given by

$$x(t) \xleftrightarrow{\text{CTFT}} 2\pi \sum_{n=-\infty}^{\infty} D_n \delta(\omega - n\omega_0). \qquad (5.62)$$

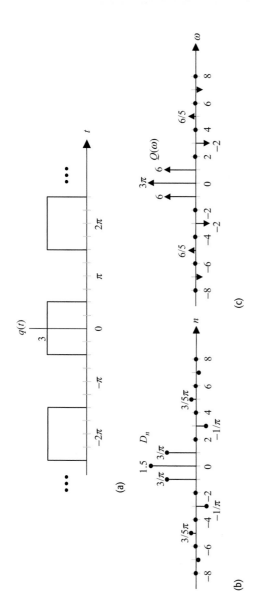

(a)

(b)

Fig. 5.13. Alternative representations for the periodic function considered in Example 5.24. (a) A periodic rectangular wavefunction $q(t)$, (b) CTFS coefficients D_n for $q(t)$, and (c) the CTFT $Q(\omega)$ of $q(t)$.

(c)

Equation (5.62) provides us with an alternative method for calculating the CTFT of periodic signals using the exponential CTFS. We illustrate the procedure in Examples 5.24 and 5.25.

Example 5.24

Calculate the CTFT representation of the periodic waveform $q(t)$ shown in Fig. 5.13(a).

Solution

The waveform $q(t)$ is a special case of the rectangular wave $x(t)$ considered in Example 4.14 with $\tau = \pi$ and $T = 2\pi$. Mathematically,

$$q(t) = 3x(t) \text{ with duty cycle } \tau/T = 1/2.$$

Using Eq. (4.49), the CTFS coefficients of $q(t)$ are given by

$$D_n = \frac{3}{2} \operatorname{sinc}\left(\frac{n}{2}\right)$$

or

$$D_n = \frac{3}{2} \operatorname{sinc}\left(\frac{n}{2}\right) = \begin{cases} \dfrac{3}{2} & n = 0 \\[2mm] 0 & n = 2k \neq 0 \\[2mm] \dfrac{3}{n\pi} & n = 4k + 1 \\[2mm] -\dfrac{3}{n\pi} & n = 4k + 3. \end{cases}$$

Substituting $\omega_0 = 1$ in Eq. (5.62) results in the following expression for the CTFT:

$$q(t) \xleftrightarrow{\text{CTFT}} 2\pi \sum_{n=-\infty}^{\infty} D_n \delta(\omega - n).$$

(a)

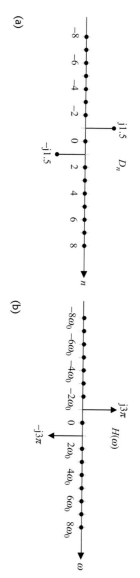

Fig. 5.14. Alternative representations for the sine wave considered in Example 5.25. (a) CTFS coefficients D_n; (b) CTFT representation $H(\omega)$.

(b)

The CTFS coefficients D_n and the CTFT $Q(\omega)$ of the periodic rectangular wave are plotted in Figs. 5.13(b) and (c).

Example 5.25

Calculate the CTFT for the periodic sine wave $h(t) = 3\sin(\omega_0 t)$.

Solution

To obtain the CTFS representation of the periodic sine wave, we expand $\sin(\omega_0 t)$ using Euler's identity. The resulting expression is as follows:

$$h(t) = 3\sin(\omega_0 t) = \frac{3}{2j}[e^{j\omega_0 t} - e^{-j\omega_0 t}],$$

which yields the following values for the exponential CTFS coefficients:

$$D_n = \begin{cases} -j1.5 & n = 1 \\ j1.5 & n = -1 \\ 0 & \text{otherwise.} \end{cases}$$

Based on Eq. (5.62), the CTFT of a periodic sine wave is given by

$$H(\omega) = 2\pi \sum_{n=-\infty}^{\infty} D_n \delta(\omega - n\omega_0) = j3\pi[\delta(\omega + \omega_0) - \delta(\omega - \omega_0)].$$

The CTFS coefficients and the CTFT for a periodic sine wave are plotted in Fig. 5.14. The above result is the same as derived in Example 5.18, with a scaling factor of 3.

5.8 CTFS coefficients as samples of CTFT

In Section 5.7, we presented a method of calculating the CTFT of a periodic signal from the CTFS representation. In this section, we solve the converse problem of calculating the CTFS coefficients from the CTFT.

Consider a time-limited aperiodic function $x(t)$, whose CTFT $X(\omega)$ is known. By following the procedure used in Section 5.1, we construct several repetitions of $x(t)$ uniformly spaced from each other with a duration of T_0. The process is illustrated in Fig. 5.1, where $x(t)$ is the aperiodic signal plotted in Fig. 5.1(a). Its periodic extension $\tilde{x}_T(t)$ is shown in Fig. 5.1(b). Using Eq. (5.3), the exponential CTFS coefficients of the periodic extension are given by

$$\tilde{D}_n = \frac{1}{T_0} \int_{\langle T_0 \rangle} x(t) e^{-jn\omega_0 t} dt = \frac{1}{T_0} \int_{-T_0/2}^{T_0/2} x(t) e^{-jn\omega_0 t} dt.$$

Since $\tilde{x}_T(t) = x(t)$ within the range $-T_0 \leq t \leq T_0$, the above expression reduces to

$$\tilde{D}_n = \frac{1}{T_0} \int_{-T_0/2}^{T_0/2} x(t) e^{-jn\omega_0 t} dt = \frac{1}{T_0} \int_{-\infty}^{\infty} x(t) e^{-jn\omega_0 t} dt = \frac{1}{T_0} X(\omega)|_{\omega = n\omega_0}, \quad (5.63)$$

which is the relationship between the CTFT of the aperiodic signal $x(t)$ and the CTFS coefficients of its periodic extension $\tilde{x}_T(t)$. In other words, we can derive the exponential CTFS coefficients of a periodic signal with period T_0 from the CTFT using the following steps.

(1) Compute the CTFT $X(\omega)$ of the aperiodic signal $x(t)$ obtained from one period of $\tilde{x}_T(t)$ as

$$x(t) = \begin{cases} \tilde{x}_T(t) & -T_0/2 \leq t \leq T_0/2 \\ 0 & \text{elsewhere.} \end{cases}$$

(2) The exponential CTFS coefficients D_n of the periodic signal $\tilde{x}_T(t)$ are given by

$$D_n = \frac{1}{T_0} X(\omega)|_{\omega = n\omega_0},$$

where ω_0 denotes the fundamental frequency of the periodic signal $\tilde{x}_T(t)$ and is given by $\omega_0 = 2\pi/T_0$.

Example 5.26

Calculate the exponential CTFS coefficients of the periodic signal $\tilde{x}_T(t)$ shown in Fig. 5.13(a).

Solution

Step 1 The aperiodic signal representing one period of $\tilde{x}_T(t)$ is given by

$$x(t) = 3 \operatorname{rect}\left(\frac{t}{\pi}\right).$$

Using Table 5.2, the CTFT of the rectangular gate function is given by

$$3 \operatorname{rect}\left(\frac{t}{\pi}\right) \xrightarrow{\text{CTFT}} 3\pi \operatorname{sinc}\left(\frac{\omega}{2}\right).$$

5.9 LTIC systems analysis using CTFT

Step 2 The exponential CTFS coefficients D_n of the periodic signal $\tilde{x}_T(t)$ are obtained from Eq. (5.63) as

$$D_n = \frac{1}{T_0} X(\omega)|_{\omega=n\omega_0} \quad \text{with} \quad T_0 = 2\pi \quad \text{and} \quad \omega_0 = 1.$$

The above expression simplifies to

$$D_n = \frac{3}{2} \operatorname{sinc}\left(\frac{n}{2}\right) = \frac{3}{n\pi} \sin\left(\frac{n\pi}{2}\right).$$

5.9.1 Transfer function of an LTIC system

In Chapters 2 and 3, we showed that an LTIC system can be modeled either by a linear, constant-coefficient differential equation or by its impulse response $h(t)$. A third representation for an LTIC system is obtained by taking the CTFT of the impulse response:

$$h(t) \xleftrightarrow{\text{CTFT}} H(\omega).$$

The CTFT $H(\omega)$ is referred to as the *Fourier transfer function* of the LTIC system and provides meaningful insights into the behavior of the system. The impulse response relates the output response $y(t)$ of an LTIC system to its input $x(t)$ using

$$y(t) = h(t) * x(t).$$

Calculating the CTFT of both sides of the equation, we obtain

$$Y(\omega) = H(\omega)X(\omega), \tag{5.64}$$

where $Y(\omega)$ and $X(\omega)$ are the respective CTFTs of the output response $y(t)$ and the input signal $x(t)$. Equation (5.64) provides an alternative definition for the transfer function as the ratio of the CTFT of the output response and the CTFT of the input signal. Mathematically, the transfer function $H(\omega)$ is given by

$$H(\omega) = \frac{Y(\omega)}{X(\omega)}. \tag{5.65}$$

Transfer function of an LTIC system

It was mentioned in Section 3.1 that, for an LTIC system, the relationship between the applied input $x(t)$ and output $y(t)$ can be described using a constant-coefficient differential equation of the following form:

$$\sum_{k=0}^{n} a_k \frac{d^k y}{dt^k} = \sum_{k=0}^{m} b_k \frac{d^k x}{dt^k}.$$

From the time-differentiation property of the CTFT, we know that

$$\frac{d^n x}{dt^n} \xleftrightarrow{\text{CTFT}} (j\omega)^n X(\omega). \tag{5.66}$$

Calculating the CTFT of both sides of Eq. (5.66) and applying the time-differentiation property, we obtain

$$\sum_{k=0}^{n} a_k (j\omega)^k Y(\omega) = \sum_{k=0}^{m} b_k (j\omega)^k X(\omega)$$

or

$$H(\omega) = \frac{Y(\omega)}{X(\omega)} = \frac{\displaystyle\sum_{k=0}^{n} b_k (j\omega)^k}{\displaystyle\sum_{k=0}^{m} a_k (j\omega)^k}. \tag{5.67}$$

Given one representation for an LTIC system, it is straightforward to derive the remaining two representations based on the CTFT and its properties. We illustrate the procedure through the following examples.

Example 5.27

Consider an LTIC system whose input–output relationship is modeled by the following third-order differential equation:

$$\frac{d^3 y}{dt^3} + 6\frac{d^2 y}{dt^2} + 11\frac{dy}{dt} + 6y(t) = 2\frac{dx}{dt} + 3x(t). \tag{5.68}$$

Calculate the transfer function $H(\omega)$ and the impulse response $h(t)$ for the LTIC system.

Solution

Using the time-differentiation property for the CTFT, we know that

$$\frac{d^n x}{dt^n} \xleftarrow{\text{CTFT}} (j\omega)^n X(\omega).$$

Taking the CTFT of both sides of Eq. (5.68) and applying the time-differentiation property yields

$$(j\omega)^3 Y(\omega) + 6(j\omega)^2 Y(\omega) + 11(j\omega)Y(\omega) + 6Y(\omega) = 2(j\omega)X(\omega) + 3X(\omega).$$

Making $Y(\omega)$ common on the left-hand side of the above expression, we obtain

$$[(j\omega)^3 + 6(j\omega)^2 + 11(j\omega) + 6]Y(\omega) = [2(j\omega) + 3]X(\omega).$$

Based on Eq. (5.67), the transfer function is therefore given by

$$H(\omega) = \frac{Y(\omega)}{X(\omega)} = \frac{2(j\omega) + 3}{(j\omega)^3 + 6(j\omega)^2 + 11(j\omega) + 6}. \tag{5.69}$$

The impulse response $h(t)$ is obtained by taking the inverse CTFT of Eq. (5.69). Factorizing the denominator, Eq. (5.69) is expressed as

$$H(\omega) = \frac{2(j\omega) + 3}{(1 + j\omega)(2 + j\omega)(3 + j\omega)},$$

which, by partial fraction expansion, reduces to

$$H(\omega) = \frac{1}{2(1 + j\omega)} + \frac{1}{(2 + j\omega)} - \frac{3}{2(3 + j\omega)}.$$

Taking the inverse CTFT:

$$h(t) = \left(\frac{1}{2}e^{-t} + e^{-2t} - \frac{3}{2}e^{-3t}\right)u(t). \qquad (5.70)$$

Equations (5.68)–(5.70) provide three equivalent representations of the LTIC system.

Example 5.28

Consider an LTIC system with the following impulse response function:

$$h(t) = \text{rect}\left(\frac{t}{\tau}\right) = \begin{cases} 1 & |t| \leq \tau/2 \\ 0 & |t| > \tau/2. \end{cases}$$

Calculate the transfer function $H(\omega)$ and the input–output relationship for the LTIC system.

Solution

From Table 5.2, we obtain the following transfer function:

$$H(\omega) = \tau \, \text{sinc}\left(\frac{\omega\tau}{2\pi}\right) = \frac{2}{\omega}\sin\left(\frac{\omega\tau}{2}\right). \qquad (5.71)$$

In other words,

$$\frac{Y(\omega)}{X(\omega)} = \frac{2}{\omega}\sin\left(\frac{\omega\tau}{2}\right),$$

which is expressed as

$$j\omega Y(\omega) = j2\sin\left(\frac{\omega\tau}{2}\right)X(\omega)$$

or

$$j\omega Y(\omega) = e^{j\omega\tau/2}X(\omega) - e^{-j\omega\tau/2}X(\omega).$$

Taking the inverse CTFT of both sides, we obtain

$$\frac{dy}{dt} = x\left(t + \frac{\tau}{2}\right) - x\left(t - \frac{\tau}{2}\right). \qquad (5.72)$$

5.9.2 Response of LTIC systems to periodic signals

In Section 4.7.2, we derived the output response of an LTIC system, shown in Fig. 5.15, of the following periodic signal:

$$x(t) = \sum_{n=-\infty}^{\infty} D_n e^{jn\omega_0 t} \qquad (5.73)$$

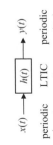

$x(t) \longrightarrow \boxed{h(t)} \longrightarrow y(t)$

periodic   LTIC   periodic
input   system   output

Fig. 5.15. Response of an LTIC system to a periodic input.

as

$$y(t) = \sum_{n=-\infty}^{\infty} D_n \mathrm{e}^{jn\omega_0 t} H(\omega)|_{\omega = n\omega_0}, \qquad (5.74)$$

where $H(\omega)$ is the CTFT of the impulse response $h(t)$ of the system and is referred to as the transfer function of the LTIC system. Corollary 4.1 is a special case of Eq. (5.74), where the input is a sinusoidal signal and the impulse response $h(t)$ is real-valued. In such cases, the output $y(t)$ can be expressed as follows:

$$k_1 \exp(j\omega_0 t) \to A_1 k_1 \exp(j\omega_0 t + j\phi_1). \qquad (5.75)$$

$$k_1 \sin(\omega_0 t) \to A_1 k_1 \sin(\omega_0 t + \phi_1), \qquad (5.76)$$

and

$$k_1 \cos(\omega_0 t) \to A_1 k_1 \cos(\omega_0 t + \phi_1), \qquad (5.77)$$

where A_1 and ϕ_1 are the magnitude and phase of $H(\omega)$ evaluated at $\omega = \omega_0$. Equations (5.73)–(5.77) can be derived directly by using the CTFT. We now prove Eq. (5.74).

Proof

The CTFT of a periodic signal $x(t)$ is given by

$$x(t) \xleftrightarrow{\text{CTFT}} 2\pi \sum_{n=-\infty}^{\infty} D_n \delta(\omega - n\omega_0).$$

Using the convolution property, the output of an LTIC with transfer function $H(\omega)$ is given by

$$Y(\omega) = 2\pi \sum_{n=-\infty}^{\infty} D_n \delta(\omega - n\omega_0) H(\omega) = 2\pi \sum_{n=-\infty}^{\infty} D_n \delta(\omega - n\omega_0) H(n\omega_0).$$

Taking the inverse CTFT of the above equation yields

$$y(t) = \sum_{n=-\infty}^{\infty} D_n \Im^{-1}\{2\pi \delta(\omega - n\omega_0)\} H(n\omega_0) = \sum_{n=-\infty}^{\infty} D_n H(n\omega_0) \mathrm{e}^{jn\omega_0 t},$$

which proves Eq. (5.74).

Example 5.29

Consider an LTIC system with impulse response given by

$$h(t) = \frac{10}{\pi} \operatorname{sinc}\left(\frac{10t}{\pi}\right), \qquad (5.78)$$

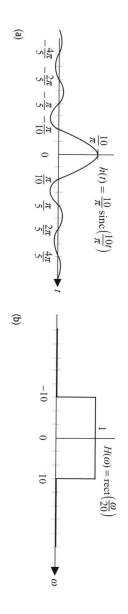

$$h(t) = \frac{10}{\pi}\,\text{sinc}\left(\frac{10t}{\pi}\right)$$

(a)

$$H(\omega) = \text{rect}\left(\frac{\omega}{20}\right)$$

(b)

Fig. 5.16. LTIC system considered in Example 5.29.
(a) Impulse response $h(t)$;
(b) transfer function $H(\omega)$.

sketched as a function of time t in Fig. 5.16(a). Determine the output response of the system for the following inputs:

(i) $x_1(t) = \sin(5t)$;
(ii) $x_2(t) = \sin(15t)$;
(iii) $x_3(t) = \sin(8t) + \sin(20t)$.

Solution

Calculating the CTFT of Eq. (5.78), the transfer function $H(\omega)$ is given by

$$H(\omega) = \text{rect}\left(\frac{\omega}{20}\right). \qquad (5.79)$$

The magnitude spectrum of the LTIC system is plotted in Fig. 5.16(b). The phase of the LTIC system is zero for all frequencies.

(i) Input $x_1(t) = \sin(5t)$. The CTFT of the input signal $x_1(t)$ is given by

$$X_1(\omega) = \frac{\pi}{j}[\delta(\omega - 5) - \delta(\omega + 5)].$$

The CTFT $Y_1(\omega)$ of the output signal is obtained by multiplying $X_1(\omega)$ by $H(\omega)$ and is given by

$$Y_1(\omega) = X_1(\omega)H(\omega) = \frac{\pi}{j}\delta(\omega - 5)H(\omega) - \frac{\pi}{j}\delta(\omega + 5)H(\omega).$$

Using the multiplication property of the impulse function, we have

$$Y_1(\omega) = \frac{\pi}{j}\delta(\omega - 5)H(5) - \frac{\pi}{j}\delta(\omega + 5)H(-5).$$

Since $H(\pm 5) = 1$, the CTFT $Y_1(\omega)$ of the output signal is given by

$$Y_1(\omega) = \frac{\pi}{j}\delta(\omega - 5) - \frac{\pi}{j}\delta(\omega - 5).$$

Taking the inverse CTFT, the output is given by

$$y_1(t) = \sin(5t).$$

The CTFT $Y_1(\omega)$ of the output signal can also be obtained by graphical multiplication, as shown in Fig. 5.17(a), where the magnitude spectrum of the transfer function $H(\omega)$ is shown as a dashed line. Since the magnitude of the transfer function $H(\omega)$ is one at the location of the two impulses contained in the CTFT of the input signal, the CTFT $Y_1(\omega)$ of the output signal is identical to the CTFT

Fig. 5.17. Frequency interpretation of the output response of an LTIC system. Response of the LTIC system (transfer function shown as a dashed line) to:
(a) $x_1(t) = \sin(5t)$;
(b) $x_2(t) = \sin(15t)$;
(c) $x_3(t) = \sin(8t) + \sin(20t)$.

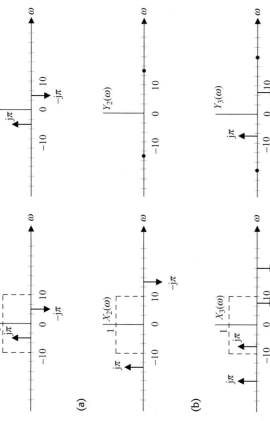

(a)

(b)

(c)

of the input signal. By calculating the inverse CTFT, we obtain the output as
$$y_1(t) = x_1(t) = \sin(5t).$$

(ii) Input $x_2(t) = \sin(15t)$. The CTFT of the input signal $x_2(t)$ is given by
$$X_2(\omega) = \frac{\pi}{j}[\delta(\omega - 15) - \delta(\omega + 15)].$$

The CTFT $Y_2(\omega)$ of the output signal is obtained by multiplying $X_1(\omega)$ by $H(\omega)$ and is given by
$$Y_2(\omega) = X_2(\omega)H(\omega) = \frac{\pi}{j}\delta(\omega - 15)H(\omega) - \frac{\pi}{j}\delta(\omega + 15)H(\omega).$$

Using the multiplication property of the impulse function, we have
$$Y_2(\omega) = X_1(\omega)H(\omega) = \frac{\pi}{j}\delta(\omega - 15)H(15) - \frac{\pi}{j}\delta(\omega + 15)H(-15).$$

Since $H(\pm 5) = 0$, the CTFT $Y_2(\omega)$ of the output signal is given by
$$Y_2(\omega) = 0.$$

Taking the inverse CTFT, the output is $y_2(t) = 0$.

As in part (i), the CTFT $Y_2(\omega)$ of the output signal can be obtained by graphical multiplication shown in Fig. 5.17(b). Since the magnitude of the transfer function $H(\omega)$ is zero at the location of the two impulses contained in the CTFT of the input signal, the two impulses are blocked from the output of the LTIC system. The CTFT $Y_1(\omega)$ of the output signal is zero, which results in $y_2(t) = 0$.

(iii) Input $x_3(t) = \sin(8t) + \sin(20t)$. Taking the CTFT of the input $x_3(t)$ yields

$$X_3(\omega) = \frac{\pi}{j}[\delta(\omega - 8) - \delta(\omega + 8)] + \frac{\pi}{j}[\delta(\omega - 20) - \delta(\omega + 20)].$$

By following the procedure used in part (i), the CTFT $Y_3(\omega)$ of the output signal is given by

$$Y_3(\omega) = \left[\frac{\pi}{j}\delta(\omega - 8)H(8) - \frac{\pi}{j}\delta(\omega + 8)H(-8)\right]$$
$$+ \left[\frac{\pi}{j}\delta(\omega - 20)H(20) - \frac{\pi}{j}\delta(\omega + 20)H(-20)\right].$$

The input signal consists of four impulse functions with two impulses located at $\omega = \pm 8$ and two located at $\omega = \pm 20$. The magnitude of the transfer function at frequencies $\omega = \pm 8$ is one, therefore the two impulse functions $\delta(\omega - 8)$ and $\delta(\omega + 8)$ are unaffected. The magnitude of the transfer function at frequencies $\omega = \pm 20$ is zero, therefore the two impulses $\delta(\omega - 20)$ and $\delta(\omega + 20)$ are eliminated from the output. The CTFT of the output signal therefore consists of only two impulse functions located at $(\omega = \pm 8)$, and is given by

$$Y_3(\omega) = \left[\frac{\pi}{j}\delta(\omega - 8) - \frac{\pi}{j}\delta(\omega + 8)\right],$$

which has the inverse CTFT of

$$y_3(t) = \sin(8t).$$

In signal processing, the LTIC system with $h(t) = (10/\pi)\,\text{sinc}(10t/\pi)$ is referred to as an ideal low-pass filter since it eliminates high-frequency components and leaves the low-frequency components unaffected. In this example, all input frequency components with frequencies greater than $\omega > 10$ are eliminated. Any input components with lower frequencies ($\omega < 10$) appear unaffected in the output of the LTIC system. The frequency ($\omega = 10$) is referred to as the *cutoff frequency* of the ideal low-pass filter.

5.9.3 Response of an LTIC system to quasi-periodic signals

The response of an LTIC system to ideal periodic signals is given by Eqs. (5.73)–(5.77). In practice, however, it is difficult to produce ideal periodic signals of infinite duration. Most practical signals start at $t = 0$ and are of finite duration. In this section, we calculate the output of an LTIC system for input signals that are not completely periodic. We refer to such signals as quasi-periodic signals.

Example 5.30

Consider the RC series circuit shown in Fig. 5.18. Determine the overall and steady state values of the output of the RC series circuit if the input signal is given by $x(t) = \sin(3t)u(t)$. Assume that the capacitor is uncharged at $t = 0$.

Solution

The CTFT of the input signal $x(t)$ is given by

$$X(\omega) = \frac{\pi}{2j}[\delta(\omega - 3) - \delta(\omega + 3)] + \frac{3}{9 - \omega^2}.$$

From the theory of electrical circuits, the transfer function of the RC series circuit is given by

$$H(\omega) = \frac{1/j\omega C}{R + 1/j\omega C} = \frac{1}{1 + j\omega CR}.$$

Substituting the value of the product $CR = 0.5$ yields

$$H(\omega) = \frac{1}{1 + j0.5\omega}.$$

By multiplying the CTFT of the input signal by the transfer function, the CTFT of the output $y(t)$ is given by

$$Y(\omega) = \left\{ \frac{\pi}{2j}[\delta(\omega - 3) - \delta(\omega + 3)] + \frac{3}{9 - \omega^2} \right\} \times \frac{1}{1 + j0.5\omega}.$$

Solving the above expression results in the following:

$$Y(\omega) = \frac{\pi}{2j}\left[\frac{\delta(\omega - 3)}{1 + j1.5} - \frac{\delta(\omega + 3)}{1 - j1.5} \right] + \frac{3}{(9 - \omega^2)(1 + j0.5\omega)}.$$

Taking the inverse CTFT of the above expression (see Problem 5.10) yields the following value for the output signal:

$$y(t) = \underbrace{\frac{2}{\sqrt{13}}\sin(3t - 56°)u(t)}_{\text{steady state value}} + \underbrace{\frac{6}{13}e^{-2t}u(t)}_{\text{transient value}}.$$

An alternative way of obtaining the steady state value of the output of the RC series circuit is suggested in Corollary 4.1. Expressed in terms of the given input, Corollary 4.1 states

$$\underbrace{\sin(3t)u(t)}_{x(t)} \longrightarrow \underbrace{A_1\sin(3t + \phi_1)u(t)}_{y(t)},$$

where A_1 and ϕ_1 are, respectively, the magnitude and phase of the transfer function at $\omega = 3$. The values of A_1 and ϕ_1 are given by

$$A_1 = |H(3)| = \left| \frac{1}{1 + j0.5(3)} \right| = \frac{2}{\sqrt{13}} \quad \text{and}$$

$$\phi_1 = <H(3) = < \left(\frac{1}{1 + j0.5(3)} \right) = -\tan^{-1}(1.5) = -56°.$$

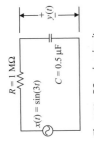

$R = 1\,\mathrm{M}\Omega$

$x(t) = \sin(3t)$

$C = 0.5\,\mu\mathrm{F}$

$+\ y(t)\ -$

Fig. 5.18. RC series circuit considered in Example 5.30.

Substituting the values of A_1 and ϕ_1 in Corollary 4.1, the steady state value of the output is given by

$$y_{ss}(t) = \frac{2}{\sqrt{13}} \sin(3t - 56°)u(t).$$

For sinusoidal signals, Corollary 4.1 provides a simpler approach of determining the steady state output.

5.9.4 Gain and phase responses

The Fourier transfer function $H(\omega)$ provides a complete description of the LTIC system. In many applications, the graphical plots of $|H(\omega)|$ and $< H(\omega)$ versus frequency ω are used to analyze the characteristics of the LTIC system. The magnitude spectrum $|H(\omega)|$ response function is also referred to as the *gain response* of the system, while the phase spectrum $<H(\omega)$ is referred to as the *phase response* of the system. Below, we provide an example to illustrate the procedure involved in plotting the magnitude and phase spectra. We also introduce Bode plots, where a logarithmic scale is used for the frequency ω-axis.

Example 5.31

Consider an LTIC system with the impulse response $h(t) = 1.25e^{-0.6t}$ $\sin(0.8t)u(t)$. Plot the gain and phase responses of the LTIC system.

Solution

The transfer function $H(\omega)$ of the LTIC system is given by

$$H(\omega) = \Im\{1.25e^{-0.6t} \sin(0.8t)u(t)\} = 1.25 \times \frac{0.8}{(0.6 + j\omega)^2 + 0.8^2}$$

The magnitude and phase spectra are as follows:

magnitude spectrum $\quad |H(\omega)| = \dfrac{1}{\sqrt{(1 - \omega^2)^2 + (1.2\omega)^2}}$

$$= \frac{1}{\sqrt{1 - 0.56\omega^2 + \omega^4}};$$

phase spectrum $\quad <H(\omega) = -\tan^{-1}\left(\dfrac{1.2\omega}{1 - \omega^2}\right).$

The magnitude spectrum $|H(\omega)| = \dfrac{1}{1 - \omega^2 + j1.2\omega}$.

Figure 5.19(a) plots the magnitude spectrum and Fig. 5.19(b) plots the phase spectrum of the LTIC system. Figure 5.19(a) illustrates that the magnitude $|H(\omega)| = 1$ for $\omega = 0$. As the frequency ω increases, the magnitude $|H(\omega)|$ drops and approaches zero at very high frequencies. From Fig. 5.19(b), we observe that the phase $<H(\omega)$ is zero at $\omega = 0$. At high frequencies, the phase $<H(\omega)$ converges to $-\pi$ radians, or $-180°$.

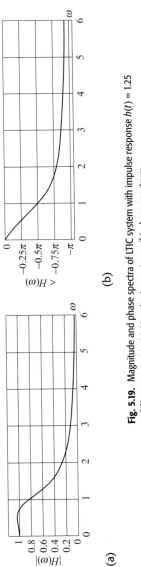

(a)

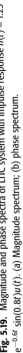

(b)

Fig. 5.19. Magnitude and phase spectra of LTIC system with impulse response $h(t) = 1.25$ $e^{-0.6t}\sin(0.8t)u(t)$. (a) Magnitude spectrum; (b) phase spectrum.

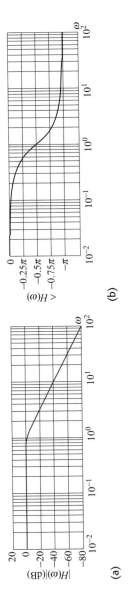

(a)

(b)

Fig. 5.20. Bode plots for the LTIC system considered in Example 5.31. (a) Magnitude plot; (b) phase plot.

Bode plots In Bode plots, the magnitude $|H(\omega)|$ in decibels and phase $<H(\omega)$ are plotted as functions of frequency ω using a logarithmic scale. Use of a logarithmic scale, with base 10, on the frequency ω-axis offers two main advantages.

(1) Compared to a linear scale, the use of a logarithmic scale allows a wider range of frequencies to be plotted, with the lower frequencies represented at a higher resolution.

(2) The asymptotic approximations of the magnitude and phase spectra can easily be sketched graphically by hand.

Figure 5.20 illustrates the Bode plots of the LTIC system considered in Example 5.31 using a logarithmic scale on the frequency axis. Figure 5.20(a) shows the magnitude Bode plot, where the magnitude $|H(\omega)|$ is expressed in decibels (dB) as $20\ \log_{10}|H(\omega)|$ and plotted as a function of $\log_{10}(\omega)$. Figure 5.20(b) shows the phase Bode plot, where the phase $<H(\omega)$ is plotted as a function of $\log_{10}(\omega)$.

5.10 MATLAB exercises

In this section, we will consider two applications of MATLAB. First, we illustrate the procedure for calculating the CTFT of a CT signal $x(t)$ using MATLAB. In our explanation, we consider an example, $x(t) = 4\cos(10\pi t)$, and write the appropriate MATLAB commands for the example at each step. Second, we list the procedure for plotting the Bode plots in MATLAB.

5.10.1 CTFT using MATLAB

Step 1 Sampling In order to manipulate the CT signals on a digital computer, the CT signals must be discretized. This is normally achieved through a process called sampling. In reality, sampling is followed by quantization, but because of the high resolution supported by MATLAB, we can neglect quantization without any appreciable loss of accuracy, at least for our purposes here. Sampling converts a CT signal $x(t)$ into an equivalent DT signal $x[k]$. To prevent any loss of information and for $x[k]$ to be an exact representation of $x(t)$, the sampling rate ω_s must be greater than at least twice the maximum frequency ω_{max} present in the signal $x(t)$, i.e.

$$\omega_s \geq 2\omega_{max}. \qquad (5.80)$$

This is referred to as the *Nyquist criterion*. We will consider sampling in depth in Chapter 9, but the information presented above is sufficient for the following discussion.

The CTFT of the periodic cosine signal is given by (see Table 5.2)

$$4\cos(10\pi t) \xleftrightarrow{\text{CTFT}} 4\pi[\delta(\omega - 10\pi) + \delta(\omega + 10\pi)]; \qquad (5.81)$$

hence, the maximum frequency in $x(t)$ is given by $\omega_{max} = 10\pi$ radians/s. Based on the Nyquist criterion, the lower bound for the sampling rate is given by

$$\omega_s \geq 20\pi \text{ radians/s}.$$

We choose a sampling rate that is 20 times the Nyquist rate, i.e. $\omega_s = 400\pi$ radians/s. The sampling interval T_s is given by

$$T_s = \frac{2\pi}{\omega_s} = 5 \text{ ms.} \qquad (5.83)$$

Selecting a time interval from -1 to 1 second to plot the sinusoidal wave, the number N of samples in $x[k]$ is 401. The MATLAB command that computes $x[k]$ is therefore given by

```
> t = -1:0.005:1;          % define time instants
> x = 4*cos(10*pi*t);      % samples of cosine wave
> subplot(221); plot(t,x)  % for CT plot
> subplot(222); stem(t,x)  % for DT plot
```

The subplots are plotted in Fig. 5.21(a) and (b) and provide a fairly accurate representation of the cosine wave.

Step 2 Fast Fourier transform In MATLAB, numeric computation of the CTFT is performed by using a fast implementation referred to as the fast Fourier transform (FFT). At this time, we will simply name the function without

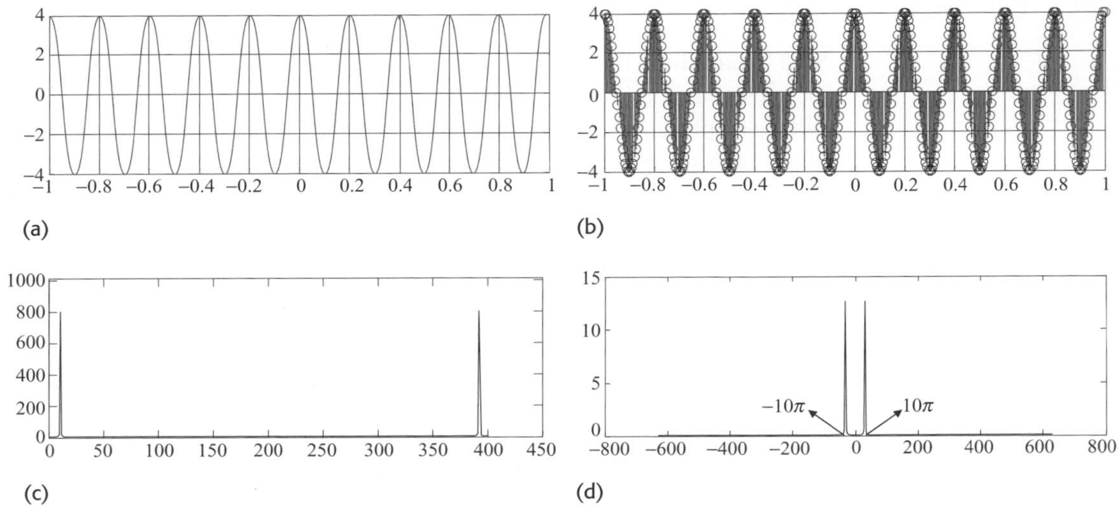

(a)

(b)

(c)

(d)

Fig. 5.21. MATLAB subplots for the time and frequency domain representations of $x(t) = 4\cos(10\pi t)$. (a) CT plot for $x(t)$; (b) DT plot for $x(t)$; (c) uncompensated CTFT of $x(t)$; (d) CTFT of $x(t)$.

worrying about its implementation. The function that evaluates FFT is fft (all lower-case letters). The MATLAB command for calculating fft is

```
> y = fft(x);              % fft computes CTFT
> subplot(223); plot(abs(y));   % abs calculates magnitude
```

The subplot of y is plotted in Fig. 5.21(c). There are two differences between y (output of the fft function) and the CTFT pair,

$$4\cos(10\pi t) \overset{\text{CTFT}}{\longleftrightarrow} 4\pi[\delta(\omega - 10\pi) + \delta(\omega + 10\pi)].$$

By looking at the peak value of the magnitude spectrum $|y|$, we note that the magnitude is not given by 4π as the CTFT pair suggests. Also, the x-axis represents the number of points instead of the appropriate frequency range ω. In steps (3) and (4), we compensate for these differences without going into the details of why the differences occur. The differences between the output of fft and CTFT will be discussed in Chapter 11.

Step 3   Compensation   Scale the magnitude of y by multiplying it by π times the sampling rate (πT_s). In our example, T_s is 5 ms. The following MATLAB command performs the scaling:

```
> z = pi*0.005*y;      % scale the magnitude of y
```

We also center z about an integer index of zero. This is accomplished by fftshift.

```
> z = fftshift(z);      % centre the CTFT about w = 0
```

Step 4 **Frequency axis** For a sequence $x[k]$ of length N with a sampling frequency ω_s, the $\mathtt{fft}$ function $y = \mathtt{fft(x)}$ produces the CTFT of $x(t)$ at N equispaced points within the frequency interval $[0, \omega_s]$. The resolution $\Delta\omega$ in the frequency domain is, therefore, given by $\Delta\omega = \omega_s/(N-1)$. After centering, performed by the $\mathtt{fftshift}$ function, the limits of the interval are changed to $[-\omega_s/2, \omega_s/2]$. The MATLAB commands to compute the appropriate values for the ω-axis are given by

```
> dw = 400* pi / 400;
> w = -400* pi/2:dw:400* pi/2;      % axis;
                                     % calculates frequency
> subplot(224); plot(w,abs(z));     % magnitude spectrum
```

The subplot of the CTFT is plotted in Fig. 5.21(d). By inspection, it is confirmed that it does correspond to the CTFT pair in Eq. (5.81).

The phase spectrum of the CTFT can be plotted using the $\mathtt{angle}$ function. For our example, the MATLAB command to plot the angle is given by

```
> subplot(224); plot(w,angle(z));       % phase spectrum
```

The above command replaces the magnitude spectrum in $\mathtt{subplot(224)}$ by the phase spectrum. For the given signal, $x(t) = 4\cos(10\pi t)$, the phase spectrum is zero for all frequencies ω. The MATLAB code for calculating the CTFT of a cosine wave is provided below in a function called $\mathtt{myctft}$.

```
function [w,z] = myctft
% MYCTFT: computes CTFT of 4*cos(10*pi*t)
% Usage: [w,z] = myctft

% compute 4 cos(10*pi*t) in time domain

A = 4;                % amplitude of cosine wave
w0 = 10*pi;           % maximum frequency in signal
ws = 20*w0;           % sampling rate
Ts = 2*pi/ws;         % sampling interval
t = -1:Ts:1;          % define time instants
x = A*cos(w0*t);      % samples of cosine wave

% compute the CTFT

y = fft(x);           % fft computes CTFT
z = pi*Ts*y;          % scale the magnitude of y
z = fftshift(z);      % centre CTFT about w = 0

% compute the frequency axis

w = -ws/2:ws/length(z):ws/2-ws/length(z);

% plots

subplot(211); plot(t,x)        % CT plot of cos(w0*t)
subplot(212); plot(w,abs(z))   % CTFT plot of cos(w0*t)
% end
```

To calculate the inverse CTFT, we replace the function `fft` with `ifft` and reverse the order of the instructions. The MATLAB code to compute the inverse CTFT is provided in a second function called `myinvctft`:

```
function [t,x] = myinvctft(w,z)
% MYINVCTFT: computes inverse CTFT of y known at
%            frequencies w
% Usage: [t,x] = myinvctft(w,z)
% compute the inverse CTFT
x = ifftshift(z);
x = ifft(x);                        % inverse fft
% compute the time instants
ws = w(length(w)) - w(1);           % sampling rate
Ts = 2*pi/ws;                       % sampling interval
t = Ts*[-floor(length(w))/2:floor(length(w))/2-1];  % sampling instants
% amplify signal by 1/(pi*Ts)
x = x/Ts;
% plots
subplot(211); plot(w,abs(z))        % CTFT plot of cos(w0*t)
subplot(212); plot(t,real(x))       % CT plot of cos(w0*t)
% end
```

5.10.2 Bode plots

MATLAB provides the bode function to sketch the Bode plot. To illustrate the application of the bode function, consider the LTIC system of Example 5.31. The system transfer function is given by

$$H(\omega) = 1.25 \times \frac{0.8}{(0.6 + j\omega)^2 + 0.8^2}.$$

In order to avoid a complex-valued representation, MATLAB expresses the Fourier transfer function in terms of the Laplace variable $s = j\omega$. In Chapter 6, we will show that the independent variable s represents the entire complex plane and leads to the generalization of the Fourier transfer function into an alternative transfer function, referred to as the Laplace transfer function. Substituting ($s = j\omega$) in $H(\omega)$ results in the following expression for the transfer function:

$$H(s) = \frac{1}{(0.6 + s)^2 + 0.8^2} = \frac{1}{s^2 + 1.2s + 1}.$$

Given $H(s)$, the Bode plots are obtained in MATLAB using the following instructions:

```
> clear;                        % clear the MATLAB environment
> num_coeff = [1];              % coefficients of the numerator
                                % in decreasing powers of s
> denom_coeff = [1 1.2 1];      % coefficients of the denominator
                                % in decreasing powers of s
> sys = tf(num_coeff,denom_coeff);  % specify the transfer function
> bode(sys,{0.01,100});        % sketch the Bode plots
```

In the above set of MATLAB instructions, we have used two new functions: tf and bode. The built-in function tf specifies the LTIC system $H(s)$ in terms of the coefficients of the polynomials of s in the numerator and denominator. Since the numerator $N(s) = 1$, the coefficients of the numerator are given by num_coeff = 1. The denominator $D(s) = s^2 + 1.2s + 1$. The coefficients of the denominator are given by denom_coeff = [1 1.2 1].

The built-in function bode sketches the Bode plots. It accepts two input arguments. The first input argument sys in used to represent the LTIC system, while the second input argument {0.01,100} specifies the frequency range, 0.01 radians/s to 100 radians/s, used to sketch the Bode plots. In setting the values for the frequency range, we use the curly parenthesis. Since the square parenthesis [0.01,100] represents only two frequencies, $\omega = 0.01$ and $\omega = 100$, it will result in the wrong plots. The second argument is optional. If unspecified, MATLAB uses a default scheme to determine the frequency range for the Bode plots.

5.11 Summary

In this chapter, we introduced the frequency representations for CT aperiodic signals. These frequency decompositions are referred to as the CTFT, which for a signal $x(t)$ is defined by the following two equations:

CTFT synthesis equation
$$x(t) = \frac{1}{2\pi} \int_{-\infty}^{\infty} X(\omega)e^{j\omega t}\, d\omega;$$

CTFT analysis equation
$$X(\omega) = \int_{-\infty}^{\infty} x(t)e^{-j\omega t}\, dt.$$

Collectively, the synthesis and analysis equations form the CTFT pair, which is denoted by

$$x(t) \xleftrightarrow{\text{CTFT}} X(\omega).$$

In Section 5.1, we derived the synthesis and analysis equations by expressing the CTFT as a limiting case of the CTFS. Several important CTFT pairs were calculated in Section 5.2. The results are listed in Table 5.2, and their magnitude and phase spectra of the CTFT are plotted in Table 5.3. In Section 5.3, we presented the partial fraction method for calculating the inverse CTFT. In Section 5.4, we covered the following symmetry properties of the CTFT.

(1) The CTFT $X(\omega)$ of a *real-valued* signal $x(t)$ is Hermitian symmetrical, i.e. $X(\omega) = X^*(-\omega)$. Due to the Hermitian symmetry property, the magnitude spectrum $|X(\omega)|$ is an even function of ω, while the phase spectrum $\angle X(\omega)$ is an odd function of ω.

(2) The CTFT $X(\omega)$ of a *real-valued and even* signal $x(t)$ is also real-valued and even, i.e. $\text{Re}\{X(\omega)\} = \text{Re}\{X(-\omega)\}$ and $\text{Im}\{X(\omega)\} = 0$.

(3) The CTFT $X(\omega)$ of a *real-valued and odd* signal $x(t)$ is also pure imaginary and odd, i.e. $\text{Re}\{X(\omega)\} = 0$ and $\text{Im}\{X(\omega)\} = -\text{Im}\{X(-\omega)\}$.

Section 5.5 considered the transformation properties of the CTFT, which are summarized as follows.

(1) The linearity property states that the CTFT of a linear combination of aperiodic signals is given by the same linear combination of the CTFT of the individual aperiodic signals.

(2) If an aperiodic signal is *time-scaled*, the CTFT is inversely time-scaled.

(3) A time shift of t_0 in the aperiodic signal does not affect the magnitude of the CTFT. However, the phase changes by an additive factor of ωt_0. This property is referred to as the *time-shifting* property.

(4) A frequency shift of ω_0 in the aperiodic signal does not affect the magnitude of the signal in the time domain. However, the phase of the signal in the time domain changes by an additive factor of ωt_0. This property is referred to as the *frequency-shifting* property.

(5) The CTFT of a *time-differentiated* periodic signal is obtained by *multiplying* the CTFT of the original signal by a factor of $j\omega$.

(6) The CTFT of a *time-integrated* periodic signal is obtained by *dividing* the CTFT of the original signal by a factor of $j\omega$ with a scaled impulse function at $\omega = 0$.

(7) The *duality property* states that there is symmetry between the time waveform and its frequency-domain representation such that the two functions in a CTFT pair are dual with respect to each other. Given an arbitrary time-domain waveform $x(t)$ and its CTFT waveform $X(\omega)$, for example, a second CTFT pair exists with the time-domain representation $X(t)$, having the same waveform as $X(\omega)$, and the CTFT $2\pi x(-\omega)$ in the frequency domain.

(8) Convolution in the time domain is equivalent to multiplication of the CTFT in the frequency domain, and vice versa. The convolution property leads to an alternative approach for evaluating the output response of an LTIC system to any arbitrary input.

Table 5.1. Magnitude $|X(\omega)|$ and phase $<X(\omega)$ for the CTFT of $x(t) = \exp(-3t)u(t)$ in Example 5.1

ω (radians/s)	$-\infty$	-1000	-100	-10	-1	0	1	10	100	∞
Magnitude: $|X(\omega)|$	0	0.001	0.01	0.096	0.316	0.333	0.316	0.096	0.01	0
Phase: $<X(\omega)$	$\pi/2$	1.57	1.54	1.28	0.32	0	-0.32	-1.28	-1.54	$-\pi/2$

The notation $a \in R^+$ implies that a is real-valued within the range $-\infty < a < \infty$.

Solution

(i) Based on the definition of the CTFT, Eq. (5.10), we obtain

$$X_1(\omega) = \Im\{e^{-at}u(t)\} = \int_{-\infty}^{\infty} e^{-at}u(t)e^{-j\omega t}\,dt = \int_{0}^{\infty} e^{-(a+j\omega)t}\,dt$$

$$= -\frac{1}{(a+j\omega)}\left[e^{-(a+j\omega)t}\right]_0^{\infty} = -\frac{1}{(a+j\omega)}\left[\lim_{t\to\infty} e^{-(a+j\omega)t} - 1\right],$$

where the term

$$\lim_{t\to\infty} e^{-(a+j\omega)t} = \lim_{t\to\infty} e^{-at} \cdot \lim_{t\to\infty} e^{-j\omega t} = 0 \cdot \lim_{t\to\infty} e^{-j\omega t} = 0.$$

Therefore,

$$X_1(\omega) = \frac{1}{a+j\omega}.$$

The magnitude and phase of $X_1(\omega)$ are given by

magnitude $\quad |X_1(\omega)| = \left|\frac{1}{a+j\omega}\right| = \frac{1}{\sqrt{a^2+\omega^2}};$

phase $\quad <X_1(\omega) = <\frac{1}{a+j\omega} = <1 - <(a+j\omega) = -\tan^{-1}\left(\frac{\omega}{a}\right).$

Table 5.1 lists the amplitude and phase of $X(\omega)$ for several values of ω with $a = 3$. The exponentially decaying function $x_1(t)$ and its magnitude and phase spectra are plotted in Fig. 5.3.

(ii) Based on the definition of the CTFT, Eq. (5.10), we obtain

$$X_2(\omega) = \Im\{e^{-a|t|}\} = \int_{-\infty}^{\infty} e^{-a|t|}e^{-j\omega t}\,dt$$

$$= \int_{-\infty}^{\infty} \underbrace{e^{-a|t|}\cos(\omega t)dt}_{\text{even function}} - j\int_{-\infty}^{\infty} \underbrace{e^{-a|t|}\sin(\omega t)dt}_{\text{odd function}}.$$

Since the integral of an odd function with limits $[-L, L]$ is zero, the above

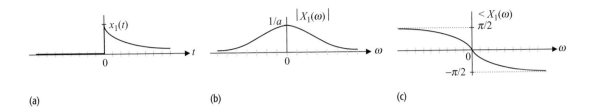

Fig. 5.3. CTFT of the causal decaying exponential function $x(t) = e^{-at}u(t)$. (a) $x(t)$; (b) magnitude spectrum; (c) phase spectrum.

equation reduces to

$$X_2(\omega) = \int_{-\infty}^{\infty} e^{-a|t|} \cos(\omega t)\mathrm{d}t = 2 \int_0^{\infty} e^{-at} \cos(\omega t)\mathrm{d}t$$

$$= \frac{2}{a^2 + \omega^2}[-ae^{-at}\cos(\omega t) + \omega e^{-at}\sin(\omega t)]_0^{\infty} = \frac{2a}{a^2 + \omega^2}.$$

Since $X_2(\omega)$ is positive real-valued, the magnitude and phase of $X_2(\omega)$ are given by

magnitude $\qquad\qquad |X_2(\omega)| = \left|\dfrac{2a}{a^2 + \omega^2}\right| = \dfrac{2a}{a^2 + \omega^2}.$

phase $\qquad\qquad\qquad <X_2(\omega) = 0.$

The non-causal exponentially decaying function $x_2(t)$ and its magnitude and phase spectra are plotted in Fig. 5.4.

We note from Example 5.1 that the magnitude spectrum is symmetric along the vertical axis while the phase spectrum is symmetric about the origin. The magnitude spectrum is, therefore, an even function of ω, while the phase spectrum is an odd function of ω. This is a consequence of the symmetry properties observed by real-valued functions. The symmetry properties are discussed in detail in Section 5.3.

Fig. 5.4. CTFT of the causal decaying exponential function $x_2(t) = \exp(-a|t|)$. (a) $x_2(t)$; (b) Magnitude spectrum; (c) phase spectrum for $a > 0$.

Example 5.2
Calculate the CTFT of a constant function $x(t) = 1$.

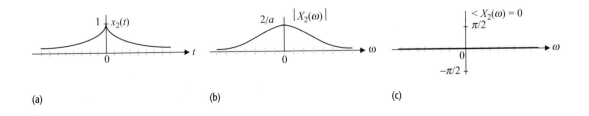

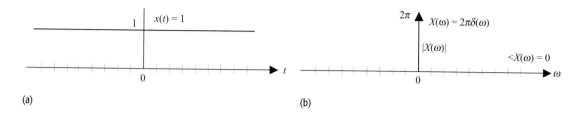

(a)   (b)

Fig. 5.5. CTFT of a constant function. (a) Constant function, $x(t) = 1$; (b) its CTFT, $X(\omega) = 2\pi\delta(\omega)$.

Solution

Based on the definition of the CTFT, Eq. (5.10), we obtain

$$X(\omega) = \Im\{1\} = \int_{-\infty}^{\infty} e^{-j\omega t} dt. \tag{5.14}$$

It can be shown that (see Problem 5.10)

$$\int_{-\infty}^{\infty} e^{j\omega t} dt = 2\pi\,\delta(\omega). \tag{5.15}$$

Substituting ω by $-\omega$ on both sides of Eq. (5.15), we obtain

$$\int_{-\infty}^{\infty} e^{-j\omega t} dt = 2\pi\,\delta(-\omega) = 2\pi\,\delta(\omega),$$

which results in

$$X(\omega) = \int_{-\infty}^{\infty} e^{-j\omega t} dt = 2\pi\,\delta(\omega).$$

In other words,

$$1 \xrightarrow{\text{CTFT}} 2\pi\,\delta(\omega). \tag{5.16}$$

The magnitude spectrum of a constant function $x(t) = 1$ therefore consists of an impulse function with area 2π located at the origin, $\omega = 0$, in the frequency domain. The magnitude spectrum is plotted in Fig. 5.5(b). The phase is zero for all frequencies ($\infty \le \omega \le -\infty$).

Example 5.3

The CTFT of an aperiodic function $g(t)$ is given by $G(\omega) = 2\pi\,\delta(\omega)$. Determine the aperiodic function $g(t)$.

Solution

Based on the CTFT synthesis equation, Eq. (5.9), we obtain

$$g(t) = \Im^{-1}\{2\pi\,\delta(\omega)\} = \frac{1}{2\pi} \int_{-\infty}^{\infty} 2\pi\,\delta(\omega)e^{j\omega t} d\omega = \int_{-\infty}^{\infty} \delta(\omega)d\omega = 1.$$

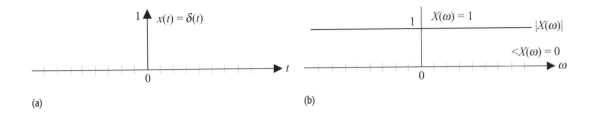

(a)   (b)

Fig. 5.6. CTFT of an impulse function. (a) Impulse function, $x(t) = \delta(t)$; (b) its CTFT, $X(\omega) = 1$.

In other words,

$$1 \xleftarrow{\text{CTFT}} 2\pi \delta(\omega). \tag{5.17}$$

Combining the results in Examples 5.2 and 5.3, we obtain the CTFT pair:

$$1 \xleftrightarrow{\text{CTFT}} 2\pi \delta(\omega). \tag{5.18}$$

Example 5.4
Determine the Fourier transform of the impulse function $x(t) = \delta(t)$.

Solution
Based on the definition of the CTFT, Eq. (5.10), we obtain

$$X(\omega) = \Im\{\delta(t)\} = \int_{-\infty}^{\infty} \delta(t)e^{-j\omega t}\,dt = \int_{-\infty}^{\infty} \delta(t)\,dt = 1.$$

Therefore,

$$\delta(t) \xrightarrow{\text{CTFT}} 1.$$

The CTFT of the impulse function located at the origin ($t = 0$) is a constant. The magnitude spectrum is shown in Fig. 5.6. The phase spectrum is zero for all frequencies ω.

Example 5.5
The CTFT of an aperiodic function $g(t)$ is given by $G(\omega) = 1$. Determine the aperiodic function $g(t)$.

Solution
Based on the CTFT analysis equation, Eq. (5.10), we obtain

$$g(t) = \Im^{-1}\{1\} = \frac{1}{2\pi} \int_{-\infty}^{\infty} 1 \cdot e^{j\omega t}\,dt. \tag{5.19}$$

By interchanging the role of ω and t in Eq. (5.15), we obtain

$$\int_{-\infty}^{\infty} e^{j\omega t} d\omega = 2\pi \delta(t).$$

Substituting the above relationship in Eq. (5.19) yields

$$g(t) = \frac{1}{2\pi} \int_{-\infty}^{\infty} e^{j\omega t} dt = \frac{1}{2\pi} \times 2\pi \delta(t) = \delta(t).$$

Therefore,

$$\delta(t) \xleftarrow{\text{CTFT}} 1. \qquad (5.20)$$

Combining the results derived in Examples 5.4 and 5.5, we can form the CTFT pair:

$$\delta(t) \xleftrightarrow{\text{CTFT}} 1. \qquad (5.21)$$

In Example 5.5, we proved that the inverse CTFT of $G(\omega) = 1$ is given by the impulse function $g(t) = \delta(t)$. In Example 5.4, we showed the converse: that the CTFT of $g(t) = \delta(t)$ is $G(\omega) = 1$. Likewise, in Examples 5.2 and 5.3, we established the CTFT pair,

$$1 \xleftrightarrow{\text{CTFT}} 2\pi \delta(\omega),$$

by computing the forward and inverse CTFT. Since the CTFT pair is unique, it is sufficient to compute either the CTFT or its inverse. Once the CTFT is derived, its inverse is established automatically, and vice versa. In the remaining examples, we form the CTFT pair by deriving either the forward CTFT or its inverse.

A second observation made from the CTFT pairs given in Eqs. (5.18) and (5.21),

$$1 \xleftrightarrow{\text{CTFT}} 2\pi \delta(\omega) \quad \text{and} \quad \delta(t) \xleftrightarrow{\text{CTFT}} 1,$$

is that the CTFT exhibits a duality property. The CTFT of a constant is the impulse function, while the CTFT of an impulse function is a constant. A factor of 2π is also introduced. We revisit the duality property in Section 5.5.

Example 5.6

Calculate the CTFT of the rectangular function $f(t)$ shown in Fig. 5.7(a).

Solution

Based on the definition of the CTFT, Eq. (5.10), we obtain

$$F(\omega) = \Im \{\text{rect}\,(t/\tau)\} = \int_{-\tau/2}^{\tau/2} 1 \cdot e^{-j\omega t} dt = \left[\frac{e^{-j\omega t}}{-j\omega}\right]_{-\tau/2}^{\tau/2},$$

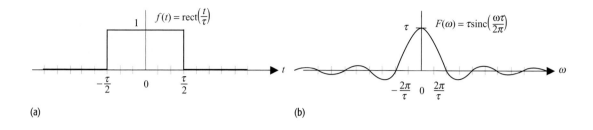

$f(t) = \text{rect}\left(\frac{t}{\tau}\right)$

$F(\omega) = \tau\text{sinc}\left(\frac{\omega\tau}{2\pi}\right)$

(a)

(b)

Fig. 5.7. CTFT of the rectangular function. (a) Rectangular function; (b) its CTFT given by the sinc function.

which simplifies to

$$F(\omega) = -\frac{1}{j\omega}[e^{-j\omega t}]_{-\tau/2}^{\tau/2} = -\frac{1}{j\omega}[e^{-j\omega\tau/2} - e^{j\omega\tau/2}] = -\frac{1}{j\omega}\left[-2j\sin\left(\frac{\omega\tau}{2}\right)\right]$$

or

$$F(\omega) = \frac{2}{\omega}\sin\left(\frac{\omega\tau}{2}\right) = \tau\,\text{sinc}\left(\frac{\omega\tau}{2\pi}\right).$$

The Fourier transform $F(\omega)$ is plotted in Fig. 5.7(b). The CTFT pair for a rectangular function is given by

$$\text{rect}\left(\frac{t}{\tau}\right) \xleftarrow{\text{CTFT}} \tau\,\text{sinc}\left(\frac{\omega\tau}{2\pi}\right). \tag{5.22}$$

Example 5.7

Determine the aperiodic function $g(t)$ whose CTFT $G(\omega)$ is the rectangular function shown in Fig. 5.8(a).

Solution

From Fig. 5.8(a), we observe that

$$G(\omega) = \begin{cases} 1 & |\omega| \leq W \\ 0 & |\omega| > W. \end{cases}$$

Based on the CTFT synthesis equation, Eq. (5.9), we obtain

$$g(t) = \Im^{-1}\left\{\text{rect}\left(\frac{\omega}{2W}\right)\right\} = \frac{1}{2\pi}\int_{-W}^{W} 1 \cdot e^{j\omega t}\,d\omega = \frac{1}{2\pi}\left[\frac{e^{j\omega t}}{jt}\right]_{-W}^{W}, \tag{5.23}$$

which simplifies to

$$g(t) = \frac{1}{j2\pi t}[e^{jWt} - e^{-jWt}] = \frac{1}{j2\pi t}[2j\sin(Wt)] = \frac{\sin(Wt)}{\pi t}$$

$$= \frac{W}{\pi}\text{sinc}\left(\frac{W}{\pi}t\right).$$

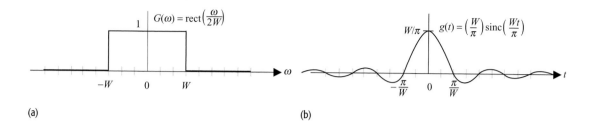

(a)   (b)

Fig. 5.8. Inverse CTFT of the rectangular function. (a) Frequency domain representation $G(\omega) =$ rect$(\omega/2W)$; (b) its inverse CTFT given by the sinc function.

The aperiodic function $g(t)$ and its CTFT are plotted in Fig. 5.8. Example 5.7 establishes the following CTFT pair:

$$\frac{W}{\pi}\text{sinc}\left(\frac{W}{\pi}t\right) \overset{\text{CTFT}}{\longleftrightarrow} \text{rect}\left(\frac{\omega}{2W}\right) = \begin{cases} 1 & |\omega| \leq W \\ 0 & |\omega| > W. \end{cases} \quad (5.24)$$

Example 5.8
Determine the signal $x(t)$ whose CTFT is a frequency-shifted impulse function $X(\omega) = \delta(\omega - \omega_0)$.

Solution
Based on the CTFT synthesis equation, Eq. (5.9), we obtain

$$x(t) = \Im^{-1}\{\delta(\omega - \omega_0)\} = \frac{1}{2\pi}\int_{-\infty}^{\infty}\delta(\omega - \omega_0)e^{j\omega t}d\omega$$

$$= \frac{1}{2\pi}e^{j\omega_0 t}\int_{-\infty}^{\infty}\delta(\omega - \omega_0)d\omega = \frac{1}{2\pi}e^{j\omega_0 t}.$$

Example 5.8 proves the following CTFT pair:

$$e^{j\omega_0 t} \overset{\text{CTFT}}{\longleftrightarrow} 2\pi\,\delta(\omega - \omega_0). \quad (5.25)$$

Substituting ω_0 by $-\omega_0$ in Eq. (5.25), we obtain another CTFT pair:

$$e^{-j\omega_0 t} \overset{\text{CTFT}}{\longleftrightarrow} 2\pi\,\delta(\omega + \omega_0). \quad (5.26)$$

In Examples 5.1 to 5.8, we evaluated several CTFT pairs for some elementary time functions. Table 5.2 lists the CTFTs for additional time functions. In practice, a graphical plot of the CTFT helps to understand the frequency properties of the function. In Table 5.3, we illustrate the frequency responses of several functions by plotting their magnitude and phase spectra. In the plots, the magnitude spectra are shown as solid lines and the phase spectra are shown as dashed lines. In certain cases, the values of the corresponding phases are zero for all frequencies, and in these cases the phase spectra are not plotted.

Table 5.2. CTFT pairs for elementary CT signals

CT signals	Time domain $x(t) = \dfrac{1}{2\pi}\displaystyle\int_{-\infty}^{\infty} X(\omega)e^{j\omega t}\, dt$	Frequency domain $X(\omega) = \displaystyle\int_{-\infty}^{\infty} x(t)e^{-j\omega t}\, dt$	Comments
(1) Constant	1	$2\pi\,\delta(\omega)$	
(2) Impulse function	$\delta(t)$	1	
(3) Unit step function	$u(t)$	$\pi\,\delta(\omega) + \dfrac{1}{j\omega}$	
(4) Causal decaying exponential function	$e^{-at}u(t)$	$\dfrac{1}{a + j\omega}$	$a > 0$
(5) Two-sided decaying exponential function	$e^{-a\lvert t\rvert}$	$\dfrac{2a}{a^2 + \omega^2}$	$a > 0$
(6) First-order time-rising causal decaying exponential function	$te^{-at}u(t)$	$\dfrac{1}{(a + j\omega)^2}$	$a > 0$
(7) Nth-order time-rising causal decaying exponential function	$t^n e^{-at}u(t)$	$\dfrac{n!}{(a + j\omega)^{n+1}}$	$a > 0$
(8) Sign function	$\mathrm{sgn}(t) = \begin{cases} 1 & t > 0 \\ -1 & t < 0 \end{cases}$	$\dfrac{2}{j\omega}$	
(9) Complex exponential	$e^{j\omega_0 t}$	$2\pi\,\delta(\omega - \omega_0)$	
(10) Periodic cosine function	$\cos(\omega_0 t)$	$\pi[\delta(\omega - \omega_0) + \delta(\omega + \omega_0)]$	
(11) Periodic sine function	$\sin(\omega_0 t)$	$\dfrac{\pi}{j}[\delta(\omega - \omega_0) - \delta(\omega + \omega_0)]$	
(12) Causal cosine function	$\cos(\omega_0 t)u(t)$	$\dfrac{\pi}{2}[\delta(\omega - \omega_0) + \delta(\omega + \omega_0)] + \dfrac{j\omega}{\omega_0^2 - \omega^2}$	
(13) Causal sine function	$\sin(\omega_0 t)u(t)$	$\dfrac{\pi}{2j}[\delta(\omega - \omega_0) - \delta(\omega + \omega_0)] + \dfrac{\omega_0}{\omega_0^2 - \omega^2}$	
(14) Causal decaying exponential cosine function	$e^{-at}\cos(\omega_0 t)u(t)$	$\dfrac{a + j\omega}{(a + j\omega)^2 + \omega_0^2}$	$a > 0$
(15) Causal decaying exponential sine function	$e^{-at}\sin(\omega_0 t)u(t)$	$\dfrac{\omega_0}{(a + j\omega)^2 + \omega_0^2}$	$a > 0$
(16) Rectangular function	$\mathrm{rect}\left(\dfrac{t}{\tau}\right) = \begin{cases} 1 & \lvert t\rvert \leq \tau/2 \\ 0 & \lvert t\rvert > \tau/2 \end{cases}$	$\tau\,\mathrm{sinc}\left(\dfrac{\omega\tau}{2\pi}\right)$	$\tau \neq 0$
(17) Sinc function	$\dfrac{W}{\pi}\mathrm{sinc}\left(\dfrac{Wt}{\pi}\right)$	$\mathrm{rect}\left(\dfrac{\omega}{2W}\right) = \begin{cases} 1 & \lvert\omega\rvert \leq W \\ 0 & \lvert\omega\rvert > W \end{cases}$	
(18) Triangular function	$\Delta\left(\dfrac{t}{\tau}\right) = \begin{cases} 1 - \dfrac{\lvert t\rvert}{\tau} & \lvert t\rvert \leq \tau \\ 0 & \text{otherwise} \end{cases}$	$\tau\,\mathrm{sinc}^2\left(\dfrac{\omega\tau}{2\pi}\right)$	$\tau > 0$
(19) Impulse train	$\displaystyle\sum_{k=-\infty}^{\infty} \delta(t - kT_0)$	$\omega_0 \displaystyle\sum_{m=-\infty}^{\infty} \delta(\omega - m\omega_0)$	angular frequency $\omega_0 = 2\pi/T_0$
(20) Gaussian function	$e^{-t^2/2\sigma^2}$	$\sigma\sqrt{2\pi}\,e^{-\sigma^2\omega^2/2}$	

(9) The *Parseval's theorem* states that the total energy in a function is the same in the time and frequency domains. Therefore, the energy in a function can be obtained either in the time domain by calculating the energy per unit time and integrating it over all time, or in the frequency domain by calculating the energy per unit frequency and integrating over all frequencies.

In Section 5.6 we derived the following condition for the existence of the CTFT of the signal $x(t)$:

$$\int_{-\infty}^{\infty} |x(t)|dt < \infty,$$

while in Sections 5.7 and 5.8 we discussed the relationship between the CTFS and CTFT of periodic signals. In particular, the CTFT of a periodic signal $x(t)$ is obtained by the relationship

$$x(t) \xleftrightarrow{\text{CTFT}} 2\pi \sum_{n=-\infty}^{\infty} D_n \delta(\omega - n\omega_0),$$

where D_n denotes the exponential CTFS coefficients and ω_0 is the fundamental frequency. Conversely, the CTFS of a periodic signal is obtained by sampling the CTFT of one period of the periodic signal at frequencies $\omega = n\omega_0$. Section 5.9 showed that the three representations (linear, constant-coefficient differential equation; impulse response; and transfer function) for LTIC systems are equivalent. Given one representation, it is straightforward to derive the remaining two representations based on the CTFT and its properties. The transfer function $H(\omega)$ plays an important role in the analysis of LTIC systems, and is typically the preferred model for representing LTIC systems. In Section 5.10, we concluded the chapter by showing the steps involved in computing the CTFT of a CT signal using MATLAB.

Problems

5.1 For each of the following CT functions, calculate the expression for the CTFT directly by using Eq. (5.10). Compare the CTFT with the corresponding entry in Table 5.2 to confirm the validity of your result.
(a) $x_1(t) = \Delta\left(\frac{t}{\tau}\right) = (1 - |t|/\tau)[u(t + \tau) - u(t - \tau)]$;
(b) $x_2(t) = t^4 e^{-at} u(t)$, with $a \in \Re^+$;
(c) $x_3(t) = e^{-at} \cos(\omega_0 t)u(t)$, with $a, \omega_0 \in \Re^+$;
(d) $x_4(t) = e^{-t^2/2\sigma^2}$, with $\sigma \in \Re$.

5.2 Calculate the CTFT of the functions shown in Figs. P5.2 (a)–(e).

5.3 Three functions $x_1(t)$, $x_2(t)$, and $x_3(t)$ have an identical magnitude spectrum $|X(\omega)|$ but different phase spectra denoted, respectively, by

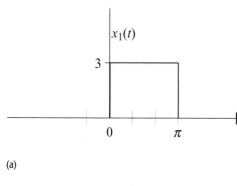

(a)

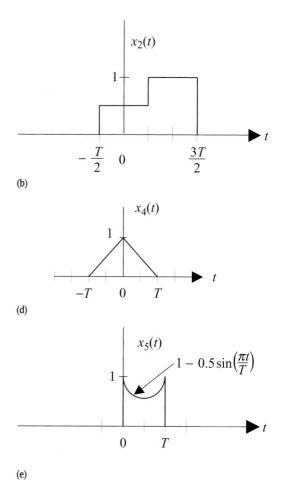

(b)

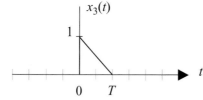

(c)

(d)

(e)

$<X_1(\omega)$, $<X_2(\omega)$, and $<X_3(\omega)$; magnitude and phase plots are shown in Figs. P5.3(a)–(d). By representing the CTFTs as $X_p(\omega) = |X(\omega)| \exp(\mathrm{j} < X_n(\omega))$, for $p = 1, 2$, and 3, and calculating the inverse CTFT, determine the functions $x_1(t)$, $x_2(t)$, and $x_3(t)$.

5.4 Using the partial fraction method, calculate the inverse Fourier transform of the following functions:

(a) $X_1(\omega) = \dfrac{(1 + \mathrm{j}\omega)}{(2 + \mathrm{j}\omega)(3 + \mathrm{j}\omega)}$;

(b) $X_2(\omega) = \dfrac{2 - \mathrm{j}\omega}{(1 + \mathrm{j}\omega)(2 + \mathrm{j}\omega)(3 + \mathrm{j}\omega)}$;

(c) $X_3(\omega) = \dfrac{2 - \mathrm{j}\omega}{(1 + \mathrm{j}\omega)(2 + \mathrm{j}\omega)^2(3 + \mathrm{j}\omega)}$;

(d) $X_4(\omega) = \dfrac{1}{(1 + \mathrm{j}\omega)(2 + 2\mathrm{j}\omega + (\mathrm{j}\omega)^2)}$;

(e) $X_5(\omega) = \dfrac{1}{(1 + \mathrm{j}\omega)^2(2 + 2\mathrm{j}\omega + (\mathrm{j}\omega)^2)^2}$.

Fig. P5.3. Amplitude and phase spectra of the three functions in Problem 5.3.

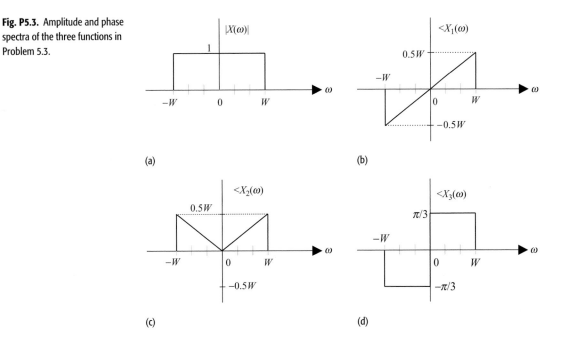

(a)

(b)

(c)

(d)

5.5 Prove the following identity:

$$\int_{-\infty}^{\infty} e^{j\omega t}\, dt = 2\pi\, \delta(\omega).$$

[Hint: Show that the integral on the left-hand side is a generalized function that satisfies Eq. (1.47) presented in Chapter 1.]

5.6 Show that the CTFT $X(\omega)$ of a real-valued even function $x(t)$ is also real and even. In other words, that $\text{Re}\{X(\omega)\} = \text{Re}\{X(-\omega)\}$ and $\text{Im}\{X(\omega)\} = 0$.

5.7 Show that the CTFT $X(\omega)$ of a real-valued odd function $x(t)$ is imaginary and odd. In other words, that $\text{Re}\{X(\omega)\} = 0$ and $\text{Im}\{X(\omega)\} = -\text{Im}\{X(-\omega)\}$.

5.8 Using the Hermitian property, determine if the time-domain functions corresponding to following CTFTs are real-valued or complex-valued. If a time-domain function is real-valued, determine if it has even or odd symmetry.

(a) $X_1(\omega) = \dfrac{5}{2 + j(\omega - 5)}$;

(b) $X_2(\omega) = \cos\left(2\omega + \dfrac{\pi}{6}\right)$;

(c) $X_3(\omega) = \dfrac{5\sin[4(\omega - \pi)]}{(\omega - \pi)}$;

(d) $X_4(\omega) = (3 + j2)\delta(\omega - 10) + (1 - j2)\delta(\omega + 10)$;

(e) $X_5(\omega) = \dfrac{1}{(1 + j\omega)(3 + j\omega)^2(5 + \omega^2)}$.

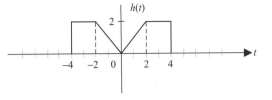

5.9 Using Table 5.2 and the properties of the CTFT, calculate the CTFT of the following functions:

(a) $x_1(t) = 5 + 3\cos(10t) - 7e^{-2t}\sin(3t)u(t)$;

(b) $x_2(t) = \dfrac{1}{\pi t}$;

(c) $x_3(t) = t^2 e^{-4|t-5|}$;

(d) $x_4(t) = 5\dfrac{\sin(3\pi t)\sin(5\pi t)}{t^2}$;

(e) $x_4(t) = 4\dfrac{\sin(3\pi t)}{t} * \dfrac{d}{dt}\left[\dfrac{\sin(4\pi t)}{t}\right]$.

5.10 Using Table 5.2 and the linearity property, show that the CTFT of the function

$$x(t) = \left[\frac{6}{13}e^{-2t} - \frac{6}{13}\cos(3t) + \frac{4}{13}\sin(3t)\right]u(t)$$

is given by

$$X(\omega) = \frac{6}{(9-\omega^2)(2+j\omega)}$$
$$-\frac{\pi}{13}[(3+j2)\delta(\omega-3) + (3-j2)\delta(\omega+3)].$$

5.11 Prove the following time-scaling property (see Eq. (5.45)) of the CTFT:

$$x(at) \xleftrightarrow{\text{CTFT}} \frac{1}{|a|}X\left(\frac{\omega}{a}\right), \quad \text{for } a \in \Re \quad \text{and} \quad a \neq 0.$$

5.12 Using the time-scaling property and the results in Example 5.12, calculate the CTFT of the function $h(t)$ shown in Fig. P5.12.

5.13 Prove the following frequency-shifting property (see Eq. (5.49)) of the CTFT:

$$h(t) = e^{j\omega_0 t}x(t) \xleftrightarrow{\text{CTFT}} X(\omega-\omega_0), \quad \text{for } \omega_0 \in \Re.$$

5.14 Prove the following time-integration property (see Eq. (5.53)) of the CTFT:

$$\int_{-\infty}^{t} x(\tau)d\tau \xleftrightarrow{\text{CTFT}} \frac{X(\omega)}{j\omega} + \pi X(0)\delta(\omega).$$

5.15 Assume that for the CTFT pair $x(t) \xleftrightarrow{\text{CTFT}} X(\omega)$, the CTFT is given by the triangular function

$$X(\omega) = \Delta\left(\frac{\omega}{3}\right) = \begin{cases} 1 - \dfrac{|\omega|}{3} & |\omega| \leq 3 \\ 0 & \text{elsewhere.} \end{cases}$$

Using the CTFT properties (listed in Table 5.4), derive the CTFT for the following set of functions:

(a) $e^{-j5t} x(2t)$;   (d) $x^2(t)$;

(b) $t^2 x(t)$;   (e) $x(t) * x(t)$;

(c) $(t+5)\dfrac{dx}{dt}$;   (f) $\cos(\omega_0 t) x(t)$ with $\omega_0 = 3/2, 3,$ and 6.

5.16 Using the transform pairs in Table 5.2 and the properties of the CTFT, calculate the inverse Fourier transform of the functions in Problem 5.8.

5.17 For each of the following functions, (i) draw a rough sketch of the function, and (ii) determine if the CTFT exists by evaluating Eq. (5.59):

(a) $x_1(t) = e^{-a|t|}$, with $a \in \Re^+$;

(b) $x_2(t) = e^{-at} \cos(\omega_0 t) u(t)$, with $a, \omega_0 \in \Re^+$;

(c) $x_3(t) = t^4 e^{-at} u(t)$, with $a \in \Re^+$;

(d) $x_4(t) = \sin(\ln(t)) u(t)$;

(e) $x_5(t) = \dfrac{1}{t}$;

(f) $x_6(t) = \cos\left(\dfrac{\pi}{2t}\right)$;

(g) $x_7(t) = e^{-t^2/2\sigma^2}$, with $\sigma \in \Re$.

5.18 Using the exponential CTFS representations (calculated in Problem 4.11), calculate the CTFT for the periodic signals shown in Fig. P4.6.

5.19 Determine the CTFS coefficients for the periodic functions shown in Fig. P4.6 from the CTFTs calculated in Problem 5.2.

5.20 Determine (i) the transfer function, and (ii) the impulse response for the LTIC systems whose input–output relationships are represented by the following linear, constant-coefficient differential equations. Assume zero initial conditions in each case.

(a) $\dfrac{d^3 y}{dt^3} + 6\dfrac{d^2 y}{dt^2} + 11\dfrac{dy}{dt} + 6y(t) = x(t)$.

(b) $\dfrac{d^2 y}{dt^2} + 3\dfrac{dy}{dt} + 2y(t) = x(t)$.

(c) $\dfrac{d^2 y}{dt^2} + 2\dfrac{dy}{dt} + y(t) = x(t)$.

(d) $\dfrac{d^2 y}{dt^2} + 6\dfrac{dy}{dt} + 8y(t) = \dfrac{dx}{dt} + 4x(t)$.

(e) $\dfrac{d^3 y}{dt^3} + 8\dfrac{d^2 y}{dt^2} + 19\dfrac{dy}{dt} + 12y(t) = x(t)$.

Fig. P5.22. (a) RC circuit system; (b) input signal.

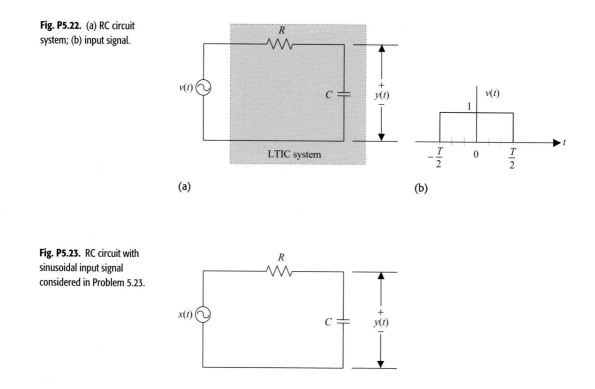

(a)   (b)

Fig. P5.23. RC circuit with sinusoidal input signal considered in Problem 5.23.

5.21 Consider the LTIC systems with the following input–output pairs:
(a) $x(t) = e^{-2t}u(t)$ and $y(t) = 5e^{-2t}u(t)$;
(b) $x(t) = e^{-2t}u(t)$ and $y(t) = 3e^{-2(t-4)}u(t-4)$;
(c) $x(t) = e^{-2t}u(t)$ and $y(t) = t^3 e^{-2t}u(t)$;
(d) $x(t) = e^{-2t}u(t)$ and $y(t) = e^{-t}u(t) + e^{-3t}u(t)$.

For each of the above systems, determine (i) the transfer function, (ii) the impulse response function, and (iii) the input–output relationship using linear constant-coefficient differential equations.

5.22 Determine the transfer function of the system shown in Fig. P5.22(a). Calculate the output of the system for the input signal shown in Fig. P5.22(b).

5.23 Using the convolution property of the CTFT, calculate the output of the system shown in Fig. P5.23 for the input signals (i) $x_1(t) = \cos(\omega_0 t)$, and (ii) $x_2(t) = \sin(\omega_0 t)$.

5.24 Sketch the gain and phase responses for the LTIC systems in Problem 5.20.

5.25 Sketch the gain and phase responses for the LTIC systems in Problem 5.21.

5.26 Show that if the transfer function $H(\omega)$ of a system is Hermitian symmetric (i.e. its impulse response $h(t)$ is real-valued), the outputs of the system to cosine and sine inputs are as follows:

$$\cos(\omega_0 t) \xrightarrow[\text{Hermitian Symmetric } H(\omega)]{} |H(\omega_0)| \cos(\omega_0 t + <H(\omega_0))$$

and

$$\sin(\omega_0 t) \xrightarrow[\text{Hermitian Symmetric } H(\omega)]{} |H(\omega_0)| \sin(\omega_0 t + <H(\omega_0)).$$

5.27 Using the results in Problem 5.26, verify the answers in Problem 5.23.

5.28 Using the results derived in Section 5.9.2 and the linearity property of the CTFT, calculate the output of the system shown in Fig. P5.23 for the following input signals. Assume that $R = 1$ MΩ and $C = 0.1$ μF.
  (i) $x_1(t) = \sin(3t)$;
  (ii) $x_2(t) = \cos(3t) - 5\sin(6t + 30°)$;
  (iii) $x_3(t) = \cos(2t) + \sin(2000t)$;
  (iv) $x_4(t) = e^{j3t} + e^{j2000t}$.

5.29 Suppose the CT signal

$$x(t) = e^{-t}u(t)$$

is applied as input to a causal LTIC system modeled by the impulse response

$$h(t) = e^{-2t}u(t)$$

Calculate the resulting output $y(t)$ using:
  (a) direct convolution;
  (b) transfer function $H(\omega)$;
  (c) differential equation.

5.30 The periodic signals shown in Figs. P4.6(a)–(e) are applied to the following LTIC systems:

  (i) $H_1(\omega) = \begin{cases} 1 & |\omega| \le \dfrac{4}{T} \\ 0 & \text{elsewhere;} \end{cases}$

  (ii) $H_2(\omega) = \begin{cases} 1 & \dfrac{4}{T} \le |\omega| \le \dfrac{8}{T} \\ 0 & \text{elsewhere.} \end{cases}$

Sketch the magnitude and phase spectra of the CTFT of the resulting outputs.

5.31 The transfer function of two LTIC systems are given by
  (i) $H_1(\omega) = \dfrac{20 - j\omega}{20 + j\omega}$;
  (ii) $H_2(\omega) = \begin{cases} 1 & |\omega| \ge 20 \\ 0 & \text{elsewhere.} \end{cases}$

(a) By sketching the magnitude spectrum of each of the LTIC systems, comment on the frequency properties of the two systems. Classify the two systems as a lowpass, highpass, bandpass, or an allpass filter. Recall that a lowpass filter blocks high-frequency components; a highpass filter blocks low-frequency components; a bandpass filters blocks frequency components within a certain band of frequencies; while an allpass filters allows all frequency components to be passed on to the output.

(b) Determine the impulse response for each of the two LTIC systems.

5.32 Sketch the gain and phase responses for the three LTIC systems given below:

(a) $h_1(t) = 2te^{-t}u(t)$;

(b) $h_2(t) = u(t)$;

(c) $h_3(t) = -2\delta(t) + 5e^{-2t}u(t)$.

For each of the three systems, show that the input signal $x(t) = \cos t$ produces the same output response. How can this result be explained?

5.33 (MATLAB exercise) By making modifications to the myctft function listed in Section 5.10, sketch the magnitude and phase spectra of the following signals:

(i) $x_1(t) = \sin(5\pi t)$ for $-2 \le t \le 2$ with sampling rate $\omega_s = 200\pi$ samples/s;

(ii) $x_2(t) = \sin(8\pi t) + \sin(20\pi t)$ for $-1.25 \le t \le 1.25$ with sampling rate $\omega_s = 1000\pi$ samples/s.

5.34 (MATLAB exercise) Compute the CTFTs of the CT functions specified in Problem 5.1. By plotting the magnitude and phase spectra, compare your computed result with the analytical expressions listed in Tables 5.2 and 5.3.

5.35 (MATLAB exercise) Compute the output response $y(t)$ for Problem 5.29 by computing the CTFT for $x(t)$ and $h(t)$, multiplying the CTFTs and then taking the inverse CTFT of the result.

5.36 (MATLAB exercise) Sketch the magnitude and phase Bode plots for the LTIC systems specified in Problems 5.20 and 5.21.

6 Laplace transform

In Chapters 4 and 5, we introduced the continuous-time Fourier series (CTFS) for CT periodic signals and the continuous-time Fourier transform (CTFT) for both CT periodic and aperiodic signals. These frequency representations provide a useful tool for determining the output of an LTIC system. Unfortunately, the CTFT is not defined for all aperiodic signals. In cases where the CTFT does not exist, an alternative procedure, based on the *Laplace transform*, is used to analyze the LTIC systems. Even for the CT signals for which the CTFT exists, the Laplace transforms are always real-valued, rational functions of the independent variable s provided that the CT functions are real. The CTFTs are complex-valued in most cases. Therefore, using the Laplace transform simplifies algebraic manipulations and leads to important flow diagram representations of the CT systems from which the hardware implementations of the CT systems are derived. Finally, the CTFT can only be applied to stable LTIC systems for which the impulse response is absolutely integrable. Since the Laplace transform exists for both stable and unstable LTIC systems, it can be used to analyze a broader range of LTIC systems.

The difference between the CTFT and the Laplace transform lies in the choice of the basis functions used in the two representations. The CTFT expands an aperiodic signal as a linear combination of complex exponential functions $e^{j\omega t}$, which are referred to as its basis functions. The Laplace transform uses e^{st} as the basis functions, where the independent Laplace variable s is complex and is given by $s = \sigma + j\omega$. The Laplace transform is, therefore, a generalization of the CTFT, since the independent variable s can take any value in the complex s-plane and is not simply restricted to the imaginary $j\omega$-axis, as is the case for the CTFT. In this chapter, we will cover the Laplace transform and its applications in the analysis of LTIC systems. To illustrate the usefulness of the Laplace transforms in signal processing, some real-world applications are presented in Chapter 8.

Chapter 6 is organized as follows. Section 6.1 defines the bilateral, or two-sided, Laplace transform and provides several examples to illustrate the steps involved in its computation. The bilateral Laplace transform is used for non-causal and causal signals. For causal signals, the bilateral Laplace transform simplifies to the one-sided, or unilateral, Laplace transform, which is covered in

Section 6.2. Section 6.3 computes the time-domain representation of a Laplace-transformed signal, while Section 6.4 considers the properties of the Laplace transform. Sections 6.5 to 6.9 propose several applications of the Laplace transform, ranging from solving differential equations (Section 6.5), evaluating the location of poles and zeros (Section 6.6), determining the causality and stability of LTIC systems from their Laplace transfer functions (Sections 6.7 and 6.8), and analyzing the outputs of LTIC systems (Section 6.9). Section 6.10 presents the cascaded, parallel, and feedback configurations for interconnecting LTI systems, and Section 6.11 concludes the chapter.

6.1 Analytical development

In Section 5.1, the CTFT pair, $x(t) \xleftrightarrow{\text{CTFT}} X(j\omega)$, was defined as follows:

CTFT synthesis equation $\qquad x(t) = \frac{1}{2\pi} \int\limits_{-\infty}^{\infty} X(j\omega) e^{j\omega t} \, d\omega;$ $\qquad$ (6.1)

CTFT analysis equation $\qquad X(j\omega) = \int\limits_{-\infty}^{\infty} x(t) e^{-j\omega t} \, dt.$ $\qquad$ (6.2)

In Eqs. (6.1) and (6.2), the CTFT of $x(t)$ is expressed as $X(j\omega)$, instead of the earlier notation $X(\omega)$, to emphasize that the CTFT is computed on the imaginary $j\omega$-axis in the complex s-plane. For a CT signal $x(t)$, the expression for the bilateral Laplace transform is derived by considering the CTFT of the modified version, $x(t)e^{-\sigma t}$, of the signal. Based on Eq. (6.2), the CTFT of the modified signal $x(t)e^{-\sigma t}$ is given by

$$\Im\{x(t)e^{-\sigma t}\} = \int\limits_{-\infty}^{\infty} x(t) e^{-\sigma t} e^{-j\omega t} \, dt, \qquad (6.3)$$

which reduces to

$$\Im\{x(t)e^{-\sigma t}\} = \int\limits_{-\infty}^{\infty} x(t) e^{-(\sigma + j\omega) t} \, dt$$

$$= X(\sigma + j\omega). \qquad (6.4)$$

Substituting $s = \sigma + j\omega$ in Eq. (6.4) leads to the following definition for the bilateral Laplace transform:[†]

Laplace analysis equation $\qquad X(s) = \Im\{x(t)e^{-\sigma t}\} = \int\limits_{-\infty}^{\infty} x(t) e^{-st} \, dt.$ $\qquad$ (6.5)

[†] The Laplace transform was discovered originally by Leonhard Euler (1707–1783), a prolific Swiss mathematician and physicist. However, it is named in honor of another mathematician and astronomer, Pierre-Simon Laplace (1749–1827), who used the transform in his work on probability theory.

To derive the synthesis equation for the bilateral Laplace transform, consider the inverse transform of the CTFT pair, $x(t)e^{-\sigma t} \xleftrightarrow{\text{CTFT}} X(\sigma + j\omega) = X(s)$. Based on Eq. (6.1), we obtain

$$x(t)e^{-\sigma t} = \frac{1}{2\pi} \int\limits_{-\infty}^{\infty} X(s)e^{j\omega t} d\omega. \tag{6.6}$$

Multiplying both sides of Eq. (6.6) by $e^{\sigma t}$ and changing the integral variable ω to s using the relationship $s = \sigma + j\omega$ yields

Laplace synthesis equation $x(t) = \dfrac{1}{2\pi j} \displaystyle\int\limits_{\sigma-j\infty}^{\sigma-j\infty} X(s)e^{st} ds. \tag{6.7}$

Solving Eq. (6.7) involves the use of contour integration and is seldom used in the computation of the inverse Laplace transform. In Section 6.3, we will consider an alternative approach based on the partial fraction expansion to evaluate the inverse Laplace transform. Collectively, Eqs. (6.5) and (6.7) form the bilateral Laplace transform pair, which is denoted by

$$x(t) \xleftrightarrow{\text{L}} X(s). \tag{6.8}$$

To illustrate the steps involved in computing the Laplace transform, we consider the following examples.

Example 6.1
Calculate the bilateral Laplace transform of the decaying exponential function: $x(t) = e^{-at}u(t)$.

Solution
Substituting $x(t) = e^{-at}u(t)$ in Eq. (6.5), we obtain

$$X(s) = \int\limits_{-\infty}^{\infty} e^{-at}u(t)e^{-st} dt = \int\limits_{0}^{\infty} e^{-(s+a)t} dt = -\frac{1}{(s+a)}e^{-(s+a)t} \Big|_{0}^{\infty}.$$

At the lower limit, $t \to 0$, $e^{-(s+a)t} = 1$. At the upper limit, $t \to \infty$, $e^{-(s+a)t} = 0$ if $\text{Re}\{s+a\} > 0$ or $\text{Re}\{s\} > -a$. If $\text{Re}\{s\} \leq -a$, then the value of $e^{-(s+a)t}$ is infinite at the upper limit, $t \to \infty$. Therefore,

$$X(s) = \begin{cases} \dfrac{1}{(s+a)} & \text{for } \text{Re}\{s\} > -a \\ \text{undefined} & \text{for } \text{Re}\{s\} \leq -a. \end{cases}$$

The set of values of s over which the bilateral Laplace transform is defined is referred to as the region of convergence (ROC). Assuming a to be a real number, the ROC is given by $\text{Re}\{s\} > -a$ for the Laplace transform of the decaying exponential function, $x(t) = e^{-at}u(t)$. Figure 6.1 highlights the ROC by shading the appropriate area in the complex s-plane.

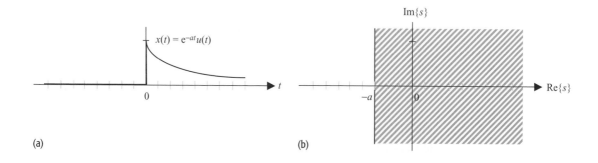

(a)　　　　　　　　　　　　　　　　　　　　　　(b)

Fig. 6.1. (a) Exponential decaying function $x(t) = e^{-at}u(t)$; (b) its associated ROC, $\text{Re}\{s\} > -a$, over which the bilateral Laplace transform exists.

Example 6.1 shows that the bilateral Laplace transform of the decaying exponential function $x(t) = e^{-at}u(t)$ will converge to a finite value $X(s) = 1/(s+a)$ within the ROC ($\text{Re}\{s\} > -a$). In other words, the bilateral Laplace transform of $x(t) = e^{-at}u(t)$ exists for all values of a within the specified ROC. No restriction is imposed on the value of a for the existence of the Laplace transform. On the other hand, the CTFT of the decaying exponential function exists only for $a > 0$. For $a < 0$, the exponential function $x(t) = e^{-at}u(t)$ is not absolutely integrable, and hence its CTFT does not exist. This is an important distinction between the CTFT and the bilateral Laplace transform. The CTFT exists for a limited number of absolutely integrable functions. By associating an ROC with the bilateral Laplace transform, we can evaluate the Laplace transform for a much larger set of functions.

Example 6.2

Calculate the bilateral Laplace transform of the non-causal exponential function $g(t) = -e^{-at}u(-t)$.

Solution

Substituting $g(t) = -e^{-at}u(-t)$ in Eq. (6.5), we obtain

$$G(s) = \int_{-\infty}^{\infty} -e^{-at}u(-t)e^{-st}\,dt = -\int_{-\infty}^{0} e^{-(s+a)t}\,dt = \frac{1}{(s+a)}e^{-(s+a)t}\Big|_{-\infty}^{0}.$$

At the upper limit, $t \to 0$, $e^{-(s+a)t} = 1$. At the lower limit, $t \to -\infty$, $e^{-(s+a)t}$ is finite only if $\text{Re}\{s+a\} < 0$, where it equals zero. The bilateral Laplace transform is therefore given by

$$G(s) = \begin{cases} \dfrac{1}{(s+a)} & \text{for } \text{Re}\{s\} < -a \\ \text{undefined} & \text{for } \text{Re}\{s\} \geq -a. \end{cases}$$

Figure 6.2 illustrates the ROC, $\text{Re}\{s\} < -a$, for the bilateral Laplace transform of $g(t) = -e^{-at}u(-t)$.

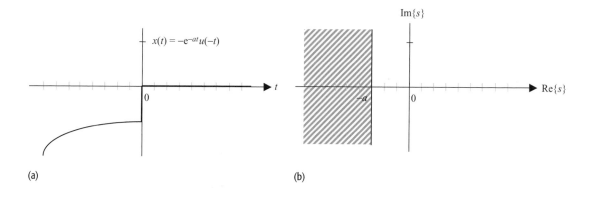

Fig. 6.2. (a) Non-causal decaying function $x(t) = -\mathrm{e}^{-at}u(-t)$; (b) its associated ROC, $\mathrm{Re}\{s\} < -a$, over which the bilateral Laplace transform exists.

In Examples 6.1 and 6.2, we have proved the following Laplace transform pairs:

$$\mathrm{e}^{-at}u(t) \overset{L}{\longleftrightarrow} \frac{1}{(s+a)} \qquad \text{with ROC: } \mathrm{Re}\{s\} > -a$$

and

$$-\mathrm{e}^{-at}u(-t) \overset{L}{\longleftrightarrow} \frac{1}{(s+a)} \qquad \text{with ROC: } \mathrm{Re}\{s\} < -a.$$

Although the algebraic expressions for the bilateral Laplace transforms are the same for the two functions, the ROCs are different. This implies that a bilateral Laplace transform is completely specified only if the algebraic expression and the ROC are both specified. This is illustrated further in Example 6.3.

Example 6.3

Calculate the inverse Laplace transform of the function $H(s) = 1/(s+a)$.

Solution

From Examples 6.1 and 6.2, we know that

$$\mathrm{e}^{-at}u(t) \overset{L}{\longleftrightarrow} \frac{1}{(s+a)} \qquad \text{with ROC: } \mathrm{Re}\{s\} > -a$$

and

$$-\mathrm{e}^{-at}u(-t) \overset{L}{\longleftrightarrow} \frac{1}{(s+a)} \qquad \text{with ROC: } \mathrm{Re}\{s\} < -a.$$

Therefore, the inverse bilateral Laplace transform is either $h(t) = \mathrm{e}^{-at}u(t)$ or $h(t) = -\mathrm{e}^{-at}u(-t)$. If we want to determine a unique inverse, we need to specify the ROC associated with the Laplace transform. If the ROC is specified as $\mathrm{Re}\{s\} > -a$, then the inverse Laplace transform $h(t) = \mathrm{e}^{-at}u(t)$. On the other hand, if the ROC is $\mathrm{Re}\{s\} > -a$, then $h(t) = \mathrm{e}^{-at}u(t)$.

The need to specify the ROC is also evident from the synthesis equation, Eq. (6.7), of the Laplace transform. To evaluate the inverse Laplace transform using Eq. (6.7), a straight line, parallel to the $j\omega$-axis, corresponding to all points s satisfying $\mathrm{Re}\{s\} = \sigma$ within the ROC, is used as the contour of integration. The complex integral, therefore, cannot be computed without having prior knowledge of the ROC.

6.2 Unilateral Laplace transform

In Section 6.1, we introduced the bilateral Laplace transform that is used to analyze both causal and non-causal LTIC systems. In signal processing, most physical systems and signals are causal. Applying the causality condition, the bilateral Laplace transform reduces to a simpler version of the Laplace transform. The Laplace transform for causal signals and systems is referred to as the unilateral Laplace transform and is defined as follows:

$$X(s) = L\{x(t)\} = \int\limits_{0^-}^{\infty} x(t)e^{-st}\,dt, \tag{6.9}$$

where the initial conditions of the system are incorporated by the lower limit ($t = 0^-$). In our subsequent discussions, we will mostly use the unilateral Laplace transform. For simplicity, we will omit the term "unilateral," therefore the Laplace transform implies the unilateral Laplace transform. When we refer to the bilateral Laplace transform, the term "bilateral" will be explicitly stated. To clarify further the differences between the unilateral and bilateral Laplace transform, we summarize the major points.

(1) The unilateral Laplace transform simplifies the analysis of causal LTIC systems. However, it cannot analyze non-causal systems directly. Since most physical systems are naturally causal, we will use the unilateral Laplace transform in our computations. The bilateral transform will be used only to analyze non-causal systems.
(2) For causal signals and systems, the unilateral and bilateral Laplace transforms are the same.
(3) The synthesis equation used for calculating the inverse of the unilateral Laplace transform is the same as Eq. (6.7) used for evaluating the inverse of the bilateral transform.

Example 6.4
Calculate the unilateral Laplace transform for the following functions:

(i) unit impulse function, $x_1(t) = \delta(t)$;
(ii) unit step function, $x_2(t) = u(t)$;
(iii) shifted gate function,

$$x_3(t) = \begin{cases} 1 & 2 \leq t \leq 4 \\ 0 & \text{otherwise}; \end{cases}$$

(iv) causal complex exponential function, $x_4(t) = e^{-j\omega_0 t} u(t)$;

(v) causal sine function, $x_5(t) = \sin(\omega_0 t)u(t)$;

(vi) causal ramp function, $x_6(t) = tu(t)$;

(vii) $x_7(t) = \begin{cases} 2t & 0 \le t \le 1 \\ 2 & 1 \le t \le 2 \\ 0 & \text{otherwise.} \end{cases}$

Solution

(i) Unit impulse function. Substituting $x_1(t) = \delta(t)$ in Eq. (6.9) yields

$$X_1(s) = \int_{0^-}^{\infty} \delta(t) e^{-st} dt.$$

Since $\delta(t) e^{-st} = \delta(t)$, the above equation reduces to

$$X_1(s) = \int_{0^-}^{\infty} \delta(t) dt = 1.$$

The Laplace transform for an impulse function is given by

$$\delta(t) \overset{L}{\longleftrightarrow} 1 \quad \text{with ROC: entire s-plane.}$$

(ii) Unit step function. Substituting $x_2(t) = u(t)$ in Eq. (6.9) yields

$$X_2(s) = \int_{0^-}^{\infty} u(t) e^{-st} dt.$$

For $\text{Re}\{s\} > 0$, the above integral reduces to

$$X_2(s) = \int_{0^-}^{\infty} e^{-st} dt = -\frac{1}{s} e^{-st} \Big|_{0}^{\infty} = 1.$$

The Laplace transform for a unit step function is given by

$$u(t) \overset{L}{\longleftrightarrow} \frac{1}{s} \quad \text{with ROC: } \text{Re}\{s\} > 0.$$

(iii) Shifted gate function. Substituting $x_3(t)$ in Eq. (6.9) yields

$$X_3(s) = \int_{2}^{4} e^{-st} dt = -\frac{1}{s} e^{-st} \Big|_{2}^{4} = \frac{1}{s}(e^{-2s} - e^{-4s}).$$

Clearly, the above expression for the Laplace transform is not valid for $s = 0$. The value of the Laplace transform for $s = 0$ is obtained by substituting $s = 0$ in Eq. (6.9). The resulting expression is given by

$$X_3(s) = \int_{0^-}^{\infty} x_3(t) dt = \int_{2}^{4} 1 \cdot dt = t\big|_{2}^{4} = 2, \quad \text{for } s = 0.$$

The Laplace transform for the shifted gate function is therefore given by

$$
X_3(s) = \begin{cases} 2 & \text{for } s = 0 \\ \dfrac{1}{s}(e^{-2s} - e^{-4s}) & \text{for } s \neq 0. \end{cases}
$$

The associated ROC is the entire s-plane.

　(iv) Causal complex exponential function. From Example 6.1, we know that

$$
e^{-at}u(t) \stackrel{L}{\longleftrightarrow} \frac{1}{(s+a)} \qquad \text{with ROC: } \text{Re}\{s\} > -\text{Re}\{a\}.
$$

Substituting $a = j\omega_0$, we obtain

$$
e^{-j\omega_0 t}u(t) \stackrel{L}{\longleftrightarrow} \frac{1}{(s+j\omega_0)} \qquad \text{with ROC: } \text{Re}\{s\} > 0.
$$

　(v) Causal sine function. By expanding $\sin(\omega_0 t) = [\exp(j\omega_0 t) - \exp(-j\omega_0 t)]/2j$, the Laplace transform for the causal sine function is given by

$$
X_5(s) = \frac{1}{2j}\int_{0^-}^{\infty} [e^{j\omega_0 t} - e^{-j\omega_0 t}]e^{-st}\,dt = \frac{1}{2j}\int_{0^-}^{\infty} e^{-(s-j\omega_0)t}\,dt - \frac{1}{2j}\int_{0^-}^{\infty} e^{-(s+j\omega_0)t}\,dt.
$$

Both integrals are finite for $\text{Re}\{s \pm j\omega_0\} > 0$ or $\text{Re}\{s\} > 0$. The Laplace transform is given by

$$
X_5(s) = \frac{1}{2j}\left[\frac{1}{s-j\omega_0}\right] - \frac{1}{2j}\left[\frac{1}{s+j\omega_0}\right] = \frac{\omega_0}{s^2 + \omega_0^2}.
$$

In other words, the Laplace transform pair is given by

$$
\sin(\omega_0 t)u(t) \stackrel{L}{\longleftrightarrow} \frac{\omega_0}{s^2 + \omega_0^2} \qquad \text{with ROC: } \text{Re}\{s\} > 0.
$$

　(vi) Causal ramp function. Substituting $x_6(t) = tu(t)$ in Eq. (6.9) yields

$$
X_6(s) = \int_{0^-}^{\infty} tu(t)e^{-st}\,dt = \int_{0}^{\infty} te^{-st}\,dt = \frac{te^{-st}}{(-s)}\bigg|_0^{\infty} - \frac{e^{-st}}{(-s)^2}\bigg|_0^{\infty},
$$

which, on simplification, leads to the following Laplace transform pair:

$$
tu(t) \stackrel{L}{\longleftrightarrow} \frac{1}{s^2} \qquad \text{with ROC: } \text{Re}\{s\} > 0.
$$

　(vii) Substituting

$$
x_7(t) = \begin{cases} 2t & 0 \leq t \leq 1 \\ 2 & 1 \leq t \leq 2 \\ 0 & \text{otherwise} \end{cases}
$$

in Eq. (6.9) leads to the following Laplace transform:

$$
X_7(s) = 2\int_{0^-}^{1} te^{-st}\,dt + 2\int_{1}^{2} e^{-st}\,dt = 2\left[\frac{te^{-st}}{-s} - \frac{e^{-st}}{(-s)^2}\right]_{0^-}^{1} + 2\left[\frac{e^{-st}}{-s}\right]_{1}^{2}.
$$

Clearly, the above integral is not defined for $s = 0$. For $s \neq 0$, the above expression reduces to

$$X_7(s) = 2\left[\frac{e^{-s}}{-s} - \frac{e^{-s}}{(-s)^2} + \frac{1}{(-s)^2}\right] + 2\left[\frac{e^{-2s}}{-s} - \frac{e^{-s}}{-s}\right]$$

$$= \frac{2}{s^2}[1 - e^{-s} - se^{-2s}].$$

For $s = 0$, the Laplace transform is given by

$$X_7(s) = \int_{0-}^{\infty} x_7(t)dt = \int_0^1 2t \cdot dt + \int_1^2 2 \cdot dt = t^2|_0^1 + 2t|_1^2 = 3.$$

The Laplace transform pair is therefore given by

$$X_7(s) = \begin{cases} 3 & \text{for } s = 0 \\ \dfrac{2}{s^2}[1 - e^{-s} - se^{-2s}] & \text{for } s \neq 0. \end{cases}$$

The associated ROC is the entire s-plane.

6.2.1 Relationship between Fourier and Laplace transforms

Comparing Eq. (6.2) with Eq. (6.5), the CTFT can be related to the bilateral Laplace transform as follows:

$$X(j\omega) = \int_{-\infty}^{\infty} x(t)e^{-j\omega t}\,dt = X(s)|_{s=j\omega}. \tag{6.10}$$

Since, for causal functions, the bilateral and unilateral Laplace transforms are the same, the above relationship is also valid for the unilateral Laplace transform for causal functions. Equation (6.8) proves that the CTFT is a special case of the Laplace transform when $s = j\omega$, i.e. $\sigma = 0$. In other words, the CTFT is the Laplace transform computed along the imaginary $j\omega$-axis in the s-plane. Table 6.1 lists the Laplace transforms for several commonly used functions. To compare the results with the corresponding CTFTs, we also include the CTFTs for the same functions in the second column of Table 6.1. When the function is causal and its CTFT exists, it is observed that the CTFT can be obtained from the Laplace transform by substituting $s = j\omega$. An alternative condition for the existence of the CTFT is, therefore, the inclusion of the $j\omega$-axis within the ROC of the Laplace transform. If the ROC does not contain the $j\omega$-axis, the substitution $s = j\omega$ cannot be made and the CTFT does not exist. For example, the ROC $\text{Re}\{s\} > 0$ for the unit step function $x(t) = u(t)$ does not contain the $j\omega$-axis. Based on our earlier reasoning, its CTFT should not exist. This appears to be in violation with the second entry in Table 6.1, where the CTFT of the unit step function is listed as follows:

$$u(t) \xleftrightarrow{\text{CTFT}} \pi\delta(\omega) + 1/j\omega.$$

Table 6.1. CTFT and Laplace transform pairs for several causal CT signals

CT signals $x(t)$	CTFT $$X(j\omega) = \int_{-\infty}^{\infty} x(t)e^{-j\omega t}\,dt$$	Laplace transform $$X(s) = \int_{-\infty}^{\infty} x(t)e^{-st}\,dt$$
(1) Impulse function $x(t) = \delta(t)$	1	1 ROC: entire s-plane
(2) Unit step function $x(t) = u(t)$	$\pi\delta(\omega) + \dfrac{1}{j\omega}$	$\dfrac{1}{s}$ ROC: $\mathrm{Re}\{s\} > 0$
(3) Causal gate function $x(t) = u(t) - u(t-a)$	$(1 - e^{-ja\omega})\left(\pi\delta(\omega) + \dfrac{1}{j\omega}\right)$	$\dfrac{1}{s}(1 - e^{-as})$ ROC: $\mathrm{Re}\{s\} > 0$
(4) Causal decaying exponential function $x(t) = e^{-at}u(t)$	$\dfrac{1}{a + j\omega}$	$\dfrac{1}{a + s}$ ROC: $\mathrm{Re}\{s\} > -a$
(5) Causal ramp function $x(t) = tu(t)$	does not exist	$\dfrac{1}{s^2}$ ROC: $\mathrm{Re}\{s\} > 0$
(6) Higher-order causal ramp function $x(t) = t^n u(t)$	does not exist	$\dfrac{n!}{s^{n+1}}$ ROC: $\mathrm{Re}\{s\} > 0$
(7) First-order time-rising causal decaying exponential function $x(t) = te^{-at}u(t)$	$\dfrac{1}{(a + j\omega)^2}$ provided $a > 0$.	$\dfrac{1}{(a + s)^2}$ ROC: $\mathrm{Re}\{s\} > -a$
(8) Higher-order time-rising causal decaying exponential function $x(t) = t^n e^{-at}u(t)$	$\dfrac{n!}{(a + j\omega)^{n+1}}$ provided $a > 0$	$\dfrac{n!}{(a + s)^{n+1}}$ ROC: $\mathrm{Re}\{s\} > -a$
(9) Causal cosine wave $x(t) = \cos(\omega_0 t)u(t)$	$\pi[\delta(\omega - \omega_0) + \delta(\omega + \omega_0)]$ $+ \dfrac{j\omega}{\omega_0^2 - \omega^2}$	$\dfrac{s}{\omega_0^2 + s^2}$ ROC: $\mathrm{Re}\{s\} > 0$
(10) Causal sine wave $x(t) = \sin(\omega_0 t)u(t)$	$\dfrac{\pi}{2j}[\delta(\omega - \omega_0) - \delta(\omega + \omega_0)]$ $+ \dfrac{\omega_0}{\omega_0^2 - \omega^2}$	$\dfrac{\omega_0}{\omega_0^2 + s^2}$ ROC: $\mathrm{Re}\{s\} > 0$
(11) Squared causal cosine wave $x(t) = \cos^2(\omega_0 t)u(t)$	$\dfrac{\pi}{2}[\delta(\omega) + \delta(\omega - 2\omega_0) + \delta(\omega + 2\omega_0)]$ $+ \dfrac{1}{j2\omega} + \dfrac{j\omega}{2(4\omega_0^2 - \omega^2)}$	$\dfrac{(2\omega_0^2 + s^2)}{s(4\omega_0^2 + s^2)}$ ROC: $\mathrm{Re}\{s\} > 0$
(12) Squared causal sine wave $x(t) = \sin^2(\omega_0 t)u(t)$	$\dfrac{\pi}{2}[\delta(\omega) - \delta(\omega - 2\omega_0) - \delta(\omega + 2\omega_0)]$ $+ \dfrac{1}{j2\omega} - \dfrac{j\omega}{2(4\omega_0^2 - \omega^2)}$	$\dfrac{2\omega_0^2}{s(4\omega_0^2 + s^2)}$ ROC: $\mathrm{Re}\{s\} > 0$
(13) Causal decaying exponential cosine function $x(t) = \exp(-at)\cos(\omega_0 t)u(t)$	$\dfrac{a + j\omega}{(a + j\omega)^2 + \omega_0^2}$ provided $a > 0$	$\dfrac{a + s}{(a + s)^2 + \omega_0^2}$ ROC: $\mathrm{Re}\{s\} > -a$
(14) Causal decaying exponential sine function $x(t) = \exp(-at)\sin(\omega_0 t)u(t)$	$\dfrac{\omega_0}{(a + j\omega)^2 + \omega_0^2}$ provided $a > 0$	$\dfrac{\omega_0}{(a + s)^2 + \omega_0^2}$ ROC: $\mathrm{Re}\{s\} > -a$

The above argument is not true because the CTFT of the unit step function contains a discontinuity at $\omega = 0$ due to the presence of the impulse function $\delta(\omega)$. Therefore, the CTFT violates the condition for the existence of CTFT. In such cases, the CTFT is not derived from its definition but is expressed using the impulse function, which is not a mathematical function in the strict sense. It is therefore natural to expect Eq. (6.10) to be invalid. Likewise, the ROC for the Laplace transform of the sine wave, cosine wave, squared cosine wave, and squared sine wave do not contain the $j\omega$-axis, and Eq. (6.10) is also not valid in these cases.

6.2.2 Region of convergence

As a side note to our discussion, we observe that the Laplace transform is guaranteed to exist at all points within the ROC. For example, consider the causal sine wave $h(t) = \sin(4t)u(t)$. We are interested in calculating the values of the Laplace transform at two points, $s_1 = 2 + j3$ and $s_2 = j3$ in the complex s-plane. Since s_1 lies within the ROC, $\mathrm{Re}\{s\} > 0$, the value of the Laplace transform at s_1 is given by

$$H(2 + j3) = \frac{4}{(2 + j3)^2 + 4^2} = \frac{4}{4 + j12 - 9 + 16} = \frac{4}{11 + j12},$$

which, as expected, is a finite value. The point $s_2 = j3$ lies outside the ROC. However, the Laplace transform is not necessarily infinite at s_2. In fact, the Laplace transform of the causal sine wave $h(t) = \sin(4t)u(t)$ is finite for all values of s on the imaginary axis except at $s = \pm j4$. The value of the Laplace transform at s_2 is given by

$$H(j3) = \frac{4}{(j3)^2 + 4^2} = -\frac{4}{5}.$$

Since the Laplace transform is not defined at two points ($s = \pm j4$) on the imaginary axis, the entire imaginary axis is excluded from the ROC. In short, if a point lies on the boundary of the ROC, it is possible that the Laplace transform exists, though the point may not be included in the ROC.

6.2.3 Spectra for the Laplace transform

In Chapter 5, the magnitude and phase spectra of the CTFT provided meaningful insights into the frequency properties of the reference function. Except for one difference, the magnitude and phase spectra of the Laplace transform (collectively referred to as the Laplace spectra) can be plotted in a similar way. Since the Laplace variable s is a complex variable, the Laplace spectra are plotted with respect to a 2D complex plane with $\mathrm{Re}\{s\} = \sigma$ and $\mathrm{Im}\{s\} = \omega$ being the two independent axes. For the magnitude spectrum, the magnitude of the Laplace transform is plotted along the z-axis within the ROC defined in the 2D complex plane. Likewise, for the phase spectrum, the phase of the

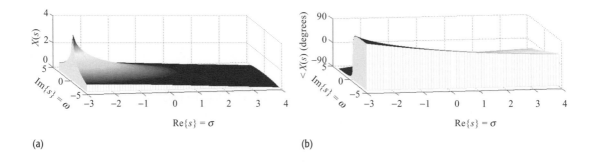

(a) (b)

Fig. 6.3. Laplace spectra for $x(t) = \mathrm{e}^{-3t} u(t)$. (a) Laplace magnitude spectrum; (b) Laplace phase spectrum.

Laplace transform is plotted along the vertical z-axis within the ROC. Both Laplace magnitude and phase spectra are, therefore, 3D plots. To illustrate the steps involved in plotting the magnitude and phase spectra, we consider the following example.

Example 6.5

Plot the Laplace spectra of the decaying exponential function $x(t) = \mathrm{e}^{-3t} u(t)$.

Solution

Based on entry (4) of Table 6.1, the Laplace transform of the decaying exponential function is given by

$$X(s) = \frac{1}{(s+3)} = \frac{1}{(\sigma + \mathrm{j}\omega + 3)} \quad \text{with ROC: } \sigma = \mathrm{Re}(s) > -3.$$

The Laplace spectra are therefore given by

magnitude spectrum $\quad |X(s)| = \dfrac{1}{\sqrt{(\sigma+3)^2 + \omega^2}};$

phase spectrum $\quad <X(s) = -\tan^{-1}\dfrac{\omega}{(\sigma+3)}.$

The magnitude and phase spectra are plotted with respect to the 2D complex s-plane in Fig. 6.3. To obtain the CTFT spectra, we can simply slice out the 2D plot along the $\mathrm{Re}\{s\} = \sigma = 0$ axis from the Laplace spectra. Figure 6.4 shows the resulting CTFT magnitude and phase spectra. These are identical to the CTFT spectra obtained directly from the CTFT and plotted in Fig. 6.4.

Fig. 6.4. CTFT spectra for $x(t) = \mathrm{e}^{-3t} u(t)$ obtained by extracting the 2D plot along the $\mathrm{Re}\{s\} = \sigma = 0$ axis in Fig. 6.3. (a) CTFT magnitude spectrum; (b) CTFT phase spectrum.

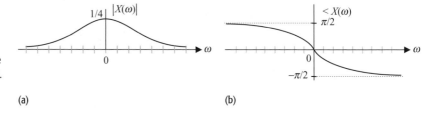

(a) (b)

6.3 Inverse Laplace transform

Evaluation of the inverse Laplace transform is an important step in the analysis of LTIC systems. The inverse Laplace transform can be calculated directly by solving the complex integral in the synthesis equation, Eq. (6.7). This approach involves contour integration, which is beyond the scope of this text. In cases where the Laplace transform takes the following rational form:

$$X(s) = \frac{N(s)}{D(s)} = \frac{b_m s^m + b_{m-1} s^{m-1} + b_{m-2} s^{m-2} + \cdots + b_1 s + b_0}{s^n + a_{n-1} s^{n-1} + a_{n-2} s^{n-2} + \cdots + a_1 s + a_0}, \quad (6.11)$$

an alternative approach based on partial fraction expansion is commonly used. The approach eliminates the need for computing Eq. (6.7) and consists of the following steps.

(1) Calculate the roots of the characteristic equation of the rational fraction, Eq. (6.11). The characteristic equation is obtained by equating the denominator $D(s)$ in Eq. (6.11) to zero, i.e.

$$D(s) = s^n + a_{n-1} s^{n-1} + a_{n-2} s^{n-2} + \cdots + a_1 s + a_0 = 0. \quad (6.12)$$

For an nth-order characteristic equation, there will be n first-order roots. Depending on the value of the coefficients $\{a_l\}$, $0 \le l \le (n-1)$, roots $\{p_r\}$, $1 \le r \le n$, of the characteristic equation may be real-valued and/or complex-valued. Assuming that roots are real-valued and do not repeat, the Laplace transform $X(s)$ is represented as

$$X(s) = \frac{N(s)}{(s-p_1)(s-p_2)\cdots(s-p_{n-1})(s-p_n)}. \quad (6.13)$$

(2) Using Heaviside's partial fraction expansion formula, explained in Appendix D, decompose $X(s)$ into a summation of the first- or second-order fractions. If no roots are repeated, $X(s)$ is decomposed as follows:

$$X(s) = \frac{k_1}{(s-p_1)} + \frac{k_2}{(s-p_2)} + \cdots + \frac{k_{n-1}}{(s-p_{n-1})} + \frac{k_n}{(s-p_n)}, \quad (6.14)$$

where the coefficients $\{k_r\}$, $1 \le r \le n$, are obtained from

$$k_r = \left[(s-p_r)\frac{N(s)}{D(s)}\right]_{s=p_r}. \quad (6.15)$$

If there are repeated or complex roots, $X(s)$ takes a slightly different form. See Appendix D for more details.
(3) From Table 6.1,

$$e^{p_r t} u(t) \overset{L}{\longleftrightarrow} \frac{1}{(s-p_r)} \quad \text{with ROC: Re}\{s\} > p_r.$$

Using the above transform pair, the inverse Laplace transform of $X(s)$ is given by

$$x(t) = k_1 e^{P_1 t} u(t) + k_2 e^{P_2 t} u(t) + \cdots + k_{n-1} e^{P_{n-1} t} u(t) + k_n e^{P_n t} u(t)$$

$$= \sum_{r=1}^{n} k_r e^{P_r t} u(t). \tag{6.16}$$

To illustrate the aforementioned procedure (steps (1) to (3)) for evaluating the inverse Laplace transform using the partial fraction expansion, we consider the following examples.

Example 6.6

Calculate the inverse Laplace transform of a right-sided sequence with transfer function

$$G(s) = \frac{7s - 6}{(s^2 - s - 6)}.$$

Solution

The characteristic equation of $G(s)$ is given by $s^2 - s - 6 = 0$, which has two roots at $s = 3$ and -2. Using the partial fraction expansion, the Laplace transform $G(s)$ is expressed as

$$G(s) = \frac{7s - 6}{(s + 2)(s - 3)} \equiv \frac{k_1}{(s + 2)} + \frac{k_2}{(s - 3)}.$$

The coefficients of the partial fractions k_1 and k_2 are given by

$$k_1 = \left[(s + 2) \frac{(7s - 6)}{(s + 2)(s - 3)} \right]_{s=-2} = \left[\frac{(7s - 6)}{(s - 3)} \right]_{s=2} = 4$$

and

$$k_2 = \left[(s - 3) \frac{(7s - 6)}{(s + 2)(s - 3)} \right]_{s=3} = \left[\frac{(7s - 6)}{(s + 2)} \right]_{s=3} = 3.$$

The partial fraction expansion of the Laplace transform $G(s)$ is therefore given by

$$G(s) = \frac{4}{(s + 2)} + \frac{3}{(s - 3)}.$$

Using entry (4) of Table 6.1, the inverse Laplace transform is

$$g(t) = (4e^{-2t} + 3e^{3t})u(t).$$

Example 6.7

Calculate the inverse Laplace transform of right-sided sequences with the following transfer functions:

(i) $X_1(s) = \dfrac{s + 3}{s(s + 1)(s + 2)};$

(ii) $X_2(s) = \dfrac{s + 5}{s^3 + 5s^2 + 17s + 13}.$

Solution

(i) In $X_1(s)$, the characteristic equation is already factorized. In terms of the partial fractions, $X_1(s)$ can be expressed as follows:

$$X_1(s) = \frac{s+3}{s(s+1)(s+2)} \equiv \frac{k_1}{s} + \frac{k_2}{(s+1)} + \frac{k_3}{(s+2)},$$

where the partial fraction coefficients k_1, k_2, and k_3 are given by

$$k_1 = \left[s \frac{(s+3)}{s(s+1)(s+2)} \right]_{s=0} = \left[\frac{(s+3)}{(s+1)(s+2)} \right]_{s=0} = \frac{3}{2},$$

$$k_2 = \left[(s+1) \frac{(s+3)}{s(s+1)(s+2)} \right]_{s=-1} = \left[\frac{(s+3)}{s(s+2)} \right]_{s=-1} = -2,$$

and

$$k_3 = \left[(s+2) \frac{(s+3)}{s(s+1)(s+2)} \right]_{s=-2} = \left[\frac{(s+3)}{s(s+1)} \right]_{s=-2} = \frac{1}{2}.$$

The partial fraction expansion of the Laplace transform $X_1(s)$ is given by

$$X_1(s) = \frac{s+3}{s(s+1)(s+2)} \equiv \frac{3}{2s} - \frac{2}{(s+1)} + \frac{1}{2(s+2)},$$

which leads to the following inverse Laplace transform:

$$x_1(t) = \left(\frac{3}{2} - 2e^{-t} + \frac{1}{2}e^{-2t} \right) u(t).$$

(ii) The characteristic equation of $X_2(s)$ is given by

$$D(s) = s^3 + 5s^2 + 17s + 13 = 0,$$

which has three roots at $s = -1$ and $-2 \pm j3$. The partial fraction expansion of $X_2(s)$ is given by

$$X_2(s) = \frac{s+5}{(s+1)(s+2+j3)(s+2-j3)} \equiv \frac{k_1}{(s+1)} + \frac{k_2s + k_3}{(s^2+4s+13)}.$$

The partial fraction coefficient k_1 is calculated to be

$$k_1 = \left[(s+1) \frac{(s+5)}{(s+1)(s^2+4s+13)} \right]_{s=-1} = \left[\frac{(s+5)}{(s^2+4s+13)} \right]_{s=-1} = \frac{2}{5}.$$

To compute coefficients k_2 and k_3, we substitute $k_1 = 2/5$ in $X_2(s)$ and expand

$$\frac{s+5}{(s+1)(s^2+4s+13)} \equiv \frac{2}{5(s+1)} + \frac{k_2s+k_3}{(s^2+4s+13)}$$

as

$$s+5 \equiv 0.4(s^2+4s+13) + (k_2s+k_3)(s+1).$$

Comparing the coefficients of s^2 on both sides of the above expression yields

$$k_2 + 0.4 = 0 \implies k_2 = -0.4.$$

Similarly, comparing the coefficients of s yields

$$k_2 + k_3 + 1.6 = 1 \implies k_3 = -0.2.$$

The partial fraction expansion of $X_2(s)$ reduces to

$$X_2(s) = \frac{2}{5(s+1)} - 0.2 \frac{2s+1}{(s+2)^2 + 9},$$

which is expressed as

$$X_2(s) = \frac{2}{5(s+1)} - 0.2 \frac{2(s+2)}{(s+2)^2 + 9} + 0.2 \frac{3}{(s+2)^2 + 9}.$$

Based on entries (4) and (13) in Table 6.1, the inverse Laplace transform is given by

$$x_1(t) = (0.4e^{-t} - 0.4e^{-2t}\cos(3t) + 0.2e^{-2t}\sin(3t))u(t).$$

6.4 Properties of the Laplace transform

The unilateral and bilateral Laplace transforms have several interesting properties, which are used in the analysis of signals and systems. These properties are similar to the properties of the CTFT covered in Section 5.4. In this section, we discuss several of these properties, including their proofs and applications, through a series of examples. A complete listing of the properties is provided in Table 6.2. In most cases, we prove the properties for the unilateral Laplace transform. The proof for the bilateral Laplace transform follows along similar lines and is not included to avoid repetition.

6.4.1 Linearity

If $x_1(t)$ and $x_2(t)$ are two arbitrary functions with the following Laplace transform pairs:

$$x_1(t) \overset{\text{L}}{\longleftrightarrow} X_1(s) \quad \text{with ROC: } R_1$$

and

$$x_2(t) \overset{\text{L}}{\longleftrightarrow} X_2(s) \quad \text{with ROC: } R_2,$$

then

$$a_1 x_1(t) + a_2 x_2(t) \overset{\text{L}}{\longleftrightarrow} a_1 X_1(s) + a_2 X_2(s) \quad \text{with ROC: at least } R_1 \cap R_2$$

$$(6.17)$$

for both unilateral and bilateral Laplace transforms.

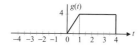

Fig. 6.5. Causal function $g(t)$ considered in Example 6.8.

Proof

Calculating the Laplace transform of $\{a_1x_1(t)+a_1x_2(t)\}$ using Eq. (6.9) yields

$$L\{a_1x_1(t)+a_2x_2(t)\} = \int_{0-}^{\infty}\{a_1x_1(t)+a_2x_2(t)\}e^{-st}\,dt$$

$$= \int_{0-}^{\infty}a_1x_1(t)e^{-st}\,dt + \int_{0-}^{\infty}a_2x_2(t)e^{-st}\,dt$$

$$= a_1\int_{0-}^{\infty}x_1(t)e^{-st}\,dt + a_2\int_{0-}^{\infty}x_2(t)e^{-st}\,dt$$

$$= a_1X_1(s)+a_2X_2(s),$$

which proves Eq. (6.17).

By definition of the ROC, the Laplace transform $X_1(s)$ is finite within the specified region R_1. Similarly, $X_2(s)$ is finite within its ROC R_2. Therefore, the linear combination $a_1X_1(s)+a_2X_2(s)$ must at least be finite in region R that represents the intersection of the two regions i.e. $R = R_1 \cap R_2$. If there is no common region between R_1 and R_2, then the Laplace transform of $\{a_1x_1(t)+a_1x_2(t)\}$ does not exist. Due to the cancellation of certain terms in $a_1X_1(s)+a_2X_2(s)$, it is also possible that the overall ROC of the linear combination is larger than $R_1 \cap R_2$. To illustrate the application of the linearity property, we consider the following example.

Example 6.8

Calculate the Laplace transform of the causal function $g(t)$ shown in Fig. 6.5.

Solution

The causal function $g(t)$ is expressed as the linear combination

$$g(t) = 4x_3(t)+2x_7(t),$$

where the CT functions $x_3(t)$ and $x_7(t)$ are defined in Example 6.4. Based on the results of Example 6.4, the Laplace transforms for $x_3(t)$ and $x_7(t)$ are given by

$$X_3(s) = \begin{cases} 2 & \text{for } s = 0 \\ \dfrac{1}{s}[e^{-2s}-e^{-4s}] & \text{for } s \neq 0 \end{cases} \quad \text{with ROC: entire s-plane}$$

and

$$X_7(s) = \begin{cases} 3 & \text{for } s = 0 \\ \dfrac{2}{s^2}[1-e^{-s}-se^{-2s}] & \text{for } s \neq 0. \end{cases} \quad \text{with ROC: entire s-plane}$$

Applying the linearity property, the Laplace transform of $g(t)$ is given by

$$G(s) = \begin{cases} 4(2) + 2(3) & \text{for } s = 0 \\ \dfrac{4}{s}(e^{-2s} - e^{-4s}) + \dfrac{4}{s^2}[1 - e^{-s} - se^{-2s}] & \text{for } s \neq 0, \end{cases}$$

which reduces to

$$G(s) = \begin{cases} 14 & \text{for } s = 0 \\ \dfrac{4}{s^2}[1 - e^{-s} - se^{-4s}] & \text{for } s \neq 0. \end{cases}$$

Note that the ROC of $G(s)$ is the intersection of the individual regions R_1 and R_2. The overall ROC R is, therefore, the entire s-plane.

6.4.2 Time scaling

If $x(t) \xleftrightarrow{\text{L}} X(s)$ with ROC:R, then the Laplace transform of the scaled signal $x(at)$, where $a \in \Re^+$, $a > 0$, is given by

$$x(at) \xleftrightarrow{\text{L}} \frac{1}{|a|} X\left(\frac{s}{a}\right) \quad \text{with ROC: } aR \tag{6.18}$$

for both unilateral and bilateral Laplace transforms. For the unilateral Laplace transform, the value of a must be greater than zero. If $a < 0$, the scaled signal $x(at)$ will be non-causal such that its unilateral Laplace transform will not exist.

Proof
By Eq. (6.9), the Laplace transform of the time-scaled signal $x(at)$ is given by

$$L\{x(at)\} = \int_{0-}^{\infty} x(at)e^{-st}\,dt$$

$$= \int_{0-}^{\infty} x(\tau)e^{-s\tau/a}\frac{d\tau}{a} \quad \text{(by substituting } \tau = at\text{)}$$

$$= \frac{1}{a} X\left(\frac{s}{a}\right),$$

which proves Eq. (6.18) for $a > 0$. To prove that the ROC of the Laplace transform of the time-scaled signal is aR, note that the values of $X(s)$ are finite within region R. For $X(s/a)$, the new region over which $X(s/a)$ is finite will transform to aR.

Example 6.9
Calculate the Laplace transform of the function $h(t)$ shown in Fig. 6.6.

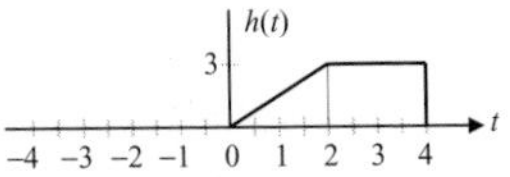

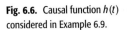

Fig. 6.6. Causal function $h(t)$ considered in Example 6.9.

Solution
In terms of the causal function $x_7(t)$, signal $h(t)$ is expressed as

$$h(t) = 1.5x_7\left(\frac{t}{2}\right) \quad \text{or} \quad h(t) = 1.5x_7(0.5t).$$

In Example 6.4, the Laplace transform of $x_7(t)$ is given by

$$X_7(s) = \begin{cases} 3 & \text{for } s = 0 \\ \dfrac{2}{s^2}[1 - e^{-s} - se^{-2s}] & \text{for } s \neq 0, \end{cases}$$

with the entire s-plane as the ROC.

Using the time-scaling property the Laplace transform of $h(t)$ is given by

$$h(t) = 1.5x_7(0.5t) \overset{L}{\longleftrightarrow} \frac{1}{0.5}(1.5)X_7\left(\frac{s}{0.5}\right) = 3X_7(2s) \quad \text{with ROC: } 2R_1,$$

which reduces to

$$H(s) = \begin{cases} 9 & \text{for } s = 0 \\ \dfrac{1.5}{s^2}[1 - e^{-2s} - 2se^{-4s}] & \text{for } s \neq 0. \end{cases}$$

The ROC associated with $H(s)$ is the entire s-plane.

6.4.3 Time shifting

If $x(t) \overset{L}{\longleftrightarrow} X(s)$ with ROC:R, then the Laplace transform of the time-shifted signal is

$$x(t - t_0) \overset{L}{\longleftrightarrow} e^{-st_0}X(s) \quad \text{with ROC: } R \qquad (6.19)$$

for both unilateral and bilateral Laplace transforms. For the unilateral Laplace transform, the value of t_0 should be carefully selected such that the time-shifted signal $x(t - t_0)$ remains causal. There is no such restriction for the bilateral Laplace transform. Also, it may be noted that time shifting a signal does not change the ROC of its Laplace transform.

Proof
By Eq. (6.9), the Laplace transform of the time-shifted signal $x(t - t_0)$ is given by

$$L\{x(t - t_0)\} = \int_{-\infty}^{\infty} x(t - t_0)e^{-st}\,dt$$

$$= e^{-st_0}\int_{-\infty}^{\infty} x(\tau)e^{-s\tau}\,d\tau \quad \text{(by substituting } \tau = t - t_0)$$

$$= e^{-st_0}X(s),$$

which proves the time-shifting property, Eq. (6.19). The Laplace transform of the time-shifted signal $x(t - t_0)$ is a product of two terms: $\exp(-st_0)$ and $X(s)$. For finite values of s and t_0, the first term is always finite. Therefore, the ROC of the Laplace transform of the time-shifted signal is the same as $X(s)$.

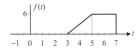

Fig. 6.7. Waveform f(t) used in
Example 6.10.

Example 6.10

Calculate the Laplace transform of the causal function $f(t)$ shown in Fig. 6.7.

Solution

In terms of the waveform $h(t)$ shown in Fig. 6.6, $f(t)$ is expressed as follows:

$$f(t) = 2h(t-3).$$

In Example 6.9, the Laplace transform of $h(t)$ is given by

$$H(s) = \begin{cases} 9 & \text{for } s = 0 \\ \dfrac{1.5}{s^2}[1 - e^{-2s} - 2se^{-4s}] & \text{for } s \neq 0, \end{cases}$$

with the entire s-plane as the ROC. Using the time-shifting property, the Laplace transform of $f(t)$ is

$$f(t) = 2h(t-3) \xleftrightarrow{\text{L}} 2e^{-3s}H(s) \quad \text{with ROC: } R,$$

which results in the following Laplace transform for $f(t)$:

$$H(s) = \begin{cases} [18e^{-3s}]_{s=0} = 18 & \text{for } s = 0 \\ \dfrac{3}{s^2}[e^{-3s} - e^{-5s} - 2se^{-7s}] & \text{for } s \neq 0, \end{cases}$$

with the entire s-plane as the ROC.

6.4.4 Shifting in the s-domain

If $x(t) \xleftrightarrow{\text{L}} X(s)$ with ROC: R, then the Laplace transform of

$$e^{s_0 t} x(t) \xleftrightarrow{\text{L}} X(s - s_0) \quad \text{with ROC: } R + \text{Re}\{s_0\} \qquad (6.20)$$

for both unilateral and bilateral Laplace transforms. Shifting a signal in the complex s-domain by s_0 causes the ROC to shift by $\text{Re}\{s_0\}$. Although the amount of shift s_0 can be complex, the shift in the ROC is always a real number. In other words, the ROC is always shifted along the horizontal axis, irrespective of the value of the imaginary component in s_0.

The shifting property can be proved directly from Eq. (6.9) by considering the CTFT of the signal $\exp(s_0 t)x(t)$. The proof is left as an exercise for the reader (see Problem 6.6).

Example 6.11

Using the Laplace transform pair

$$u(t) \xleftrightarrow{\text{L}} \frac{1}{s} \quad \text{with ROC: } \text{Re}\{s\} > 0,$$

calculate the Laplace transform of (i) $x_1(t) = \cos(\omega_0 t)u(t)$ and (ii) $x_2(t) = \sin(\omega_0 t)u(t)$.

Solution

Using the above Laplace transform pair and s-shifting property, the Laplace transforms of $\exp(j\omega_0 t)u(t)$ and $\exp(-j\omega_0 t)\,u(t)$ are given by

$$e^{j\omega_0 t}u(t) \xleftrightarrow{\text{L}} \frac{1}{(s-j\omega_0)} \quad \text{with ROC: } \text{Re}\{s\} > 0$$

and

$$e^{-j\omega_0 t}u(t) \xleftrightarrow{\text{L}} \frac{1}{(s+j\omega_0)} \quad \text{with ROC: } \text{Re}\{s\} > 0.$$

(i) To calculate the Laplace transform of $x_1(t) = \cos(\omega_0 t)\,u(t)$, we add the above transform pairs to obtain

$$e^{j\omega_0 t}u(t) + e^{-j\omega_0 t}u(t) \xleftrightarrow{\text{L}} \frac{1}{(s-j\omega_0)} + \frac{1}{(s+j\omega_0)} \quad \text{with ROC: } \text{Re}\{s\} > 0,$$

which reduces to

$$\cos(\omega_0 t)u(t) \xleftrightarrow{\text{L}} \frac{s}{s^2 + \omega_0^2} \quad \text{with ROC: } \text{Re}\{s\} > 0.$$

(ii) To evaluate the Laplace transform of $x_2(t) = \sin(\omega_0 t)u(t)$, we take the difference of the above transform pairs to obtain

$$e^{j\omega_0 t}u(t) - e^{-j\omega_0 t}u(t) \xleftrightarrow{\text{L}} \frac{1}{(s-j\omega_0)} - \frac{1}{(s+j\omega_0)} \quad \text{with ROC: } \text{Re}\{s\} > 0,$$

which simplifies to

$$\sin(\omega_0 t)u(t) \xleftrightarrow{\text{L}} \frac{\omega_0}{s^2 + \omega_0^2} \quad \text{with ROC: } \text{Re}\{s\} > 0.$$

6.4.5 Time differentiation

If $x(t) \xleftrightarrow{\text{L}} X(s)$ with ROC:R, then the Laplace transform of

$$\frac{dx}{dt} \xleftrightarrow{\text{L}} sX(s) - x(0^-) \quad \text{with ROC: } R. \tag{6.21}$$

Note that if the function $x(t)$ is causal, $x(0^-) = 0$.

Proof

By Eq. (6.9), the Laplace transform of the derivative dx/dt is given by

$$\text{L}\left\{\frac{dx}{dt}\right\} = \int_{0^-}^{\infty} \frac{dx}{dt} e^{-st}\,dt.$$

Applying integration by parts on the right-hand side of the equation yields

$$\text{L}\left\{\frac{dx}{dt}\right\} = \underbrace{x(t)e^{-st}\Big|_{0^-}^{\infty}}_{A} - (-s)\underbrace{\int_{0^-}^{\infty} x(t)e^{-st}\,dt}_{X(s)}.$$

Considering the first term, denoted by A, we note that for the upper limit, $t \to \infty$, the value of A is zero due to the decaying exponential term. For the lower limit, $t \to 0^-$, A equals $x(0^-)$. The above equation therefore reduces to

$$L\left\{\frac{dx}{dt}\right\} = -x(0^-) + sX(s).$$

Corollary 6.1 By repeatedly applying the differentiation property n times, it is straightforward to prove that

$$\frac{d^n x}{dt^n} \overset{L}{\longleftrightarrow} s^n X(s) - s^{n-1}x(0^-) - \cdots - sx^{(n-2)}(0^-)$$
$$- x^{(n-1)}(0^-) \quad \text{with ROC: } R, \quad (6.22)$$

where $x^{(k)}$ denotes the kth derivative of $x(t)$, i.e. $x^{(k)} = d^k x/dt^k$.

Example 6.12

Based on the Laplace transform pair

$$u(t) \overset{L}{\longleftrightarrow} \frac{1}{s} \quad \text{with ROC: } \text{Re}\{s\} > 0,$$

calculate the Laplace transform for the impulse function $x(t) = \delta(t)$.

Solution

Based on entry (2) of Table 6.1, we know that

$$\frac{du(t)}{dt} \overset{L}{\longleftrightarrow} s \cdot \frac{1}{s} - u(t)\Big|_{t=0} \quad \text{with ROC: } \text{Re}\{s\} > 0.$$

Using the time-differentiation property, the Laplace transform of the first derivative of $u(t)$ is given by

$$\frac{du(t)}{dt} \overset{L}{\longleftrightarrow} s \cdot \frac{1}{s} - u(t)|_{t=0^-} \quad \text{with ROC: } \text{Re}\{s\} > 0.$$

Knowing that $du/dt = \delta(t)$ and $u(0^-) = 0$, we obtain

$$\delta(t) \overset{L}{\longleftrightarrow} 1 \quad \text{with ROC: } \text{Re}\{s\} > 0.$$

6.4.6 Time integration

If $x(t) \overset{L}{\longleftrightarrow} X(s)$ with ROC: R, then

unilateral Laplace transform $\quad \displaystyle\int_{0^-}^{t} x(\tau)d\tau \overset{L}{\longleftrightarrow} \frac{X(s)}{s}$

$$\text{with ROC: } R \cap \text{Re}\{s\} > 0; \quad (6.23)$$

bilateral Laplace transform $\quad \displaystyle\int_{-\infty}^{t} x(\tau)d\tau \overset{L}{\longleftrightarrow} \frac{X(s)}{s} + \frac{1}{s}\int_{-\infty}^{0^-} x(\tau)d\tau$

$$\text{with ROC: } R \cap \text{Re}\{s\} > 0. \quad (6.24)$$

Table 6.2. Properties of the Laplace transform

The corresponding properties of the CTFT are also listed in the table for comparison

Properties in the time domain	CTFT $$X(j\omega) = \int_{-\infty}^{\infty} x(t)e^{-j\omega t}\,dt$$	Laplace transform $$X(s) = \int_{-\infty}^{\infty} x(t)e^{-st}\,dt$$
Linearity $a_1 x_1(t) + a_2 x_2(t)$	$a_1 X_1(\omega) + a_2 X_2(\omega)$	$a_1 X_1(s) + a_2 X_2(s)$ ROC: at least $R_1 \cap R_2$
Time scaling $x(at)$	$\dfrac{1}{|a|} X\left(\dfrac{\omega}{a}\right)$	$\dfrac{1}{|a|} X\left(\dfrac{s}{a}\right)$ with ROC: aR
Time shifting $x(t - t_0)$	$e^{-j\omega_0 t} X(\omega)$	$e^{-st_0} X(s)$ with ROC: R
Frequency/s-domain shifting $x(t)e^{j\omega_0 t}$ or $x(t)e^{s_0 t}$	$X(\omega - \omega_0)$	$X(s - s_0)$ with ROC: $R + Re\{s_0\}$
Time differentiation dx/dt	$j\omega X(\omega)$	$sX(s) - x(0^-)$ with ROC: R
Time integration $\displaystyle\int_{-\infty}^{t} x(\tau)d\tau$	$\dfrac{X(\omega)}{j\omega} + \pi X(0)\delta(\omega)$	$\dfrac{X(s)}{s}$ with ROC: $R \cap Re\{s\} > 0$
Frequency/s-domain differentiation $(-t)x(t)$	$-j dX/d\omega$	dX/ds
Duality $X(t)$	$2\pi x(\omega)$	not applicable
Time convolution $x_1(t) * x_2(t)$	$X_1(\omega)X_2(\omega)$	$X_1(s)X_2(s)$ ROC includes $R_1 \cap R_2$
Frequency/s-domain convolution $x_1(t)x_2(t)$	$\dfrac{1}{2\pi} X_1(\omega) * X_2(\omega)$	$\dfrac{1}{2\pi} X_1(s) * X_2(s)$ ROC includes $R_1 \cap R_2$
Parseval's relationship	$\displaystyle\int_{-\infty}^{\infty} |x(t)|^2 dt = \dfrac{1}{2\pi} \int_{-\infty}^{\infty} |X(\omega)|^2 d\omega$	not applicable
Initial value $x(0^+)$ if it exists	$\dfrac{1}{2\pi} \displaystyle\int_{-\infty}^{\infty} X(\omega)d\omega$	$\lim\limits_{s \to \infty} sX(s)$ provided $s = \infty$ is included in the ROC of $sX(s)$
Final value $x(\infty)$ if it exists	not applicable	$\lim\limits_{s \to 0} sX(s)$ provided $s = 0$ is included in the ROC of $sX(s)$

The proof of the time-integration property is left as an exercise for the reader (see Problem 6.7).

Example 6.13
Given the Laplace transform pair
$$\cos(\omega_0 t)u(t) \xrightarrow{\text{ L }} \frac{s}{(s^2 + \omega_0^2)} \quad \text{with ROC: } \operatorname{Re}\{s\} > 0,$$
derive the unilateral Laplace transform of $\sin(\omega_0 t)u(t)$.

Solution
By applying the time-integration property to the aforementioned unilateral Laplace transform pair yields
$$\int_{0^-}^{t} \cos(\omega_0 \tau)u(\tau)\mathrm{d}\tau \xrightarrow{\text{ L }} \frac{1}{s}\frac{s}{(s^2 + \omega_0^2)} \quad \text{with ROC: } \operatorname{Re}\{s\} > 0,$$
where the left-hand side of the transform pair is given by
$$\int_{0^-}^{t} \cos(\omega_0 \tau)u(\tau)\mathrm{d}\tau = \int_{0}^{t} \cos(\omega_0 \tau)\mathrm{d}\tau = \left.\frac{\sin(\omega_0 \tau)}{\omega_0}\right|_{0}^{t} = \frac{1}{\omega_0}\sin(\omega_0 t).$$
Substituting the value of the integral in the transform pair, we obtain
$$\sin(\omega_0 t)u(t) \xrightarrow{\text{ L }} \frac{\omega_0}{(s^2 + \omega_0^2)} \quad \text{with ROC: } \operatorname{Re}\{s\} > 0,$$

6.4.7 Time and s-plane convolution

If $x_1(t)$ and $x_2(t)$ are two arbitrary functions with the following Laplace transform pairs:
$$x_1(t) \xrightarrow{\text{ L }} X_1(s) \quad \text{with ROC: } R_1 \quad \text{and} \quad x_2(t) \xrightarrow{\text{ L }} X_2(s) \quad \text{with ROC: } R_2,$$
then the convolution property states that

time convolution $\qquad x_1(t) * x_2(t) \xrightarrow{\text{ L }} X_1(s)X_2(s)$

$$\text{containing at least ROC: } R_1 \cap R_2; \quad (6.25)$$

s-plane convolution $\quad x_1(t)x_2(t) \xrightarrow{\text{ L }} \frac{1}{2\pi \mathrm{j}}[X_1(s) * X_2(s)]$

$$\text{containing at least ROC: } R_1 \cap R_2. \quad (6.26)$$

The convolution property is valid for both unilateral (for causal signals) and bilateral (for non-causal signals) Laplace transforms. The overall ROC of the convolved signals may be larger than the intersection of regions R_1 and R_2 because of possible cancellation of poles in the products.

Proof
Consider the Laplace transform of $x_1(t) * x_2(t)$:
$$L\{x_1(t) * x_2(t)\} = \int_{0^-}^{\infty} [x_1(t) * x_2(t)]\mathrm{e}^{-st}\mathrm{d}t = \int_{0^-}^{\infty}\left[\int_{-\infty}^{\infty} x_1(\tau)x_2(t - \tau)\mathrm{d}\tau\right]\mathrm{e}^{-st}\mathrm{d}t.$$

Interchanging the order of integration, we get

$$L\{x_1(t) * x_2(t)\} = \int\limits_{-\infty}^{\infty} x_1(\tau) \left[\int\limits_{0^-}^{\infty} x_2(t-\tau)e^{-st}dt \right] d\tau.$$

By noting that the inner integration $\int x_2(t-\tau) \exp(-st)dt = X_2(s) \exp(-s\tau)$, the above integral simplifies to

$$L\{x_1(t) * x_2(t)\} = X_2(s) \int\limits_{-\infty}^{\infty} x_1(\tau)e^{-s\tau}d\tau = X_2(s)X_1(s),$$

which proves Eq. (6.25). The s-plane convolution property may be proved in a similar fashion.

Like the CTFT convolution property discussed in Section 5.5.8, the Laplace time-convolution property provides us with an alternative approach to calculate the output $y(t)$ when a CT signal $x(t)$ is applied at the input of an LTIC system with the impulse response $h(t)$. In Chapter 3, we proved that the zero-state output response $y(t)$ is obtained by convolving the input signal $x(t)$ with the impulse response $h(t)$, i.e. $y(t) = h(t) * x(t)$. Using the time-convolution property, the Laplace transform $Y(s)$ of the resulting output $y(t)$ is given by

$$y(t) = x(t) * h(t) \xrightarrow{\text{L}} Y(s) = X(s)H(s),$$

where $X(s)$ and $H(s)$ are the Laplace transforms of the input signal $x(t)$ and the impulse response $h(t)$ of the LTIC systems. In other words, the Laplace transform of the output signal is obtained by multiplying the Laplace transforms of the input signal and the impulse response. The procedure for calculating the output $y(t)$ of an LTI system in the complex s-domain, therefore, consists of the following four steps.

(1) Calculate the Laplace transform $X(s)$ of the input signal $x(t)$. If the input signal and the impulse response are both causal functions, then the unilateral Laplace transform is used. If either of the two functions is non-causal, the bilateral Laplace transform must be used.
(2) Calculate the Laplace transform $H(s)$ of the impulse response $h(t)$ of the LTIC system. The Laplace transform $H(s)$ is referred to as the Laplace transfer function of the LTIC system and provides a meaningful insight into the behavior of the system.
(3) Based on the convolution property, the Laplace transform $Y(s)$ of the output response $y(t)$ is given by the product of the Laplace transforms of the input signal and the impulse response of the LTIC systems. Mathematically, this implies that $Y(s) = X(s)H(s)$.
(4) Calculate the output response $y(t)$ in the time domain by taking the inverse Laplace transform of $Y(s)$ obtained in step (3).

Since the Laplace-transform-based approach for calculating the output response of an LTIC system does not involve integration, it is preferred over the time-domain approaches.

Example 6.14

In Example 3.6, we showed that in response to the input signal $x(t) = e^{-t}u(t)$, the LTIC system with the impulse response $h(t) = e^{-2t}u(t)$ produces the following output:

$$y(t) = (e^{-t} - e^{-2t})u(t).$$

Example 5.21 derived the result using the CTFT. We now derive the result using the Laplace transform.

Solution

Since the input signal and impulse response are both causal functions, we take the unilateral Laplace transform of both signals. Based on Table 6.1, the resulting transform pairs are given by

$$x(t) = e^{-t}u(t) \overset{L}{\longleftrightarrow} X(s) = \frac{1}{(s+1)} \quad \text{with ROC: Re}\{s\} > -1$$

and

$$h(t) = e^{-2t}u(t) \overset{L}{\longleftrightarrow} H(s) = \frac{1}{(s+2)} \quad \text{with ROC: Re}\{s\} > -2.$$

Based on the time-convolution property, the Laplace transform $Y(s)$ of the resulting output $y(t)$ is given by

$$y(t) = h(t) * x(t) \overset{L}{\longleftrightarrow} Y(s) = \frac{1}{(s+1)(s+2)} \quad \text{with ROC: Re}\{s\} > -1,$$

where the ROC of the Laplace transform of the output is obtained by taking the intersection of the regions $\text{Re}\{s\} > -1$ and $\text{Re}\{s\} > -2$, associated with the applied input and the impulse response. Using partial fraction expansion, $Y(s)$ may be expressed as follows:

$$Y(s) = \underbrace{\frac{1}{(s+1)}}_{\text{ROC}:\,\text{Re}\{s\}>-1} - \underbrace{\frac{1}{(s+2)}}_{\text{ROC}:\,\text{Re}\{s\}>-2}.$$

Taking the inverse Laplace transform of the individual terms on the right-hand side of this equation yields

$$y(t) = (e^{-t} - e^{-2t})u(t),$$

which is the same as the result produced by direct convolution and the approach based on the CTFT time-convolution property.

6.4.8 Initial- and final-value theorems

If $x(t) \overset{L}{\longleftrightarrow} X(s)$ with ROC:R, then

initial-value theorem $x(0^+) = \lim\limits_{t \to 0^+} x(t) = \lim\limits_{s \to \infty} sX(s)$ provided $x(0^+)$ exists;

$$(6.27)$$

final-value theorem $x(\infty) = \lim\limits_{t \to \infty} x(t) = \lim\limits_{s \to 0} sX(s)$ provided $x(\infty)$ exists.

$$(6.28)$$

The initial-value theorem is valid only for the unilateral Laplace transform as it requires the reference signal $x(t)$ to be zero for $t < 0$. In addition, $x(t)$ should not contain an impulse function or any other higher-order discontinuities at $t = 0$. The second constraint is required to ensure a unique value of $x(t)$ at $t = 0^+$. However, the final-value theorem may be used with either the unilateral or bilateral Laplace transform. The proof of these theorems is left as an exercise for the reader (see Problems 6.8 and 6.9).

Example 6.15

Calculate the initial and final values of the functions $x_1(t)$, $x_2(t)$, and $x_3(t)$, whose Laplace transforms are specified below:

(i) $X_1(s) = \dfrac{s+3}{s(s+1)(s+2)}$ with ROC R_1: $\text{Re}\{s\} > 0$;

(ii) $X_2(s) = \dfrac{s+5}{s^3 + 5s^2 + 17s + 13}$ with ROC R_2: $\text{Re}\{s\} > -1$;

(iii) $X_3(s) = \dfrac{5}{s^2 + 25}$ with ROC R_3: $\text{Re}\{s\} > 0$.

Solution

(i) Applying the initial-value theorem, Eq. (6.27), to $X_1(s)$, we obtain

$$x_1(0^+) = \lim_{t \to 0^+} x_1(t) = \lim_{s \to \infty} sX_1(s) = \lim_{s \to \infty} \frac{s(s+3)}{s(s+1)(s+2)}$$

$$= \lim_{s \to \infty} \frac{(s+3)}{(s+1)(s+2)} = 0.$$

Applying the final-value theorem, Eq. (6.28), to $X_1(s)$ yields

$$x_1(\infty) = \lim_{t \to \infty} x_1(t) = \lim_{s \to 0} sX_1(s) = \lim_{s \to 0} \frac{s(s+3)}{s(s+1)(s+2)}$$

$$= \lim_{s \to 0} \frac{(s+3)}{(s+1)(s+2)} = 1.5.$$

These initial and final values of $x(t)$ can be verified from the following inverse Laplace transform of $X_1(s)$ derived in Example 6.7(i):

$$x_1(t) = (1.5 - 2e^{-t} + 0.5e^{-2t})u(t).$$

(ii) Applying the initial-value theorem, Eq. (6.27), to $X_2(s)$, we obtain

$$x_2(0^+) = \lim_{t \to 0^+} x_2(t) = \lim_{s \to \infty} s X_2(s) = \lim_{s \to \infty} \frac{s(s + 5)}{s^3 + 5s^2 + 17s + 13}$$

$$= \lim_{s \to \infty} \frac{2}{6s} = 0.$$

Applying the final-value theorem, Eq. (6.28), to $X_2(s)$ yields

$$x_2(\infty) = \lim_{t \to \infty} x_2(t) = \lim_{s \to 0} s X_2(s) = \lim_{s \to 0} \frac{s(s + 5)}{s^3 + 5s^2 + 17s + 13} = 0.$$

The initial and final values of $x(t)$ can be verified from the following inverse Laplace transform of $X_1(s)$ derived in Example 6.7(ii):

$$x_1(t) = (0.4e^{-t} - 0.4e^{-2t} \cos(3t) + 0.2e^{-2t} \sin(3t))u(t).$$

(iii) Applying the initial-value theorem, Eq. (6.27), to $X_3(s)$, we obtain

$$x_3(0^+) = \lim_{t \to 0^+} x_3(t) = \lim_{s \to \infty} s X_3(s) = \lim_{s \to \infty} \frac{5s}{s^2 + 25} = \lim_{s \to \infty} \frac{5}{2s} = 0.$$

Applying the final-value theorem, Eq. (6.28), to $X_3(s)$ yields

$$x_3(\infty) = \lim_{t \to \infty} x_3(t) = \lim_{s \to 0} s X_3(s) = \lim_{s \to 0} \frac{5s}{s^2 + 25} = 0.$$

To confirm the initial and final values obtained in (iii), we determine these values directly from the inverse transform of $X_3(s) = 5/(s^2 + 25)$. From Table 6.1, the inverse Laplace transform of $X_3(s)$ is given by $x_3(t) = \sin(5t)u(t)$. Substituting $t = 0^+$, the initial value $x_3(0^+) = 0$, which verifies the value determined from the initial-value theorem. Applying the limit $t \to \infty$ to $x_3(t)$, the final value of $x_3(t)$ cannot be determined due to the oscillatory behavior of the sinusoidal wave. As a result, the final-value theorem provides an erroneous answer. The discrepancy between the result obtained from the final-value theorem and the actual value $x_3(\infty)$ occurs because the point $s = 0$ is not included in the ROC of $s X_3(s)$ R_3: $\text{Re}\{s\} > 0$. As such, the expression for the Laplace transform $s X_3(s)$ is not valid for $s = 0$. In such cases, the final-value theorem cannot be used to determine the value of the function as $t \to \infty$. Similarly, the point $s = \infty$ must be present within the ROC of $s X_3(s)$ to apply the initial-value theorem.

6.5 Solution of differential equations

An important application of the Laplace transform is to solve linear, constant-coefficient differential equations. In Section 3.1, we used a time-domain approach to obtain the zero-input, zero-state, and overall solution of differential equations. In this section, we discuss an alternative approach based on the Laplace transform. We illustrate the steps involved in the Laplace-transform-based approach through Examples 6.16 and 6.17.

Example 6.16

In Example 3.2, we calculated the output voltage $y(t)$ across resistor $R = 5\,\Omega$ of an RC series circuit, which is modeled by the linear, constant-coefficient differential equation

$$\frac{dy}{dt} + 4y(t) = \frac{dx}{dt} \tag{6.29}$$

for an initial condition $y(0^-) = 2$ V and a sinusoidal voltage $x(t) = \sin(2t)u(t)$ applied at the input of the RC circuit. Repeat Example 3.2 using the Laplace-transform-based approach.

Solution

Overall response To compute the overall response of the RC circuit, we take the Laplace transform of each term on both sides of Eq. (6.29). The Laplace transform $X(s)$ of the input signal $x(t)$ is given by

$$X(s) = L\{x(t)\} = L\{\sin(2t)u(t)\} = \frac{2}{s^2 + 4}.$$

Using the time-differentiation property,

$$L\left\{\frac{dx}{dt}\right\} = sX(s) - x(0^-) = \frac{2s}{s^2 + 4}.$$

Expressed in terms of the Laplace transform pair, $y(t) \xleftrightarrow{\;L\;} Y(s)$, the transform of the first derivative of $y(t)$ is given by

$$L\left\{\frac{dy}{dt}\right\} = sY(s) - y(0^-) = sY(s) - 2.$$

Taking the Laplace transform of Eq. (6.29) and substituting the above values yields

$$[sY(s) - 2] + 4Y(s) = \frac{2s}{s^2 + 4}. \tag{6.30}$$

Rearranging and collecting the terms corresponding to $Y(s)$ on the left-hand side of the equation results in the following:

$$[s + 4]Y(s) = 2 + \frac{2s}{s^2 + 4}$$

or

$$Y(s) = \frac{2s^2 + 2s + 8}{(s + 4)(s^2 + 4)} \equiv \frac{A}{(s + 4)} + \frac{Bs + C}{(s^2 + 4)}, \tag{6.31}$$

where Eq. (6.31) is obtained by the partial fraction expansion. The partial fraction coefficient A is given by

$$A = \left[(s + 4)\frac{2s^2 + 2s + 8}{(s + 4)(s^2 + 4)}\right]_{s=-4} = \frac{32}{20} = 1.6.$$

To obtain the values of the partial fraction coefficients B and C, we multiply both sides of Eq. (6.31) by $(s + 4)(s^2 + 4)$ and substitute $A = 1.6$. The resulting expression is as follows:

$$
\begin{aligned}
2s^2 + 2s + 8 &= A(s^2 + 4) + (s + 4)(Bs + C) \\
&= (A + B)s^2 + (4B + C)s + 4(A + C) \\
&= (1.6 + B)s^2 + (4B + C)s + 4(1.6 + C).
\end{aligned}
$$

Comparing the coefficients of s^2, we obtain $1.6 + B = 2$, or $B = 0.4$. Similarly, comparing the coefficients of s gives $4B + C = 2$, or $C = 0.4$. The expression for $Y(s)$ is, therefore, given by

$$
Y(s) = \frac{1.6}{(s + 4)} + \frac{0.4s + 0.4}{(s^2 + 4)} = \frac{1.6}{(s + 4)} + 0.4\frac{s}{(s^2 + 4)} + 0.2\frac{2}{(s^2 + 4)},
$$

which has the following inverse Laplace transform:

$$
y(t) = [1.6e^{-4t} + 0.4\cos(2t) + 0.2\sin(2t)]u(t).
$$

The aforementioned value of the overall output signal is same as the solution derived in Eq. (3.10) using the time-domain approach. We now proceed with the calculation of the zero-input response $y_{zi}(t)$ and zero-state response $y_{zs}(t)$.

Zero-input response To obtain the zero-input response $y_{zi}(t)$, we assume that the value of input $x(t) = 0$ in Eq. (6.29), i.e.

$$
\frac{dy_{zi}}{dt} + 4y_{zi}(t) = 0.
$$

Taking the Laplace transform of the above equation and substituting:

$$
L\left\{\frac{dy_{zi}}{dt}\right\} = sY_{zi}(s) - y_{zi}(0^-) = sY_{zi}(s) - 2,
$$

gives

$$
[s + 4]Y_{zi}(s) = 2,
$$

which reduces to

$$
Y_{zi}(s) = \frac{2}{s + 4}.
$$

Taking the inverse Laplace transform results in the following expression for the zero-input response:

$$
y_{zi}(t) = 2e^{-4t}u(t),
$$

which is same as the result derived in Example 3.2.

Zero-state response To obtain the zero-state response, we assume that the initial condition $y_{zs}(0^-) = 0$. This changes the value of the Laplace transform of the first derivative of $y(t)$ as follows:

$$
L\left\{\frac{dy_{zs}}{dt}\right\} = sY_{zs}(s) - y_{zs}(0^-) = sY_{zs}(s).
$$

Taking the Laplace transform of Eq. (6.29) yields

$$(s+4)Y_{zs}(s) = \frac{2s}{s^2+4}.$$

Using the partial fraction expansion, the above equation is expressed as follows:

$$Y_{zs}(s) = \frac{2s}{(s+4)(s^2+4)} \equiv -\frac{0.4}{(s+4)} + \frac{0.4s+0.4}{(s^2+4)}.$$

Taking the inverse Laplace transform, the zero-state response is given by

$$y(t) = [-0.4e^{-4t} + 0.4\cos(2t) + 0.2\sin(2t)]u(t),$$

which is same as the result derived in Example 3.2.

We also know from Chapter 3 that the overall response $y(t)$ is the sum of the zero-input response $y_{zi}(t)$ and the zero-state response $y_{zs}(t)$. This is easily verifiable for the above results.

Example 6.17

In Example 3.3, the following differential equation

$$\frac{d^2w}{dt^2} + 7\frac{dw}{dt} + 12w(t) = 12x(t) \tag{6.32}$$

was used to model the RLC series circuit shown in Fig. 3.1. Determine the zero-input, zero-state, and overall response of the system produced by the input $x(t) = 2e^{-t}u(t)$ given the initial conditions, $w(0^-) = 5$ V and $\dot{w}(0^-) = 0$.

Solution

Overall response The Laplace transforms of the individual terms in Eq. (6.32) are given by

$$X(s) = L\{x(t)\} = L\{2e^{-t}u(t)\} = \frac{2}{s+1},$$

$$W(s) = L\{w(t)\},$$

$$L\left\{\frac{dw}{dt}\right\} = sW(s) - w(0^-) = sW(s) - 5,$$

and

$$L\left\{\frac{d^2w}{dt^2}\right\} = s^2W(s) - sw(0^-) - \dot{w}(0^-) = s^2W(s) - 5s.$$

Taking the Laplace transform of both sides of Eq. (6.32) and substituting the above values yields

$$[s^2W(s) - 5s] + 7[sW(s) - 5] + 12W(s) = \frac{24}{s+1}$$

or

$$[s^2 + 7s + 12]W(s) = 5s + 35 + \frac{24}{s+1} = \frac{5s^2 + 40s + 59}{s+1},$$

which reduces to

$$W(s) = \frac{5s^2 + 40s + 59}{(s+1)\,(s^2 + 7s + 12)} = \frac{5s^2 + 40s + 59}{(s+1)(s+3)(s+4)}.$$

Taking the partial fraction expansion, we obtain

$$\frac{5s^2 + 40s + 59}{(s+1)(s+3)(s+4)} \equiv \frac{k_1}{(s+1)} + \frac{k_2}{(s+3)} + \frac{k_3}{(s+4)},$$

where the partial fraction coefficients are given by

$$k_1 = \left[(s+1)\frac{5s^2 + 40s + 59}{(s+1)(s+3)(s+4)}\right]_{s=-1} = \frac{5 - 40 + 59}{(2)(3)} = 4,$$

$$k_2 = \left[(s+3)\frac{5s^2 + 40s + 59}{(s+1)(s+3)(s+4)}\right]_{s=-3} = \frac{45 - 120 + 59}{(-2)(1)} = 8,$$

and

$$k_3 = \left[(s+4)\frac{5s^2 + 40s + 59}{(s+1)(s+3)(s+4)}\right]_{s=-4} = \frac{80 - 160 + 59}{(-3)(-1)} = -7.$$

Substituting the values of the partial fraction coefficients k_1, k_2, and k_3, we obtain

$$W(s) \equiv \frac{4}{(s+1)} + \frac{8}{(s+3)} - \frac{7}{(s+4)}.$$

Calculating the inverse Laplace transform of both sides, we obtain the output signal as follows:

$$w(t) \equiv [4e^{-t} + 8e^{-3t} - 7e^{-4t}]u(t).$$

Zero-input response To calculate the zero-input output, the input signal is assumed to be zero. Equation (6.32) reduces to

$$\frac{d^2 w_{zi}}{dt^2} + 7\frac{dw_{zi}}{dt} + 12w_{zi}(t) = 0, \tag{6.33}$$

with initial conditions $w(0^-) = 5$ and $\dot{w}(0^-) = 0$. Calculating the Laplace transform of Eq. 6.33 yields

$$[s^2 W_{zi}(s) - 5s] + 7[s W_{zi}(s) - 5] + 12W_{zi}(s) = 0$$

or

$$W_{zi}(s) = \frac{5s + 35}{s^2 + 7s + 12}.$$

Using the partial fraction expansion, the above equation is expressed as follows:

$$W_{zi}(s) = \frac{5s + 35}{s^2 + 7s + 12} \equiv \frac{20}{s+3} - \frac{15}{s+4}.$$

Taking the inverse Laplace transform, the zero-input response is given by

$$w_{zi}(t) \equiv [20e^{-3t} - 15e^{-4t}]u(t).$$

Zero-state response To calculate the zero-state output, the initial conditions are assumed to be zero, i.e. $w(0^-) = 0$ and $\dot{w}(0^-) = 0$. Taking the Laplace transform Eq. (6.31) and applying zero initial conditions yields

$$s^2 W_{zs}(s) + 7s\, W_{zs}(s) + 12 W_{zs}(s) = \frac{24}{s+1}$$

or

$$W_{zs}(s) = \frac{24}{(s+1)(s^2 + 7s + 12)}.$$

Using the partial fraction expansion, we obtain

$$W_{zs}(s) = \frac{24}{(s+1)(s+3)(s+4)} \equiv \frac{4}{(s+1)} - \frac{12}{(s+3)} + \frac{8}{(s+4)}.$$

Taking the inverse Laplace transform, the zero-state response of the system is given by

$$w_{zs}(t) \equiv [4e^{-t} - 12e^{-3t} + 8e^{-4t}]u(t).$$

The overall, zero-input, and zero-state responses calculated in the Laplace domain are the same as the results computed in Example 3.3 using the time-domain approach.

A direct consequence of solving a linear, constant-coefficient differential equation is the evaluation of the Laplace transfer function $H(s)$ for the LTIC system. The Laplace transfer function is defined as the ratio of the Laplace transform $Y(s)$ of the output signal $y(t)$ to the Laplace transform $X(s)$ of the input signal $x(t)$. Mathematically,

$$H(s) = \frac{Y(s)}{X(s)}, \qquad (6.34)$$

which is obtained by taking the Laplace transform of the differential equation and solving for $H(s)$, as defined in Eq. (6.34). The above procedure provides an algebraic expression for the Laplace transfer function. Its ROC is obtained by observing whether the LTIC is causal or non-causal. Given the algebraic expression and the ROC, the inverse Laplace transform of the Laplace transfer function $H(s)$ leads to the impulse response $h(t)$ of the LTIC system. The Laplace transfer function is also useful for analyzing the stability of the LTIC systems, which is considered in Sections 6.6 and 6.7.

6.6 Characteristic equation, zeros, and poles

In this section, we will define the key concepts related to the stability of LTIC systems. Although these concepts can be applied to general LTIC systems, we will assume a system with a rational transfer function $H(s)$ of the following

For such right-sided functions, it is straightforward to show that the ROC R of its transfer function $H(s)$ will be of the form $\text{Re}\{s\} > \sigma_0$, containing the right side of the s-plane. Consider, for example, the bilateral Laplace transform pairs

$$e^{-at}u(t) \overset{L}{\longleftrightarrow} \frac{1}{s+a} \quad \text{with ROC: } \text{Re}\{s\} > -a$$

and

$$-e^{-at}u(-t) \overset{L}{\longleftrightarrow} \frac{1}{s+a} \quad \text{with ROC: } \text{Re}\{s\} < -a.$$

The first function $e^{-at}u(t)$ is right sided and its ROC: $\text{Re}\{s\} > -a$ occupies the right side of the s-plane. On the other hand, the second function $-e^{-at}u(-t)$ is left sided, and its ROC: $\text{Re}\{s\} < -a$ occupies the left side of the s-plane. Based on the above observations and the Laplace transform pairs listed in Table 6.1, we state the following properties for the ROC.

Property 1 The ROC consists of 2D strips that are parallel to the imaginary $j\omega$-axis.

Property 2 For a right-sided function, the ROC takes the form $\text{Re}\{s\} > \sigma_0$ and consists of the right side of the complex s-plane.

Property 3 For a left-sided function, the ROC takes the form $\text{Re}\{s\} < \sigma_0$ and consists of most of the left side of the complex s-plane.

Property 4 For a finite duration function, the ROC consists of the entire s-plane except for the possible deletion of the point $s = 0$.

Property 5 For a double-sided function, the ROC takes the form $\sigma_1 < \text{Re}\{s\} < \sigma_2$ and is a confined strip within the complex s-plane.

Property 6 The ROC of a rational transfer function does not contain any pole.

Combining Property 6 with the causality constraint ($\text{Re}\{s\} > \sigma_0$) discussed earlier in the section, we obtain the following condition for a causal LTIC system.

Property 7 The ROC R for a *right-sided* LTIC system with the rational transfer function $H(s)$ is given by R: $\text{Re}\{s\} > \text{Re}\{p_r\}$, where p_r is the location of the rightmost pole among the n poles determined using Eq. (6.36).

Since the impulse response of a causal system is a right-sided function, the ROC of a causal system satisfies Property 7. The converse of Property 7 leads to Property 8 for a left-sided sequence.

Property 8 The ROC R for a *left-sided* function with the rational transfer function $H(s)$ is given by R: $\text{Re}\{s\} < \text{Re}\{p_l\}$ where p_l is the leftmost pole among the n poles determined using Eq. (6.36).

To illustrate the application of the properties of the ROC, we consider the following example.

Example 6.19

Consider the LTIC systems in Example 6.18(i) and (ii). Calculate the impulse response if the specified LTIC systems are causal. Repeat for non-causal systems.

Solution

(i) Using the partial fraction expansion, $H_1(s)$ can be expressed as follows:

$$H_1(s) = \frac{(s+4)(s+5)}{s^2(s+2)(s-2)} \equiv \frac{k_1}{s} + \frac{k_2}{s^2} + \frac{k_3}{(s+2)} + \frac{k_4}{(s-2)},$$

where

$$k_1 = \left[\frac{d}{ds}\left(\frac{(s+4)(s+5)}{(s+2)(s-2)}\right)\right]_{s=0} \equiv \left[\frac{2s+9}{s^2-4} - \frac{(s+4)(s+5)}{(s^2-4)^2}2s\right]_{s=0} = -\frac{9}{4},$$

$$k_2 = \left[s^2\frac{(s+4)(s+5)}{s^2(s+2)(s-2)}\right]_{s=0} \equiv \frac{(4)(5)}{2(-2)} = -5,$$

$$k_3 = \left[(s+2)\frac{(s+4)(s+5)}{s^2(s+2)(s-2)}\right]_{s=-2} \equiv \frac{(2)(3)}{4(-4)} = -\frac{3}{8},$$

and

$$k_4 = \left[(s-2)\frac{(s+4)(s+5)}{s^2(s+2)(s-2)}\right]_{s=2} \equiv \frac{(6)(7)}{4(4)} = \frac{21}{8}.$$

Therefore,

$$H_1(s) \equiv -\frac{9}{4s} - \frac{5}{s^2} - \frac{3}{8(s+2)} + \frac{21}{8(s-2)}.$$

If $H_1(s)$ represents a causal LTIC system, then its ROC, based on Property 7, is given by R_c: $\text{Re}\{s\} > 2$. Based on the linearity property, the overall ROC R_c is only possible if the ROCs for the individual terms in $H_1(s)$ are given by

$$H_1(s) = \underbrace{-\frac{9}{4s}}_{\text{ROC:Re}\{s\}>0} - \underbrace{\frac{5}{s^2}}_{\text{ROC:Re}\{s\}>0} - \underbrace{\frac{3}{8(s+2)}}_{\text{ROC:Re}\{s\}>-2} + \underbrace{\frac{21}{8(s-2)}}_{\text{ROC:Re}\{s\}>2}.$$

By calculating the inverse Laplace transform, the impulse response for a causal LTIC system is obtained as follows:

$$h_1(t) = \left[-\frac{9}{4} - 5t - \frac{3}{8}e^{-2t} + \frac{21}{8}e^{2t}\right]u(t).$$

If $H_1(s)$ represents a non-causal system, then its ROC can have three different values: $\text{Re}\{s\} < -2$; $-2 < \text{Re}\{s\} < 0$; or $0 < \text{Re}\{s\} < 2$ in the s-plane. Selecting $\text{Re}\{s\} < -2$ as the ROC will lead to a left-sided signal. The remaining two choices will lead to a double-sided signal. Assuming that we select the overall ROC to be R_{nc}: $\text{Re}\{s\} < -2$, the ROCs for the individual terms in $H_1(s)$ are given by

$$H_1(s) = -\underbrace{\frac{9}{4s}}_{\text{ROC:Re\{s\}<0}} - \underbrace{\frac{5}{s^2}}_{\text{ROC:Re\{s\}<0}} - \underbrace{\frac{3}{8(s+2)}}_{\text{ROC:Re\{s\}<-2}} + \underbrace{\frac{21}{8(s-2)}}_{\text{ROC:Re\{s\}<2}}.$$

Taking the inverse Laplace transform, the impulse response for a non-causal LTIC system is given by

$$h_1(t) = \left[\frac{9}{4} + 5t + \frac{3}{8}e^{-2t} - \frac{21}{8}e^{2t}\right]u(-t).$$

(ii) Using the partial fraction expansion, $H_2(s)$ may be expressed as follows:

$$H_2(s) = \frac{3}{10(s+1)} - \frac{3s-1}{10(s^2+4s+13)}.$$

If $H_2(s)$ represents a causal system, then its ROC is given by R_c: $\text{Re}\{s\} > -1$. The ROCs associated with the individual terms in $H_2(s)$ are given by

$$H_2(s) = \underbrace{\frac{3}{10(s+1)}}_{\text{ROC:Re\{s\}>-1}} - \underbrace{\frac{3s-1}{10(s^2+4s+13)}}_{\text{ROC:Re\{s\}>-2}}.$$

Taking the inverse Laplace transform, the impulse response for a causal LTIC system is given by

$$h_2(t) = \left[\frac{3}{10}e^{-t} - \frac{3}{10}e^{-2t}\cos(3t) + \frac{7}{30}e^{-2t}\sin(3t)\right]u(t).$$

If $H_2(s)$ represents a non-causal system, then several different choices of ROC are possible. One possible choice is given by R_{nc}: $\text{Re}\{s\} < -2$. The ROCs associated with the individual terms in $H_2(s)$ are given by

$$H_2(s) = \underbrace{\frac{3}{10(s+1)}}_{\text{ROC:Re\{s\}<-1}} - \underbrace{\frac{3s-1}{10(s^2+4s+13)}}_{\text{ROC:Re\{s\}<-2}}.$$

Taking the inverse Laplace transform, the impulse response for a causal LTIC system is given by

$$h_2(t) = \left[-\frac{3}{10}e^{-t} + \frac{3}{10}e^{-2t}\cos(3t) - \frac{7}{30}e^{-2t}\sin(3t)\right]u(-t).$$

6.8 Stable and causal LTIC systems

In Section 3.7.3, we showed that the impulse response $h(t)$ of a BIBO stable system satisfies the condition

$$\int_{-\infty}^{\infty} |h(t)|dt < \infty. \tag{6.39}$$

In this section, we derive an equivalent condition to determine the stability of an LTIC system modeled with a rational Laplace transfer function $H(s)$ given in Eq. (6.35). Since we are mostly interested in causal systems, we assume that the Laplace transfer function $H(s)$ corresponds to a right-sided system. The poles of a system with a transfer function as given in Eq. (6.35) can be calculated by solving the characteristic equation, Eq. (6.36). Three types of poles are possible. Out of the n possible poles, assume that there are L poles at $s = 0$, K real poles at $s = -\sigma_k$, $1 \leq k \leq K$, and M pairs of complex-conjugate poles at $s = -\alpha_m \pm j\omega_m$, $1 \leq m \leq M$, such that $L + K + 2M = n$. In terms of its poles, the transfer function, Eq. (6.35), is given by

$$H(s) = \frac{N(s)}{D(s)} = \frac{N(s)}{s^L \prod\limits_{k=1}^{K}(s + \sigma_k) \prod\limits_{m=1}^{M}\left(s^2 + 2\alpha_m s + \left(\alpha_m^2 + \omega_m^2\right)\right)}. \tag{6.40}$$

From Table 6.1, the repeated roots at $s = 0$ correspond to the following term in the time domain:

$$\frac{1}{n!}t^n u(t) \overset{L}{\longleftrightarrow} \frac{1}{s^n}. \tag{6.41}$$

Since term $t^n u(t)$ is unbounded as $t \to \infty$, a stable LTIC system will not contain such unstable terms. Therefore, we assume that $L = 0$. The partial fraction expansion of Eq. (6.40) with $L = 0$ results in the following expression:

$$H(s) = \frac{A_1}{(s + \sigma_1)} + \cdots + \frac{A_K}{(s + \sigma_K)} + \frac{B_1 s + C_1}{\left(s^2 + 2\alpha_1 s + \left(\alpha_1^2 + \omega_1^2\right)\right)} + \cdots$$

$$+ \frac{B_M s + C_M}{\left(s^2 + 2\alpha_M s + \left(\alpha_M^2 + \omega_M^2\right)\right)}, \tag{6.42}$$

where $\{A_k, B_m, C_m\}$ are the partial fraction coefficients. Calculating the inverse Laplace transform of Eq. (6.42) and assuming a causal system, we obtain the following expression for the impulse response $h(t)$ of the LTIC system:

$$h(t) = \sum_{k=1}^{K} \underbrace{A_k e^{-\sigma_k t} u(t)}_{h_k(t)} + \sum_{m=1}^{M} \underbrace{r_m e^{-\alpha_m t} \cos(\omega_m t + \theta_m) u(t)}_{h_m(t)}, \tag{6.43}$$

where we have expressed the terms with conjugate poles in the polar format. Constants $\{r_m, \theta_m\}$ are determined from the values of the partial fraction coefficients $\{B_m, C_m\}$ and α_m.

In Eq. (6.43), we have two types of terms on the right-hand side of the equation. Summation I consists of K real exponential functions of the type $h_k(t) = A_k \exp(-\sigma_k t)u(t)$. Depending upon the value of σ_k, each of these functions $h_k(t)$ may have a constant, decaying exponential or a rising exponential waveform.

Summation II consists of exponentially modulated sinusoidal functions of the type $h_m(t) = r_m \exp(-\alpha_m t) \cos(\omega_m t + \theta_m)u(t)$. The stability characteristic

Fig. 6.9. Nature of the shape of the terms $h_k(t)$ and $h_m(t)$ for different sets of values for σ_k and α_m. For real-valued coefficients b_n in $D(s)$, the complex poles occur as complex-conjugate pairs.

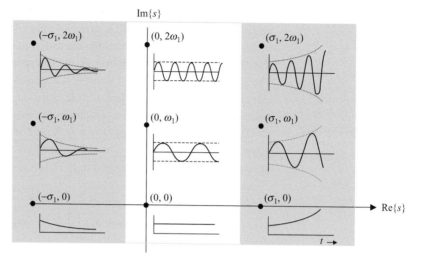

of the functions $h_m(t)$ included in the second summation depends upon the value of α_m. To illustrate the effect of the values of σ_k and α_m on the stability of the LTIC system, Fig. 6.9 plots the shape of the waveforms in the time domain corresponding to terms $h_k(t)$ and $h_m(t)$ for different sets of values for σ_k and α_m. The three plots along the real axis, Re$\{s\}$, at coordinates $(-\sigma_1, 0)$, $(0, 0)$, and $(\sigma_1, 0)$ represent the terms $h_k(t)$ in summation I. For $H(s)$ to correspond to a stable LTIC system, each of the terms in Eq. (6.43) should satisfy the stability condition, Eq. (6.39). Clearly, terms $h_k(t) = A_k \exp(-\sigma_k t)u(t)$ are stable if $\sigma_k > 0$, where terms $h_k(t)$ would correspond to decaying exponential functions. In the three cases plotted along the real axis in Fig. 6.9, this is observed by the impulse response $h_k(t)$ at coordinate $(-\sigma_1, 0)$. In other words, summation I will be stable if the value of σ_k in term $h_k(t) = A_k \exp(-\sigma_k t)u(t)$ is positive. The real roots $s = -\sigma_k$, for $1 \leq k \leq K$, must therefore lie in the left-half s-plane for summation I to be stable.

Similarly, term $h_m(t) = r_m \exp(-\alpha_m t) \cos(\omega_m t + \theta_m)u(t)$ in summation II is stable if $\alpha_m > 0$, where $h_m(t)$ would correspond to a decaying sinusoidal waveform. This is evident from the remaining six coordinates selected in Fig. 6.9. If the value of α_m in term $h_m(t) = r_m \exp(-\alpha_m t) \cos(\omega_m t + \theta_m)u(t)$ is set to a negative value, corresponding to the two impulse responses $h_m(t)$ at coordinates (α_1, ω_1) and $(\alpha_1, 2\omega_1)$, term $h_m(t)$ corresponds to an unstable waveform. Only when the value of α_m is set to be positive, corresponding to the waveforms at coordinates $(-\alpha_1, \omega_1)$ and $(-\alpha_1, 2\omega_1)$, is term $h_m(t)$ stable. This implies that the location of the complex poles $s = -\alpha_m \pm j\omega_m$, $1 \leq m \leq M$, should also lie in the left-half s-plane for the LTIC system to be stable. Based on the above discussion, we state the following conditions for the stability of the LTIC systems with causal implementation for the impulse responses.

Property 9 A causal LTIC system with n poles $\{p_l\}$, $1 \leq l \leq n$, will be absolutely BIBO stable if and only if the real part of all poles are non-zero negative numbers, i.e. if

$$\text{Re}\{p_l\} < 0 \text{ for all } l. \tag{6.44}$$

Equation (6.44) states that a causal LTIC system will be absolutely BIBO stable if and only if all of its poles lie in the left half of the s-plane, (i.e. to the left of the $j\omega$-axis). In other words, a causal LTIC system will be absolutely BIBO stable and causal if the ROC occupies the entire right half of the s-plane including the $j\omega$-axis.

We illustrate the application of the stability condition in Eq. (6.44) in Example 6.20.

Example 6.20

In Example 6.18, we considered the following LTIC systems:

(i) $H_1(s) = \dfrac{(s+4)(s+5)}{s^2(s+2)(s-2)}$;

(ii) $H_2(s) = \dfrac{(s+4)}{s^3 + 5s^2 + 17s + 13}$;

(iii) $H_3(s) = \dfrac{1}{e^s + 10}$.

Assuming that the systems are causal, determine if the systems are BIBO stable.

Solution

(i) The LTIC system with transfer function $H_1(s)$ has four poles located at $s = -2, 0, 0, 2$. Since all the poles do not lie in the left half of the s-plane, the transfer function does not represent an absolutely BIBO stable and causal system. The impulse response of the causal implementation of the LTIC system was calculated in Example 6.19. It can be easily verified that the time-domain stability condition, Eq. (6.39), is not satisfied because of the rising exponential function $21/8 \exp(2t)u(t)$ and the ramp function $5t$, which have infinite areas.

(ii) The LTIC system with transfer function $H_2(s)$ has three poles located at $s = -1, -2 \pm j3$. Since all the poles lie in the left-half s-plane, the transfer function represents an absolutely BIBO stable and causal system. The impulse response of the causal implementation of the LTIC system was calculated in Example 6.19. It can be easily verified that the time-domain stability condition, Eq. (6.39), is satisfied as all terms are decaying exponential functions with finite areas.

(iii) The LTIC system with transfer function $H_3(s)$ has multiple poles located at $s = -2.3 + j(2m + 1)\pi$, for $m = 0, \pm 1, \pm 2, \ldots$ Since all the poles lie in the left-half s-plane, the transfer function represents an absolutely BIBO stable and causal system.

6.8.1 Marginal stability

In our previous discussion, we considered absolutely stable and unstable systems. An absolutely stable system has all the poles in the left half of the complex s-plane. A causal implementation of such a system is stable in the sense that as long as the input is bounded, the system produces a bounded output. On the contrary, an absolutely unstable system has one or more poles in the right half of the complex s-plane. The impulse response of a causal implementation of such a system includes a growing exponential function, making the system unstable. An intermediate case arises when a system has unrepeated poles on the imaginary jω-axis. The remaining poles are in the left half of the complex s-plane. Such a system is referred to as a marginally stable system. The condition for marginally stable system is stated below.

Property 10 An LTIC system, with K unrepeated poles $s_k = j\omega_k$, $1 \leq k \leq K$, on the imaginary jω-axis and all remaining poles in the left-half s-plane, is stable for all bounded input signals that do not include complex exponential terms of the form $\exp(-j\omega_k t)$, for $1 \leq k \leq K$. If the poles on the imaginary jω-axis are repeated, then the LTIC system is unstable.

The following example demonstrates that a marginally stable system becomes unstable if the input signal includes a complex exponential $\exp(-j\omega_0 t)$ with frequency ω_0 corresponding to coordinate $s = j\omega_0$ of the location of the pole of the system on the imaginary jω-axis in the complex s-plane.

Example 6.21
Consider an LTIC system with transfer function

$$H(s) = \frac{25}{s^2 + 25}$$

representing a marginally stable system. Determine the output of the LTIC system for the following inputs:

(i) $x_1(t) = u(t)$;
(ii) $x_2(t) = \sin(5t)u(t)$.

Solution
(i) Taking the Laplace transform of the input gives $X_1(s) = 1/s$. The Laplace transform of the output is given by

$$Y_1(s) = H(s)X_1(s) = \frac{25}{s(s^2 + 25)} \equiv \frac{1}{s} - \frac{s}{(s^2 + 25)}.$$

Taking the inverse Laplace transform gives the following value of the output in the time domain:

$$y_1(t) = (1 - \cos(5t))u(t).$$

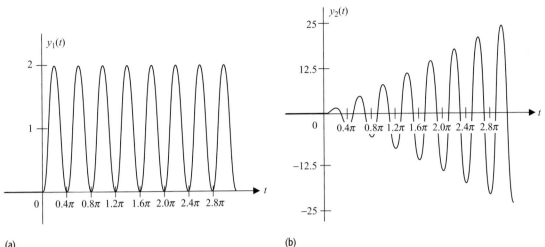

(a)   (b)

Fig. 6.10. Waveforms of the output signals produced by a marginally stable system resulting from (a) $x_1(t) = u(t)$ and (b) $x_2(t) = \sin(5t)u(t)$, as considered in Example 6.21.

As expected for a marginally stable system, the output $y_1(t)$ produced by a bounded input $x_1(t) = u(t)$ in the above expression is bounded for all time t. Figure 6.10(a) plots the bounded output $y_1(t)$ as a function of time t.

(ii) Taking the Laplace transform of the input gives $X_2(s) = 5/s^2 + 25$. The Laplace transform of the output is given by

$$Y_2(s) = H(s)X_2(s) = \frac{125}{(s^2 + 25)^2}.3$$

Using the transform pair

$$\frac{1}{2a^3}(\sin(at) - at\cos(at))u(t) \overset{L}{\longleftrightarrow} \frac{1}{(s^2 + a^2)^2},$$

the output $y_2(t)$ in the time domain is given by

$$y_2(t) = 0.5(\sin(5t) - 5t\cos(5t))u(t).$$

1In part (ii), a sinusoidal signal $\sin(5t) = (\exp(j5t) - \exp(-j5t))/2j$ is applied at the input of a marginally stable system with poles located at $s = \pm j5$ on the imaginary $j\omega$-axis. Note that the fundamental frequency ($\omega_0 = 5$) of the sinusoidal input is the same as the location ($s = \pm j5$) of the poles in the complex s-plane. In such cases, Property 6.10 states that the resulting output $y_2(t)$ will be unbounded. The second term $-5t\cos(5t)u(t)$ indeed makes the output unbounded. This is illustrated in Fig. 6.10(b), where $y_2(t)$ is plotted as a function of time t.

6.8.2 Improving stability using zeros

To conclude our discussion on stability, let us consider an LTIC system with transfer function given by

$$H_{ap}(s) = \frac{(s - a - jb)}{(s + a - jb)}, \tag{6.45}$$

Fig. 6.11. Locations of poles
"×" and zeros "o" of an allpass
system in the complex s-plane.
The ROC of the causal
implementation of the allpass
system is highlighted by the
shaded region.

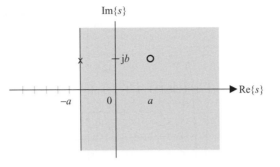

Fig. 6.11. Locations of poles
"×" and zeros "o" of an allpass
system in the complex s-plane.
The ROC of the causal
implementation of the allpass
system is highlighted by the
shaded region.

having a pole at $s = (-a + jb)$ and a zero at $s = (a + jb)$. As shown in
Fig. 6.11, the locations of the pole and zero are symmetric about the
imaginary jω-axis in the complex s-plane. Clearly, a causal implementation of the transfer function $H(s)$ of the system will be stable as its ROC:
Re$\{s\} > -a$ includes the imaginary jω-axis. The CTFT of the LTIC system is
evaluated as

$$H_{ap}(j\omega) = H_{ap}(s)|_{s=j\omega} = \frac{(j\omega - a - jb)}{(j\omega + a - jb)}, \qquad (6.46)$$

with the CTFT spectra as follows:

$$\text{magnitude spectrum} \quad |H_{ap}(j\omega)| = \frac{\sqrt{(-a)^2 + (\omega - b)^2}}{\sqrt{(a)^2 + (\omega - b)^2}} = 1; \qquad (6.47)$$

$$\text{phase spectrum} \quad <H_{ap}(j\omega) = \tan^{-1}\left(\frac{\omega - b}{-a}\right) - \tan^{-1}\left(\frac{\omega - b}{a}\right). \qquad (6.48)$$

Such a system is referred to as an *allpass system*, since it allows all frequencies
present in the input signal to pass through the system without any attenuation.
Of course, the phase of the input signal is affected, but in most applications we
are more concerned about the magnitude of the signal.

An allpass system specified in Eq. (6.45) is frequently used to stabilize an
unstable system. Consider an LTIC system with the transfer function

$$H(s) = \frac{H_1(s)}{(s - a - jb)}, \qquad (6.49)$$

where the component $H_1(s)$ is assumed to have all poles in the left half of
the s-plane and is, therefore, stable. A causal implementation of the transfer
function $H(s)$ is unstable because of the existence of the term $(s - a - jb)$
in the denominator. This term results in a pole at $s = (a + jb)$ and introduces
instability into the system. Such a system can be made stable by cascading it

with an allpass system that has a zero at the location of the unstable pole. The transfer function of the overall cascaded system is given by

$$H_{\text{overall}}(s) = H(s)H_{\text{ap}}(s) = \frac{H_1(s)}{(s + a - jb)}, \qquad (6.50)$$

which is stable because the unstable pole at $s = (a + jb)$ is canceled by the zero of the allpass system. The new pole at $s = (-a + jb)$ lies in the left-half s-plane and satisfies the stability requirements. Note that the magnitude response of the overall cascaded system is the product of the magnitude responses of the unstable and allpass systems, and is given by

$$|H_{\text{overall}}(j\omega)| = |H(j\omega)||H_{\text{ap}}(j\omega)| = |H(j\omega)|, \qquad (6.51)$$

since $|H_{ap}(j\omega)| = 1$. Hence, by cascading an unstable system with an allpass system, which has a zero at the location of the unstable pole, we have stabilized the system without affecting its magnitude response. The only change in the system is in its phase. Such a *pole–zero cancellation* approach is frequently used in applications where information is contained in the magnitude of the signal and the phase is relatively unimportant. One such application is the amplitude modulation system described in Section 2.1.3, which is used for radio communications.

6.9 LTIC systems analysis using Laplace transform

In Section 6.4.7, we showed that the output response of an LTIC system could be computed using the convolution property in the complex s-plane. This eliminates the need to compute the computationally intense convolution integral in the time domain. Below, we provide another example for calculating the output using the Laplace transform. Our motivation in reintroducing this topic is to compare the Laplace-transform-based analysis technique with the CTFT-based approach.

Example 6.22
In Example 5.26, we determined the overall and steady state values of the output of the RC series circuit with the CTFT transfer function

$$H(\omega) = \frac{1/j\omega C}{R + 1/j\omega C} = \frac{1}{1 + j\omega CR}$$

and constant $CR = 0.5$ for the input signal $x(t) = \sin(3t)u(t)$. For simplicity, we assumed that the capacitor is uncharged at $t = 0$. Here we solve the problem in Example 5.26 using the Laplace transform.

Solution

The Laplace transform of the input signal $x(t)$ is given by

$$X(s) = L\{\sin(3t)u(t)\} = \frac{3}{s^2 + 9}.$$

The Laplace transfer function of the RC series circuit is given by

$$H(s) = H(\omega)|_{j\omega=s} = \frac{1}{1 + sCR}.$$

Substituting the value of the product $CR = 0.5$ yields

$$H(s) = \frac{1}{1 + 0.5\,s} = \frac{2}{s + 2}.$$

The Laplace transform $Y(s)$ of the output signal is given by

$$Y(s) = H(s)X(s) = \frac{6}{(s + 2)(s^2 + 9)} \equiv \frac{6}{13(s + 2)} - \frac{6s - 12}{13(s^2 + 9)}$$

or

$$Y(s) = \frac{6}{13(s + 2)} - \frac{6}{13}\frac{s}{(s^2 + 9)} + \frac{4}{13}\frac{3}{(s^2 + 9)}.$$

Taking the inverse transform leads to the following expression for the overall output in the time domain:

$$y(t) = \left[\frac{6}{13}e^{-2t} - \frac{6}{13}\cos(3t) + \frac{4}{13}\sin(3t)\right]u(t)$$

$$= \left[\frac{6}{13}e^{-2t} + \frac{2}{\sqrt{13}}\sin(3t - 56°)\right]u(t).$$

The steady state value of the output is computed by applying the limit $t \to \infty$ to the overall output:

$$y_{ss}(t) = \lim_{t \leftarrow \infty} y(t) = \frac{2}{\sqrt{13}}\sin(3t - 56°)u(t).$$

In Chapters 5 and 6, we presented two frequency-domain approaches to analyze CT signals and systems. The CTFT-based approach introduced in Chapter 5 uses the real frequency ω, whereas the Laplace-transform-based approach uses the complex frequency σ. Both approaches have advantages. Depending upon the application under consideration, the appropriate transform is selected.

Comparing Example 6.22 with Example 5.26, the Laplace transform appears to be a more convenient tool for the transient analysis. For the steady state analysis, the Laplace transform does not seem to offer any advantage over the CTFT. The transient analysis is very important for applications in control systems, including process control and guided missiles. In signal processing applications, such as audio, image, and video processing, the transients are generally ignored. In such applications, the CTFT is sufficient to analyze the steady state response. This is precisely why most signal processing literature uses the CTFT, while the control systems literature uses the Laplace transform. Important applications of the Laplace transforms such as analysis of the spring damper system and the modeling of the human immune system are presented in Chapter 8.

6.10 Block diagram representations

In the preceding discussion, we considered relatively elementary LTIC systems described by linear, constant-coefficient differential equations. Most practical structures are more complex, consisting of a combination of several LTIC systems. In this section, we analyze the cascaded, parallel, and feedback configurations used to synthesize larger systems.

6.10.1 Cascaded configuration

A series or cascaded configuration between two systems is illustrated in Fig. 6.12(a). The output of the first system $H_1(s)$ is applied as input to the second system $H_2(s)$. Assuming that the Laplace transform of the input $x(t)$, applied to the first system, is given by $X(s)$, the Laplace transform $W(s)$ of the output $w(t)$ of the first system is given by

$$w(t) = x(t) * h_1(t) \overset{\text{L}}{\longleftrightarrow} W(s) = X(s)H_1(s). \qquad (6.52)$$

The resulting signal $w(t)$ is applied as input to the second system $H_2(s)$, which leads to the following overall output:

$$y(t) = w(t) * h_2(t) \overset{\text{L}}{\longleftrightarrow} Y(s) = W(s)H_2(s). \qquad (6.53)$$

Substituting the value of $w(t)$ from Eq. (6.52), Eq. (6.53) reduces to

$$y(t) = x(t) * h_1(t) * h_2(t) \overset{\text{L}}{\longleftrightarrow} Y(s) = W(s)H_1(s)H_2(s). \qquad (6.54)$$

In other words, the cascaded configuration is equivalent to a single LTIC system with transfer function

Fig. 6.12. Cascaded configuration for connecting LTIC systems: (a) cascaded connection; (b) its equivalent single system.

$$h(t) = h_1(t) * h_2(t) \overset{\text{L}}{\longleftrightarrow} H(s) = H_1(s)H_2(s). \qquad (6.55)$$

The system $H(s)$ equivalent to the cascaded configuration is shown in Fig. 6.12(b).

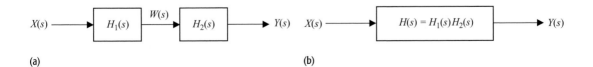

(a)   (b)

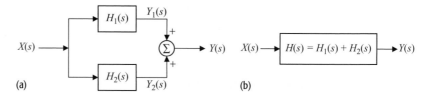

Fig. 6.13. Parallel configuration for connecting LTIC systems: (a) parallel connection; (b) its equivalent single system.

6.10.2 Parallel configuration

The parallel configuration between two systems is illustrated in Fig. 6.13(a). A single input $x(t)$ is applied simultaneously to the two systems. The overall output $y(t)$ is obtained by adding the individual outputs $y_1(t)$ and $y_2(t)$ of the two systems. The individual outputs of the two systems are given by

$$\text{system (1)} \quad y_1(t) = x(t) * h_1(t) \xleftrightarrow{\text{L}} Y_1(s) = X(s)H_1(s); \quad (6.56)$$

$$\text{system (2)} \quad y_2(t) = x(t) * h_2(t) \xleftrightarrow{\text{L}} Y_2(s) = X(s)H_2(s). \quad (6.57)$$

Combining the two outputs, the overall output $y(t)$ is given by

$$y(t) = y_1(t) + y_2(t) \xleftrightarrow{\text{L}} Y(s) = Y_1(s) + Y_2(s). \quad (6.58)$$

Substituting Eqs. (6.56) and (6.57) into the above equation yields

$$y(t) = x(t) * [h_1(t) + h_2(t)] \xleftrightarrow{\text{L}} Y(s) = X(s)[H_1(s) + H_2(s)]. \quad (6.59)$$

In other words, the parallel configuration is equivalent to a single LTIC system with transfer function

$$h(t) = h_1(t) + h_2(t) \xleftrightarrow{\text{L}} H(s) = H_1(s) + H_2(s). \quad (6.60)$$

The parallel configuration and its equivalent single-stage system are shown in Fig. 6.13.

6.10.3 Feedback configuration

The feedback connection between two systems is shown in Fig. 6.14(a). In a feedback system, the overall output $y(t)$ is applied at the input of the second system $H_2(s)$. The output $w(t)$ of the second system is fed back into the input of the overall system through an adder. In terms of the applied input $x(t)$ and $w(t)$, the output of the adder is given by

$$E(s) = X(s) - W(s). \quad (6.61)$$

The outputs of the two LTIC systems are given by

$$\text{system (1)} \qquad Y(s) = E(s)H_1(s); \quad (6.62)$$

$$\text{system (2)} \qquad W(s) = Y(s)H_2(s). \quad (6.63)$$

Fig. 6.14. Feedback
configuration for connecting
LTIC systems: (a) feedback
connection; (b) its equivalent
single system.

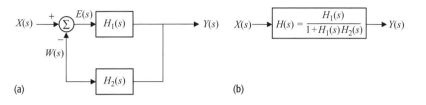

Substituting the value of $E(s) = Y(s)/H_1(s)$ from Eq. (6.62) and $W(s)$ from Eq. (6.63) into Eq. (6.61) yields

$$Y(s) = H_1(s)[X(s) - H_2(s)Y(s)]. \qquad (6.64)$$

Rearranging terms containing $Y(s)$, we obtain

$$[1 + H_1(s)H_2(s)]Y(s) = H_1(s)X(s),$$

which leads to the following transfer function for the feedback system:

$$H(s) = \frac{Y(s)}{X(s)} = \frac{H_1(s)}{1 + H_1(s)H_2(s)}. \qquad (6.65)$$

The feedback configuration and its equivalent single system are shown in Fig. 6.14.

Example 6.23

Determine (i) the impulse response and (ii) the transfer function of the interconnected systems shown in Figs. 6.15(a)–(c).

Solution

(a) To calculate the overall impulse response, we proceed in the Laplace domain. The transfer function $H_1(s)$ of the cascaded systems shown in the lower branch of the system in Fig.6.15(a) is given by

$$H_1(s) = L\{\delta(t - 1)\}H(s) = e^{-s}H(s).$$

The overall transfer function $H_a(s)$ is therefore given by

$$H_a(s) = H(s) + H_1(s) = (1 + e^{-s})H(s).$$

Taking the inverse of the above transfer function gives the impulse response:

$$h_a(t) = h(t) + h(t - 1).$$

(b) The system in Fig. 6.15(b) is the feedback configuration with transfer functions $H_1(s) = 1$ and $H_2(s) = L\{\alpha\delta(t - T)\} = \alpha e^{-Ts}$. Substituting the values of $H_1(s)$ and $H_2(s)$ into Eq. (6.65) yields

$$H_b(s) = \frac{1}{1 + \alpha e^{-Ts}}.$$

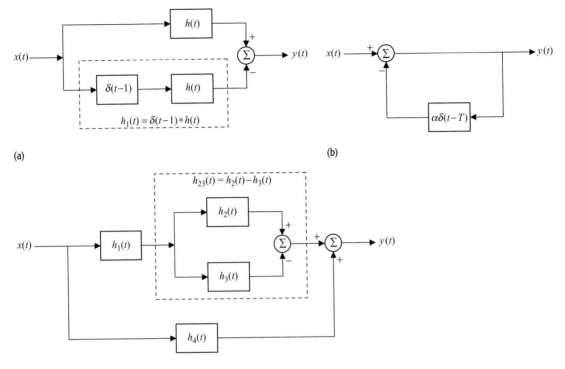

(a)

(b)

(c)

Fig. 6.15. Interconnections between LTIC systems. Parts (a)–(c) correspond to parts (a)–(c) of Example 6.23.

Since $H_b(s)$ is not a rational function of s, the inverse Laplace transform is evaluated from the definition in Eq. (6.7), which involves contour integration.

(c) The transfer function of the parallel configuration shown in the dashed box is given by

$$H_{23}(s) = H_2(s) - H_3(s).$$

In terms of $H_{23}(s)$, the transfer function $H_{123}(s)$ of the top path is given by

$$H_{123}(s) = H_1(s)H_{23}(s).$$

Substituting the value of $H_{23}(s)$, the above expression reduces to

$$H_{123}(s) = H_1(s)[H_2(s) - H_3(s)].$$

The overall transfer function of the system in Fig. 6.15(c) is given by

$$H_c(s) = H_{123}(s) + H_4(s)$$

or

$$H_c(s) = H_1(s)[H_2(s) - H_3(s)] + H_4(s).$$

Taking the inverse Laplace transform of the above equation leads to the follow-ing expression for the overall impulse response:

$$h_c(t) = h_1(t) * h_2(t) - h_1(t) * h_3(t) + h_4(t).$$

6.11 Summary

In this chapter, we introduced the bilateral and unilateral Laplace transforms used for the analysis of LTIC signals and systems. The Laplace transforms are a generalization of the CTFT, where the independent Laplace variable, $s = \sigma + j\omega$, can take any value in the complex s-plane and is not simply restricted to the $j\omega$-axis, as is the case for the CTFT. The values of s for which the Laplace transforms converge constitute the region of convergence (ROC) of the Laplace transforms. In Section 6.2, we derived the unilateral Laplace transforms and the associated ROCs for a number of elementary CT signals; these transform pairs are listed in Table 6.1. Direct computation of the inverse Laplace transform involves contour integration, which is difficult to compute analytically. For Laplace transforms, which take a rational form, the inverse can be easily determined using the partial fraction approach cov-ered in Section 6.3. The properties of the Laplace transform are covered in Section 6.4 and listed in Table 6.2. In particular, we covered the linearity, scaling, shifting, differentiation, integration, and convolution properties, as summarized below.

(1) The linearity property implies that the Laplace transform of a linear com-bination of signals is obtained by taking the same linear combination in the complex s-domain. In other words,

$$a_1 x_1(t) + a_2 x_2(t) \overset{L}{\longleftrightarrow} a_1 X_1(s) + a_2 X_2(s) \quad \text{with ROC: at least } R_1 \cap R_2.$$

(2) Scaling a signal by a factor of a in the time domain is equivalent to scaling its Laplace transform by a factor of $1/a$ in the s-domain; i.e.

$$x(at) \overset{L}{\longleftrightarrow} \frac{1}{|a|} X\left(\frac{s}{a}\right) \quad \text{with ROC: } aR.$$

(3) Shifting a signal in the time domain is equivalent to multiplication by a complex exponential in the s-domain. Mathematically, the time-shifting property is expressed as follows:

$$x(t - t_0) \overset{L}{\longleftrightarrow} e^{-st_0} X(s) \quad \text{with ROC: } R.$$

(4) The converse of the time-shifting property is also true. In other words, shifting a signal in the s-domain is equivalent to multiplication by a complex exponential in the time domain:

$$e^{s_0 t} x(t) \overset{L}{\longleftrightarrow} X(s - s_0) \quad \text{with ROC: } R + \text{Re}\{s_0\}.$$

(5) Differentiation in the time domain is equivalent to multiplication by s in the complex s-domain. This is referred to as the time-differentiation property and is expressed as follows:

$$\frac{dx}{dt} \overset{L}{\longleftrightarrow} sX(s) - x(0^-) \quad \text{with ROC: } R.$$

(6) Integration in the time domain is equivalent to division by s in the complex s-domain. This is referred to as the time-integration property and is expressed as follows:

$$\text{unilateral Laplace transform} \quad \int_{0^-}^{t} x(\tau)d\tau \overset{L}{\longleftrightarrow} \frac{X(s)}{s}$$

$$\text{with ROC: } R \cap \text{Re}\{s\} > 0;$$

$$\text{bilateral Laplace transform} \quad \int_{-\infty}^{t} x(\tau)d\tau \overset{L}{\longleftrightarrow} \frac{X(s)}{s} + \frac{1}{s}\int_{-\infty}^{0^-} x(\tau)d\tau$$

$$\text{with ROC: } R \cap \text{Re}\{s\} > 0.$$

(7) The convolution property states that convolution in the time domain is equivalent to multiplication in the s-domain, and vice versa. Mathematically, the convolution property is stated as follows:

$$\text{time convolution} \quad x_1(t) * x_2(t) \overset{L}{\longleftrightarrow} X_1(s)X_2(s)$$

$$\text{containing at least ROC: } R_1 \cap R_2;$$

$$\text{s-plane convolution} \quad x_1(t)x_2(t) \overset{L}{\longleftrightarrow} \frac{1}{2\pi j}[X_1(s) * X_2(s)]$$

$$\text{containing at least ROC: } R_1 \cap R_2.$$

(8) The initial- and final-value theorems provide us with an alternative approach for calculating the limits of a CT function $x(t)$ as $t \to 0$ and $t \to \infty$ from the following expressions:

$$\text{initial-value theorem} \quad x(0^+) = \lim_{t \to 0^+} x(t) = \lim_{s \to \infty} sX(s)$$

$$\text{provided } x(0^+) \text{ exists;}$$

$$\text{final-value theorem} \quad x(\infty) = \lim_{t \to \infty} x(t) = \lim_{s \to 0} sX(s)$$

$$\text{provided } x(\infty) \text{ exists.}$$

The initial-value theorem is valid for the unilateral Laplace transform, while the final-value theorem is valid for both unilateral and bilateral transforms.

Sections 6.5 to 6.9 discussed various applications of the Laplace transform. The time-differentiation property is used in Section 6.5 to solve linear, constant-coefficient differential equations. Section 6.6 uses the properties of the ROC associated with the Laplace transform with an emphasis on causal systems. Sections 6.7 and 6.8 define the stability of the causal LTIC systems in terms of the poles and zeros of its transfer function. The key points are summarized below.

(1) The causal implementation of an absolutely BIBO stable system must have all of its poles in the left half of the complex s-plane.

(2) If even a single pole lies in the right half of the s-plane, the causal implementation of the system is unstable.

(3) If no pole lies in the right half of the s-plane, but one or more first-order poles lie on the imaginary $j\omega$-axis, the LTIC system is referred to as a marginally stable system.

(4) An unstable system may be transformed into a stable system by cascading the unstable system with an allpass system, which has zeros at the locations of the unstable poles.

Section 6.9 described an analysis technique based on the Laplace transform to calculate the output of an LTIC system. We showed that the Laplace-transform-based analysis approach is suitable for studying the transient response of the systems. The CTFT-based approach is appropriate for analyzing the steady state response of the system.

Finally, Section 6.10 discussed the cascaded, parallel, and feedback configurations used to interconnect two LTIC systems. If two systems with impulse responses $h_1(t)$ and $h_2(t)$ are connected, the overall impulse response and the corresponding transfer functions are as follows:

cascaded configuration $\qquad h(t) = h_1(t) * h_2(t) \overset{L}{\longleftrightarrow} H(s) = H_1(s)H_2(s);$

parallel configuration $\qquad h(t) = h_1(t) + h_2(t) \overset{L}{\longleftrightarrow} H(s) = H_1(s) + H_2(s);$

feedback configuration $\qquad H(s) = \dfrac{H_1(s)}{1 + H_1(s)H_2(s)}.$

A practical system comprises multiple LTIC systems interconnected with a combination of cascaded, parallel, and feedback configurations.

Problems

6.1 Using the definition in Eq. (6.5), calculate the bilateral Laplace transform and the associated ROC for the following CT functions:

(a) $x(t) = e^{-5t}u(t) + e^{4t}u(-t);$ $\qquad$ (d) $x(t) = e^{-3|t|}\cos(5t);$

(b) $x(t) = e^{-3|t|};$ $\qquad$ (e) $x(t) = e^{7t}\cos(9t)u(-t);$

(c) $x(t) = t^2\cos(10t)u(-t);$ $\qquad$ (f) $x(t) = \begin{cases} 1 - |t| & 0 \leq |t| \leq 1 \\ 0 & \text{otherwise.} \end{cases}$

6.2 Using Eq. (6.9), calculate the unilateral Laplace transform and the associated ROC for the following CT functions:

(a) $x(t) = t^5 u(t);$ $\qquad$ (d) $x(t) = e^{-3t}\cos(9t)u(t);$

(b) $x(t) = \sin(6t)u(t);$ $\qquad$ (e) $x(t) = t^2\cos(10t)u(t);$

(c) $x(t) = \cos^2(6t)u(t);$ $\qquad$ (f) $x(t) = \begin{cases} 1 - |t| & 0 \leq t \leq 1 \\ 0 & \text{otherwise.} \end{cases}$

6.3 Using the partial fraction expansion approach, calculate the inverse Laplace transform for the following rational functions of s:

(a) $X(s) = \dfrac{s^2 + 2s + 1}{(s+1)(s^2 + 5s + 6)}$;   ROC : Re$\{s\} > -1$;

(b) $X(s) = \dfrac{s^2 + 2s + 1}{(s+1)(s^2 + 5s + 6)}$;   ROC : Re$\{s\} < -3$;

(c) $X(s) = \dfrac{s^2 + 3s - 4}{(s+1)(s^2 + 5s + 6)}$;   ROC : Re$\{s\} > -1$;

(d) $X(s) = \dfrac{s^2 + 3s - 4}{(s+1)(s^2 + 5s + 6)}$;   ROC : Re$\{s\} < -3$;

(e) $X(s) = \dfrac{s^2 + 1}{s(s+1)(s^2 + 2s + 17)}$;   ROC : Re$\{s\} > 0$;

(f) $X(s) = \dfrac{s + 1}{(s+2)^2(s^2 + 7s + 12)}$;   ROC : Re$\{s\} > -2$;

(g) $X(s) = \dfrac{s^2 - 2s + 1}{(s+1)^3(s^2 + 16)}$;   ROC : Re$\{s\} < -1$.

6.4 The Laplace transforms of two CT signals $x_1(t)$ and $x_2(t)$ are given by the following expressions:

$$x_1(t) \overset{L}{\longleftrightarrow} \frac{s}{s^2 + 5s + 6} \quad \text{with ROC}(R_1) : \text{Re}\{s\} > -2$$

and

$$x_2(t) \overset{L}{\longleftrightarrow} \frac{1}{s^2 + 5s + 6} \quad \text{with ROC}(R_2) : \text{Re}\{s\} > -2.$$

Determine the Laplace transform and the associated ROC R of the combined signal $x_1(t) + 2x_2(t)$. Explain how the ROC R of the combined signal exceeds the intersection $(R_1 \cap R_2)$ of the individual ROCs R_1 and R_2.

6.5 Calculate the time-domain representation of the bilateral Laplace transform

$$X(s) = \frac{s^2}{(s^2 - 1)(s^2 - 4s + 5)(s^2 + 4s + 5)}$$

if the ROC R is specified as follows:

(a) R : Re$\{s\} < -2$;   (d) R : $1 <$ Re$\{s\} < 2$;

(b) R : $-2 <$ Re$\{s\} < -1$;   (e) R : Re$\{s\} > 2$.

(c) R : $-1 <$ Re$\{s\} < 1$;

6.6 Prove the frequency-shifting property, Eq. (6.20), as stated in Section 6.4.4.

6.7 Prove the time-integration property for the unilateral and bilateral Laplace transform as stated in Section 6.4.6.

6.8 Prove the initial-value theorem, Eq. (6.27), as stated in Section 6.4.8.

6.9 Prove the final-value theorem, Eq. (6.28), as stated in Section 6.4.8.

6.10 Using the transform pairs in Table 6.1 and the properties of the Laplace transform, prove the following Laplace transform pairs:

(a) $t\cos(\omega_0 t)u(t) \xleftrightarrow{\text{L}} \dfrac{s^2 - \omega_0^2}{(s^2 + a^2)^2}$;

(b) $t\sin(\omega_0 t)u(t) \xleftrightarrow{\text{L}} \dfrac{2\omega_0 s}{(s^2 + a^2)^2}$;

(c) $\dfrac{1}{2a^3}(\sin(at) - at\cos(at))u(t) \xleftrightarrow{\text{L}} \dfrac{1}{(s^2 + a^2)^2}$.

6.11 Express the Laplace transform and the associated ROC for the following functions in terms of the Laplace transform $X(s)$ with ROC R_x of the CT function $x(t)$:

(a) $\cos(10t)x(t)$;

(b) $e^{-5t}x(4t - 3)$;

(c) $(t - 4)^4 \dfrac{\mathrm{d}}{\mathrm{d}t}[x(t - 4)]$;

(d) $[x(t) + 2]^2$;

(e) $\displaystyle\int_{-\infty}^{t} e^{-\alpha s_0} x(\alpha)\,\mathrm{d}\alpha$.

6.12 Using the initial- and final-value theorems, calculate the initial and final values of the causal CT functions with the following unilateral Laplace transforms. In each case, first determine the ROC to see if the initial value exists.

(a) $X(s) = \dfrac{s}{s^2 + 7s + 1}$;

(b) $X(s) = \dfrac{s}{s^2 + 5s - 4}$;

(c) $X(s) = \dfrac{s^2 + 9}{s^2 - 25}$;

(d) $X(s) = \dfrac{s^2 + 2s + 1}{s^2 + 3s + 4}$;

(e) $X(s) = e^{-5s}\dfrac{s^2 + 4}{s(s + 1)(s + 2)(s + 3)}$.

6.13 Solve the following initial-value differential equations using the Laplace transform method:

(a) $\dfrac{\mathrm{d}^2 y}{\mathrm{d}t^2} + 3\dfrac{\mathrm{d}y}{\mathrm{d}t} + 2y(t) = \delta(t);\quad y(0^-) = \dot{y}(0^-) = 0$;

(b) $\dfrac{\mathrm{d}^2 y}{\mathrm{d}t^2} + 4\dfrac{\mathrm{d}y}{\mathrm{d}t} + 4y(t) = u(t);\quad y(0^-) = \dot{y}(0^-) = 0$;

(c) $\dfrac{\mathrm{d}^2 y}{\mathrm{d}t^2} + 6\dfrac{\mathrm{d}y}{\mathrm{d}t} + 8y(t) = te^{-3t}u(t);\quad y(0^-) = \dot{y}(0^-) = 1$;

(d) $\dfrac{\mathrm{d}^3 y}{\mathrm{d}t^3} + 8\dfrac{\mathrm{d}^2 y}{\mathrm{d}t^2} + 19\dfrac{\mathrm{d}y}{\mathrm{d}t} + 12y(t) = tu(t)$;

$y(0^-) = 1; \dot{y}(0^-) = \ddot{y}(0^-) = 0$;

(e) $\dfrac{\mathrm{d}^4 y}{\mathrm{d}t^4} + 2\dfrac{\mathrm{d}^2 y}{\mathrm{d}t^2} + y(t) = u(t);\quad y(0^-) = \dot{y}(0^-) = \ddot{y}(0^-) = \dddot{y}(0^-) = 0$.

6.14 Determine (i) the Laplace transfer function, (ii) the impulse response function, and (iii) the input–output relationship (in the form of a linear constant-coefficient differential equation) for the causal LTIC systems

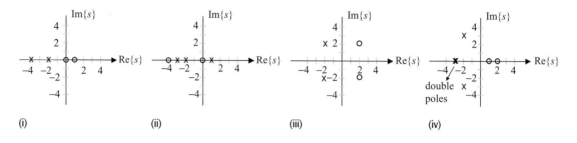

(i)　　　　　　　　(ii)　　　　　　　　(iii)　　　　　　　　(iv)

Fig. P6.17. Pole – zero plots for Problem 6.17.

with the following input–output pairs:

(a) $x(t) = 4u(t)$　　　and　　$y(t) = tu(t) + e^{-2t}u(t)$;

(b) $x(t) = e^{-2t}u(t)$　and　$y(t) = 3e^{-2(t-4)}u(t-4)$;

(c) $x(t) = tu(t)$　　　and　　$y(t) = [t^2 - 3e^{-4t}]u(t)$;

(d) $x(t) = e^{-2t}u(t)$　and　$y(t) = e^{-t}u(t) + e^{-3t}u(t)$;

(e) $x(t) = e^{-3t}u(t)$　and　$y(t) = e^{t}u(-t) + e^{-3t}u(t)$.

6.15 Sketch the location of the poles and zeros for the following transfer functions, and determine if the corresponding causal systems are stable, unstable, or marginally stable:

(a) $H(s) = \dfrac{s^2 + 1}{s^2 + 2s + 1}$;

(b) $H(s) = \dfrac{2s + 5}{s^2 + s - 6}$;

(c) $H(s) = \dfrac{3s + 10}{s^2 + 9s + 18}$;

(d) $H(s) = \dfrac{s + 2}{s^2 + 9}$;

(e) $H(s) = \dfrac{s^2 + 3s + 2}{s^3 + 3s^2 + 2s}$.

6.16 Without explicitly calculating the output, determine if the LTIC system with the transfer function

$$H(s) = \frac{s^2 + 1}{(s + 5)(s^2 + 4)(s^2 + 9)(s^2 + 4s + 5)}$$

produces a bounded output for the following set of inputs:

(a) $x(t) = e^{-j2t}u(t)$;

(b) $x(t) = [e^{-(1+j4)t} + e^{-(2+j5)t}]u(t)$;

(c) $x(t) = [\cos(t) + \sin(4t)]u(t)$;

(d) $x(t) = [\cos(2t) + \sin(3t)]u(t)$;

(e) $x(t) = [e^{-(1+j2)t}\sin(3t)]u(t)$.

6.17 The pole–zero plots of four causal LTIC systems are shown in Fig. P6.17. Determine if the LTIC systems are stable. Also determine the transfer function $H(s)$ for each system. Assume that $H(4) = 1$ in all cases, and the poles and zeros are all located at integer coordinates in the s-plane.

6.18 Determine the transfer functions of all possible non-causal implementations of the LTIC systems considered in Fig. P6.17. Specify which transfer functions represent stable systems.

6.19 The inverse of an LTIC system is defined as the system that when cascaded with the original system results in an overall transfer function of unity. Without calculating the transfer functions, determine the pole–zero plots of the inverse systems associated with the LTIC systems whose pole–zero plots are specified in Fig. P6.17.

6.20 An LTIC system has an impulse response $h(t)$ with the Laplace transfer function $H(s)$, which satisfies the following properties:
(a) the impulse response $h(t)$ is even and real-valued;

(b) the area enclosed by the impulse response is 8, i.e. $\int\limits_{-\infty}^{\infty} h(t)\mathrm{d}t = 8$;

(c) the Laplace transfer function $H(s)$ has four poles but no zeros;
(d) the Laplace transfer function $H(s)$ has a complex pole at $s = 0.5\exp(\mathrm{j}\pi/4)$.
Determine the Laplace transfer function $H(s)$ and the associated ROC.

6.21 Consider the RLC series circuit shown in Fig. 3.1. The relationship between the input voltage $x(t)$ and the output voltage $w(t)$ is given by the following differential equation:
$$\frac{\mathrm{d}^2 w}{\mathrm{d}t^2} + \frac{R}{L}\frac{\mathrm{d}w}{\mathrm{d}t} + \frac{1}{L}w(t) = \frac{1}{LC}x(t).$$
By determining the locations of the poles of the transfer function describing the RLC series circuit, show that the causal implementation of the RLC circuit is always stable for positive values ($R > 0$, $L > 0$, and $C > 0$) of the passive components.

6.22 Given the transfer function
$$H(s) = \frac{s^2 - s - 6}{(s^2 + 3s + 1)(s^2 + 7s + 12)}$$
(a) determine all possible choices for the ROC;
(b) determine the impulse response of a causal implementation of the transfer function $H(s)$;
(c) determine the left-sided impulse response with the specified transfer function $H(s)$;
(d) determine all possible choices of double-sided impulse responses having the specified transfer function $H(s)$.
(e) Which of the four impulse responses obtained in (b)–(d) are stable?

6.23 Repeat Problem 6.22 for the following transfer function:
$$H(s) = \frac{s^2 - 5s - 84}{(s^2 - 2s - 35)(s^2 + 9s + 20)}.$$

6.24 For most practical applications, we are interested in implementing a causal and stable system. The causal implementations of some of the transfer functions specified in Problem 6.15 are not stable. For each such transfer function, specify an allpass system that may be cascaded in a series

Fig. P6.25. Interconnected systems specified in Problem 6.25.

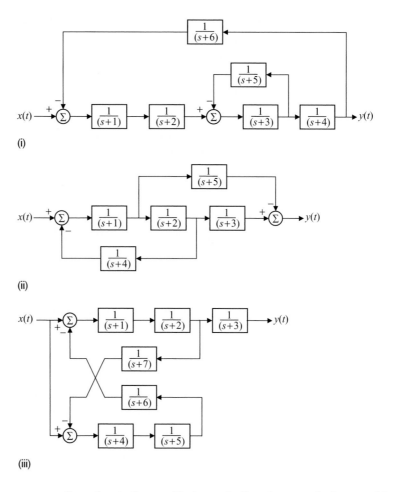

(i)

(ii)

(iii)

configuration to the specified transfer function to make its causal implementation stable.

6.25 Determine the overall transfer function for the three interconnected systems shown in Fig. P6.25.

6.26 Using the function residue available in MATLAB toolboxes, calculate the partial fraction coefficients for the transfer functions considered in Problem 6.3.

6.27 Using the functions tf and bode available in the MATLAB control system toolbox, plot the frequency characteristics of the systems with transfer functions considered in Problem 6.15.

6.28 Repeat Problem 6.27 using the function freqs available in the MATLAB signal processing toolbox.

6.29 Using the functions tf and impulse available in the MATLAB control system toolbox, calculate the impulse response of the systems with transfer functions considered in Problem 6.15.

6.30 (a) Using the MATLAB function `roots`, calculate the location of poles and zeros of the following transfer functions:

(i) $H_1(s) = \dfrac{s^2 - 5s - 84}{s^4 + 7s^3 - 33s^2 - 355s - 700}$;

(ii) $H_2(s) = \dfrac{s^2 - 19s + 84}{s^4 + 7s^3 - 33s^2 - 355s - 700}$;

(iii) $H_3(s) = \dfrac{s^3 + 20s^2 + 15s + 61}{s^4 + 5s^3 + 31s^2 + 125s + 150}$;

(iv) $H_4(s) = \dfrac{s^3 - 10s^2 + 25s + 7}{s^6 + 6s^5 + 42s^4 + 48s^3 + 288s^2 + 96s + 544}$;

(v) $H_5(s) = \dfrac{s^2 + 3s + 7}{s^3 + (6 - j7)s^2 + (11 - j28)s + (6 - j21)}$.

(b) From the location of poles and zeros in the s-plane, determine if the systems are (i) absolutely stable, (ii) marginally stable, or (iii) unstable.

7 Continuous-time filters

A common requirement in signal processing is to modify the frequency contents of a continuous-time (CT) signal in a predefined manner. In communication systems, for example, noise and interference from the neighboring channels corrupt the information-bearing signal transmitted via a communication channel, such as a telephone line. By exploiting the differences between the frequency characteristics of the transmitted signal and the channel noise, a linear time-invariant system (LTI) system can be designed to compensate for the distortion introduced during the transmission. Such an LTI system is referred to as a frequency-selective filter, which processes the received signal to eliminate the high-frequency components introduced by the channel interference and noise from the low-frequency components constituting the information-bearing signal. The range of frequencies eliminated from the CT signal applied at the input of the filter is referred to as the stop band of the filter, while the range of frequencies that is left relatively unaffected by the filter constitute the pass band of the filter.

Graphic equalizers used in stereo sound systems provide another application for the continuous-time (CT) filters. A graphic equalizer consists of a combination of CT filters, each tuned to a different band of frequencies. By selectively amplifying or attenuating the frequencies within the operational bands of the constituent filters, a graphic equalizer maintains sound consistency within dissimilar acoustic environments and spaces. The operation of a graphic equalizer is somewhat different from that of a frequency-selective filter used in our earlier example of the communication system since it amplifies or attenuates selected frequency components of the input signal. A frequency-selective filter, on the other hand, attempts to eliminate the frequency components completely within the stop band of the filter.

This chapter focuses on the design of CT filters. We are particularly interested in the frequency-selective filters that are categorized in four different categories (lowpass, highpass, bandpass, and bandstop) in Section 7.1. Practical approximations to the frequency characteristics of the ideal frequency-selective filters are presented in Section 7.2, where acceptable levels of distortion is tolerated within the pass and stop bands of the ideal filters. Section 7.3 designs three realizable implementations of an ideal lowpass filter. These implementations are referred to as the Butterworth, Chebyshev, and elliptic filters. Section 7.4 transforms the frequency characteristics of the highpass, bandpass, and bandstop filters in terms of the characteristics of the lowpass filters. These transformations are exploited to design the highpass, bandpass, and bandstop filters. Finally, the chapter is concluded with a summary of important concepts in Section 7.5.

7.1 Filter classification

An ideal frequency-selective filter is a system that passes a prespecified range of frequency components without any attenuation but completely rejects the remaining frequency components. As discussed earlier, the range of input frequencies that is left unaffected by the filter is referred to as the pass band of the filter, while the range of input frequencies that are blocked from the output is referred to as the stop band of the filter. In terms of the magnitude spectrum, the absolute value of the transfer function $|H(\omega)|$ of the frequency filter, therefore, toggles between the values of A and zero as a function of frequency ω. The gain $|H(\omega)|$ is A, typically set to one, within the pass band, while $|H(\omega)|$ is zero within the stop band. Depending upon the range of frequencies within the pass and stop bands, an ideal frequency-selective filter is categorized in four different categories. These categories are defined in the following discussion.

7.1.1 Lowpass filters

The transfer function $H_{\mathrm{lp}}(\omega)$ of an ideal lowpass filter is defined as follows:

$$H_{\mathrm{lp}}(\omega) = \begin{cases} A & |\omega| \leq \omega_{\mathrm{c}} \\ 0 & |\omega| > \omega_{\mathrm{c}}, \end{cases} \tag{7.1}$$

where ω_{c} is referred to as the cut-off frequency of the filter. The pass band of the lowpass filter is given by $|\omega| \leq \omega_{\mathrm{c}}$, while the stop band of the lowpass filter is given by $\omega_{\mathrm{c}} < |\omega| < \infty$. The frequency characteristics of an ideal lowpass filter are plotted in Fig. 7.1(a), where we observe that the magnitude $|H_{\mathrm{lp}}(\omega)|$ toggles between the values of A within the pass band and zero within the stop band. The phase $<H_{\mathrm{lp}}(\omega)$ of an ideal lowpass filter is zero for all frequencies.

7.1.2 Highpass filters

The transfer function $H_{\mathrm{hp}}(\omega)$ of an ideal highpass filter is defined as follows:

$$H_{\mathrm{hp}}(\omega) = \begin{cases} 0 & |\omega| \leq \omega_{\mathrm{c}} \\ A & |\omega| > \omega_{\mathrm{c}}, \end{cases} \tag{7.2}$$

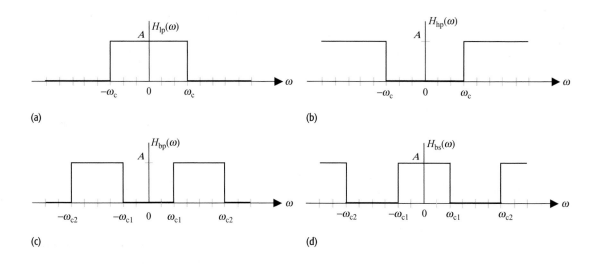

(a)

(a)

(b)

(c)

(d)

Fig. 7.1. Magnitude spectra of ideal frequency-selective filters. (a) Lowpass filter; (b) highpass filter; (c) bandpass filter; (d) bandstop filter.

where ω_c is the cut-off frequency of the filter. In other words, the transfer function of an ideal highpass filter $H_{hp}(\omega)$ is related to the transfer function of an ideal lowpass filter $H_{lp}(\omega)$ by the following relationship:

$$H_{hp}(\omega) = A - H_{lp}(\omega). \tag{7.3}$$

The pass band of the lowpass filter is given by $\omega_c < |\omega| < \infty$, while the stop band of the lowpass filter is given by $|\omega| \leq \omega_c$. The frequency characteristics of an ideal lowpass filter are plotted in Fig. 7.1(b). As was the case for the ideal lowpass filter, the phase $<H_{hp}(\omega)$ of an ideal highpass filter is zero for all frequencies.

7.1.3 Bandpass filters

The transfer function $H_{bp}(\omega)$ of an ideal bandpass filter is defined as follows:

$$H_{bp}(\omega) = \begin{cases} A & \omega_{c1} \leq |\omega| \leq \omega_{c2} \\ 0 & |\omega| < \omega_{c1} \text{ and } \omega_{c2} < |\omega| < \infty, \end{cases} \tag{7.4}$$

where ω_{c1} and ω_{c2} are collectively referred to as the cut-off frequencies of the ideal bandpass filter. The lower frequency ω_{c1} is referred to as the lower cut off, while the higher frequency ω_{c2} is referred to as the higher cut off. Unlike the highpass filter, the bandpass filter has a finite bandwidth as it only allows a range of frequencies ($\omega_{c1} \leq \omega \leq \omega_{c2}$) to be passed through the filter.

7.1.4 Bandstop filters

The transfer function $H_{bs}(\omega)$ of an ideal bandstop filter is defined as follows:

$$H_{bp}(\omega) = \begin{cases} 0 & \omega_{c1} \leq |\omega| \leq \omega_{c2} \\ A & |\omega| < \omega_{c1} \text{ and } \omega_{c2} < |\omega| < \infty, \end{cases} \tag{7.5}$$

where ω_{c1} and ω_{c2} are, respectively, referred to as the lower cut-off and higher cut-off frequencies of the ideal bandstop filter. A bandstop filter can be implemented from a bandpass filter using the following relationship:

$$H_{bs}(\omega) = A - H_{bp}(\omega). \tag{7.6}$$

The ideal bandstop filter is the converse of the ideal bandpass filter as it eliminates a certain range of frequencies ($\omega_{c1} \leq \omega \leq \omega_{c2}$) from the input signal.

In the above discussion, we used the transfer function to categorize different types of frequency selective filters. Example 7.1 derives the impulse response for ideal lowpass and highpass filters.

Example 7.1

Determine the impulse response of an ideal lowpass filter and an ideal highpass filter. In each case, assume a gain of A within the pass band and a cut-off frequency of ω_c.

Solution

Taking the inverse CTFT of Eq. (7.1), we obtain

$$h_{lp}(t) = \Im^{-1}\{H_{lp}(\omega)\} = \frac{1}{2\pi} \int_{-\omega_c}^{\omega_c} A \cdot e^{j\omega t}\, d\omega = \left.\frac{Ae^{j\omega t}}{j2\pi t}\right|_{\omega_c}^{\omega_c}$$

$$= \frac{A}{j2\pi t}[e^{j\omega_c t} - e^{-j\omega_c t}],$$

which reduces to

$$h_{lp}(t) = \frac{2jA\sin(\omega_c t)}{j2\pi t} = \frac{\omega_c A}{\pi}\mathrm{sinc}\left(\frac{\omega_c t}{\pi}\right). \tag{7.7}$$

To derive the impulse response $h_{hp}(t)$ of the ideal highpass filter, we take the inverse CTFT of Eq. (7.3). The resulting relationship is given by

$$h_{hp}(t) = A\delta(t) - h_{lp}(t) = A\delta(t) - \frac{\omega_c A}{\pi}\mathrm{sinc}\left(\frac{\omega_c t}{\pi}\right). \tag{7.8}$$

Fig. 7.2. Impulse responses $h(t)$ of: (a) ideal lowpass filter and (b) ideal highpass filter derived in Example 7.1.

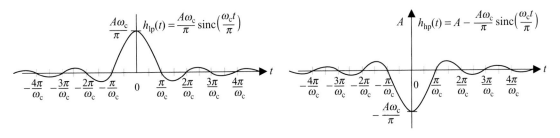

(a)

(b)

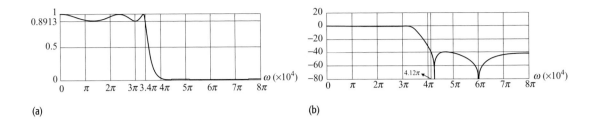

(a)

(b)

Fig. 7.4. Magnitude spectrum of the practical lowpass filter in Example 7.2 using (a) a linear scale and (b) a decibel scale along the y-axis.

uses a linear scale for the magnitude. Figure 7.4(b) uses a decibel scale to plot the magnitude spectrum.

Expressed on a linear scale, the pass-band ripple δ_p is given by $10^{-1/20}$ or 0.8913. From Fig. 7.4(a), we observe that the pass-band frequency ω_p corresponding to $|H(\omega)| = 0.8913$ is given by $3.4\pi \times 10^4$ radians/s. Therefore, the pass band is specified by $|\omega| \leq 3.4\pi \times 10^4$ radians/s.

To determine the stop band, we use Fig. 7.4(b), which uses a decibel scale $20 \times \log_{10}|H(\omega)|$ to plot the magnitude spectrum. Figure 7.4(b) shows that the smallest frequency for which the magnitude spectrum equals a gain of 40 dB is given by $4.12\pi \times 10^4$ radians/s. The stop band is therefore specified by $|\omega| > 4.12\pi \times 10^4$ radians/s.

Based on the aforementioned results, it is straightforward to derive the transition band as follows:

$$3.4\pi \times 10^4 < |\omega| < 4.12\pi \times 10^4 \text{ radians/s.}$$

7.2.1 Cut-off frequency

An important parameter in the design of CT filters is the cut-off frequency ω_c of the filter, which is defined as the frequency at which the gain of the filter drops to 0.7071 times its maximum value. Assuming a gain of unity within the pass band, the gain at the cut-off frequency ω_c is given by 0.7071 or -3 dB on a logarithmic scale. Since the cut-off frequency lies typically within the transitional band of the filter, therefore

$$\omega_\mathrm{p} \leq \omega_\mathrm{c} \leq \omega_\mathrm{s} \tag{7.11}$$

for a lowpass filter. Note that the equality $\omega_\mathrm{p} = \omega_\mathrm{c} = \omega_\mathrm{s}$ implies a transitional band of zero bandwidth and is valid only for ideal filters.

As a side note to our discussion, we observe that in this chapter we only consider positive values of frequencies ω in plotting the magnitude spectrum. The majority of our designs are based on real-valued impulse responses, which lead to frequency spectra that satisfy the Hermitian symmetry. Exploiting the even symmetry for the magnitude spectrum, it is therefore sufficient to specify the magnitude spectrum only for positive frequencies in such cases. The pass-band, stop-band, and cut-off frequencies are also specified by positive values, though their counter-negative values exist for all three parameters.

Example 7.3

Determine the cut-off frequency for the lowpass filter specified in Example 7.2.

Solution

Based on the magnitude spectrum, we note that the maximum gain of the filter is given by 1 or 0 dB. At the cut off frequency ω_c,

$$|H(\omega_c)| = 0.7071 \times 1 = 0.7071,$$

which implies that

$$\left| \frac{5.018 \times 10^3 (j\omega_c)^4 + 2.682 \times 10^{14} (j\omega_c)^2 - 1.026 \times 10^4 (j\omega_c) + 3.196 \times 10^{24}}{(j\omega_c)^5 + 9.863 \times 10^4 (j\omega_c)^4 + 2.107 \times 10^{10} (j\omega_c)^3 + 1.376 \times 10^{15} (j\omega_c)^2 + 1.026 \times 10^{20} (j\omega_c) + 3.196 \times 10^{24}} \right|$$
$$= 0.7071.$$

The above equality can be solved for ω_c using numerical techniques in MATLAB. The value of the cut-off frequency is given by $\omega_c = 3.462\pi \times 10^4$ radians/s. Note that the cut-off frequency lies within the transitional band in between the pass and stop bands of the lowpass filter as derived in Example 7.2.

7.3 Design of CT lowpass filters

To begin our discussion of the design of CT filters, we consider a prototype or normalized lowpass filter, defined as a lowpass filter, with a cut-off frequency of $\omega_c = 1$ radians/s. The remaining specifications for the pass and stop bands of the normalized lowpass filter are assumed to be given by

pass band $(0 \leq |\omega| \leq \omega_p$ radians/s) $\qquad 1 - \delta_p \leq |H(\omega)| \leq 1 + \delta_p;$ (7.12)

stop band $(|\omega| > \omega_s$ radians/s)| $\qquad\qquad\qquad H(\omega)| \leq \delta_s,$ (7.13)

with $\omega_p \leq \omega_c \leq \omega_s$. Using the transfer function of the normalized lowpass filter, it is straightforward to implement any of the more complicated CT filters. Section 7.4 considers the frequency transformations used to convert a lowpass filter into another category of frequency-selective filters.

There are several specialized implementations such as Butterworth, Type I Chebyshev, Type II Chebysev, and elliptic filters, which may be used to design a normalized lowpass filter. Figure 7.5 shows representative characteristics of these implementations, where we observe that the Butterworth filter (Fig. 7.5(a)) has a monotonic transfer function such that the gain decreases monotonically from its maximum value of unity at $\omega = 0$ along the positive frequency axis. The magnitude spectrum of the Butterworth filter has negligible ripples within the pass and stop bands, but has a relatively lower fall off leading to a wide transitional band. By allowing some ripples in either the pass or stop band, the Type I and Type II Chebyshev filters incorporate a sharper fall off. The Type I Chebyshev filter constitutes ripples within the pass band, while the

Type II Chebyshev filter allows for the stop-band ripples. Compared with the Butterworth filter, both Type I and Type II Chebyshev filters have narrower transitional bands. The elliptic filters allow for the sharpest fall off by incorporating ripples in both the pass and stop bands of the filter. The elliptic filters have the narrowest transitional band. To compare the transitional bands, Fig. 7.5 plots the magnitude spectra resulting from the Butterworth, Type I Chebyshev, Type II Chebysev, and elliptic filters with the same order N.

Figure 7.5 confirms our earlier observations that the Butterworth filter (Fig. 7.5(a)) has the widest transitional band. Both the Type I and Type II Chebyshev filters (Figs. 7.5(b) and (c)) have roughly equal transitional bands, which are narrower than the transitional band of the Butterworth filter. The elliptic filter (Fig. 7.5(d)) has the narrowest transitional band but includes ripples in both the pass and stop bands.

We now consider the design techniques for the four specialized implementations with a brief explanation of the MATLAB library functions useful for computing the transfer functions of the implementations.

7.3.1 Butterworth filters

The frequency characteristics of an Nth-order lowpass Butterworth filter are given by

$$|H(\omega)| = \frac{1}{\sqrt{1 + \left(\dfrac{\omega}{\omega_c}\right)^{2N}}}, \tag{7.14}$$

where ω_c is the cut-off frequency of the filter. Substituting $\omega_c = 1$ for the normalized implementation, the transfer function of the normalized lowpass Butterworth filter of order N is given by

$$|H(\omega)| = \frac{1}{\sqrt{1 + \omega^{2N}}}. \tag{7.15}$$

To derive the Laplace transfer function $H(s)$ of the normalized Butterworth filter, we use the following relationship:

$$|H(\omega)|^2 = H(s)H(-s)|_{s=j\omega}. \tag{7.16}$$

Substituting $\omega = s/j$, Eq. (7.16) reduces to

$$H(s)H(-s) = |H(s/j)|^2. \tag{7.17}$$

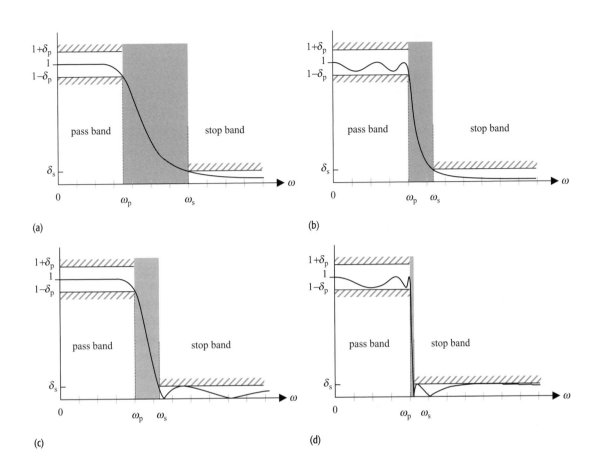

Fig. 7.5. Frequency characteristics of standard implementations of lowpass filters of order N.
(a) Butterworth filter; (b) Type-I Chebyshev filter; (c) Type-II Chebyshev filter; (d) elliptic filter.

Further substituting $H(s/j)$ from Eq. (7.15) leads to the following expression:

$$H(s)H(-s) = \frac{1}{1 + \left(\frac{s}{j}\right)^{2N}}, \qquad (7.18)$$

where the denominator represents the characteristic function for $H(s)H(-s)$. The poles of $H(s)H(-s)$ occur at

$$\left(\frac{s}{j}\right)^{2N} = -1 = e^{j(2n-1)\pi} \qquad (7.19)$$

or

$$s = j\exp\left[j\frac{(2n-1)\pi}{2N}\right] = \exp\left[j\frac{\pi}{2} + j\frac{(2n-1)\pi}{2N}\right] \qquad (7.20)$$

for $0 \leq n \leq 2N-1$. It is clear that the $2N$ poles for $H(s)H(-s)$, specified in Eq. (7.20), are evenly distributed along the unit circle in the complex s-plane. Of these, N poles would lie in the left half of the s-plane, while the remaining N poles would be in the right half of the s-plane. To ensure a causal and

Table 7.1. Location of the $2N$ poles for $H(s)H(-s)$ in Example 7.4 for $N=7$

n	0	1	2	3	4	5	6	7	8	9	10	11	12	13
p_n	$e^{j3\pi/7}$	$e^{j4\pi/7}$	$e^{j5\pi/7}$	$e^{j6\pi/7}$	$e^{j\pi}$	$e^{-j6\pi/7}$	$e^{-j5\pi/7}$	$e^{-j4\pi/7}$	$e^{-j3\pi/7}$	$e^{-j2\pi/7}$	$e^{-j\pi/7}$	1	$e^{j\pi/7}$	$e^{j2\pi/7}$

stable implementation, the transfer function $H(s)$ of the normalized lowpass Butterworth filter is determined from the N poles lying in the left half of the s-plane and is given by

$$H(s) = \frac{1}{\displaystyle\prod_{n=1}^{N}(s - p_n)}, \tag{7.21}$$

where p_n, for $1 \leq n \leq N$, denotes the location of the poles in the left-half s-plane.

Example 7.4
Determine the Laplace transfer function $H(s)$ for the normalized Butterworth filter with cut-off frequency $\omega_c = 1$ and order $N = 7$.

Solution
Using Eq. (7.20), the poles of $H(s)H(-s)$ are given by

$$s = \exp\left[j\frac{\pi}{2} + j\frac{(2n-1)\pi}{14}\right]$$

for $0 \leq n \leq 13$. Substituting different values of n, the locations of the poles are specified by Table 7.1. Figure 7.6 plots the locations of the poles for $H(s)H(-s)$ in the complex s-plane. Allocating the poles located in the left-half s-plane ($1 \leq n \leq 7$), the Laplace transfer function $H(s)$ of the Butterworth filter is given by

$$H(s) = \frac{1}{(s - e^{j4\pi/7})(s - e^{j5\pi/7})(s - e^{j6\pi/7})(s - e^{j\pi})(s - e^{-j6\pi/7})(s - e^{-j5\pi/7})(s - e^{-j4\pi/7})},$$

which simplifies to

$$H(s) = \frac{1}{(s+1)[(s - e^{j4\pi/7})(s - e^{-j4\pi/7})][(s - e^{j5\pi/7})(s - e^{-j5\pi/7})][(s - e^{j6\pi/7})(s - e^{-j6\pi/7})]}$$

or

$$H(s) = \frac{1}{(s+1)(s^2 + 0.4450\,s + 1)(s^2 + 1.2470\,s + 1)(s^2 + 1.8019\,s + 1)}.$$

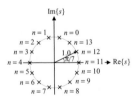

Fig. 7.6. Location of the poles for $H(s)H(-s)$ in the complex s-plane for $N = 7$. The poles lying in the left-half s-plane are allocated to the Butterworth filter.

In Example 7.4, we observed that the locations of poles for the normalized Butterworth filter are complex. Since the poles occur in complex-conjugate pairs, the coefficients of the Laplace transfer function for the normalized

Table 7.2. Denominator $D(s)$ for transfer function $H(s)$ of the Butterworth filter

N	$D(s)$
1	$(s + 1)$
2	$(s^2 + 1.414s + 1)$
3	$(s + 1)(s^2 + s + 1)$
4	$(s^2 + 0.7654s + 1)(s^2 + 1.8478s + 1)$
5	$(s + 1)(s^2 + 0.6810\,s + 1)(s^2 + 1.6810\,s + 1)$
6	$(s^2 + 0.5176s + 1)(s^2 + 1.4142s + 1)(s^2 + 1.9319s + 1)$
7	$(s + 1)(s^2 + 0.4450\,s + 1)(s^2 + 1.2470\,s + 1)(s^2 + 1.8019\,s + 1)$
8	$(s^2 + 0.3902s + 1)(s^2 + 1.1111s + 1)(s^2 + 1.6629s + 1)(s^2 + 1.9616s + 1)$
9	$(s + 1)(s^2 + 0.3473\,s + 1)(s^2 + s + 1)(s^2 + 1.5321\,s + 1)(s^2 + 1.8794\,s + 1)$
10	$(s^2 + 0.3129\,s + 1)(s^2 + 0.9080s + 1)(s^2 + 1.4142s + 1)(s^2 + 1.7820s + 1)(s^2 + 1.9754\,s + 1)$

Butterworth filter are all real-valued. In general, Eq. (7.21) can be simplified as follows:

$$H(s) = \frac{1}{D(s)} = \frac{1}{s^N + a_{N-1}s^{N-1} + \cdots + a_1 s + 1} \qquad (7.22)$$

and represents the transfer function of the normalized Butterworth filter of order N.

Repeating Example 7.4 for different orders ($1 \leq N \leq 10$), the transfer functions $H(s)$ of the resulting normalized Butterworth filters can be similarly computed. Since the numerator of the transfer function is always unity, Table 7.2 lists the polynomials for the denominator $D(s)$ for $1 \leq N \leq 10$.

7.3.1.1 Design steps for the lowpass Butterworth filter

In this section, we will design a Butterworth lowpass filter based on the specifications illustrated in Fig. 7.3(a). Mathematically, the specifications can be expressed as follows:

pass band ($0 \leq |\omega| \leq \omega_p$ radians/s) $\quad 1 - \delta_p \leq |H(\omega)| \leq 1 + \delta_p;$ $\qquad (7.23)$

stop band ($|\omega| > \omega_s$ radians/s) $\qquad\qquad\qquad |H(\omega)| \leq \delta_s.$ $\qquad\qquad\qquad (7.24)$

At times, Eq. (7.23) is also expressed in terms of the pass-band ripple as $20 \log_{10}\delta_p$ dB. Similarly, Eq. (7.24) is expressed in terms of the stop-band ripple as $20 \log_{10}\delta_s$ dB. The design of the Butterworth filter consists of the following steps, which we refer to as Algorithm 7.3.1.1.

Step 1 Determine the order N of the Butterworth filter. To determine the order N of the filter, we calculate the gain of the filter at the corner frequencies $\omega = \omega_p$

and $\omega = \omega_s$. Using Eq. (7.15), the two gains are given by

pass-band corner frequency $(\omega = \omega_p)$ $\quad |H(\omega_p)|^2 = \dfrac{1}{1 + (\omega_p/\omega_c)^{2N}} = (1 - \delta_p)^2;$

$$(7.25)$$

stop-band corner frequency $(\omega = \omega_s)$ $\quad |H(\omega_s)|^2 = \dfrac{1}{1 + (\omega_s/\omega_c)^{2N}} = (\delta_s)^2.$

$$(7.26)$$

Equations (7.25) and (7.26) can alternatively be expressed as follows:

$$(\omega_p/\omega_c)^{2N} = \frac{1}{(1 - \delta_p)^2} - 1 \tag{7.27}$$

and

$$(\omega_s/\omega_c)^{2N} = \frac{1}{(\delta_s)^2} - 1. \tag{7.28}$$

Dividing Eq. (7.27) by Eq. (7.28) and simplifying in terms of N, we obtain the following expression:

$$N = \frac{1}{2} \times \frac{\ln(G_p/G_s)}{\ln(\omega_p/\omega_s)}, \tag{7.29}$$

where the gain terms are given by

$$G_p = \frac{1}{(1 - \delta_p)^2} - 1 \quad \text{and} \quad G_s = \frac{1}{(\delta_s)^2} - 1. \tag{7.30}$$

Step 2 Using Table 7.2 or otherwise determine the transfer function for the normalized Butterworth filter of order N. The transfer function for the normalized Butterworth filter is denoted by $H(S)$ with the Laplace variable S capitalized to indicate the normalized domain.

Step 3 Determine the cut-off frequency ω_c of the Butterworth filter using either of the following two relationships:

pass-band constraint $\quad\quad\quad\quad \omega_c = \dfrac{\omega_p}{(G_p)^{1/2N}};$ $\quad\quad\quad$ (7.31)

stop-band constraint $\quad\quad\quad\quad \omega_c = \dfrac{\omega_s}{(G_s)^{1/2N}}.$ $\quad\quad\quad$ (7.32)

If Eq. (7.31) is used to compute the cut-off frequency, then the Butterworth filter will satisfy the pass-band constraint exactly. Similarly, the stop-band constraint will be satisfied exactly if Eq. (7.32) is used to determine the cut-off frequency.

Step 4 Determine the transfer function $H(s)$ of the required lowpass filter from the transfer function for the normalized Butterworth filter $H(S)$, obtained in Step 2, and the cut-off frequency ω_c, using the following transformation:

$$H(s) = H(S)|_{S=s/\omega_c}.$$

Note that the transformation $S = s/\omega_c$ represents scaling in the Laplace domain. It is therefore clear that the normalized cut-off frequency of 1 radian/s used in the normalized Butterworth filter is transformed to a value of ω_c as required in Step 3.

Step 5 Sketch the magnitude spectrum from the transfer function $H(s)$ determined in Step 4. Confirm that the transfer function satisfies the initial design specifications.

Examples 7.5 and 7.6 illustrate the application of the design algorithm.

Example 7.5
Design a Butterworth lowpass filter with the following specifications:

pass band $(0 \leq |\omega| \leq 5$ radians/s$)$ $0.8 \leq |H(\omega)| \leq 1$;

stop band $(|\omega| > 20$ radians/s$)$ $|H(\omega)| \leq 0.20$.

Solution
Using Step 1 of Algorithm 7.3.1.1, the gain terms G_p and G_s are given by

$$G_p = \frac{1}{(1 - \delta_p)^2} - 1 = \frac{1}{0.8^2} - 1 = 0.5625$$

and

$$G_s = \frac{1}{(\delta_s)^2} - 1 = \frac{1}{0.2^2} - 1 = 24.$$

Using Eq. (7.29), the order of the Butterworth filter is given by

$$N = \frac{1}{2} \times \frac{\ln(G_p/G_s)}{\ln(\omega_p/\omega_s)} = \frac{1}{2} \times \frac{\ln(0.5625/24)}{\ln(5/20)} = 1.3538.$$

We round off the order of the filter to the higher integer value as $N = 2$.

Using Step 2 of Algorithm 7.3.1.1, the transfer function $H(S)$ of the normalized Butterworth filter with a cut-off frequency of 1 radian/s is given by

$$H(S) = \frac{1}{S^2 + 1.414S + 1}.$$

Using the pass-band constraint, Eq. (7.31), in Step 3 of Algorithm 7.3.1.1, the cut-off frequency of the required Butterworth filter is given by

$$\omega_c = \frac{\omega_p}{(G_p)^{1/2N}} = \frac{5}{(0.5625)^{1/4}} = 5.7735 \text{ radians/s.}$$

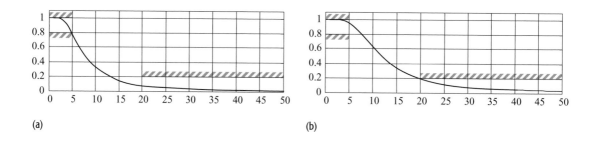

(a)   (b)

Fig. 7.7. Magnitude spectra of the Butterworth lowpass filters, designed in Example 7.5, as a function of ω. Part (a) satisfies the constraint at the pass-band corner frequency, while part (b) satisfies the magnitude constraint at the stop-band corner frequency.

Using Step 4 of Algorithm 7.3.1.1, the transfer function $H(s)$ of the required Butterworth filter is obtained by the following transformation:

$$H(s) = H(S)|_{S=s/\omega_c} = \left.\frac{1}{S^2 + 1.414S + 1}\right|_{S=s/5.7735},$$

which simplifies to

$$H(s) = \frac{1}{(s/5.7735)^2 + 1.414s/5.7735 + 1} = \frac{33.3333}{s^2 + 8.1637s + 33.3333}.$$

Step 5 plots the magnitude spectrum of the Butterworth filter. The CTFT transfer function of the Butterworth filter is given by

$$H(\omega) = H(s)|_{s=j\omega} = \frac{33.3333}{(j\omega)^2 + 8.1637(j\omega) + 33.3333}.$$

The magnitude spectrum $|H(\omega)|$ is plotted in Fig. 7.7(a) with the specifications shown by the shaded lines. We observe that the design specifications are indeed satisfied by the magnitude spectrum.

Alternative implementation An alternative implementation of the aforementioned Butterworth filter can be obtained by using the stop-band constraint, Eq. (7.32), in Step 3 of Algorithm 7.3.1.1. The cut-off frequency of the alternative implementation of the Butterworth filter is given by

$$\omega_c = \frac{\omega_s}{(G_s)^{1/2N}} = \frac{20}{(24)^{1/4}} = 9.0360 \text{ radians/s.}$$

Using Step 4 of Algorithm 7.3.1.1, the transfer function $H(s)$ of the alternative implementation is obtained by the following transformation:

$$H(s) = H(S)|_{S=s/\omega_c} = \left.\frac{1}{S^2 + 1.414S + 1}\right|_{S=s/9.0360},$$

which simplifies to

$$H(s) = \frac{1}{(s/9.0360)^2 + 1.414s/9.0360 + 1} = \frac{81.6497}{s^2 + 12.7769s + 81.6497}.$$

Step 5 plots the magnitude spectrum of the alternative implementation of the Butterworth filter in Fig. 7.7(b), which satisfies the initial design specifications.

Example 7.6
Design a lowpass Butterworth filter with the following specifications:

pass band $(0 \leq |\omega| \leq 50$ radians/s$)$ -1 dB $\leq 20 \log_{10}|H(\omega)| \leq 0$;

stop band $(|\omega| > 100$ radians/s$)$ $20 \log_{10}|H(\omega)| \leq -15$ dB.

Solution
Expressed on a linear scale, the pass-band gain is given by $(1 - \delta_p) = 10^{-1/20} = 0.8913$. Similarly, the stop-band gain is given by $\delta_s = 10^{-15/20} = 0.1778$.

Using Step 1 of Algorithm 7.3.1.1, the gain terms G_p and G_s are given by

$$G_p = \frac{1}{(1 - \delta_p)^2} - 1 = \frac{1}{0.8913^2} - 1 = 0.2588$$

and

$$G_s = \frac{1}{(\delta_s)^2} - 1 = \frac{1}{0.1778^2} - 1 = 30.6327.$$

The order N of the Butterworth filter is obtained using Eq. (7.29) as follows:

$$N = \frac{1}{2} \times \frac{\ln(G_p/G_s)}{\ln(\omega_p/\omega_s)} = \frac{1}{2} \times \frac{\ln(0.2588/30.6327)}{\ln(50/100)} = 3.4435.$$

We round off the order of the filter to the higher integer value as $N = 4$.

Using Step 2 of Algorithm 7.3.1.1, the transfer function $H(S)$ of the normalized Butterworth filter with a cut-off frequency of 1 radian/s is given by

$$H(S) = \frac{1}{(S^2 + 0.7654S + 1)(S^2 + 1.8478S + 1)}.$$

Using the pass-band constraint, Eq. (7.31), in Step 3 of Algorithm 7.3.1.1, the cut-off frequency of the required Butterworth filter is given by

$$\omega_c = \frac{\omega_p}{(G_p)^{1/2N}} = \frac{50}{(0.2588)^{1/8}} = 59.2038 \text{ radians/s.}$$

Using Step 4 of Algorithm 7.3.1.1, the transfer function $H(s)$ of the required Butterworth filter is obtained by the following transformation:

$$H(s) = H(S)|_{S=s/\omega_c} = \frac{1}{(S^2 + 0.7654S + 1)(S^2 + 1.8478S + 1)}\bigg|_{S=s/59.2038},$$

which simplifies to

$$H(s) = \frac{(3.5051 \times 10^3)^2}{(s^2 + 45.3146s + 3.5051 \times 10^3)(s^2 + 109.396s + 3.5051 \times 10^3)}$$

or

$$H(s) = \frac{1.2286 \times 10^7}{s^4 + 154.7106 \, s^3 + 1.1976 \times 10^4 s^2 + 5.4228 \times 10^5 s + 1.2286 \times 10^7}.$$

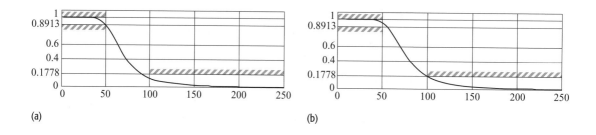

(a)   (b)

Fig. 7.8. Magnitude spectra of the Butterworth lowpass filters, designed in Example 7.6, as a function of ω. Part (a) satisfies the constraint at the pass-band corner frequency, while part (b) satisfies the magnitude constraint at the stop-band corner frequency.

Step 5 plots the magnitude spectrum of the Butterworth filter. The CTFT transfer function of the Butterworth filter is given by

$$H(\omega) = H(s)|_{s=j\omega}$$

$$= \frac{1.2286 \times 10^7}{(j\omega)^4 + 154.7106\,(j\omega)^3 + 1.1976 \times 10^4 (j\omega)^2 + 5.4228 \times 10^5 (j\omega) + 1.2286 \times 10^7}.$$

The magnitude spectrum $|H(\omega)|$ is plotted in Fig. 7.8(a), where the labels on the y-axis are chosen to correspond to the specified gains for the filter. We observe that the design specifications are satisfied by the magnitude spectrum.

Alternative implementation An alternative implementation of the aforementioned Butterworth filter can be obtained by using the stop-band constraint, Eq. (7.32), in Step 3 of Algorithm 7.3.1.1. The cut-off frequency of the alternative implementation of the Butterworth filter is given by

$$\omega_c = \frac{\omega_s}{(G_s)^{1/2N}} = \frac{100}{(30.6327)^{1/4}} = 65.1969 \text{ radians/s.}$$

Using Step 4 of Algorithm 7.3.1.1, the transfer function $H(s)$ of the alternative implementation is obtained by the following transformation:

$$H(s) = H(S)|_{S=s/\omega_c} = \left.\frac{1}{(S^2 + 0.7654S + 1)(S^2 + 1.8478S + 1)}\right|_{S=s/65.1969},$$

which simplifies to

$$H(s) = \frac{(4.2506 \times 10^3)^2}{(s^2 + 49.9017s + 4.2506 \times 10^3)(s^2 + 120.4708s + 4.2506 \times 10^3)}$$

or

$$H(s) = \frac{1.8068 \times 10^7}{s^4 + 170.3725\,s^3 + 1.4513 \times 10^4 s^2 + 7.2419 \times 10^5 s + 1.8068 \times 10^7}.$$

Step 5 plots the magnitude spectrum of the alternative implementation of the Butterworth filter in Fig. 7.8(b), which satisfies the initial design specifications.

7.3.1.2 Butterworth filter design using MATLAB

MATLAB incorporates a number of functions to implement the design algorithm for the Butterworth filter specified in Section 7.3.1.1. The order N and the cut-off wc frequency for the filter in Step 1 of Algorithm 7.3.1.1 can be determined using the library function buttord, which has the following calling syntax:

```
>> [N,wc] = buttord(wp,ws,Rp,Rs,'s');
```

where wp is the corner frequency of the pass band, ws is the corner frequency of the stop band, Rp is the permissible ripple in the pass band in decibels, and Rs is the permissible attenuation in the stop band in decibels. The last argument 's' specifies that a CT filter in the Laplace domain is to be designed. In determining the cut-off frequency, MATLAB uses the stop-band constraint, Eq. (7.32).

Having determined the order and the cut-off frequency, the coefficients of the numerator and denominator polynomials of the Butterworth filter can be determined using the library function butter with the following calling syntax:

```
>> [num,den] = butter(N,wc,'s');
```

where num is a vector containing the coefficients of the numerator and den is a vector containing the coefficients of the denominator in decreasing powers of s.

Finally, the transfer function $H(s)$ can be determined using the library function tf as follows:

```
>> H = tf(num,den).
```

For Example 7.5, the MATLAB commands for designing the Butterworth filter are given by

```
>> wp=5; ws=20; Rp=1.9382; Rs=13.9794;
                              % specify design parameters
                              % Rp = -20*log10(0.8)
                                % = 1.9382dB
                              % Rs = -20*log10(0.2)
                                % = 13.9794dB
>> [N,wc]=buttord (wp,ws,Rp,Rs,'s');
                              % determine order and
                              % cut-off freq
>> [num,den]=butter (N,wc,'s');
                              % determine num and denom
                              % coeff.
>> Ht = tf(num,den);         % determine transfer
                              % function
>> [H,w] = freqs(num,den);   % determine magnitude
                              % spectrum
>> plot(w,abs(H));           % plot magnitude spectrum
```

Stepwise implementation of the above code returns the following values for different variables:

```
Instruction II:   N = 2; wc = 9.0360;
Instruction III: num = [0 0 81.6497]; den = [1.0000
     12.7789 81.6497];
Instruction IV:   Ht = 1/(s^2 + 12.78s + 81.65);
```

The magnitude spectrum is the same as that given in Fig. 7.7(b).

7.3.2 Type I Chebyshev filters

Butterworth filters have a relatively low roll off in the transitional band, which leads to a large transitional bandwidth. Type I Chebyshev filters reduce the bandwidth of the transitional band by using an approximating function, referred to as the Type I Chebyshev polynomial, with a magnitude response that has ripples within the pass band. We start with the definition of the Chebyshev polynomial.

7.3.2.1 Type I Chebyshev polynomial

The Nth-order Type I Chebyshev polynomial is defined as

$$T_N(\omega) = \begin{cases} \cos(N \cos^{-1}(\omega)) & |\omega| \le 1 \\ \cosh(N \cosh^{-1}(\omega)) & |\omega| > 1, \end{cases} \tag{7.33}$$

where $\cosh(x)$ denotes the hyperbolic cosine function, which is given by

$$\cosh(x) = \cos(jx) = \frac{e^x + e^{-x}}{2}. \tag{7.34}$$

Starting from the initial values of $T_0(\omega) = 1$ and $T_1(\omega) = \omega$, the higher orders of the Type I Chebyshev polynomial can be recursively generated using the following expression:

$$T_n(\omega) = 2\omega T_{n-1}(\omega) - T_{n-2}(\omega). \tag{7.35}$$

Table 7.3 lists the Chebyshev polynomial for different values of n within the range $0 \le n \le 10$.

Using Eq. (7.33), the roots of the Type I Chebyshev polynomial $T_N(\omega)$ can be derived as follows:

$$\omega_n = \cos\left[\frac{(2n+1)\pi}{2N}\right], \tag{7.36}$$

for $0 \le n \le N - 1$.

Table 7.3. Chebyshev polynomial $T_N(\omega)$ for different values of N

N	$T_N(\omega)$
0	1
1	ω
2	$2\omega^2 - 1$
3	$4\omega^3 - 3\omega$
4	$8\omega^4 - 8\omega^2 + 1$
5	$16\omega^5 - 20\omega^3 + 5\omega$
6	$32\omega^6 - 48\omega^4 + 18\omega^2 - 1$
7	$64\omega^7 - 112\omega^5 + 56\omega^3 - 7\omega$
8	$128\omega^8 - 256\omega^6 + 160\omega^4 - 32\omega^2 + 1$
9	$256\omega^9 - 576\omega^7 + 432\omega^5 - 120\omega^3 + 9\omega$
10	$512\omega^{10} - 1280\omega^8 + 1120\omega^6 - 400\omega^4 + 50\omega^2 - 1$

7.3.2.2 Type I Chebyshev filter

The frequency characteristics of the Type I Chebyshev filter of order N are defined as follows:

$$|H(\omega)| = \frac{1}{\sqrt{1 + \varepsilon^2 T_N^2(\omega/\omega_p)}}, \tag{7.37}$$

where ω_p is the pass-band corner frequency and ε is the ripple control parameter that adjusts the magnitude of the ripple within the pass band. Substituting $\omega_p = 1$, the frequency characteristics of the normalized Type I Chebyshev filter of order N are expressed in terms of the Chebyshev polynomial as follows:

$$|H(\omega)| = \frac{1}{\sqrt{1 + \varepsilon^2 T_N^2(\omega)}}. \tag{7.38}$$

Based on Eqs. (7.35) and (7.38), we make the following observations for the frequency characteristics of the normalized Type I Chebyshev filter.

(1) For $\omega = 0$, the Chebyshev polynomial $T_N(\omega)$ has a value of ± 1 or 0. This can be shown by substituting $\omega = 0$ in Eq. (7.33), which yields

$$T_N(0) = \cos(N \cos^{-1}(0)) = \cos\left(\frac{N(2n+1)\pi}{2}\right) = \begin{cases} \pm 1 & N \text{ is even} \\ 0 & N \text{ is odd.} \end{cases} \tag{7.39}$$

Equation (7.37) implies that the dc component $|H(0)|$ of the Type I Chebyshev filter is given by

$$|H(0)| = \begin{cases} \dfrac{1}{\sqrt{1 + \varepsilon^2}} & N \text{ is even} \\ 1 & N \text{ is odd.} \end{cases} \tag{7.40}$$

(2) For $\omega = 1$ radian/s, the value of the Chebyshev polynomial $T_N(\omega)$ is given by

$$T_N(1) = \cos(N \cos^{-1}(1)) = \cos(2nN\pi) = 1. \tag{7.41}$$

Therefore, the magnitude $|H(\omega)|$ of the normalized Type I Chebyshev filter at $\omega = 1$ radian/s is given by

$$|H(1)| = \frac{1}{\sqrt{1 + \varepsilon^2}}, \tag{7.42}$$

irrespective of the order N of the normalized Chebyshev filter.

(3) For large values of ω within the stop band, the magnitude response of the normalized Type I Chebyshev filter can be approximated by

$$|H(\omega)| \approx \frac{1}{\varepsilon T_N(\omega)}, \tag{7.43}$$

since $\varepsilon T_N(\omega) \gg 1$. If $N \gg 1$, then a second approximation can be made by ignoring the lower degree terms in $T_N(\omega)$ and using the approximation $T_N(\omega) \approx 2^{N-1}\omega^N$. Equation (7.43) is therefore simplified as follows:

$$|H(\omega)| \approx \frac{1}{\varepsilon} \times \frac{1}{2^{N-1}\omega^N}. \tag{7.44}$$

(4) Since

$$H(s)H(-s)|_{s=j\omega} = |H(\omega)|^2,$$

$H(s)H(-s)$ can be derived from Eq. (7.38) as follows:

$$H(s)H(-s) = \frac{1}{1 + \varepsilon^2 T_N^2(s/j)}. \tag{7.45}$$

The $2N$ poles of $H(s)H(-s)$ are obtained by solving the characteristic equation,

$$1 + \varepsilon^2 T_N^2(s/j) = 0, \tag{7.46}$$

and are given by

$$\begin{aligned} s_n = {} & \sin\left(\frac{2n-1}{2N}\pi\right) \sinh\left(\frac{1}{N}\sinh^{-1}\left(\frac{1}{\varepsilon}\right)\right) \\ & + j\cos\left(\frac{2n-1}{2N}\pi\right) \cosh\left(\frac{1}{N}\sinh^{-1}\left(\frac{1}{\varepsilon}\right)\right) \end{aligned} \tag{7.47}$$

for $1 \leq n \leq 2N-1$. To derive a stable implementation of the normalized Type I Chebyshev filter, the N poles in the left-hand s-plane are included in the Laplace transfer function of $H(s)$. From Eq. (7.45), it is clear that there are no zeros for the normalized Type I Chebyshev filter.

Properties (1)–(4) are used to derive the design algorithm for the Type I Chebyshev filter, which is explained in the following.

7.3.2.3 Design steps for the lowpass filter

In this section, we will design a lowpass Type I Chebyshev filter based on the following specifications:

pass band $(0 \leq |\omega| \leq \omega_p$ radians/s) $\quad 1 - \delta_p \leq |H(\omega)| \leq 1 + \delta_p$;

stop band $(|\omega| > \omega_s$ radians/s) $\quad\quad\quad |H(\omega)| \leq \delta_s$.

Since the Type I Chebyshev filter is designed in terms of its normalized version, Eq. (7.37), we normalize the aforementioned specifications by the pass-band corner frequency ω_p. The normalized specifications are as follows:

pass band $(0 \leq |\omega| \leq 1)$ $\quad 1 - \delta_p \leq |H(\omega)| \leq 1 + \delta_p$;

stop band $(|\omega| > \omega_s/\omega_p)$ $\quad\quad |H(\omega)| \leq \delta_s$.

Step 1 Determine the value of the ripple control factor ε. Equation (7.42) computes the value of the ripple control factor ε:

$$\varepsilon = \sqrt{G_p} \quad \text{with} \quad G_p = \frac{1}{(1 - \delta_p)^2} - 1. \tag{7.48}$$

Step 2 Calculate the order N of the normalized Chebyshev polynomial. The gain at the normalized stop-band corner frequency ω_s/ω_p is obtained from Eq. (7.38) as

$$|H(\omega_s/\omega_p)|^2 = \frac{1}{1 + \varepsilon^2 T_N^2(\omega_s/\omega_p)} = (\delta_s)^2. \tag{7.49}$$

Substituting the value of the Chebyshev polynomial $T_N(\omega)$ from Eq. (7.33) and simplifying the resulting equation, we obtain

$$N = \frac{\cosh^{-1}[(G_s/G_p)^{0.5}]}{\cosh^{-1}[\omega_s/\omega_p]}, \tag{7.50}$$

where the gain terms G_p and G_s are given by

$$G_p = \frac{1}{(1 - \delta_p)^2} - 1 \quad \text{with} \quad G_s = \frac{1}{(\delta_s)^2} - 1. \tag{7.51}$$

Step 3 Determine the location of the $2N$ poles of $H(S)H(-S)$ using Eq. (7.47). To derive a stable implementation for the normalized Type I Chebyshev filter $H(S)$, the N poles lying in the left-half s-plane are selected to derive the transfer function $H(S)$. If required, a constant gain term K is also multiplied with $H(S)$

such that the gain $|H(0)|$ of the normalized Type I Chebyshev filter is unity at $\omega = 0$.

Step 4 Derive the transfer function $H(s)$ of the required lowpass filter from the transfer function $H(S)$ of the normalized Type I Chebyshev filter, obtained in Step 3, using the following transformation:

$$H(s) = H(S)|_{S=s/\omega_p}. \qquad (7.52)$$

Step 5 Sketch the magnitude spectrum from the transfer function $H(s)$ determined in Step 4. Confirm that the transfer function satisfies the initial design specifications.

Example 7.7
Repeat Example 7.6 using the Type I Chebyshev filter.

Solution
For the given specifications, Example 7.6 calculates the pass-band and stop-band gain on a linear scale as $(1 - \delta_p) = 0.8913$ and $\delta_s = 10^{-15/20} = 0.1778$ with the gain terms given by $G_p = 0.2588$ and $G_s = 30.6327$.

Step 1 determines the value of the ripple control factor ε:

$$\varepsilon = \sqrt{G_p} = \sqrt{0.2588} = 0.5087.$$

Step 2 determines the order N of the Chebyshev polynomial:

$$N = \frac{\cosh^{-1}[(30.6327/0.2588)^{0.5}]}{\cosh^{-1}[100/50]} = 2.3371.$$

We round off N to the closest higher integer, $N = 3$.

Step 3 determines the location of the six poles of $H(S)H(-S)$:

$$[-0.2471 + j0.9660, \quad -0.2471 - j0.9660, \quad 0.2471 + j0.9660,$$
$$0.2471 - j0.9660, \quad 0.4943, \quad -0.4943].$$

The three poles lying in the left-half s-plane are included in the transfer function $H(S)$ of the normalized Type I Chebyshev filter. These poles are located at

$$[-0.2471 + j0.9660, \quad -0.2471 - j0.9660, \quad -0.4943].$$

The transfer function for the normalized Type I Chebyshev filter is therefore given by

$$H(S) = \frac{K}{(S + 0.2471 + j0.9660)(S + 0.2471 - j0.9660)(S + 0.4943)},$$

which simplifies to

$$H(S) = \frac{K}{S^3 + 0.9885S^2 + 1.2386S + 0.4914}.$$

Fig. 7.9. Magnitude spectrum of the Type I Chebyshev lowpass filter designed in Example 7.7.

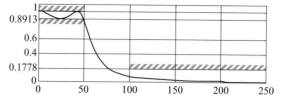

Since $|H(\omega)|$ at $\omega = 0$ is $K/0.4914$, K is set to 0.4914 to make the dc gain equal to unity. The new transfer function with unity gain at $\omega = 0$ is given by

$$H(S) = \frac{0.4914}{S^3 + 0.9885S^2 + 1.2386S + 0.4914}.$$

Step 4 transforms the normalized Type I Chebyshev filter using the following relationship:

$$H(s) = H(S)|_{S=s/50} = \frac{0.4914}{(s/50)^3 + 0.9885(s/50)^2 + 1.2386(s/50) + 0.4914}$$

or

$$H(s) = \frac{6.1425 \times 10^4}{s^3 + 49.425s^2 + 3.0965 \times 10^3 s + 6.1425 \times 10^4},$$

which is the transfer function of the required lowpass filter.

The magnitude spectrum of the Type I Chebyshev filter is plotted in Fig. 7.9. It is observed that Fig. 7.9 satisfies the initial design specifications.

Examples 7.6 and 7.7 used the Butterworth and Type I Chebyshev implementations to design a lowpass filter based on the same specifications. Comparing the magnitude spectra (Figs. 7.8 and 7.9) for the resulting filters, we note that the Butterworth filter has a monotonic gain with negligible ripples in the pass and stop bands. By introducing pass-band ripples, the Type I Chebyshev implementation is able to satisfy the design specifications with a lower order N for the lowpass filter, thus reducing the complexity of the filter. However, savings in the complexity are achieved at the expense of ripples, which are added to the the pass band of the frequency characteristics of the Type I Chebyshev filter.

7.3.2.4 Type I Chebyshev filter design using MATLAB

MATLAB uses the `cheb1ord` and `cheby1` functions to implement the Type I Chebyshev filter. The `cheb1ord` function determines the order N of the Type I Chebyshev filter from the pass-band corner frequency wp, stop-band corner frequency ws, pass-band attenuation rp, and the stop-band attenuation rs. In terms of the filter specifications, Eqs. (7.23) and (7.24), the values of the pass-band attenuation rp and the stop-band attenuation rs are given by

$$rp = 20 \times \log_{10}(\delta_p) \quad \text{and} \quad rs = 20 \times \log_{10}(\delta_s).$$

The `cheb1ord` also returns wn, another design parameter referred to as the Chebyshev natural frequency to use with `cheby1` to achieve the design specifications. The syntax for `cheb1ord` is given by

```
>> [N,wn] = cheb1ord(wp,ws,rp,rs,'s');
```

To determine the coefficients of the numerator and denominator of the Type I Chebyshev filter, MATLAB uses the `cheb1` function with the following syntax:

```
>> [num,den] = cheby1(N,rp,wn,'s');
```

The transfer function $H(s)$ can be determined using the library function `tf` as follows:

```
>> H = tf(num,den);
```

For Example 7.7, the MATLAB commands for designing the Butterworth filter are given by

```
>> wp=50; ws=100; rp=1; rs=15;
                                % specify design parameters
>> [N,wn] = cheb1ord (wp,ws,rp,rs,'s');
                                % determine order and
                                % natural freq
>> [num,den] = cheby1 (N,rp,wn,'s');
                                % determine num and denom
                                % coeff.
>> Ht = tf(num,den);            % determine transfer
                                % function
>> [H,w] = freqs(num,den);      % determine magnitude
                                % spectrum
>> plot(w,abs(H));              % plot magnitude spectrum
```

Stepwise implementation of the above code returns the following values for different variables:

```
Instruction II:   N = 3; wn = 50;
Instruction III: num = [0 0 0 61413.3]; den =
    [1.0000 49.417 3096 61413.3];
Instruction IV: Ht = 61413.3/ (s^3 + 49.417s^2 + 3096s
    + 61413.3);
```

The magnitude spectrum is the same as that given in Fig. 7.9.

7.3.3 Type II Chebyshev filters

The Type II Chebyshev filters, or the inverse Chebyshev filters, are monotonic within the pass band and introduce ripples in the stop band. Such an implementation is preferred over the Type I Chebyshev filter in applications where a constant gain is desired within the pass band.

The frequency characteristics of the Type II Chebyshev filter are given by

$$|H(\omega)| = \frac{1}{\sqrt{1 + \left[\varepsilon^2 T_N^2(\omega_s/\omega)\right]^{-1}}} = \sqrt{\frac{\varepsilon^2 T_N^2(\omega_s/\omega)}{1 + \varepsilon^2 T_N^2(\omega_s/\omega)}}, \qquad (7.53)$$

where ω_s is the lower corner frequency of the stop band. To derive the normalized version of the Type II Chebyshev filter, we set $\omega_s = 1$ in Eq. (7.53) leading to the following expression for the frequency characteristics of the normalized Type II Chebyshev filter:

$$|H(\omega)| = \frac{1}{\sqrt{1 + \left[\varepsilon^2 T_N^2(1/\omega)\right]^{-1}}} = \sqrt{\frac{\varepsilon^2 T_N^2(1/\omega)}{1 + \varepsilon^2 T_N^2(1/\omega)}}. \qquad (7.54)$$

In the following section, we list the steps involved in the design of the Type II Chebyshev filter.

7.3.3.1 Design steps for the lowpass filter

The design of the lowpass Type II Chebyshev filter is based on the following specifications:

pass band $(0 \leq |\omega| \leq \omega_p$ radians/s$)$ $1 - \delta_p \leq |H(\omega)| \leq 1 + \delta_p$;

stop band $(|\omega| > \omega_s$ radians/s$)$ $|H(\omega)| \leq \delta_s$.

Normalizing the specifications with the stop-band corner frequency ω_s, we obtain

pass band $(0 \leq |\omega| \leq \omega_p/\omega_s)$ $1 - \delta_p \leq |H(\omega)| \leq 1 + \delta_p$;

stop band $(|\omega| > 1)$ $|H(\omega)| \leq \delta_s$.

Step 1 Compute the value of the ripple factor by setting the normalized frequency $\omega = 1$ in Eq. (7.54). Since the Type II Chebyshev filter is normalized with respect to ω_s, the normalized frequency $\omega = 1$ corresponds to ω_s and the filter gain $H(1) = \delta_s$. Substituting $H(1) = \delta_s$ in Eq. (7.54), we obtain

$$|H(1)| = \sqrt{\frac{\varepsilon^2}{1 + \varepsilon^2}} = \delta_s,$$

which simplifies to

$$\varepsilon = \frac{1}{\sqrt{G_s}}, \qquad (7.55)$$

with the gain term specified in Eq. (7.51).

Step 2 Compute the order N of the Type II Chebyshev filter. To derive an expression for the order N, we compute the gain $|H(\omega)|$ at the normalized pass-band corner frequency ω_p/ω_s. Substituting $|H(\omega)| = (1 - \delta_p)$ at $\omega = \omega_p/\omega_s$, we obtain

$$\frac{\varepsilon^2 T_N^2(\omega_s/\omega_p)}{1 + \varepsilon^2 T_N^2(\omega_s/\omega_p)} = (1 - \delta_p)^2.$$

Substituting the value of the Chebyshev polynomial from Eq. (7.33) and simplifying the resulting expression with respect to N yields

$$N = \frac{\cosh^{-1}[(G_s/G_p)^{0.5}]}{\cosh^{-1}[\omega_s/\omega_p]}, \tag{7.56}$$

where the gain terms G_p and G_s are defined in Eq. (7.51). Note that the expression for the order of the filter for the Type II Chebyshev filter is the same as the corresponding expression, Eq. (7.50), for the Type I Chebyshev filter.

Step 3 Determine the location of the poles and zeros of the transfer function $H(S)$ of the normalized Type II Chebyshev filter. Substituting

$$H(s)H(-s)|_{s=j\omega} = |H(\omega)|^2,$$

the Laplace transfer function for the normalized Type II Chebyshev filter is given by

$$H(s)H(-s) = \frac{\varepsilon^2 T_N^2(j/s)}{1 + \varepsilon^2 T_N^2(j/s)}. \tag{7.57}$$

The poles of $H(s)H(-s)$ are obtained by solving for the roots of the characteristic equation,

$$1 + \varepsilon^2 T_N^2(j/s) = 0. \tag{7.58}$$

Comparing with the characteristic equation for $H(s)H(-s)$ of the Type I Chebyshev filter, Eq. (7.46), we note that (s/j) in the Chebyshev polynomial of Eq. (7.46) is replaced by (j/s) in Eq. (7.58). This implies that the poles of the normalized Type II Chebyshev filter are simply the inverse of the poles of the Type I Chebyshev filter. Hence, the location of the poles for the normalized Type II Chebyshev filter can be computed by determining the locations of the poles for the normalized Type I Chebyshev filter and then taking the inverse.

The zeros of $H(s)H(-s)$ are obtaining by solving

$$T_N^2(j/s) = 0. \tag{7.59}$$

The zeros of $H(s)H(-s)$ are therefore the inverse of the roots of the Chebyshev polynomial $T_N(\omega) = T_N(s/j)$, which are given by

$$\omega = \cos\left[\frac{(2n+1)\pi}{2N}\right].$$

The zeros of $H(s)$ are therefore given by

$$s = \frac{j}{\cos\left(\dfrac{(2n+1)\pi}{2N}\right)} \qquad (7.60)$$

for $0 \leq n \leq N-1$. The poles and zeros are used to evaluate the transfer function $H(S)$ for the normalized Type II Chebyshev filter. If required, a constant gain term K is also multiplied by $H(S)$ such that the gain $|H(0)|$ of the normalized Type II Chebyshev filter is unity at $\omega = 0$.

Step 4 Derive the transfer function $H(s)$ of the required lowpass filter from the transfer function $H(S)$ of the normalized Type II Chebyshev filter, obtained in Step 3, using the following transformation:

$$H(s) = H(S)|_{S=s/\omega_s}. \qquad (7.61)$$

Step 5 Sketch the magnitude spectrum from the transfer function $H(s)$ determined in Step 4. Confirm that the transfer function satisfies the initial design specifications.

Example 7.8
Repeat Example 7.6 using the Type II Chebyshev filter.

Solution
As calculated in Example 7.6, the pass-band and stop-band gain are $(1 - \delta_p) = 0.8913$ and $\delta_s = 10^{-15/20} = 0.1778$. The gain terms are also calculated as $G_p = 0.2588$ and $G_s = 30.6327$.

Step 1 determines the value of the ripple control factor ε:

$$\varepsilon = \frac{1}{\sqrt{G_s}} = \frac{1}{\sqrt{30.6327}} = 0.1807.$$

Step 2 determines the order N of the Chebyshev polynomial:

$$N = \frac{\cosh^{-1}[(30.6327/0.2588)^{0.5}]}{\cosh^{-1}[100/50]} = 2.3371.$$

We round off N to the closest higher integer, $N = 3$.

Step 3 determines the location of the poles and zeros of $H(S)H(-S)$. We first determine the location of poles for the Type I Chebyshev filter with $\varepsilon = 0.1807$ and $N = 3$. Using Eq. (7.47), the location of poles for $H(s)H(-s)$ of the Type I Chebyshev filter is given by

$$[-0.4468 + j1.1614, \quad -0.4468 - j1.1614, \quad 0.4468 + j1.1614, \quad 0.4468$$
$$-j1.1614, \quad 0.8935, \quad -0.8935].$$

Selecting the poles located in the left-half s-plane, we obtain

$$[-0.4468 + j1.1614, \quad -0.4468 + j1.1614, \quad -0.8935].$$

The poles of the normalized Type II Chebyshev filter are located at the inverse of the above locations and are given by

$$[-0.2885 - j0.7501, \quad -0.2885 + j0.7501, \quad -1.1192].$$

The zeros of the normalized Chebyshev Type II filter are computed using Eq. (7.60) and are given by

$$[-j1.1547, \quad +j1.1547, \quad \infty].$$

The zero at $s = \infty$ is neglected. The transfer function for the normalized Type II Chebyshev filter is given by

$$H(S) = \frac{K(S + j1.1547)(S - j1.1547)}{(S + 0.2885 + j0.7501)(S + 0.2885 - j0.7501)(S + 1.1192)},$$

which simplifies to

$$H(S) = \frac{K(S^2 + 1.3333)}{S^3 + 1.6962S^2 + 1.2917S + 0.7229}.$$

Since $|H(\omega)|$ at $\omega = 0$ is $1.3333/0.7229 = 1.8444$, K is set to $1/1.8444 = 0.5422$ to make the dc gain equal to unity. The new transfer function with unity gain at $\omega = 0$ is given by

$$H(S) = \frac{0.5422(S^2 + 1.3333)}{S^3 + 1.6962S^2 + 1.2917S + 0.7229}.$$

Step 4 normalizes $H(S)$ based on the following transformation:

$$H(s) = H(S)|_{S=s/100} = \frac{0.5422((s/100)^2 + 1.3333)}{(s/100)^3 + 1.6962(s/100)^2 + 1.2917(s/100) + 0.7229},$$

which simplifies to

$$H(s) = \frac{54.22(s^2 + 1.3333 \times 10^4)}{s^3 + 1.6962 \times 10^2 s^2 + 1.2917 \times 10^4 s + 0.7229 \times 10^6}.$$

Step 5 plots the magnitude spectrum, which is shown in Fig. 7.10. As expected, the frequency characteristics in Fig. 7.10 have a monotonic gain within the pass band and ripples within the stop band. Also, it is noted that the magnitude spectrum $|H(\omega)| = 0$ between the frequencies of $\omega = 100$ and $\omega = 150$ radians/s. This zero value corresponds to the location of the complex zeros in $H(s)$. Setting

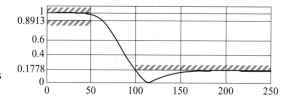

Fig. 7.10. Magnitude spectrum of the Type II Chebyshev lowpass filter designed in Example 7.8.

the numerator of $H(s)$ equal to zero, we get two zeros at $s = \pm j115.4686$, which lead to a zero magnitude at a frequency of $\omega = 115.4686$.

7.3.3.2 Type II Chebyshev filter design using MATLAB

MATLAB provides the `cheb2ord` and `cheby2` functions to implement the Type II Chebyshev filter. The usage of these functions is the same as the `cheb1ord` and `cheby1` functions for the Type I Chebyshev filter except for the `cheby2` function, for which the stop-band constraints (stop-band ripple `rs` and stop-band corner frequency `ws`) are specified. The code for Example 7.8 is as follows:

```
>> wp=50; ws=100; rp=1; rs=15;
                                % specify design parameters
>> [N,wn] = cheb2ord (wp,ws,rp,rs,'s');
                                % determine order and
                                % natural freq
>> [num,den] = cheby2(N,rs,ws,'s');
                                % determine num and denom
                                % coeff.
>> Ht = tf(num,den);           % determine transfer
                                % function
>> [H,w] = freqs(num,den);     % determine magnitude
                                % spectrum
>> plot(w,abs(H));             % plot magnitude spectrum
```

Stepwise implementation of the above code returns the following values for different variables:

```
Instruction II:   N = 3; wn = 78.6980;
Instruction III:  num = [0 54.212 0 722835];
   den = [1.0000 169.63 12917 722835];
Instruction IV: Ht = (54.21s^2 + 722800) /(s^3 + 169.6s^2
   + 12920s + 722800);
```

The magnitude spectrum is the same as that given in Fig. 7.10.

7.3.4 Elliptic filters

Elliptic filters, also referred to as Cauer filters, include both pass-band and stop-band ripples. Consequently, elliptic filters can achieve a very narrow bandwidth for the transition band. The frequency characteristics of the elliptic filter are given by

$$|H(\omega)| = \frac{1}{\sqrt{1 + \varepsilon^2 U_N^2(\omega/\omega_p)}}, \tag{7.62}$$

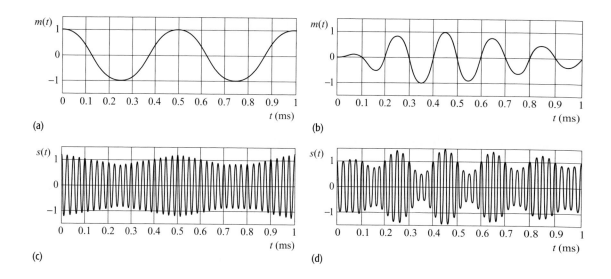

Fig. 8.2. Amplitude modulation in time domain for two different modulating signals: (a) pure sinusoidal signal with fundamental frequency of 2 kHz; (b) synthetic audio signal. The modulated signal for the pure sinusoidal signal is shown in (c) and that for real speech is shown in (d).

Figure 8.2 shows the results of amplitude modulation for two different modulating signals: a pure sinusoidal signal with the fundamental frequency of 2 kHz is plotted in Fig. 8.2(a) and a synthetic audio signal is plotted in Fig. 8.2(b). Both signals are amplitude modulated with the carrier signal $\cos(\omega_c t + \phi_c)$ having a fundamental frequency of $f_c = 40$ kHz and an epoch of $\phi_c = 0$ radians. In the case of the pure sinusoidal signal, the modulation index k is selected to have a value of 0.2, while for the real audio signal the modulation index is set to 0.7. The results of amplitude modulation are shown in Fig. 8.2(c) for the pure sinusoidal signal and in Fig 8.2(d) for the audio signal. In both cases, we observe that the amplitude of the carrier signal is adjusted according to the magnitude of the modulating signal. In other words, the modulating signal acts as an envelope and controls the amplitude of the carrier.

To illustrate the effect of modulation on the frequency content of the modulating signal, we use the CTFT. Equation (8.1) is expressed as follows:

$$s(t) = A\cos(\omega_c t + \phi_c) + Akm(t)\cos(\omega_c t + \phi_c). \tag{8.2}$$

Without loss of generality, we set $A = 1$ and $\phi_c = 0$. Using the multiplication property for the CTFT, we obtain

$$S(\omega) = \pi[\delta(\omega - \omega_c) + \delta(\omega + \omega_c)] + \frac{1}{2}k[M(\omega - \omega_c) + M(\omega + \omega_c)]. \tag{8.3}$$

Equation (8.3) proves that the spectrum of the modulated signal $s(t)$ is the sum of three components: the scaled spectrum of the carrier signal, the scaled replica of the modulating signal $m(t)$ shifted to $+\omega_c$, and the scaled replica of the modulating signal $m(t)$ shifted to $-\omega_c$. This result is illustrated in Fig. 8.3(c) for the baseband signal $m(t)$, which is band-limited to ω_{max} and has the spectrum shown in Fig. 8.3(a). The two replicas of the CTFT $M(\omega)$ of the modulating signal in Fig. 8.3(c) are referred to as the side bands of the AM signal.

Table 7.4. Comparison of the different implementations of a lowpass filter

Type of filter	Order	Pass band	Transition band	Stop band
Butterworth	highest order (4)	monotonic gain	widest width	monotonic gain
		either pass or stop bands specs are met; the other is overdesigned		
Type I Chebyshev	moderate order (3)	ripples are present; exact specs are met	narrow width	monotonic gain; overdesigned specs
Type II Chebyshev	moderate order (3); same as Type I	montonic gain; overdesigned specs	narrow width; similar to Type I	ripples are present; exact specs are met
Elliptic	lowest order (2)	ripples are present; exact specs are met	narrowest width	ripples are present; exact specs are met

Using MATLAB,
$\psi[0.25] = \text{ellipke}(0.25^2) = 1.5962$, $\psi[0.9958] = \text{ellipke}(0.9958^2) = 3.7830$, $\psi[0.0085] = \text{ellipke}(0.0085^2) = 1.5708$, and $\psi[0.8660] = \text{ellipke}(0.8660^2) = 2.1564$. The value of N is given by

$$N = \frac{1.5962 \times 3.7830}{1.5708 \times 2.1564} = 1.7827.$$

Rounding off to the nearest higher integer, the order N of the filter equals 2.

Examples 7.6 to 7.9 designed a lowpass filter for the same specifications based on four different implementations derived from the Butterworth, Type I Chebyshev, Type II Chebyshev, and elliptic filters. Table 7.4 compares the properties of these four implementations with respect to the frequency responses within the pass, transition, and stop bands.

In terms of the complexity of the implementations, the elliptic filters provide the lowest order at the expense of equiripple gains in both the pass and stop bands. The Chebyshev filters provide monotonic gain in either the pass or stop band, but increase the order of the implementation. The Butterworth filters provide monotonic gains of maximally flat nature in both the pass and stop bands. However, the Butterworth filters are of the highest order and have the widest transition bandwidth.

Another factor considered in choice of implementation is the phase response of the filter. Generally, ripples add non-linearity to the phase responses. Therefore, the elliptic filter may not be the best choice in applications where a linear phase is important.

7.3.4.1 Elliptic filter design using MATLAB

MATLAB provides the `ellipord` and `ellip` functions to implement the elliptic filters. The usage of these functions is similar to the `cheb1ord` and

Fig. 7.11. Magnitude spectrum of the elliptic lowpass filter designed in Example 7.9.

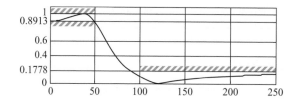

`cheby1` functions used to design Type I Chebyshev filters. The code to implement an elliptic filter for Example 7.9 is as follows:

```
>> wp=50; ws=100; rp=1; rs=15;
                           % specify design parameters
>> [N,wn] = ellipord (wp,ws,rp,rs,'s');
                           % determine order and
                           % natural freq
>> [num,den] = ellip(N,rp,rs,wn,'s');
                           % determine num and denom
                           % coeff.
>> Ht = tf(num,den);       % determine transfer
                           % function
>> [H,w] = freqs(num,den); % determine magnitude
                           % spectrum
>> plot(w,abs(H));         % plot magnitude spectrum
```

Stepwise implementation of the above code returns the following values for different variables:

```
Instruction II:    N = 2; wn = 50;
Instruction III:    num = [0.1778 0 2639.66];
   den = [1.0000 48.384 2961.75];
Instruction IV: Ht = (0.1778s^2 + 2640)/(s^2 + 48.38s
   + 2962);
```

The magnitude spectrum is plotted in Fig. 7.11.

7.4 Frequency transformations

In Section 7.3, we designed a collection of specialized CT lowpass filters. In this section, we consider the design techniques for the remaining three categories (highpass, bandpass, and bandstop filters) of CT filters. A common approach for designing CT filters is to convert the desired specifications into the specifications of a normalized or prototype lowpass filter using a frequency transformation that maps the required frequency-selective filter into a lowpass filter. Based on the transformed specifications, a normalized lowpass filter is designed using the techniques covered in Section 7.3. The transfer function $H(S)$ of the normalized lowpass filter is then transformed back into the original frequency domain. Transformation for converting a lowpass filter to a highpass filter is considered next, followed by the lowpass to bandpass, and lowpass to bandstop transformations.

7.4.1 Lowpass to highpass filter

The transformation that converts a lowpass filter with the transfer function $H(S)$ into a highpass filter with transfer function $H(s)$ is given by

$$S = \frac{\xi_p}{s}, \qquad (7.67)$$

where $S = \sigma + j\omega$ represents the lowpass domain and $s = \gamma + j\xi$ represents the highpass domain. The frequency $\xi = \xi_p$ represents the pass-band corner frequency for the highpass filter. In terms of the CTFT domain, Eq. (7.67) can be expressed as follows:

$$\omega = -\frac{\xi_p}{\xi} \quad \text{or} \quad \xi = -\frac{\xi_p}{\omega}. \qquad (7.68)$$

Figure 7.12 shows the effect of applying the frequency transformation in Eq. (7.68) to the specifications of a highpass filter. Equation (7.68) maps the highpass specifications in the range $-\infty < \xi \le 0$ to the specifications of a lowpass filter in the range $0 \le \omega < \infty$. Similarly, the highpass specifications for the positive range of frequencies $(0 < \xi \le \infty)$ are mapped to the lowpass specifications within the range $-\infty \le \omega < 0$. Since the magnitude spectra are symmetrical about the y-axis, the change from positive ξ frequencies to negative ω frequencies does not affect the nature of the filter in the entire domain.

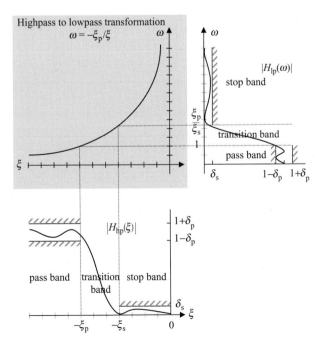

Fig. 7.12. Highpass to lowpass transformation.

From Fig. 7.12, it is clear that Eq. (7.68), or alternatively Eq. (7.67), represents a highpass to lowpass transformation. We now exploit this transformation to design a highpass filter.

Example 7.10

Design a highpass Butterworth filter with the following specifications:

stop band $(0 \leq |\xi| \leq 50$ radians/s) $\quad 20\log_{10}|H(\xi)| \leq -15\,\text{dB}$;

pass band $(|\xi| > 100$ radians/s) $\quad -1\,\text{dB} \leq 20\log_{10}|H(\xi)| \leq 0$.

Solution

Using Eq. (7.67) with $\xi_p = 100$ radians/s to transform the specifications from the domain $s = \gamma + j\xi$ of the highpass filter to the domain $S = \sigma + j\omega$ of the lowpass filter, we obtain

stop band $(2 < |\omega| \leq \infty$ radians/s) $\quad 20\log_{10}|H(\omega)| \leq 15\,\text{dB}$;

pass band $(|\omega| < 1$ radian/s) $\quad -1\,\text{dB} \leq 20\log_{10}|H(\omega)| \leq 0$.

The above specifications are used to design a normalized lowpass Butterworth filter. Expressed on a linear scale, the pass-band and stop-band gains are given by

$$(1 - \delta_p) = 10^{-1/20} = 0.8913 \quad \text{and} \quad \delta_s = 10^{-15/20} = 0.1778.$$

The gain terms G_p and G_s are given by

$$G_p = \frac{1}{(1 - \delta_p)^2} - 1 = \frac{1}{0.8913^2} - 1 = 0.2588$$

and

$$G_s = \frac{1}{(\delta_s)^2} - 1 = \frac{1}{0.1778^2} - 1 = 30.6327.$$

The order N of the Butterworth filter is obtained using Eq. (7.29) as follows:

$$N = \frac{1}{2} \times \frac{\ln(G_p/G_s)}{\ln(\xi_p/\xi_s)} = \frac{1}{2} \times \frac{\ln(0.2588/30.6327)}{\ln(1/2)} = 3.4435.$$

We round off the order of the filter to the higher integer value as $N = 4$.

Using the stop-band constraint, Eq. (7.32), the cut-off frequency of the required Butterworth filter is given by

$$\omega_c = \frac{\omega_s}{(G_s)^{1/2N}} = \frac{2}{(30.6327)^{1/8}} = 1.3039 \text{ radians/s.}$$

Fig. 7.13. Magnitude spectrum
of the Butterworth highpass
filter designed in Example 7.10.

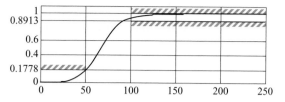

The poles of the lowpass filter are located at

$$S = \omega_c \exp\left[j\frac{\pi}{2} + j\frac{(2n-1)\pi}{8}\right]$$

for $1 \leq n \leq 4$. Substituting different values of n yields

$$S = [-0.4990 + j1.2047 \quad -1.2047 + j0.4990 \quad -1.2047$$
$$-j0.4990 \quad -0.4990 - j1.2047].$$

The transfer function of the lowpass filter is given by

$$H(S) = \frac{K}{(S+0.4490-j1.2047)(S+0.4490+j1.2047)(S+1.2047-j0.4990)(S+1.2047+j0.4990)}$$

or

$$H(S) = \frac{K}{S^4 + 3.4074S^3 + 5.8050S^2 + 5.7934S + 2.8909}.$$

To ensure a dc gain of unity for the lowpass filter, we set $K = 2.8909$. The
transfer function of a unity gain lowpass filter is given by

$$H(S) = \frac{2.8909}{S^4 + 3.4074S^3 + 5.8050S^2 + 5.7934S + 2.8909}.$$

To derive the transfer function of the required highpass filter, we use Eq. (7.67)
with $\xi_p = 100$ radians/s. The transfer function of the highpass filter is given by

$$H(s) = H(S)|_{S=100/s}$$
$$= \frac{2.8909}{(100/s)^4 + 3.4074(100/s)^3 + 5.8050(100/s)^2 + 5.7934(100/s) + 2.8909}$$

or

$$H(s) = \frac{s^4}{s^4 + 2.004 \times 10^2 s^3 + 2.008 \times 10^4 s^2 + 1.179 \times 10^6 s + 3.459 \times 10^7}.$$

The magnitude spectrum of the highpass filter is given in Fig. 7.13, which
confirms that the given specifications are satisfied.

7.4.1.1 MATLAB code for designing highpass filters

The MATLAB code for the design of the highpass filter required in
Example 7.10 using the Butterworth, Type I Chebyshev, Type II Chebyshev,

and elliptic implementations is included below. In each case, MATLAB auto-matically designs the highpass filter. No explicit transformations are needed.

```
>> % Matlab code for designing highpass filter
>> wp=100; ws=50; Rp=1; Rs=15;
                                    % design
                                    % specifications
>>                                  % for Butterworth
                                    % filter
>> [N, wc] = buttord(wp,ws,Rp,Rs,'s');
                                    % determine order
                                    % and cut off
>> [num1,den1] = butter(N,wc,'high','s');
                                    % determine
                                    % transfer
                                    % function
>> H1 = tf(num1,den1);
>> %%%%%                            % Type I Chebyshev
                                    % filter
>> [N, wn] = cheb1ord(wp,ws,Rp,Rs,'s');
>> [num2,den2] = cheby1(N,Rp,wn,'high','s');
>> H2 = tf(num2,den2);
>> %%%%%                            % Type II Chebyshev
                                    % filter
>> [N,wn] = cheb2ord(wp,ws,Rp,Rs,'s') ;
>> [num3,den3] = cheby2(N,Rs,wn,'high','s') ;
>> H3 = tf(num3,den3);
>> %%%%%                            % Elliptic filter
>> [N,wn] = ellipord(wp,ws,Rp,Rs,'s') ;
>> [num4,den4] = ellip(N,Rp,Rs,wn,'high','s') ;
>> H4 = tf(num4,den4);
```

In the above code, note that wp > ws. Also, an additional argument of 'high' is included in the design statements for different filters, which is used to specify a highpass filter. The aforementioned MATLAB code results in the following transfer functions for the different implementations:

Butterworth

$$H(s) = \frac{s^4}{s^4 + 2.004 \times 10^2 s^3 + 2.008 \times 10^4 s^2 + 1.179 \times 10^6 s + 3.459 \times 10^7};$$

Type I Chebyshev $$H(s) = \frac{s^3}{s^3 + 252.1 s^2 + 2.012 \times 10^4 s + 2.035 \times 10^6};$$

Type II Chebyshev $$H(s) = \frac{s^3 + 3.027 \times 10^3 s}{s^3 + 113.5 s^2 + 9.473 \times 10^3 s + 3.548 \times 10^5};$$

elliptic $$H(s) = \frac{0.8903 s^2 + 1501}{s^2 + 81.68 s + 8441}.$$

The transfer function for the Butterworth filter is the same as that derived analytically in Example 7.10.

7.4.2 Lowpass to bandpass filter

The transformation that converts a lowpass filter with the transfer function $H(S)$ into a bandpass filter with transfer function $H(s)$ is given by

$$S = \frac{s^2 + \xi_{p1}\xi_{p2}}{s(\xi_{p2} - \xi_{p1})},\tag{7.69}$$

where $S = \sigma + j\omega$ represents the lowpass domain and $s = \gamma + j\xi$ represents the bandpass domain. The frequency $\xi = \xi_{p1}$ and ξ_{p2} represents the two pass-band corner frequencies for the bandpass filter with $\xi_{p2} > \xi_{p1}$. In terms of the CTFT variables ω and ξ, Eq. (7.69) can be expressed as follows:

$$\omega = \frac{\xi^2 - \xi_{p1}\xi_{p2}}{\xi(\xi_{p2} - \xi_{p1})}.\tag{7.70}$$

From Eq. (7.70), it can be shown that the pass-band corner frequencies ξ_{p1} and $-\xi_{p2}$ of the bandpass filter are both mapped in the lowpass domain to $\omega = -1$, whereas the pass-band corner frequencies $-\xi_{p1}$ and ξ_{p2} are mapped to $\omega = 1$. Also, the pass band $\xi_{p1} \leq |\xi| = \xi_{p2}$ of the bandpass filter is mapped to the pass band $-1 \leq |\xi| \leq 1$ of the lowpass filter. These results can be confirmed by substituting different values for the bandpass domain frequencies ξ and evaluating the corresponding lowpass domain frequencies.

Considering the stop-band corner frequencies of the bandpass filter, Eq. (7.70) can be used to show that the stop-band corner frequency $\pm\xi_{s1}$ is mapped to

$$\omega_{s1} = \left| \frac{\xi_{p1}\xi_{p2} - \xi_{s1}^2}{\xi_{s1}(\xi_{p2} - \xi_{p1})} \right|,\tag{7.71}$$

and that the stop-band corner frequency $\pm\xi_{s2}$ is mapped to

$$\omega_{s2} = \left| \frac{\xi_{p1}\xi_{p2} - \xi_{s2}^2}{\xi_{s2}(\xi_{p2} - \xi_{p1})} \right|.\tag{7.72}$$

As a lower value for the stop-band frequency for the lowpass filter leads to more stringent requirements, the stop-band corner frequency for the lowpass filter is selected from the minimum of the two values computed in Eqs. (7.71) and (7.72). Mathematically, this implies that

$$\omega_s = \min(\omega_{s1}, \omega_{s2}).\tag{7.73}$$

Example 7.11 designs a bandpass filter.

Example 7.11

Design a bandpass Butterworth filter with the following specifications:

stop band I $(0 \leq \mid\xi\mid \leq 50$ radians/s)	$20\log_{10}\mid H(\xi)\mid \leq -20$ dB;
pass band $(100 \leq \mid\xi\mid \leq 200$ radians/s)	-2 dB $\leq 20\log_{10}\mid H(\xi)\mid \leq 0$;
stop band II $(\mid\xi\mid \geq 380$ radians/s)	$20\log_{10}\mid H(\xi)\mid \leq -20$ dB.

Solution

For $\xi_{p1} = 100$ radians/s and $\xi_{p2} = 200$ radians/s, Eq. (7.70) becomes

$$\omega = \frac{\xi^2 - 2 \times 10^4}{100\xi},$$

to transform the specifications from the domain $s = \gamma + j\xi$ of the bandpass filter to the domain $S = \sigma + j\omega$ of the lowpass filter. The specifications for the normalized lowpass filter are given by

pass band $(0 \leq \mid\omega\mid < 1$ radian/s) $-2 \leq 20\log_{10}\mid H(\omega)\mid \leq 0$;

stop band $(\mid\omega\mid \geq \min(3.2737, 3.5)$ radians/s $20\log_{10}\mid H(\omega)\mid \leq -20$.

The above specifications are used to design a normalized lowpass Butterworth filter. Expressed on a linear scale, the pass-band and stop-band gains are given by

$$(1 - \delta_p) = 10^{-2/20} = 0.7943 \quad \text{and} \quad \delta_s = 10^{-20/20} = 0.1.$$

The gain terms G_p and G_s are given by

$$G_p = \frac{1}{(1 - \delta_p)^2} - 1 = \frac{1}{0.7943^2} - 1 = 0.5850$$

and

$$G_s = \frac{1}{(\delta_s)^2} - 1 = \frac{1}{0.1^2} - 1 = 99.$$

The order N of the Butterworth filter is obtained using Eq. (7.29) as follows:

$$N = \frac{1}{2} \times \frac{\ln(G_p/G_s)}{\ln(\xi_p/\xi_s)} = \frac{1}{2} \times \frac{\ln(0.5850/99)}{\ln(1/3.2737)} = 2.1232.$$

We round off the order of the filter to the higher integer value as $N = 3$.

Using the stop-band constraint, Eq. (7.31), the cut-off frequency of the lowpass Butterworth filter is given by

$$\omega_c = \frac{\omega_s}{(G_s)^{1/2N}} = \frac{3.2737}{(99)^{1/6}} = 1.5221 \text{ radians/s.}$$

The poles of the lowpass filter are located at

$$S = \omega_c \exp\left[j\frac{\pi}{2} + j\frac{(2n-1)\pi}{6}\right]$$

Fig. 7.14. Magnitude spectrum of the Butterworth bandpass filter designed in Example 7.11.

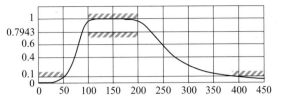

for $1 \leq n \leq 3$. Substituting different values of n yields

$$S = [-0.7610 + \text{j}1.3182 \quad -0.7610 - \text{j}1.3182 \quad -1.5221].$$

The transfer function of the lowpass filter is given by

$$H(S) = \frac{K}{(S + 0.7610 + \text{j}1.3182)(S + 0.7610 + \text{j}1.3182)(S + 1.5221)}$$

or

$$H(S) = \frac{K}{S^3 + 3.0442 S^2 + 4.6336 S + 3.5264}.$$

To ensure a dc gain of unity for the lowpass filter, we set $K = 3.5364$. The transfer function of the unity gain lowpass filter is given by

$$H(S) = \frac{3.5264}{S^3 + 3.0442 S^2 + 4.6336 S + 3.5264}.$$

To derive the transfer function of the required bandpass filter, we use Eq. (7.69) with $\xi_{p1} = 100$ radians/s and $\xi_{p2} = 200$ radians/s. The transformation is given by

$$S = \frac{s^2 + 2 \times 10^4}{100\,s},$$

from which the transfer function of the bandpass filter is calculated as follows:

$$H(s) = H(S)\big|_{S = \frac{s^2+2\times10^4}{100\,s}}$$

$$= \frac{3.5264}{\left[\dfrac{s^2 + 2 \times 10^4}{100\,s}\right]^3 + 3.0442 \left[\dfrac{s^2 + 2 \times 10^4}{100\,s}\right]^2 + 4.6336 \left[\dfrac{s^2 + 2 \times 10^4}{100\,s}\right] + 3.5264},$$

which reduces to

$$H(s) = \frac{3.5264 \times 10^6 s^3}{s^6 + 3.0442 \times 10^2 s^5 + 1.0633 \times 10^5 s^4 + 1.5703 \times 10^7 s^3 + 2.1267 \times 10^9 s^2 + 1.2177 \times 10^{11} s + 8 \times 10^{12}}.$$

The magnitude spectrum of the bandpass filter is given in Fig. 7.14, which confirms that the given specifications for the bandpass filter are satisfied.

7.4.2.1 MATLAB code for designing bandpass filters

The MATLAB code for the design of the bandpass filter required in Example 7.11 using the Butterworth, Type I Chebyshev, Type II Chebyshev, and elliptic implementations is as follows:

```
>> % MATLAB code for designing bandpass filter
>> wp=[100 200]; ws=[50 380]; Rp=2;        % Specifications
   Rs=20;
>> % Butterworth filter
>> [N, wc] = buttord(wp,ws,Rp,Rs,'s');
>> [num1,den1] = butter(N,wc,'s');
>> H1 = tf(num1,den1);
>> % Type I Chebyshev filter
>> [N, wn] = cheb1ord(wp,ws,Rp,Rs,'s');
>> [num2,den2] = cheby1(N,Rp,wn,'s');
>> H2 = tf(num2,den2);
>> % Type II Chebyshev filter
>> [N,wn] = cheb2ord(wp,ws,Rp,Rs,'s');
>> [num3,den3] = cheby2(N,Rs,wn,'s');
>> H3 = tf(num3,den3);
>> % Elliptic filter
>> [N,wn] = ellipord(wp,ws,Rp,Rs,'s');
>> [num4,den4] = ellip(N,Rp,Rs,wn,'s');
>> H4 = tf(num4,den4);
```

The type of filter is specified by the dimensions of the pass-band and stop-band frequency vectors. Since wp and ws are both vectors, MATLAB knows that either a bandpass or bandstop filter is being designed. From the range of the values within wp and ws, MATLAB is also able to differentiate whether a bandpass or a bandstop filter is being specified. In the above example, since the range (50–380 Hz) of frequencies specified within the stop-band frequency vector ws exceeds the range (100–200 Hz) specified within the pass-band frequency vector wp, MATLAB is able to make the final determination that a bandpass filter is being designed. For a bandstop filter, the converse would hold true.

The aforementioned MATLAB code produces bandpass filters with the following transfer functions:

Butterworth $$H(s) = \frac{3.526 \times 10^6 s^4}{s^6 + 304.4 s^5 + 1.063 \times 10^5 s^4 + 1.57 \times 10^7 s^3 + 2.127 \times 10^9 s^2 + 1.218 \times 10^{11} s + 8 \times 10^{12}};$$

Type I Chebyshev $$H(s) = \frac{6.538 \times 10^3 s^2}{s^4 + 80.38 s^3 + 4.8231 \times 10^4 s^2 + 1.608 \times 10^6 s + 4 \times 10^8};$$

Type II Chebyshev $$H(s) = \frac{0.1 s^4 + 1.801 \times 10^4 s^2 + 4 \times 10^7}{s^4 + 1.588 \times 10^2 s^3 + 5.4010 \times 10^4 s^2 + 3.176 \times 10^6 s + 4 \times 10^8};$$

elliptic $$H(s) = \frac{0.1 s^4 + 1.101 \times 10^4 s^2 + 4 \times 10^7}{s^4 + 74.67 s^3 + 4.8819 \times 10^4 s^2 + 1.493 \times 10^6 s + 4 \times 10^8}.$$

Note that the transfer function for the bandpass Butterworth filter is the same as that derived by hand in Example 7.9.

7.4.3 Lowpass to bandstop filter

The transformation to convert a lowpass filter with the transfer function $H(S)$ into a bandstop filter with transfer function $H(s)$ is given by the following expression:

$$S = \frac{s(\xi_{p2} - \xi_{p1})}{s^2 + \xi_{p1}\xi_{p2}}, \tag{7.74}$$

where $S = \sigma + j\omega$ represents the lowpass domain and $s = \gamma + j\xi$ represents the bandstop domain. The frequency $\xi = \xi_{p1}$ and ξ_{p2} represents the two pass-band corner frequencies for the bandpass filter with $\xi_{p2} > \xi_{p1}$. Note that the transformation in Eq. (7.74) is the inverse of the lowpass to bandpass transformation specified in Eq. (7.69).

In terms of the CTFT domain, Eq. (7.74) can be expressed as follows:

$$\omega = \frac{\xi(\xi_{p2} - \xi_{p1})}{\xi_{p1}\xi_{p2} - \xi^2}, \tag{7.75}$$

which can be used to confirm that Eq. (7.74) is indeed a lowpass to bandstop transformation.

As for the bandpass filter, Eq. (7.75) leads to two different values of the stop-band frequencies,

$$\omega_{s1} = \left| \frac{\xi_{s1}(\xi_{p2} - \xi_{p1})}{\xi_{p1}\xi_{p2} - \xi_{s1}^2} \right| \quad \text{and} \quad \omega_{s2} = \left| \frac{\xi_{s2}(\xi_{p2} - \xi_{p1})}{\xi_{p1}\xi_{p2} - \xi_{s2}^2} \right| \tag{7.76}$$

for the lowpass filter. The smaller of the two values is selected as the stop-band corner frequency for the normalized lowpass filter. Example 7.12 designs a bandstop filter.

Example 7.12
Design a bandstop Butterworth filter with the following specifications:

pass band I ($0 \le |\xi| \le 100$ radians/s) $-2\,\text{dB} \le 20\log_{10}|H(\xi)| \le 0$;

stop band ($150 \le |\xi| \le 250$ radians/s) $20\log_{10}|H(\xi)| \le -20\,\text{dB}$;

pass band II ($|\xi| \ge 370$ radians/s) $-2\,\text{dB} \le 20\log_{10}|H(\xi)| \le 0$.

Solution
For $\xi_{p1} = 100$ radians/s and $\xi_{p2} = 370$ radians/s, Eq. (7.70) becomes

$$\omega = \frac{270\xi}{3.7 \times 10^4 - \xi^2},$$

to transform the specifications from the domain $s = \gamma + j\xi$ of the bandstop filter to the domain $S = \sigma + j\omega$ of the lowpass filter. The specifications for the

The aforementioned MATLAB code produces the following transfer functions for the four filters:

Butterworth $H(s) = \dfrac{s^6 + 1.125 \times 10^5 s^4 + 4.219 \times 10^9 s^2 + 5.273 \times 10^{13}}{s^6 + 430.2 s^5 + 2.05 \times 10^5 s^4 + 4.221 \times 10^7 s^3 + 7.688 \times 10^9 s^2 + 6.049 \times 10^{11} s + 5.273 \times 10^{13}}$;

Type I Chebyshev $H(s) = \dfrac{0.7943 s^4 + 5.957 \times 10^4 s^2 + 1.117 \times 10^9}{s^4 + 262.4 s^3 + 1.627 \times 10^5 s^2 + 9.839 \times 10^6 s + 1.406 \times 10^9}$;

Type II Chebyshev $H(s) = \dfrac{s^4 + 8.015 \times 10^4 s^2 + 1.406 \times 10^9}{s^4 + 304.5 s^3 + 1.265 \times 10^5 s^2 + 1.142 \times 10^6 s + 1.406 \times 10^9}$;

elliptic $H(s) = \dfrac{0.7943 s^4 + 6.776 \times 10^4 s^2 + 1.117 \times 10^9}{s^4 + 227.5 s^3 + 1.568 \times 10^5 s^2 + 8.53 \times 10^6 s + 1.406 \times 10^9}$.

7.5 Summary

Chapter 7 defines the CT filters as LTI systems used to transform the frequency characteristics of the CT signals in a predefined manner. Based on the magnitude spectrum $|H(\omega)|$, Section 7.1 classifies the frequency-selective filters into four different categories.

(1) An ideal lowpass filter removes frequency components above the cut-off frequency ω_c from the input signal, while retaining the lower frequency components $\omega \leq \omega_c$.
(2) An ideal highpass filter is the converse of the lowpass filter since it removes frequency components below the cut-off frequency ω_c from the input signal, while retaining the higher frequency components $\omega \leq \omega_c$.
(3) An ideal bandpass filter retains a selected range of frequency components between the lower cut-off frequency ω_{c1} and the upper cutoff frequency ω_{c2} of the filter. All other frequency components are eliminated from the input signal.
(4) A bandstop filter is the converse of the bandpass filter and rejects all frequency components between the lower cut-off frequency ω_{c1} and the upper cut-off frequency ω_{c2} of the filter. All other frequency components are retained at the output of the bandstop filter.

The ideal frequency filters are not physically realizable. Section 7.2 introduces practical implementations of the ideal filters obtained by introducing ripples in the pass and stop bands. A transition band is also included to eliminate the sharp transition between the pass and stop bands.

In Section 7.3, we considered the design of practical lowpass filters. We presented four implementations of practical filters: Butterworth, Type I Chebyshev, Type II Chebyshev, and elliptic filters, for which the design algorithms were covered. The Butterworth filters provide a maximally flat gain within the pass band but have a higher-order N than the Chebyshev and elliptic filters designed with the same specifications. By introducing ripples within the pass band, Type I Chebyshev filters reduce the required order N of the designed filter. Alternatively, Type II Chebyshev filters introduce ripples within the stop band to reduce the order N of the filter. The elliptic filters allow ripples in both the pass and stop bands to derive a filter with the lowest order N among the four implementations. MATLAB instructions to design the four implementations are also presented in Section 7.3.

In Section 7.4, we covered three transformations for converting a highpass filter to a lowpass filter, a bandpass to a lowpass filter, and a bandstop to a lowpass filter. Using these transformations, we were able to map the specifications of any type of the frequency-selective filters in terms of a normalized lowpass filter. After designing the normalized lowpass filter using the design algorithms covered in Section 7.3, the transfer function of the lowpass filter is transformed back into the original domain of the frequency-selective filter.

Problems

7.1 Determine the impulse response of an ideal bandpass filter and an ideal bandstop filter. In each case, assume a gain of A within the pass bands and cut off frequencies of ω_{c1} and ω_{c2}.

7.2 Derive and sketch the location of the poles for the Butterworth filters of orders $N = 12$ and 13 in the complex s-plane.

7.3 Show that a lowpass Butterworth filter with an odd value of order N will always have at least one pole on the real axis in the complex s-plane.

7.4 Show that all complex poles of the lowpass Butterworth filter occur in conjugate pairs.

7.5 Show that the Nth -order Type I Chebyshev polynomial $T_N(\omega)$ has N simple roots in the interval $[-1, 1]$, which are given by

$$\omega_n = \cos\left[\frac{(2n+1)\pi}{2N}\right] \qquad 0 \le n \le N - 1.$$

7.6 Show that the roots of the characteristic equation

$$1 + \varepsilon^2 T_N^2(j/s) = 0$$

for the Type II Chebyshev filter are the inverse of the roots of the characteristic equation

$$1 + \varepsilon^2 T_N^2(s/j) = 0$$

for the Type I Chebyshev filter.

7.7 Design a Butterworth lowpass filter for the following specifications:

pass band $(0 \le |\omega| \le 10 \text{ radians/s})$ $0.9 \le |H(\omega)| \le 1$;

stop band $(|\omega| > 20 \text{ radians/s})$ $|H(\omega)| \le 0.10,$

by enforcing the pass-band requirements. Repeat for the stop-band requirements. Sketch the magnitude spectrum and confirm that the magnitude spectrum satisfies the design specifications.

7.8 Repeat Problem 7.7 for the following specifications:

pass band $(0 \leq |\omega| \leq 50$ radians/s$)$ $-1 \leq 20 \log_{10}|H(\omega)| \leq 0$;

stop band $(|\omega| > 65$ radians/s$)$ $20 \log_{10}|H(\omega)| \leq -25$.

7.9 Repeat (a) Problem 7.7 and (b) Problem 7.8 for the Type I Chebyshev filter.

7.10 Repeat (a) Problem 7.7 and (b) Problem 7.8 for the Type II Chebyshev filter.

7.11 Determine the order of the elliptic filters for the specifications included in (a) Problem 7.7 and (b) Problem 7.8.

7.12 Using the results in Problems 7.7–7.11, compare the implementation complexity of the Butterworth, Type I Chebyshev, Type II Chebyshev, and elliptic filters for the specifications included in (a) Problem 7.7 and (b) Problem 7.8.

7.13 By selecting the corner frequencies of the pass and stop bands, show that the transformation

$$S = \frac{s(\xi_{p2} - \xi_{p1})}{s^2 + \xi_{p1}\xi_{p2}}$$

maps a normalized lowpass filter into a bandstop filter.

7.14 Design a Butterworth highpass filter for the following specifications:

stop band $(0 \leq |\omega| \leq 15$ radians/s$)$ $|H(\omega)| \leq 0.15$;

pass band $(|\omega| > 30$ radians/s$)$ $0.85 \leq |H(\omega)| \leq 1$.

Sketch the magnitude spectrum and confirm that it satisfies the design specifications.

7.15 Repeat Problem 7.14 for the Type I Chebyshev filter.

7.16 Repeat Problem 7.14 for the Type II Chebyshev filter.

7.17 Design a Butterworth bandpass filter for the following specifications:

stop band I $(0 \leq |\xi| \leq 75$ radians/s$)$ $20 \log_{10}|H(\omega)| \leq -15$;

pass band $(100 \leq |\xi| \leq 150$ radians/s$)$ $-1 \leq 20 \log_{10}|H(\omega)| \leq 0$;

stop band II $(|\xi| \geq 175$ radians/s$)$ $20 \log_{10}|H(\omega)| \leq -15$.

Sketch the magnitude spectrum and confirm that it satisfies the design specifications.

7.18 Repeat Problem 7.17 for the Type I Chebyshev filter.

7.19 Repeat Problem 7.17 for the Type II Chebyshev filter.

7.20 Design a Butterworth bandstop filter for the following specifications:

pass band I $(0 \leq |\xi| \leq 25 \text{ radians/s})$ 　 $-4 \leq 20 \log_{10}|H(\omega)| \leq 0$;

stop band $(100 \leq |\xi| \leq 250 \text{ radians/s})$ 　 $20 \log_{10}|H(\omega)| \leq -20$;

pass band II $(|\xi| \geq 325 \text{ radians/s})$ 　 $-4 \leq 20 \log_{10}|H(\omega)| \leq 0$.

Sketch the magnitude spectrum and confirm that it satisfies the design specifications.

7.21 Repeat Problem 7.20 for the Type I Chebyshev filter.

7.22 Repeat Problem 7.20 for the Type II Chebyshev filter.

7.23 Determine the transfer function of the four implementations: (a) Butterworth, (b) Type I Chebyshev, (c) Type II Chebyshev, and (d) elliptic, of the lowpass filter specified in Problem 7.7 using MATLAB. Plot the frequency characteristics and confirm that the specifications are satisfied by the designed implementations.

7.24 Determine the transfer function of the four implementations: (a) Butterworth, (b) Type I Chebyshev, (c) Type II Chebyshev, and (d) elliptic, of the lowpass filter specified in Problem 7.8 using MATLAB. Plot the frequency characteristics and confirm that the specifications are satisfied by the designed implementations.

7.25 Determine the transfer function of the four implementations: (a) Butterworth, (b) Type I Chebyshev, (c) Type II Chebyshev, and (d) elliptic, of the highpass filter specified in Problem 7.14 using MATLAB. Plot the frequency characteristics and confirm that the specifications are satisfied by the designed implementations.

7.26 Determine the transfer function of the four implementations (a) Butterworth, (b) Type I Chebyshev, (c) Type II Chebyshev, and (d) elliptic, of the bandpass filter specified in Problem 7.17 using MATLAB. Plot the frequency characteristics and confirm that the specifications are satisfied by the designed implementations.

7.27 Determine the transfer function of the four implementations: (a) Butterworth, (b) Type I Chebyshev, (c) Type II Chebyshev, and (d) elliptic, of the bandstop filter specified in Problem 7.20 using MATLAB. Plot the frequency characteristics and confirm that the specifications are satisfied by the designed implementations.

8 Case studies for CT systems

Several aspects of continuous-time (CT) systems were covered in Chapters 3–7. Among the concepts covered, we used the convolution integral in Chapter 3 to determine the output $y(t)$ of a linear time-invariant, continuous-time (LTIC) system from the input signal $x(t)$ and the impulse response $h(t)$. Chapters 4 and 5 defined the frequency representations, namely the CT Fourier transform (CTFT) and the CT Fourier series (CTFS) and evaluated the output signal $y(t)$ of the LTIC system in the frequency domain. The CTFT was also used to estimate the frequency characteristics of the LTIC system by plotting the magnitude and phase spectra. Chapter 6 introduced the Laplace transform widely used as an alternative for the CTFT in control systems, where the analysis of the transient response is of paramount importance. Chapter 7 presented techniques for designing LTIC systems based on the specified frequency characteristics. When an LTIC system is described in terms of its frequency characteristics, it is referred to as a frequency-selective filter. Design techniques for four types of analog filters, lowpass filters, highpass filters, bandpass filters, and bandstop filters, were also covered in Chapter 7. In this chapter, we provide applications for the LTIC systems. Our goal is to illustrate how the tools developed in the earlier chapters can be utilized in real-world applications.

The organization of this chapter is as follows. Section 8.1 considers analog communication systems. In particular, we illustrate the use of amplitude modulation in communication systems for transmitting information to the receivers. Based on the CTFT, spectral analysis of the process of modulation provides insight into the performance of the communication systems. Section 8.2 introduces a spring damper system and shows how the Laplace transform is useful in analyzing the stability of the system. Section 8.3 analyzes the armature-controlled, direct current (dc) motor by deriving its impulse response and transfer function. The immune system of humans is considered in Section 8.4. Analytical models for the immune system are considered and later analyzed using the simulink toolbox available in MATLAB.

Fig. 8.1. Schematic diagram
modeling the process of
amplitude modulation.

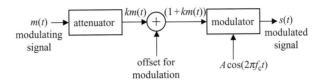

8.1 Amplitude modulation of baseband signals

Section 2.1.3 introduced modulation as a frequency-shifting operation where
the frequency contents of the information-bearing signal are moved to a higher
frequency range. Modulation leads to two main advantages in communications.
First, since the length of the antenna is inversely proportional to the frequency
of the information signal, transmitting information bearing low-frequency
baseband signals directly leads to antennas with impractical lengths. By shifting the frequency content of the information signal to a higher frequency range,
the length of the antenna is considerably reduced. Secondly, modulation leads
to frequency division multiplexing (FDM), where multiple signals are coupled
together by shifting them to a range of different frequencies and are then transmitted simultaneously. This provides considerable savings in the transmission
time and the power consumed by the communication systems. In this section,
we consider a common form of modulation, referred to as amplitude modulation
(AM), used frequently in radio communications.

Amplitude modulation is a popular technique used for broadcasting radio
stations within a local community. In North America, a frequency range of 520
to 1710 kHz is assigned to the AM stations. Typically, each station occupies a
bandwidth of 10 kHz. To limit the range of transmission to a few kilometers,
the transmitted power for a station ranges from 0.1 to 50 kW, such that the same
AM band can be reused by another community without interference. In this
section, we use the CTFT to analyze AM-based communication systems.

A schematic diagram of an AM system is shown in Fig. 8.1, where $m(t)$
represents a baseband signal with non-zero frequency components within the
range $-\omega_{\max} \leq \omega \leq \omega_{\max}$. The output of the AM system is given by

$$s(t) = A[1 + km(t)]\cos(\omega_c t + \phi_c). \qquad (8.1)$$

The multiplication term $A\cos(\omega_c t + \phi_c)$ represents the sinusoidal carrier,
whose amplitude is denoted by A, and the radian frequency is given by
$\omega_c = 2\pi f_c$.[†] The constant phase term ϕ_c is referred to as the epoch of the carrier,
while the factor k is referred to as the modulation index, which is adjusted such
that the intermediate signal $(1 + km(t))$ is always positive for all $t \geq 0$.

[†] Note that ω_c represents the fundamental frequency of the sinusoidal carrier signal $c(t)$, and
should not be confused with ω_c, used to denote the cut-off frequency of the CT filter.

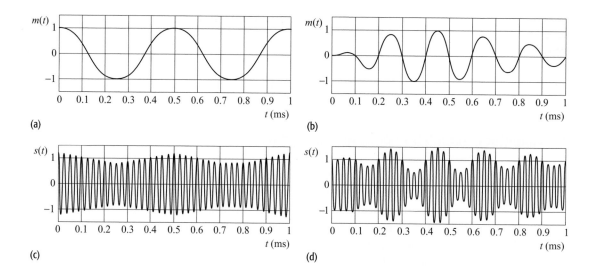

Fig. 8.2. Amplitude modulation in time domain for two different modulating signals: (a) pure sinusoidal signal with fundamental frequency of 2 kHz; (b) synthetic audio signal. The modulated signal for the pure sinusoidal signal is shown in (c) and that for real speech is shown in (d).

Figure 8.2 shows the results of amplitude modulation for two different modulating signals: a pure sinusoidal signal with the fundamental frequency of 2 kHz is plotted in Fig. 8.2(a) and a synthetic audio signal is plotted in Fig. 8.2(b). Both signals are amplitude modulated with the carrier signal $\cos(\omega_c t + \phi_c)$ having a fundamental frequency of $f_c = 40$ kHz and an epoch of $\phi_c = 0$ radians. In the case of the pure sinusoidal signal, the modulation index k is selected to have a value of 0.2, while for the real audio signal the modulation index is set to 0.7. The results of amplitude modulation are shown in Fig. 8.2(c) for the pure sinusoidal signal and in Fig 8.2(d) for the audio signal. In both cases, we observe that the amplitude of the carrier signal is adjusted according to the magnitude of the modulating signal. In other words, the modulating signal acts as an envelope and controls the amplitude of the carrier.

To illustrate the effect of modulation on the frequency content of the modulating signal, we use the CTFT. Equation (8.1) is expressed as follows:

$$s(t) = A\cos(\omega_c t + \phi_c) + Akm(t)\cos(\omega_c t + \phi_c). \tag{8.2}$$

Without loss of generality, we set $A = 1$ and $\phi_c = 0$. Using the multiplication property for the CTFT, we obtain

$$S(\omega) = \pi[\delta(\omega - \omega_c) + \delta(\omega + \omega_c)] + \frac{1}{2}k[M(\omega - \omega_c) + M(\omega + \omega_c)]. \tag{8.3}$$

Equation (8.3) proves that the spectrum of the modulated signal $s(t)$ is the sum of three components: the scaled spectrum of the carrier signal, the scaled replica of the modulating signal $m(t)$ shifted to $+\omega_c$, and the scaled replica of the modulating signal $m(t)$ shifted to $-\omega_c$. This result is illustrated in Fig. 8.3(c) for the baseband signal $m(t)$, which is band-limited to $\omega_{\max}$ and has the spectrum shown in Fig. 8.3(a). The two replicas of the CTFT $M(\omega)$ of the modulating signal in Fig. 8.3(c) are referred to as the side bands of the AM signal.

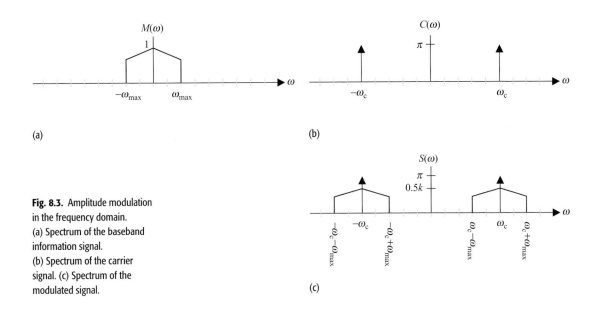

Fig. 8.3. Amplitude modulation in the frequency domain. (a) Spectrum of the baseband information signal. (b) Spectrum of the carrier signal. (c) Spectrum of the modulated signal.

We now consider the extraction of the information signal $x(t)$ from the modulated signal $s(t)$. This procedure is referred to as demodulation, which is explained in Sections 8.1.1 and 8.1.2.

8.1.1 Synchronous demodulation

The objective of demodulation is to reconstruct $m(t)$ from $s(t)$. Analyzing the spectrum $S(\omega)$ of the modulated signal $s(t)$, the following method extracts the information-bearing signal $m(t)$ from $s(t)$.

(1) Frequency shift the modulated signal $s(t)$ by ω_c (or $-\omega_c$). If the modulated signal is frequency-shifted by ω_c, one of the side bands is shifted to zero frequency, while the second side band is shifted to $2\omega_c$. Conversely, if the modulated signal is frequency-shifted by $-\omega_c$, the two side bands are shifted to zero and $-2\omega_c$.

(2) In order to remove the side band shifted to the non-zero frequency, the result obtained in Step (1) is passed through a lowpass filter having a pass band of $(-\omega_{max} \leq \omega \leq \omega_{max})$. The output of the lowpass filter consists of a scaled version of the modulating signal and an impulse at $\omega = 0$. The impulse represents the dc component and is removed by subtracting a constant value in the time domain as shown in Step (3).

(3) A constant voltage equal to the dc component is subtracted from the output of the lowpass signal.

Step (1) can be performed by multiplying the AM signal $s(t)$ by the demodulating carrier $\cos(\omega_c t)$ having the same fundamental frequency and phase as the

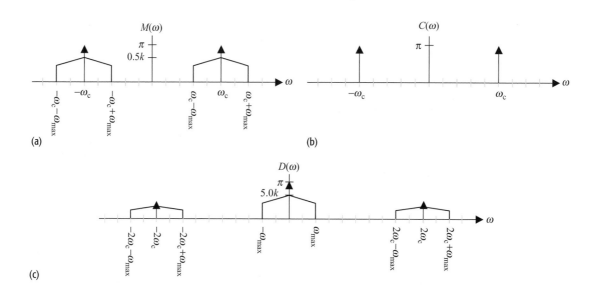

Fig. 8.4. Demodulation in the frequency domain. (a) Spectrum of the modulated signal. (b) Spectrum of the carrier signal. (c) Spectrum of the demodulated signal.

modulating carrier. In the time domain, the result of the multiplication is given by

$$d(t) = s(t)c(t) = [\cos(\omega_c t) + km(t)\cos(\omega_c t)]\cos(\omega_c t), \qquad (8.4)$$

which is expressed as

$$d(t) = s(t)c(t) = \underbrace{\frac{1}{2}[1 + km(t)]}_{d_{\text{low}}(t)} + \underbrace{\frac{1}{2}[1 + km(t)]\cos(2\omega_c t)}_{d_{\text{high}}(t)}. \qquad (8.5)$$

Equation (8.5) shows that the demodulated signal $d(t)$ has two components. The first component $d_{\text{low}}(t)$ is the low-frequency component, which consists of a constant factor of $1/2$ and a scaled replica of the modulated signal. The second component $d_{\text{high}}(t)$ is the higher-frequency component and can be filtered out, as explained next. Taking the CTFT of Eq. (8.5) yields

$$D(\omega) = \frac{1}{2\pi}[S(\omega) * C(\omega)] = \underbrace{\left[\pi\delta(\omega) + \frac{k}{2}M(\omega)\right]}_{d_{\text{low}}(t)}$$

$$+ \underbrace{\left[\frac{\pi}{2}\delta(\omega - 2\omega_c) + \frac{k}{4}M(\omega - 2\omega_c)\right] + \left[\frac{\pi}{2}\delta(\omega + 2\omega_c) + \frac{k}{4}M(\omega + 2\omega_c)\right]}_{d_{\text{high}}(t)},$$

$$(8.6)$$

which is plotted in Fig. 8.4(c). Recall that Fig. 8.4(a) represents the spectrum of the modulated signal $m(t)$ and that Fig. 8.4(b) represents the spectrum of the carrier signal $c(t)$. By filtering $d(t)$ with a lowpass filter having a pass band of

$-\omega_c \leq \omega \leq \omega_c$, the lowpass component $d_{\text{low}}(t)$ is extracted. The information signal $m(t)$ is then obtained from $d_{\text{low}}(t)$ using the following relationship:

$$m(t) = 2[d_{\text{low}}(t) - 1]. \tag{8.7}$$

8.1.2 Synchronous demodulation with non-zero epochs

In synchronous demodulation, the epoch ϕ_c of the modulating carrier is assumed to be identical to the epoch of the demodulating carrier. In practice, perfect synchronization between the carriers is not possible, which leads to distortion in the signal reconstructed from demodulation. To illustrate the effect of distortion introduced by unsynchronized carriers, consider the following modulated signal:

$$s(t) = A\cos(\omega_c t + \phi_c) + Akm(t)\cos(\omega_c t + \phi_c), \tag{8.8}$$

as derived in Eq. (8.2). Assume that the demodulator carrier is given by

$$c_2(t) = A\cos(\omega_c t + \theta_c(t)), \tag{8.9}$$

which has a time-varying epoch $\theta_c(t) \neq \phi_c$. Using $c_2(t)$, the demodulated signal is given by

$$d(t) = s(t)c_2(t) = [A\cos(\omega_c t + \phi_c) + Akm(t)\cos(\omega_c t + \phi_c)]\cos(\omega_c t + \theta_c(t)), \tag{8.10}$$

which simplifies to

$$d(t) = \underbrace{\frac{A}{2}[1 + km(t)]\cos(\phi_c - \theta_c(t))}_{d_{\text{low}}(t)} + \underbrace{\frac{A}{2}[1 + km(t)]\cos(2\omega_c t + \phi_c + \theta_c(t))}_{d_{\text{high}}(t)}. \tag{8.11}$$

Equation (8.11) illustrates that the demodulated signal contains a low-frequency component $d_{\text{low}}(t)$ and a higher-frequency component $d_{\text{high}}(t)$. By passing the demodulated signal through a lowpass filter, the higher-frequency component is removed. The output of the lowpass filter is given by

$$y(t) = \frac{A}{2}[1 + km(t)]\cos(\phi_c - \theta_c(t)). \tag{8.12}$$

Even after eliminating the dc component, the reconstructed signal has the following form:

$$y(t) = \frac{A}{2}km(t)\cos(\phi_c - \theta_c(t)), \tag{8.13}$$

where distortion is caused by the factor of $\cos(\phi_c - \theta_c(t))$. Since the epoch $\theta_c(t)$ is time-varying, it is difficult to eliminate the distortion. To reconstruct $x(t)$ precisely, the phase difference between the carrier signals used at the modulator and demodulator must be kept equal to zero over time. In other words,

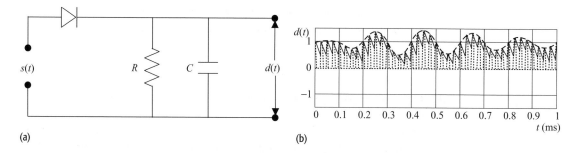

(a)

(b)

Fig. 8.5. Asynchronous AM demodulation. (a) RC parallel circuit coupled with a diode to implement the envelope detector. (b) Reconstructed signal $d(t)$ is shown as a solid line. For comparison, the information component $[1 + km(t)]$ is shown as the envelope of the AM signal (dashed line).

perfect synchronization between the modulator and demodulator is essential to retrieve the information signal $m(t)$. For this reason, the aforementioned modulation scheme based on multiplying the modulated signal by the carrier signal and lowpass filtering is referred to as synchronous demodulation. Although synchronous demodulation is an elegant way of retrieving the information signal $m(t)$, the demodulator has a high implementation cost due to the synchronization required between the two carriers. An alternative scheme, which does not require synchronization of the modulating and demodulating carriers, is referred to as asynchronous demodulation, which is considered in the following section.

8.1.3 Asynchronous demodulation

In amplitude modulation, the information-bearing signal $m(t)$ modulates the magnitude of the carrier signal $c(t)$. This is illustrated in Figs. 8.2(c) and (d), where the envelope of the amplitude modulated signal follows the information component $[1 + km(t)]$. In asynchronous demodulation, we reconstruct the information signal $m(t)$ by tracking the envelope of the modulated signal.

Figure 8.5(a) shows a parallel RC circuit used to reconstruct the information-bearing signal $m(t)$ from the amplitude modulated signal $s(t)$ applied at the input of the RC circuit. The diode acts as a half-wave rectifier removing the negative values from the modulated signal, while the capacitor C tracks the envelope of the AM signal by charging to the peak of the sinusoidal carrier during the positive transition of the signal. During the negative transitions of the carrier, the capacitor discharges slightly, but is again recharged by the next positive transition. The process is illustrated in Fig. 8.5(b), where the demodulated signal is represented by a solid line. We observe that the demodulated signal closely follows the envelope of the modulated signal and is a good approximation of the information-bearing signal.

8.2 Mechanical spring damper system

The spring damping system, considered in Section 2.1.5, is a classic example of a second-order system; the schematic diagram for such a system is shown in

Fig. 8.6. The input–output relationship of the spring damping system is given by Eq. (2.16), which for convenience of reference is repeated below:

$$M\frac{d^2 y}{dt} + r\frac{dy}{dt} + ky(t) = x(t). \tag{8.14}$$

In Eq. (8.14), M is the mass attached to the spring, r is the frictional coefficient, k is the spring constant, $x(t)$ is the force applied to pull the mass, and $y(t)$ is the displacement of the mass caused by the force. In this section, we analyze the spring damping system using the methods discussed in Chapters 5 and 6. Using the Laplace transform, the transfer function determines the stability of the system.

8.2.1 Transfer function

Taking the Laplace transform of both sides of Eq. (8.14) and assuming zero initial conditions, we obtain

$$(Ms^2 + rs + k)Y(s) = X(s), \tag{8.15}$$

which results in the following transfer function:

$$H(s) = \frac{Y(s)}{X(s)} = \frac{1}{(Ms^2 + rs + k)}. \tag{8.16}$$

Alternatively, Eq. (8.16) can be expressed as follows:

$$H(s) = \frac{1/M}{s^2 + (r/M)s + k/M} = \frac{1/M}{s^2 + 2\xi_n \omega_n s + \omega_n^2}, \tag{8.17}$$

where

$$\omega_n = \sqrt{\frac{k}{M}} \quad \text{and} \quad \xi_n = \frac{r}{2\sqrt{kM}}.$$

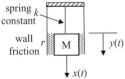

spring k constant

wall friction r

The characteristic equation of the mechanical spring damping system is given by

$$s^2 + 2\xi_n \omega_n s + \omega_n^2 = 0, \tag{8.18}$$

which has two poles at

$$p_1 = -\xi_n \omega_n + \omega_n \sqrt{\xi_n^2 - 1} \quad \text{and} \quad p_2 = -\xi_n \omega_n - \omega_n \sqrt{\xi_n^2 - 1}. \tag{8.19}$$

Depending on the value of ξ_n, the poles p_1 and p_2 may lie in different locations within the s-plane. If $\xi_n = 1$, poles p_1 and p_2 are real-valued and identical. If $\xi_n > 1$, poles p_1 and p_2 are real-valued but not equal. Finally, if $\xi_n < 1$, poles p_1 and p_2 are complex conjugates of each other. In the following, we calculate the impulse response $h(t)$ of the spring damping system for three sets of values of ξ_n and show that the characteristics of the system depend on the value of ξ_n.

Fig. 8.6. Mechanical spring damping system.

Case 1 ($\xi_n = 1$) For $\xi_n = 1$, Eq. (8.17) reduces to

$$H(s) = \frac{1/M}{s^2 + 2\omega_n s + \omega_n^2} = \frac{1/M}{(s + \omega_n)^2}, \tag{8.20}$$

with repeated roots at $s = -\omega_n, -\omega_n$. Taking the inverse transform, the impulse response is given by

$$h(t) = \frac{1}{M} t e^{-\omega_n t} u(t). \tag{8.21}$$

Case 2 ($\xi_n > 1$) For $\xi_n > 1$, the poles p_1 and p_2 of the spring damping system are real-valued and given by

$$p_1 = -\xi_n \omega_n + \omega_n \sqrt{\xi_n^2 - 1} \quad \text{and} \quad p_2 = -\xi_n \omega_n - \omega_n \sqrt{\xi_n^2 - 1}. \tag{8.22}$$

The transfer function of the spring damping system can be expressed as follows:

$$H(s) = \frac{1/M}{s^2 + 2\xi_n \omega_n s + \omega_n^2} = \frac{1/M}{(s - p_1)(s - p_2)}, \tag{8.23}$$

which leads to the impulse response

$$\begin{aligned} h(t) &= \frac{1}{M} \frac{1}{(p_1 - p_2)} [e^{p_1 t} - e^{p_2 t}] u(t) = \frac{1}{2\omega_n M \sqrt{\xi_n^2 - 1}} e^{-\xi_n \omega_n t} \\ &\quad \times \left[e^{\omega_n \sqrt{\xi_n^2 - 1}\, t} - e^{-\omega_n \sqrt{\xi_n^2 - 1}\, t} \right] u(t). \end{aligned} \tag{8.24}$$

Case 3 ($\xi_n < 1$) For $\xi_n < 1$, the poles p_1 and p_2 of the spring damping system are complex and are given by

$$p_1 = -\xi_n \omega_n + j\omega_n \sqrt{1 - \xi_n^2} \quad \text{and} \quad p_2 = -\xi_n \omega_n - j\omega_n \sqrt{1 - \xi_n^2}. \tag{8.25}$$

By repeating the procedure for Case 2, the impulse response of the spring damping system is given by

$$H(s) = \frac{1/M}{s^2 + 2\xi_n \omega_n s + \omega_n^2} = \frac{1/M}{(s - p_1)(s - p_2)}, \tag{8.26}$$

which leads to the impulse response

$$h(t) = \frac{1}{\omega_n M \sqrt{1 - \xi_n^2}} e^{-\xi_n \omega_n t} \sin \left[\omega_n \sqrt{1 - \xi_n^2}\, t \right] u(t). \tag{8.27}$$

Figure 8.7 shows the impulse response of the spring damping system for the three cases considered earlier. We set $M = 10$ and $\omega_n = 0.3$ radians/s in each case. For Case 1 with $\xi_n = 1$, the impulse response decreases monotonically approaching the steady state value of zero. Such systems are referred to as critically damped systems.

For Case 2 with $\xi_n = 4$, the impulse response of the spring damping system again approaches the steady state value of zero. Initially, the deviation from the steady state value is smaller than that of the critically damped system, but the

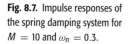

Fig. 8.7. Impulse responses of the spring damping system for $M = 10$ and $\omega_n = 0.3$.

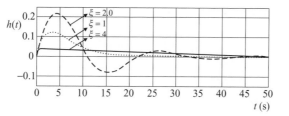

overall duration over which the steady state value is achieved is much longer. Such systems are referred to as overdamped systems.

For Case 3 with $\xi_n = 0.2$, the spring acts as a flexible system. The impulse response approaches the steady state value of zero after several oscillations. Such systems are referred to as underdamped systems. Since the fundamental frequency ω_n is 0.3 radians/s, the period of oscillation is given by

$$T = \frac{2\pi}{\omega_n} = \frac{2\pi}{0.3} = 21.95 \text{ seconds.} \qquad (8.28)$$

Based on Eq. (8.28), parameter ω_n is referred to as the fundamental frequency of the spring damping system. Since parameter ξ_n determines the level of damping, it is referred to as the damping constant.

8.3 Armature-controlled dc motor

Electrical motors form an integral component of most electrical and mechanical devices such as automobiles, ac generators, and power supplies. Broadly speaking, electrical motors can be classified into two categories: direct current (dc) motors and alternating current (ac) motors. Within each category, there are additional subclassifications covering different applications. In this section, we analyze the armature-controlled dc motor by deriving its transfer function and impulse response.

Figure 8.8(a) shows an armature-controlled dc motor, in which an armature, consisting of several copper conducting coils, is placed within a magnetic field generated by a permanent or an electrical magnet. A voltage applied across the armature results in a flow of current through the armature circuit. Interaction between the electrical and magnetic fields causes the armature to rotate, the direction of rotation being determined by the following empirical rule, derived by Faraday.

Extend the thumb, index finger, and middle finger of the right hand such that the three are mutually orthogonal to each other. If the index finger points in the direction of the current and the middle finger in the direction of the magnetic field, then the thumb points in the direction of motion of the armature.

Fig. 8.8. Armature-controlled dc motor. (a) Cross-section; (b) schematic representation.

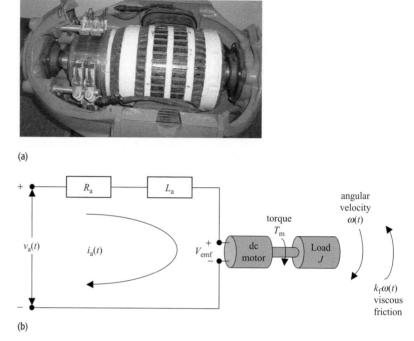

(a)

(b)

8.3.1 Mathematical model

The linear model of the armature-controlled dc motor is shown in Fig. 8.8(b), where a load J is coupled to the armature through a shaft. Rotation of the armature of the dc motor causes the desired motion in the attached load J. Moving a conductor within a magnetic field also generates a back electromagnetic field (emf) to be induced in the dc motor. The back emf results in an opposing emf voltage, which is denoted by V_{emf} in Fig. 8.8(b). In the following analysis, we decompose the motors into three components: armature, motor, and load. The equations for the three components are presented below.

Armature circuit Applying Kirchhoff's voltage law to the armature circuit, we obtain

$$L_a \frac{di_a}{dt} + R_a i_a + \underbrace{k_f \omega(t)}_{V_{emf}(t)} = v_a(t), \tag{8.29}$$

where $v_a(t)$ denotes the armature voltage and $i_a(t)$ denotes the armature current. The electrical components of the armature circuit are given by L_a and R_a, where L_a denotes the self inductance of the armature and R_a denotes the self resistance of the armature. The emf voltage $V_{emf}(t)$ is approximated by the product of the feedback factor k_f and the angular velocity $\omega(t)$.

Motor circuit The torque T_m, induced by the applied voltage across the armature, is given by

$$T_m = k_m i_a(t), \tag{8.30}$$

where k_m is referred to as the motor or armature constant and $i_a(t)$ is the armature current. The armature constant k_m depends on the physical properties of the dc motor such as the strength of the magnetic field and the density of the armature coil.

Load The load component of the dc motor is obtained by applying Newton's third law of motion, which states that the sum of the applied and reactive forces is zero. The applied forces are the torques around the motor shaft. The reactive force causes acceleration of the armature and equals the product of the inertial load J and the derivative of the angular rate $\omega(t)$. In other words,

$$\sum_p T_p = J \frac{d\omega}{dt}, \tag{8.31}$$

where J denotes the inertia of the rotor. There are three different torques, i.e. $p = 3$, observed at the shaft: (i) motor torque T_m represented by Eq. (8.30); (ii) frictional torque T_f given by $r\omega(t)$, r being the frictional constant; and (iii) load disturbance torque T_d. In other words, Eq. (8.31) can be expressed as follows:

$$J \frac{d\omega}{dt} = T_m - r\omega(t) - T_d. \tag{8.32}$$

Since the angular velocity $\omega(t)$ is related to the shaft position $\theta(t)$ by the following expression:

$$\omega(t) = \frac{d\theta}{dt}, \tag{8.33}$$

Eq. (8.32) can be expressed as follows:

$$J \frac{d^2\theta}{dt^2} + r \frac{d\theta}{dt} = T_m - T_d = T_L, \tag{8.34}$$

where T_L denotes the difference between the motor torque T_m and the load disturbance torque T_d.

8.3.2 Transfer function

The dc motor shown in Fig. 8.8 is modeled as a linear time-invariant (LTI) system with the armature voltage $v_a(t)$ considered as the input signal and the shaft position $\theta(t)$ as the output signal. We now derive the transfer function of the linearized model.

Number of antigens At any given time, the total number of antigens present in the human body depends on three factors: (i) external antigens entering the human body from outside; (ii) reproduced antigens produced within the human body by the already existing antigens; and (iii) destroyed antigens that are eradicated by the antibodies. The net change in the number of antigens is modeled by the following equation:

$$\frac{da}{dt} = \alpha a(t) - \eta b(t) + g(t), \tag{8.49}$$

where α denotes the reproduction rate at which the antigens are multiplying within the human body and η is the destruction rate at which the antigens are being destroyed by the antibodies.

Number of lymphocytes Assuming that the number of lymphocytes is proportional to the number of antigens, the number of lymphocytes present within the human body is given by

$$l(t) = \beta a(t), \tag{8.50}$$

where β is the proportionality constant relating the number of lymphocytes to the number of antigens. The value of β generally depends on many factors, including the health of the patient and external stimuli. In general, β varies with time in a non-linear fashion. For simplicity, however, we can assume that β is a constant.

Number of plasma cells The change in the number of plasma cells is proportional to the number of lymphocytes $l(t)$. Typically, there is a delay of τ seconds between the instant that the antigens are detected and the instant that the plasma cells are generated. Therefore, the number of plasma cells depends on $l(t - \tau)$, where the proportionality constant is assumed to be unity. Also, a large portion of plasma cells die due to aging. The number of plasma cells at any time t can therefore be expressed as follows:

$$\frac{dp}{dt} = l(t - \tau) - \gamma p(t), \tag{8.51}$$

where γ denotes the rate at which the plasma cells die due to aging.

Number of antibodies The number of antibodies depends on three factors: (i) new antibodies being generated by the human body (the rate of generation μ of the new antibodies is proportional to the number of plasma cells in the human body); (ii) destroyed antibodies lost to the antigens (the rate of destruction σ of such antibodies is proportional to the number of existing antigens); and (iii) dead antibodies lost to aging. We assume that the antibodies die at the rate of λ because of aging. Combining the three factors, the number of antibodies at any time t is given by

$$\frac{db}{dt} = \mu p(t) - \sigma a(t) - \lambda b(t). \tag{8.52}$$

8.4.2 Transfer function

Equations (8.49)–(8.52) describe the linearized model used to analyze the human immune system. To develop the transfer function, we take the Laplace transform of Eqs. (8.49)–(8.52). The resulting expressions can be expressed as follows:

number of antigens $\qquad A(s) = \dfrac{1}{(s - \alpha)} [G(s) - \eta B(s)]; \qquad$ (8.53)

number of lymphocytes $\qquad L(s) = \beta A(s); \qquad$ (8.54)

number of plasma cells $\qquad P(s) = \dfrac{e^{-\tau s}}{(s + \gamma)} L(s); \qquad$ (8.55)

number of antibodies $\qquad B(s) = \dfrac{1}{(s + \lambda)} [\mu P(s) - \sigma A(s)]. \qquad$ (8.56)

In Eqs. (8.53)–(8.56), variables $A(s)$, $G(s)$, $L(s)$, $P(s)$, and $B(s)$ are, respectively, the Laplace transforms of the number of antigens $a(t)$ present within the human body, the number of antigens $g(t)$ entering the human body, the number of lymphocytes $l(t)$ within the blood, the total number of plasma cells $p(t)$ within the human body, and the number of antibodies $b(t)$ in the blood. Assuming the number of antigens $g(t)$ entering the human body to be the input and the number of antibodies $b(t)$ produced to be the output, the human immune system can be modeled by the schematic diagram shown in Fig. 8.11(a). Figure 8.11(b) is the simplified version of Fig. 8.11(a), which yields the following transfer function for the human immune system:

$$T(s) = \frac{M(s)}{(1 + \eta M(s))} = \frac{\mu \beta e^{-\tau s} - \sigma(s + \gamma)}{(s - \alpha)(s + \lambda)(s + \gamma) + \eta[\mu \beta e^{-\tau s} - \sigma(s + \gamma)]}. \tag{8.57}$$

8.4.3 System simulations

The simplified model of the human immune system is still a fairly complex system to be analyzed analytically. The characteristic equation of the human immune system is not a polynomial of s, therefore evaluation of its poles is difficult. In this section, we simulate the human immune system using the simulink toolbox available in MATLAB.

8.4.3.1 Simulation 1

In simulink, a system is simulated using a block diagram where the subblocks represent different subsystems. Figure 8.12 shows the simulink representation of

Fig. 8.11. Schematic models for the immune response system. (a) Detailed model; (b) simplified model.

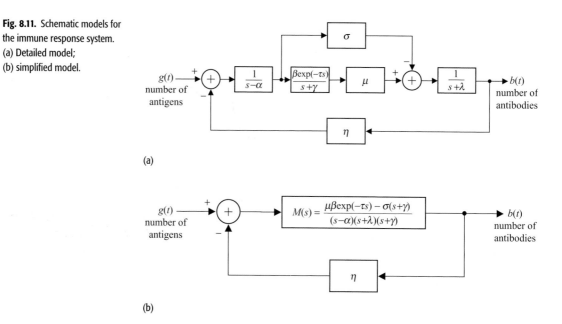

(a)

(b)

the human immune system shown in Fig. 8.11. We have assumed a hypothetical case with the values of the proportionality constants given by

$$\alpha = 0.1, \quad \beta = 0.5, \quad \gamma = 0.1, \quad \mu = 0.5, \quad \tau = 0.2, \quad \lambda = 0.1,$$
$$\sigma = 0.1, \quad \text{and} \quad \eta = 0.5.$$

The proportionality constants α, γ, σ, and λ related to the antigens are deliberately kept smaller than the proportionality constants β, η, and μ related to the antibodies for quick recovery from the infection. The input signal $g(t)$ modeling the number of antigens entering the human body is approximated by a pulse and is shown in Fig. 8.13(a). The duration of the pulse is 0.5 s, implying that the antigens keep entering the human body at a constant level for the first 0.5 s. The outputs $a(t)$, $p(t)$, and $b(t)$ are monitored by the simulated scope available in simulink. The output of the scope is shown in Fig. 8.13(b), where we observe

Fig. 8.12. Simulink model for Simulation 1 modeling the immune response system of humans.

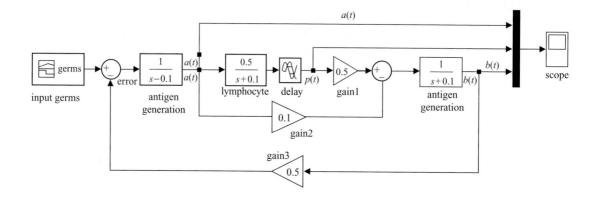

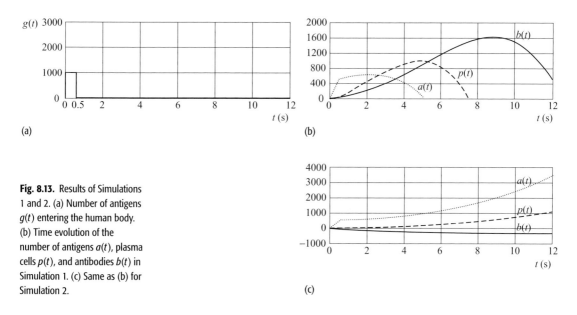

Fig. 8.13. Results of Simulations 1 and 2. (a) Number of antigens $g(t)$ entering the human body. (b) Time evolution of the number of antigens $a(t)$, plasma cells $p(t)$, and antibodies $b(t)$ in Simulation 1. (c) Same as (b) for Simulation 2.

that the number of antigens increases linearly for the initial duration of 0.5 s. Since the human body generates lymphocytes with a delay τ, which is 0.2 s in our simulation, the number of plasma cells $p(t)$ starts rising with a delay of 0.2 s. After 0.5 s, no external antigens enter the human body. However, new antigens are being reproduced by the already existing antigens present inside the human body. As a result, the number of antigens $a(t)$ keeps rising, even after 0.5 s. After roughly 3 s, the respective strengths of lymphocytes and plasma cells is high enough to impact the overall population of the antigens. The number of antigens $a(t)$ starts decreasing after 3 s. After 5.3 s, all antigens in the body are destroyed. At this time, the body stops producing any further plasma cells. After this stage, the number of plasma cells $p(t)$ starts decreasing, as some of these cells die naturally due to aging. As the number of plasma cells decreases, the number of antibodies $b(t)$ also decreases and after 10 s no antibodies are present in the simulation.

8.4.3.2 Simulation 2

Simulation 1 models successful eradication of the antigens. Let us now consider the other extreme, where the antigens are lethal such that the human immune system is unable to terminate the infection. The proportionality constants β and μ related to the antibodies have lower values than those specified in Simulation 1. Also, the delay τ between the instant when the antigens are detected to the instant when antibodies are produced is increased to 1 s. The simulated values of the constants are given by

$$\alpha = 0.1, \quad \beta = 0.1, \quad \gamma = 0.1, \quad \mu = 0.3, \quad \tau = 1, \quad \lambda = 0.1, \quad \sigma = 0.1,$$
$$\text{and} \quad \eta = 0.5.$$

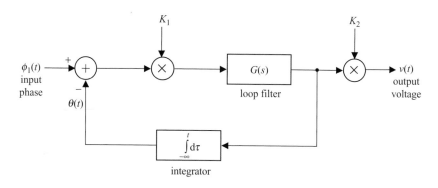

Fig. P8.11. Block diagram representation of a phase-locked loop.

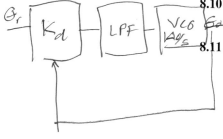

8.7 By integrating the impulse response $h(t)$ of the armature-controlled dc motor, derive Eq. (8.42) for $\xi_n = 1$.

8.8 Assume that the inductance L_a of the induction motor, shown in Fig. 8.8(b), is zero. Determine the transfer function $H(s)$ and impulse response $h(t)$ for the modified model. Based on the location of the poles, comment on the stability of the induction motor.

8.9 Repeat Problem 8.7 for Eq. (8.43) with $\xi_n > 1$ and Eq. (8.44) with $\xi_n < 1$.

8.10 Based on Eqs. (8.53)–(8.56), derive the expression for the transfer function $H(s)$ of the human immune system shown in Eq. (8.57).

8.11 In order to achieve synchronization between the modulating and demodulating carriers, a special circuit referred to as a phase-locked loop (PLL) is commonly used in communications. The block diagram representing the PLL is shown in Fig. P8.11.

Show that the transfer function of the PLL is given by

$$\frac{V(s)}{\phi(s)} = K_1 K_2 \frac{s G(s)}{s + K_1 G(s)},$$

where K_1 and K_2 are gain constants and $G(s)$ is the transfer function of a loop filter. Specify the condition under which the PLL acts as an ideal differentiator. In other words, derive the expression for $G(s)$ when the transfer function of the PLL equals Ks, with K being a constant.

8.12 Repeat the simulink simulation for the human immune system for the following values of the proportionality constants:

$$\alpha = 0.3, \quad \beta = 0.1, \quad \gamma = 0.25, \quad \mu = 0.6, \quad \tau = 1, \quad \lambda = 0.1,$$

$$\sigma = 0.4, \quad \text{and} \quad \eta = 0.2$$

Sketch the time evolution of the antigens, plasma cells, and antibodies.

PART III

Discrete-time signals and systems

9 Sampling and quantization

Part II of the book covered techniques for the analysis of continuous-time (CT) signals and systems. In Part III, we consider the corresponding analysis techniques for discrete-time (DT) sequences and systems. A DT sequence may occur naturally. Examples are the one-dimensional (1D) hourly measurements $x[k]$ made with an electronic thermometer, or the two-dimensional (2D) image $x[m, n]$ recorded with a digital camera, as illustrated earlier in Fig. 1.1. Alternatively, a DT sequence may be derived from a CT signal by a process known as sampling. A widely used procedure for processing CT signals consists of transforming these signals into DT sequences by sampling, processing the resulting DT sequences with DT systems, and converting the DT outputs back into the CT domain. This concept of DT processing of CT signals is illustrated by the schematic diagram shown in Fig. 9.1. Here, the input CT signal $x(t)$ is converted to a DT sequence $x[k]$ by the *sampling* module, also referred to as the A/D converter. The DT sequence is then processed by the *DT system* module. Finally, the output $y[k]$ of the DT module is converted back into the CT domain by the *reconstruction* module. The reconstruction module is also referred to as the D/A converter. Although the intermediate waveforms, $x[k]$ and $y[k]$, are DT sequences, the overall shaded block may be considered as a CT system since it accepts a CT signal $x(t)$ at its input and produces a CT output $y(t)$. If the internal working of the shaded block is hidden, one would interpret that the overall operation of Fig. 9.1 results from a CT system.

In practice, a CT signal can either be processed by using a full CT setup, in which the individual modules are themselves CT systems (as explained in Chapters 3–8), or by using a CT–DT hybrid setup (as shown in Fig. 9.1). Both approaches have advantages and disadvantages. The primary advantage of CT signal processing is its higher speed as DT systems are not as fast as their counterparts in the CT domain due to limits on the sampling rate of the A/D converter and the clock rate of the processor used to implement the DT systems. In spite of its limitation in speed, there are important advantages with DT signal processing, such as improved flexibility, self-calibration, and data-logging. Whereas CT systems have a limited performance range, DT systems are more

Fig. 9.1. Processing CT signals
using DT systems.

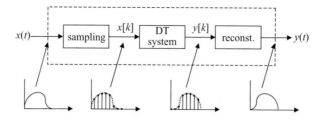

flexible and can be reprogrammed such that the same hardware can be used in a
variety of different applications. In addition, the characteristics of CT systems
tend to vary with changes in the operating conditions and with age. The DT
systems have no such problems as the digital hardware used to implement these
systems does not drift with age or with changes in the operating conditions and,
therefore, can be self-calibrated easily. Digital signals, obtained by quantizing
DT sequences, are less sensitive to noise and interference than analog signals
and are widely used in communication systems. Finally, the data available from
the DT systems can be stored in a digital server so that the performance of the
system can be monitored over a long period of time. In summary, the advantages of the DT system outweigh their limitations in most applications. Until
the late 1980s, most signal processing applications were implemented with CT
systems constructed with analog components such as resistors, capacitors, and
operational amplifiers. With the recent availability of cheap digital hardware,
it is a common practice now to perform signal processing in the DT domain
based on the hybrid setup shown in Fig. 9.1.

   Although, a CT–DT hybrid setup similar to Fig. 9.1 is advantageous in many
applications, care should be taken during the design stage. For example, during
the sampling process some loss of information is generally inevitable. Consequently, if the system is not designed properly, the performance of a CT–DT
hybrid setup may degrade significantly as compared with a CT setup. In this
chapter, we focus on the analysis of the sampling process and the converse
step of reconstructing a CT signal from its DT version. In addition, we also
analyze the process of quantization for converting an analog signal to a digital signal. Both time-domain and frequency-domain analyses are used where
appropriate.

   The organization of Chapter 9 is as follows. Section 9.1 introduces the
impulse-train sampling process and derives a necessary condition, referred to as
the sampling theorem, under which a CT signal can be perfectly reconstructed
from its sampled DT version. We observe that violating the sampling theorem
leads to distortion or aliasing in the frequency domain. Section 9.2 introduces
the practical implementations for impulse-train sampling. These implementations are referred to as pulse-train sampling and zero-order hold.

   In Section 9.3, we introduce another discretization process called quantization, which, in conjunction with sampling, converts a CT signal into a digital
signal. In Section 9.4, we present an application of sampling and quantization

used in recording music on a compact disc (CD). Finally, Section 9.5 concludes our discussion with a summary of the key concepts introduced in the chapter.

9.1 Ideal impulse-train sampling

In this section, we consider sampling of a CT signal $x(t)$ with a bounded CTFT $X(\omega)$ such that

$$X(\omega) = 0 \quad \text{for} \quad |\omega| > 2\pi\beta. \tag{9.1}$$

A CT signal $x(t)$ satisfying Eq. (9.1) is referred to as a baseband signal, which is band-limited to $2\pi\beta$ radians/s or β Hz. In the following discussion, we prove that a baseband signal $x(t)$ can be transformed into a DT sequence $x[k]$ with no loss of information if the sampling interval T_s satisfies the criterion that $T_s \leq 1/2\beta$.

To derive the DT version of the baseband signal $x(t)$, we multiply $x(t)$ by an impulse train:

$$s(t) = \sum_{k=-\infty}^{\infty} \delta(t - kT_s), \tag{9.2}$$

where T_s denotes the separation between two consecutive impulses and is called the sampling interval. Another related parameter is the sampling rate ω_s, with units of radians/s, which is defined as follows:

$$\omega_s = \frac{2\pi}{T_s}. \tag{9.3}$$

Mathematically, the resulting sampled signal, $x_s(t) = x(t) \cdot s(t)$, is given by

$$x_s(t) = x(t) \sum_{k=-\infty}^{\infty} \delta(t - kT_s) = \sum_{k=-\infty}^{\infty} x(kT_s)\delta(t - kT_s). \tag{9.4}$$

Figure 9.2 illustrates the time-domain representation of the process of the impulse-train sampling. Figure 9.2(a) shows the time-varying waveform representing the baseband signal $x(t)$. In Figs. 9.2(b) and (c), we plot the sampled signal $x_s(t)$ for two different values of the sampling interval. In Fig. 9.2(b), the sampling interval $T_s = T$ and the sampled signal $x_s(t)$ provides a fairly good approximation of $x(t)$. In Fig. 9.2(c), the sampling interval T_s is increased to $2T$. With T_s set to a larger value, the separation between the adjacent samples in $x_s(t)$ increases. Compared to Fig. 9.2(b), the sampled signal in Fig. 9.2(c) provides a coarser representation of $x(t)$. The choice of T_s therefore determines how accurately the sampled signal $x_s(t)$ represents the original CT signal $x(t)$. To determine the optimal value of T_s, we consider the effect of sampling in the frequency domain.

Fig. 9.2. Time-domain illustration of sampling as a product of the band-limited signal and an impulse train. (a) Original signal $x(t)$; (b) sampled signal $x_s(t)$ with sampling interval $T_s = T$; (c) sampled signal $x_s(t)$ with sampling interval $T_s = 2T$.

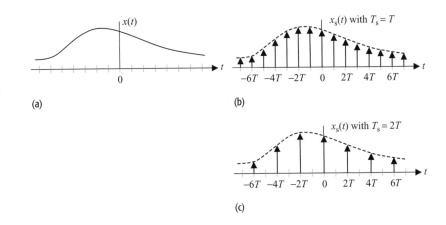

Calculating the CTFT of Eq. (9.4), the CTFT $X_s(\omega)$ of the sampled signal $x_s(t)$ is given by

$$
X_s(\omega) = \Im \left\{ x(t) \sum_{k=-\infty}^{\infty} \delta(t - kT_s) \right\} = \frac{1}{2\pi} F\{x(t)\} * \Im \left\{ \sum_{k=-\infty}^{\infty} \delta(t - kT_s) \right\}
$$

$$
= \frac{1}{2\pi} \left[X(\omega) * \frac{2\pi}{T_s} \sum_{m=-\infty}^{\infty} \delta \left(\omega - \frac{2m\pi}{T_s} \right) \right] = \frac{1}{T_s} \sum_{m=-\infty}^{\infty} X \left(\omega - \frac{2m\pi}{T_s} \right)
$$

(9.5)

where $*$ denotes the CT convolution operator. In deriving Eq. (9.5), we used the following CTFT pair:

$$
\sum_{k=-\infty}^{\infty} \delta(t - kT_s) \overset{\text{CTFT}}{\longleftrightarrow} \frac{2\pi}{T_s} \sum_{m=-\infty}^{\infty} \delta \left(\omega - \frac{2m\pi}{T_s} \right)
$$

Fig. 9.3. Frequency-domain illustration of the impulse-train sampling. (a) Spectrum $X(\omega)$ of the original signal $x(t)$; (b) spectrum $X_s(\omega)$ of the sampled signal $x_s(t)$ with sampling rate $\omega_s \geq 4\pi\beta$; (c) spectrum $X_s(\omega)$ of the sampled signal $x_s(t)$ with sampling rate $\omega_s < 4\pi\beta$.

based on entry (19) of Table 5.2. Equation (9.5) implies that the spectrum $X_s(\omega)$ of the sampled signal $x_s(t)$ is a periodic extension, consisting of the shifted replicas of the spectrum $X(\omega)$ of the original baseband signal $x(t)$. Figure 9.3 illustrates the frequency-domain interpretation of Eq. (9.5). The spectrum of the original signal $x(t)$ is assumed to be an arbitrary trapezoidal waveform and is shown in Fig. 9.3(a). The spectrum $X_s(\omega)$ of the sampled signal $x_s(t)$ is plotted

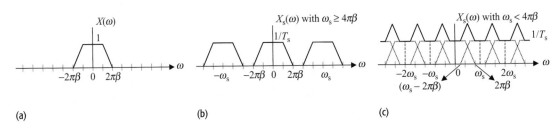

in Figs. 9.3(c) and (d) for the following two cases:

case I $\omega_s \geq 4\pi\beta$;

case II $\omega_s < 4\pi\beta$.

When the sampling rate $\omega_s \geq 4\pi\beta$, no overlap exists between consecutive replicas in $X_s(\omega)$. However, as the sampling rate ω_s is decreased such that $\omega_s < 4\pi\beta$, adjacent replicas overlap with each other. The overlapping of replicas is referred to as aliasing, which distorts the spectrum of the original baseband signal $x(t)$ such that $x(t)$ cannot be reconstructed from its samples. To prevent aliasing, the sampling rate $\omega_s \geq 4\pi\beta$. This condition is referred to as the *sampling theorem* and is stated in the following.[†]

Sampling theorem   *A baseband signal $x(t)$, band-limited to $2\pi\beta$ radians/s, can be reconstructed accurately from its samples $x(kT)$ if the sampling rate ω_s, in radians/s, satisfies the following condition:*

$$\omega_s \geq 4\pi\beta. \tag{9.6a}$$

Alternatively, the sampling theorem may be expressed in terms of the sampling rate $f_s = \omega_s/2\pi$ in samples/s, or the sampling interval T_s. To prevent aliasing,

sampling rate (samples/s) $f_s \geq 2\beta$;   (9.6b)

or

sampling interval $T_s \leq 1/2\beta$.   (9.6c)

The minimum sampling rate f_s (Hz) required for perfect reconstruction of the original band-limited signal is referred to as the Nyquist rate.

The sampling theorem is applicable for *baseband* signals, where the signal contains low-frequency components within the range $0 - \beta$ Hz. In some applications, such as communications, we come across bandpass signals that also contain a band of frequencies, but the occupied frequency range lies within the band $\beta_2 - \beta_1$ Hz with $\beta_1 \neq 0$. In these cases, although the maximum frequency of β_2 Hz implies the Nyquist sampling rate of $2\beta_2$ *Hz* it is possible to achieve perfect reconstruction with a lower sampling rate (see Problem 9.8).

[†] The sampling theorem was known in various forms in the mathematics literature before its application in signal processing, which started much later, in the 1950s. Several people developed independently or contributed towards its development. Notable contributions, however, were made by E. T. Whittaker (1873–1956), Harry Nyquist (1889–1976), Karl Küpfmüller (1897–1977), V. A. Kotelnikov (1908–2005), Claude Shannon (1916–2001), and I. Someya.

9.1.1 Reconstruction of a band-limited signal from its samples

Figure 9.3(b) illustrates that the CTFT $X_s(\omega)$ of the sampled signal $x_s(t)$ is a periodic extension of the CTFT of the original signal $x(t)$. By eliminating the replicas in $X_s(\omega)$, we should be able to reconstruct $x(t)$. This is accomplished by applying the sampled signal $x_s(t)$ to the input of an ideal lowpass filter (LPF) with the following transfer function:

$$H(\omega) = \begin{cases} T_s & |\omega| \leq \omega_s/2 \\ 0 & \text{elsewhere.} \end{cases} \tag{9.7}$$

The CTFT $Y(\omega)$ of the output $y(t)$ of the LPF is given by $Y(\omega) = X_s(\omega)H(\omega)$, and therefore all shifted replicas at frequencies $\omega > \omega_s/2$ are eliminated. All frequency components within the pass band $\omega \leq \omega_s/2$ of the LPF are amplified by a factor of T_s to compensate for the attenuation of $1/T_s$ introduced during sampling. The process of reconstructing $x(t)$ from its samples in the frequency domain is illustrated in Fig. 9.4. We now proceed to analyze the reconstruction process in the time domain.

According to the convolution property, multiplication in the frequency domain transforms to convolution in the time domain. The output $y(t)$ of the lowpass filter is therefore the convolution of its impulse response $h(t)$ with the sampled signal $x_s(t)$. Based on entry (17) of Table 5.2, the impulse response of an ideal lowpass filter with the transfer function given in Eq. (9.7) is given by

$$h(t) = \text{sinc}\left(\frac{\omega_s t}{2\pi}\right). \tag{9.8}$$

Convolving the impulse response $h(t)$ with the sampled signal, $x_s(t) = \sum_{k=-\infty}^{\infty} x(kT_s)\delta(t - kT_s)$ yields

Fig. 9.4. Reconstruction of the original baseband signal x(t) by ideal lowpass filtering. (a) Spectrum of the sampled signal $x_s(t)$; (b) transfer function $H(\omega)$ of the lowpass filter; (c) spectrum of the reconstructed signal x(t).

$$y(t) = \text{sinc}\left(\frac{\omega_s t}{2\pi}\right) * \sum_{k=-\infty}^{\infty} x(kT_s)\delta(t - kT_s), \tag{9.9}$$

which reduces to

$$y(t) = \sum_{k=-\infty}^{\infty} x(kT_s)\left[\text{sinc}\left(\frac{\omega_s t}{2\pi}\right) * \delta(t - kT_s)\right] \tag{9.10}$$

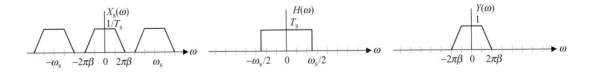

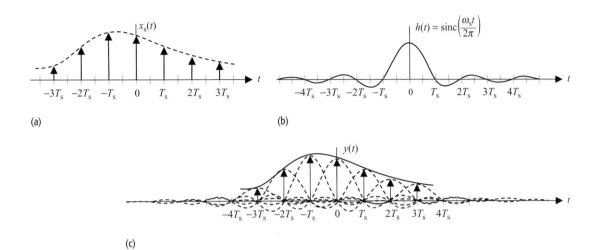

(a)

(b)

(c)

Fig. 9.5. Reconstruction of the band-limited signal in the time domain. (a) Sampled signal $x_s(t)$; (b) impulse response $h(t)$ of the lowpass filter; (c) reconstructed signal $x(t)$ obtained by convolving $x_s(t)$ with $h(t)$.

or

$$y(t) = \sum_{k=-\infty}^{\infty} x(kT_s) \left[\text{sinc} \left(\frac{\omega_s(t - kT_s)}{2\pi} \right) \right]. \qquad (9.11)$$

Equation (9.11) implies that the original signal $x(t)$ is reconstructed by adding a series of time-shifted sinc functions, whose amplitudes are scaled according to the values of the samples at the center location of the sinc functions. The sinc functions in Eq. (9.11) are called the interpolating functions and the overall process is referred to as the band-limited interpolation. The time-domain interpretation of the reconstruction of the original band-limited signal $x(t)$ is illustrated in Fig. 9.5. At $t = kT_s$, only the kth sinc function, with amplitude $x(kT_s)$, is non-zero. The remaining sinc functions are all zero. The value of the reconstructed signal at $t = kT_s$ is therefore given by $x(kT_s)$. In other words, the values of the reconstructed signal at the sampling instants are given by the respective samples. The values in between two samples are interpolated using a linear combination of the time-shifted sinc functions.

Example 9.1

Consider the following sinusoidal signal with the fundamental frequency f_0 of 4 kHz:

$$g(t) = 5 \cos(2\pi f_0 t) = 5 \cos(8000\pi t).$$

(i) The sinusoidal signal is sampled at a sampling rate f_s of 6000 samples/s and reconstructed with an ideal LPF with the following transfer function:

$$H_1(\omega) = \begin{cases} 1/6000 & |\omega| \le 6000\pi \\ 0 & \text{elsewhere.} \end{cases}$$

Determine the reconstructed signal.

(ii) Repeat (i) for a sampling rate f_s of 12 000 samples/s and an ideal LPF with the following transfer function:

$$H_2(\omega) = \begin{cases} 1/12\,000 & |\omega| \leq 12\,000\pi \\ 0 & \text{elsewhere.} \end{cases}$$

Solution

(i) The CTFT $G(\omega)$ of the sinusoidal signal $g(t)$ is given by

$$G(\omega) = 5\pi\,[\delta(\omega - 8000\pi) + \delta(\omega + 8000\pi)].$$

Using Eq. (9.4), the CTFT $G_s(\omega)$ of the sampled signal with a sampling rate $\omega_s = 2\pi(6000)$ radians/s ($T_s = 1/6000$ s) is expressed as follows:

$$G_s(\omega) = 6000 \sum_{m=-\infty}^{\infty} G(\omega - 2\pi m(6000)) = 6000 \sum_{m=-\infty}^{\infty} G(\omega - 12\,000 m\pi).$$

Substituting the value of $G(\omega)$ in the above expression yields

$$G_s(\omega) = 6000 \sum_{m=-\infty}^{\infty} 5\pi\,[\delta(\omega - 8000\pi - 12\,000\,m\pi)$$
$$+ \delta(\omega + 8000\pi - 12\,000\,m\pi)]$$

$$= 6000(5\pi)\left[\cdots + \underbrace{\delta(\omega + 16\,000\pi) + \delta(\omega + 32\,000\pi)}_{m=-2} \right.$$

$$+ \underbrace{\delta(\omega + 4000\pi) + \delta(\omega + 20\,000\pi)}_{m=-1}$$

$$+ \underbrace{\delta(\omega - 8000\pi) + \delta(\omega + 8000\pi)}_{m=0} + \underbrace{\delta(\omega - 20\,000\pi) + \delta(\omega - 4000\pi)}_{m=1}$$

$$\left. + \underbrace{\delta(\omega - 32\,000\pi) + \delta(\omega - 16\,000\pi)}_{m=2} + \cdots \right].$$

When the sampled signal is passed through the ideal LPF with transfer function $H_1(\omega)$, all frequency components $|\omega| > 6000\pi$ radians/s) are eliminated from the output. The CTFT $Y(\omega)$ of the output $y(t)$ of the LPF is given by

$$Y(\omega) = H_1(\omega)G_s(\omega) = \frac{1}{6000} \cdot 6000(5\pi)[\delta(\omega + 4000\pi) + \delta(\omega - 4000\pi)].$$

Calculating the inverse CTFT, the reconstructed signal is given by $y(t) = 5\cos(4000\pi t)$.

Fig. 9.6. Sampling and reconstruction of a sinusoidal signal $g(t) = 5 \cos(8000\pi t)$ at a sampling rate of 6000 samples/s. CTFTs of: (a) the sinusoidal signal $g(t)$; (b) the impulse train $s(t)$; (c) the sampled signal $g_s(t)$; and (d) the signal reconstructed with an ideal LPF $H_1(\omega)$ with a cut-off frequency of 6000π radians/s.

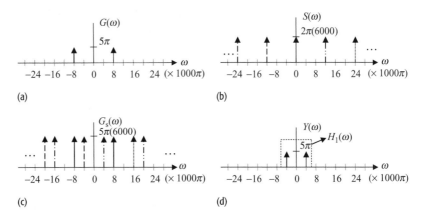

The graphical representation of the sampling and reconstruction of the sinusoidal signal in the frequency domain is illustrated in Fig. 9.6. The CTFTs of the sinusoidal signal $g(t)$ and the impulse train $s(t)$ are plotted, respectively, in Fig. 9.6(a) and Fig. 9.6(b). Since the CTFT $S(\omega)$ of $s(t)$ consists of several impulses, the CTFT $G_s(\omega)$ of the sampled signal $g_s(t)$ is obtained by convolving the CTFT $G(\omega)$ of the sinusoidal signal $g(t)$ separately with each impulse in $G_s(\omega)$ and then applying the principle of superposition. To emphasize the results of individual convolutions, a different pattern is used in Fig. 9.6(b) for each impulse in $S(\omega)$. For example, the impulse $\delta(\omega)$ located at origin in $S(\omega)$ is shown in Fig. 9.6(b) by a solid line. Convolving $G(\omega)$ with $\delta(\omega)$ results in two impulses located at $\omega = \pm 8000\pi$, which are shown in Fig. 9.6(c) by solid lines. Similarly for the other impulses in $S(\omega)$.

The output $y(t)$ is obtained by applying $G_s(\omega)$ to the input of an ideal LPF with a cut-off frequency of 6000π radians/s. Clearly, only the two impulses at $\omega = \pm 4000\pi$, corresponding to the sinusoidal signal $\cos(4000\pi t)$, lie within the pass band of the lowpass filter. The remaining impulses are eliminated from the output. This results in an output, $y(t) = \cos(4000\pi t)$, which is different from the original signal.

(ii) The CTFT $G_s(\omega)$ of the sampled signal with $\omega_s = 2\pi(12\,000)$ radians/s ($T_s = 1/12\,000$ s) is given by

$$G_s(\omega) = 12\,000 \sum_{m=-\infty}^{\infty} G(\omega - 2\pi m(12\,000))$$

$$= 12\,000 \sum_{m=-\infty}^{\infty} G(\omega - 24\,000m\pi).$$

Substituting the value of the CTFT $G(\omega) = 5\pi[\delta(\omega - 8000\pi) + \delta(\omega + 8000\pi)]$ in the above equation, we obtain

$$G_s(\omega) = 12\,000 \sum_{m=-\infty}^{\infty} 5\pi[\delta(\omega - 8000\pi - 24\,000m\pi)$$
$$+ \delta(\omega + 8000\pi - 24\,000m\pi)]$$

$$= 12\,000(5\pi) \left[\cdots + \underbrace{\delta(\omega + 40\,000\pi) + \delta(\omega + 56\,000\pi)}_{m=-2} \right.$$
$$+ \underbrace{\delta(\omega + 16\,000\pi) + \delta(\omega + 32\,000\pi)}_{m=-1}$$
$$+ \underbrace{\delta(\omega - 8000\pi) + \delta(\omega + 8000\pi)}_{m=0} + \underbrace{\delta(\omega - 32\,000\pi) + \delta(\omega - 16\,000\pi)}_{m=1}$$
$$\left. + \underbrace{\delta(\omega - 56\,000\pi) + \delta(\omega - 40\,000\pi)}_{m=2} + \cdots \right].$$

To reconstruct the original sinusoidal signal, the sampled signal is passed through an ideal LPF $H_2(\omega)$. The frequency components outside the pass-band range $|\omega| \leq 12\,000\pi$ radians/s are eliminated from the ouput. The CTFT $Y(\omega)$ of the output $y(t)$ of the LPF is therefore given by

$$Y(\omega) = 5\pi[\delta(\omega + 8000\pi) + \delta(\omega - 8000\pi)],$$

which results in the reconstructed signal

$$y(t) = 5\cos(8000\pi t).$$

The graphical interpretation of the aforementioned sampling and reconstruction process is illustrated in Fig. 9.7.

As the signal $g(t)$ is a sinusoidal signal with frequency 4 kHz, the Nyquist sampling rate is 8 kHz. In part (i), the sampling rate (6 kHz) is lower than the Nyquist rate, and consequently the reconstructed signal is different from the original signal due to the aliasing effect. In part (ii), the sampling rate is higher than the Nyquist rate, and as a result the original sinusoidal signal is accurately reconstructed.

9.1.2 Aliasing in sampled sinusoidal signals

As demonstrated in Example 9.1, undersampling of a baseband signal at a sampling rate less than the Nyquist rate leads to aliasing. Under such conditions, perfect reconstruction of the baseband signal is not possible from its samples. In this section, we consider undersampling of a sinusoidal signal

$$x(t) = \cos(2\pi f_0 t)$$

with a fundamental frequency of f_0 Hz. The sampling rate f_s, in samples/s, is assumed to be less than the Nyquist rate of $2f_0$, i.e. $f_s < 2f_0$. We show

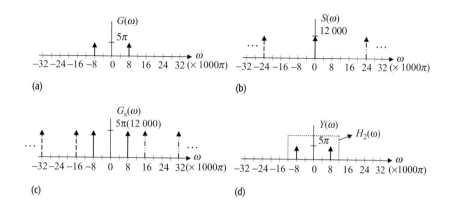

Fig. 9.7. Sampling and reconstruction of a sinusoidal signal $g(t) = 5\cos(8000\pi t)$ at a sampling rate of 12 000 samples/s. CTFTs of: (a) the sinusoidal signal $g(t)$; (b) the impulse train $s(t)$; (c) the sampled signal $g_s(t)$; and (d) the signal reconstructed with an ideal LPF $H_2(\omega)$ with a cut-off frequency of 12 000π radians/s.

that the reconstructed signal is sinusoidal but with a different fundamental frequency.

Using Eq. (9.4), the CTFT $X_s(\omega)$ of the sampled sinusoidal signal $x_s(t)$ is given by

$$X_s(\omega) = f_s \sum_{m=-\infty}^{\infty} X(\omega - 2m\pi f_s). \qquad (9.12)$$

In Eq. (9.12), we substitute the CTFT, $X(\omega) = \pi[\delta(\omega - 2\pi f_0) + \delta(\omega + 2\pi f_0)]$, of the sinusoidal signal $x(t)$. The resulting expression is as follows:

$$X_s(\omega) = \pi f_s \sum_{m=-\infty}^{\infty} \delta(\omega + 2\pi(f_0 - mf_s)) + \pi f_s \sum_{k=-\infty}^{\infty} \delta(\omega - 2\pi(f_0 + kf_s)). \qquad (9.13)$$

To reconstruct $x(t)$, the sampled signal $x_s(t)$ is filtered with an ideal LPF with transfer function

$$H(\omega) = \begin{cases} T_s & |\omega| \le \pi f_s \\ 0 & \text{elsewhere.} \end{cases} \qquad (9.14)$$

Within the pass band $|\omega| \le \pi f_s$ of the LPF, the input frequency components are amplified by a factor of T_s or $1/f_s$. All frequency components within the stop band $|\omega| > \pi f_s$ are eliminated from the reconstructed signal $y(t)$. In addition, the CTFT of the reconstructed signal $y(t)$ satisfies the following properties.

(1) The CTFT $Y(\omega)$ consists of impulses located at frequencies $\omega = -2\pi(f_0 - mf_s)$ and $\omega = 2\pi(f_0 + kf_s)$, where m and k are integers such that $|(f_0 - mf_s)| \le f_s/2$ and $|(f_0 + kf_s)| \le f_s/2$. Since the two conditions are satisfied only for $m = -k$, the locations of the impulses are given by $\omega = \pm 2\pi(f_0 - mf_s)$.
(2) If $|(f_0 - mf_s)| \le f_s/2$, then $|(f_0 - (m+1)f_s)| > f_s/2$ and $|(f_0 - (m-1)f_s)| > f_s/2$. Combined with (1), this implies that only two impulses at $\omega = \pm 2\pi(f_0 - mf_s)$ will be present in $Y(\omega)$.
(3) Each impulse in $Y(\omega)$ will have a magnitude (enclosed area) of π.

Based on properties (1)–(3) listed above, the spectrum of the reconstructed signal is given by

$$Y(\omega) = \pi[\delta(\omega + 2\pi(f_0 - mf_s)) + \delta(\omega - 2\pi(f_0 - mf_s))]. \qquad (9.15)$$

Calculating the inverse CTFT of Eq. (9.15) leads to the following sinusoidal signal:

$$y(t) = \cos(2\pi(f_0 - mf_s)t), \qquad (9.16)$$

where m is an integer such that $|(f_0 - mf_s)| \leq f_s/2$.

Lemma 9.1 If a sinusoidal signal $x(t) = \cos(2\pi f_0 t)$ is undersampled such that the sampling rate $f_s < 2f_0$, then the signal reconstructed with an ideal LPF, with pass band $|\omega| \leq \pi f_s$, is another sinusoidal signal

$$y(t) = \cos(2\pi(f_0 - mf_s)t),$$

where m is a positive integer satisfying the condition $|(f_0 - mf_s)| < f_s/2$.

In Example 9.1(i), for example, the fundamental frequency $f_0 = 4000$ Hz and the sampling rate $f_s = 6000$ samples/s is less than the Nyquist rate. Selecting $m = 1$, the reconstructed signal $y(t)$ is given by

$$y(t) = \cos(2\pi(f_0 - mf_s)t) = \cos(2\pi(4000 - 6000)t) = \cos(4000\pi t).$$

The result obtained from Lemma 9.1 is in agreement with the expression derived in Example 9.1(i).

Example 9.2

A signal generator produces a sinusoidal tone $x(t) = \cos(2\pi f_0 t)$ with fundamental frequency f_0 between 1 Hz and 1000 kHz. The signal is sampled with a sampling rate $f_s = 6000$ samples/s and is reconstructed using an ideal LPF with a cut-off frequency $\omega_c = \pi f_s = 6000\pi$ radians/s. Determine the reconstructed signal for $f_0 = 500$ Hz, 2.5 kHz, 2.8 kHz, 3.2 kHz, 3.5 kHz, 7 kHz, 10 kHz, 20 kHz, and 1000 kHz.

Solution

Table 9.1 lists the reconstructed signals obtained by applying Lemma 9.1. The sampling frequency f_s in the top three entries of Table 9.1 satisfies the sampling theorem. Therefore, the original signal is reconstructed without any distortion. In the remaining entries, the sampling theorem is violated. Lemma 9.1 is used to determine the fundamental frequency of the reconstructed sinusoidal signal, which is different from that of the original signal due to aliasing. The reconstructed signals are tabulated in entries (4)–(9) of Table 9.1. An interesting observation is that the reconstructed signals for the sinusoidal waveforms $x(t) = \cos(5600\pi t)$ and $x(t) = \cos(6400\pi t)$, listed in entries (3)–(4)

Table 9.1. Signals reconstructed from samples of a sinusoidal tone $x(t) = \cos(2\pi f_0 t)$ for different values of the fundamental frequency f_0; the sampling frequency f_s is kept constant at 6000 samples/s

	Funadmental frequency (f_0)	Original signal	$\|(f_0 - mf_s)\| < f_s/2$	Reconstructed signal	Comments
(1)	500 Hz	$\cos(1000\pi t)$	$f_s > 2f_0$	$\cos(1000\pi t)$	no aliasing
(2)	2.5 kHz	$\cos(5000\pi t)$	$f_s > 2f_0$	$\cos(5000\pi t)$	no aliasing
(3)	2.8 kHz	$\cos(5600\pi t)$	$f_s > 2f_0$	$\cos(5600\pi t)$	no aliasing
(4)	3.2 kHz	$\cos(6400\pi t)$	$\|3200 - 1 \times 6000\|$	$\cos(5600\pi t)$	aliasing
(5)	3.5 kHz	$\cos(7000\pi t)$	$\|3500 - 1 \times 6000\|$	$\cos(5000\pi t)$	aliasing
(6)	7 kHz	$\cos(14000\pi t)$	$\|7000 - 1 \times 6000\|$	$\cos(2000\pi t)$	aliasing
(7)	10 kHz	$\cos(20000\pi t)$	$\|10000 - 2 \times 6000\|$	$\cos(4000\pi t)$	aliasing
(8)	20 kHz	$\cos(40000\pi t)$	$\|20000 - 3 \times 6000\|$	$\cos(4000\pi t)$	aliasing
(9)	1000 kHz	$\cos(2 \times 10^6 \pi t)$	$\|10^6 - 167 \times 6000\|$	$\cos(4000\pi t)$	aliasing

of Table 9.1, are identical. Similarly, the reconstructed signals for the sinusoidal waveforms $x(t) = \cos(5000\pi t)$ and $x(t) = \cos(7000\pi t)$, listed in entries (2) and (5) of Table 9.1, are also identical. Finally, the reconstructed signals for the sinusoidal waveforms $x(t) = \cos(20\,000\pi t)$, $x(t) = \cos(40\,000\pi t)$, and $x(t) = \cos(2 \times 10^6 \pi t)$, listed in entries (7)–(9) of Table 9.1, are the same. The identical waveforms are the consequences of aliasing.

9.2 Practical approaches to sampling

Section 9.1 introduced the impulse-train sampling used to derive the DT version of a band-limited CT signal. In practice, impulses are difficult to generate and are often approximated by narrow rectangular pulses. The resulting approach is referred to as pulse-train sampling, which is discussed in Section 9.2.1. A second practical implementation, referred to as the zero-order hold, is discussed in Section 9.2.2.

9.2.1 Pulse-train sampling

In pulse-train sampling, the impulse train $s(t)$ is approximated by a rectangular pulse train of the form

$$r(t) = \sum_{k=-\infty}^{\infty} p_1(t - kT_s) = \left[p_1(t) * \sum_{k=-\infty}^{\infty} \delta(t - kT_s) \right], \qquad (9.17)$$

where $p_1(t)$ represents a rectangular pulse of duration $\tau \ll T_s$, which is given by

$$p_1(t) = \text{rect}\left(\frac{t}{\tau}\right).$$

Fig. 9.8. Time-domain illustration of the pulse-train sampling of a CT signal. (a) Original signal $x(t)$; (b) pulse train $r(t)$; (c) sampled signal $x_s(t) = x(t)r(t)$.

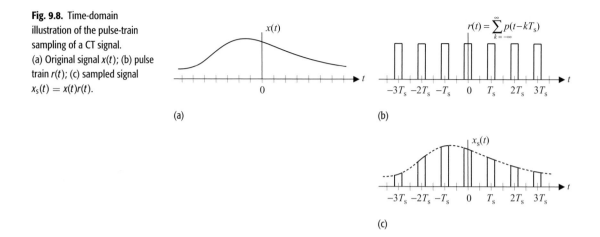

(a)

(b)

(c)

As in impulse-train sampling, the sampled signal $x_s(t)$ is obtained by multiplying the reference signal $x(t)$ by $r(t)$ such that

$$x_s(t) = x(t)r(t) = x(t)\left[p_1(t) * \sum_{k=-\infty}^{\infty} \delta(t - kT_s) \right]. \qquad (9.18)$$

Based on Eq. (9.18), the time-domain representation of the process of pulse-train sampling is shown in Fig. 9.8. The sampled signal, shown in Fig. 9.8(c), consists of several pulses of duration τ. The magnitude of the rectangular pulses in $x_s(t)$ follows the reference signal $x(t)$ within the duration of the pulses.

To analyze the process in the frequency domain, we consider the CTFS expansion of the periodic pulse train. The exponential CTFS representation of $r(t)$ is given by(see Example 9.14)

$$r(t) = \sum_{n=-\infty}^{\infty} D_n e^{jn\omega_s t} \quad \text{with} \quad D_n = \frac{\omega_s \tau}{2\pi} \operatorname{sinc}\left(\frac{n\omega_s \tau}{2\pi}\right), \qquad (9.19)$$

where ω_s is the sampling rate in radians/s and is given by $\omega_s = 2\pi f_s = 2\pi/T_s$. the CTFT of $r(t)$ is given by

$$R(\omega) = 2\pi \sum_{n=-\infty}^{\infty} D_n \delta(\omega - n\omega_s) \quad \text{with} \quad D_n = \frac{\omega_s \tau}{2\pi} \operatorname{sinc}\left(\frac{n\omega_s \tau}{2\pi}\right). \qquad (9.20)$$

Based on Eq. (9.18), the CTFT $X_s(\omega)$ of sampled signal $x_s(t)$ is given by

$$X_s(\omega) = \frac{1}{2\pi} X(\omega) * R(\omega). \qquad (9.21a)$$

Substituting the value of $R(\omega)$ from Eq. (9.20) yields

$$X_s(\omega) = \frac{1}{2\pi} X(\omega) * R(\omega) = \sum_{n=-\infty}^{\infty} D_n X(\omega - n\omega_s). \qquad (9.21b)$$

Fig. 9.9. Frequency-domain illustration of the pulse-train sampling of a CT signal. Spectrum of (a) the original signal $x(t)$; (b) the pulse train $r(t)$; (c) the sampled signal $x_s(t) = x(t)r(t)$.

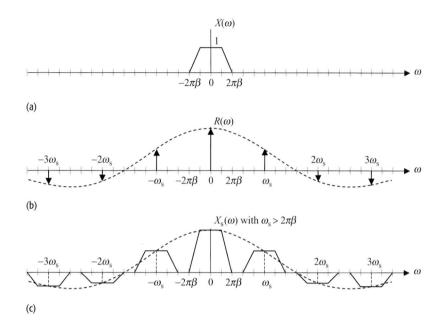

Based on Eq. (9.21b), Fig. 9.9 illustrates the frequency-domain interpretation of the pulse-train sampling. The spectrum $X(\omega)$ of the original signal $x(t)$ is shown in Fig. 9.9(a), while the spectrum $R(\omega)$ of the pulse train $r(t)$ is shown in Fig. 9.9(b). The spectrum $X_s(\omega)$ of the sampled signal $x_s(t)$ is obtained by convolving $X(\omega)$ with $R(\omega)$. As shown in Fig. 9.9(c), $X_s(\omega)$ consists of several shifted replicas of $X(\omega)$ attenuated with a factor of D_n. Compared to the impulse-train sampling, the spectra of the two sampled signals are identical except for a varying attenuation factor of D_n introduced by the pulse-train sampling.

Reconstruction of the original signal $x(t)$ from the pulse-train sampled signal $x_s(t)$ is achieved by filtering $x_s(t)$ with an ideal LPF having a cut-off frequency $\omega_c = \omega_s/2$ and a gain of $1/D_0$ in the pass band. The LPF eliminates all shifted replicas present at frequencies $|\omega| > \omega_s/2$. This leaves a single replica at $\omega = 0$, which is the same as the CTFT of the original signal. For perfect reconstruction, pulse-train sampling should not introduce any aliasing. To prevent aliasing between different replicas, the sampling rate f_s must satisfy the sampling theorem, i.e. $\omega_s = 2\pi f_s \geq 4\pi\beta$.

9.2.2 Zero-order hold

A second practical implementation of sampling is achieved by the sample-and-hold circuit, which samples the band-limited input signal $x(t)$ at discrete time $(t = kT_s)$ and maintains the sampled value for the next T_s seconds. To prevent aliasing, the sampling interval T_s must satisfy the sampling theorem. This

Fig. 9.10. Time-domain illustration of the zero-order hold operation for a CT signal. (a) Original signal $x(t)$; (b) zero-order hold output $x_s(t)$.

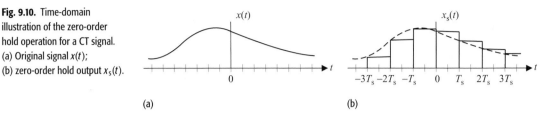

(a)   (b)

zero-order hold operation is illustrated in Fig. 9.10. Unlike the pulse-train sampling, the amplitude of the sampled signal is maintained constant for T_s seconds until the next sample is taken.

For mathematical analysis, the zero-order hold operation can be modeled by the following expression:

$$x_s(t) = \sum_{k=-\infty}^{\infty} x(kT_s) p_2(t - kT_s) \tag{9.22a}$$

or

$$x_s(t) = p_2(t) * \sum_{k=-\infty}^{\infty} x(kT_s)\delta(t - kT_s) = p_2(t) * \left[x(t) \sum_{k=-\infty}^{\infty} \delta(t - kT_s) \right], \tag{9.22b}$$

where $p_2(t)$ represents a rectangular pulse given by

$$p_2(t) = \text{rect}\left(\frac{t - 0.5\,T_s}{T_s} \right). \tag{9.23}$$

Equation (9.22b) models the zero-hold operation and is different from Eq. (9.18) in two ways. First, the duration of the pulse $p_2(t)$ in Eq. (9.22b) is the same as the sampling interval T_s, whereas the duration of the pulse $p_1(t)$ is much smaller than T_s in pulse-train sampling. Secondly, the order of operation in the sampled signal $x_s(t)$ is different from that used in the corresponding sampled signal in pulse-train sampling. In Eq. (9.22b), the sampled signal $x_s(t)$ is obtained by convolving $p_2(t)$ with a periodic impulse train, which is scaled by the values of the reference signal at the location of the impulse functions. In Eq. (9.18), on the other hand, $x_s(t)$ is obtained by multiplying the original signal directly by the periodic pulse train $r(t)$.

The CTFT of Eq. (9.22b) is given by

$$X_s(\omega) = P_2(\omega) \cdot \frac{1}{2\pi} \left[X(\omega) * \frac{2\pi}{T_s} \sum_{k=-\infty}^{\infty} \delta\left(\omega - \frac{2k\pi}{T_s} \right) \right], \tag{9.24}$$

where $P_2(\omega)$ denotes the CTFT of the rectangular pulse $p_2(t)$. Based on entry (16) of Table 5.2, the CTFT of $p_2(t)$ is given by the following transform pair:

$$\text{rect}\left(\frac{t - 0.5\,T_s}{T_s} \right) \overset{\text{CTFT}}{\longleftrightarrow} T_s \, \text{sinc}\left(\frac{\omega T_s}{2\pi} \right) e^{-j0.5\,\omega T_s}.$$

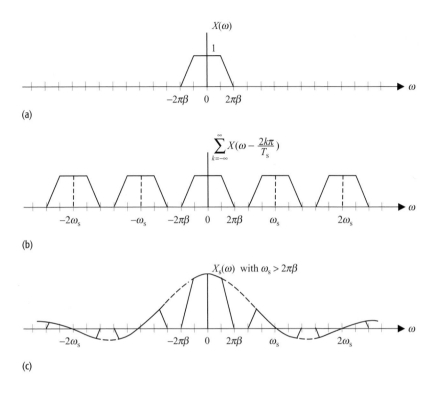

Fig. 9.11. Frequency-domain illustration of the zero-order hold operation for a CT signal. CTFTs of the: (a) original signal $x(t)$; (b) periodic replicas; and (c) the sampled signal $x_s(t)$.

Substituting the value of $P_2(\omega)$, Eq. (9.23) can be expressed as follows:

$$X_s(\omega) = e^{-j0.5\,\omega T_s}\,\operatorname{sinc}\left(\frac{\omega T_s}{2\pi}\right) \cdot \sum_{k=-\infty}^{\infty} X\left(\omega - \frac{2k\pi}{T_s}\right). \qquad (9.25)$$

Based on Eq. (9.25), Fig. 9.11 illustrates the frequency-domain interpretation of the zero-hold operation. The spectrum $X_s(\omega)$ of the sampled signal is shown in Fig. 9.11(c), which contains scaled replicas of the CTFT of the original base-band signal. Unlike the pulse-train sampling, some distortion in the amplitude is introduced in the central replica located at $\omega = 0$. This distortion can be minimized by increasing the width of the main lobe of the sinc function in Eq. (9.25). Since the width of the main lobe is given by $2\pi/T_s$, it is equivalent to reducing the sampling interval T_s.

To recover the original CT signal, the sampled signal is filtered with an LPF having a cut-off frequency $\omega_c = \omega_s/2$. Due to the amplitude distortion introduced in the central replica, ideal lowpass filtering recovers an approximate version of the original CT signal. For perfect reconstruction, the filter with the transfer function given by

$$H(\omega) = \begin{cases} \dfrac{1}{\operatorname{sinc}(\omega T_s/2\pi)} & |\omega| \leq \omega_s/2 \\ 0 & \text{elsewhere} \end{cases} \qquad (9.26)$$

Fig. 9.12. Input–output
relationship of an L-level
quantizer used to discretize the
sample values $x[kT_s]$ of a DT
sequence $x[k]$. (a) Uniform
quantizer; (b) non-uniform
quantizer.

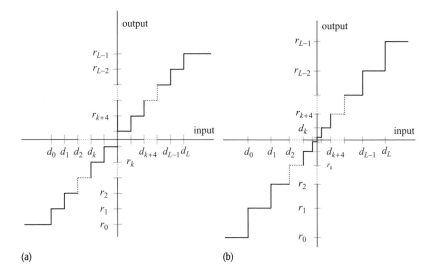

(a)   (b)

is used. The above filter is referred to as the compensation, or anti-imaging,
filter. Filtering $X_s(\omega)$ with the anti-imaging filter introduces a linear phase $-\omega T_s$
corresponding to the exponential term $\exp(-j\omega T_s)$. Inclusion of a linear phase
in the frequency domain is equivalent to a delay in the time domain and is
therefore harmless and not considered as a distortion.

9.3 Quantization

The process of sampling, discussed in Sections 9.1 and 9.2, converts a CT signal
$x(t)$ into a DT sequence $x[k]$, with each sample representing the amplitude of
the CT signal $x(t)$ at a particular instant $t = kT_s$. The amplitude $x[kT_s]$ of a
sample in $x[k]$ can still have an infinite number of possible values. To produce
a true digital sequence, each sample in $x[k]$ is approximated to a finite set
of values. The last step is referred to as *quantization* and is the focus of our
discussion in this section.

9.3.1 Uniform and non-uniform quantization

Figure 9.12(a) illustrates the input–output relationship for an L-level uniform
quantizer. The peak-to-peak range of the input sequence $x[k]$ is divided uniformly into $(L + 1)$ quantization levels $\{d_0, d_1, \ldots, d_L\}$ such that the separation $\Delta = (d_{m+1} - d_m)$ is the same between any two consecutive levels. The
separation Δ between two quantization levels is referred to as the *quantile interval* or quantization step size. For a given input, the output of the quantizer is
calculated from the following relationship:

$$y[k] = r_m = \frac{1}{2}[d_m + d_{m+1}] \quad \text{for} \quad d_m \leq x[k] < d_{m+1} \quad \text{and} \quad 0 \leq m < L.$$

$$(9.27)$$

In other words, the quantized value of the input lying within the levels d_m and d_{m+1} is given by r_m, which equals $0.5(d_m + d_{m+1})$. The quantization levels $\{d_0, d_1, \ldots, d_L\}$ are referred to as the *decision levels*, while the output levels $\{r_0, r_1, \ldots, r_{L-1}\}$ are referred to as the *reconstruction levels*.

Equation (9.27) approximates the analog sample values by using a finite number of quantization levels. The approximation introduces a distortion, which is referred to as the quantization error. The peak value of the quantization error is one-half of the quantile interval in the positive or negative direction.

The quantizer illustrated in Fig. 9.12(a) is called a uniform quantizer because the quantization levels are uniformly distributed between the minimum and maximum ranges of the input sequence. In most practical applications, the distribution of the amplitude of the input sequence is skewed towards low values. In speech communication, for example, low speech volumes dominate the sequence most of the time. Large-amplitude values are extremely rare and typically occupy only 15% to 25% of the communication time. A uniform quantizer will be wasteful, with most of the quantization levels rarely used. In such applications, we use non-uniform quantization, which provides fine quantization at frequently occurring lower volumes and coarse quantization at higher volumes. The input–output relationship of a non-uniform quantizer is shown in Fig. 9.12(b). The quantile interval is small at low values of the input sequence and large at high values of the sequence.

Example 9.3

Consider an audio recording system where the microphone generates a CT voltage signal within the range $[-1, 1]$ volts. Calculate the decision and reconstruction levels for an eight-level uniform quantizer.

Solution

For an $L = 8$ level quantizer with peak-to-peak range of $[-1, 1]$ volts, the quantile interval Δ is given by

$$\Delta = \frac{1 - (-1)}{8} = 0.25 \, \text{V}.$$

Starting with the minimum voltage of -1 V, the decision levels d_m are uniformly distributed between -1 V and 1 V. In other words,

$$d_m = -1 + m\Delta \quad \text{for} \quad 0 \leq m \leq L.$$

Substituting different values of m, we obtain

$$d_m = -1 \, \text{V}, -0.75 \, \text{V}, -0.5 \, \text{V}, -0.25 \, \text{V}, 0 \, \text{V}, 0.25 \, \text{V},$$
$$0.50 \, \text{V}, 0.75 \, \text{V}, 1 \, \text{V}.$$

Fig. 9.13. Derivation of a PCM sequence from a CT signal $x(t)$. The original CT signal $x(t)$ is shown by the dotted line, while the PCM sequence is shown as a stem plot.

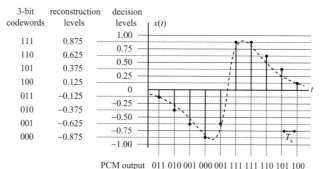

PCM output 011 010 001 000 001 111 111 110 101 100

Using Eq. (9.27), the reconstruction levels r_m are given by

$$r_m = -0.875 \text{ V}, -0.625 \text{ V}, -0.375 \text{ V}, -0.125 \text{ V}, 0.125 \text{ V},$$
$$0.375 \text{ V}, 0.625 \text{ V}, 0.875 \text{ V}.$$

The maximum quantization error is one-half of the quantile interval Δ and is given by 0.125 V.

9.3.1.1 Pulse code modulation

Pulse code modulation (PCM) is the analog-to-digital conversion of a CT signal, where the quantized samples of the CT signal are represented by finite-length digital words. The essential features of PCM are illustrated in Fig. 9.13, where a CT signal, with a peak-to-peak range of ± 1 V, is sampled and quantized by an eight-level uniform quantizer. As derived in Example 9.3, the decision levels d_m are located at $[-1 \text{ V}, -0.75 \text{ V}, -0.5 \text{ V}, -0.25 \text{ V}, 0 \text{ V}, 0.25 \text{ V}, 0.50 \text{ V}, 0.75 \text{ V}, 1$ V], while the corresponding reconstruction levels r_m are located at $[-0.875 \text{ V}, -0.625 \text{ V}, -0.375 \text{ V}, -0.125 \text{ V}, 0.125 \text{ V}, 0.375 \text{ V}, 0.625 \text{ V}, 0.875 \text{ V}]$. Since there are eight reconstruction levels, each quantized sample can be encoded by a minimum of $\ell = \log_2(L) = \log_2(8) = 3$-bit word. We assign the 3-bit word 000 to the reconstruction level $r_0 = -0.875$ V, 001 to the reconstruction level $r_1 = -0.625$, and so on for the remaining reconstruction levels as shown in Fig. 9.13. The PCM representation of the waveform $x(t)$ shown in Fig. 9.13 is therefore given by the following bits:

$$[011 \; 010 \; 001 \; 000 \; 001 \; 111 \; 111 \; 110 \; 101 \; 100],$$

where the final output is parsed in terms of 3-bit codewords.

9.3.2 Fidelity of quantized signal

In Table 9.2, we list the sampling frequency, the total number of quantization levels, and the resulting raw (uncompressed) data rate for a number of commercial audio applications. Low-fidelity applications, for example the telephone

Table 9.2. Raw data rates for digital audio used in commercial applications

Applications	Bandwidth (Hz)	Sampling rate (samples/s)	Quantization levels (L)	Raw data rate (bytes/s)
Telephone	200–3400	8000	2^8	8000
AM radio		11 025	2^8	11 025
FM radio (stereo)		22 050	2^{16}	88 200
CD (stereo)	20–20 000	44 100	2^{16}	176 400
Digital audio tape (stereo)	20–20 000	48 000	2^{16}	192 000

and the AM radio, are sampled at a relatively low sampling rate followed by a coarse quantizer to generate the PCM sequence. The quality of the reconstructed audio is moderate in such applications. In high-fidelity applications, for example the FM radio, compact disc (CD), and digital audio tape (DAT), the sampling rate is much higher to ensure accurate reconstruction of the high-frequency components. The number of levels in the quantizer is also increased to 2^{16} to reduce the effect of the quantization error. Two channels, one for the right speaker and the other for the left speaker, are transmitted for high-fidelity applications. Compared to a single channel, the data rate is effectively doubled with the transmission of two channels. The CD and DAT provide excellent audio quality and are generally recognized as world standards for achieving fidelity of audio reproduction that surpasses any other existing technique. In the following section, we discuss the CD digital audio system in more detail.

9.4 Compact discs

The compact disc (CD) digital audio system was defined jointly in 1979 by the Sony Corporation of Japan and the Philips Corporation of the Netherlands. The most important component of the CD digital audio system is an optical disc about 120 mm in diameter, which is used as the storage medium for recording data. The optical disc is referred to as the compact disc (CD) and stores about 10^{10} bits of data in the form of minute pits. To read data, the CD is optically scanned with a laser.

Before music can be recorded on a CD, it is preprocessed and converted into PCM data. The schematic diagram of the preprocessing stage for a single music channel is illustrated in Fig. 9.14(a). Each channel of the music signal is amplified and applied at the input of a lowpass filter (LPF), referred to as the anti-aliasing filter. Since the human ear is only sensitive to frequency components within the range 20 Hz–20 kHz, the anti-aliasing filter limits the bandwidth of the input channel to 20 kHz. Following the anti-aliasing filter is the PCM system, which converts the CT music channel into binary data. The sampling rate used in

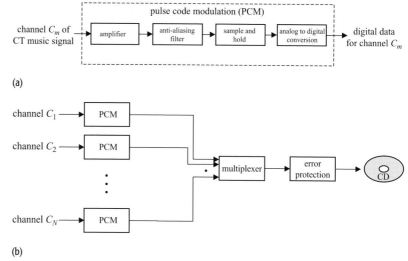

(a)

(b)

the sample-and-hold circuit is 44.1 ksamples/s, which exceeds the Nyquist rate
by a margin of 4.1 ksamples/s. The additional margin reduces the complexity
of the anti-aliasing filter by allowing a fair transition bandwidth between the
pass and stop bands of the filter. The audio samples obtained from the sample-
and-hold circuit are quantized using 2^{16}-level uniform quantization. Finally,
each quantized sample is encoded with a 16-bit codeword, which results in a
raw data rate of $(44\,100 \text{ samples/s} \times 16 \text{ bits/sample}) = 705.6$ kbits per second
(kbps) or $705.6/8 = 88.2$ kBytes per second (kBps).

For high-fidelity performance, several channels of the music signal are
recorded on a CD. For commonly used stereo systems, only two channels
corresponding to the left and right speakers are recorded. Many home theatre
systems now record a much higher number of channels to simulate surround
sound and other audio effects. Each channel of the music signal is preprocessed by the system illustrated in Fig. 9.14(a) and converted into raw PCM
data. Figure 9.14(b) illustrates the multiplexing stage, where data streams from
different channels are interleaved together into a single continuous bit stream.
The final step in the multiplexing stage is an error control scheme, which adds
an additional layer of protection to the music data. Any scanning errors that
were introduced whilst data were being read out from the CD are concealed
by the error control scheme. The output of the error control circuit is stored on
the CD. To record more music on a single CD, PCM data may be compressed
using an audio compression standard such as MP3.

A CD player reverses each step illustrated in Fig. 9.14. Data read from the
CD is checked for possible scanning errors. After correcting or concealing
the detected errors, the data streams for the individual channels are derived from
the interleaved bit stream. By following the reconstruction procedure outlined
in Section 9.1.1, each data stream is used to reconstruct the corresponding music

channel. The reconstructed channels are played simultaneously to simulate the effect of real audio.

Example 9.4

Consider a digital monochrome CCD camera that records an image $x[m, n]$ at a resolution of 800×1200 picture elements (pixels). In other words, each image consists of $800 \times 1200 = 0.96 \times 10^6$ pixels. Assuming that the human visual system cannot distinguish between more than 200 different shades of gray, determine how many bytes are required to store a single image. If the CCD camera has 32 million bytes of memory space to store images, how many images can be saved simultaneously in the camera?

Solution

An image pixel can have 200 different shades of gray. The number of bits required to represent the intensity value of each pixel is given by $\lceil \log_2(200) \rceil$ or $\lceil 7.64 \rceil$ or 8 bits;

$$\text{space required to save one image}$$
$$= 0.96 \times 10^6 \text{ pixels } \times 8 \text{ bits/pixel}$$
$$= 7.68 \times 10^6 \text{ bits or } 0.96 \times 10^6 \text{ bytes.}$$

Since the disc space for storing images is 32×10^6 bytes,

$$\text{number of images that can be stored simultaneously}$$
$$= 32 \times 10^6 \text{ bytes}/0.96 \times 10^6 \text{ bytes}$$
$$= 33.$$

9.5 Summary

In this chapter, we introduced the principle of sampling that is used to transform a CT baseband signal into an equivalent DT sequence. Section 9.1 discussed the ideal impulse-train sampling, where a periodic impulse train is multiplied by a CT baseband signal, resulting in a sequence of equally spaced samples at the location of the impulses ($t = kT_s$). In the frequency domain, the spectrum of the sampled signal consists of several shifted replicas of the spectrum of the original signal. We observe that the original CT signal is recoverable from its DT version by ideal lowpass filtering if the sampling rate $f_s = 1/T_s$ is greater than twice the highest frequency present in the baseband signal. This condition is referred to as the sampling theorem. Violating the sampling theorem distorts the spectrum of the original baseband signal; a phenomenon known as aliasing.

In practice, impulses are difficult to generate and are often approximated by narrow rectangular pulses. This leads to a more practical approach to sampling, covered in Section 9.2, in which a periodic rectangular pulse train is multiplied by the CT baseband signal to produce the sampled signal. Compared with the ideal impulse train sampling, the spectra of the two sampled signals are identical, except that the shifted replicas in the spectrum of the pulse-train are attenuated by a sinc function. Reconstruction of the original signal in rectangular pulse-train sampling is also achieved by lowpass filtering the sampled signal. A second practical implementation of sampling uses a zero-order hold circuit to sample the CT signal; this is covered in Section 9.2.2.

To encode a CT signal into a digital waveform, Section 9.3 introduces the process of quantization, in which the values of the samples are approximated to a finite set of levels. This involves replacing the exact sample value with the closest level defined by the L-level quantizer. In uniform quantization, the quantization levels are distributed uniformly between the maximum and minimum ranges of the input sequence. A uniform quantizer results in high quantization error in most practical applications, where the distribution of the sample values is skewed towards low amplitudes. In such cases, most of the quantization levels in the uniform quantizer are rarely used. A non-uniform quantizer reduces the overall quantization error by providing finer quantization at frequently occurring lower amplitudes and coarser quantization at less frequent higher amplitudes.

Sampling is used in a number of important applications. Section 9.4 introduces the compact disc (CD) and illustrates how sampling and quantization are used to convert an analog music signal into binary data, which can be stored on a CD. Since digital signals are less sensitive to distortion and interference than analog signals, the audio CD provides excellent audio quality that surpasses most analog storage mechanisms.

Problems

9.1 For the following CT signals, calculate the maximum sampling period T_s that produces no aliasing:

(a) $x_1(t) = 5 \, \text{sinc}(200t)$;

(b) $x_2(t) = 5 \, \text{sinc}(200t) + 8 \sin(100\pi t)$;

(c) $x_3(t) = 5 \, \text{sinc}(200t) \sin(100\pi t)$;

(d) $x_4(t) = 5 \, \text{sinc}(200t) * \sin(100\pi t)$, where $*$ denotes the CT convolution operation.

9.2 A famous theorem known as the uncertainty principle states that a baseband signal cannot be time-limited. By calculating the inverse CTFT of the following baseband signals, show that the uncertainty principle is indeed satisfied by the following signals (assume that ω_0 and W are real, positive constants):

(a) $X_1(\omega) = \text{rect}\left(\dfrac{\omega}{2W}\right)e^{-j2\omega}$;

(b) $X_2(\omega) = \begin{cases} 1 & |\omega| \leq W \\ 0 & \text{elsewhere}; \end{cases}$

(c) $X_3(\omega) = \text{rect}\left(\dfrac{\omega - \omega_0}{2W}\right) + \text{rect}\left(\dfrac{\omega + \omega_0}{2W}\right);$

(d) $X_4(\omega) = u(\omega - \omega_0) - u(\omega - 2\omega_0).$

9.3 The converse of the uncertainty principle, explained in Problem 9.2, is also true. In other words, a time-limited signal cannot be band-limited. By calculating the CTFT of the following time-limited signals, show that the converse of the uncertainty principle is indeed true (assume that τ, T, and α are real, positive constants):

(a) $x_1(t) = \cos(\omega_0 t)[u(t + T) - u(t - T)];$

(b) $x_2(t) = \text{rect}\left(\dfrac{t}{\tau}\right) * \text{rect}\left(\dfrac{t}{\tau}\right)$

$\qquad\qquad\qquad\qquad\qquad$ ($*$ denotes the CT convolution operator);

(c) $x_3(t) = e^{-\alpha|t|} \, \text{rect}\left(\dfrac{t}{\tau}\right);$

(d) $x_4(t) = \delta(t - 5) + \delta(t + 5).$

9.4 The CT signal $x(t) = v_1(t) \, v_2(t)$ is sampled with an ideal impulse train:

$$s(t) = \sum_{k=-\infty}^{\infty} \delta(t - kT_{\text{s}}).$$

(a) Assuming that $v_1(t)$ and $v_2(t)$ are two baseband signals band-limited to 200 Hz and 450 Hz, respectively, compute the minimum value of the sampling rate f_{s} that does not introduce any aliasing.

(b) Repeat part (a) if the waveforms for $v_1(t)$ and $v_2(t)$ are given by the following expression:

$$v_1(t) = \text{sinc}(600t) \quad \text{and} \quad v_2(t) = \text{sinc}(1000t).$$

(c) Assuming that a sampling interval of $T_{\text{s}} = 2\,\text{ms}$ is used to sample $x(t) = v_1(t)v_2(t)$ specified in part (b), sketch the spectrum of the sampled signal. Can $x(t)$ be accurately recovered from the sampled signal?

(d) Repeat part (c) for a sampling interval of $T_{\text{s}} = 0.1\,\text{ms}$.

9.5 The CT signal $x(t) = \sin(400\pi t) + 2\cos(150\pi t)$ is sampled with an ideal impulse train. Sketch the CTFT of the sampled signal for the following values of the sampling rate:

(a) $f_{\text{s}} = 100$ samples/s;

(b) $f_{\text{s}} = 200$ samples/s;

(c) $f_{\text{s}} = 400$ samples/s;

(d) $f_{\text{s}} = 500$ samples/s.

In each case, calculate the reconstructed signal using an ideal LPF with the transfer function given in Eq. (9.7) and a cut-off frequency of $\omega_{\text{s}}/2 = \pi f_{\text{s}}$.

9.6 Consider the following CT signal:

$$x(t) = \begin{cases} 0.25(3 - |t|) & 0 \leq |t| \leq 3 \\ 0 & \text{otherwise.} \end{cases}$$

(a) Calculate the CTFT $X(\omega)$. Determine the bandwidth of the signal and the ideal Nyquist sampling rate.

(b) If the bandwidth is infinite, approximate the bandwidth as β Hz, such that

$$|X(\omega)| < 0.01 \max|X(\omega)| \quad \text{for} \quad |\omega| > 2\pi\beta$$

and recalculate a practical Nyquist sampling rate.

(c) Discretize $x(t)$ using a sampling interval of $T_s = 1$ s. Plot the resulting DT sequence $x[k]$ corresponding to the duration $-5 \leq t \leq 5$.

(d) Quantize the signal $x[k]$ obtained in (c) with the uniform quantizer derived in Example 9.3. Plot the quantization error with respect to k. What is the maximum value of the quantization error?

(e) Repeat (d) using a uniform quantizer with $L = 16$ reconstruction levels defined within the dynamic range $[-1, 1]$. Plot the quantization error with respect to k. What is the maximum value of the quantization error? Compare the plot with your answer obtained in (d).

9.7 Show that the CTFS representation of the rectangular pulse train $r(t)$ as defined in Eq. (9.17) is given by Eq. (9.19).

9.8 The spectrum of a CT signal $x(t)$ satisfies the following conditions:

$$X(\omega) = 0 \quad \text{for} \quad |\omega| < \omega_1 \quad \text{or} \quad |\omega| > \omega_2 \quad \text{with} \quad \omega_2 > \omega_1 > 0.$$

In other words, the CTFT $X(\omega)$ of $x(t)$ is non-zero only within the range of frequencies $\omega_1 \leq |\omega| \leq \omega_2$. Such a signal is referred to as a bandpass signal.

(a) Show that the bandpass signal $x(t)$ can be sampled with an ideal impulse train at a rate less than the Nyquist rate of $2(\omega_2/2\pi)$ samples/s and can be perfectly reconstructed with a bandpass filter with the following transfer function:

$$H_{bp}(\omega) = \begin{cases} p & \omega_\ell \leq |\omega| \leq \omega_u \\ 0 & \text{elsewhere.} \end{cases}$$

(b) Determine the minimum sampling rate for which perfect reconstruction is possible.

(c) Compute the values of parameters p, ω_ℓ, and ω_u used to specify the transfer function of the bandpass filter.

9.9 An alternative to the bandpass sampling procedure introduced in Problem 9.8 is the system illustrated in Fig. P9.9. For a real-valued bandpass signal $x(t)$ with the spectrum shown in Fig. P9.9(a), the

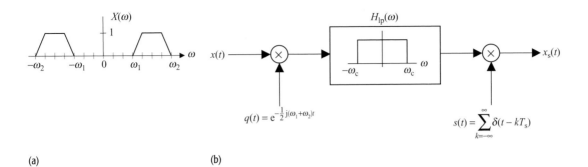

(a)   (b)

Fig. P. 9.9. (a) Spectrum of a bandpass signal $x(t)$; (b) ideal sampling of a bandpass baseband signal.

cut-off frequency of the ideal LPF $H_{lp}(\omega)$ in Fig. P9.9(b) is given by $\omega_c = 0.5(\omega_2 - \omega_1)$.

(a) Sketch the spectrum of the sampled signal $x_s(t)$.

(b) Determine the maximum value of the sampling interval T_s that introduces no aliasing. Compare this sampling interval with that obtained from the Nyquist rate.

(c) Implement a reconstruction system to recover $x(t)$ from the sampled signal $x_s(t)$.

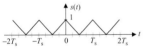

Fig. P. 9.10. Sawtooth function used in sawtooth wave sampling.

9.10 An alternative to ideal impulse train sampling is sawtooth wave sampling. Here, a CT signal $x(t)$ is multiplied with a periodic sawtooth wave $s(t)$ (shown in Fig. P9.10). Denote the resulting signal by $z(t) = x(t) * s(t)$.

(a) Derive an expression for the CTFT $Z(\omega)$ of the signal $z(t)$ in terms of the CTFT of the original signal $x(t)$.

(b) Assuming that the CTFT of the original signal $x(t)$ is shown in Fig. 9.3(a), sketch the spectrum of the CTFT of the signal $z(t)$.

(c) Based on your answer to part (b), can $x(t)$ be reconstructed from $z(t)$? If yes, state the conditions under which $x(t)$ may be reconstructed. Sketch the block diagram of the reconstruction system including the specifications of any filters used.

(d) By comparing the CTFTs, state how $z(t)$ relates to the sampled signal $x_s(t)$ obtained by ideal impulse train sampling.

9.11 Repeat Problem 9.10 with an alternating sign impulse train,

$$s(t) = \sum_{k=-\infty}^{\infty} (-1)^k \delta(t - kT_s),$$

as the sampling signal.

9.12 Repeat Problem 9.10 with the periodic signal,

$$s(t) = \sum_{k=-\infty}^{\infty} [\delta(t - kT_s) + \delta(t - \Delta - kT_s)],$$

as the sampling signal.

9.13 A CT band-limited signal $x(t)$ is sampled at its Nyquist rate f_s and transmitted over a band-limited channel modeled with the transfer function

$$H_{ch}(\omega) = \begin{cases} 1 & 4\pi f_s \leq |\omega| \leq 8\pi f_s \\ 0 & \text{otherwise.} \end{cases}$$

Let the signal received at the end of the channel be $x_{ch}(t)$. Determine the reconstruction system that recovers the CT signal $x(t)$ from $x_{ch}(t)$.

9.14 If the quantization noise needs to be limited to $\pm p\%$ of the peak-to-peak value of the input signal, show that the number of bits in each PCM word must satisfy the following inequality:

$$n \geq 3.32 \log_{10}\left(\frac{50}{p}\right).$$

9.15 A voice signal with a bandwidth of 4 kHz and an amplitude range of ± 20 mV is converted to digital data using a PCM system.
  (a) Determine the maximum sampling interval T_s that can be used to sample the voice signal.
  (b) If the PCM system has an accuracy of $\pm 5\%$ during the quantization step, determine the length of the codewords in bits.
  (c) Determine the data rate in bps (bits/s) of the resulting PCM sequence.

9.16 A baseband signal with a bandwidth of 100 kHz and an amplitude range of ± 1 V is to be transmitted through a channel which is constrained to a maximum transmission speed of 2 Mbps. Your task is to design a uniform quantizer that introduces minimum quantization error. Determine the maximum number of levels L in the uniform quantizer. What is the maximum distortion introduced by the uniform quantizer? Assume the Nyquist rate for sampling.

9.17 Consider the input–output relationship of an ideal sampling system given by

$$x_s(t) = x(t) \sum_{k=-\infty}^{\infty} \delta(t - kT_s) = \sum_{k=-\infty}^{\infty} x(kT_s)\delta(t - kT_s).$$

Determine if the ideal sampling system is (i) linear, (ii) time-invariant, (iii) memoryless, (iv) causal, (v) stable, and (vi) invertible.

9.18 Consider the input–output relationship of a DT quantizer with L decision levels, given by

$$y[k] = Q\{x[k]\} = \frac{1}{2}[d_m + d_{m+1}] \quad \text{for} \quad d_m \leq x[k] < d_{m+1} \quad \text{and}$$
$$0 \leq m < L.$$

Determine if the DT quantizer is (i) linear, (ii) time-invariant, (iii) memoryless, (iv) causal, (v) stable, and (vi) invertible.

9.19 Consider a digital mp3 player that has 1024×10^6 bytes of memory. Assume that the audio clips stored in the player have an average duration of five minutes.

(a) Assuming a sampling rate of $44\,100$ samples/s and 16 bits/sample/ channel quantization, determine the average storage space required (without any form of compression) to store a stereo (i.e. two-channel) audio clip.

(b) Assume that the audio clips are stored in the mp3 format, which reduces the audio file size to roughly one-eighth of its original size. Calculate the storage space required to store an mp3-compressed audio clip.

(c) How many mp3-compressed audio files can be stored in the mp3 player?

9.20 Consider a digital color camera with a resolution of 2560×1920 pixels.

(a) Calculate the storage space required to store an image in the camera without any compression. Assume three color channels and quantization of 8 bit/pixel/channel.

(b) Assume that the images are stored in the camera in the JPEG format, which reduces an image to roughly one-tenth of its original size. Calculate the storage space required to store a JPEG-compressed image.

(c) If the camera has 512×10^6 bytes of memory, determine the number of JPEG-compressed images that can be stored in the camera.

10 Time-domain analysis of discrete-time systems

An important subset of discrete-time (DT) systems satisfies the linearity and time-invariance properties, discussed in Chapter 2. Such DT systems are referred to as linear, time-invariant, discrete-time (LTID) systems. In this chapter, we will develop techniques for analyzing LTID systems. As was the case for the LTIC systems discussed in Part II, we are primarily interested in calculating the output response $y[k]$ of an LTID system to a DT sequence $x[k]$ applied at the input of the system.

In the time domain, an LTID system is modeled either with a linear, constant-coefficient difference equation or with its impulse response $h[k]$. Section 10.1 covers linear, constant-coefficient difference equations and develops numerical techniques for solving such equations. Section 10.2 defines the impulse response $h[k]$ as the output of an LTID system to an unit impulse function $\delta[k]$ applied at the input of the system and shows how the impulse response can be derived from a linear, constant-coefficient difference equation. Section 10.3 proves that any arbitrary DT sequence can be represented as a linear combination of time-shifted DT impulse functions. This development leads to a second approach for calculating the output $y[k]$ based on convolving the applied input sequence $x[k]$ with the impulse response $h[k]$ in the DT domain. The resulting operation is referred to as the convolution sum and is defined in Section 10.4. Section 10.5 introduces two graphical methods for calculating the convolution sum, and Section 10.6 lists several important properties of the convolution sum. A special case of convolution sum, referred to as the periodic or circular convolution, occurs when the two operands are periodic sequences. Section 10.7 develops techniques for computing the periodic convolution and shows how it may be used to compute the linear convolution. In Section 10.8, we revisit the causality, stability, and invertibility properties of LTID systems and express these properties in terms of the impulse response $h[k]$. MATLAB instructions for computing the convolution sum are listed in Section 10.9. The chapter is concluded in Section 10.10 with a summary of the important concepts covered in the chapter.

10.1 Finite-difference equation representation of LTID systems

As discussed in Section 3.1, an LTIC system can be modeled using a linear, constant-coefficient differential equation. Likewise, the input–output relationship of a linear DT system can be described using a difference equation, which takes the following form:

$$y[k + n] + a_{n-1}y[k + n - 1] + \cdots + a_0 y[k]$$
$$= b_m x[k + m] + b_{m-1}x[k + m - 1] + \cdots + b_0 x[k], \qquad (10.1)$$

where $x[k]$ denotes the input sequence and $y[k]$ denotes the resulting output sequence, and coefficients a_r (for $0 \leq r \leq n - 1$), and b_r (for $0 \leq r \leq m$) are parameters that characterize the DT system. The coefficients a_r and b_r are constants if the DT system is also time-invariant. For causal signals and systems analysis, the following n initial (or ancillary) conditions must be specified in order to obtain the solution of the nth-order difference equation in Eq. (10.1):

$$y[-1], y[-2], \ldots, y[-n].$$

We now consider an iterative procedure for solving linear, constant-coefficient difference equations.

Example 10.1

The DT sequence $x[k] = 2ku[k]$ is applied at the input of a DT system described by the following difference equation:

$$y[k + 1] - 0.4y[k] = x[k].$$

By iterating the difference equation from the ancillary condition $y[-1] = 4$, compute the output response $y[k]$ of the DT system for $0 \leq k \leq 5$.

Solution
Express $y[k + 1] - 0.4y[k] = x[k]$ as follows:

$$y[k] = 0.4y[k - 1] + x[k - 1]$$
$$= 0.4y[k - 1] + 2(k - 1)\, u(k - 1) \quad \{\because x[k] = 2k\, u[k]\},$$

which can alternatively be expressed as

$$y[k] = \begin{cases} 0.4y[k - 1] & k = 0 \\ 0.4y[k - 1] + 2(k - 1) & k \geq 1. \end{cases}$$

Fig. 10.1. Input and output sequences for Example 10.1. (a) Input sequence $x[k]$; (b) output sequence $y[k]$.

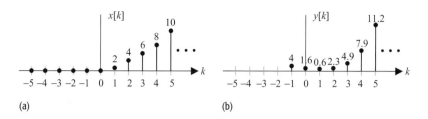

(a)   (b)

By iterating from $k = 0$, the output response is computed as follows:

$$y[0] = 0.4y[-1] = 1.6,$$
$$y[1] = 0.4y[0] + 2 \times 0 = 0.64,$$
$$y[2] = 0.4y[1] + 2 \times 1 = 2.256,$$
$$y[3] = 0.4y[2] + 2 \times 2 = 4.902,$$
$$y[4] = 0.4y[3] + 2 \times 3 = 7.961,$$
$$y[5] = 0.4y[4] + 2 \times 4 = 11.184.$$

Additional values of the output sequence for $k > 5$ can be similarly evaluated from further iterations with respect to k. The input and output sequences are plotted in Fig. 10.1 for $0 \leq k \leq 5$.

In Chapter 3, we showed that the output response of a CT system, represented by the differential equation in Eq. (3.1), can be decomposed into two components: the zero-state response and the zero-input response. This is also valid for the DT systems represented by the difference equation in Eq. (10.1). The output response $y[k]$ can be expressed as

$$y[k] = \underbrace{y_{zi}[k]}_{\text{zero-input response}} + \underbrace{y_{zs}[k]}_{\text{zero-state response}}, \qquad (10.2)$$

where $y_{zi}[k]$ denotes the *zero-input response* (or the *natural response*) of the system and $y_{zs}[k]$ denotes the *zero-state response* (or the *forced response*) of the DT system.

The zero-input component $y_{zi}[k]$ for a DT system is the response produced by the system because of the initial conditions, and is not due to any external input. To calculate the zero-input component $y_{zi}[k]$, we assume that the applied input sequence $x[k] = 0$. On the other hand, the zero-state response $y_{zs}[k]$ arises due to the input sequence and does not depend on the initial conditions of the system. To calculate the zero-state response $y_{zs}[k]$, the initial conditions are assumed to be zero. Based on Eq. (10.2), a DT system represented by Eq. (10.1) can be considered as an incrementally linear system (see Section 2.2.1) where the additive offset is caused by the initial conditions (see Fig. 2.10). If the initial conditions are zero, the DT system becomes on LTID system. We now solve Example 10.1 in terms of the zero-input and zero-state components of the output.

Example 10.2

Repeat Example 10.1 to calculate (i) the zero-input response $y_{zi}[k]$, (ii) the zero-state response $y_{zs}[k]$, and (iii) the overall output response $y[k]$ for $0 \leq k \leq 5$.

Solution

(i) The zero-input response of the system is obtained by solving the following difference equation:

$$y[k + 1] - 0.4y[k] = x[k],$$

with input $x[k] = 0$ and ancillary condition $y[-1] = 4$. The difference equation reduces to

$$y_{zi}[k] = 0.4y_{zi}[k - 1],$$

with ancillary condition $y_{zi}[-1] = 4$. Iterating for $k = 0, 1, 2, 3, 4$, and 5 yields

$$y_{zi}[0] = 0.4y_{zi}[-1] = 1.6,$$
$$y_{zi}[1] = 0.4y_{zi}[0] = 0.64,$$
$$y_{zi}[2] = 0.4y_{zi}[1] = 0.256,$$
$$y_{zi}[3] = 0.4y_{zi}[2] = 0.1024,$$
$$y_{zi}[4] = 0.4y_{zi}[3] = 0.0410,$$
$$y_{zi}[5] = 0.4y_{zi}[4] = 0.0164.$$

(ii) The zero-state response of the system is calculated by solving the following difference equation:

$$y_{zs}[k] = 0.4y_{zs}[k - 1] + 2(k - 1)u[k - 1],$$

with ancillary condition $y_{zs}[-1] = 0$. Iterating the difference equation for $k = 0, 1, 2, 3, 4$, and 5 yields

$$y_{zs}[0] = 0.4y_{zs}[-1] + 2 \times (-1) \times 0 = 0,$$
$$y_{zs}[1] = 0.4y_{zs}[0] + 2 \times 0 \times 1 = 0,$$
$$y_{zs}[2] = 0.4y_{zs}[1] + 2 \times 1 \times 1 = 2,$$
$$y_{zs}[3] = 0.4y_{zs}[2] + 2 \times 2 \times 1 = 4.8,$$
$$y_{zs}[4] = 0.4y_{zs}[3] + 2 \times 3 \times 1 = 7.92,$$
$$y_{zs}[5] = 0.4y_{zs}[4] + 2 \times 4 \times 1 = 11.168.$$

(iii) Adding the zero-input and zero-state components obtained in parts (i) and (ii), yields

$$y[0] = y_{zi}[0] + y_{zs}[0] = 1.6,$$
$$y[1] = y_{zi}[1] + y_{zs}[1] = 0.64,$$
$$y[2] = y_{zi}[2] + y_{zs}[2] = 2.256,$$
$$y[3] = y_{zi}[3] + y_{zs}[3] = 4.902,$$
$$y[4] = y_{zi}[4] + y_{zs}[4] = 7.961,$$
$$y[5] = y_{zi}[5] + y_{zs}[5] = 11.184.$$

Note that the overall output response $y[k]$ is identical to the output response obtained in Example 10.1. By iterating with respect to k, additional values for the output response $y[k]$ for $k > 5$ can be computed.

In Section 10.1, we used a linear, constant-coefficient difference equation to model an LTID system. A second model is based on the impulse response $h[k]$ of a system. This alternative representation leads to a different approach for analyzing LTID systems. Section 10.2 presents this alternative approach.

10.2 Representation of sequences using Dirac delta functions

In this section, we show that any arbitrary sequence $x[k]$ may be represented as a linear combination of time-shifted, DT impulse functions. Recall that a DT impulse function is defined in Eq. (1.51) as follows:

$$\delta[k] = \begin{cases} 1 & k = 0 \\ 0 & k \neq 0. \end{cases} \tag{10.3}$$

We are interested in representing any DT sequence $x[k]$ as a linear combination of shifted impulse functions, $\delta[k - m]$, for $-\infty < m < \infty$. We illustrate the procedure using the arbitrary function $x[k]$ shown in Fig. 10.2(a). Figures 10.2(b)–(f) represent $x[k]$ as a linear combination of a series of simple functions $x_m[k]$, for $-\infty < m < \infty$. Since $x_m[k]$ is non-zero only at one location ($k = m$), it represents a scaled and time-shifted impulse function. In other words,

$$x_m[k] = x[m]\delta[k - m]. \tag{10.4}$$

Fig. 10.2. Representation of a DT sequence as a linear combination of time-shifted impulse functions. (a) Arbitrary sequence $x[k]$; (b)–(f) its decomposition using DT impulse functions.

In terms of $x_m[k]$, the DT sequence $x[k]$ is, therefore, represented by

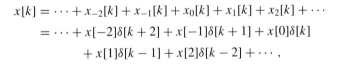

$$\begin{aligned} x[k] &= \cdots + x_{-2}[k] + x_{-1}[k] + x_0[k] + x_1[k] + x_2[k] + \cdots \\ &= \cdots + x[-2]\delta[k+2] + x[-1]\delta[k+1] + x[0]\delta[k] \\ &\quad + x[1]\delta[k-1] + x[2]\delta[k-2] + \cdots, \end{aligned}$$

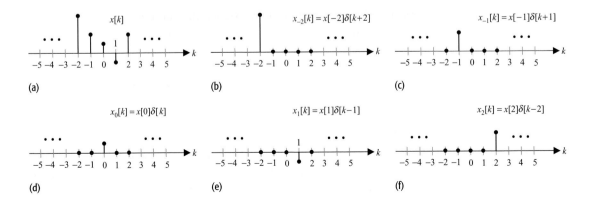

which reduces to

$$x[k] = \sum_{m=-\infty}^{\infty} x[m]\delta[k - m]. \tag{10.5}$$

Equation (10.5) provides an alternative representation of an arbitrary DT function using a linear combination of time-shifted DT impulses. In Eq. (10.5), variable m denotes the dummy variable for the summation that disappears as the summation is computed. Recall that a similar representation exists for the CT functions and is given by Eq. (3.24).

10.3 Impulse response of a system

In Section 10.1, a constant-coefficient difference equation is used to specify the input–output characteristics of an LTID system. An alternative representation of an LTID system is obtained by specifying its impulse response. In this section, we will formally define the impulse response and illustrate how the impulse response of an LTID system can be derived directly from the difference equation modeling the LTID system.

Definition 10.1 *The impulse response $h[k]$ of an LTID system is the output of the system when a unit impulse $\delta[k]$ is applied at the input of the LTID system. Following the notation introduced in Eq. (2.1b), the impulse response can be expressed as follows:*

$$\delta[k] \rightarrow h[k], \tag{10.6}$$

with zero ancillary conditions.

Note that an LTID system satisfies the linearity and the time-shifting properties. Therefore, if the input is a scaled and time-shifted impulse function $a\delta[k - k_0]$, the output, Eq. (10.6), of the DT system is also scaled by a factor of a and time-shifted by k_0, i.e.

$$a\delta[k - k_0] \rightarrow ah[k - k_0], \tag{10.7}$$

for any arbitrary constants a and k_0. Section 10.4 illustrates how Eq. (10.7) can be generalized to calculate the output of LTID systems for any arbitrary input.

Example 10.3
Consider the LTID systems with the following input–output relationships:

(i) $y[k] = x[k - 1] + 2x[k - 3];$ \hfill (10.8)
(ii) $y[k + 1] - 0.4y[k] = x[k].$ \hfill (10.9)

Calculate the impulse responses for the two LTID systems. Also, determine the output responses of the LTID systems when the input is given by $x[k] = 2\delta[k] + 3\delta[k - 1]$.

Solution

(i) The impulse response of a system is the output of the system when the input sequence $x[k] = \delta[k]$. Therefore, the impulse response $h[k]$ of system (i) can be obtained by substituting $y[k]$ by $h[k]$ and $x[k]$ by $\delta[k]$ in Eq. (10.8). In other words, the impulse response for system (i) is given by

$$h[k] = \delta[k - 1] + 2\delta[k - 3].$$

To evaluate the output response resulting from the input sequence $x[k] = 2\delta[k] + 3\delta[k - 1]$, we use the linearity and time-invariance properties of the system. The outputs resulting from the two terms $2\delta[k]$ and $3\delta[k - 1]$ in the input sequence are as follows:

$$2\delta[k] \rightarrow 2h[k] = 2\delta[k - 1] + 4\delta[k - 3]$$

and

$$3\delta[k - 1] \rightarrow 3h[k - 1] = 3\delta[k - 2] + 6\delta[k - 4].$$

Applying the superposition principle, the output $y[k]$ to input $x[k] = 2\delta[k] + 3\delta[k - 1]$ is given by

$$2\delta[k] + 3\delta[k - 1] \rightarrow 2h[k] + 3h[k - 1]$$

or

$$\begin{aligned} y[k] &= (2\delta[k - 1] + 4\delta[k - 3]) + (3\delta[k - 2] + 6\delta[k - 4]) \\ &= 2\delta[k - 1] + 3\delta[k - 2] + 4\delta[k - 3] + 6\delta[k - 4]). \end{aligned}$$

(ii) On substituting $y[k]$ by $h[k]$ and $x[k]$ by $\delta[k]$ in Eq. (10.9), the impulse response of the LTID system (ii) is represented by the following recursive equation:

$$h[k + 1] - 0.4h[k] = \delta[k]. \tag{10.10a}$$

Equation (10.10a) is a difference equation, which can be solved by substituting $k = m - 1$. The resulting equation is given by

$$h[m] = \delta[m - 1] + 0.4h[m - 1]. \tag{10.10b}$$

To solve for the delayed response $h[m - 1]$, we substitute $k = m - 2$ in Eq. (10.10b). The resulting expression is given by

$$h[m - 1] = \delta[m - 2] + 0.4h[m - 2]. \tag{10.10c}$$

Substituting the above value of $h[m- 1]$ from Eq. (10.10c) in Eq. 10.10(b) yields

$$h[m] = \delta[m - 1] + 0.4\delta[m - 2] + 0.4^2 h[m - 2].$$

The aforementioned procedure can be repeated for the delayed impulse response $h[m-2]$ on the right-hand side of the equation, then for the resulting $h[m-3]$, and so on. The final result is as follows:

$$h[m] = \delta[m-1] + 0.4\delta[m-2] + 0.4^2\delta[m-3] + 0.4^3\delta[m-4] + \cdots$$

or

$$h[m] = \sum_{\ell=1}^{\infty} 0.4^{\ell-1}\delta[m-\ell] = 0.4^{m-1}u[m-1]$$

or

$$h[k] = 0.4^{k-1}u[k-1],$$

which is the required expression for the impulse response of the system.

Next, we proceed to calculate the output of the LTID system for the input sequence $x[k] = 2\delta[k] + 3\delta[k-1]$. Because the system is linear and time-invariant, the output sequence $y[k]$ resulting from input $x[k] = 2\delta[k] + 3\delta[k-1]$ is given by

$$2\delta[k] + 3\delta[k-1] \rightarrow 2h[k] + 3h[k-1]$$

or

$$\begin{aligned} y[k] &= 2 \times 0.4^{k-1}u[k-1] + 3 \times 0.4^{k-2}u[k-2] \\ &= 2 \times 0.4^0\delta[k-1] + (2 \times 0.4^{k-1}u[k-2] + 3 \times 0.4^{k-2}u[k-2]) \\ &= 2\delta[k-1] + 3.8 \times 0.4^{k-2}u[k-2]. \end{aligned}$$

Example 10.4

The impulse response of an LTID system is given by $h[k] = 0.5^k u[k]$. Determine the output of the system for the input sequence $x[k] = \delta[k-1] + 3\delta[k-2] + 2\delta[k-6]$.

Solution

Because the system is LTID, it satisfies the linearity and time-shifting properties. The individual responses to the three terms $\delta[k-1]$, $3\delta[k-2]$, and $2\delta[k-6]$ in the input sequence $x[k]$ are given by

$$\delta[k-1] \rightarrow h[k-1] = 0.5^{k-1}u[k-1],$$
$$3\delta[k-2] \rightarrow 3h[k-2] = 3 \times 0.5^{k-2}u[k-2],$$

and

$$2\delta[k-6] \rightarrow 2h[k-6] = 2 \times 0.5^{k-6}u[k-6].$$

Fig. 10.3. (a) Impulse response $h[k]$ of the LTID system specified in Example 10.4. (b) Output $y[k]$ of the LTID system for input $x[k] = \delta[k-1] + 3\delta[k-2] + 2\delta[k-6]$.

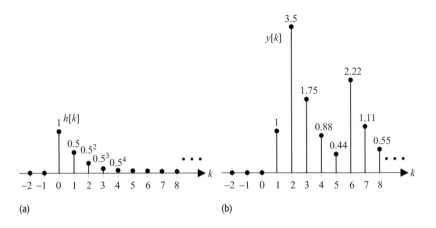

Applying the principle of superposition, the overall response to the input sequence $x[k]$ is given by

$$y[k] = h[k-1] + 3h[k-2] + 2h[k-6].$$

Substituting the value of $h[k] = 0.5^k u[k]$ results in the output response:

$$y[k] = 0.5^{k-1} u[k-1] + 3 \times 0.5^{k-2} u[k-2] + 2 \times 0.5^{k-6} u[k-6].$$

The impulse response $h[k]$ and the resulting output sequence are plotted in Figs 10.3(a) and (b).

10.4 Convolution sum

Examples 10.3 and 10.4 compute the output of an LTID system for relatively elementary input sequences $x[k]$ consisting of a few scaled and time-shifted impulses. In this section, we extend the approach to more complex input sequences.

It was shown in Eq. (10.5), which is reproduced below for clarity, that any arbitrary input sequence can be represented as a linear combination of time-shifted impulse functions as follows:

$$x[k] = \sum_{m=-\infty}^{\infty} x[m]\delta[k-m]. \qquad (10.11)$$

Note that in Eq. (10.11), $x[m]$ is a scalar representing the magnitude of the impulse $\delta[k-m]$ located at $k = m$. In terms of the impulse response $h[k]$, the output resulting from a single impulse $x[m]\delta[k-m]$ is given by

$$x[m]\delta[k-m] \longrightarrow x[m]h[k-m]. \qquad (10.12)$$

Fig. 10.4. Output response of a system to an arbitrary input sequence x[k].

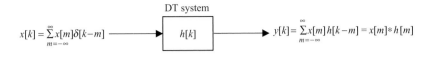

DT system

$$x[k] = \sum_{m=-\infty}^{\infty} x[m]\delta[k-m] \longrightarrow \boxed{h[k]} \longrightarrow y[k] = \sum_{m=-\infty}^{\infty} x[m]h[k-m] = x[m] * h[m]$$

Applying the principle of superposition, the overall output $y[k]$ resulting from the input sequence $x[k]$, represented by Eq. (10.11), is given by

$$\underbrace{\sum_{m=-\infty}^{\infty} x[m]\delta[k-m]}_{x[k]} \longrightarrow \underbrace{\sum_{m=-\infty}^{\infty} x[m]h[k-m]}_{y[k]}, \qquad (10.13)$$

where the summation on the right-hand side, used to compute the output response $y[k]$, is referred to as the *convolution sum*. Equation (10.13) provides us with a second approach for calculating the output $y[k]$. It states that the output $y[k]$ can be calculated by convolving the input sequence $x[k]$ with the impulse response $h[k]$ of the LTID system. Mathematically, Eq. (10.13) is expressed as follows:

$$y[k] = x[k] * h[k] = \sum_{m=-\infty}^{\infty} x[m]h[k-m], \qquad (10.14)$$

where $*$ denotes the convolution sum. Figure 10.4 illustrates the process of convolution. The convolution operation defined in Eq. (10.14) is commonly referred to as the *linear* convolution, in contrast to a special type of convolution known as *periodic* convolution, which is discussed in Section 10.6.

We now consider several examples to illustrate the steps involved in computing the convolution sum.

Example 10.5

Assuming that the impulse response of an LTID system is given by $h[k] = 0.5^k u[k]$, determine the output response $y[k]$ to the input sequence $x[k] = 0.8^k u[k]$.

Solution

Using Eq. (10.14), the output response $y[k]$ of the LTID system is given by

$$y[k] = \sum_{m=-\infty}^{\infty} x[m]h[k-m] = \sum_{m=-\infty}^{\infty} 0.8^m u[m]0.5^{k-m} u[k-m].$$

Using the values of the unit step function $u[m]$, the above summation simplifies as follows:

$$y[k] = \sum_{m=0}^{\infty} 0.8^m 0.5^{k-m} u[k-m].$$

Depending on the value of k, the output response $y[k]$ of the system may take two different forms for $k \geq 0$ or $k < 0$. We consider the two cases separately.

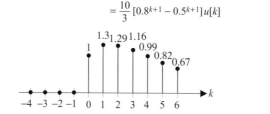

Case 1 $(k < 0)$　When $k < 0$, the unit step function $u[k - m] = 0$ within the limits of summation $(0 \le m \le \infty)$. Therefore, the output sequence $y[k] = 0$ for $k < 0$.

Case II $(k \ge 0)$　When $k \ge 0$, the unit step function $u[k - m]$ has the following values:

$$u[k - m] = \begin{cases} 1 & m \le k \\ 0 & m > k. \end{cases}$$

The output sequence $y[k]$ is therefore given by

$$y[k] = \sum_{m=0}^{k} 0.8^m 0.5^{k-m} = 0.5^k \sum_{m=0}^{k} \left(\frac{0.8}{0.5}\right)^m,$$

for $k \ge 0$. The above summation represents a geometric progression (GP) series. Using the GP series sum formula provided in Appendix A, Section A.3, the output response $y[k]$ is calculated as follows:

$$y[k] = 0.5^k \left[\frac{1 - (0.8/0.5)^{k+1}}{1 - (0.8/0.5)}\right] = \frac{10}{3}[0.8^{k+1} - 0.5^{k+1}].$$

Combining the two cases $(k < 0$ and $k \ge 0)$, the output response $y[k]$ is given by

$$y[k] = \begin{cases} 0 & k < 0 \\ \dfrac{10}{3}[0.8^{k+1} - 0.5^{k+1}] & k \ge 0 \end{cases}$$

$$= \frac{10}{3}[0.8^{k+1} - 0.5^{k+1}]u[k].$$

The output response of the system is plotted in Fig. 10.5.

　　Example 10.5 shows how to calculate the convolution sum analytically. In many situations, it is more convenient to use a graphical approach to evaluate the convolution sum. Section 10.5 describes the graphical approach.

10.5　Graphical method for evaluating the convolution sum

The graphical approach for calculating the convolution sum is similar to the graphical procedure for calculating the convolution integral for the LTIC system, discussed in Chapter 3. In the following, we highlight the main steps in calculating the convolution sum between two sequences $x[k]$ and $h[k]$.

Algorithm 10.1 Graphical procedure for computing the linear convolution

(1) Sketch the waveform for input $x[m]$ by changing the independent variable of $x[k]$ from k to m and keep the waveform for $x[m]$ fixed during steps (2)–(7).

(2) Sketch the waveform for the impulse response $h[m]$ by changing the independent variable from k to m.

(3) Reflect $h[m]$ about the vertical axis to obtain the time-inverted impulse response $h[-m]$.

(4) Shift the sequence $h[-m]$ by a selected value of k. The resulting function represents $h[k-m]$.

(5) Multiply the input sequence $x[m]$ by $h[k-m]$ and plot the product function $x[m]h[k-m]$.

(6) Calculate the summation $\sum_{m=-\infty}^{\infty} x[m]h[k-m]$.

(7) Repeat steps (4)–(6) for $-\infty \leq k \leq \infty$ to obtain the output response $y[k]$ over all time k.

The graphical approach for calculating the output response is illustrated through a series of examples.

Example 10.6

Repeat Example 10.5 with input $x[k] = 0.8^k u[k]$ and impulse response $h[k] = 0.5^k u[k]$ to determine the output of the LTID system using the graphical convolution approach.

Solution

Following steps (1)–(3) of Algorithm 10.1, the DT sequences $x[m] = 0.8^m u[m]$, $h[m] = 0.5^m u[m]$ and its time reflection $h[-m] = 0.5^{-m} u[-m]$ are plotted in Fig. 10.6. Based on step (4), the sequence $h[k-m] = h[-(m-k)]$ is obtained by shifting $h[-m]$ by k samples. To compute the output sequence, we consider two cases based on the values of k.

Case 1 For $k < 0$, the waveform $h[k-m]$ is on the left-hand side of the vertical axis. As is apparent in Fig. 10.6, step (5a), waveforms for $h[k-m]$ and $x[m]$ do not overlap. In other words, the product $x[m]h[k-m] = 0$, for $-\infty \leq m \leq \infty$, as long as $k < 0$. The output sequence $y[k]$ is therefore zero for $k < 0$.

Case 2 For $k \geq 0$, we see from Fig. 10.6, step (5b), that the non-zero parts of $h[k-m]$ and $x[m]$ overlap over the range $m = [0, k]$. Therefore,

$$y[k] = \sum_{m=0}^{k} x[m]h[k-m] = \sum_{m=0}^{k} 0.8^m 0.5^{k-m}.$$

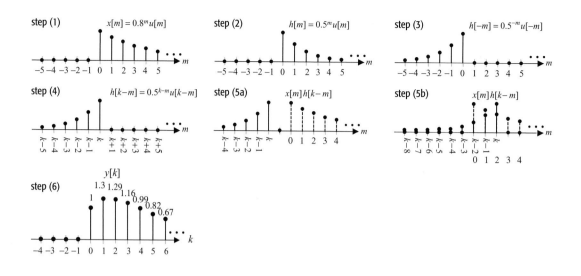

Fig. 10.6. Convolution of the input sequence $x[k]$ with the impulse response $h[k]$ in Example 10.6.

As shown in Example 10.5, the above summation simplifies to

$$y[k] = \frac{10}{3}[0.8^{k+1} - 0.5^{k+1}] \quad \text{for } k \geq 0.$$

Combining Cases 1 and 2, the overall output sequence is given by

$$y[k] = \frac{10}{3}[0.8^{k+1} - 0.5^{k+1}]u[k].$$

The final output response is plotted in Fig. 10.6, step (6).

Example 10.7

For the following DT sequences:

$$x[k] = \begin{cases} 2 & 0 \leq k \leq 2 \\ 0 & \text{otherwise} \end{cases} \quad \text{and} \quad h[k] = \begin{cases} k+1 & 0 \leq k \leq 4 \\ 0 & \text{otherwise,} \end{cases}$$

calculate the convolution sum $y[k] = x[k] * h[k]$ using the graphical approach.

Solution

Following steps (1)–(3) of Algorithm 10.1, the sequences $x[m]$, $h[m]$, and its reflection $h[-m]$ are plotted as a function of the independent variable m in Fig. 10.7, steps (1)–(3). The DT sequence $h[k - m] = h[-(m - k)]$ is obtained by shifting the time-reflected function $h[-m]$ by k. Depending on the value of k, five special cases arise. We consider these cases separately.

Case 1 For $k < 0$, we see from Fig. 10.7, step (5a), that the non-zero parts of $h[k - m]$ and $x[m]$ do not overlap. In other words, output $y[k] = 0$ for $k < 0$.

Case 2 For $0 \leq k \leq 2$, we see from Fig. 10.7, step (5b), that the non-zero parts of $h[k - m]$ and $x[m]$ overlap over the duration $m = [0, k]$. Therefore, the

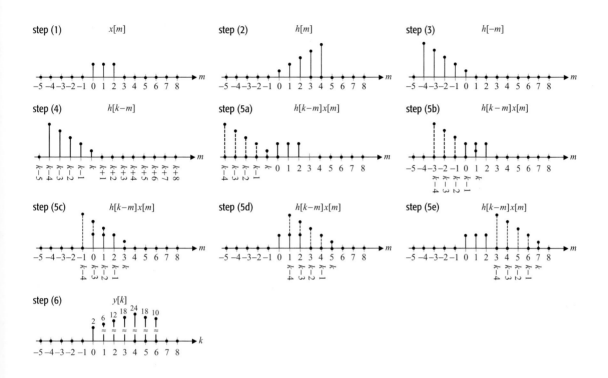

Fig. 10.7. Convolution of the input sequence $x[k]$ with the impulse response $h[k]$ in Example 10.7.

output response for $0 \le k \le 2$ is given by

$$y[k] = \sum_{m=0}^{k} x[m]h[k-m] = \sum_{m=0}^{k} 2 \times (k - m + 1) = 2(k+1)\sum_{m=0}^{k} 1 - 2\sum_{m=0}^{k} m$$

$$= 2(k+1)^2 - 2\sum_{m=1}^{k} m.$$

The summation $\sum_{m=1}^{k} m$ is an arithmetic progression (AP) series. Using the AP series summation formula provided in Appendix A, Section A.3, the output response $y[k]$ for $0 \le k \le 2$ is calculated as follows:

$$y[k] = 2(k+1)^2 - k(k+1) = k^2 + 3k + 2.$$

Case 3 For $2 \le k \le 4$, we see from Fig. 10.7, step (5c), that the non-zero part of $h[k-m]$ completely overlaps $x[m]$ over the region $m = [0, 2]$. The output response $y[k]$ for $2 \le k \le 4$ is given by

$$y[k] = \sum_{m=0}^{2} x[m]h[k-m] = \sum_{m=0}^{2} 2 \times (k - m + 1)$$

$$= 2(k+1)\sum_{m=0}^{2} 1 - 2\sum_{m=0}^{2} m = 6(k+1) - 6 = 6k.$$

Case 4 For $4 \leq k \leq 6$, we see from Fig. 10.7, step (5d), that the non-zero part of $h[k - m]$ partially overlaps $x[m]$ over the region $m = [k - 4, 2]$. The output $y[k]$ for $5 \leq k \leq 6$ is given by

$$y[k] = \sum_{m=k-4}^{2} x[m]h[k - m] = \sum_{m=k-4}^{2} 2 \times (k - m + 1)$$

$$= 2(k + 1) \sum_{m=k-4}^{2} 1 - 2 \sum_{m=k-4}^{2} m$$

$$= 2(k + 1)(7 - k) - (7 - k)(k - 2) = -k^2 + 3k + 8.$$

Case 5 For $k > 6$, we see from Fig. 10.7, step (5e), that the non-zero parts of $h[k - m]$ and $x[m]$ do not overlap. Therefore, the product $x[m]h[k - m] = 0$ for all values of m. The value of the output sequence $y[k] = 0$ for $k > 6$.

Combining the above five cases, we obtain

$$y[k] = \begin{cases} 0 & k < 0, k > 6 \\ k^2 + 3k + 2 & 0 \leq k \leq 2 \\ 6k & 2 \leq k \leq 4 \\ -k^2 + 3k + 8 & 4 \leq k \leq 6, \end{cases}$$

which is plotted in Fig. 10.7, step (6).

10.5.1 Sliding tape method

The graphical convolution approach, illustrated in Examples 10.6 and 10.7, for LTID systems is similar to the graphical convolution procedure for LTIC systems. However, sketching the figures for the time-reversed and time-shifted impulse functions may prove to be difficult in certain cases. There is a variant of the graphical method for DT convolution, known as the sliding tape method, which is convenient in cases where the convolved sequences are relatively short in length. Instead of drawing the figures in such cases, we compute the convolution sum using a table whose entries are the values of the DT sequences at different instances. We illustrate the sliding tape method in Examples 10.8 and 10.9.

Example 10.8

For the two sequences $x[k]$ and $h[k]$ defined in Example 10.7, calculate the convolution $y[k] = x[k] * h[k]$ using the sliding tape method.

Solution

The convolution of $x[k]$ and $h[k]$ using the sliding tape method is illustrated in Table 10.1. The first row represents the m-axis; the second row represents the input sequence $x[m]$; and the third row represents the impulse response

Table 10.1. Convolution of $x[k]$ and $h[k]$ using the sliding tape method for Example 10.8

m	...	−5	−4	−3	−2	−1	0	1	2	3	4	5	6	7	...	k	$y[k]$
$x[m]$							2	2	2								
$h[m]$							1	2	3	4	5						
$h[-m]$			5	4	3	2	1										
$h[-1-m]$		5	4	3	2	1										−1	0
$h[0-m]$			5	4	3	2	1									0	2
$h[1-m]$				5	4	3	2	1								1	6
$h[2-m]$					5	4	3	2	1							2	12
$h[3-m]$						5	4	3	2	1						3	18
$h[4-m]$							5	4	3	2	1					4	24
$h[5-m]$								5	4	3	2	1				5	18
$h[6-m]$									5	4	3	2	1			6	10
$h[7-m]$										5	4	3	2	1		7	0

$h[m]$ for different values of m. Following the steps involved in convolution, we generate the values for the sequence $h[k - m]$ and store the value in a row. To generate the values of $h[k - m]$, we first form the function $h[-m]$, which is obtained by time-inverting $h[m]$. The result is illustrated in the fourth row of Table 10.1. The time-reversed function $h[-m]$ is used to generate $h[k - m]$ by right-shifting $h[-m]$ by k time units. For example, the fifth row contains the values of the function $h[-1 - m] = h[-(m + 1)]$. Similarly, rows (6)–(13) contain the values of the function $h[k - m] = h[-(m - k)]$ for the range $0 \leq k \leq 7$. In order to calculate $y[k]$ for a fixed value of k, we multiply the entries in the row containing $x[m]$ by the corresponding entries contained in the row for $h[k - m]$ and then evaluate the summation:

$$y[k] = \sum_{m=-\infty}^{\infty} x[m]h[k - m].$$

For $k = -1$, we note that the non-zero entries of $x[m]$ and $h[k - m]$ do not overlap. Therefore, $y[k] = 0$ for $k = -1$. Since there is also no overlap for $k < -1$, the output $y[k] = 0$ for $k \leq -1$.

The aforementioned multiplication process is repeated for different values of k. For $k = 0$, we note that the only overlap between the non-zero values of $x[m]$ and $h[-m]$ occurs for $m = 0$. The output response is therefore given by

$$y[0] = 2 \cdot 1 = 2.$$

These values of time instant $k = 0$ and the output response $y[0] = 2$ are stored in the last two columns of row (6), corresponding to the entries of $h[0 - m]$ in Table 10.1. Similarly, for $k = 1$, we observe that the overlap between the non-zero values of $x[m]$ and $h[1 - m]$ occurs for $m = 0$ and 1. The output response is given by

$$y[0] = 2 \cdot 2 + 2 \cdot 1 = 6$$

Table 10.2. Convolution of $x[k]$ and $h[k]$ using the sliding tape method for Example 10.9

m	...	−5	−4	−3	−2	−1	0	1	2	3	4	5	6	...	k	$y[k]$
$h[m]$						3	1	−2	3	−2						
$x[m]$						−1	1	2								
$x[-m]$						2	1	−1								
$x[-3-m]$			2	1	−1										−3	0
$x[-2-m]$				2	1	−1									−2	−3
$x[-1-m]$					2	1	−1								−1	2
$x[0-m]$						2	1	−1							0	9
$x[1-m]$							2	1	−1						1	−3
$x[2-m]$								2	1	−1					2	1
$x[3-m]$									2	1	−1				3	4
$x[4-m]$										2	1	−1			4	−4
$x[5-m]$											2	1	−1		5	0

and is stored in the last column of Table 10.1. We repeat the process for increasing values of k until the overlap between $x[m]$ and $h[k - m]$ is eliminated. In Table 10.1, this occurs for $k > 7$, beyond which the output response $y[k]$ is zero.

By comparison with the result obtained in Example 10.7, we note that the output response $y[k]$ obtained using the sliding tape method is identical to the one obtained using the graphical approach.

Example 10.9

For the following pair of the input sequence $x[k]$ and impulse response $h[k]$:

$$x[k] = \begin{cases} -1 & k = -1 \\ 1 & k = 0 \\ 2 & k = 1 \\ 0 & \text{otherwise} \end{cases} \quad \text{and} \quad h[k] = \begin{cases} 3 & k = -1, 2 \\ 1 & k = 0 \\ -2 & k = 1, 3 \\ 0 & \text{otherwise,} \end{cases}$$

calculate the output response using the sliding tape method.

Solution

The output $y[k]$ can be calculated by convolving the input sequence $x[k]$ with the impulse response $h[k]$. Since convolution satisfies the distributive property, i.e.

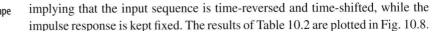

$$y[k] = x[k] * h[k] = h[k] * x[k],$$

Table 10.2 reverses the role of the input sequence $x[k]$ with that of the impulse response $h[k]$ and computes the following summation:

$$y[k] = \sum_{m=-\infty}^{\infty} h[m]x[k - m],$$

implying that the input sequence is time-reversed and time-shifted, while the impulse response is kept fixed. The results of Table 10.2 are plotted in Fig. 10.8.

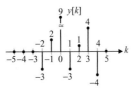

Fig. 10.8. Output response calculated using the sliding tape method in Example 10.9.

10.6 Periodic convolution

Linear convolution is used to convolve aperiodic sequences. If the convolving sequences are periodic, the result of linear convolution is unbounded. In such cases, a second type of convolution, referred to as *periodic* or *circular convolution*, is generally used.

Consider two periodic sequences $x_p[k]$ and $h_p[k]$, with identical fundamental period K_0. The subscript p denotes periodicity. The relationship for the periodic convolution between two periodic sequences is defined as follows:

$$y_p[k] = x_p[k] \otimes h_p[k] = \sum_{m=\langle K_0 \rangle} x_p[m]h_p[k-m], \qquad (10.15)$$

where the summation on the right-hand side of Eq. (10.15) is defined over one complete period K_0. In calculating the summation, we can, therefore, start from any arbitrary position (say $m = m_0$) as long as one complete period of the sequences is covered by the summation. For the lower limit $m = m_0$, the upper limit is given by $m = m_0 + K_0 - 1$. In the text, the periodic convolution is denoted by the operator $\otimes$, whereas the linear convolution is denoted by $*$.

The steps involved in calculating the periodic convolution are given in the following algorithm.

Algorithm 10.2 Graphical procedure for computing the periodic convolution

(1) Sketch the waveform for input $x_p[m]$ by changing the independent variable of $x_p[k]$ from k to m and keep the waveform for $x_p[m]$ fixed during steps (2)–(7).

(2) Sketch the waveform for the impulse response $h_p[m]$ by changing the independent variable from k to m.

(3) Reflect $h_p[m]$ about the vertical axis to obtain the time-inverted impulse response $h_p[-m]$. Set the time index $k = 0$.

(4) Shift the function $h_p[-m]$ by a selected value of k. The resulting sequence represents $h_p[k-m]$.

(5) Multiply input sequence $x_p[m]$ by $h_p[k-m]$ and plot the product function $x_p[m]h_p[k-m]$.

(6) Calculate the summation $\sum_{m=\langle K_0 \rangle} x_p[m]h_p[k-m]$ for $m = [m_0, m_0 + K_0 - 1]$ to determine $y_p[k]$ for the value of k selected in step (4).

(7) Increment k by one and repeat steps (4)–(6) till all values of k in the specified range ($0 \le k \le K_0 - 1$) are exhausted.

(8) Since $y_p[k]$ is periodic with period K_0, the values of $y_p[k]$ outside the range $0 \le k \le K_0 - 1$ are determined from the values obtained in steps (6) and (7).

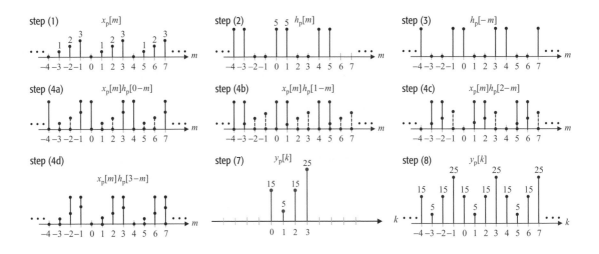

Fig. 10.9. Periodic convolution of the periodic sequences $x[k]$ and $h[k]$ in Example 10.10.

By comparing the aforementioned procedure for computing the periodic convolution with the procedure specified for evaluating the linear convolution in Section 10.5, we observe that steps (4), (6), and (7) are different in the two algorithms. In the linear convolution, the summation

$$\sum_{m=-\infty}^{\infty} x[m]h[k-m]$$

is computed within the limits $m = [-\infty, \infty]$ for different values of k in the range $-\infty \le k \le \infty$. In the periodic convolution, however, the summation is computed over one complete period, say $m = [m_0, m_0 + K_0 - 1]$ for a reduced range $(0 \le k \le K_0 - 1)$.

Example 10.10

Determine the periodic convolution between the following periodic sequences:

$$x_p[k] = k, \quad \text{for} \quad 0 \le k \le 3 \quad \text{and} \quad h_p[k] = \begin{cases} 5 & k = 0, 1 \\ 0 & k = 2, 3, \end{cases}$$

with the fundamental period $K_0 = 4$.

Solution

Following steps (1)–(3), the periodic sequences $x_p[m]$, $h_p[m]$, and its reflected version $h_p[-m]$ are plotted in Fig. 10.9, steps (1)–(3). Since the fundamental period $K_0 = 4$, we compute the result of the periodic convolution as follows:

$$y_p[k] = x_p[k] \otimes h_p[k] = \sum_{m=0}^{3} x_p[m]h_p[k-m] \tag{10.16}$$

for $0 \le k \le 3$. The DT periodic sequences $h_p[k-m]$ and $x_p[m]$ for $k = 0, 1, 2,$ and 3 are plotted, respectively, in Fig. 10.9, steps 4(a)–(d). The convolution summation, Eq. (10.16), has the following values:

$(k = 0)$ $y_p[0] = x_p[0]h_p[0] + x_p[1]h_p[-1] + x_p[2]h_p[-2] + x_p[3]h_p[-3]$

$= 0 \times 5 + 1 \times 0 + 2 \times 0 + 3 \times 5 = 15;$

$(k = 1)$ $y_p[1] = x_p[0]h_p[1] + x_p[1]h_p[0] + x_p[2]h_p[-1] + x_p[3]h_p[-2]$

$= 0 \times 0 + 1 \times 5 + 2 \times 0 + 3 \times 0 = 5;$

$(k = 2)$ $y_p[2] = x_p[0]h_p[2] + x_p[1]h_p[1] + x_p[2]h_p[0] + x_p[3]h_p[-1]$

$= 0 \times 0 + 1 \times 5 + 2 \times 5 + 3 \times 0 = 15;$

$(k = 3)$ $y_p[3] = x_p[0]h_p[3] + x_p[1]h_p[2] + x_p[2]h_p[1] + x_p[3]h_p[0]$

$= 0 \times 0 + 1 \times 0 + 2 \times 5 + 3 \times 5 = 25.$

The remaining values of $y_p[k]$ are easily determined by exploiting the periodicity property of $y_p[k]$. The output $y_p[k]$ is plotted in Fig. 10.9, step (8).

An alternative procedure for computing the periodic convolution can be obtained by calculating the limits of Eq. (10.15) for $m = 0$ to $m = K_0 - 1$. The resulting expression is given by

$$y_p[k] = \sum_{m=0}^{K_0-1} x_p[m]h_p[k - m]$$

or

$$y_p[k] = x_p[0]h_p[k] + x_p[1]h_p[k - 1] + x_p[2]h_p[k - 2]$$
$$+ \cdots + x_p[K_0 - 1]h_p[k - (K_0 - 1)],$$

for $0 \leq k \leq K_0 - 1$. Expanding the above equation in terms of the time index k yields

$$\left.\begin{aligned} y_p[0] &= x_p[0]h_p[0] + x_p[1]h_p[-1] + x_p[2]h_p[-2] + \cdots \\ &\quad + x_p[K_0 - 1]h_p[-(K_0 - 1)], \\ y_p[1] &= x_p[0]h_p[1] + x_p[1]h_p[0] + x_p[2]h_p[-1] \\ &\quad + \cdots + x_p[K_0 - 1]h_p[-(K_0 - 2)], \\ y_p[2] &= x_p[0]h_p[2] + x_p[1]h_p[1] + x_p[2]h_p[0] + \cdots \\ &\quad + x_p[K_0 - 1]h_p[-(K_0 - 3)], \\ &\vdots \\ y_p[K_0 - 1] &= x_p[0]h_p[K_0 - 1] + x_p[1]h_p[K_0 - 2] \\ &\quad + x_p[2]h_p[K_0 - 3] + \cdots + x_p[K_0 - 1]h_p[0]. \end{aligned}\right\}$$

(10.17)

Since $h_p[k]$ is periodic,

$$h_p[k] = h_p[k + K_0] \quad \text{or} \quad h_p[-k] = h_p[K_0 - k]. \tag{10.18}$$

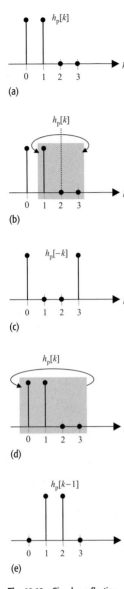

(a)

(b)

(c)

(d)

(e)

Fig. 10.10. Circular reflection and shifting for a periodic sequence. (a) Original periodic sequence $h_p[k]$. (b) Procedure to determine circularly reflected sequence $h_p[-k]$ from $h_p[k]$. (c) Circularly reflected sequence $h_p[-k]$. (d) Procedure to determine circularly shifted sequence $h_p[k-1]$ from $h_p[k]$. (e) Circularly shifted sequence $h_p[k-1]$.

Equation (10.18) is referred to as *periodic* or *circular reflection*. Before proceeding with the alternative algorithm for periodic convolution, we explain circular reflection in more detail.

Example 10.11

For the periodic sequence

$$h_p[k] = \begin{cases} 5 & k = 0, 1 \\ 0 & k = 2, 3, \end{cases}$$

with fundamental period $K_0 = 4$, determine the circularly reflected sequence $h_p[-k]$ and the circular shifted sequence $h_p[k-1]$.

Solution

Let $v_p[k]$ denote the circular reflected sequence $h_p[-k]$. Using $v_p[k] = h_p[-k] = h_p[K_0 - k]$, the values of the circularly reflected signals are given by

$k = 0$ $\qquad\qquad v_p[0] = h_p[K_0] = h_p[0] = 5;$

$k = 1$ $\qquad\qquad v_p[1] = h_p[K_0 - 1] = h_p[3] = 0;$

$k = 2$ $\qquad\qquad v_p[2] = h_p[K_0 - 2] = h_p[2] = 0;$

$k = 3$ $\qquad\qquad v_p[3] = h_p[K_0 - 3] = h_p[1] = 5.$

The original sequence $h_p[k]$ is plotted in Fig. 10.10(a), and the circularly reflected sequence $h_p[-k]$ is plotted in Fig. 10.10(c). Note that the circularly reflected signal $h_p[-k]$ can be obtained directly from $h_p[k]$ by keeping the value of $h_p[0]$ fixed and then reflecting the remaining values of $h_p[k]$ for $1 \le k \le K_0 - 1$ about $k = K_0/2$. This procedure is illustrated in Fig. 10.10(b).

Substituting $0 \le k \le K_0 - 1$, the values for the circularly shifted signal $w_p[k] = h_p[k-1]$ are obtained as follows:

$k = 0$ $\qquad\qquad w_p[0] = h_p[-1] = h_p[K_0 - 1] = 0;$

$k = 1$ $\qquad\qquad w_p[1] = h_p[0] = 5;$

$k = 2$ $\qquad\qquad v_p[2] = h_p[1] = 5;$

$k = 3$ $\qquad\qquad v_p[3] = h_p[2] = 0.$

The circularly shifted sequence $h_p[k - 1]$ is plotted in Fig. 10.10(e). The circularly shifted signal $h_p[k - 1]$ can also be obtained directly from $h_p[k]$ by shifting $h_p[k]$ towards the left by one time unit and moving the overflow value of $h_p[K_0 - 1]$ back into the sequence. This procedure is illustrated in Fig. 10.10(d).

To derive the alternative algorithm for periodic convolution, we substitute different values of k within the range $1 \le k \le K_0 - 1$ in Eq. (10.18). The resulting equations are given by

$$h_p[-1] = h_p[K_0 - 1]; h_p[-2] = h_p[K_0 - 2]; \ldots; h_p[-(K_0 - 1)] = h_p[1],$$

which are substituted in Eq. (10.17) to obtain

$$
\left.\begin{aligned}
y_{\mathrm{p}}[0] &= x_{\mathrm{p}}[0]h_{\mathrm{p}}[0] + x_{\mathrm{p}}[1]h_{\mathrm{p}}[K_0 - 1] + x_{\mathrm{p}}[2]h_{\mathrm{p}}[K_0 - 2] \\
&\quad + \cdots x_{\mathrm{p}}[K_0 - 1]h_{\mathrm{p}}[1], \\
y_{\mathrm{p}}[1] &= x_{\mathrm{p}}[0]h_{\mathrm{p}}[1] + x_{\mathrm{p}}[1]h_{\mathrm{p}}[0] + x_{\mathrm{p}}[2]h_{\mathrm{p}}[K_0 - 1] \\
&\quad + \cdots x_{\mathrm{p}}[K_0 - 1]h_{\mathrm{p}}[2], \\
y_{\mathrm{p}}[2] &= x_{\mathrm{p}}[0]h_{\mathrm{p}}[2] + x_{\mathrm{p}}[1]h_{\mathrm{p}}[1] + x_{\mathrm{p}}[2]h_{\mathrm{p}}[0] + \cdots x_{\mathrm{p}}[K_0 - 1]h_{\mathrm{p}}[3], \\
&\quad\vdots \\
y_{\mathrm{p}}[K_0 - 1] &= x_{\mathrm{p}}[0]h_{\mathrm{p}}[K_0 - 1] + x_{\mathrm{p}}[1]h_{\mathrm{p}}[K_0 - 2] \\
&\quad + x_{\mathrm{p}}[2]h_{\mathrm{p}}[K_0 - 3] + \cdots x_{\mathrm{p}}[K_0 - 1]h_{\mathrm{p}}[0].
\end{aligned}\right\}
$$

$$(10.19)$$

These expressions require values from only one period ($0 \le k \le K_0 - 1$) of the input sequence $x_{\mathrm{p}}[k]$ and the impulse response $h_{\mathrm{p}}[k]$. Therefore, we can implement the periodic convolution from a single period of the convolving functions. The main steps involved in such an implementation are listed in the following algorithm.

Algorithm 10.3 Alternative procedure for computing the periodic convolution

(1) Sketch one period of the waveform for input $x_{\mathrm{p}}[m]$ by changing the independent variable of $x_{\mathrm{p}}[k]$ from k to m within the range $0 \le k \le K_0 - 1$.

(2) Sketch one period of the waveform for the impulse response $h_{\mathrm{p}}[m]$ by changing the independent variable from k to m within the range $0 \le k \le K_0 - 1$.

(3) Reflect $h_{\mathrm{p}}[m]$ such that $h_{\mathrm{p}}[-m] = h_{\mathrm{p}}[K_0 - m]$ as defined by the circular reflection. Set $k = 0$.

(4) Using the circularly reflected function $h_{\mathrm{p}}[-m]$, determine the waveform for $h_{\mathrm{p}}[k - m] = h_{\mathrm{p}}[-(m - k)]$.

(5) Multiply the function $x_{\mathrm{p}}[m]$ by $h_{\mathrm{p}}[k - m]$ for $0 \le m \le K_0 - 1$ and plot the product function $x_{\mathrm{p}}[m]h_{\mathrm{p}}[k - m]$.

(6) Calculate the summation $\sum_{m=0}^{K_0 - 1} x_{\mathrm{p}}[m]h_{\mathrm{p}}[k - m]$ to determine $y_{\mathrm{p}}[k]$ for the value of k selected in step (4).

(7) Increment k by one and repeat steps (4)–(6) till all values of k within the range $0 \le k \le K_0 - 1$ are exhausted.

(8) Since $y_{\mathrm{p}}[k]$ is periodic with period K_0, the values of $y_{\mathrm{p}}[k]$ outside the range $0 \le k \le K_0 - 1$ are determined from the values obtained in step (7).

We illustrate the alternative implementation by repeating Example 10.10 and using the modified algorithm.

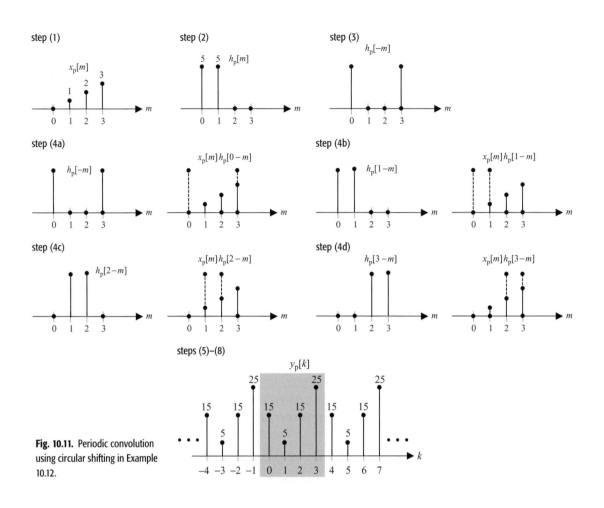

Example 10.12

Using Algorithm 10.3, determine the periodic convolution of the periodic sequences

$$x_p[k] = k \ (0 \leq k \leq 3) \quad \text{and} \quad h_p[k] = \begin{cases} 5 & k = 0, 1 \\ 0 & k = 2, 3, \end{cases}$$

with fundamental period $K_0 = 4$.

Solution

Following steps (1) and (2), the applied input and the impulse response are plotted as a function of m in Fig. 10.11, steps (1) and (2).

Following step (3), the circularly reflected impulse response $v_p[m] = h_p[-m] = h_p[K_0 - m]$ for $0 \leq m \leq 3$ is calculated as follows:

$$v_p[0] = h_p[0] = 1; \quad v_p[1] = h_p[-1] = h_p[3] = 0; \quad v_p[2] = h_p[-2]$$
$$= h_p[2] = 0; \quad \text{and} \quad v_p[3] = h_p[-3] = h_p[1] = 3.$$

For $k = 0$, the DT sequence $h_p[k - m] = h_p[-m]$. The value of the output response at $k = 0$ is given by

$$y_p[0] = \sum_{m=0}^{K_0-1} x_p[m]h_p[-m] = 0(5) + 1(0) + 2(0) + 3(5) = 15.$$

For $k = 1$, the DT sequence $h_p[k - m] = h_p[1 - m]$. The new sequence $h_p[1 - m] = h_p[-(m - 1)]$ is obtained by circularly shifting $h_p[-m]$ towards the right by one sample, with the last sample at $m = 3$ taking the place of the first sample at $m = 0$. The sequence $h_p[1 - m]$ is plotted in Fig. 10.11, step (4b). Multiplying by $h_p[m]$, the value of the output response at $k = 1$ is given by

$$y_p[1] = \sum_{m=0}^{K_0-1} x_p[m]h_p[1 - m] = 0(5) + 1(5) + 2(0) + 3(0) = 5.$$

For $k = 2$, the DT sequence $h_p[k - m] = h_p[2 - m]$. The new sequence $h_p[2 - m]$ is obtained by circularly shifting $h_p[1 - m]$ towards the right by one sample, with the last sample at $m = 3$ taking the place of the first sample at $m = 0$. The sequence $h_p[2 - m]$ is plotted in Fig. 10.11, step (4c). Multiplying by $h_p[m]$, the value of the output response at $k = 2$ is given by

$$y_p[2] = \sum_{m=0}^{K_0-1} x_p[m]h_p[2 - m] = 0(0) + 1(5) + 2(5) + 3(0) = 15.$$

For $k = 3$, the DT sequence $h_p[k - m] = h_p[3 - m]$. The new sequence $h_p[3 - m]$ is obtained by circularly shifting $h_p[2 - m]$ towards the right by one sample, with the last sample at $m = 3$ taking the place of the first sample at $m = 0$. The sequence $h_p[3 - m]$ is plotted in Fig. 10.11, step (4d). Multiplying by $h_p[m]$, the value of the output response at $k = 3$ is given by

$$y_p[3] = \sum_{m=0}^{K_0-1} x_p[m]h_p[3 - m] = 0(0) + 1(0) + 2(5) + 3(5) = 25.$$

The final output $y_p[k]$, obtained from steps (5)–(8) of Algorithm 10.3, is plotted in Fig. 10.11, Steps (5)–(8). Observe that the result is identical to that in Fig. 10.9, which was obtained using the full periodic convolution.

10.6.1 Linear convolution through periodic convolution

In this chapter, we have introduced two types of DT convolution. The linear convolution, defined in Eq. (10.14), is used to convolve aperiodic sequences, while the periodic convolution, defined in Eq. (10.15), is used for convolving periodic sequences. Definition 10.3 states a condition under which the results of the periodic and linear convolution are the same.

Definition 10.3 *Assume that $x[k]$ and $h[k]$ are two aperiodic DT sequences of finite length such that the following are true.*

(i) *The DT sequence $x[k] = 0$ outside the range $k_{\ell 1} \leq k \leq k_{u1}$. Note that it is possible for $x[k]$ to have some zero values within the range $k_{\ell 1} \leq k \leq k_{u1}$. The length K_x of $x[k]$ is given by $K_x = (k_{u1} - k_{\ell 1} + 1)$ samples.*

(ii) *The DT sequence $h[k] = 0$ outside the range $k_{\ell 2} \leq k \leq k_{u2}$. As for $x[k]$, it is possible for $h[k]$ to have intermittent zero values within the range $k_{\ell 2} \leq k \leq k_{u2}$. The length K_h of $h[k]$ is given by $K_h = k_{u2} - k_{\ell 2} + 1$ samples.*

Add the appropriate number of zeros to the two sequences $x[k]$ and $h[k]$ so that they have the same length $K_0 \geq (K_x + K_h - 1)$. The procedure of adding zeros to a sequence is referred to as zero padding. The periodic extensions of zero-padded $x[k]$ and $h[k]$ are denoted by $x_p[k]$ and $h_p[k]$, which have the same fundamental period of $K_0 \geq (K_x + K_h - 1)$. Mathematically, the single periods of $x_p[k]$ and $h_p[k]$ are defined as follows:

$$x_p[k] = \begin{cases} x[k] & k_{\ell 1} \leq k \leq k_{u1} \\ 0 & k_{u1} < k \leq K_0 + k_{\ell 1} - 1 \end{cases} \qquad (10.20a)$$

and

$$h_p[k] = \begin{cases} h[k] & k_{\ell 2} \leq k \leq k_{u2} \\ 0 & k_{u2} < k \leq K_0 + k_{\ell 2} - 1. \end{cases} \qquad (10.20b)$$

It can be shown that the linear convolution between $x[k]$ and $h[k]$ can be obtained from the periodic convolution between $x_p[k]$ and $h_p[k]$ using the following relationship:

$$x[k] * h[k] = x_p[k] \otimes h_p[k],$$

for $(k_{\ell 1} + k_{\ell 2}) \leq k \leq (k_{u1} + k_{u2})$.

Definition 10.3 provides us with an alternative algorithm for implementing the linear convolution through the periodic convolution. The advantage of the above approach lies in computationally efficient implementations of the periodic convolution, which are much faster than the implementations of the linear convolution. Chapter 12 presents one such approach using the discrete Fourier transform (DFT) to compute the periodic convolution.

Algorithm 10.4 Computing linear convolution from periodic convolution

(1) Consider two time-limited DT sequences $x[k]$ and $h[k]$. The DT sequence $x[k] = 0$ outside the range $k_{\ell 1} \leq k \leq k_{u1}$ of length $K_x = k_{u1} - k_{\ell 1} + 1$ samples. Similarly, the DT sequence $h[k] = 0$ outside the range $k_{\ell 2} \leq k \leq k_{u2}$ of length $K_h = k_{u2} - k_{\ell 2} + 1$ samples.

(2) Select an arbitrary integer $K_0 \geq K_x + K_h - 1$.

(3) Compute the periodic extension $x_p[k]$ of $x[k]$ using Eq. (10.20a).

(4) Compute the periodic extension $h_p[k]$ of $h[k]$ using Eq. 10.20b).

Fig. 10.12. Periodic convolution using circular shifting in Example 10.13.

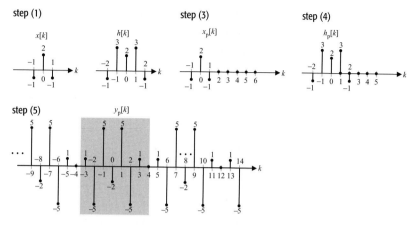

(5) Calculate the periodic convolution $y_p[k] = x_p[k] \otimes h_p[k]$. The result of the linear convolution is obtained by selecting the range $k_{\ell 1} + k_{\ell 2} \leq k \leq k_{u1} + k_{u2}$ of $y_p[k]$.

Example 10.13 illustrates the aforementioned procedure.

Example 10.13

Compute the linear convolution of the following DT sequences:

$$x[k] = \begin{cases} 2 & k = 0 \\ -1 & |k| = 1 \\ 0 & \text{otherwise} \end{cases} \quad \text{and} \quad h[k] = \begin{cases} 2 & k = 0 \\ 3 & |k| = 1 \\ -1 & |k| = 2 \\ 0 & \text{otherwise,} \end{cases}$$

using the periodic convolution method outlined in Algorithm 10.4.

Solution

The DT sequences $x[k]$ and $h[k]$ are plotted in Fig. 10.12, step (1). We observe that the length K_x of $x[k]$ is 3, while the length K_h of $h[k]$ is 5.

Based on step (2), the value of $K_0 \geq 3 + 5 - 1$ or 7. We select $K_0 = 8$.

Following step (3), we form $x_p[k]$ by padding $x[k]$ with $K_0 - K_x$ or five zeros. The resulting sequence $x_p[k]$ is shown in Fig. 10.12, step (3).

Following step (4), we form $h_p[k]$ by padding $h[k]$ with $K_0 - K_h$, or three zeros. The resulting sequence $h_p[k]$ is shown in Fig. 10.12, step (4).

Following step (5), the periodic convolution of the DT sequences $x_p[k]$ and $h_p[k]$ is performed using the sliding tape method. The final result is shown in Table 10.3, where only one period ($K_0 = 8$) of each sequence within the duration $k = [-3, 4]$ is considered.

The sliding tape approach illustrated in Table 10.3 is slightly different from that of Table 10.2. The reflection and shifting operations in Table 10.3 are based on circular reflection and circular shifting since periodic sequences are

Table 10.3. Periodic convolution of $x_p[k]$ and $h_p[k]$ in Example 10.13

m	-3	-2	-1	0	1	2	3	4	k	$y_p[k]$
$h_p[k]$	0	-1	3	2	3	-1	0	0		
$x_p[k]$	0	0	-1	2	-1	0	0	0		
$x_p[-k]$	0	0	-1	2	-1	0	0	0		
$x_p[-4-k]$	-1	0	0	0	0	0	-1	2	-4	0
$x_p[-3-k]$	2	-1	0	0	0	0	0	-1	-3	1
$x_p[-2-k]$	-1	2	-1	0	0	0	0	0	-2	-5
$x_p[-1-k]$	0	-1	2	-1	0	0	0	0	-1	5
$x_p[0-k]$	0	0	-1	2	-1	0	0	0	0	-2
$x_p[1-k]$	0	0	0	-1	2	-1	0	0	1	5
$x_p[2-k]$	0	0	0	0	-1	2	-1	0	2	-5
$x_p[3-k]$	0	0	0	0	0	-1	2	-1	3	1
$x_p[4-k]$	-1	0	0	0	0	0	-1	2	4	0

being convolved. The values of the output sequence $y_p[k]$ over one period $(-3 \leq k \leq 4)$ are listed in the right-hand column of Table 10.3.

The plot of the periodic output $y_p[k]$ is sketched in Fig. 10.12, step (5). The result of the linear convolution $y[k] = x[k] * h[k]$ is obtained by selecting one period of the periodic output $y_p[k]$ within the duration $k_{\ell 1} + k_{\ell 2} \leq k \leq k_{u1} + k_{u2}$, which equals $-3 \leq k \leq 3$.

10.7 Properties of the convolution sum

The properties of the DT linear convolution sum are similar to the properties of the CT convolution integral presented in Chapter 3. In the following, we list the properties of linear convolution for DT sequences followed by the corresponding properties for the periodic convolution.

Commutative property

$$x_1[k] * x_2[k] = x_2[k] * x_1[k]. \qquad (10.21)$$

The commutative property states that the order of the convolution operands does not affect the result of the convolution. In the context of LTID systems, the commutative property implies that the input sequence and the impulse response of the DT system may be interchanged without affecting the output response. The periodic convolution also satisfies the commutative property provided that the two sequences have the same fundamental period K_0.

Distributive property

$$x_1[k] * \{x_2[k] + x_3[k]\} = x_1[k] * x_2[k] + x_1[k] * x_3[k]. \qquad (10.22)$$

The distributive property states that convolution is a linear operation with respect to addition. The periodic convolution also satisfies the distributive property provided that the three sequences have the same fundamental period K_0.

Associative property

$$x_1[k] * \{x_2[k] * x_3[k]\} = \{x_1[k] * x_2[k]\} * x_3[k]. \qquad (10.23)$$

This property states that changing the order of the linear convolution operands does not affect the result of the linear convolution. The periodic convolution also satisfies the associative property provided that the three sequences have the same fundamental period K_0.

Shift property If $x_1[k] * x_2[k] = g[k]$, then

$$x_1[k - k_1] * x_2[k - k_2] = g[k - k_1 - k_2] \qquad (10.24)$$

for any arbitrary integer constants k_1 and k_2. In other words, if the two operands of the linear convolution sum are shifted then the result of the convolution sum is shifted in time by a duration that is the sum of the individual time shifts introduced in the operands. The periodic convolution satisfies the shift property with respect to the circular shift operation.

Length of convolution Let the non-zero lengths of the convolution operands $x_1[k]$ and $x_2[k]$ be denoted by K_1 and K_2 time units, respectively. It can be shown that the non-zero length of the linear convolution $(x_1[k] * x_2[k])$ is $K_1 + K_2 - 1$ time units. The periodic convolution does not satisfy the length property. The circular convolution of two periodic sequences with fundamental period K_0 is also of length K_0.

Convolution with impulse function

$$x_1[k] * \delta[k - k_0] = x_1[k - k_0]. \qquad (10.25)$$

In other words, convolving a DT sequence with a unit impulse function whose origin is located at $k = k_0$ shifts the DT sequence by k_0 time units. Since periodic convolution is defined in terms of periodic sequences and the impulse function is not a periodic sequence, Eq. (10.25) is not valid for the periodic convolution.

Convolution with unit step function

$$x_1[k] * u[k] = \sum_{m=-\infty}^{\infty} x[m]u[k - m] = \sum_{m=-\infty}^{k} x[m]. \qquad (10.26)$$

Equation (10.26) states that convolving a DT sequence $x[k]$ with a unit step function produces the running sum of the original sequence $x[k]$ as a function of time k. Since periodic convolution is defined in terms of periodic sequences and the unit step function is not periodic, Eq. (10.26) is not valid for the periodic convolution.

Causal functions If one of the sequences is causal, the expression for linear convolution, Eq. (10.14), can be written in a simpler form. For example, if $h[k] = 0$ for $k < 0$, the convolution sum $y[k]$ in Eq. (10.14) is expressed as follows:

$$y[k] = x[k] * h[k] = \sum_{m=-\infty}^{\infty} h[m]x[k-m]$$

$$= \sum_{m=0}^{\infty} h[m]x[k-m]. \qquad (10.27a)$$

However, if $h[k]$ is both causal and time-limited, i.e. if $h[k] = 0$ for $k < 0$ and $k > K$, then the convolution sum is expressed as follows:

$$y[k] = \sum_{m=0}^{K} h[m]x[k-m]. \qquad (10.27b)$$

Since periodic convolution is defined in terms of periodic sequences, which are not causal, Eqs. (10.27a) and (10.27b) are not valid for the periodic convolution.

Example 10.14

Simplify the following expressions using the properties of the discrete-time convolution:

(i) $(x[k] + 2\delta[k-1]) * \delta[k-2]$,
(ii) $(x[k-1] - 3\delta[k+1]) * (\delta[k-2] + u[k-1])$,

where $x[k]$ is an arbitrary function and $\delta[k]$ is the unit impulse function.

Solution
(i) Applying the distributive property,

$$(x[k] + 2\delta[k-1]) * \delta[k-2] = \underbrace{x[k] * \delta[k-2]}_{\text{term I}} + \underbrace{2\delta[k-1] * \delta[k-2]}_{\text{term II}}.$$

In both terms I and II, convolution with an impulse function is involved. Equation (10.25) yields

$$\text{term I} = x[k] * \delta[k-2] = x[k-2]$$

and

$$\text{term II} = 2\delta[k-1] * \delta[k-2] = 2\delta[k-3].$$

The simplified expression for (i) is as follows:

$$(x[k] + 2\delta[k-1]) * \delta[k-2] = x[k-2] + 2\delta[k-3].$$

(ii) Applying the distributive property,

$$(x[k-1] - 3\delta[k+1]) * (\delta[k-2] + u[k-1])$$

$$= \underbrace{x[k-1] * \delta[k-2]}_{\text{term I}} - \underbrace{3\delta[k+1] * \delta[k-2]}_{\text{term II}}$$

$$+ \underbrace{x[k-1] * u[k-1]}_{\text{term III}} - \underbrace{3\delta[k+1] * u[k-1]}_{\text{term IV}}.$$

Terms I, II, and IV involve convolution with an impulse function. Equation (10.24) yields

$$\text{term I} = x[k-1] * \delta[k-2] = x[k-3],$$
$$\text{term II} = 3\delta[k+1] * \delta[k-2] = 3\delta[k-1],$$

and

$$\text{term IV} = 3\delta[k+1] * u[k-1] = 3u[k].$$

Term III involves convolution with a unit step function. We express term III as follows:

$$\text{term III} = x[k-1] * u[k-1] = (\delta[k-1] * x[k]) * (u[k] * \delta[k-1])$$
$$= (x[k] * u[k]) * (\delta[k-1] * \delta[k-1]) = (x[k] * u[k]) * \delta[k-2].$$

Using Eq. (10.26) we can further simplify term III to obtain

$$\text{term III} = (x[k] * u[k]) * \delta[k-2] = \left(\sum_{m=-\infty}^{k} x[m] \right) * \delta[k-2]$$

$$= \sum_{m=-\infty}^{k-2} x[m].$$

The simplified expression for (ii) is given by

$$(x[k-1] - 3\delta[k+1]) * (\delta[k-2] + u[k-1])$$

$$= x[k-3] - 3\delta[k-1] + 3u[k] + \sum_{m=-\infty}^{k-2} x[m].$$

10.8 Impulse response of LTID systems

In Section 2.2, we considered several properties of DT systems. Since the characteristics of an LTID system is completely specified by its impulse response, it is logical to assume that its properties can also be completely determined from its impulse response. In this section, we express some of the basic properties of the LTID systems defined in Section 2.2 in terms of the impulse response of the LTID systems. We consider the memory, causality, stability, and invertibility properties for the LTID systems.

10.8.1 Memoryless LTID systems

A DT system is said to be memoryless if its output $y[k]$ at time instant $k = k_0$ depends only on the value of the applied input sequence $x[k]$ at the same time instant $k = k_0$. In other words, a memoryless LTID system typically has the input–output relationship of the following form:

$$y[k] = ax[k],$$

where a is a constant. By substituting $x[k] = \delta[k]$, the impulse response $h[k]$ of a memoryless system can be expressed as

$$h[k] = a\delta[k]. \tag{10.28}$$

> An LTID system will be memoryless if and only if its impulse response $h[k] = a\delta[k]$. Equivalently, an LTID system is memoryless if and only if $h[k] = 0$ for $k \neq 0$.

10.8.2 Causal LTID systems

A DT system is said to be causal if the output at time instant $k = k_0$ depends only on the value of the applied input sequence $x[k]$ at and before the time instant $k = k_0$. Using the reasoning similar to that given in Section 3.7.2 for the CT system, the following can be stated.

> An LTID system will be causal if and only if its impulse response $h[k] = 0$ for $k < 0$.

10.8.3 Stable LTID systems

A DT system is BIBO stable if an arbitrary bounded input sequence always produces a bounded output sequence. Consider a bounded sequence $x[k]$ with $|x[k]| < B_x$, for all k, applied as the input to an LTID system with impulse response $h[k]$. The magnitude of the output $y[k]$ is given by

$$|y[k]| = \left| \sum_{m=-\infty}^{\infty} h[m]x[k-m] \right|.$$

Using the traingle inequality, we can say that the output is bounded by the following limit:

$$|y[k]| \leq \sum_{m=-\infty}^{\infty} |h[m]||x[k-m]|.$$

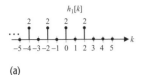

(a)

$h_2[k]$

(b)

$h_3[k]$

(c)

Fig. 10.13. Impulse responses for systems considered in Example 10.15.

Since $|x[k]| < B_x$, the above inequality reduces to

$$|y[k]| \le B_x \sum_{m=-\infty}^{\infty} |h[m]|.$$

It is clear from the above expression that for the output $y[k]$ to be bounded (i.e. $|y[k]| < \infty$), the summation $\sum_{m=-\infty}^{\infty} |h[m]|$ needs to be bounded. The stability condition can therefore be stated as follows.

If the impulse response $h[k]$ of an LTID system satisfies the following condition:

$$\sum_{k=-\infty}^{\infty} |h[k]| < \infty, \tag{10.29}$$

the LTID system is BIBO stable.

Example 10.15
Determine which of the LTID systems with impulse responses, shown in Figs 10.13(a)–(c), are memoryless, causal, and stable.

Solution
(a) Memoryless: since $h_1[k] \neq 0$ for $k \neq 0$, the DT system in Fig. 10.13(a) is not memoryless. In fact, the impulse response $h_1[k]$ extends to $-\infty$, therefore this system has an infinite memory.

Causality: since $h_1[k] \neq 0$ for all $k < 0$, the system is not causal.

Stability: using Eq. (10.29),

$$\sum_{k=-\infty}^{\infty} |h_1[k]| = \sum_{k=-\infty}^{2} |h_1[k]| = \sum_{\substack{k=-\infty \\ k \text{ is even}}}^{2} 2 = \infty.$$

Therefore, the system is not stable.

(b) Memoryless: since $h_2[k] \neq 0$ for $k \neq 0$, the DT system in Fig. 10.13(b) is not memoryless. The impulse response $h_2[k]$ has a finite memory of two time units.

Causality: since $h_2[k] = 0$ for all $k < 0$, the system is causal.

Stability: using Eq. (10.29),

$$\sum_{k=-\infty}^{\infty} |h_2[k]| = \sum_{k=0}^{2} |h_2[k]| = 3 + 2 + 1 = 6.$$

Therefore, the system is BIBO stable.

(c) Memoryless: since $h_3[k] = 0$ for $k \neq 0$, the DT system in Fig. 10.13(c) is memoryless.

Causality: since $h_3[k] = 0$ for all $k < 0$, the system is causal. Also note that all memoryless systems are causal.

Stability: using Eq. (10.29),

$$\sum_{k=-\infty}^{\infty} |h_3[k]| = |h_3[0]| = 5.$$

Therefore, the system is BIBO stable.

10.8.4 Invertible LTID systems

Consider an LTID system with impulse response $h[k]$. The output $y_1[k]$ of the system for an input sequence $x[k]$ is given by $y_1[k] = x[k] * h[k]$. To check its invertibility property, we cascade a second LTID system with impulse response $h_i[k]$ in series with the original system. The output of the second system is given by

$$y_2[k] = y_1[k] * h_i[k] = (x[k] * h[k]) * h_i[k]$$
$$= x[k] * (h[k] * h_i[k]),$$

based on the associative property.

For the second system to be an inverse of the original system, the final output $y_2[k]$ should be the same as $x[k]$, the input to the first LTID system. This is possible only if

$$h[k] * h_i[k] = \delta[k]. \tag{10.30}$$

The existence of $h_i[k]$ proves that an LTID system is invertible. At times, it is difficult to determine the inverse system $h_i[k]$ in the time domain. In Chapter 11, when we introduce the discrete Fourier transform, we will revisit the topic and illustrate how the impulse response of the inverse system can be evaluated with relative ease in the frequency domain.

Example 10.16

Determine which of the following systems is invertible:

(i) $h[k] = \delta[k - 3]$;
(ii) $h[k] = \delta[k] + \delta[k - 1]$.

Solution

(i) Because $\delta[k - 3] * \delta[k + 3] = \delta[k]$, system (i) is invertible. The impulse response $h_i[k]$ of the inverse of system (i) is given by

$$h_i[k] = \delta[k + 3].$$

(ii) It is difficult to calculate the impulse response of the inverse system in the time domain. Using the DTFT introduced in Chapter 11, we can show that the impulse response of the inverse of system (ii) is given by

$$h_i[k] = \sum_{m=0}^{\infty} (-1)^m \delta[k-m] = \delta[k] - \delta[k-1] + \delta[k-2] - \delta[k-3] \pm \cdots$$

We can show indirectly that $h_i[k]$ is indeed the impulse response of the inverse of system (ii) by proving that $h[k] * h_i[k] = \delta[k]$:

$$\begin{aligned}
h[k] * h_i[k] &= (\delta[k] + \delta[k-1]) * h_i[k] = h_i[k] + h_i[k-1] \\
&= (\delta[k] - \delta[k-1] + \delta[k-2] - \delta[k-3] \pm \cdots) + (\delta[k-1] \\
&\quad - \delta[k-2] + \delta[k-3] - \delta[k-4] \pm \cdots) \\
&= \delta[k].
\end{aligned}$$

10.9 Experiments with MATLAB

MATLAB provides several functions (also referred to as M-files) for processing DT signals and LTID systems. In this section, we will focus on the MATLAB implementations of the difference equations with known ancillary conditions, convolution of two DT signals, and deconvolution.

10.9.1 Difference equations

Consider the following linear, constant-coefficient difference equation:

$$\begin{aligned}
y[k+n] &+ a_{n-1} y[k+n-1] + \cdots + a_0 y[k] \\
&= b_m x[k+m] + b_{m-1} x[k+m-1] + \cdots + b_0 x[k], \quad (10.31)
\end{aligned}$$

which models the relationship between the input sequence $x[k]$ and the output response $y[k]$ of an LTID system. The ancillary conditions $y[-1], y[-2], \ldots,$ $y[-n]$ are also specified.

To solve the difference equation, MATLAB provides a built-in function `filter` with the syntax

```
>> [y] = filter(B,A,X,Zi);
```

In terms of the difference equation, Eq. (10.31), the input variables B and A are defined as follows:

$$\text{A} = [1, a_{n-1}, \ldots, a_0] \quad \text{and} \quad \text{B} = [b_m, b_{m-1}, \ldots, b_0],$$

while X is the vector containing the values of the input sequence and Zi denotes the initial conditions of the delays used to implement the difference equation. The initial conditions used by the `filter` function are not the past values of the output $y[k]$ but a modified version of these values. The initial conditions used by MATLAB can be obtained by using another built-in function, `filtic`. The calling syntax for the `filtic` function is as follows:

```
>> [Zi] = filtic(B,A,yinitial);
```

For an n-order difference equation, the input variable `yinitial` is set to

$$\text{yinitial} = [y[-1], y[-2], \ldots, y[-n]].$$

To illustrate the usage of the built-in function `filter`, let us repeat Example 10.1 using MATLAB.

Example 10.17

The DT sequence $x[k] = 2ku[k]$ is applied at the input of an LTID system described by the following difference equation:

$$y[k+1] - 0.4\, y[k] = x[k],$$

with the ancillary condition $y[-1] = 4$. Compute the output response $y[k]$ of the LTID system for $0 \leq k \leq 50$ using MATLAB.

Solution

The MATLAB code used to solve the difference equation is listed below. The explanation follows each instruction in the form of comments.

```
>> k = [0:50];           % time index k = [0, 1,
                         % ...50]
>> X = 2*k.*(k>=1);      % Input signal
>> A = [1 -0.4];         % Coefficients with y[k]
>> B = [0 1];            % Coefficients with x[k]
>> Zi = filtic(B,A,4);   % Initial condition
>> Y = filter(B,A,X,Zi); % Calculate output
```

The output response is stored in the vector Y. Printing the first six values of the output response yields

```
Y = [1.6 0.6400 2.2560 4.9024 7.9610 11.1844],
```

which corresponds to the values of the output response $y[k]$ for the duration $0 \leq k \leq 5$. Comparing with the numerical solution obtained in Example 10.1, we observe that the two results are identical.

Next we proceed with a second-order difference equation.

Example 10.18

The DT sequence $x[k] = 0.5^k u[k]$ is applied at the input of an LTID system described by the following second-order difference equation:

$$y[k+2] + y[k+1] + 0.25y[k] = x[k+2],$$

with ancillary conditions $y[-1] = 1$ and $y[-2] = -2$. Compute the output response $y[k]$ of the LTID system for $0 \leq k \leq 50$ using MATLAB.

Solution

The MATLAB code used to solve the difference equation is listed below. The explanation follows each instruction in the form of comments.

```
>> k = [0:50];              % time index k = [0, 1,
                            %  ...50]
>> X = 0.5.^k.*(k>=0);      % Input signal
>> A = [1 1 0.25];          % Coefficients with y[k]
>> B = [1 0 0];             % Coefficients with x[k]
>> Zi = filtic(B,A,[1 -2]); % Initial condition
>> Y = filter(B,A,X,Zi);    % Calculate output
```

The output response is stored in the vector Y. Printing the first six values of the output response yields

$$Y = [0.5000 \quad -0.2500 \quad 0.3750 \quad -0.1875 \quad 0.1563$$
$$-0.0781].$$

To confirm if the MATLAB code is correct, we also compute the values of the output response in the range $0 \leq k \leq 5$. We express $y[k+2] + y[k+1] + 0.25y[k] = x[k+2]$ as follows:

$$y[k] = -y[k-1] - 0.25y[k-2] + x[k],$$

with ancillary conditions $y[-1] = 1$ and $y[-2] = -2$. Solving the difference equation iteratively yields

$$y[0] = -y[-1] - 0.25y[-2] + x[0] = -1 - 0.25(-2) + 1 = 0.5,$$
$$y[1] = -y[0] - 0.25y[-1] + x[1] = -0.5 - 0.25(1) + 0.5 = -0.25,$$
$$y[2] = -y[1] - 0.25y[0] + x[2] = -(-0.25) - 0.25(0.5) + 0.25 = 0.375,$$
$$y[3] = -y[2] - 0.25y[1] + x[3] = -0.375 - 0.25(-0.25) + 0.125$$
$$= -0.1875,$$
$$y[4] = -y[3] - 0.25y[2] + x[4] = -(-0.1875) - 0.25(0.375) + 0.0625$$
$$= 0.1563,$$

and

$$y[5] = -y[4] - 0.25y[3] + x[2] = -0.1563 - 0.25(-0.1875) + 0.031\,25$$
$$= -0.0782,$$

which are the same as the values computed using MATLAB.

The expressions for the initial conditions for the higher-order difference equations are more complex. Fortunately, most systems are causal with zero ancillary conditions. The initial conditions Zi are zero in such cases.

10.9.2 Convolution

Consider two time-limited DT sequences $x_1[k]$ and $x_2[k]$, where $x_1[k] \neq 0$ within the range $k_{\ell 1} \leq k \leq k_{u1}$ and $x_2[k] \neq 0$ within the range $k_{\ell 2} \leq k \leq k_{u2}$. The length K_1 of the DT sequence $x_1[k]$ is given by $K_1 = k_{u1} - k_{\ell 1} + 1$ samples, while the length K_2 of the DT sequence $x_2[k]$ is $K_2 = k_{u2} - k_{\ell 2} + 1$ samples. In MATLAB, two vectors are required to represent each DT signal. The first vector contains the sample values, while the second vector stores the time indices corresponding to the sample values. For example, the following DT sequence:

$$x[k] = \begin{cases} -1 & k = -1 \\ 1 & k = 0 \\ 2 & k = 1 \\ 0 & \text{otherwise} \end{cases}$$

has the following MATLAB representation:

```
>> kx = [-1 0 1];      % time indices where x is nonzero
>> x = [-1 1 2];       % Sample values for DT sequence x
```

To perform DT convolution, MATLAB provides a built-in function `conv`. We illustrate its usage by repeating Example 10.9 with MATLAB.

Example 10.19

Consider the following two DT sequences $x[k]$ and $h[k]$ specified in Example 10.9:

$$x[k] = \begin{cases} -1 & k = -1 \\ 1 & k = 0 \\ 2 & k = 1 \\ 0 & \text{otherwise} \end{cases} \quad \text{and} \quad h[k] = \begin{cases} 3 & k = -1, 2 \\ 1 & k = 0 \\ -2 & k = 1, 3 \\ 0 & \text{otherwise.} \end{cases}$$

Compute the convolution $y[k] = x[k] * h[k]$ using MATLAB.

Solution

The MATLAB code used to convolve the two functions is given below. As before, the explanation follows each instruction in the form of comments.

```
>> kx = [-1 0 1];       % time indices where x is nonzero
>> x = [-1 1 2];        % Sample values for DT sequence x
>> kh = [-1 0 1 2 3];   % time indices where y is nonzero
>> h = [3 1 -2 3 -2];   % Sample values for DT sequence y
>> y = conv(x,h);       % Convolve x with h
>> ky = kx(1)+kh(1):kx(length(kx))+kh(length(kh));
        % ky= time indices for y
```

In the above instructions, note that MATLAB does not calculate the indices of the result of convolution. These indices have to be calculated separately based on the observation that we made on the starting and last indices of the convolved result.

The computed values of y are given by

$$y = [-3\ 2\ 9\ -3\ 1\ 4\ -4],$$

with the computed indices

$$ky = [-2\ -1\ 0\ 1\ 2\ 3\ 4].$$

Note that the above result is the same as the one obtained in Example 10.9.

The function deconv performs the inverse of the convolution sum. Given a DT input sequence x and the output sequence y, for example, the impulse response h can be determined using the following instructions:

```
>> h2 = deconv(y,x);                 % Deconvolve x out of y
>> kh2 = ky(1)-kx(1):ky(length(ky))-kx(length(kx));
                                      % kh2 = indices for h2
```

Note that h2 has the same sample values and indices kh2 as those of h.

10.10 Summary

In this chapter, we developed analytical techniques for LTID systems. We saw that the output sequence $y[k]$ of an LTID system can be calculated analytically in the time domain using two different methods. In Section 10.1, we determined the output of a DT system by solving a linear, constant-coefficient difference equation. The solution of such a difference equation can be expressed as a sum of two components: the zero-input response and the zero-state response. The zero-input response is the output produced by the DT system because of the initial conditions. For most DT systems, the zero-input response decays to zero with increasing time. The zero-state response results from the input sequence. The overall output of a DT system is the sum of the zero-input response and the zero-state response. A DT system, of the form shown in Eq. (10.1), will be an LTID system if all initial conditions are zero. In other words, the zero-input response of an LTID system is always zero.

An alternative representation for determining the output of an LTID system is based on the impulse response of the system. In Section 10.3, we defined the impulse response $h[k]$ as the output of an LTID system when a unit impulse $\delta[k]$ is applied at the input of the system. In Section 10.4, we proved that the output $y[k]$ of an LTID system could be obtained by convolving the input sequence $x[k]$ with its impulse response $h[k]$. The resulting convolution sum can either be solved analytically or by using a graphical approach. The graphical approach was illustrated through several examples in Section 10.5. In discrete time, the convolution of two periodic functions is also defined and is known as periodic or circular convolution. The periodic convolution is discussed in Section 10.6, where we mentioned that the linear convolution may be efficiently calculated through periodic convolution. The convolution sum satisfies the commutative, distributive, associative, and time-shifting properties.

(1) The commutative property states that the order of the convolution operands does not affect the result of the convolution.
(2) The distributive property states that convolution is a linear operation with respect to addition.
(3) The associative property is an extension of the commutative property to more than two convolution operands. It states that changing the order of the convolution operands does not affect the result of the convolution sum.
(4) The time-shifting property states that if the two operands of the convolution sum are shifted in time then the result of the convolution sum is shifted by a duration that is the sum of the individual time shifts introduced in the convolution operands.
(5) If the lengths of the two functions are K_1 and K_2 samples, the convolution sum of these two functions will have a length of $K_1 + K_2 - 1$ samples.
(6) Convolving a sequence with a unit DT impulse function with the origin at $k = k_0$ shifts the sequence by k_0 time units.
(7) Convolving a sequence with a unit DT step function produces the running sum of the original sequence as a function of time k.

Finally, in Section 10.8, we expressed the memoryless, causality, stability, and invertibility properties of an LTID system in terms of its impulse response.

(1) An LTID system will be memoryless if and only if its impulse response $h[k] = 0$ for $k \neq 0$.
(2) An LTID system will be causal if and only if its impulse response $h[k] = 0$ for $k < 0$.
(3) The impulse response $h[k]$ of a (BIBO) stable LTID system is absolutely summable, i.e.

$$\sum_{k=-\infty}^{\infty} |h[k]| < \infty.$$

(4) An LTID system will be invertible if there exists another LTID system with impulse response $h_i[k]$ such that $h[k] * h_i[k] = \delta[k]$. The system with the impulse response $h_i[k]$ is the inverse system.

In the next chapter, we consider the frequency representations of DT sequences and systems.

Problems

10.1 Consider the input sequence $x[k] = 2u[k]$ applied to a DT system modeled with the following input–output relationship:

$$y[k + 1] - 2y[k] = x[k],$$

and ancillary condition $y[-1] = 2$.

(a) Determine the response $y[k]$ by iterating the difference equation for $0 \le k \le 5$.

(b) Determine the zero-input response $y_{zi}[k]$ for $0 \le k \le 5$.

(c) Calculate the zero-state response $y_{zs}[k]$ for $0 \le k \le 5$.

(d) Verify that $y[k] = y_{zi}[k] + y_{zs}[k]$.

10.2 Repeat Problem 10.1 for the applied input $x[k] = 0.5^k u[k]$ and the input–output relationship

$$y[k + 2] - y[k + 1] + 0.5y[k] = x[k],$$

with ancillary conditions $y[-1] = 0$ and $y[-2] = 1$.

10.3 Repeat Problem 10.1 for the applied input $x[k] = (-1)^k u[k]$ and the input–output relationship

$$y[k + 2] - 0.75y[k + 1] + 0.125y[k] = x[k],$$

with ancillary conditions $y[-1] = 1$ and $y[-2] = -1$.

10.4 Show that the convolution of two sequences $a^k u[k]$ and $b^k u[k]$ is given by

$$(a^k u[k]) * (b^k u[k]) = \begin{cases} (k + 1)a^k u[k] & a = b \\ \dfrac{1}{a - b}(a^{k+1} - b^{k+1})u[k] & a \ne b. \end{cases}$$

10.5 Calculate the convolution $(x_1[k] * x_2[k])$ for the following pairs of sequences:

(a) $x_1[k] = u[k + 2] - u[k - 3]$, $x_2[k] = u[k + 4] - u[k - 5]$;

(b) $x_1[k] = 0.5^k u[k]$, $x_2[k] = 0.8^k u[k - 5]$;

(c) $x_1[k] = 7^k u[-k + 2]$, $x_2[k] = 0.4^k u[k - 4]$;

(d) $x_1[k] = 0.6^k u[k]$, $x_2[k] = \sin(\pi k/2)u[-k]$;

(e) $x_1[k] = 0.5^{|k|}$, $x_2[k] = 0.8^{|k|}$.

10.6 For the following pairs of sequences:

(a) $x[k] = \begin{cases} k & 0 \le k \le 3 \\ 0 & \text{otherwise} \end{cases}$ and $h[k] = \begin{cases} 2 & -1 \le k \le 2 \\ 0 & \text{otherwise}; \end{cases}$

(b) $x[k] = \begin{cases} |k| & |k| \le 2 \\ 0 & \text{otherwise} \end{cases}$ and $h[k] = \begin{cases} 2^{-k} & 0 \le k \le 3 \\ 0 & \text{otherwise}, \end{cases}$

calculate the DT convolution $y[k] = x[k] * h[k]$ using (i) the graphical approach and (ii) the sliding tape method.

10.7 Using the sliding tape method and the following equation:

$$y[k] = \sum_{m=-\infty}^{\infty} h[m]x[k - m],$$

calculate the convolution of the sequences in Example 10.8 and show that the convolution output is identical to that obtained in Example 10.8.

10.8 Using the sliding tape method and the following equation:

$$y[k] = \sum_{m=-\infty}^{\infty} x[m]h[k - m],$$

calculate the convolution of the sequences in Example 10.9 and show that the convolution output is identical to that obtained in Example 10.9.

10.9 The linear convolution between two sequences $x[k]$ and $h[k]$ of lengths K_1 and K_2, respectively, can be performed using periodic convolution by considering periodic extensions of the two zero-padded sequences. Calculate the linear convolution of the sequences defined in Example 10.8 using the periodic convolution approach with the fundamental period K_0 set to 10. Repeat for K_0 set to 13.

10.10 Repeat Example 10.7 using the periodic convolution approach with K set to 10.

10.11 Repeat Example 10.7 using the periodic convolution approach with K set to 15.

10.12 Repeat Example 10.12 with K set to 8.

10.13 Calculate the unit step response of the DT systems with the following impulse responses:
(a) $h[k] = u[k + 7] - u[k - 8]$;
(b) $h[k] = 0.4^k u[k]$;
(c) $h[k] = 2^k u[-k]$;
(d) $h[k] = 0.6^{|k|}$;
(e) $h[k] = \sum_{m=-\infty}^{\infty} (-1)^m \delta(k - 2m)$.

10.14 Simplify the following expressions using the properties of discrete-time convolution:
(a) $(x[k] + 2\delta[k - 1]) * \delta[k - 2]$;
(b) $(x[k] + 2\delta[k - 1]) * (\delta[k + 1] + \delta[k - 2])$;
(c) $(x[k] - u[k - 1]) * \delta[k - 2]$;
(d) $(x[k] - x[k - 1]) * u[k]$,
where $x[k]$ is an arbitrary function, $\delta[k]$ is the unit impulse function, and $u[k]$ is the unit step function.

10.15 Prove Definition 10.3 by expanding the right-hand side of the periodic convolution and showing it to be equal to the left-hand side.

10.16 Prove the time-shifting property stated in Eq. (10.24).

10.17 Show that the linear convolution $y[k]$ of a time-limited DT sequence $x_1[k]$ that is non-zero only within the range $k_{\ell 1} \leq k \leq k_{u1}$ with another time-limited DT sequence $x_2[k]$ that is non-zero only within the range

$k_{\ell 2} \le k \le k_{u2}$ is time-limited, and is non-zero only within the range $k_{\ell 1} + k_{\ell 2} \le k \le k_{u1} + k_{u2}$.

10.18 For each of the following impulse responses, determine if the DT system is (i) memoryless; (ii) causal; and (iii) stable:

(a) $h[k] = u[k + 7] - u[k - 8]$;

(b) $h[k] = \sin\left(\frac{\pi k}{8}\right) u[k]$;

(c) $h[k] = 6^k u[-k]$;

(d) $h[k] = 0.9^{|k|}$;

(e) $h[k] = \displaystyle\sum_{m=-\infty}^{\infty} (-1)^m \delta(k - 2m)$.

10.19 Determine which of the following pair of impulse responses correspond to inverse systems:

(a) $h_1[k] = u[-k - 1]$, $h_2[k] = \delta[k - 1] - \delta[k]$;

(b) $h_1[k] = 0.5^k u[k]$, $h_2[k] = \delta[k] - 0.5\delta[k - 1]$;

(c) $h_1[k] = 0.8^k k u[k]$, $h_2[k] = 0.8\delta[k - 1] - 2\delta[k]$
                                            $+ 1.25\delta[k + 1]$;

(d) $h_1[k] = k u[k]$, $h_2[k] = \delta[k + 1] - 2\delta[k] + \delta[k - 1]$;

(e) $h_1[k] = (k + 1)0.8^k u[k]$, $h_2[k] = \delta[k] - 1.6\delta[k - 1]$
                                            $+ 0.64\delta[k - 2]$.

10.20 Repeat Problems 10.1–10.3 to compute the first 50 samples of the output response using the `filter` and `filtic` functions available in MATLAB.

10.21 Repeat Problem 10.5 using the `conv` function available in MATLAB. For a sequence with infinite length, you may truncate the sequence when the value of the sequence is less than 0.1% of its maximum value.

10.22 The MATLAB function `impz` can be used to determine the impulse response of an LTID system from its difference equation representation. Determine the first 50 samples of the impulse response of the LTID systems with the difference equations specified in Problems 10.1–10.3.

11 Discrete-time Fourier series and transform

In Chapter 10, we developed analysis techniques for LTID systems based on the convolution sum by representing the input sequence $x[k]$ as a linear combination of time-shifted unit impulse functions. In this chapter, we introduce frequency-domain representations for DT sequences and LTID systems based on weighted superpositions of complex exponential functions. For periodic sequences, the resulting representation is referred to as the discrete-time Fourier series (DTFS), while for aperiodic sequences the representation is called the discrete-time Fourier transform (DTFT). We exploit the properties of the discrete-time Fourier series and Fourier transform to develop alternative techniques for analyzing DT sequences. The derivations of these results closely parallel the development of the CT Fourier series (CTFS) and CT Fourier transform (CTFT) as presented in Chapters 4 and 5.

The organization of this chapter is as follows. In Section 11.1, we introduce the exponential form of the DTFS and illustrate the procedure used to calculate the DTFS coefficients through a series of examples. The DTFT provides frequency representations for aperiodic sequences and is presented in Section 11.2. Section 11.3 defines the condition for the existence of the DTFT, and Section 11.4 extends the scope of the DTFT to represent periodic sequences. Section 11.5 lists the properties of the DTFT and DTFS, including the time-convolution property, which states that the convolution of two DT sequences in the time domain is equivalent to the multiplication of the DTFTs of the two sequences in the frequency domain. The convolution property provides us with an alternative technique to compute the output response of the LTID system. The DTFT of the impulse response is referred to as the transfer function, which is covered in Section 11.6. Section 11.7 defines the magnitude and phase spectra for LTID systems, and Section 11.8 relates the CTFT and DTFT of periodic and aperiodic waveforms to each other. Finally, the chapter is concluded in Section 11.9 with a summary of important concepts covered in the chapter.

11.1 Discrete-time Fourier series

In Example 4.4, we proved that the set of complex exponential functions $\exp(jn\omega_0 t)$, $n \in Z$, defines an orthonormal set of functions over the interval $t = (t_0, t_0 + T_0)$ with duration $T_0 = 2\pi/\omega_0$. This orthonormal set of exponentials was used to derive the CT Fourier series. In the same spirit, we now show that the discrete-time (DT) complex exponential sequences form an orthonormal set in the DT domain and are used to derive the DTFS. We start with the definition of the orthonormal sequences.

Definition 11.1 *Two sequences $p[k]$ and $q[k]$ are said to be orthogonal over interval $k = [k_1, k_2]$ if*

$$\text{orthogonality property} \quad \sum_{k=k_1}^{k_2} p[k]q^*[k] = \sum_{k=k_1}^{k_2} p^*[k]q[k] = 0, \; p[k] \neq q[k],$$

$$(11.1)$$

where the superscript $*$ denotes complex conjugation. In addition to Eq. (11.1), both signals $p[k]$ and $q[k]$ must also satisfy the following unit magnitude property to satisfy the orthonormality condition:

$$\text{unit magnitude property} \quad \sum_{k=k_1}^{k_2} p[k]p^*[k] = \sum_{k=k_1}^{k_2} q[k]q^*[k] = 1. \quad (11.2)$$

Definition 11.2 *A set comprising an arbitrary number of N functions, say $\{p_1[k], p_2[k], \ldots, p_N[k]\}$, is mutually orthogonal over interval $k = [k_1, k_2]$ if*

$$\sum_{k=k_1}^{k_2} p_m[k]p_n^*[k] = \begin{cases} E_n \neq 0 & m = n \\ 0 & m \neq n, \end{cases} \quad (11.3)$$

for $1 \leq m, n \leq N$. In addition, if $E_n = 1$ for all n, the orthogonal set is referred to as an orthonormal set.

Based on Definitions 11.1 and 11.2, we show that the DT complex sequences form an orthogonal set.

Proposition 11.1 *The set of discrete-time complex exponential sequences $\{\exp(jn\Omega_0 k), n \in Z\}$, is orthogonal over the interval $[r, r + K_0 - 1]$, where the duration $K_0 = 2\pi/\Omega_0$ and r is an arbitrary integer.*

Proof
Consider the following summation:

$$\sum_{k=r}^{r+K_0-1} e^{jm\Omega_0 k} e^{-jn\Omega_0 k} = \sum_{k=r}^{r+K_0-1} e^{j(m-n)\Omega_0 k}.$$

Substituting $p = k - r$ to make the lower limit of the summation equal to zero yields

$$\sum_{k=r}^{r+K_0-1} e^{j(m-n)\Omega_0 k} = \sum_{p=0}^{K_0-1} e^{j(m-n)\Omega_0(p+r)} = e^{j(m-n)\Omega_0 r} \sum_{p=0}^{K_0-1} e^{j(m-n)\Omega_0 p}.$$

The above summation is solved for two different cases, $m = n$ and $m \neq n$.

Case I For $m = n$, the summation reduces to

$$e^{j(m-n)\Omega_0 r} \sum_{p=0}^{K_0-1} e^{j(m-n)\Omega_0 p} = 1 \cdot \sum_{p=0}^{K_0-1} 1 = K_0.$$

Case II For $m \neq n$, the summation forms a GP series and is simplified as follows:

$$e^{j(m-n)\Omega_0 r} \sum_{p=0}^{K_0-1} e^{j(m-n)\Omega_0 p} = e^{j(m-n)\Omega_0 r} \left[\frac{1 - e^{j(m-n)\Omega_0 K_0}}{1 - e^{j(m-n)\Omega_0}} \right].$$

Because $\Omega_0 K_0 = 2\pi$ and indices m and n are integers, the exponential term in the numerator is given by

$$e^{j(m-n)\Omega_0 K_0} = e^{j(m-n)2\pi} = 1.$$

Therefore, for $m \neq n$ the summation reduces to

$$e^{j(m-n)\Omega_0 r} \sum_{p=0}^{K_0-1} e^{j(m-n)\Omega_0 p} = e^{j(m-n)\Omega_0 r} \left[\frac{1 - 1}{1 - e^{j(m-n)\Omega_0}} \right] = 0.$$

Combining the results of cases I and II, we obtain

$$\sum_{k=r}^{r+K_0-1} e^{jm\Omega_0 k} e^{-jn\Omega_0 k} = \begin{cases} K_0 & \text{if } m = n \\ 0 & \text{if } m \neq n. \end{cases}$$

In other words, the set of DT complex exponential sequences $\{\exp(jn\Omega_0 k), n \in Z\}$ is orthogonal over the specified interval $[r, r + K_0 - 1]$.

An important difference between the DT and CT complex exponential functions lies in the frequency–periodicity property of the DT exponential sequences. Since

$$e^{jn(\Omega_0 + 2\pi)k} = e^{jn\Omega_0 k} e^{jn2\pi k} = e^{jn\Omega_0 k},$$

the exponential sequence $\exp(jn\Omega_0 k)$ is identical to $\exp(jn(\Omega_0 + 2\pi)k)$. This is in contrast to the CT exponentials, where $\exp(jn\omega_0 t)$ is different from $\exp(jn(\omega_0 + 2\pi)t)$. The following example illustrates the frequency periodicity for the DT sinusoidal signals. Using the Euler property, a DT complex exponential $\exp(jn\Omega_0 k)$ can be expressed as follows:

$$e^{jn\Omega_0 k} = \cos(n\Omega_0 k) + j\sin(n\Omega_0 k).$$

Example 11.1 shows that both the real and imaginary components of the complex exponential satisfy the frequency–periodicity property; therefore, the DT complex exponential should also satisfy the frequency–periodicity property.

Example 11.1

Consider a CT sinusoidal function with a fundamental frequency of 1.4 Hz, i.e.

$$x(t) = \cos(2.8\pi t + \phi),$$

where ϕ is the constant phase. Sample the function with a sampling rate of 1 sample/s and determine the fundamental frequency of the resulting DT sequence.

Solution

In the time domain, the DT sequence is obtained by sampling $x(t)$ at $t = kT$. Since the sampling interval $T = 1$ s,

$$x[k] = x(kT) = \cos(2.8\pi k + \phi),$$

which is periodic with a period $\Omega_1 = 2.8\pi$ radians/s. Because the CT signal $x(t)$ is a sinusoid with a fundamental frequency of 1.4 Hz, the minimum sampling rate, required to avoid aliasing, is given by 2.8 samples/s. Since the sampling rate of 1 samples/s is less than the Nyquist sampling rate, aliasing is introduced due to sampling. Based on Lemma 9.1, the reconstructed signal is given by

$$y(t) = \cos(2\pi(1.4 - 1)t) = \cos(0.8\pi t + \phi).$$

Substituting $t = kT$, the DT representation of the reconstructed signal is given by

$$y[k] = \cos(0.8\pi k + \phi),$$

which is periodic with a period $\Omega_2 = 0.8\pi$ radians/s. From the above analysis, it is clear that the DT sequences $x[k] = \cos(2.8\pi k + \phi)$ and $y[k] = \cos(0.8\pi k + \phi)$ are identical. This is because the difference in the fundamental frequencies Ω_1 and Ω_2 is 2π.

Proposition 11.2 *A discrete-time periodic function $x[k]$ with period K_0 can be expressed as a superposition of DT complex exponentials as follows:*

$$x[k] = \sum_{n=<K_0>} D_n e^{jn\Omega_0 k}, \tag{11.4}$$

where Ω_0 is the fundamental frequency, given by $\Omega_0 = 2\pi/K_0$, and the discrete-time Fourier series (DTFS) coefficients D_n for $1 \leq n \leq K_0$ are given by

$$D_n = \frac{1}{K_0} \sum_{k=\langle K_0 \rangle} x[k] e^{-jn\Omega_0 k}. \tag{11.5}$$

In Eq. (11.5), the limit of $k = \langle K_0 \rangle$ implies that the sum can be taken over any K_0 consecutive samples of $x[k]$. Unless otherwise specified, we would consider the range $0 \leq k \leq K_0 - 1$ in our derivations.

Proof

To verify the DTFS, we expand the right-hand side of Eq. (11.4) by substituting the value of D_n from Eq. (11.5). With K_0 consecutive exponentials in the range $0 \leq n \leq k_0 - 1$, the resulting expression is given by

$$\sum_{n=0}^{K_0-1} D_n e^{jn\Omega_0 k} = \sum_{n=0}^{K_0-1} \left[\frac{1}{K_0} \sum_{m=0}^{K_0-1} x[m] e^{-jn\Omega_0 m} \right] e^{jn\Omega_0 k}.$$

Interchanging the order of the summation yields

$$\sum_{n=0}^{K_0-1} D_n e^{jn\Omega_0 k} = \frac{1}{K_0} \sum_{m=0}^{K_0-1} x[m] \left[\sum_{n=0}^{K_0-1} e^{jn\Omega_0(k-m)} \right]. \tag{11.6}$$

From Proposition 11.1, we have

$$\sum_{n=0}^{K_0-1} e^{jn\Omega_0(k-m)} = \begin{cases} K_0 & \text{if } k = m \\ 0 & \text{if } k \neq m. \end{cases}$$

The right-hand side of Eq. (11.6) reduces to

$$\sum_{n=0}^{K_0-1} D_n e^{jn\Omega_0 k} = \frac{1}{K_0} \sum_{m=0}^{K_0-1} x[m] K_0 \delta[m - k] = \frac{1}{K_0} K_0 x[k] = x[k],$$

and therefore proves Proposition 11.2.

Examples 11.2–11.5 calculate the DTFS for selected DT periodic sequences.

Example 11.2

Determine the DTFS coefficients of the following periodic sequence:

$$h[k] = \begin{cases} 1 & |k| \leq N \\ 0 & N + 1 \leq k \leq K_0 - N - 1, \end{cases} \tag{11.7}$$

with a fundamental period $K_0 > (2N + 1)$.

Solution

With K_0 consecutive samples in the range $-N \leq k \leq K_0 - N - 1$, Eq. (11.5) reduces to

$$D_n = \frac{1}{K_0} \sum_{k=-N}^{N} 1 \cdot e^{-jn\Omega_0 k} + \frac{1}{K_0} \sum_{k=N+1}^{K_0-N-1} 0 \cdot e^{-jn\Omega_0 k} = \frac{1}{K_0} \sum_{k=-N}^{N} e^{-jn\Omega_0 k}.$$

Table 11.1. Values of D_n for $0 \leq n \leq 9$ in Example 11.2

n	0	1	2	3	4	5	6	7	8	9
D_n	0.300	0.262	0.162	0.038	−0.062	−0.100	−0.062	0.038	0.162	0.262

Fig. 11.1. (a) DT periodic sequence $h[k]$; (b) its DTFS coefficients calculated in Example 11.2.

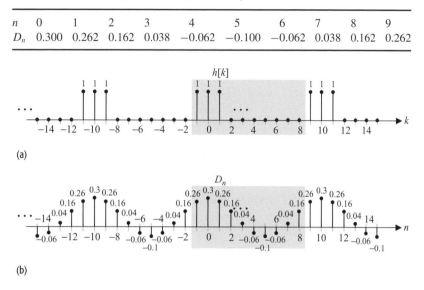

(a)

(b)

The summation represents a GP series and simplifies as follows:

$$D_n = \frac{1}{K_0} \left[e^{jn\Omega_0 N} \frac{1 - e^{-jn\Omega_0(2N+1)}}{1 - e^{-jn\Omega_0}} \right]$$

$$= \frac{1}{K_0} \left[e^{jn\Omega_0 N} \underbrace{\frac{e^{-jn\Omega_0(2N+1)/2}}{e^{-jn\Omega_0/2}}}_{=1} \frac{e^{jn\Omega_0(2N+1)/2} - e^{-jn\Omega_0(2N+1)/2}}{e^{jn\Omega_0/2} - e^{-jn\Omega_0/2}} \right]$$

$$= \frac{1}{K_0} \left[\frac{\sin\left(\frac{2N+1}{2} n\Omega_0\right)}{\sin\left(\frac{1}{2} n\Omega_0\right)} \right].$$

Substituting the value of the fundamental frequency $\Omega_0 = 2\pi/K_0$ yields

$$D_n = \frac{1}{K_0} \left[\frac{\sin\left(\frac{2N+1}{K_0} n\pi\right)}{\sin\left(\frac{1}{K_0} n\pi\right)} \right], \tag{11.8}$$

which represents a DT sinc function.

As a special case, we plot the values of the coefficients D_n for $N = 1$ and $K_0 = 10$ in Fig. 11.1. The expression for the DTFS coefficients is given by

$$D_n = \frac{1}{10} \left[\frac{\sin(0.3n\pi)}{\sin(0.1n\pi)} \right],$$

with the values for $0 \leq n \leq 9$ given in Table 11.1.

The value of the DTFS coefficient D_0 is calculated using L'Hôpital's rule as follows:

$$D_0 = \lim_{n \to 0} \frac{1}{10} \left[\frac{\sin(0.3n\pi)}{\sin(0.1n\pi)} \right] = \lim_{n \to 0} \frac{1}{10} \left[\frac{(0.3\pi)\cos(0.3n\pi)}{(0.1\pi)\cos(0.1n\pi)} \right] = 0.3.$$

In Fig. 11.1(b), we observe that the DTFS coefficients are periodic with a period of 10, which is the same as the fundamental period of the original sequence $h[k]$. One such period is highlighted in Fig. 11.1(b).

11.1.1 Periodicity of DTFS coefficients

In Example 11.2, we noted that the DTFS coefficients D_n of a periodic sequence are themselves periodic with a period of K_0. In Proposition 11.3, we show that this is true for any DT periodic sequence.

Proposition 11.3 *The DTFS coefficients D_n of a periodic sequence $x[k]$, with a period of K_0, are themselves periodic with a period of K_0. In other words,*

$$D_n = D_{n+mK_0} \quad for\ m \in Z. \tag{11.9}$$

Proof
By definition, the DTFS coefficients are expressed as follows:

$$D_{n+mK_0} = \frac{1}{K_0} \sum_{k=\langle K_0 \rangle} x[k] e^{-j(n+mK_0)\Omega_0 k}$$

$$= \frac{1}{K_0} \sum_{k=\langle K_0 \rangle} x[k] e^{-jn\Omega_0 k} e^{-jm\Omega_0 K_0 k}$$

where the exponential term $\exp(-jm\Omega_0 K_0 k) = \exp(-j2m\pi k) = 1$. The above expression reduces to

$$D_{n+mK_0} = \frac{1}{K_0} \sum_{k=\langle K_0 \rangle} x[k] e^{-jn\Omega_0 k},$$

which, by definition, is D_n.

In the following examples, we calculate the DTFS coefficients D_n over one period ($n = \langle K_0 \rangle$) and exploit the periodicity property to obtain the DTFS coefficients outside this range.

Example 11.3
Determine the DTFS coefficients of the periodic DT sequence $x[k]$ with one fundamental period defined as

$$x[k] = 0.5^k u[k], \quad 0 \le k \le 14. \tag{11.10}$$

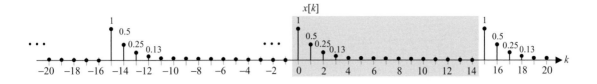

Fig. 11.2. Periodic DT sequence defined in Example 11.3.

Solution

The DT sequence $x[k]$ is plotted in Fig. 11.2. Since its period $K_0 = 15$, the fundamental frequency $\Omega_0 = 2\pi/15$. The DTFS coefficients D_n are given by

$$D_n = \frac{1}{15} \sum_{k=0}^{14} 0.5^k e^{-jn\Omega_0 k} = \frac{1}{15} \sum_{k=0}^{14} (0.5 e^{-jn\Omega_0})^k,$$

which is a GP series that simplifies to

$$D_n = \frac{1}{15} \cdot \frac{1 - (0.5 e^{-jn\Omega_0})^{15}}{1 - 0.5 e^{-jn\Omega_0}} = \frac{1}{15} \cdot \frac{1 - 0.5^{15} e^{-j15n\Omega_0}}{1 - 0.5 e^{-jn\Omega_0}}.$$

Since $\Omega_0 = 2\pi/15$, the exponential term in the numerator, $\exp(-j15n\Omega_0) = \exp(-j2n\pi) = 1$. Expanding the exponential term in the denominator as $\exp(-jn\Omega_0) = \cos(n\Omega_0) - j\sin(n\Omega_0)$, the DTFS coefficients are given by

$$D_n = \frac{1}{15} \cdot \frac{1 - 0.5^{15}}{1 - 0.5\cos(n\Omega_0) + j0.5\sin(n\Omega_0)}$$

$$\approx \frac{1}{15} \cdot \frac{1}{1 - 0.5\cos(n\Omega_0) + j0.5\sin(n\Omega_0)}. \tag{11.11}$$

As the DTFS coefficients are complex, we determine the magnitude and phase of the coefficients as follows:

magnitude $$|D_n| = \frac{1}{15} \cdot \frac{1}{\sqrt{(1 - 0.5\cos(n\Omega_0))^2 + (0.5\sin(n\Omega_0))^2}}$$

$$= \frac{1}{15} \cdot \frac{1}{\sqrt{1.25 - \cos(n\Omega_0)}}; \tag{11.12}$$

phase $$<D_n = -\tan^{-1}\left[\frac{0.5\sin(n\Omega_0)}{1 - 0.5\cos(n\Omega_0)}\right], \tag{11.13}$$

where $\Omega_0 = 2\pi/15$. The magnitude and phase spectra of the DTFS coefficients are plotted in Figs. 11.3(a) and (b), in which one period of D_n is highlighted by a shaded region.

Example 11.4

Determine the DTFS coefficients of the following periodic function:

$$x[k] = A e^{j((2\pi m/N)k + \theta)}, \tag{11.14}$$

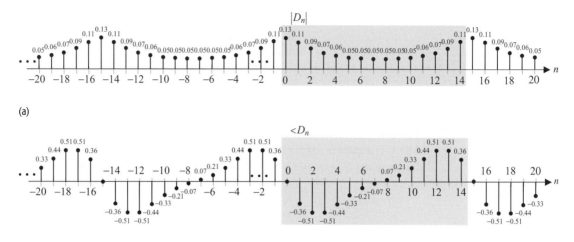

(a)

(b)

Fig. 11.3. (a) Magnitude spectrum and (b) phase spectrum of the DTFS coefficients in Example 11.3.

where the greatest common divisor between the fundamental period N and the integer constant m is one.

Solution

We first show that the DT sequence $x[k]$ is periodic and determine its fundamental period. It was mentioned in Proposition 1.1 that a DT complex exponential sequence $x[k] = \exp(j(\Omega_0 k + \theta))$ is periodic if $2\pi/\Omega_0$ is a rational number. In this case, $2\pi/\Omega_0 = N/m$, which is a rational number as m, K and N are all integers. In other words, the sequence $x[k]$ is periodic. Using Eq. (1.8), the fundamental period of $x[k]$ is calculated to be

$$K_0 = (2\pi/\Omega_0)p = pN/m,$$

where p is the smallest integer that results in an integer value for K_0. Note that the fraction N/m represents a rational number, which cannot be reduced further since the greatest common divisor between m and N is given to be one. Selecting $p = m$, the fundamental period is obtained as $K_0 = N$.

To compute the DTFS coefficients, we express $x[k]$ as follows:

$$x[k] = Ae^{j\theta}e^{j\Omega_0 mk}$$

and compare this expression with Eq. (11.4). For $0 \leq n \leq K_0 - 1$, we observe that

$$D_n = \begin{cases} Ae^{j\theta} & \text{if } n = m \\ 0 & \text{if } n \neq m. \end{cases} \tag{11.15}$$

As a special case, we consider $A = 2$, $K_0 = 6$, $m = 5$, and $\theta = \pi/4$. The magnitude and phase spectra for the selected values are shown in Figs. 11.4(a) and (b), where we have used the periodicity property of the DTFS

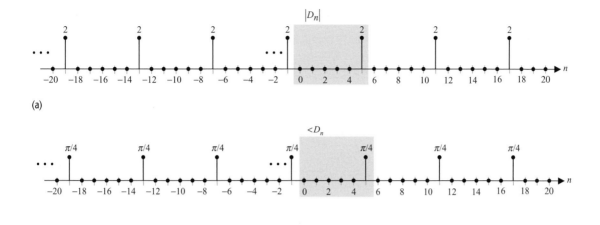

(a)

(b)

Fig. 11.4. (a) Magnitude spectrum and (b) phase spectrum of the DTFS coefficients in Example 11.4.

coefficients to plot the values of the coefficients outside the duration $0 \le n \le (K_0 - 1)$.

Substituting $\theta = 0$ in Example 11.4 results in Corollary 11.1.

Corollary 11.1 The DTFS coefficients corresponding to the complex exponential sequence $x[k] = A \exp(j2\pi mk/K_0)$ with the fundamental period K_0 are given by

$$D_n = \begin{cases} A & \text{if } n = m, m \pm K_0, m \pm 2K_0, \dots \\ 0 & \text{elsewhere,} \end{cases} \qquad (11.16)$$

provided the greatest common divisor between the m and K_0 is one.

Example 11.5
Determine the DTFS coefficients of the following sinusoidal sequence:

$$y[k] = B \sin\left(\frac{2\pi m}{K_0}k + \theta\right), \qquad (11.17)$$

where the greatest common divisor between integers m and K_0 is one. The phase component θ is constant with respect to time.

Solution
Using Proposition 1.1, it is straightforward to show that the sinusoidal sequence $y[k]$ is periodic with fundamental period K_0. The fundamental frequency is given by $\Omega_0 = 2\pi/K_0$.

Based on Eq. (11.5), and noting that $\Omega_0 = 2\pi/K_0$, the DTFS coefficients are given by

$$
\begin{aligned}
D_n &= \frac{1}{K_0} \sum_{k=<K_0>} B \sin(m\Omega_0 k + \theta) \cdot e^{-jn\Omega_0 k} \\
&= \frac{1}{K_0} \sum_{k=<K_0>} B \left[\frac{e^{j(m\Omega_0 k+\theta)} - e^{j(m\Omega_0 k+\theta)}}{2j} \right] \cdot e^{-jn\Omega_0 k} \\
&= -j\frac{B}{2K_0} e^{j\theta} \underbrace{\sum_{k=\langle K_0 \rangle} e^{j(m-n)\Omega_0 k}}_{\text{summation I}} + j\frac{B}{2K_0} e^{-j\theta} \underbrace{\sum_{k=\langle K_0 \rangle} e^{-j(m+n)\Omega_0 k}}_{\text{summation II}}.
\end{aligned}
$$

In proving Proposition 11.2, we used the following summation:

$$
\sum_{n=0}^{K_0-1} e^{jn\Omega_0(k-m)} = \begin{cases} K_0 & \text{if } k = m \\ 0 & \text{if } k \neq m. \end{cases}
$$

Therefore, summations I and II are given by

$$
\text{I} = \sum_{k=\langle K_0 \rangle} e^{j(m-n)\Omega_0 k} = \begin{cases} K_0 & \text{if } n = m \\ 0 & \text{if } n \neq m; \end{cases}
$$

$$
\text{II} = \sum_{k=\langle K_0 \rangle} e^{-j(m+n)\Omega_0 k} = \begin{cases} K_0 & \text{if } n = -m \\ 0 & \text{if } n \neq -m, \end{cases}
$$

which results in the following values for the DTFS coefficients:

$$
D_n = \begin{cases} -j\dfrac{B}{2} e^{j\theta} & \text{for } n = m \\[2mm] j\dfrac{B}{2} e^{-j\theta} & \text{for } n = -m \\[2mm] 0 & \text{elsewhere,} \end{cases} \tag{11.18}
$$

within one period $(-m \leq n \leq (K_0 - m - 1))$.

As a special case, let us consider the DTFS for the following discrete sinusoidal sequence:

$$
y[k] = 3 \sin\left(\frac{2\pi}{7} k + \frac{\pi}{4} \right),
$$

which has a fundamental period of $K_0 = 7$. Substituting $B = 3$, $m = 1$, and $\theta = \pi/4$ into Eq. (11.18), we obtain

$$
D_n = \begin{cases} -j\dfrac{3}{2} e^{j\frac{\pi}{4}} & \text{for } n = 1 \\[2mm] j\dfrac{3}{2} e^{-j\frac{\pi}{4}} & \text{for } n = -1 \\[2mm] 0 & \text{elsewhere,} \end{cases} \tag{11.19}
$$

for $-1 \leq n \leq 5$. The magnitude and phase spectra for the sinusoidal sequence are shown in Figs. 11.5(a) and (b).

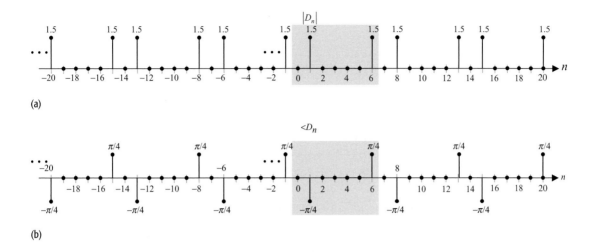

(a)

(b)

Fig. 11.5. (a) Magnitude spectrum and (b) phase spectrum of the DTFS coefficients in Example 11.5.

Corollary 11.2 The DTFS coefficients of the sinusoidal sequence $x[k] = B \sin(2\pi mk/K_0)$ are given by

$$
D_n = \begin{cases}
-j\dfrac{B}{2} & \text{for } n = m, m \pm K_0, m \pm 2K_0, \ldots \\[2mm]
j\dfrac{B}{2} & \text{for } n = -m, -m \pm K_0, -m \pm 2K_0, \ldots \\[2mm]
0 & \text{elsewhere},
\end{cases} \tag{11.20}
$$

provided that the greatest common divisor between integers m and K_0 is one.

11.2 Fourier transform for aperiodic functions

In Section 11.1, we used the exponential DTFS to derive the frequency representations for periodic sequences. In this section, we consider the frequency representations for aperiodic sequences. The resulting representation is called the DT Fourier transform (DTFT).

Figure 11.6(a) shows the waveform of an aperiodic sequence $x[k]$, which is zero outside the range $M_1 \le k \le M_2$. Such a sequence is referred to as a time-limited sequence having a length of $M_2 - M_1 + 1$ samples. As was the case for the CTFT, we consider periodic repetitions of $x[k]$ uniformly spaced with a duration of K_0 between each other; $K_0 \ge (M_2 - M_1 + 1)$ such that the adjacent replicas of $x[k]$ do not overlap with each other. The resulting sequence is referred to as the periodic extension of $x[k]$ and is denoted by $\tilde{x}_{K_0}[k]$. If we increase the value of K_0, in the limit, we obtain

$$
\lim_{K_0 \to \infty} \tilde{x}_{K_0}[k] = x[k]. \tag{11.21}
$$

Fig. 11.6. (a) Time-limited
sequence $x[k]$; (b) its periodic
extension.

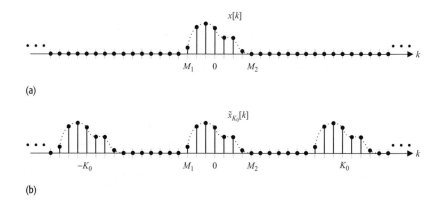

(a)

(b)

Since $\tilde{x}_{K_0}[k]$ is periodic with fundamental period K_0 (or fundamental frequency Ω_0), we can express it using the DTFS as follows:

$$\tilde{x}_{K_0}[k] = \sum_{n=\langle K_0\rangle} D_n e^{jn\Omega_0 k}, \tag{11.22}$$

where the DTFS coefficients D_n are given by

$$D_n = \frac{1}{K_0}\sum_{k=\langle K_0\rangle} \tilde{x}_{K_0}[k]e^{-jn\Omega_0 k},$$

for $1 \le n \le K_0$. Using Eq. (11.21), the above equation can be expressed as follows:

$$D_n = \lim_{K_0\to\infty}\frac{1}{K_0}\sum_{k=-\infty}^{\infty} x[k]e^{-jn\Omega_0 k} \tag{11.23}$$

for $1 \le n \le K_0$. Let us now define a new function $X(\Omega)$, which is continuous with respect to the independent variable Ω:

$$X(\Omega) = \sum_{k=-\infty}^{\infty} x[k]e^{-j\Omega k}. \tag{11.24}$$

In Eq. (11.24), the independent variable Ω is continuous in the range $-\infty \le \Omega \le \infty$. In terms of $X(\Omega)$, Eq. (11.23) can be expressed as follows:

$$D_n = \lim_{K_0\to\infty}\frac{1}{K_0}X(n\Omega_0). \tag{11.25}$$

The function $X(n\Omega_0)$ is obtained by sampling $X(\Omega)$ at discrete points $\Omega = n\Omega_0$.

Given the DTFS coefficients D_n of $\tilde{x}_{K_0}[k]$, the aperiodic sequence $x[k]$ can be obtained by substituting the values of D_n in Eq. (11.22) and solving for $M_1 \le k \le M_2$. The resulting expression is given by

$$x[k] = \lim_{K_0\to\infty}\tilde{x}_{K_0}[k] = \lim_{K_0\to\infty}\sum_{n=\langle K_0\rangle}\frac{1}{K_0}X(n\Omega_0)e^{jn\Omega_0 k}. \tag{11.26a}$$

In the limit $K_0 \to \infty$, the angular frequency Ω_0 takes a very small value, say $\Delta\Omega$, with the fundamental period $K_0 = 2\pi/\Delta\Omega$. In the limit $K_0 \to \infty$, Eq. (11.26a) can, therefore, be expressed as follows:

$$x[k] = \lim_{\Delta\Omega \to 0} \sum_{n=\langle K_0 \rangle} \frac{1}{2\pi} X(n\Delta\Omega) e^{jnk\Delta\Omega} \Delta\Omega. \qquad (11.26b)$$

Substituting $\Omega = n\Delta\Omega$ and applying the limit $\Delta\Omega \to 0$, Eq. (11.26b) reduces to the following integral:

$$x[k] = \frac{1}{2\pi} \int_{\langle 2\pi \rangle} X(\Omega) e^{jk\Omega} d\Omega. \qquad (11.27)$$

In Eq. (11.27), the limits of integration are derived by evaluating the duration $n = \langle K_0 \rangle$ in terms of Ω as follows:

$$\Omega = \langle n\Delta\Omega \rangle|_{n=\langle K_0 \rangle} = \left\langle n\left(\frac{2\pi}{K_0}\right)\right\rangle\Big|_{n=\langle K_0 \rangle} = \langle 2\pi \rangle,$$

implying that any frequency range of 2π may be used to solve the integral in Eq. (11.27). Collectively, Eq. (11.24), in conjunction with Eq. (11.27), is referred to as the DTFT pair.

Definition 11.3 *The DTFT pair for an aperiodic sequence $x[k]$ is given by*

DTFT synthesis equation $\qquad x[k] = \dfrac{1}{2\pi} \displaystyle\int_{\langle 2\pi \rangle} X(\Omega) e^{jk\Omega} d\Omega;$ $\qquad (11.28a)$

DTFT analysis equation $\qquad X(\Omega) = \displaystyle\sum_{k=-\infty}^{\infty} x[k] e^{-j\Omega k}.$ $\qquad (11.28b)$

In the subsequent discussion, we will denote the DTFT pair as follows:

$$x[k] \xleftrightarrow{\text{DTFT}} X(\Omega). \qquad (11.28c)$$

Example 11.6
Calculate the Fourier transform of the following functions:

(i)   unit impulse sequence, $x_1[k] = \delta[k]$;

(ii) gate sequence, $x_2[k] = \text{rect}\left(\dfrac{k}{2N+1}\right) = \begin{cases} 1 & |k| \leq N \\ 0 & \text{elsewhere}; \end{cases}$

(iii) decaying exponential sequence, $x_3[k] = p^k u[k]$ with $|p| < 1$.

Solution
(i) By definition,

$$X_1(\Omega) = \sum_{k=-\infty}^{\infty} \delta[k] e^{-j\Omega k} = e^{-j\Omega k}|_{k=0} = 1.$$

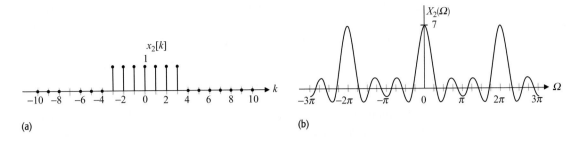

Fig. 11.7. (a) Rectangular
sequence $x_2[k]$ with a width of
seven samples. (b) DTFT of the
rectangular sequence derived in
Example 11.6(ii).

(ii) By definition,

$$X_2(\Omega) = \sum_{k=-\infty}^{\infty} x_2[k] e^{-j\Omega k} = \sum_{k=-N}^{N} 1 \cdot e^{-j\Omega k}.$$

The summation represents a GP series with $\exp(-j\Omega)$ as the ratio between two consecutive terms. The GP series simplifies to

$$X_2(\Omega) = (e^{-j\Omega})^{-N} \frac{1 - (e^{-j\Omega})^{2N+1}}{1 - e^{-j\Omega}}$$

$$= (e^{-j\Omega})^{-N} \frac{(e^{-j\Omega})^{(2N+1)/2}}{(e^{-j\Omega})^{1/2}} \frac{(e^{-j\Omega})^{-(2N+1)/2} - (e^{-j\Omega})^{(2N+1)/2}}{(e^{-j\Omega})^{-1/2} - (e^{-j\Omega})^{1/2}}$$

$$= \frac{e^{j(2N+1)\Omega/2} - e^{-j(2N-1)\Omega/2}}{e^{j\Omega/2} - e^{-j\Omega/2}} = \frac{\sin\left(\dfrac{2N+1}{2}\Omega\right)}{\sin\left(\dfrac{1}{2}\Omega\right)}.$$

As a special case, we assume $N = 3$ and plot the rectangular sequence $x_2[k]$ and its DTFT $X_2(\Omega)$ in Fig. 11.7.

(iii) By definition,

$$X_3(\Omega) = \sum_{k=-\infty}^{\infty} p^k u[k] e^{-j\Omega k} = \sum_{k=0}^{\infty} (p e^{-j\Omega})^k.$$

The summation represents a GP series, which can be simplified to

$$X_3(\Omega) = \frac{1}{1 - p e^{-j\Omega}} = \frac{1}{1 - p \cos\Omega + j p \sin\Omega}.$$

The DTFT $X_3(\Omega)$ is a complex-valued function of the angular frequency Ω. Its magnitude and phase spectra are determined below:

magnitude spectrum $$|X_3(\Omega)| = \frac{|1|}{|1 - p \cos\Omega + j p \sin\Omega|}$$

$$= \frac{1}{\sqrt{1 - 2p \cos\Omega + p^2}};$$

phase spectrum $$\angle X_3(\Omega) = \angle 1 - \angle(1 - p \cos\Omega + j p \sin\Omega)$$

$$= -\tan^{-1}\left(\frac{p \sin\Omega}{1 - p \cos\Omega}\right).$$

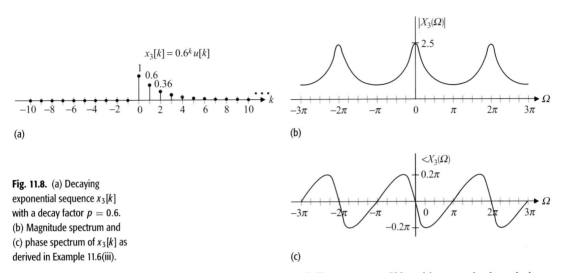

Fig. 11.8. (a) Decaying exponential sequence $x_3[k]$ with a decay factor $p = 0.6$. (b) Magnitude spectrum and (c) phase spectrum of $x_3[k]$ as derived in Example 11.6(iii).

As a special case, we plot the DT sequence $x_3[k]$ and its magnitude and phase spectra for $p = 0.6$ in Figs. 11.8(a)–(c).

In Example 11.6, we calculated the DTFTs for three different sequences and observed that all three DTFTs are periodic with period $\Omega_0 = 2\pi$. This property is referred to as the frequency–periodicity property and is satisfied by all DTFTs. In Section 11.4, we present a mathematical proof verifying the frequency–periodicity property.

Example 11.7

Calculate the DT sequences for the following DTFTs:

(i) $X_1(\Omega) = 2\pi \displaystyle\sum_{m=-\infty}^{\infty} \delta(\Omega - 2m\pi)$;

(ii) $X_2(\Omega) = 2\pi \displaystyle\sum_{m=-\infty}^{\infty} \delta(\Omega - \Omega_0 - 2m\pi)$.

Solution

(i) Using the synthesis equation, Eq. (11.28a), the inverse DTFT of $X_1(\Omega)$ is given by

$$x_1[k] = \frac{1}{2\pi} \int_{(2\pi)} X_1(\Omega) e^{jk\Omega} d\Omega = \frac{1}{2\pi} \int_{-\pi}^{\pi} 2\pi \sum_{m=-\infty}^{\infty} \delta(\Omega - 2m\pi) e^{jk\Omega} d\Omega$$

$$= \int_{-\pi}^{\pi} \sum_{m=-\infty}^{\infty} \delta(\Omega - 2m\pi) \underbrace{e^{jk2m\pi}}_{=1} d\Omega \quad [\because \delta(\Omega - \theta) f(\Omega) = \delta(\Omega - \theta) f(\theta)]$$

$$= \int_{-\pi}^{\pi} \sum_{m=-\infty}^{\infty} \delta(\Omega - 2m\pi) d\Omega.$$

The integral on the right-hand side of the above equation includes several impulse functions located at $\Omega = 0, \pm 2\pi, \pm 4\pi, \ldots$ Only the impulse function located at $\Omega = 0$ falls in the frequency range $\Omega = [-\pi, \pi]$. Therefore, $x_1[k]$ can be simplified as follows:

$$x_1[k] = \int_{-\pi}^{\pi} \delta(\Omega)\mathrm{d}\Omega = 1.$$

(ii) Using the synthesis equation, (11.28a), the inverse DTFT of $X_2(\Omega)$ is given by

$$x_1[k] = \frac{1}{2\pi} \int_{(2\pi)} X_2(\Omega)\mathrm{e}^{\mathrm{j}k\Omega}\mathrm{d}\Omega = \frac{1}{2\pi}\int_{-\pi}^{\pi} 2\pi \sum_{m=-\infty}^{\infty} \delta(\Omega - \Omega_0 - 2m\pi)\mathrm{e}^{\mathrm{j}k\Omega}\mathrm{d}\Omega.$$

$$= \int_{-\pi}^{\pi} \sum_{m=-\infty}^{\infty} \delta(\Omega - \Omega_0 - 2m\pi)\mathrm{e}^{\mathrm{j}k(\Omega_0-2m\pi)}\,\mathrm{d}\Omega$$

$$= \int_{-\pi}^{\pi} \sum_{m=-\infty}^{\infty} \delta(\Omega - \Omega_0 - 2m\pi)\mathrm{e}^{\mathrm{j}k\Omega_0}\underbrace{\mathrm{e}^{\mathrm{j}k2m\pi}}_{=1}\,\mathrm{d}\Omega$$

$$= \mathrm{e}^{\mathrm{j}k\Omega_0}\int_{-\pi}^{\pi} \sum_{m=-\infty}^{\infty} \delta(\Omega - \Omega_0 - 2m\pi)\mathrm{d}\Omega.$$

The integral on the right-hand side of the above equation includes several impulse functions located at $\Omega = \Omega_0 + 2m\pi$. Only one of these infinite number of impulse functions will be present in the frequency range $\Omega = [-\pi, \pi]$. Therefore, the integral will have a vaue of unity and the function $x_2[k]$ can be simplified as follows:

$$x_2[k] = \mathrm{e}^{\mathrm{j}k\Omega_0}.$$

Table 11.2 lists the DTFT and DTFS representations for several DT sequences. In situations where a DT sequence is aperiodic, the DTFS representation is not possible and therefore not included in the table. The DTFT of the periodic sequences is determined from its DTFS representation and is covered in Section 11.4.

Table 11.3 plots the DTFT for several DT sequences. In situations where a DT sequence or its DTFT is complex, we plot both the magnitude and phase components. The magnitude component is shown using a bold line, and the phase component is shown using a dashed line.

Example 11.7 illustrates the calculation of a DT function from its DTFT using Eq. (11.28a). In many cases, it may be easier to calculate a DT function from its DTFT using the partial fraction expansion and the DTFT pairs listed in Table 11.2. This procedure is explained in more detail in

Table 11.2. DTFTs and DTFSs for elementary DT sequences
Note that the DTFS does not exist for aperiodic sequences

Sequence: $x[k]$	DTFS: $D_n = \dfrac{1}{K_0} \displaystyle\sum_{k=\langle K_0\rangle} x[k]e^{-jn\Omega_0 k}$	DTFT: $X(\Omega) = \displaystyle\sum_{k=-\infty}^{\infty} x[k]e^{-j\Omega k}$
(1) $x[k] = 1$	$D_n = 1$	$X(\Omega) = 2\pi \displaystyle\sum_{m=-\infty}^{\infty} \delta(\Omega - 2m\pi)$
(2) $x[k] = \delta[k]$	does not exist	$X(\Omega) = 1$
(3) $x[k] = \delta[k - k_0]$	does not exist	$X(\Omega) = e^{-j\Omega k_0}$
(4) $x[k] = \displaystyle\sum_{m=-\infty}^{\infty} \delta(k - mK_0)$	$D_n = \dfrac{1}{K_0}$ for all n	$X(\Omega) = \dfrac{2\pi}{K_0} \displaystyle\sum_{m=-\infty}^{\infty} \delta\left(\Omega - \dfrac{2m\pi}{K_0}\right)$
(5) $x[k] = u[k]$	does not exist	$X(\Omega) = \pi \displaystyle\sum_{m=-\infty}^{\infty} \delta(\Omega - 2m\pi) + \dfrac{1}{1 - e^{-j\Omega}}$
(6) $x[k] = p^k u[k]$ with $|p| < 1$	does not exist	$X(\Omega) = \dfrac{1}{1 - pe^{-j\Omega}}$
(7) First-order time-rising decaying exponential $x[k] = (k + 1)p^k u[k]$, with $|p| < 1$.	does not exist	$X(\Omega) = \dfrac{1}{(1 - pe^{-j\Omega})^2}$
(8) Complex exponential (periodic) $x[k] = e^{jk\Omega_0}$ $K_0 = 2\pi p/\Omega_0$	$D_n = \begin{cases} 1 & n = p \pm rK_0 \\ 0 & \text{elsewhere} \end{cases}$ for $-\infty < r < \infty$	$X(\Omega) = 2\pi \displaystyle\sum_{m=-\infty}^{\infty} \delta(\Omega - \Omega_0 - 2m\pi)$
(9) Complex exponential (aperiodic) $x[k] = e^{jk\Omega_0}$, $2\pi/\Omega_0 \neq$ rational	does not exist	$X(\Omega) = 2\pi \displaystyle\sum_{m=-\infty}^{\infty} \delta(\Omega - \Omega_0 - 2m\pi)$
(10) Cosine (periodic) $x[k] = \cos(\Omega_0 k)$ $K_0 = 2\pi p/\Omega_0$	$D_n = \begin{cases} \dfrac{1}{2} & n = \pm p \pm rK_0 \\ 0 & \text{elsewhere} \end{cases}$ for $-\infty < r < \infty$	$X(\Omega) = \pi \displaystyle\sum_{m=-\infty}^{\infty} \delta(\Omega + \Omega_0 - 2m\pi)$ $+ \pi \displaystyle\sum_{m=-\infty}^{\infty} \delta(\Omega - \Omega_0 - 2m\pi)$
(11) Cosine (aperiodic) $x[k] = \cos(\Omega_0 k)$, $2\pi/\Omega_0 \neq$ rational	does not exist	$X(\Omega) = \pi \displaystyle\sum_{m=-\infty}^{\infty} \delta(\Omega + \Omega_0 - 2m\pi)$ $+ \pi \displaystyle\sum_{m=-\infty}^{\infty} \delta(\Omega - \Omega_0 - 2m\pi)$
(12) Sine (periodic) $x[k] = \sin(\Omega_0 k)$ $K_0 = 2\pi p/\Omega_0$	$D_n = \begin{cases} \dfrac{1}{2j} & n = \pm p \pm rK_0 \\ 0 & \text{elsewhere} \end{cases}$ for $-\infty < r < \infty$	$X(\Omega) = j\pi \displaystyle\sum_{m=-\infty}^{\infty} \delta(\Omega + \Omega_0 - 2m\pi)$ $- j\pi \displaystyle\sum_{m=-\infty}^{\infty} \delta(\Omega - \Omega_0 - 2m\pi)$
(13) Sine (aperiodic) $x[k] = \sin(\Omega_0 k)$, $2\pi/\Omega_0 \neq$ rational	does not exist	$X(\Omega) = j\pi \displaystyle\sum_{m=-\infty}^{\infty} \delta(\Omega + \Omega_0 - 2m\pi)$ $- j\pi \displaystyle\sum_{m=-\infty}^{\infty} \delta(\Omega - \Omega_0 - 2m\pi)$

(cont.)

Table 11.2. (*cont.*)

Sequence: $x[k]$	DTFS: $D_n = \dfrac{1}{K_0}\displaystyle\sum_{k=\langle K_0\rangle} x[k]e^{-jn\Omega_0 k}$	DTFT: $X(\Omega) = \displaystyle\sum_{k=-\infty}^{\infty} x[k]e^{-j\Omega k}$
(14) Rectangular (periodic) $x[k] = \begin{cases} 1 & \|k\| \le N \\ 0 & N < \|k\| \le K_0/2 \end{cases}$ $x[k] = x[k + K_0]$	$D_n = \begin{cases} (2N+1)/K_0 & k = rK_0 \\ \dfrac{1}{K_0}\left[\dfrac{\sin\left(\frac{2N+1}{K_0}n\pi\right)}{\sin\left(\frac{1}{K_0}n\pi\right)}\right] & \text{elsewhere} \end{cases}$	$X(\Omega) = 2\pi \displaystyle\sum_{n=-\infty}^{\infty} D_n\delta\left(\Omega - \dfrac{2n\pi}{K_0}\right)$
(15) Rectangular (aperiodic) $x[k] = \begin{cases} 1 & \|k\| \le N \\ 0 & \text{elsewhere} \end{cases}$	does not exist	$X(\Omega) = \dfrac{\sin\left(\dfrac{2N+1}{2}\Omega\right)}{\sin\left(\dfrac{1}{2}\Omega\right)}$
(16) Sinc $x[k] = \dfrac{W}{\pi}\,\text{sinc}\left(\dfrac{Wk}{\pi}\right) =$ $\dfrac{\sin(Wk)}{\pi k}$ for $0 < W < \pi$	does not exist	$X(\Omega) = \begin{cases} 1 & \|\Omega\| \le W \\ 0 & W < \|\Omega\| \le \pi \end{cases}$ $X(\Omega) = X(\Omega + 2\pi)$
(17) Arbitrary periodic sequence with period K_0 $x[k] = \displaystyle\sum_{n=\langle K_0\rangle} D_n e^{jn\Omega_0 k}$	$D_n = \dfrac{1}{K_0}\displaystyle\sum_{k=\langle K_0\rangle} x[k]e^{-jn\Omega_0 k}$	$X(\Omega) = 2\pi \displaystyle\sum_{n=-\infty}^{\infty} D_n\delta\left(\Omega - \dfrac{2n\pi}{K_0}\right)$

Appendix D, and has been used later in this chapter in solving Examples 11.15 and 11.18.

11.3 Existence of the DTFT

Definition 11.4 *The DTFT $X(\Omega)$ of a DT sequence $x[k]$ is said to exist if*

$$|X(\Omega)| < \infty, \quad \text{for} - \infty < \Omega < \infty. \tag{11.29}$$

The above definition for the existence of the DTFT satisfies our intuition that a valid function should be finite for all values of the independent variable. Substituting the value of the DTFT $X(\Omega)$ from Eq. (11.28b), Eq. (11.29) can be expressed as follows:

$$\left| \sum_{k=-\infty}^{\infty} x[k]e^{-j\Omega k} \right| < \infty,$$

which is satisfied if

$$\sum_{k=-\infty}^{\infty} |x[k]| \cdot |e^{-j\Omega k}| < \infty.$$

Table 11.3. DTFT spectra for elementary DT sequences

Sequence	Time-domain waveform	Magnitude and phase spectra

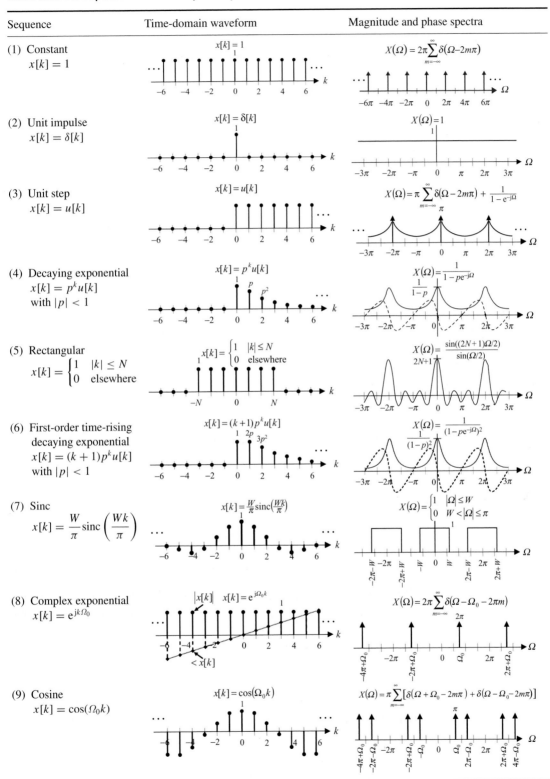

(1) Constant
$x[k] = 1$

$x[k] = 1$

$X(\Omega) = 2\pi \sum_{m=-\infty}^{\infty} \delta(\Omega - 2m\pi)$

(2) Unit impulse
$x[k] = \delta[k]$

$x[k] = \delta[k]$

$X(\Omega) = 1$

(3) Unit step
$x[k] = u[k]$

$x[k] = u[k]$

$X(\Omega) = \pi \sum_{m=-\infty}^{\infty} \delta(\Omega - 2m\pi) + \dfrac{1}{1 - e^{-j\Omega}}$

(4) Decaying exponential
$x[k] = p^k u[k]$
with $|p| < 1$

$x[k] = p^k u[k]$

$X(\Omega) = \dfrac{1}{1 - pe^{-j\Omega}}$

(5) Rectangular
$x[k] = \begin{cases} 1 & |k| \leq N \\ 0 & \text{elsewhere} \end{cases}$

$x[k] = \begin{cases} 1 & |k| \leq N \\ 0 & \text{elsewhere} \end{cases}$

$X(\Omega) = \dfrac{\sin((2N+1)\Omega/2)}{\sin(\Omega/2)}$

(6) First-order time-rising
decaying exponential
$x[k] = (k+1)p^k u[k]$
with $|p| < 1$

$x[k] = (k+1)p^k u[k]$

$X(\Omega) = \dfrac{1}{(1 - pe^{-j\Omega})^2}$

(7) Sinc
$x[k] = \dfrac{W}{\pi} \mathrm{sinc}\left(\dfrac{Wk}{\pi}\right)$

$x[k] = \dfrac{W}{\pi} \mathrm{sinc}\left(\dfrac{Wk}{\pi}\right)$

$X(\Omega) = \begin{cases} 1 & |\Omega| \leq W \\ 0 & W < |\Omega| \leq \pi \end{cases}$

(8) Complex exponential
$x[k] = e^{jk\Omega_0}$

$|x[k]| \quad x[k] = e^{j\Omega_0 k}$
$< x[k]$

$X(\Omega) = 2\pi \sum_{m=-\infty}^{\infty} \delta(\Omega - \Omega_0 - 2\pi m)$

(9) Cosine
$x[k] = \cos(\Omega_0 k)$

$x[k] = \cos(\Omega_0 k)$

$X(\Omega) = \pi \sum_{m=-\infty}^{\infty} [\delta(\Omega + \Omega_0 - 2m\pi) + \delta(\Omega - \Omega_0 - 2m\pi)]$

From the Euler's formula, we know that $|\exp(-j\Omega k)| = 1$. Therefore, an alternative expression to verify the existence of the DTFT is given by

$$\sum_{k=-\infty}^{\infty} |x[k]| < \infty.$$

Condition for the existence of DTFT The DTFT $X(\Omega)$ of a DT sequence $x[k]$ exists if

$$\sum_{k=-\infty}^{\infty} |x[k]| < \infty. \tag{11.30}$$

Equation (11.30) is a sufficient condition to verify the existence of the DTFT.

Example 11.8

Determine if the DTFTs exist for the following functions:

(i) causal exponential function, $x_1[k] = p^k u[k]$.
(ii) cosine waveform, $x_2[k] = \cos(\Omega_0 k)$.

Solution

(i) Equation (11.30) yields

$$\sum_{k=-\infty}^{\infty} |x_1[k]| = \sum_{k=-\infty}^{\infty} |p^k u[k]| = \sum_{k=0}^{\infty} |p^k| = \begin{cases} \dfrac{1}{1-p} & 0 < |p| < 1 \\ \infty & |p| \geq 1. \end{cases}$$

Therefore, the DTFT of the exponential sequence $x_1[k] = p^k u[k]$ exists if $0 < |p| < 1$. Under such a condition, $x_1[k]$ is a decaying exponential sequence with the summation in Eq. (11.30) having a finite value.

(ii) Equation (11.30) yields

$$\sum_{k=-\infty}^{\infty} |x_2[k]| = \sum_{k=-\infty}^{\infty} |\cos(\Omega_0 k)| \to \infty.$$

Therefore, the DTFT does not exist for the cosine waveform. However, this appears to be in violation of Table 11.2, which lists the following DTFT pair for the cosine sequence:

$$\cos(\Omega_0 k) \xleftarrow{\text{DTFT}} \pi \sum_{m=-\infty}^{\infty} [\delta(\Omega + \Omega_0 - 2\pi m) + \delta(\Omega - \Omega_0 - 2\pi m)].$$

Looking closely at the above DTFT pair, we note that the DTFT $X(\Omega)$ of the cosine function consists of continuous impulse functions at discrete frequencies $\Omega = (\pm\Omega_0 - 2\pi m)$, for $-\infty < m < \infty$. Since the magnitude of a continuous impulse function is infinite, $|X(\Omega)|$ is infinite at the location of the impulses. The infinite magnitude of the impulses in the DTFT $X(\Omega)$ leads to the violation of the existence condition stated in Eq. (11.30).

Example 11.8 has introduced a confusing situation for the cosine sequence. We proved that the condition of the existence of the DTFT is violated by the cosine waveform, yet its DTFT can be expressed mathematically. A similar behavior is exhibited by most periodic sequences. So how do we determine the DTFT for a periodic sequence? We cannot use the definition of the DTFT, Eq. (11.28b), since the procedure will lead to infinite DTFT values. In such cases, an alternative procedure based on the DTFS is used; this is explained in Section 11.4.

11.4 DTFT of periodic functions

Consider a periodic function $x[k]$ with fundamental period K_0. The DTFS representation for $x[k]$ is given by

$$x[k] = \sum_{n=\langle K_0 \rangle} D_n e^{jn\Omega_0 k}, \tag{11.31}$$

where $\Omega_0 = 2\pi/K_0$ and the DTFS coefficients are given by

$$D_n = \frac{1}{K_0} \sum_{k=\langle K_0 \rangle} x[k] e^{-jn\Omega_0 k}. \tag{11.32}$$

Calculating the DTFT of both sides of Eq. (11.31), we obtain

$$X(\Omega) = \Im \left\{ \sum_{n=\langle K_0 \rangle} D_n e^{jn\Omega_0 k} \right\}.$$

Since the DTFT satisfies the linearity property, the above equation can be expressed as follows:

$$X(\Omega) = \sum_{n=\langle K_0 \rangle} D_n \Im\{e^{jn\Omega_0 k}\}, \tag{11.33}$$

where the DTFT of the complex exponential sequence is given by

$$\Im\{e^{jn\Omega_0 k}\} = 2\pi \sum_{m=-\infty}^{\infty} \delta(\Omega - n\Omega_0 - 2\pi m).$$

Using the above value for the DTFT of the complex exponential, Eq. (11.33) takes the following form:

$$X(\Omega) = \sum_{n=\langle K_0 \rangle} D_n 2\pi \sum_{m=-\infty}^{\infty} \delta(\Omega - n\Omega_0 - 2\pi m).$$

By changing the order of summation in the above equation and substituting $\Omega_0 = 2\pi/K_0$, we have

$$X(\Omega) = 2\pi \sum_{m=-\infty}^{\infty} \sum_{n=\langle K_0 \rangle} D_n \delta\left(\Omega - \frac{2n\pi}{K_0} - 2\pi m\right).$$

Since the DTFT is periodic with a period of 2π, we determine the DTFT in the range $\Omega = [0, 2\pi]$ and use the periodicity property to determine the DTFT values outside the specified range. Taking $n = 0, 1, 2, \ldots, K_0 - 1$ and $m = 0$, the following terms of $X(\Omega)$ lie within the range $\Omega = [0, 2\pi]$:

$$X(\Omega) = 2\pi D_0 \delta(\Omega) + 2\pi D_1 \delta\left(\Omega - \frac{2\pi}{K_0}\right) + 2\pi D_2 \delta\left(\Omega - \frac{4\pi}{K_0}\right)$$

$$+ \cdots + 2\pi D_{K_0-1}\delta\left(\Omega - \frac{2(K_0-1)\pi}{K_0}\right) \qquad (11.34a)$$

or

$$X(\Omega) = 2\pi \sum_{n=\langle K_0 \rangle} D_n \delta\left(\Omega - \frac{2n\pi}{K_0}\right), \qquad (11.34b)$$

for $0 \le \Omega \le 2\pi$. Since $X(\Omega)$ is periodic, Eq. (11.34b) can also be expressed as follows:

$$X(\Omega) = 2\pi \sum_{n=-\infty}^{\infty} D_n \delta\left(\Omega - \frac{2n\pi}{K_0}\right), \qquad (11.35)$$

which is the DTFT of the periodic sequence $x[k]$ for the entire Ω-axis. The values of the DTFS coefficients lying outside the range $0 \le n \le (K_0 - 1)$ are evaluated from Eq. (11.9) to be

$$D_n = D_{n+mK_0} \quad \text{for } m \in Z.$$

Definition 11.5 *The DTFT $X(\Omega)$ of a periodic sequence $x[k]$ is given by*

$$X(\Omega) = 2\pi \sum_{n=-\infty}^{\infty} D_n \delta\left(\Omega - \frac{2n\pi}{K_0}\right), \qquad (11.36a)$$

where D_n are the DTFS coefficients of $x[k]$. The DTFS coefficients are given by

$$D_n = \frac{1}{K_0} \sum_{k=\langle K_0 \rangle} x[k] e^{-jn\Omega_0 k} \qquad (11.36b)$$

for $0 \le n \le K_0 - 1$ and the values outside the range are evaluated from the following periodicity relationship:

$$D_n = D_{n+mK_0} \quad \text{for } m \in Z. \qquad (11.36c)$$

Example 11.9
Calculate the DTFT of the following periodic sequences:

(i) $x_1[k] = k$ for $0 \le k \le 3$, with the fundamental period $K_0 = 4$;

(ii) $x_2[k] = \begin{cases} 5 & k = 0, 1 \\ 0 & k = 2, 3, \end{cases}$ with the fundamental period $K_0 = 4$;

(iii) $x_3[k] = 0.5^k$ for $0 \le k \le 14$, with the fundamental period $K_0 = 15$;

(iv) $x_4[k] = 3 \sin\left(\frac{2\pi}{7}k + \frac{\pi}{4}\right)$, with the fundamental period $K_0 = 7$.

Fig. 11.9. DTFT of the periodic sequence $x_1[k] = k$, $0 \leq k \leq 3$, with fundamental period $K_0 = 4$. (a) Magnitude spectrum; (b) phase spectrum.

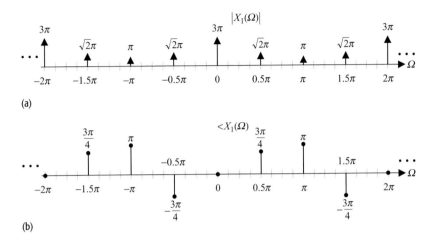

(a)

(b)

Solution

(i) Using Eq. (11.36a), the DTFT of $x_1[k]$ is given by

$$X_1(\Omega) = 2\pi \sum_{n=-\infty}^{\infty} D_n \delta\left(\Omega - \frac{2n\pi}{4}\right) = 2\pi \sum_{n=-\infty}^{\infty} D_n \delta\left(\Omega - \frac{n\pi}{2}\right).$$

Substituting $\Omega_0 = 2\pi/K_0 = \pi/2$ in Eq. (11.36b), the DTFS coefficients D_n for $x_1[k]$ are given by

$$D_n = \frac{1}{4}\sum_{k=0}^{3} k e^{-jn\pi k/2} = \frac{1}{4}[e^{-jn\pi/2} + 2e^{-jn\pi} + 3e^{-j3n\pi/2}].$$

For $0 \leq n \leq 3$, the values of the DTFS coefficients are as follows:

$n = 0$ $D_0 = \dfrac{1}{4}[1 + 2 \cdot 1 + 3 \cdot 1] = \dfrac{3}{2};$

$n = 1$ $D_1 = \dfrac{1}{4}[e^{-j\pi/2} + 2 \cdot e^{-j\pi} + 3 \cdot e^{-j3\pi/2}]$

                $= \dfrac{1}{4}[-j + 2(-1) + 3(j)] = -\dfrac{1}{2}[1 - j];$

$n = 2$ $D_2 = \dfrac{1}{4}[e^{-j\pi} + 2 \cdot e^{-j2\pi} + 3 \cdot e^{-j3\pi}]$

                $= \dfrac{1}{4}[-1 + 2(1) + 3(-1)] = -\dfrac{1}{2};$

$n = 3$ $D_3 = \dfrac{1}{4}[e^{-j3\pi/2} + 2 \cdot e^{-j3\pi} + 3 \cdot e^{-j9\pi/2}]$

                $= \dfrac{1}{4}[j + 2(-1) + 3(-j)] = -\dfrac{1}{2}[1 + j].$

The values of the DTFS coefficients that lie outside the range $0 \leq n \leq 3$ can be obtained by using the periodicity property $D_{n+4} = D_n$.

Since $X_1(\Omega)$ is a complex-valued function, its magnitude and phase spectra are plotted separately in Figs. 11.9(a) and (b). The area enclosed by the impulse

Fig. 11.10. DTFT of the periodic sequence $x_2[k]$, with fundamental period $K_0 = 4$. (a) Magnitude spectrum; (b) phase spectrum.

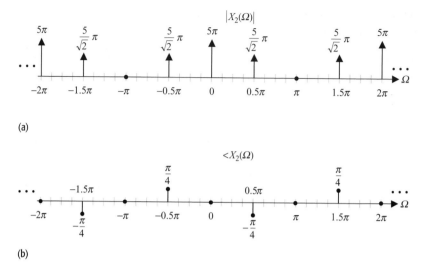

(a)

(b)

functions in the magnitude spectrum is given by $2\pi D_n$ and is indicated at the top of each impulse in Fig. 11.9(a).

(ii) Using Eq. (11.36a), the DTFT of $x_2[k]$ is given by

$$X_2(\Omega) = 2\pi \sum_{n=-\infty}^{\infty} D_n \delta\left(\Omega - \frac{2n\pi}{4}\right) = 2\pi \sum_{n=-\infty}^{\infty} D_n \delta\left(\Omega - \frac{n\pi}{2}\right).$$

Substituting $\Omega_0 = 2\pi/K_0 = \pi/2$ in Eq. (11.36b), the DTFS coefficients D_n are as follows:

$$D_n = \frac{1}{4}\sum_{k=0}^{1} 5e^{-jn\pi k/2} = \frac{1}{4}[5 + 5e^{-j\pi n/2}] = \frac{5}{2}e^{-j\pi n/4} \cos\left(\frac{\pi n}{4}\right).$$

For $0 \le n \le 3$, the values of the DTFS coefficients are as follows:

$$n = 0\,(\Omega = 0) \qquad D_0 = \frac{5}{2} \text{ with } |D_0| = \frac{5}{2}, \; <D_0 = 0;$$

$$n = 1\,(\Omega = 0.5\pi) \quad D_1 = \frac{5}{2\sqrt{2}}e^{-j\pi/4} \text{ with } |D_1| = \frac{5}{2\sqrt{2}}, \; <D_1 = -\frac{\pi}{4};$$

$$n = 2\,(\Omega = \pi) \qquad D_2 = 0 \text{ with } |D_2| = 0, \; <D_2 = 0;$$

$$n = 3\,(\Omega = 1.5\pi) \quad D_3 = -\frac{5}{2\sqrt{2}}e^{-j3\pi/4} \text{ with } |D_3| = \frac{5}{2\sqrt{2}},$$

$$<D_3 = \pi - \frac{3\pi}{4} = \frac{\pi}{4}.$$

The magnitude and phase spectra are plotted separately in Figs. 11.10(a) and (b), where the values of the DTFS coefficients lying outside $0 \le n \le 3$ are obtained using the periodicity property $D_{n+4} = D_n$.

(iii) Using Eq. (11.36a), the DTFT of $x_3[k]$ is given by

$$X_3(\Omega) = 2\pi \sum_{n=-\infty}^{\infty} D_n \delta\left(\Omega - \frac{2n\pi}{15}\right).$$

Table 11.4. Values of $|D_n|$ and $<D_n$ for $0 \leq n \leq 14$ in Example 11.9(iii)

The radian frequency Ω corresponding to each value of n is given in the second row

n	0	1	2	3	4	5	6	7	8	9	10	11	12	13	14
Ω	0	$2\pi/15$	$4\pi/15$	$6\pi/15$	$8\pi/15$	$10\pi/15$	$12\pi/15$	$14\pi/15$	$16\pi/15$	$18\pi/15$	$20\pi/15$	$22\pi/15$	$24\pi/15$	$26\pi/15$	$28\pi/15$
$|D_n|$	0.133	0.115	0.088	0.069	0.057	0.050	0.047	0.045	0.045	0.047	0.050	0.057	0.069	0.088	0.115
$<D_n$	0	-0.11π	-0.16π	-0.16π	-0.14π	-0.11π	-0.07π	-0.02π	0.02π	0.07π	0.11π	0.14π	0.16π	0.16π	0.11π

The DTFS coefficients of $x_3[k]$ are computed in Example 11.3. Substituting $\Omega_0 = 2\pi/K_0 = 2\pi/15$ in Eqs. (11.11)–(11.13), we obtain

$$D_n = \frac{1}{15} \left(\frac{1}{1 - 0.5 \cos\left(\dfrac{2n\pi}{15}\right) + j0.5 \sin\left(\dfrac{2n\pi}{15}\right)} \right),$$

where the magnitude component is given by

$$|D_n| = \frac{1}{15} \left(\frac{1}{\sqrt{1.25 - \cos\left(\dfrac{2n\pi}{15}\right)}} \right)$$

and the phase component is given by

$$<D_n = -\tan^{-1} \left[\frac{0.5 \sin\left(\dfrac{2n\pi}{15}\right)}{1 - 0.5 \cos\left(\dfrac{2n\pi}{15}\right)} \right].$$

The magnitude and phase components of the DTFS coefficients D_n for $0 \leq n \leq 14$ are given in Table 11.4.

The values of the DTFS coefficients, lying outside $0 \leq n \leq 14$ are obtained using the periodicity property $D_{n+14} = D_n$. The magnitude and phase of $X_3(\Omega)$ are plotted in Fig. 11.11.

(iv) In Example 11.5, the DTFS coefficients D_n of $x_4[k]$ are computed and are given by Eq. (11.19), which is reproduced here:

$$D_n = \begin{cases} -j1.5e^{j\pi/4;} & \text{for } n = 1 \\ j1.5e^{-j\pi/4;} & \text{for } n = -1 \quad \text{for } -1 \leq n \leq 5 \\ 0 & \text{elsewhere.} \end{cases}$$

The values of the DTFS coefficients lying outside $-1 \leq n \leq 5$ are obtained using the periodicity property $D_{n+7} = D_n$. Using Eq. (11.36a), the DTFT of

Fig. 11.11. DTFT of the periodic sequence $x_3[k] = 0.5^k u[k]$, $0 \le k \le 14$, with fundamental period $K_0 = 15$. (a) Magnitude spectrum; (b) phase spectrum.

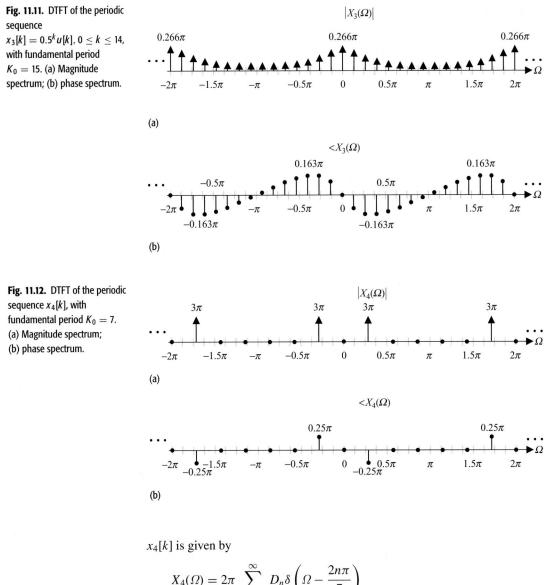

Fig. 11.12. DTFT of the periodic sequence $x_4[k]$, with fundamental period $K_0 = 7$. (a) Magnitude spectrum; (b) phase spectrum.

$x_4[k]$ is given by

$$X_4(\Omega) = 2\pi \sum_{n=-\infty}^{\infty} D_n \delta\left(\Omega - \frac{2n\pi}{7}\right)$$

$$= 2\pi \sum_{\substack{n=-\infty \\ n=7m+1}}^{\infty} D_1 \delta\left(\Omega - \frac{2n\pi}{7}\right) + 2\pi \sum_{\substack{n=-\infty \\ n=7m-1}}^{\infty} D_{-1} \delta\left(\Omega - \frac{2n\pi}{7}\right)$$

$$= -j3\pi e^{j(\pi/4)} \sum_{m=-\infty}^{\infty} \delta\left(\Omega - \frac{2\pi}{7} - 2m\pi\right)$$

$$+ j3\pi e^{-j(\pi/4)} \sum_{m=-\infty}^{\infty} \delta\left(\Omega + \frac{2\pi}{7} - 2m\pi\right).$$

The magnitude and phase of $X_4(\Omega)$ are plotted in Fig. 11.12.

11.5 Properties of the DTFT and the DTFS

In this section, we present the properties of the DTFT. These properties are similar to the properties for the CTFT discussed in Chapter 5. In most cases, we do not explicitly state the DTFS properties, but a list of the DTFS properties is included in Table 11.5.

11.5.1 Periodicity

DTFT The DTFT $X(\Omega)$ of an arbitrary DT sequence $x[k]$ is periodic with a period $\Omega_0 = 2\pi$. Mathematically,

$$X(\Omega) = X(\Omega + 2\pi). \tag{11.37}$$

DTFS The DTFS coefficients D_n of a periodic sequence $x[n]$ with period K_0 are periodic with respect to the coefficient number n and has a period K_0. In other words,

$$D_n = D_{n+mK_0}, \tag{11.38}$$

for $0 \leq n \leq K_0 - 1$ and $-\infty < m < \infty$. Recall that the coefficient number $n = K_0$ corresponds to the frequency $\Omega_n = 2\pi n/K_0 = 2\pi$. Therefore, the frequency–periodicity property of the DTFS and DTFT are in fact the same.

11.5.2 Hermitian symmetry

The DTFT $X(\Omega)$ of a real-valued sequence $x[k]$ satisfies

$$X(-\Omega) = X^*(\Omega), \tag{11.39a}$$

where $X^*(\Omega)$ denotes the complex conjugate of $X(\Omega)$. By expressing the DTFT $X(\Omega)$ in terms of its real and imaginary components,

$$X(\Omega) = \text{Re}\{X(\Omega)\} + j\,\text{Im}\{X(\Omega)\},$$

Eq. (11.39a) can be expressed as follows:

$$\text{Re}\{X(-\Omega)\} + j\,\text{Im}\{X(-\Omega)\} = \text{Re}\{X(\Omega)\} - j\,\text{Im}\{X(\Omega)\}.$$

Separating the real and imaginary components yields

$$\text{Re}\{X(-\Omega)\} = \text{Re}\{X(\Omega)\} \quad \text{and} \quad \text{Im}\{X(-\Omega)\} = -\text{Im}\{X(\Omega)\}, \tag{11.39b}$$

which implies that the real component $\text{Re}\{X(\Omega)\}$ of the DTFT $X(\Omega)$ of a real-valued sequence $x[k]$ is an even function of frequency Ω and that its imaginary component $\text{Im}\{X(\Omega)\}$ is an odd function of Ω. In terms of the magnitude and phase spectra of the DTFT $X(\Omega)$, the Hermitian symmetry property can be expressed as follows:

$$|X(-\Omega)| = |X(\Omega)| \quad \text{and} \quad <X(-\Omega) = - <X(\Omega), \qquad (11.39\text{c})$$

implying that the magnitude spectrum is even and that the phase spectrum is odd.

As extensions of the Hermitian symmetry properties, we consider the special cases when: (a) $x[k]$ is real-valued and even and (b) $x[k]$ is imaginary-valued and odd.

Case 1 If $x[k]$ is both real-valued and even, then its DTFT $X(\Omega)$ is also real-valued and even, with the imaginary component $\text{Im}\{X(\Omega)\} = 0$. In other words,

$$\text{Re}\{X(-\Omega)\} = \text{Re}\{X(\Omega)\} \quad \text{and} \quad \text{Im}\{X(-\Omega)\} = 0. \qquad (11.39\text{d})$$

Case 2 If $x[k]$ is both imaginary-valued and odd, then its DTFT $X(\Omega)$ is also imaginary-valued and odd, with the real component $\text{Re}\{X(\Omega)\} = 0$. In other words,

$$\text{Re}\{X(-\Omega)\} = 0 \quad \text{and} \quad \text{Im}\{X(-\Omega)\} = -\text{Im}\{X(\Omega)\}. \qquad (11.39\text{e})$$

11.5.3 Linearity

Like the CTFT, both the DTFT and DTFS satisfy the linearity property.

DTFT If $x_1[k]$ and $x_2[k]$ are two DT sequences with the following DTFT pairs:

$$x_1[k] \xleftrightarrow{\text{DTFT}} X_1(\Omega) \quad \text{and} \quad x_2[k] \xleftrightarrow{\text{DTFT}} X_2(\Omega),$$

then the linearity property states that

$$a_1 x_1[k] + a_2 x_2[k] \xleftrightarrow{\text{DTFT}} a_1 X_1(\Omega) + a_2 X_2(\Omega), \qquad (11.40\text{a})$$

for any arbitrary constants a_1 and a_2, which may be complex-valued.

DTFS If $x_1[k]$ and $x_2[k]$ are two periodic DT sequences with the same fundamental period K_0 and the following DTFS pairs:

$$x_1[k] \xleftrightarrow{\text{DTFS}} D_n^{x_1} \quad \text{and} \quad x_2[k] \xleftrightarrow{\text{DTFS}} D_n^{x_2},$$

then the DTFS coefficients of the periodic DT sequence $x_3[k] = a_1 x_1[k] + a_2 x_2[k]$, which also has a period of K_0, are given by

$$\underbrace{a_1 x_1[k] + a_2 x_2[k]}_{x_3[k]} \xleftrightarrow{\text{DTFS}} \underbrace{a_1 D_n^{x_1} + a_2 D_n^{x_2}}_{D_n^{x_3}}, \qquad (11.40\text{b})$$

for any arbitrary constants a_1 and a_2, which may be complex-valued.

11.5.4 Time scaling

The time-scaling property of the CTFT, defined in Section 5.4.2, states that if a CT function $x(t)$ is time-compressed by a factor of a $(a \neq 0)$, its CTFT $X(\omega)$ is expanded in the frequency domain by a factor of a, and vice versa. For the DTFT, the time-scaling property has a limited scope, as illustrated in the following.

Decimation Since decimation is an irreversible nature of the decimation operation, the DTFT of $x[k]$ and the decimated sequence $y[k] = x[mk]$ are not related to each other.

Interpolation In the DT domain, interpolation is defined in Chapter 1 as follows:

$$x^{(m)}[k] = \begin{cases} x\left[\dfrac{k}{m}\right] & \text{if } k \text{ is a multiple of integer } m \\ 0 & \text{otherwise.} \end{cases} \tag{11.41a}$$

The interpolated sequence $x^{(m)}[k]$ inserts $(m-1)$ zeros in between adjacent samples of the DT sequence $x[k]$. The time-scaling property for the interpolated sequence $x^{(m)}[k]$ is given as follows.
     If

$$x[k] \xleftrightarrow{\text{DTFT}} X(\Omega),$$

then the DTFT $X^{(m)}(\Omega)$ of $x^{(m)}[k]$ is given by

$$X^{(m)}(\Omega) = X(m\Omega), \tag{11.41b}$$

for $2 \leq m < \infty$. Equation (11.41b) shows that interpolation in time results in compression in the frequency domain. To demonstrate the application of the interpolation property, consider the DTFT of a rectangular sequence:

$$x[k] = \text{rect}\left(\frac{k}{3}\right) \xleftrightarrow{\text{DTFT}} \frac{\sin(3.5\Omega)}{\sin(0.5\Omega)}.$$

Using the interpolation property, the DTFT of the interpolated function $x^{(2)}[k]$ for $m = 2$ is given by

$$x^{(2)}[k] \xleftrightarrow{\text{DTFT}} X(2\Omega) = \frac{\sin(7\Omega)}{\sin(\Omega)}.$$

The functions $x[k]$ and $x^{(2)}[k]$ and their DTFTs are shown graphically in Fig. 11.13.

11.5.5 Time shifting

The time-shifting operation delays or advances the reference sequence in time. Given a signal $x[k]$, the time-shifted signal is given by $x[k - k_0]$, where k_0 is an integer. If the value of the shift k_0 is positive, the reference sequence $x[k]$

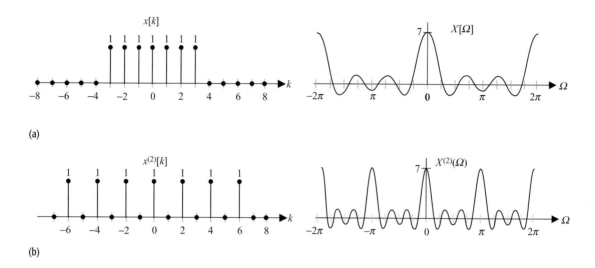

(a)

(b)

Fig. 11.13. Time-scaling property. (a) DTFT pair for a rectangular sequence $x[k]$ with a length of seven samples. (b) DTFT pair for $x^{(2)}[k]$ obtained by interpolating $x[k]$ by a factor of $m = 2$.

is delayed and shifted towards the right-hand side of the k-axis. On the other hand, if the value of the shift k_0 is negative, sequence $x[k]$ advances and is shifted towards the left-hand side of the k-axis. The DTFT of the time-shifted sequence $x[k-k_0]$ is related to the DTFT of the reference sequence $x[k]$ using the following time-shifting property.

If

$$x[k] \xleftarrow{\text{DTFT}} X(\Omega)$$

then

$$x[k - k_0] \xleftarrow{\text{DTFT}} e^{-jk_0\Omega}X(\Omega), \qquad (11.42)$$

for integer values of k_0.

Example 11.10

Using the time-shifting property, calculate the DTFT of the following sequence:

$$x[k] = \begin{cases} 0.75 & (3 \leq k \leq 9) \\ 0.5^{k-12} & (12 \leq k < \infty) \\ 0 & \text{elsewhere.} \end{cases}$$

Solution

The DT sequence $x[k]$, plotted in Fig. 11.14, can be expressed as a linear combination of: (i) a time-shifted gate or rectangular sequence, denoted by $x_2[k]$ in Example 11.6, (ii), as follows:

$$x_2[k] = \text{rect}\left(\frac{k}{2N+1}\right) = \begin{cases} 1 & |k| \leq N \\ 0 & \text{elsewhere,} \end{cases}$$

Fig. 11.14. DT sequence $x[k]$
used in Example 11.10.

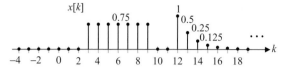

with $N = 3$; and (ii) a time-shifted decaying exponential sequence, denoted by $x_3[k]$ in Example 11.6, (iii), as follows:

$$x_3[k] = p^k u[k],$$

with decay factor $p = 0.5$. In terms of $x_2[k]$ and $x_3[k]$, the expression for $x[k]$ is given by

$$x[k] = 0.75x_2[k - 6] + x_3[k - 12].$$

Using the linearity and time-shifting properties, the DTFT $X(\Omega)$ of $x[k]$ is given by

$$X(\Omega) = 0.75e^{-j6\Omega}X_2(\Omega) + e^{-j12\Omega}X_3(\Omega).$$

From the results in Example 11.6, the DTFTs for the sequences $x_2[k]$ and $x_3[k]$ are given by

$$X_2(\Omega) = \frac{\sin(3.5\Omega)}{\sin(0.5\Omega)} \quad \text{and} \quad X_3(\Omega) = \frac{1}{1 - 0.5e^{-j\Omega}}.$$

Substituting the values of the DTFTs results in the following:

$$X(\Omega) = 0.75e^{-j6\Omega}\frac{\sin(3.5\Omega)}{\sin(0.5\Omega)} + e^{-j12\Omega}\frac{1}{1 - 0.5e^{-j\Omega}}.$$

11.5.6 Frequency shifting

In the time-shifting property, we observed the change in the DTFT when the DT sequence $x[k]$ is shifted in the time domain. The frequency-shifting property addresses the converse problem of how shifting the DTFT $X(\Omega)$ in the frequency domain affects the sequence $x[k]$ in the time domain.

If

$$x[k] \xleftrightarrow{\text{DTFT}} X(\Omega)$$

then

$$x[k]e^{j\Omega_0 k} \xleftrightarrow{\text{DTFT}} X(\Omega - \Omega_0), \tag{11.43}$$

for $0 \leq \Omega_0 < 2\pi$.

Example 11.11
Using the frequency-shifting property, calculate the DTFT of $x[k] = \cos(\Omega_0 k)\cos(\Omega_1 k)$ with $(\Omega_0 + \Omega_1) < \pi$.

Solution

Using Table 11.2, the DTFT of $\cos(\Omega_0 k)$ is given by

$$\cos(\Omega_0 k) \xleftrightarrow{\text{DTFT}} \pi \sum_{m=-\infty}^{\infty} [\delta(\Omega + \Omega_0 - 2m\pi) + \delta(\Omega - \Omega_0 - 2m\pi)].$$

Using the frequency-shifting property,

$$\cos\ (\Omega_0 k) e^{j\Omega_1 k} \xleftrightarrow{\text{DTFT}} \pi \sum_{m=-\infty}^{\infty} [\delta(\Omega + \Omega_0 - \Omega_1 - 2m\pi)$$
$$+ \delta(\Omega - \Omega_0 - \Omega_1 - 2m\pi)]$$

and

$$\cos(\Omega_0 k) e^{-j\Omega_1 k} \xleftrightarrow{\text{DTFT}} \pi \sum_{m=-\infty}^{\infty} [\delta(\Omega + \Omega_0 + \Omega_1 - 2m\pi)$$
$$+ \delta(\Omega - \Omega_0 + \Omega_1 - 2m\pi)].$$

Adding the two DTFT pairs and noting that $[\exp(j\Omega_1 k) + \exp(-j\Omega_1 k)] = 2\cos(\Omega_1 k)$, we obtain

$$\cos(\Omega_0 k) \cos(\Omega_1 k) \xleftrightarrow{\text{DTFT}} \frac{\pi}{2} \sum_{m=-\infty}^{\infty} [\delta(\Omega + \Omega_0 - \Omega_1 - 2m\pi)$$
$$+ \delta(\Omega - \Omega_0 - \Omega_1 - 2m\pi)$$
$$+ \delta(\Omega + \Omega_0 + \Omega_1 - 2m\pi)$$
$$+ \delta(\Omega - \Omega_0 + \Omega_1 - 2m\pi)].$$

The above DTFT can also be obtained by expressing

$$2\cos(\Omega_0 k) \cos(\Omega_1 k) = \cos[(\Omega_0 + \Omega_1)k] + \cos[(\Omega_0 - \Omega_1)k]$$

and calculating the DTFT of the right-hand side of the above expression.

11.5.7 Time differencing

The time differencing in the DT domain is the counterpart of differentiation in the CT domain. The time-differencing property is stated as follows.
   If

$$x[k] \xleftrightarrow{\text{DTFT}} X(\Omega)$$

then

$$x[k] - x[k-1] \xleftrightarrow{\text{DTFT}} [1 - e^{-j\Omega}] X(\Omega). \tag{11.44}$$

The proof of Eq. (11.44) follows directly from the application of the time-shifting property.

Example 11.12

Based on the DTFT of the unit step $u[k]$ and the time-shifting property, calculate the DTFT of $x[k] = \delta[k]$.

Solution

Using Table 11.2, the DTFT of the unit step function is given by

$$u[k] \xleftrightarrow{\text{DTFT}} \pi \sum_{m=-\infty}^{\infty} \delta(\Omega - 2m\pi) + \frac{1}{1 - e^{-j\Omega}}.$$

Applying the time-differencing property yields

$$u[k] - u[k-1] \xleftrightarrow{\text{DTFT}} (1 - e^{-j\Omega}) \left[\pi \sum_{m=-\infty}^{\infty} \delta(\Omega - 2m\pi) + \frac{1}{1 - e^{-j\Omega}} \right],$$

which reduces to

$$u[k] - u[k-1] \xleftrightarrow{\text{DTFT}} 1 + \pi \sum_{m=-\infty}^{\infty} \delta(\Omega - 2m\pi)(1 - e^{-j\Omega})|_{\Omega=2m\pi}.$$

Since $[1 - \exp(-j2m\pi)] = 0$, the above DTFT pair reduces to

$$\delta[k] \xleftrightarrow{\text{DTFT}} 1.$$

11.5.8 Differentiation in frequency

If

$$x[k] \xleftrightarrow{\text{DTFT}} X(\Omega)$$

then

$$-jkx[k] \xleftrightarrow{\text{DTFT}} \frac{dX}{d\Omega}. \tag{11.45}$$

Example 11.13

Based on the DTFT of the exponential decaying function and the frequency differentiation property, calculate the DTFT of $x[k] = (k+1)p^k u[k]$.

Solution

In Table 11.2, the DTFT of the exponential decaying function is given as

$$p^k u[k] \xleftrightarrow{\text{DTFT}} \frac{1}{1 - pe^{-j\Omega}}.$$

Using the frequency-differentiation property, we obtain

$$(-jk)p^k u[k] \xleftrightarrow{\text{DTFT}} \frac{d}{d\Omega} \left[\frac{1}{1 - pe^{-j\Omega}} \right] = \frac{-jpe^{-j\Omega}}{(1 - pe^{-j\Omega})^2}$$

or

$$kp^k u[k] \xleftrightarrow{\text{DTFT}} j\frac{d}{d\Omega} \left[\frac{1}{1 - pe^{-j\Omega}} \right] = \frac{pe^{-j\Omega}}{(1 - pe^{-j\Omega})^2}.$$

Adding the DTFT pairs for $p^k u[k]$ and $kp^k u[k]$ yields

$$(k+1)p^k u[k] \xleftrightarrow{\text{DTFT}} \frac{1}{1 - pe^{-j\Omega}} + \frac{pe^{-j\Omega}}{(1 - pe^{-j\Omega})^2} = \frac{1}{(1 - pe^{-j\Omega})^2}.$$

11.5.9 Time summation

The time summation in the DT domain is the counterpart of integration in the CT domain. The time-summation property is defined as follows.
     If

$$x[k] \xleftrightarrow{\text{DTFT}} X(\Omega)$$

then

$$\sum_{n=-\infty}^{k} x[n] \xleftrightarrow{\text{DTFT}} \frac{1}{(1 - e^{-j\Omega})} X(\Omega) + \pi X(0) \sum_{m=-\infty}^{\infty} \delta(\Omega - 2\pi m). \quad (11.46)$$

Example 11.14

Based on the DTFT of the unit impulse sequence and the time-summation property, calculate the DTFT of the unit step sequence.

Solution

Using Table 11.2, the DTFT of the unit impulse sequence is given by

$$\delta[k] \xleftrightarrow{\text{DTFT}} 1.$$

Using the time-summation property, we obtain

$$\sum_{n=-\infty}^{k} \delta[n] \xleftrightarrow{\text{DTFT}} \frac{1}{1 - e^{-j\Omega}} \cdot 1 + \pi \cdot 1 \sum_{m=-\infty}^{\infty} \delta(\Omega - 2\pi m),$$

which yields

$$u[k] \xleftrightarrow{\text{DTFT}} \frac{1}{1 - e^{-j\Omega}} + \pi \sum_{m=-\infty}^{\infty} \delta(\Omega - 2\pi m).$$

11.5.10 Time convolution

In Section 10.5, we showed that the output response of an LTID system is obtained by convolving the input sequence with the impulse response of the system. Sometimes the resulting convolution sum is difficult to solve analytically in the time domain. The convolution property provides us with an alternative approach, based on the DTFT, of calculating the output response. Below we state the convolution property and explain its application in calculating the output response of an LTID system using an example.
     If $x_1[k]$ and $x_2[k]$ are two DT sequences with the following DTFT pairs:

$$x_1[k] \xleftrightarrow{\text{DTFT}} X_1(\Omega) \quad \text{and} \quad x_2[k] \xleftrightarrow{\text{DTFT}} X_2(\Omega),$$

then the time-convolution property states that

$$x_1[k] * x_2[k] \xleftrightarrow{\text{DTFT}} X_1(\Omega)X_2(\Omega). \tag{11.47}$$

In other words, the convolution between two DT sequences in the time domain is equivalent to multiplication of the DTFTs of the two functions in the frequency domain. Note that the CTFT also has a similar property, as stated in Section 5.4.8.

Equation (11.47) provides us with an alternative technique for calculating the convolution sum using the DTFT. Expressed in terms of the following DTFT pairs:

$$x[k] \xleftrightarrow{\text{DTFT}} X(\Omega), \; h[k] \xleftrightarrow{\text{DTFT}} H(\Omega), \quad \text{and} \quad y[k] \xleftrightarrow{\text{DTFT}} Y(\Omega),$$

the output sequence $y[k]$ can be expressed in terms of the impulse response $h[k]$ and the input sequence $x[k]$ as follows:

$$y[k] = x[k] * h[k] \xleftrightarrow{\text{DTFT}} Y(\Omega) = X(\Omega)H(\Omega). \tag{11.48}$$

In other words, the DTFT of the output sequence is obtained by multiplying the DTFTs of the input sequence and the impulse response. The procedure for evaluating the output $y[k]$ of an LTID system in the frequency domain therefore consists of the following four steps.

(1) Calculate the DTFT $X(\Omega)$ of the input signal $x[k]$.
(2) Calculate the DTFT $H(\Omega)$ of the impulse response $h[k]$ of the LTID system. The DTFT $H(\Omega)$ is referred to as the transfer function of the LTID system and provides a meaningful insight into the behavior of the system.
(3) Based on the convolution property, the DTFT of the output $y[k]$ is given by $Y(\Omega) = H(\Omega)X(\Omega)$.
(4) The output $y[k]$ in the time domain is obtained by calculating the inverse DTFT of $Y(\Omega)$ obtained in step (3).

Since the DTFTs are periodic with period $\Omega = 2\pi$, steps (1)–(4) can be applied only to the frequency range $[-\pi \leq \Omega \leq \pi]$.

Example 11.15

The exponential decaying sequence $x[k] = a^k u[k], 0 < a < 1$, is applied at the input of an LTID system with the impulse response $h[k] = b^k u[k], 0 < b < 1$. Using the DTFT approach, calculate the output of the system.

Solution

Based on Table 11.2, the DTFTs for the input sequence and the impulse response are given by

$$x[k] \xleftrightarrow{\text{DTFT}} \frac{1}{1 - ae^{-j\Omega}} \quad \text{and} \quad h[k] \xleftrightarrow{\text{DTFT}} \frac{1}{1 - be^{-j\Omega}}.$$

The DTFT $Y(\Omega)$ of the output signal is therefore calculated as follows:

$$y[k] = x[k] * h[k] \xrightarrow{\text{DTFT}} Y(\Omega) = \frac{1}{(1 - ae^{-j\Omega})(1 - be^{-j\Omega})}.$$

The inverse of the DTFT $Y(\Omega)$ takes two different forms depending on the values of a and b:

$$Y(\Omega) = \begin{cases} \dfrac{1}{(1 - ae^{-j\Omega})^2} & a = b \\[3mm] \dfrac{1}{(1 - ae^{-j\Omega})(1 - be^{-j\Omega})} & a \neq b. \end{cases}$$

We consider the two cases separately.

Case 1 $(a = b)$ The inverse DTFT follows directly from Table 11.2 as follows:

$$y[k] = (k + 1)a^k u[k + 1].$$

Case 2 $(a \neq b)$ Using partial fraction expansion, the DTFT $Y(\Omega)$ is expressed as follows:

$$Y(\Omega) = \frac{A}{1 - ae^{-j\Omega}} + \frac{B}{1 - be^{-j\Omega}}, \tag{11.49}$$

where the partial fraction coefficients are given by

$$A = \left. \frac{1}{1 - be^{-j\Omega}} \right|_{ae^{-j\Omega} = 1} = \frac{a}{a - b}$$

and

$$B = \left. \frac{1}{1 - ae^{-j\Omega}} \right|_{be^{-j\Omega} = 1} = -\frac{b}{a - b}.$$

Substituting the values of A and B in Eq. (11.49) and calculating the inverse DTFT yields

$$y[k] = \frac{1}{a - b}[a^{k+1} - b^{k+1}]u[k].$$

Combining case 1 with case 2, we obtain

$$y[k] = \begin{cases} (k + 1)a^k u[k] & a = b \\[2mm] \dfrac{1}{a - b}[a^{k+1} - b^{k+1}]u[k] & a \neq b. \end{cases} \tag{11.50}$$

11.5.11 Periodic convolution

The time-convolution property, defined by Eq. (11.47), is used to calculate the output of convolving aperiodic sequences. In Section 10.6, we defined the

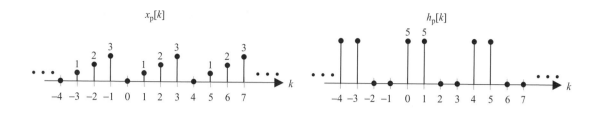

Fig. 11.15. Periodic sequences $x_p[k]$ and $h_p[k]$ used in Example 11.16.

periodic, or circular, convolution to convolve periodic sequences. We now show how the periodic convolution can be calculated using the DTFS.

If $x_1[k]$ and $x_2[k]$ are two DT periodic sequences with the same fundamental period K_0 and the following DTFS pairs:

$$x_1[k] \overset{\text{DTFS}}{\longleftrightarrow} D_n^{x_1} \quad \text{and} \quad x_2[k] \overset{\text{DTFS}}{\longleftrightarrow} D_n^{x_2},$$

then the periodic convolution property states that

$$x_1[k] \otimes x_2[k] \overset{\text{DTFS}}{\longleftrightarrow} K_0 D_n^{x_1} D_n^{x_2}. \tag{11.51}$$

We illustrate the application of the periodic convolution property by revisiting Example 10.10.

Example 11.16

In Example 10.10, we calculated the periodic convolution $y_p[k]$ of the two periodic sequences $x_p[k]$ and $h_p[k]$, defined over one period ($K_0 = 4$) as $x_p[k] = k, 0 \leq k \leq 3$, and $h_p[k] = 5, 0 \leq k \leq 1$, in the time domain. Repeat Example 10.10 using the periodic convolution property.

Solution

The periodic sequences $x_p[k]$ and $h_p[k]$ are shown in Fig. 11.15. In part (i) of Example 11.9, we calculated the DTFS coefficients of $x_p[k]$ as follows:

$$D_0^{x_p} = \frac{3}{2}, \quad D_1^{x_p} = -\frac{1}{2}[1-j], \quad D_2^{x_p} = -\frac{1}{2}, \quad \text{and} \quad D_3^{x_p} = -\frac{1}{2}[1+j].$$

Similarly, in part (ii) of Example 11.9 we calculated the DTFS coefficients of $h_p[k]$:

$$D_0^{h_p} = \frac{5}{2}, \quad D_1^{h_p} = \frac{5}{4}[1-j], \quad D_2^{h_p} = 0, \quad \text{and} \quad D_3^{h_p} = \frac{5}{4}[1+j].$$

Using the periodic convolution property, the DTFS coefficients of $y_p[k]$ are

$$D_0^{y_p} = K_0 D_0^{x_p} D_0^{h_p} = 15;$$
$$D_1^{y_p} = K_0 D_1^{x_p} D_1^{h_p} = j5;$$
$$D_2^{y_p} = K_0 D_2^{x_p} D_2^{h_p} = 0;$$
$$D_3^{y_p} = K_0 D_3^{x_p} D_3^{h_p} = -j5.$$

Calculating the inverse DTFS, the DT sequence $y_p[k]$ is given by

$$y_p[k] = \sum_{n=0}^{3} D_n^{y_p} e^{-j\frac{2\pi}{4}nk}$$

$$= \left[15 + j5 \cdot e^{-j\frac{\pi}{2}k} + 0 \cdot e^{-j\pi k} - j5 \cdot e^{-j\frac{3\pi}{2}k} \right].$$

Calculating the values of $y_p[k]$ within one period ($0 \le k \le 3$) yields

$k = 0$ $y_p[0] = 15 + j5 - j5 = 15;$

$k = 1$ $y_p[1] = 15 + j5e^{-j\frac{\pi}{2}} - j5e^{-j\frac{3\pi}{2}} = 15 - 5 - 5 = 5;$

$k = 2$ $y_p[2] = 15 + j5e^{-j\pi} - j5e^{-j3\pi} = 15 - j5 + j5 = 15;$

$k = 3$ $y_p[3] = 15 + j5e^{-j\frac{3\pi}{2}} - j5e^{-j\frac{9\pi}{2}} = 15 + 5 + 5 = 25.$

The above result is identical to the result obtained in Example 10.10.

Example 11.16 shows how periodic convolution can be calculated using the DTFS periodic-convolution property. A more computationally efficient approach of calculating the periodic convolution is based on the discrete Fourier transform (DFT). The theory of DFT will be presented in Chapter 12.

11.5.12 Frequency convolution

The time-convolution property (see Section 11.5.10) states that the convolution between two DT sequences in the time domain is equivalent to the multiplication of the DTFTs of the two sequences in the frequency domain. The converse of the time-convolution property is also true, and is referred to as the frequency-convolution property.

If $x_1[k]$ and $x_2[k]$ are two DT sequences with the following DTFT pairs:

$$x_1[k] \xrightarrow{\text{DTFT}} X_1(\Omega) \quad \text{and} \quad x_2[k] \xrightarrow{\text{DTFT}} X_2(\Omega),$$

then the frequency-convolution property states that

$$x_1[k]x_2[k] \xrightarrow{\text{DTFT}} \frac{1}{2\pi} \int_{\langle 2\pi \rangle} X_1(\theta)X_2(\Omega - \theta)d\theta. \tag{11.52}$$

The limits of integration in Eq. (11.52) are given by $\langle \Omega = 2\pi \rangle$, which implies that any range of 2π may be chosen during the integration.

The frequency-convolution property is widely used in digital communications systems, where it is commonly referred to as the modulation property.

11.5.13 Parseval's theorem

If $x[k]$ is an energy signal and $x[k] \xleftrightarrow{\text{DTFT}} X(\Omega)$, the energy of the DT signal $x[k]$ is given by

$$E_x = \sum_{k=-\infty}^{\infty} |x[k]|^2 = \frac{1}{2\pi} \int_{\langle 2\pi \rangle} |X(\Omega)|^2 d\Omega. \qquad (11.53)$$

Parseval's theorem states that the DTFT is a lossless transform as there is no loss of energy if a signal is transformed to the frequency domain.

Example 11.17
Using Parseval's theorem, evaluate the following integral:

$$\int_{\langle 2\pi \rangle} \left[\frac{\sin\left(\dfrac{2N+1}{2}\Omega\right)}{\sin\left(\dfrac{1}{2}\Omega\right)} \right]^2 d\Omega.$$

Solution
Since

$$\text{rect}\left(\frac{k}{2N+1}\right) \xleftrightarrow{\text{DTFT}} \frac{\sin\left(\dfrac{2N+1}{2}\Omega\right)}{\sin\left(\dfrac{1}{2}\Omega\right)},$$

Eq. (11.53) computes the area enclosed by the squared DT sinc function within one period $\Omega = \langle 2\pi \rangle$ as follows:

$$\frac{1}{2\pi} \int_{\langle 2\pi \rangle} \left[\frac{\sin\left(\dfrac{2N+1}{2}\Omega\right)}{\sin\left(\dfrac{1}{2}\Omega\right)} \right]^2 d\Omega = \sum_{k=-\infty}^{\infty} \left| \text{rect}\left[\frac{k}{2N+1}\right] \right|^2,$$

where

$$\text{rect}\left[\frac{k}{2N+1}\right] = \begin{cases} 1 & |k| \leq N \\ 0 & \text{elsewhere.} \end{cases}$$

Simplifying the summation on the right-hand side of this equation yields

$$\int_{\langle 2\pi \rangle} \left[\frac{\sin\left(\dfrac{2N+1}{2}\Omega\right)}{\sin\left(\dfrac{1}{2}\Omega\right)} \right]^2 d\Omega = 2\pi(2N+1).$$

We have presented several properties in Sections 11.5.1–11.5.12. Table 11.5 lists the properties of the DTFS and Table 11.6 lists the properties of the DTFT.

Table 11.5. Properties of the DTFS: sequences $x[k]$, $x_1[k]$, and $x_2[k]$ are periodic with a period of K_0

Properties	Time domain	Frequency domain	Comments
	$x[k] = \sum_{n=\langle K_0 \rangle} D_n e^{jn\Omega_0 k}$	$D_n = \frac{1}{K_0} \sum_{k=\langle K_0 \rangle} x[k] e^{-jn\Omega_0 k}$	$\Omega_0 = 2\pi/K_0$
	$x_1[k]$	$D_n^{x_1}$	$\Omega_0 = 2\pi/K_0$
	$x_2[k]$	$D_n^{x_2}$	$\Omega_0 = 2\pi/K_0$
Periodicity	$x[k]$	$D_n = D_{n+K_0}$	
Linearity	$a_1 x_1[k] + a_2 x_2[k]$	$a_1 D_n^{x_1} + a_2 D_n^{x_2}$	$a_1, a_2 \in C$
Scaling	$x^{(m)}[k]$ with period mK_0	$\frac{1}{m} D_n$	$m = 1, 2, 3, \ldots$
Time shifting	$x[k - k_0]$	$\exp\left(j\frac{2\pi k_0}{K_0} n\right) D_n$	$k_0 \in R$
Frequency shifting	$\exp\left(j\frac{2\pi n_0}{K_0} k\right) x[k]$	D_{n-n_0}	$n_0 \in R$
Time differencing	$x[k] - x[k-1]$	$\left[1 - \exp\left(j\frac{2\pi}{K_0} n\right)\right] D_n$	
Time summation	$S = \sum_{m=-\infty}^{k} x[m]$	$\dfrac{1}{1 - \exp\left(j\frac{2\pi}{K_0} n\right)} D_n$	summation S is finite only if $D_0 = 0$
Periodic convolution	$\sum_{n=\langle K_0 \rangle} x_1[n] x_2[n-k]$	$K_0 D_n^{x_1} D_n^{x_2}$	convolution over a period K_0
Frequency convolution	$x_1[k] x_2[k]$	$\sum_{m=\langle K_0 \rangle} D_m^{x_1} D_{m-n}^{x_2}$	multiplication in time domain
Parseval's relationship	$\frac{1}{K_0} \sum_{k=\langle K_0 \rangle} |x[k]|^2 = \sum_{n=\langle K_0 \rangle} |D_n|^2$		power of a periodic sequence

		Symmetry properties	
		DTFS: $D_{-n} = D_n^*$	Comments
Hermitian property	$x[k]$ is a real-valued sequence	real and imaginary components: $\begin{cases} \text{Re}\{D_{-n}\} = \text{Re}\{D_n\} \\ \text{Im}\{D_{-n}\} = -\text{Im}\{D_n\} \end{cases}$	real component is even; imaginary component is odd
		magnitude and phase spectra: $\begin{cases} |D_{-n}| = |D_n| \\ <D_{-n} = -<D_n \end{cases}$	magnitude spectrum is even; phase spectrum is odd
Real-valued and even function	$x[k]$ is an even and real-valued sequence	$\begin{cases} \text{Re}\{D_{-n}\} = \text{Re}\{D_n\} \\ \text{Im}\{D_{-n}\} = 0 \end{cases}$	DTFS is real-valued and even
Real-valued and odd function	$x[k]$ is an odd and real-valued sequence	$\begin{cases} \text{Re}\{D_{-n}\} = 0 \\ \text{Im}\{D_{-n}\} = -\text{Im}\{D_n\} \end{cases}$	DTFS is imaginary and odd

Table 11.6. Properties of the discrete-time Fourier transform (DTFT)

Transformation properties	Time domain	Frequency domain	Comments
	$x[k] = \dfrac{1}{2\pi} \displaystyle\int_{(2\pi)} X(\Omega)\mathrm{e}^{\mathrm{j}k\Omega}\mathrm{d}\Omega$	$X(\Omega) = \displaystyle\sum_{k=-\infty}^{\infty} x[k]\mathrm{e}^{-\mathrm{j}\Omega k}$	
	$x_1[k]$	$X_1(\Omega)$	
	$x_2[k]$	$X_2(\Omega)$	
Periodicity	$x[k]$	$X(\Omega) = X(\Omega + 2\pi)$	
Linearity	$a_1 x_1[k] + a_2 x_2[k]$	$a_1 X_1(\Omega) + a_2 X_2(\Omega)$	$a_1, a_2 \in C$
Scaling	$x^{(m)}[k]$	$X(m\Omega)$	$m = 1, 2, 3, \ldots$
Time shifting	$x[k - k_0]$	$\exp(-\mathrm{j}k_0\Omega)X(\Omega)$	$k_0 \in R$
Frequency shifting	$\exp(\mathrm{j}k\Omega_0)x[k]$	$X(\Omega - \Omega_0)$	$\Omega_0 \in R$
Time differencing	$x[k] - x[k - 1]$	$[1 - \exp(\mathrm{j}\Omega)]X(\Omega)$	
Time summation	$S = \displaystyle\sum_{m=-\infty}^{k} x[m]$	$\dfrac{1}{1 - \exp(\mathrm{j}\Omega)}X(\Omega) +$ $\pi X(0) \displaystyle\sum_{m=-\infty}^{\infty} \delta(\Omega - 2\pi m)$	provided summation S is finite
Time convolution	$x_1[k] * x_2[k]$	$X_1(\Omega)X_2(\Omega)$	
Periodic convolution	$x_1[k] \otimes x_2[k]$	$X_1(\Omega)X_2(\Omega)$	over period K_0
Frequency convolution	$x_1[k]x_2[k]$	$\dfrac{1}{2\pi} \displaystyle\int_{(2\pi)} X_1(\theta)X_2(\Omega - \theta)\mathrm{d}\theta$	multiplication in time domain
Parseval's relationship	$E_x = \displaystyle\sum_{k=-\infty}^{\infty} |x[k]|^2 = \dfrac{1}{2\pi} \displaystyle\int_{(2\pi)} |X(\Omega)|^2\, \mathrm{d}\Omega$		energy in a signal

Symmetry properties			
		DTFT: $X(-\Omega) = X^*(\Omega)$	
Hermitian property	$x[k]$ is a real-valued function	real and imaginary component: $\begin{cases} \mathrm{Re}\{X(-\Omega)\} = \mathrm{Re}\{X(\Omega)\} \\ \mathrm{Im}\{X(-\Omega)\} = -\mathrm{Im}\{X(\Omega)\} \end{cases}$	real component is even; imaginary component is odd
		magnitude and phase spectra: $\begin{cases} |X(-\Omega)| = |X(\Omega)| \\ {<}X(-\Omega) = -\,{<}X(\Omega) \end{cases}$	magnitude spectrum is even; phase spectrum is odd
Real-valued and even function	$x[k]$ is even and real-valued	$\begin{cases} \mathrm{Re}\{X(-\Omega)\} = \mathrm{Re}\{X(\Omega)\} \\ \mathrm{Im}\{X(\Omega)\} = 0 \end{cases}$	DTFT is real-valued and even
Real-valued and odd function	$x[k]$ is odd and real-valued	$\begin{cases} \mathrm{Re}\{X(-\Omega)\} = 0 \\ \mathrm{Im}\{X(-\Omega)\} = -\mathrm{Im}\{X(\Omega)\} \end{cases}$	DTFT is imaginary and odd

11.6 Frequency response of LTID systems

In Chapter 10, we presented two different representations to specify the input–output relationship of an LTID system. Section 10.1 used a linear, constant-coefficient difference equation, while Section 10.3 used the impulse response $h[k]$ to model an LTID system. A third representation for an LTID system is obtained by calculating the DTFT of the impulse response,

$$h[k] \xleftrightarrow{\text{DTFT}} H(\Omega).$$

The DTFT $H(\Omega)$ is referred to as the *Fourier transfer function* of the LTID system. In conjunction with the linear convolution property, the transfer function $H(\Omega)$ can be used to determine the output response $y[k]$ of the LTID system due to the input sequence $x[k]$. In the time domain, the output response $y[k]$ is given by

$$y[k] = x[k] * h[k].$$

Calculating the DTFT of both sides of the equation, we obtain

$$Y(\Omega) = X(\Omega)H(\Omega) \tag{11.54}$$

or

$$H(\Omega) = \frac{Y(\Omega)}{X(\Omega)}, \tag{11.55}$$

where $Y(\Omega)$ and $X(\Omega)$ are, respectively, the DTFTs of the output response $y[k]$ and the input signal $x[k]$. Equation (11.55) provides an alternative definition for the transfer function as the ratio of the DTFT of the output signal and the DTFT of the input signal.

Given one representation for an LTID system, it is straightforward to derive the remaining two representations based on the DTFT and its properties. In the following, we derive a formula to calculate the transfer function of an LTID system from its difference equation representation.

Consider an LTID system whose input–output relationship is given by the following difference equation:

$$y[k + n] + a_{n-1}y[k + n - 1] + \cdots + a_0 y[k] = b_m x[k + m]$$
$$+ b_{m-1}x[k + m - 1] + \cdots + b_0 x[k]. \tag{11.56}$$

Calculating the DTFT of both sides of the above equation, we obtain

$$\left\{ e^{jn\Omega} + a_{n-1}e^{j(n-1)\Omega} + \cdots + a_0 \right\} Y(\Omega) = \left\{ b_m e^{jm\Omega} + b_{m-1}e^{j(m-1)\Omega} \right.$$
$$\left. + \cdots + b_0 \right\} X(\Omega),$$

which reduces to the following transfer function:

$$H(\Omega) = \frac{Y(\Omega)}{X(\Omega)} = \frac{b_m e^{jm\Omega} + b_{m-1}e^{j(m-1)\Omega} + \cdots + b_0}{e^{jn\Omega} + a_{n-1}e^{j(n-1)\Omega} + \cdots + a_0}. \tag{11.57}$$

The impulse response $h[k]$ of the LTID system can be obtained by calculating the inverse DTFT of the transfer function $H(\Omega)$.

Fig. 11.16. Impulse response $h[k]$ of the LTID system derived in Example 11.18.

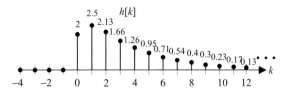

Example 11.18

The input–output relationship of an LTID system is given by the following difference equation:

$$y[k + 2] - \frac{3}{4}y[k + 1] + \frac{1}{8}y[k] = 2x[k + 2]. \qquad (11.58)$$

Determine the transfer function and the impulse response of the system.

Solution

Calculating the DTFT of Eq. (11.58) yields

$$\left\{e^{j2\Omega} - \frac{3}{4}e^{j\Omega} + \frac{1}{8}\right\} Y(\Omega) = 2e^{j2\Omega} X(\Omega),$$

which results in the following transfer function:

$$H(\Omega) = \frac{2e^{j2\Omega}}{e^{j2\Omega} - \frac{3}{4}e^{j\Omega} + \frac{1}{8}} = \frac{2}{1 - \frac{3}{4}e^{-j\Omega} + \frac{1}{8}e^{-j2\Omega}}$$

$$= \frac{2}{\left(1 - \frac{1}{2}e^{-j\Omega}\right)\left(1 - \frac{1}{4}e^{-j\Omega}\right)}.$$

To calculate the impulse response of the LTID system, we calculate the partial fraction of $H(\Omega)$ as follows:

$$H(\Omega) = \frac{4}{1 - \frac{1}{2}e^{-j\Omega}} - \frac{2}{1 - \frac{1}{4}e^{-j\Omega}}.$$

By calculating the inverse DTFT of both sides, the impulse response $h[k]$ is given by

$$h[k] = 4\left(\frac{1}{2}\right)^k u[k] - 2\left(\frac{1}{4}\right)^k u[k],$$

which is plotted in Fig. 11.16.

11.7 Magnitude and phase spectra

The Fourier transfer function $H(\Omega)$ provides a complete description of an LTID system. In most cases, $H(\Omega)$ is a complex function of the angular frequency Ω. Therefore, it is difficult to analyze the frequency characteristics of the transfer

Fig. 11.17. Gain and phase
responses of an LTID system.
The LTID system provides a gain
of $|H(\Omega)|$ to the magnitude and
a phase change of $<H(\Omega)$ to the
phase of the sinusoidal input.

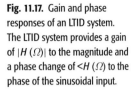

function directly from the mathematical expression. By expressing the transfer
function $H(\Omega)$ as

$$H(\Omega) = |H(\Omega)|e^{j<H(\Omega)}, \tag{11.59}$$

the LTID system is analyzed by plotting the magnitude $|H(\Omega)|$ and phase
$<H(\Omega)$ as functions of frequency Ω. The plot of the magnitude $|H(\Omega)|$ with
respect to frequency Ω is referred to as the magnitude spectrum, while the plot
of the phase $<H(\Omega)$ with respect to frequency Ω is referred to as the phase
spectrum. Collectively, magnitude and phase spectra are used to analyze the
LTID system.

A second interpretation of the magnitude and phase spectra is obtained by
considering a sinusoidal sequence $x[k] = \cos(\Omega_0 k)$ applied at the input of an
LTID system with transfer function $H(\Omega)$. The DTFT of the output of the LTID
system is given by

$$\begin{aligned}
Y(\Omega) &= \Im\{x[k]\}H(\Omega) \\
&= \pi[\delta(\Omega - \Omega_0) + \delta(\Omega + \Omega_0)]|H(\Omega)|e^{j<H(\Omega)} \\
&= \pi\left[\delta(\Omega - \Omega_0)|H(\Omega_0)|e^{j<H(\Omega_0)} + \delta(\Omega + \Omega_0)|H(-\Omega_0)|e^{j<H(-\Omega_0)}\right].
\end{aligned}$$

Assuming that the impulse response $h[k]$ of the LTID system is real-valued and
then applying the Hermitian symmetry property, we observe that the magnitude
response $|H(\Omega)|$ is an even function of Ω while the phase response $<H(\Omega)$ is
an odd function of Ω. Mathematically,

$$|H(-\Omega)| = |H(\Omega)| \quad \text{and} \quad <H(-\Omega) = - <H(\Omega).$$

The DTFT $Y(\Omega)$ of the output of the LTID system is therefore given by

$$Y(\Omega) = \pi|H(\Omega_0)|\left[\delta(\Omega - \Omega_0)e^{j<H(\Omega_0)} + \delta(\Omega + \Omega_0)e^{-j<H(\Omega_0)}\right].$$

Calculating the inverse DTFT, the output of the LTID system is given by

$$y[k] = \frac{1}{2}|H(\Omega_0)|\left[e^{j(\Omega_0 k + <H(\Omega_0))} + e^{-j(\Omega_0 k + <H(\Omega_0))}\right] \tag{11.60a}$$

or

$$y[k] = |H(\Omega_0)| \cos(\Omega_0 k + <H(\Omega_0)). \tag{11.60b}$$

Figure 11.17 is a schematic diagram of the gain and phase changes in the sinusoidal input caused by an LTID system. Computed at the fundamental frequency
$\Omega = \Omega_0$ of the sinusoidal input, the magnitude $|H(\Omega)|$ of the transfer function
determines the gain introduced by the LTID system, while the phase $<H(\Omega)$ at
$\Omega = \Omega_0$ determines the phase change in the applied sinusoidal sequence. The
magnitude $|H(\Omega)|$ and phase $<H(\Omega)$ are therefore also referred to as the gain
and phase responses of the LTID system.

Example 11.19

Plot the magnitude and phase spectra of the LTID system specified in Example 11.18.

Solution

From Example 11.18, the transfer function of the LTID system is given by

$$H(\Omega) = \frac{2}{1 - \frac{3}{4}e^{-j\Omega} + \frac{1}{8}e^{-j2\Omega}}.$$

Using Euler's formula $\exp(j\Omega) = \cos\Omega + j\sin\Omega$ and similarly for $\exp(j2\Omega)$, yields

$$H(\Omega) = \frac{2}{1 - \frac{3}{4}\cos\Omega + \frac{1}{8}\cos(2\Omega) + j\left[\frac{3}{4}\sin\Omega - \frac{1}{8}\sin(2\Omega)\right]},$$

which leads to the following expressions for the magnitude and phase responses:

$$|H(\Omega)| = \frac{2}{\sqrt{\left[1 - \frac{3}{4}\cos\Omega + \frac{1}{8}\cos(2\Omega)\right]^2 + \left[\frac{3}{4}\sin\Omega - \frac{1}{8}\sin(2\Omega)\right]^2}}$$

$$= \frac{2}{\sqrt{\frac{101}{64} - \frac{27}{16}\cos\Omega + \frac{1}{4}\cos(2\Omega)}};$$

$$<H(\Omega) = <2- <\left\{1 - \frac{3}{4}\cos\Omega + \frac{1}{8}\cos(2\Omega) + j\left[\frac{3}{4}\sin\Omega - \frac{1}{8}\sin(2\Omega)\right]\right\}$$

$$= -\tan^{-1}\left\{\frac{\frac{3}{4}\sin\Omega - \frac{1}{8}\sin(2\Omega)}{1 - \frac{3}{4}\cos\Omega + \frac{1}{8}\cos(2\Omega)}\right\}.$$

Figures 11.18(a) and (b) plot the magnitude and phase spectra in the frequency range $\Omega = [-\pi, \pi]$. Because the DTFT is periodic with period $\Omega_0 = 2\pi$, the magnitude and phase spectra at other frequencies can be calculated using the periodicity property. It is observed that the gain $|H(\Omega)|$ of the LTID system has the maximum value of 16/3 at frequency $\Omega = 0$. The gain $|H(\Omega)|$ at $\Omega = 0$ is also referred to as the dc component of the impulse response $h[k]$, and is the sum $\sum h[k]$ over the duration of the impulse response. As the frequency increases to π (or decreases to $-\pi$), the gain decreases monotonically and has a minimum value of 16/15 at $\Omega = \pm\pi$ radians/s. For LTID systems, the frequency $\Omega = \pm\pi$ radians/s corresponds to the maximum frequency. The transfer function $H(\Omega)$ represents a non-uniform amplifier as the lower-frequency components are amplified at a relatively higher scale than the high-frequency components.

The phase response $<H(\Omega)$ of the LTID system has a value of zero at $\Omega = 0$. As the frequency increases from zero, the phase decreases to its minimum

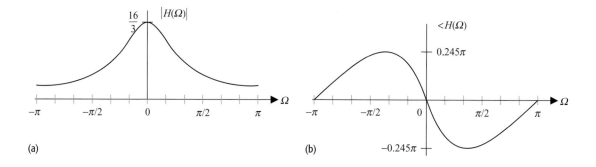

Fig. 11.18. (a) Magnitude
spectrum and (b) phase
spectrum of the LTID system
considered in Example 11.19. The
responses are shown in the
frequency range $\Omega = [-\pi, \pi]$.

value of -0.245π radians at $\Omega = 0.37\pi$ radians/s. From $\Omega = 0.37\pi$ radians/s to $\Omega = \pi$ radians/s, the phase increases and approaches zero at $\Omega = \pi$ radians/s. For negative frequencies, the phase increases to its maximum value of 0.245π radians at $\Omega = -0.37\pi$ radians/s, after which the phase decreases and approaches zero at $\Omega = -\pi$ radians/s.

It is also observed that the transfer function $H(\Omega)$ satisfies the Hermitian symmetry property stated in Eq. (11.39a). Since the impulse response $h[k]$ is a real-valued function, the magnitude spectrum $|H(\Omega)|$ is an even function of Ω and is therefore symmetric about the y-axis in Fig. 11.18(a). On the other hand, the phase spectrum $<H(\Omega)$ is an odd function of Ω and is therefore symmetric about the origin in Fig. 11.18(b). In cases where the impulse response $h[k]$ is a real-valued function, the plots in the range $\Omega = [0, \pi]$ are sufficient to represent the frequency response completely. The frequency response within the range $\Omega = [-\pi, 0]$ can then be obtained using the Hermitian symmetry property.

Example 11.20

Derive and plot the frequency responses of the LTID systems with the following impulse responses:

(i) $\quad h[k] = \dfrac{\sin(\pi k/6)}{\pi k};$ $\qquad\qquad\qquad\qquad\qquad\qquad\qquad$ (11.61)

(ii) $\quad g[k] = \delta[k] - \dfrac{\sin(\pi k/6)}{\pi k}.$ $\qquad\qquad\qquad\qquad\qquad$ (11.62)

Solution

(i) We express $h[k]$ as a sinc function as $h[k] = \dfrac{1}{6}\dfrac{\sin(\pi k/6)}{\pi k/6} = \dfrac{1}{6}\operatorname{sinc}(k/6)$.

Using Table 11.2, the transfer function is given by

$$H(\Omega) = \begin{cases} 1 & |\Omega| \leq \pi/6 \\ 0 & \pi/6 < |\Omega| \leq \pi. \end{cases} \qquad\qquad (11.63)$$

The impulse response $h[k]$ and its magnitude spectrum $|H(\Omega)|$ are plotted in Figs. 11.19(a) and (b) within the frequency range $\Omega = [0, \pi]$. Since the transfer function $H(\Omega)$ is real-valued, the phase spectrum is zero. The transfer function, Eq. (11.63), or equivalently the impulse response, Eq. (11.61), represents an

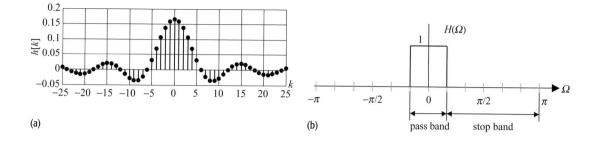

(a)   (b)   pass band   stop band

Fig. 11.19. (a) Impulse response $h[k]$ and (b) magnitude spectrum $|H(\Omega)|$ of an ideal lowpass filter specified in Example 11.20(i). The phase response is zero for all frequencies.

ideal lowpass filter since the low-frequency components in the input sequence, which lie within the range $0 \leq \Omega \leq \pi/6$, are passed through the system without attenuation. On the other hand, the higher-frequency components within the range $\pi/6 \leq \Omega \leq \pi$ are completely blocked. Lowpass filters are widely used in digital signal processing and will be considered in more detail in Chapter 14.

(ii) Expressing the impulse response $g[k]$ in terms of the impulse response $h[k]$ given in part (i), we obtain

$$g[k] = \delta[k] - h[k].$$

Using the linearity property, the transfer function of $g[k]$ is given by

$$G(\Omega) = 1 - H(\Omega).$$

Substituting the value of $H(\Omega)$ from Eq. (11.63) yields

$$G(\Omega) = \begin{cases} 0 & |\Omega| \leq \pi/6 \\ 1 & \pi/6 < |\Omega| \leq \pi. \end{cases} \tag{11.64}$$

The impulse response $g[k]$ and its magnitude spectrum $|G(\Omega)|$ are plotted in Figs. 11.20(a) and (b). It is observed that the low-frequency components within the range $0 \leq \Omega \leq \pi/6$ are completely blocked from the output, while the high-frequency components within the range $\pi/6 < \Omega \leq \pi$ are passed through the system without any attenuation. Such a system is referred to as an ideal highpass filter. Like lowpass filters, the highpass filters are also widely used in digital signal processing, and will be considered in more detail in Chapter 14.

In the previous example, we considered calculating the magnitude and phase spectra of an LTID system. The following example illustrates how the spectra may be used to calculate the output of an LTID system for elementary sinusoidal sequences.

Example 11.21

A continuous-time audio signal $x(t) = 3\,\cos(1000\pi t) + 5\,\cos(2000\pi t)$ is sampled at a sampling rate of 8000 samples/s to produce the DT sequence $x[k]$. Calculate the output signals if the DT signal $x[k]$ is applied at the input of an

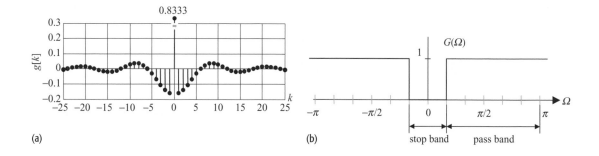

(a)   (b)
                                        stop band   pass band

Fig. 11.20. (a) Impulse response g[k] and (b) magnitude spectrum $|G(\Omega)|$ of an ideal highpass filter specified in Example 11.20(ii). The phase response is zero for all frequencies.

LTID systems with the following transfer functions:

(i) $H_1(\Omega) = \dfrac{2}{1 - \dfrac{3}{4}e^{-j\Omega} + \dfrac{1}{8}e^{-j2\Omega}}$;   (11.65)

(ii) $H_2(\Omega) = \begin{cases} 1 & |\Omega| \le \dfrac{\pi}{6} \\ 0 & \dfrac{\pi}{6} < |\Omega| \le \pi, \end{cases}$ (11.66)

(iii) $H_3(\Omega) = \begin{cases} 0 & |\Omega| \le \dfrac{\pi}{6} \\ 1 & \dfrac{\pi}{6} < |\Omega| \le \pi. \end{cases}$ (11.67)

Solution

The DT sequence $x[k]$ is given by

$$x[k] = x(kT_s) = 3\cos(1000\pi kT_s) + 5\cos(2000\pi kT_s).$$

Substituting $T_s = 1/8000$, we obtain

$$x[k] = 3\cos\left(\frac{\pi k}{8}\right) + 5\cos\left(\frac{\pi k}{4}\right),$$

which implies that $x[k]$ consist of two frequency components, $\Omega_1 = \pi/8$ and $\Omega_2 = \pi/4$. This is also apparent from the DTFT of $x[k]$, given by

$$X(\Omega) = 3\pi\left[\delta\left(\Omega - \frac{\pi}{8}\right) + \delta\left(\Omega + \frac{\pi}{8}\right)\right] + 5\pi\left[\delta\left(\Omega - \frac{\pi}{4}\right) + \delta\left(\Omega + \frac{\pi}{4}\right)\right],$$

which consists of impulses at frequencies $\Omega_1 = \pm\pi/8$ and $\Omega_2 = \pm\pi/4$.

As the DTFT is 2π-periodic, in the above equation we showed $X(\Omega)$ only in the frequency range $-\pi \le \Omega \le \pi$. This simplifies the analysis, and hence we will use the same approach to express the DTFTs in the following.

If the transfer function of an LTID system is $H(\Omega)$, the DTFT $Y(\Omega)$ of the output sequence is given by

$$\begin{aligned} Y(\Omega) &= H(\Omega)X(\Omega) \\ &= H(\Omega)\, 3\pi\left[\delta\left(\Omega - \frac{\pi}{8}\right) + \delta\left(\Omega + \frac{\pi}{8}\right)\right] + 5\pi\left[\delta\left(\Omega - \frac{\pi}{4}\right) + \delta\left(\Omega + \frac{\pi}{4}\right)\right] \\ &= 3\pi\left[\delta\left(\Omega - \frac{\pi}{8}\right)H\left(\frac{\pi}{8}\right) + \delta\left(\Omega + \frac{\pi}{8}\right)H\left(-\frac{\pi}{8}\right)\right] \\ &\quad + 5\pi\left[\delta\left(\Omega - \frac{\pi}{4}\right)H\left(\frac{\pi}{4}\right) + \delta\left(\Omega + \frac{\pi}{4}\right)H\left(-\frac{\pi}{4}\right)\right]. \end{aligned}$$

The DTFT $Y(\Omega)$ is obtained by substituting the values of the transfer function $H(\Omega)$ at frequencies $\Omega_1 = \pm\pi/8$ and $\Omega_2 = \pm\pi/4$.

(i) For the transfer function in Eq. (11.65), the values of $H_1(\Omega)$ are given by

$\Omega = \pi/8$	$H_1(\pi/8) = 4.04 - \text{j}2.03,$	$\|H_1(\pi/8)\| = 4.52,$	$<H_1(\pi/8) = -0.465$ radians;
$\Omega = -\pi/8$	$H_1(-\pi/8) = 4.04 + \text{j}2.03,$	$\|H_1(-\pi/8)\| = 4.52,$	$<H_1(-\pi/8) = 0.465$ radians;
$\Omega = \pi/4$	$H_1(\pi/4) = 2.44 - \text{j}2.11,$	$\|H_1(\pi/4)\| = 3.22,$	$<H_1(\pi/4) = -0.71$ radians;
$\Omega = -\pi/4$	$H_1(-\pi/4) = 2.44 + \text{j}2.11,$	$\|H_1(-\pi/4)\| = 3.22,$	$<H_1(-\pi/4) = 0.71$ radians.

The DTFT $Y_1(\Omega)$ of the output sequence is therefore given by

$$
\begin{aligned}
Y_1(\Omega) = 3\pi &\left[\delta\left(\Omega - \frac{\pi}{8}\right) 4.52 e^{-\text{j}0.465} + \delta\left(\Omega + \frac{\pi}{8}\right) 4.52 e^{\text{j}0.465} \right] \\
+ \pi 5 &\left[\delta\left(\Omega - \frac{\pi}{4}\right) 3.22 e^{-\text{j}0.71} + \delta\left(\Omega + \frac{\pi}{4}\right) 3.22 e^{\text{j}0.71} \right] \\
= 13.56\pi &\left[\delta\left(\Omega - \frac{\pi}{8}\right) e^{-\text{j}0.465} + \delta\left(\Omega + \frac{\pi}{8}\right) e^{\text{j}0.465} \right] \\
+ 16.10\pi &\left[\delta\left(\Omega - \frac{\pi}{4}\right) e^{-\text{j}0.71} + \delta\left(\Omega + \frac{\pi}{4}\right) e^{\text{j}0.71} \right].
\end{aligned}
$$

Calculating the inverse DTFT, the output sequence is obtained as

$$
y_1[k] = 13.56 \cos\left(\frac{\pi}{8}k - 0.465\right) + 16.10 \cos\left(\frac{\pi}{4}k - 0.71\right),
$$

where we have expressed the constant phase in radians. Expressing the constant phase in degrees yields

$$
y_1[k] = 13.56 \cos\left(\frac{\pi}{8}k - 26.67°\right) + 16.10 \cos\left(\frac{\pi}{4}k - 40.80°\right).
$$

The LTID system $H_1(\Omega)$ acts like an amplifier as the sinusoidal component $3\cos(\pi k/8)$ with fundamental frequency $\Omega_1 = \pi/8$ is amplified by a factor of 4.52, while the sinusoidal component $5\cos(\pi k/4)$ with fundamental frequency $\Omega_2 = \pi/4$ is amplified by a factor of 3.22. The difference in the gains is also apparent in the magnitude spectrum plotted in Fig. 11.18, where the low-frequency components have a higher amplification factor than that of the higher-frequency components.

(ii) For the transfer function in Eq. (11.66), the values of the transfer function $H_2(\Omega)$ at frequencies $\Omega_1 = \pm\pi/8$ and $\Omega_2 = \pm\pi/4$ are given by

$\Omega = \pi/8$	$H_2(\pi/8) = 1,$	$\|H_2(\pi/8)\| = 1,$	$<H_2(\pi/8) = 0$ radians;
$\Omega = -\pi/8$	$H_2(-\pi/8) = 1,$	$\|H_2(-\pi/8)\| = 1,$	$<H_2(-\pi/8) = 0$ radians;
$\Omega = \pi/4$	$H_2(\pi/4) = 0,$	$\|H_2(\pi/4)\| = 0,$	$<H_2(\pi/4) = 0$ radians;
$\Omega = -\pi/4$	$H_2(-\pi/4) = 0,$	$\|H_2(-\pi/4)\| = 0,$	$<H_2(-\pi/4) = 0$ radians.

The DTFT $Y_2(\Omega)$ of the output sequence is therefore given by

$$Y_2(\Omega) = 3\pi \left[\delta\left(\Omega - \frac{\pi}{8}\right) \cdot 1 + \delta\left(\Omega + \frac{\pi}{8}\right) \cdot 1 \right]$$
$$+ 5\pi \left[\delta\left(\Omega - \frac{\pi}{4}\right) \cdot 0 + \delta\left(\Omega + \frac{\pi}{4}\right) \cdot 0 \right]$$
$$= 3\pi \left[\delta\left(\Omega - \frac{\pi}{8}\right) + \delta\left(\Omega + \frac{\pi}{8}\right) \right].$$

Calculating the inverse DTFT, the output sequence is obtained as

$$y_2[k] = 3\cos\left(\frac{\pi}{8}k\right).$$

The LTID system $H_2(\Omega)$ acts like an ideal lowpass filter as the sinusoidal component $3\cos(\pi k/8)$ with low fundamental frequency $\Omega_1 = \pi/8$ is not attenuated, while the sinusoidal component $5\cos(\pi k/4)$ with high fundamental frequency $\Omega_2 = \pi/4$ is blocked from the output.

(iii) For the transfer function in Eq. (11.67), the values of the transfer function $H_3(\Omega)$ at frequencies $\Omega_1 = \pm\pi/8$ and $\Omega_2 = \pm\pi/4$ are given by

$$\Omega = \pi/8, \qquad H_3(\pi/8) = 0, \qquad |H_3(\pi/8)| = 0, \quad <H_3(\pi/8) = 0 \text{ radians;}$$
$$\Omega = -\pi/8, \quad H_3(-\pi/8) = 0, \quad |H_3(-\pi/8)| = 0, \quad <H_3(-\pi/8) = 0 \text{ radians;}$$
$$\Omega = \pi/4, \qquad H_3(\pi/4) = 1, \qquad |H_3(\pi/4)| = 1, \quad <H_3(\pi/4) = 0 \text{ radians;}$$
$$\Omega = -\pi/4, \quad H_3(-\pi/4) = 1, \quad |H_3(-\pi/4)| = 1, \quad <H_3(-\pi/4) = 0 \text{ radians.}$$

The DTFT $Y_3(\Omega)$ of the output sequence is therefore given by

$$Y_3(\Omega) = 3\pi \left[\delta\left(\Omega - \frac{\pi}{8}\right) \cdot 0 + \delta\left(\Omega + \frac{\pi}{8}\right) \cdot 0 \right]$$
$$+ 5\pi \left[\delta\left(\Omega - \frac{\pi}{4}\right) \cdot 1 + \delta\left(\Omega + \frac{\pi}{4}\right) \cdot 1 \right]$$
$$= 5\pi \left[\delta\left(\Omega - \frac{\pi}{4}\right) + \delta\left(\Omega - \frac{\pi}{4}\right) \right].$$

Calculating the inverse DTFT, the output sequence is obtained as

$$y_3[k] = 5\cos\left(\frac{\pi}{4}k\right).$$

The LTID system $H_3(\Omega)$ acts like an ideal highpass filter as the sinusoidal component $3\cos(\pi k/8)$ with lower fundamental frequency $\Omega_1 = \pi/8$ is blocked, while the sinusoidal component $5\cos(\pi k/4)$ with higher fundamental frequency $\Omega_2 = \pi/4$ is unattenuated in the output sequence.

11.8 Continuous- and discrete-time Fourier transforms

In Chapters 4, 5, and 11, we derived frequency representations for CT and DT waveforms. In particular, we considered the following four frequency representations:

(1) CTFT for CT periodic signals;
(2) CTFT for CT aperiodic signals;
(3) DTFT for DT periodic sequences;
(4) DTFT for DT aperiodic sequences.

In this section, we compare the Fourier transforms for different types of signals.

The CT periodic signals are typically represented by the CTFS,

$$\tilde{x}(t) = \sum_{n=-\infty}^{\infty} D_n e^{jn\omega_0 t},$$

where D_n denotes the CTFS coefficients and ω_0 is the fundamental frequency of the CT periodic signal. By exploiting the CTFT pair

$$e^{jn\omega_0 t} \xleftrightarrow{\text{CTFT}} 2\pi \delta(\omega - n\omega_0),$$

the CTFT for the CT periodic signals is given by

$$\tilde{x}(t) \xleftrightarrow{\text{CTFT}} X(\omega) = 2\pi \sum_{n=-\infty}^{\infty} D_n \delta(\omega - n\omega_0)$$

and consists of a train of time-shifted impulse functions. In other words, the CTFT of CT periodic signals is discrete in nature.

For CT aperiodic signals, the CTFT $X(\omega)$ is given by

$$x(t) \xleftrightarrow{\text{CTFT}} X(\omega) = \int_{-\infty}^{\infty} x(t) e^{-j\omega t} \, dt,$$

which is generally aperiodic and continuous in the frequency domain.

Similar to the CT periodic signal, the frequency representation for a DT periodic sequence is obtained by using the following DTFS:

$$\tilde{x}[k] = \sum_{n=\langle K_0 \rangle} D_n e^{jn\Omega_0 k},$$

where D_n denotes the DTFS coefficients and Ω_0 is the fundamental frequency of the DT periodic signal. We observed that the DTFS is periodic with period $K_0 = 2\pi/\Omega_0$ such that

$$D_n = D_{n+mK_0}$$

for $-\infty < m < \infty$. By exploiting the DTFT pair

$$e^{jn\Omega_0 k} \xleftrightarrow{\text{DTFT}} 2\pi \delta(\Omega - n\Omega_0),$$

the DTFT for a DT periodic sequence is given by

$$\tilde{x}[k] \xleftrightarrow{\text{DTFT}} X(\Omega) = 2\pi \sum_{n=-\infty}^{\infty} D_n \delta(\Omega - n\Omega_0).$$

We showed that the DTFT of a DT periodic sequence is discrete as it consists of several discrete time-shifted impulse functions. In addition, the DTFT of a DT periodic sequence is itself periodic in the frequency domain, with a fundamental period $\Omega_0 = 2\pi$.

Finally, the DTFT of a DT aperiodic sequence is given by

$$X(\Omega) = \sum_{k=-\infty}^{\infty} x[k] e^{-j\Omega k}.$$

Table 11.7. Fourier transforms for different types of waveforms

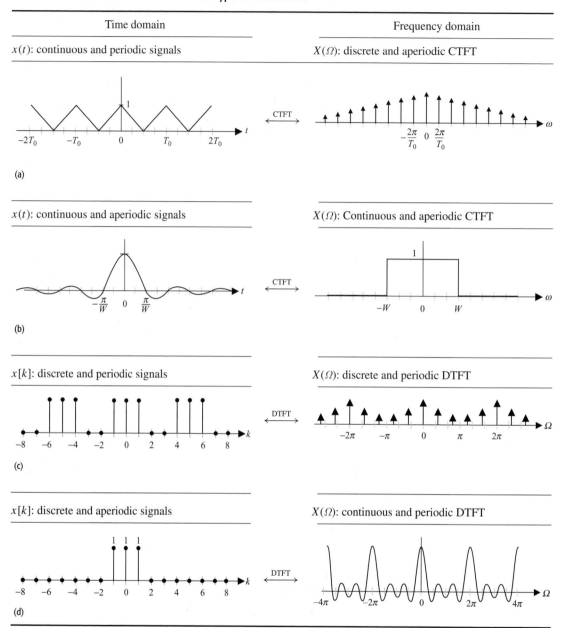

(a)

(b)

(c)

(d)

We observed that the DTFT of a DT aperiodic sequence is continuous as it is defined for all frequencies Ω. Like the DTFT of a DT periodic sequence, the DTFT of a DT aperiodic sequence is periodic in the frequency domain, with a fundamental period $\Omega_0 = 2\pi$.

The aforementioned discussion on the four types of Fourier transforms is summarized in Table 11.7, where we observe that periodicity in the time domain corresponds to discreteness in the frequency domain. The CTFT for the CT periodic signals, illustrated in row (a) of Table 11.7, and the DTFT for the DT periodic signals, illustrated in row (c), are both discrete in the frequency domain. The converse of the observation is also true, as discreteness in the time domain corresponds to periodicity in the frequency domain. The converse statement is illustrated in rows (c) and (d), where periodic and aperiodic DT sequences are considered. The DTFT for both the periodic and aperiodic DT sequences is periodic with period $\Omega_0 = 2\pi$.

When a signal is both discrete and periodic in the time domain, such as the DT periodic sequence illustrated in row (c) of Table 11.7, the DTFT is also both periodic and discrete in the frequency domain. This observation is exploited in digital signal processing. To compute the DTFT on digital computers, it is always assumed that the waveform is discrete and periodic, even when the original waveform is neither discrete nor periodic. Chapter 12 presents the theory of the discrete Fourier transform (DFT), which is a very powerful tool for computing the CTFT and DTFT.

11.9 Summary

In this chapter, we presented the frequency representation for DT sequences. For periodic sequences, we derived the DTFS, which is defined as

$$\tilde{x}[k] = \sum_{n=\langle K_0 \rangle} D_n e^{jn\Omega_0 k},$$

where Ω_0 is the fundamental frequency, given by $\Omega_0 = 2\pi/K_0$, and the discrete-time Fourier series (DTFS) coefficients D_n, for $1 \leq n \leq K_0$, are given by

$$D_n = \frac{1}{K_0} \sum_{k=\langle K_0 \rangle} x[k] e^{-jn\Omega_0 k}.$$

The DTFS coefficients of periodic sequences are themselves periodic with a period K_0 such that

$$D_n = D_{n+mK_0} \quad \text{for } m \in Z.$$

Section 11.2 derived the DTFT for an aperiodic sequence $x[k]$ as follows:

DTFT synthesis equation $\qquad x[k] = \dfrac{1}{2\pi} \int\limits_{(2\pi)} X(\Omega) e^{jk\Omega} \mathrm{d}\Omega;$

DTFT analysis equation $\quad X(\Omega) = \displaystyle\sum_{k=-\infty}^{\infty} x[k] e^{-j\Omega k},$

and showed that the DTFT is periodic in the frequency domain with a period $\Omega_0 = 2\pi$. As such, the frequencies $\Omega = 0, 2\pi, 4\pi, \ldots$ are considered as the same frequencies and are referred to as the lowest possible frequency for the DTFT. Similarly, the frequencies $\Omega = \pi, 3\pi, 5\pi, \ldots$ are the same and are referred to as the highest possible frequency for the DTFT.

Section 11.3 derived a sufficient condition for the existence of the DTFT for aperiodic DT sequences as follows:

$$\sum_{k=-\infty}^{\infty} |x_2[k]| < \infty.$$

The periodic DT sequences do not satisfy the above condition for the existence of the DTFT. Instead the DTFT of a periodic sequence is obtained by calculating the DTFT of its DTFS representation, which results in the following DTFT:

$$X(\Omega) = 2\pi \sum_{n=-\infty}^{\infty} D_n \delta\left(\Omega - \frac{2n\pi}{K_0}\right),$$

where D_n are the DTFS coefficients of the periodic sequence $x[k]$.

Section 11.4 covered the properties of the DTFT. In particular, we covered the following properties.

(1) The periodicity property states that the DTFT of any DT sequence is periodic with period 2π.
(2) The Hermitian symmetry property states that the DTFT of a real-valued sequence is Hermitian. In other words, the real component of the DTFT of a real-valued sequence is even, while the imaginary component is odd.
(3) The linearity property states that the overall DTFT of a linear combination of DT sequences is given by the same linear combination of the individual DTFTs.
(4) The time-scaling property is only applicable for time-expanded (or interpolated) sequences. It states that interpolating a sequence in the time domain compresses its DTFT in the frequency domain.
(5) The time-shifting property states that shifting a sequence in the time domain towards the right-hand side by an integer constant m is equivalent to multiplying the DTFT of the original sequence by a complex exponential $\exp(-j\Omega m)$. Similarly, shifting towards the left-hand side by

integer m is equivalent to multiplying the DTFT of the original sequence by a complex exponential $\exp(j\Omega m)$.

(6) The frequency-shifting property is the converse of the time-shifting property. It states that shifting the DTFT in the frequency domain towards the right-hand side by Ω_0 is equivalent to multiplying the original sequence by a complex exponential $\exp(j\Omega_0 m)$. Similarly, shifting the DTFT towards the left-hand side by Ω_0 is equivalent to multiplying the DTFT of the original sequence by a complex exponential $\exp(-j\Omega_0 m)$.

(7) The frequency-differentiation property states that differentiating the DTFT with respect to the frequency Ω is equivalent to multiplying the original sequence by a factor of $-jk$.

(8) Time differencing is defined as the difference between the original sequence and its time-shifted version with a shift of one sample towards the right-hand side. The time-differencing property states that time differencing a signal in the time domain is equivalent to multiplying its DTFT by a factor of $(1 - \exp(-j\Omega m))$.

(9) The time-summation property is the converse of the time-differencing property. The time-summation property states that the DTFT of the running sum of a sequence is obtained by dividing the DTFT of the original sequence by a factor of $(1 - \exp(-j\Omega m))$ and adding impulses located at multiples of 2π.

(10) The time-convolution property states that the convolution of two DT sequences is equivalent to the multiplication of the DTFTs of the two sequences in the time domain.

(11) Periodic convolution is an extension of time convolution to periodic sequences, where only single periods of the two periodic sequences are convolved. The periodic-convolution property states that the periodic convolution in the time domain is equivalent to multiplying the DTFS coefficients of the two periodic sequences by each other in the frequency domain.

(12) The frequency-convolution property states that periodic convolution of two DTFTs with period 2π is equivalent to multiplication of their sequences in the time domain.

The DTFT of the impulse response of an LTID system is referred to as the Fourier transfer function, which is generally complex-valued. The plot of the magnitude of the Fourier transfer function with respect to frequency Ω is referred to as the magnitude spectrum, while the plot of the phase of the Fourier transfer function with respect to frequency Ω is referred to as the phase spectrum. Sections 11.6 and 11.7 illustrated how the magnitude and phase spectra provide meaningful insights into the analysis of the LTID systems. In particular, we covered the ideal lowpass filter, which blocks all frequency components above a certain cut-off frequency $\Omega > \Omega_c$ in the applied input sequence. All frequency components $\Omega \leq \Omega_c$ are left unattenuated in the output response of an ideal lowpass filter.

The magnitude spectrum of an ideal lowpass filter is unity within its pass band $(\Omega \le \Omega_c)$ and zero within its stop band $(\Omega_c < \Omega \le \pi)$.

The converse of the ideal lowpass filter is the ideal highpass filter, which blocks all frequency components below a certain cut-off frequency $\Omega < \Omega_c$ in the applied input sequence. All frequency components $\Omega \ge \Omega_c$ are left unattenuated in the output response of an ideal highpass filter. The magnitude spectrum of an ideal highpass filter is unity within the pass band $(\Omega_c \le \Omega \le \pi)$ and zero within the stop band $(0 \le \Omega < \Omega_c)$.

Section 11.8 compared the Fourier representations of CT and DT periodic and aperiodic waveforms. We showed that the Fourier representations of periodic waveforms are discrete, whereas the Fourier representations of discrete waveforms are periodic.

Problems

11.1 Determine the DTFS representation for each of the following DT periodic sequences. In each case, plot the magnitude and phase of the DTFS coefficients.

(i) $x[k] = k$ for $0 \le k \le 5$ and $x[k+6] = x[k]$;

(ii) $x[k] = \begin{cases} 1 & (0 \le k \le 2) \\ 0.5 & (3 \le k \le 5) \\ 0 & (6 \le k \le 8) \end{cases}$ and $x[k+9] = x[k]$;

(iii) $x[k] = 3\sin\left(\dfrac{2\pi}{7}k + \dfrac{\pi}{4}\right)$;

(iv) $x[k] = 2e^{j\left(\frac{5\pi}{3}k + \frac{\pi}{4}\right)}$;

(v) $x[k] = \displaystyle\sum_{m=-\infty}^{\infty} \delta(k - 5m)$;

(vi) $x[k] = \cos(10\pi k/3)\cos(2\pi k/5)$;

(vii) $x[k] = |\cos(2\pi k/3)|$.

11.2 Given the following DTFS coefficients, determine the DT periodic sequence in the time domain:

(i) $D_n = \begin{cases} 1 & (0 \le k \le 2) \\ 0.5 & (3 \le k \le 5) \\ 0 & (6 \le k \le 8) \end{cases}$ and $D_{n+9} = D_n$;

(ii) $D_n = \begin{cases} 1 - j0.5 & (n = -1) \\ 1 & (n = 0) \\ 1 + j0.5 & (n = 1) \\ 0 & (2 \le n \le 5) \end{cases}$ and $D_{n+7} = D_n$;

(iii) $D_n = 1 + \dfrac{3}{4}\sin\left(\dfrac{\pi n}{8}\right)$ $(0 \le n \le 6)$ and $D_{n+7} = D_n$;

(iv) $D_n = (-1)^n$ $(0 \le n \le 7)$ and $D_{n+8} = D_n$;

(v) $D_n = e^{jn\pi/4}$ $(0 \le n \le 7)$ and $D_{n+8} = D_n$.

11.3 Determine if the following DT sequences satisfy the DTFT existence
property:
  (i) $x[k] - 2$;
  (ii) $x[k] = \begin{cases} 3 - |k| & |k| < 3 \\ 0 & \text{otherwise}; \end{cases}$
  (iii) $x[k] = k3^{-|k|}$;
  (iv) $x[k] = \alpha^k \cos(\omega_0 k) u[k]$, $|\alpha| < 1$;
  (v) $x[k] = \alpha^k \sin(\omega_0 k + \phi) u[k]$, $|\alpha| < 1$;
  (vi) $x[k] = \dfrac{\sin(\pi k/5) \sin(\pi k/7)}{\pi^2 k^2}$;
  (vii) $x[k] = \displaystyle\sum_{m=-\infty}^{\infty} \delta(k - 5m - 3)$;
  (viii) $x[k] = \begin{cases} 3 - |k| & |k| < 3 \\ 0 & |k| = 3 \end{cases}$ and $x[k + 7] = x[k]$;
  (ix) $x[k] = e^{j(0.2\pi k + 45°)}$;
  (x) $x[k] = k3^{-k} u[k] + e^{j(0.2\pi k + 45°)}$.

11.4 (a) Calculate the DTFT of the DT sequences specified in Problem 11.3.
  (b) Calculate the DTFT of the periodic DT sequences specified in
      Problem 11.1.

11.5 Given the following transform pair:

$$x_1[k] \xleftrightarrow{\text{DTFT}} X_1(\Omega) \text{ and } x_2[k] \xleftrightarrow{\text{DTFT}} X_2(\Omega),$$

express the DTFT of the following DT sequences in terms of the DTFTs
$X_1(\Omega)$ and $X_2(\Omega)$:
  (i) $(-1)^k x_1[k]$;
  (ii) $(k - 5)^2 x_2[k - 4]$;
  (iii) $ke^{-j4k} x_1[3 - k]$;
  (iv) $\displaystyle\sum_{m=-\infty}^{\infty} [x_1[k - 4m] + x_2[k - 6m]]$;
  (v) $x_1[5 - k] x_2[7 - k]$.

11.6 Calculate the DT sequences with the following DTFT representations
defined over the frequency range $-\pi \leq \Omega \leq \pi$:
  (i) $X(\Omega) = \dfrac{4e^{-j\Omega}}{1 - 5e^{-j\Omega} + 6e^{-j2\Omega}}$;
  (ii) $X(\Omega) = \dfrac{2e^{-j2\Omega}}{(1 - 4e^{-j\Omega})^2 (1 - 2e^{-j\Omega})}$;
  (iii) $X(\Omega) = 8 \sin(7\Omega) \cos(9\Omega)$;
  (iv) $X(\Omega) = \dfrac{4e^{-j4\Omega}}{10 - 6\cos\Omega}$;
  (v) $X(\Omega) = \begin{cases} 1 & 0.25\pi \leq |\Omega| < 0.75\pi \\ 0 & |\Omega| \leq 0.25\pi \text{ and } 0.75\pi \leq |\Omega| < \pi. \end{cases}$

11.7 (a) Prove the Hermitian symmetry property, Eq. (11.39a), for a real-valued DT sequence.

(b) Problem 11.6 lists the DTFTs of several sequences. Applying the Hermitian property, determine whether these sequences are real-valued.

11.8 Prove the frequency-differentiation property of the DTFT.

11.9 Prove the time-convolution property of the DTFT.

11.10 Prove the time-shifting property of the DTFT.

11.11 Given the following transfer function:

$$H(\Omega) = \frac{1}{(1 - 0.3e^{-j\Omega})(1 - 0.5e^{-j\Omega})(1 - 0.7e^{-j\Omega})},$$

(i) determine the impulse response of the LTID system;
(ii) determine the difference equation representation of the LTID system;
(iii) determine the unit step response of the LTID system by using the time-convolution property of the DTFT;
(iv) determine the unit step response of the LTID system by convolving the unit step sequence with the impulse response obtained in part (i).

11.12 Given the following difference equation:

$$y[k] + y[k-1] + \frac{1}{4}y[k-2] = x[k] - x[k-2],$$

(i) determine the transfer function representing the LTID system;
(ii) determine the impulse response of the LTID system;
(iii) determine the output of the LTID system for the input $x[k] = (1/2)^k u[k]$ using the time-convolution property;
(iv) determine the output of the LTID system by convolving the input $x[k] = (1/2)^k u[k]$ with the impulse response obtained in part (ii).

11.13 Determine the output response of the LTID systems with the specified inputs and impulse responses using Fourier transform approach:
(i) $x[k] = u[k]$ and $h[k] = 4^{-|k|}$;
(ii) $x[k] = 2^{-k}u[k]$ and $h[k] = 2^k u[-k-1]$;
(iii) $x[k] = u[k] - u[k-9]$ and $h[k] = 3^k u[-k+4]$;
(iv) $x[k] = k5^{-k}u[k]$ and $h[k] = 5^k u[-k]$;
(v) $x[k] = u[k+2] - u[-k-3]$ and $h[k] = u[k-5] - u[-k-6]$.

11.14 Given that the transfer function of an LTID system is given by

$$H(\Omega) = \frac{1}{(1 + 3e^{-j\Omega})},$$

determine and sketch the following as a function of frequency Ω over the range $-\pi \leq \Omega \leq \pi$:

(i) Re$\{H(\Omega)\}$;   (iii) $|H(\Omega)|$;

(ii) Im$\{H(\Omega)\}$;   (iv) $<H(\Omega)$.

11.15 Calculate and plot the magnitude and phase response of the LTID systems specified in Problem 11.13.

11.16 The impulse response of an LTID system is given by

$$h[k] = 3\delta[k+3] - 2\delta[k+2] + \delta[k+1] + 5\delta[k]$$
$$- \delta[k-1] - 2\delta[k-2] - 3\delta[k-3] + 4\delta[k-4].$$

Without explicitly determining the transfer function $H(\Omega)$, evaluate the following using the properties of the DTFT:

(i) $H(\Omega)|_{\Omega=0}$;

(ii) $H(\Omega)|_{\Omega=\pi}$;

(iii) $<H(\Omega)$;

(iv) $\displaystyle\int_{-\pi}^{\pi} H(\Omega)\mathrm{d}\Omega$.

(v) Determine and sketch the DT sequence with the DTFT $H(-\Omega)$.

(vi) Determine and sketch the DT sequence with the DTFT Re$\{H(\Omega)\}$.

11.17 Using Parseval's theorem, determine the following sum:

$$\sum_{k=-\infty}^{\infty} \frac{\sin(\pi k/5)\sin(\pi k/7)}{k^2}.$$

11.18 Consider an LTID system with the following impulse response:

$$h[k] = \mathrm{sinc}(3k/4).$$

Determine the output responses of the LTID system for the following inputs:

(i) $x[k] = \cos(11\pi k/16)\cos(3\pi k/16)$;

(ii) $x[k] = k$ for $0 \leq k \leq 5$ and $x[k+6] = x[k]$;

(iii) $x[k] = \begin{cases} 1 & (0 \leq k \leq 2) \\ 0.5 & (3 \leq k \leq 5) \\ 0 & (6 \leq k \leq 8) \end{cases}$ and $x[k+9] = x[k]$;

(iv) $x[k] = \displaystyle\sum_{m=-\infty}^{\infty} \delta(k-5m)$;

(v) $x[k] = \mathrm{sinc}(k/3)$.

11.19 When the DT sequence

$$x[k] = 4^{-k}u[k] + 3^{-k}u[k]$$

is applied at the input of an LTID system, the output response is given by

$$y[k] = 2 \left(\frac{1}{4}\right)^k u[k] - 4 \left(\frac{3}{4}\right)^k u[k].$$

(i) Determine the Fourier transfer function $H(\Omega)$ of the LTID system.
(ii) Determine the impulse response $h[k]$ of the LTID system.
(iii) Determine the difference equation representing the LTID system.
(iv) Determine if the system is causal.

11.20 Repeat Example 11.21 for each of the following signals, assuming that the sampling rate to discretize the CT signals is 8000 samples/s:
(i) $x_1(t) = 2 + 3\cos(400\pi t) + 7\cos(800\pi t)$;
(ii) $x_2(t) = 2\cos(4000\pi t) + 5\cos(6000\pi t)$;
(iii) $x_3(t) = 5\cos(600\pi t) + 9\cos(900\pi t) + 2\cos(3000\pi t)$;
(iv) $x_4(t) = 4\cos(600\pi t) + 6\cos(12000\pi t)$.

11.21 Repeat Example 11.21 for each of the following signals, assuming that the sampling rate to discretize the CT signals is 22 000 samples/s:
(i) $x_1(t) = 2 + 3\cos(8000\pi t) + 7\cos(18\,000\pi t)$;
(ii) $x_2(t) = 2\cos(10\,000\pi t) + 5\cos(30\,000\pi t)$;
(iii) $x_3(t) = 5\cos(600\pi t) + 9\cos(900\pi t) + 2\cos(3000\pi t)$;
(iv) $x_4(t) = 4\cos(28\,000\pi t) + 6\cos(18\,000\pi t)$.

12 Discrete Fourier transform

In Chapter 11, we introduced the discrete-time Fourier transform (DTFT) that provides us with an alternative representation for DT sequences. The DTFT transforms a DT sequence $x[k]$ into a function $X(\Omega)$ in the DTFT frequency domain Ω. The independent variable Ω is continuous and is confined to the range $-\pi \leq \Omega < \pi$. With the increased use of digital computers and specialized hardware in digital signal processing (DSP), interest has focused around transforms that are suitable for digital computations. Because of the continuous nature of Ω, direct implementation of the DTFT is not suitable on such digital devices. This chapter introduces the discrete Fourier transform (DFT), which can be computed efficiently on digital computers and other DSP boards.

The DFT is an extension of the DTFT for time-limited sequences with an additional restriction that the frequency Ω is discretized to a finite set of values given by $\Omega = 2\pi r/M$, for $0 \leq r \leq (M-1)$. The number M of the frequency samples can have any value, but is typically set equal to the length N of the time-limited sequence $x[k]$. If M is chosen to be a power of 2, then it is possible to derive highly efficient implementations of the DFT. These implementations are collectively referred to as the fast Fourier transform (FFT) and, for an M-point DFT, have a computational complexity of $O(M\log_2 M)$. This chapter discusses a popular FFT implementation and extends the theoretical DTFT results derived in Chapter 11 to the DFT.

The organization of this chapter is as follows. Section 12.1 motivates the discussion of the DFT by expressing it as a special case of the continuous-time Fourier transform (CTFT). The formal definition of the DFT is presented in Section 12.2, including its matrix-vector representation. Section 12.3 applies the DFT to estimation of the spectra of both DT and CT signals. Section 12.4 derives important properties of the DFT, while Section 12.5 uses the DFT as a tool to convolve two DT sequences in the frequency domain. A fast implementation of the DFT based on the decimation-in-time algorithm is presented in Section 12.6. Finally, Section 12.7 concludes the chapter with a summary of the important concepts.

12.1 Continuous to discrete Fourier transform

In order to motivate the discussion of the DFT, let us assume that we are interested in computing the CTFT of a CT signal $x(t)$ using a digital computer. The three main steps involved in the digital computation of the CTFT are illustrated in Fig. 12.1. The waveforms for the CT signal $x(t)$ and its CTFT $X(\omega)$, shown in Figs. 12.1(a) and (b), are arbitrarily chosen, and hence the following procedure applies to any CT signal. A brief explanation of each of the three steps is provided below.

Step 1: Analog-to-digital conversion In order to store a CT signal into a digital computer, the CT signal is digitized. This is achieved through two processes known as sampling and quantization, collectively referred to as analog-to-digital (A/D) conversion by convention. In this discussion, we only consider sampling, ignoring the distortion introduced by quantization. The CT signal $x(t)$ is sampled by multiplying it by an impulse train:

$$s_1(t) = \sum_{m=-\infty}^{\infty} \delta(t - mT_1), \qquad (12.1)$$

illustrated in Fig. 12.1(c). The sampled waveform is given by

$$x_1(t) = x(t)s_1(t) = x(t) \times \sum_{m=-\infty}^{\infty} \delta(t - mT_1) \qquad (12.2)$$

and is shown in Fig. 12.1(e). Since multiplication in the time domain is equivalent to convolution in the frequency domain, the CTFT $X_1(\omega)$ of the sampled signal $x_1(t)$ is given by

$$X_1(\omega) = \Im\left[x(t) \times \sum_{m=-\infty}^{\infty} \delta(t - mT_1)\right] = \frac{1}{2\pi}\left[X(\omega) * \frac{2\pi}{T_1}\sum_{m=-\infty}^{\infty} \delta\left(\delta - \frac{2m\pi}{T_1}\right)\right]$$

$$= \frac{1}{T_1}\sum_{m=-\infty}^{\infty} X\left(\omega - \frac{2m\pi}{T_1}\right) \qquad (12.3)$$

The above result was also derived in Eq. (9.5) of Chapter 9, and is graphically illustrated in Figs. 12.1(b), (d), and (f), where we note that the spacing between adjacent replicas of $X(\omega)$ in $X_1(\omega)$ is given by $2\pi/T_1$. Since no restriction is imposed on the bandwidth of the CT signal $x(t)$, limited aliasing may also be introduced in $X_1(\omega)$.

To derive the discretized representation of $x(t)$ from Eq. (12.3), sampling is followed by an additional step (shown in Fig. 12.1(g)), where the CT impulses are converted to the DT impulses. Equation (12.3) can now be extended to

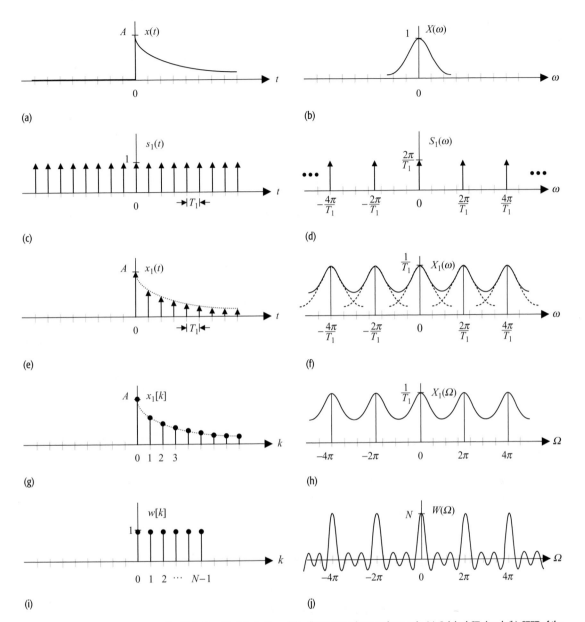

Fig. 12.1. Graphical derivation of the discrete Fourier transform pair. (a) Original CT signal. (b) CTFT of the original CT signal. (c) Impulse train sampling of CT signal. (d) CTFT of the impulse train in part (c). (e) CT sampled signal. (f) CTFT of the sampled signal in part (e). (g) DT representation of CT signal in part (a). (h) DTFT of the DT representation in part (g). (i) Rectangular windowing sequence. (j) DTFT of the rectangular window. (k) Time-limited sequence representing part (g). (l) DTFT of time-limited sequence in part (k). (m) Inverse DTFT of frequency-domain impulse train in part (n). (n) Frequency-domain impulse train. (o) Inverse DTFT of part (p). (p) DTFT representation of CT signal in part (a). (q) Inverse DFT of part (r). (r) DFT representation of CT signal in part (a).

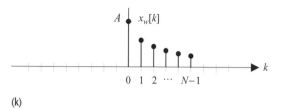

(k)

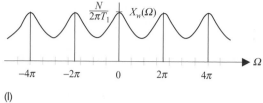

(l)

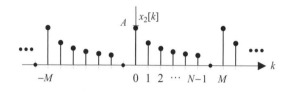

(m)

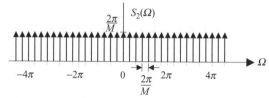

(n)

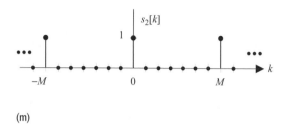

(o)

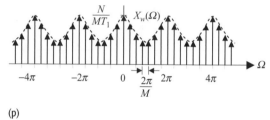

(p)

(q)

(r)

Fig. 12.1. (*cont.*)

derive the DTFT of the DT sequence $x_1[k]$ as follows:

$$x_1[k] = \sum_{m=-\infty}^{\infty} x(mT_1)\delta(t - mT_1). \tag{12.4}$$

Calculating the CTFT of both sides of Eq. (12.4) yields

$$X_1(\omega) = \sum_{m=-\infty}^{\infty} x(mT_1)e^{-j\omega mT_1}. \tag{12.5}$$

Substituting $x_1[m] = x(mT_1)$ and $\Omega = \omega T_1$ in Eq. (12.5) leads to

$$X_1(\Omega) = X_1(\omega)|_{\omega=\Omega/T_1} = \sum_{m=-\infty}^{\infty} x_1[m]e^{-jm\Omega},$$

which is the standard definition of the DTFT introduced in Chapter 11. The DTFT spectrum $X_1(\Omega)$ of $x_1[k]$ is obtained by changing the frequency axis ω of the CTFT spectrum $X_1(\omega)$ according to the relationship $\Omega = \omega T_1$. The DTFT spectrum $X_1(\Omega)$ is illustrated in Fig. 12.1(h).

Step 2: Time limitation The discretized signal $x_1[k]$ can possibly be of infinite length. Therefore, it is important to truncate the length of the discretized signal $x_1[k]$ to a finite number of samples. This is achieved by multiplying the discretized signal by a rectangular window,

$$w[k] = \begin{cases} 1 & 0 \le k \le (N-1) \\ 0 & \text{elsewhere,} \end{cases} \tag{12.6}$$

of length N. The DTFT $X_w(\Omega)$ of the time-limited signal $x_w[k] = x_1[k]w[k]$ is obtained by convolving the DTFT $X_1(\Omega)$ with the DTFT $W(\Omega)$ of the rectangular window, which is a sinc function. In terms of $X_1(\Omega)$, the DTFT $X_w(\Omega)$ of the time-limited signal is given by

$$X_w(\Omega) = \frac{1}{2\pi} \left[X_1(\Omega) \otimes \frac{\sin(0.5N\Omega)}{\sin(0.5\Omega)} e^{-j(N-1)/2} \right], \tag{12.7}$$

which is shown in Fig. 12.1(l) with its time-limited representation $x_w[k]$ plotted in Fig. 12.1(k). Symbol $\otimes$ in Eq. (12.7) denotes the circular convolution.

Step 3: Frequency sampling The DTFT $X_w(\Omega)$ of the time-limited signal $x_w[k]$ is a continuous function of Ω and must be discretized to be stored on a digital computer. This is achieved by multiplying $X_w(\Omega)$ by a frequency-domain impulse train, whose DTFT is given by

$$S_2(\Omega) = \frac{2\pi}{M} \sum_{m=-\infty}^{\infty} \delta\left(\Omega - \frac{2\pi m}{M}\right). \tag{12.8}$$

The discretized version of the DTFT $X_w(\Omega)$ is therefore expressed as follows:

$$X_2(\Omega) = X_w(\Omega)S_2(\Omega) = \frac{1}{M} \left[X_1(\Omega) \otimes \frac{\sin(0.5N\Omega)}{\sin(0.5\Omega)} e^{-j(N-1)/2} \right]$$

$$\times \sum_{m=-\infty}^{\infty} \delta\left(\Omega - \frac{2\pi m}{M}\right). \tag{12.9}$$

The DTFT $X_2(\Omega)$ is shown in Fig. 12.1(p), where the number M of frequency samples within one period ($-\pi \le \Omega \le \pi$) of $X_2(\Omega)$ depends upon the fundamental frequency $\Omega_2 = 2\pi/M$ of the impulse train $S_2(\Omega)$. Taking the inverse DTFT of Eq. (12.9), the time-domain representation $x_2[k]$ of the frequency-sampled signal $X_2(\Omega)$ is given by

$$x_2[k] = [x_w[k] * s_2[k]] = [x_1[k] \cdot w[k]] * \sum_{m=-\infty}^{\infty} \delta(k - mM), \tag{12.10}$$

and is shown in Fig. 12.1(o).

The discretized version of the DTFT $X_w(\Omega)$ is referred to as the discrete Fourier transform (DFT) and is generally represented as a function of the frequency index r corresponding to DTFT frequency $\Omega_r = 2r\pi/M$, for $0 \leq r \leq (M-1)$. To derive the expression for the DFT, we substitute $\Omega = 2r\pi/M$ in the following definition of the DTFT:

$$X_2(\Omega) = \sum_{k=0}^{N-1} x_2[k]e^{-jk\Omega}, \tag{12.11}$$

where we have assumed $x_2[k]$ to be a time-limited sequence of length N. Equation (12.11) reduces as follows:

$$X_2(\Omega_r) = \sum_{k=0}^{N-1} x_2[k]e^{-j(2\pi kr/M)}, \tag{12.12}$$

for $0 \leq r \leq (M-1)$. Equation (12.12) defines the DFT and can easily be implemented on a digital device since it converts a discrete number N of input samples in $x_2[k]$ to a discrete number M of DFT samples in $X_2(\Omega_r)$. To illustrate the discrete nature of the DFT, the DFT $X_2(\Omega_r)$ is also denoted as $X_2[r]$. The DFT spectrum $X_2[r]$ is plotted in Fig. 12.1(r).

Let us now return to the original problem of determining the CTFT $X(\omega)$ of the original CT signal $x(t)$ on a digital device. Given $X_2[r] = X_2(\Omega_r)$, it is straightforward to derive the CTFT $X(\omega)$ of the original CT signal $x(t)$ by comparing the CTFT spectrum, shown in Fig. 12.1(b), with the DFT spectrum, shown in Fig. 12.1(r). We note that one period of the DFT spectrum within the range $-(M-1)/2 \leq r \leq (M-1)/2$ (assuming M to be odd) is a fairly good approximation of the CTFT spectrum. This observation leads to the following relationship:

$$X(\omega_r) \approx \frac{MT_1}{N} X_2[r] = \frac{MT_1}{N} \sum_{k=0}^{N-1} x_2[k]e^{-j(2\pi kr/M)}, \tag{12.13}$$

where the CT frequencies $\omega_r = \Omega_r/T_1 = 2\pi r/(M \times T_1)$ for $-(M-1)/2 \leq r \leq (M-1)/2$.

Although Fig. 12.1 illustrates the validity of Eq. (12.13) by showing that the CTFT $X(\omega)$ and the DFT $X_2[r]$ are similar, there are slight variations in the two spectra. These variations result from aliasing in Step 1 and loss of samples in Step 2. If the CT signal $x(t)$ is sampled at a sampling rate less than the Nyquist limit, aliasing between adjacent replicas distorts the signal. A second distortion is introduced when the sampled sequence $x_1[k]$ is multiplied by the rectangular window $w[k]$ to limit its length to N samples. Some samples of $x_1[k]$ are lost in the process. To eliminate aliasing, the CT signal $x(t)$ should be band-limited, whereas elimination of the time-limited distortion requires $x(t)$ to be of finite length. These are contradictory requirements since a CT signal cannot be both time-limited and band-limited at the same time. As a result, at least one of the aforementioned distortions would always be present when approximating the CTFT with the DFT. This implies that Eq. (12.12) is an approximation for the CTFT $X(\omega)$ that, even at its best, only leads to a near-optimal estimation of the spectral content of the CT signal.

On the other hand, the DFT representation provides an accurate estimate of the DTFT of a time-limited sequence $x[k]$ of length N. By comparing the DFT spectrum, Fig. 12.1(h), with the DFT spectrum, Fig. 12.1(r), the relationship between the DTFT $X_2(\Omega)$ and the DFT $X_2[r]$ is derived. Except for a factor of K/M, we note that $X_2[r]$ provides samples of the DTFT at discrete frequencies $\Omega_r = 2\pi r/M$, for $0 \leq r \leq (M-1)$. The relationship between the DTFT and DFT is therefore given by

$$X_2(\Omega_r) = \frac{N}{M} X_2[r] = \frac{N}{M} \sum_{k=0}^{N-1} x_2[k] e^{-j(2\pi kr/M)} \qquad (12.14)$$

for $\Omega_r = 2\pi r/M, 0 \leq r \leq (M-1)$. We now proceed with the formal definitions for the DFT.

12.2 Discrete Fourier transform

Based on our discussion in Section 12.1, the M-point DFT and inverse DFT for a time-limited sequence $x[k]$, which is non-zero within the limits $0 \leq k \leq N - 1$, is given by

Forward DFT $X[r] = \displaystyle\sum_{k=0}^{N-1} x[k] e^{-j(2\pi kr/M)}$ for $0 \leq r \leq M - 1$; (12.15)

Inverse DFT $x[k] = \dfrac{1}{M} \displaystyle\sum_{r=0}^{M-1} X[r] e^{j(2\pi kr/M)}$ for $0 \leq k \leq N - 1$.

$$(12.16)$$

Equations (12.15) and (12.16) are also, respectively, known as DFT *analysis* and *synthesis* equations. Equation (12.15) was derived in Section 12.1. By substituting the expression for $X[r]$ from Eq. (12.15), the analysis equation, Eq. (12.16), can be formally proved, and vice versa. The formal proofs of the DFT pair are left as an exercise for the reader. In Eqs. (12.15) and (12.16), the length M of the DFT is typically set to be greater or equal to the length N of the aperiodic sequence $x[k]$. Unless otherwise stated, we assume $M = N$ in the discussion that follows. Collectively, the DFT pair is denoted as

$$x[k] \xleftrightarrow{\text{DFT}} X[r]. \qquad (12.17)$$

Examples 12.1 and 12.2 illustrate the steps involved in calculating the DFTs of aperiodic sequences.

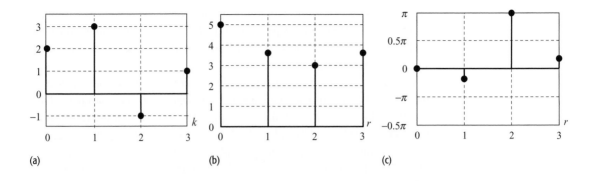

(a)   (b)   (c)

Fig. 12.2. (a) DT sequence $x[k]$; (b) magnitude spectrum and (c) phase spectrum of its DTFT $X[r]$ computed in Example 12.1.

Example 12.1

Calculate the four-point DFT of the aperiodic sequence $x[k]$ of length $N = 4$, which is defined as follows:

$$x[k] = \begin{cases} 2 & k = 0 \\ 3 & k = 1 \\ -1 & k = 2 \\ 1 & k = 3. \end{cases}$$

Solution

Using Eq. (12.15), the four-point DFT of $x[k]$ is given by

$$X[r] = \sum_{k=0}^{3} x[k] e^{-j(2\pi kr/4)}$$

$$= 2 + 3 \times e^{-j(2\pi r/4)} - 1 \times e^{-j(2\pi(2)r/4)} + 1 \times e^{-j(2\pi(3)r/4)},$$

for $0 \le r \le 3$. Substituting different values of r, we obtain

$$r = 0 \quad X[0] = 2 + 3 - 1 + 1 = 5;$$

$$r = 1 \quad X[1] = 2 + 3e^{-j(2\pi/4)} - e^{-j(2\pi(2)/4)} + e^{-j(2\pi(3)/4)}$$

$$= 2 + 3(-j) - 1(-1) + 1(j) = 3 - 2j;$$

$$r = 2 \quad X[2] = 2 + 3e^{-j(2\pi(2)/4)} - e^{-j(2\pi(2)(2)/4)} + e^{-j(2\pi(3)(2)/4)}$$

$$= 2 + 3(-1) - 1(1) + 1(-1) = -3;$$

$$r = 3 \quad X[3] = 2 + 3e^{-j(2\pi(3)/4)} - e^{-j(2\pi(2)(3)/4)} + e^{-j(2\pi(3)(3)/4)}$$

$$= 2 + 3(j) - 1(-1) + 1(-j) = 3 + j2.$$

The magnitude and phase spectra of the DFT are plotted in Figs. 12.2(b) and (c), respectively.

Example 12.2

Calculate the inverse DFT of

$$X[r] = \begin{cases} 5 & r = 0 \\ 3 - j2 & r = 1 \\ -3 & r = 2 \\ 3 + j2 & r = 3. \end{cases}$$

Solution

Using Eq. (12.13), the inverse DFT of $X[r]$ is given by

$$x[k] = \frac{1}{4} \sum_{r=0}^{3} X[r] e^{j(2\pi kr/4)} = \frac{1}{4} \left[5 + (3 - j2) \times e^{j(2\pi k/4)} - 3 \times e^{j(2\pi(2)k/4)} \right.$$

$$\left. + (3 + j2) \times e^{j(2\pi(3)k/4)} \right],$$

for $0 \le k \le 3$. On substituting different values of k, we obtain

$$x[0] = \frac{1}{4}[5 + (3 - j2) - 3 + (3 + j2)] = 2;$$

$$x[1] = \frac{1}{4}\left[5 + (3 - j2)e^{j(2\pi/4)} - 3e^{j(2\pi(2)/4)} + (3 + j2)e^{j(2\pi(3)/4)}\right]$$

$$= \frac{1}{4}[5 + (3 - j2)(j) - 3(-1) + (3 + j2)(-j)] = 3;$$

$$x[2] = \frac{1}{4}\left[5 + (3 - j2)e^{j(2\pi(2)/4)} - 3e^{j(2\pi(2)(2)/4)} + (3 + j2)e^{j(2\pi(3)(2)/4)}\right]$$

$$= \frac{1}{4}[5 + (3 - j2)(-1) - 3(1) + (3 + j2)(-1)] = -1;$$

$$x[3] = \frac{1}{4}\left[5 + (3 - j2)e^{j(2\pi(3)/4)} - 3e^{j(2\pi(2)(3)/4)} + (3 + j2)e^{j(2\pi(3)(3)/4)}\right]$$

$$= \frac{1}{4}[5 + (3 - j2)(-j) - 3(-1) + (3 + j2)(j)] = 1.$$

Examples 12.1 and 12.2 prove the following DFT pair:

$$x[k] = \begin{cases} 2 & k = 0 \\ 3 & k = 1 \\ -1 & k = 2 \\ 1 & k = 3 \end{cases} \xleftrightarrow{\text{DFT}} X[r] = \begin{cases} 5 & r = 0 \\ 3 - j2 & r = 1 \\ -3 & r = 2 \\ 3 + j2 & r = 3, \end{cases}$$

where both the DT sequence $x[k]$ and its DFT $X[r]$ have length $N = 4$.

Example 12.3

Calculate the N-point DFT of the aperiodic sequence $x[k]$ of length N, which is defined as follows:

$$x[k] = \begin{cases} 1 & 0 \le k \le N_1 - 1 \\ 0 & N_1 \le k \le N. \end{cases}$$

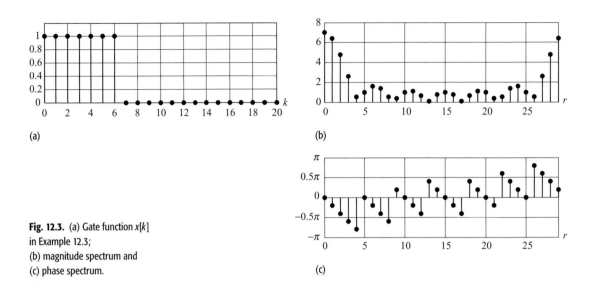

Fig. 12.3. (a) Gate function $x[k]$ in Example 12.3; (b) magnitude spectrum and (c) phase spectrum.

Solution

Using Eq. (12.15), the DFT of $x[k]$ is given by

$$X[r] = \sum_{k=0}^{N-1} x[k] e^{-j(2\pi kr/N)} = \sum_{k=0}^{N_1-1} 1 \cdot e^{-j(2\pi kr/N)}$$

$$+ \sum_{k=N_1}^{N-1} 0 \cdot e^{-j(2\pi kr/N)} = \sum_{k=0}^{N_1-1} e^{-j(2\pi kr/N)},$$

for $0 \le r \le (N-1)$. The right-hand side of this equation represents a GP series, which can be simplified as follows:

$$X[r] = \sum_{k=0}^{N_1-1} e^{-j(2\pi kr/N)} = \begin{cases} N_1 & r = 0 \\ \dfrac{1 - e^{-j(2\pi r N_1/N)}}{1 - e^{-j(2\pi r/N)}} & r \ne 0 \end{cases}$$

$$= \begin{cases} N_1 & r = 0 \\ e^{-j(\pi r(N_1-1)/N)} \dfrac{\sin(\pi r N_1/N)}{\sin(\pi r/N)} & r \ne 0. \end{cases}$$

Since $X[r]$ is a complex-valued function, its magnitude and phase components are given by

$$r = 0 \qquad |X[r]| = N_1 \quad \text{and} \quad <X[r] = 0;$$

$$r \ne 0 \quad |X[r]| = \frac{\sin(\pi r N_1/N)}{\sin(\pi r/N)}$$

$$<X[r] = -\frac{\pi r(N_1 - 1)}{N} + <\sin(\pi r N_1/N) - <\sin(\pi r/N).$$

The magnitude and phase spectra for $N_1 = 7$ and length $N = 30$ are shown in Figs. 12.3(b) and (c).

12.2.1 DFT as matrix multiplication

An alternative representation for computing the DFT is obtained by expanding Eq. (12.15) in terms of the time and frequency indices (k, r). For $N = M$, the resulting equations are expressed as follows:

$$
\left.
\begin{aligned}
X[0] &= x[0] + x[1] + x[2] + \cdots + x[N-1], \\
X[1] &= x[0] + x[1]e^{-j(2\pi/N)} + x[2]e^{-j(4\pi/N)} \\
&\quad + \cdots + x[N-1]e^{-j(2(N-1)\pi/N)}, \\
X[2] &= x[0] + x[1]e^{-j(4\pi/N)} + x[2]e^{-j(8\pi/N)} \\
&\quad + \cdots + x[N-1]e^{-j(4(N-1)\pi/N)}, \\
&\;\;\vdots \\
X[N-1] &= x[0] + x[1]e^{-j(2(N-1)\pi/N)} + x[2]e^{-j(4(N-1)\pi/N)} \\
&\quad + \cdots + x[N-1]e^{-j(2(N-1)(N-1)\pi/N)}.
\end{aligned}
\right\}
$$

(12.18)

In the matrix-vector format they are given by

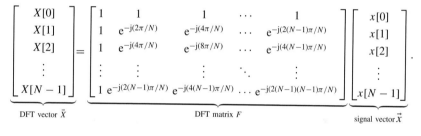

$$
\underbrace{\begin{bmatrix} X[0] \\ X[1] \\ X[2] \\ \vdots \\ X[N-1] \end{bmatrix}}_{\text{DFT vector } \vec{X}}
=
\underbrace{\begin{bmatrix}
1 & 1 & 1 & \cdots & 1 \\
1 & e^{-j(2\pi/N)} & e^{-j(4\pi/N)} & \cdots & e^{-j(2(N-1)\pi/N)} \\
1 & e^{-j(4\pi/N)} & e^{-j(8\pi/N)} & \cdots & e^{-j(4(N-1)\pi/N)} \\
\vdots & \vdots & \vdots & \ddots & \vdots \\
1 & e^{-j(2(N-1)\pi/N)} & e^{-j(4(N-1)\pi/N)} & \cdots & e^{-j(2(N-1)(N-1)\pi/N)}
\end{bmatrix}}_{\text{DFT matrix } F}
\underbrace{\begin{bmatrix} x[0] \\ x[1] \\ x[2] \\ \vdots \\ x[N-1] \end{bmatrix}}_{\text{signal vector } \vec{x}}.
$$

(12.19)

Equation (12.19) shows that the DFT coefficients $X[r]$ can be computed by left-multiplying the DT sequence $x[k]$, arranged in a column vector $\vec{x}$ in ascending order with respect to the time index k, by the DFT matrix F.

Similarly, the expression for the inverse DFT given in Eq. (12.16) can be expressed as follows:

$$
\underbrace{\begin{bmatrix} x[0] \\ x[1] \\ x[2] \\ \vdots \\ x[N-1] \end{bmatrix}}_{\text{signal vector } x}
= \frac{1}{N}
\underbrace{\begin{bmatrix}
1 & 1 & 1 & \cdots & 1 \\
1 & e^{j(2\pi/N)} & e^{j(4\pi/N)} & \cdots & e^{j(2(N-1)\pi/N)} \\
1 & e^{j(4\pi/N)} & e^{j(8\pi/N)} & \cdots & e^{j(4(N-1)\pi/N)} \\
\vdots & \vdots & \vdots & \ddots & \vdots \\
1 & e^{j(2(N-1)\pi/N)} & e^{j(4(N-1)\pi/N)} & \cdots & e^{j(2(N-1)(N-1)\pi/N)}
\end{bmatrix}}_{\text{DFT matrix } G = F^{-1}}
\underbrace{\begin{bmatrix} X[0] \\ X[1] \\ X[2] \\ \vdots \\ X[N-1] \end{bmatrix}}_{\text{DFT vector } x},
$$

(12.20)

which implies that the DT sequence $x[k]$ can be obtained by left-multiplying the DFT coefficients $X[r]$, arranged in a column vector $\vec{X}$ in ascending order with respect to the DFT coefficient index r, by the inverse DFT matrix G. It is straightforward to show that $G \times F = F \times G = I_N$, where I_N is the identity matrix of order N.

Example 12.4 repeats Example 12.1 using the matrix-vector representation for the DFT.

Example 12.4

Calculate the four-point DFT of the aperiodic signal $x[k]$ considered in Example 12.1.

Solution

Arranging the values of the DT sequence in the signal vector x, we obtain

$$x = [2 \quad 3 \quad -1 \quad 1]^{\mathrm{T}},$$

where superscript T represents the transpose operation for a vector. Using Eq. (12.19), we obtain

$$
\begin{bmatrix} X[0] \\ X[1] \\ X[2] \\ X[3] \end{bmatrix} = \underbrace{\begin{bmatrix} 1 & 1 & 1 & 1 \\ 1 & e^{-j(2\pi/N)} & e^{-j(4\pi/N)} & e^{-j(6\pi/N)} \\ 1 & e^{-j(4\pi/N)} & e^{-j(8\pi/N)} & e^{-j(12\pi/N)} \\ 1 & e^{-j(6\pi/N)} & e^{-j(12\pi/N)} & e^{-j(18\pi/N)} \end{bmatrix}}_{\text{DFT matrix: } F} \begin{bmatrix} x[0] \\ x[1] \\ x[2] \\ x[3] \end{bmatrix}
$$

$$
= \underbrace{\begin{bmatrix} 1 & 1 & 1 & 1 \\ 1 & e^{-j(2\pi/4)} & e^{-j(4\pi/4)} & e^{-j(6\pi/4)} \\ 1 & e^{-j(4\pi/4)} & e^{-j(8\pi/4)} & e^{-j(12\pi/4)} \\ 1 & e^{-j(6\pi/4)} & e^{-j(12\pi/4)} & e^{-j(18\pi/4)} \end{bmatrix}}_{\text{DFT matrix: } F} \begin{bmatrix} 2 \\ 3 \\ -1 \\ 1 \end{bmatrix} = \begin{bmatrix} 5 \\ 3-j2 \\ -3 \\ 3+j2 \end{bmatrix}.
$$

The above values for the DFT coefficients are the same as the ones obtained in Example 12.1.

Example 12.5

Calculate the inverse DFT of $X[r]$ considered in Example 12.2.

Solution

Arranging the values of the DFT coefficients in the DFT vector X, we obtain

$$X = [5 \quad 3-j2 \quad -3 \quad 3+j2]^{\mathrm{T}}.$$

Using Eq. (12.20), the DFT vector X is given by

$$
\begin{bmatrix} x[0] \\ x[1] \\ x[2] \\ x[3] \end{bmatrix} = \frac{1}{4} \begin{bmatrix} 1 & 1 & 1 & 1 \\ 1 & e^{j(2\pi/N)} & e^{j(4\pi/N)} & e^{j(6\pi/N)} \\ 1 & e^{j(4\pi/N)} & e^{j(8\pi/N)} & e^{j(12\pi/N)} \\ 1 & e^{j(6\pi/N)} & e^{j(12\pi/N)} & e^{j(18\pi/N)} \end{bmatrix} \begin{bmatrix} X[0] \\ X[1] \\ X[2] \\ X[3] \end{bmatrix}
$$

$$
= \frac{1}{4} \begin{bmatrix} 1 & 1 & 1 & 1 \\ 1 & e^{j(2\pi/4)} & e^{j(4\pi/4)} & e^{j(6\pi/4)} \\ 1 & e^{j(4\pi/4)} & e^{j(8\pi/4)} & e^{j(12\pi/4)} \\ 1 & e^{j(6\pi/4)} & e^{j(12\pi/4)} & e^{j(18\pi/4)} \end{bmatrix} \begin{bmatrix} 5 \\ 3-j2 \\ -3 \\ 3+j2 \end{bmatrix} = \frac{1}{4} \begin{bmatrix} 8 \\ 12 \\ -4 \\ 4 \end{bmatrix} = \begin{bmatrix} 2 \\ 3 \\ -1 \\ 1 \end{bmatrix}.
$$

The above values for the DT sequence $x[k]$ are the same as the ones obtained in Example 12.2.

12.2.2 DFT basis functions

The matrix-vector representation of the DFT derived in Section 12.2.1 can be used to determine the set of basis functions for the DFT representation. Expressing Eq. (12.20) in the following format:

$$
\begin{bmatrix} x[0] \\ x[1] \\ x[2] \\ \vdots \\ x[N-1] \end{bmatrix} = \frac{1}{N} X[0] \begin{bmatrix} 1 \\ 1 \\ 1 \\ \vdots \\ 1 \end{bmatrix} + \frac{1}{N} X[1] \begin{bmatrix} 1 \\ e^{j(2\pi/N)} \\ e^{j(4\pi/N)} \\ \vdots \\ e^{j(2(N-1)\pi/N)} \end{bmatrix} + \frac{1}{N} X[2]
$$

$$
\begin{bmatrix} 1 \\ e^{j(4\pi/N)} \\ e^{j(8\pi/N)} \\ \vdots \\ e^{j(4(N-1)\pi/N)} \end{bmatrix} + \cdots \frac{1}{N} X[N-1] \begin{bmatrix} 1 \\ e^{j(2(N-1)\pi/N)} \\ e^{j(4(N-1)\pi/N)} \\ \vdots \\ e^{j(2(N-1)(N-1)\pi/N)} \end{bmatrix}, \tag{12.21}
$$

it is clear that the basis functions for the N-point DFT are given by the following set of N vectors:

$$
F_r = \frac{1}{N} \begin{bmatrix} 1 & \exp\left(\dfrac{j2\pi r}{N}\right) & \exp\left(\dfrac{j4\pi r}{N}\right) & \cdots & \exp\left(\dfrac{j2(N-1)\pi r}{N}\right) \end{bmatrix}^T,
$$

for $0 \leq r \leq (N-1)$. Equation (12.21) illustrates that the DFT represents a DT sequence as a linear combination of complex exponentials, which are weighted by the corresponding DFT coefficients. Such a representation is useful for the analysis of LTID systems.

As an example, Fig. 12.4 plots the real and imaginary components of the basis vectors for the eight-point DFT of length $N = 8$. From Fig. 12.4(a), we observe that the real components of the basis vectors correspond to a cosine function sampled at different sampling rates. Similarly, the imaginary components of the basis vectors correspond to a sine function sampled at different sampling

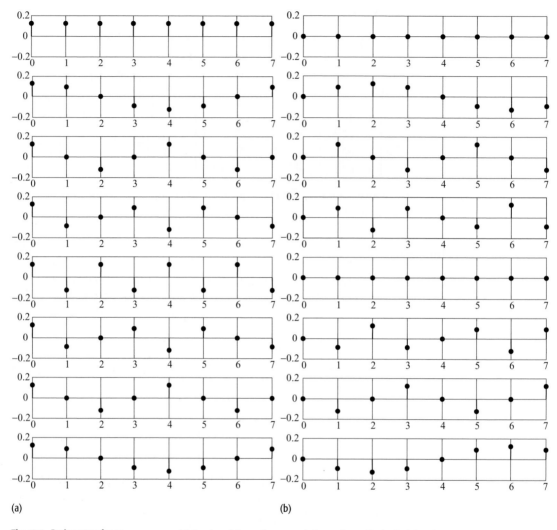

(a)   (b)

Fig. 12.4. Basis vectors for an eight-point DFT. (a) Real components; (b) imaginary components.

rates. This should not be surprising, since Euler's identity expands a complex exponential as a complex sum of cosine and sine terms.

We now proceed with the estimation of the spectral content of both DT and CT signals using the DFT.

12.3 Spectrum analysis using the DFT

In this section, we illustrate how the DFT can be used to estimate the spectral content of the CT and DT signals. Examples 12.6–12.8 deal with the CT signals, while Examples 12.9 and 12.10 deal with the DT sequences.

Example 12.6

Using the DFT, estimate the frequency characteristics of the decaying exponential signal $g(t) = \exp(-0.5t)u(t)$. Plot the magnitude and phase spectra.

Solution

Following the procedure outlined in Section 12.1, the three steps involved in computing the CTFT are listed below.

Step 1: Impulse-train sampling Based on Table 5.1, the CTFT of the decaying exponential is given by

$$g(t) = e^{-0.5t}u(t) \xleftrightarrow{\text{CTFT}} G(\omega) = \frac{1}{0.5 + j\omega}.$$

This CTFT pair implies that the bandwidth of $g(t)$ is infinite. Ideally speaking, the sampling theorem can never be satisfied for the decaying exponential signal. However, we exploit the fact that the magnitude $|G(\omega)|$ of the CTFT decreases monotonically with higher frequencies and we neglect any frequency components at which the magnitude falls below a certain threshold η. Selecting the value of $\eta = 0.01 \times |G(\omega)|_{\text{max}}$, the threshold frequency B is given by

$$\left| \frac{1}{0.5 + j2\pi B} \right| \leq 0.01 \times |G(\omega)|_{\text{max}}.$$

Since the maximum value of the magnitude $|G(\omega)|$ is 2 at $\omega = 0$, the above expression reduces to

$$\sqrt{0.25 + (2\pi B)^2} \geq 50,$$

or $B \geq 7.95$ Hz. The Nyquist sampling rate f_1 is therefore given by

$$f_1 \geq 2 \times 7.95 = 15.90 \text{ samples/s}.$$

Selecting a sampling rate of $f_1 = 20$ samples/s, or a sampling interval $T_1 = 1/20 = 0.05$ s, the DT approximation of the decaying exponential is given by

$$g[k] = g(kT_1) = e^{-0.5kT_1}u[k] = e^{-0.025k}u[k].$$

Since there is a discontinuity in the CT signal $g(t)$ at $t = 0$ with $g(0^-) = 0$ and $g(0^+) = 1$, the value of $g[k]$ at $k = 0$ is set to $g[0] = 0.5$. based on Eq. (1.1).

Step 2: Time-limitation To truncate the length of $g[k]$, we apply a rectangular window of length $N = 201$ samples. The truncated sequence is given by

$$g_w[k] = e^{-0.025k}(u[k] - u[k - 201]) = \begin{cases} e^{-0.025k} & 0 \leq k \leq 200 \\ 0 & \text{elsewhere.} \end{cases}$$

The subscript w in $g_w[k]$ denotes the truncated version of $g[k]$ obtained by multiplying by the window function $w[k]$. Note that the truncated sequence $g_w[k]$ is a fairly good approximation of $g[k]$, as the peak magnitude of the truncated samples is given by 0.0066 and occurs at $k = 201$. This is only 0.66% of the peak value of the complex exponential $g[k]$.

Step 3: DFT computation The DFT of the truncated DT sequence $g_w[k]$ can now be computed directly from Eq. (12.16). MATLAB provides a built-in function fft, which has the calling syntax of

```
>> G = fft(g);
```

where g is the signal vector containing the values of the DT sequence $g_w[k]$ and G is the computed DFT. Both g and G have a length of N, implying that an N-point DFT is being taken. The built-in function fft computes the DFT within the frequency range $0 \leq r \leq (N-1)$. Since the DFT is periodic, we can obtain the DFT within the frequency range $-(N-1)/2 \leq r \leq (N-1)/2$ by a circular shift of the DFT coefficients. In MATLAB, this is accomplished by the fftshift function.

Having computed the DFT, we use Eq. (12.12) to estimate the CTFT of the original CT decaying exponential signal $g(t)$. The MATLAB code for computing the CTFT is as follows:

```
>> f1 = 20;                 % set sampling rate
>> t1 = 1/f1;               % set sampling interval
>> N = 201; k = 0:N-1;      % set length of DT sequence to
                            % N = 201
>> g = exp(-0.025*k);       % compute the DT sequence
>> g(1) = 0.5;               % initialize the first sample
>> G = fft(g);              % determine the 201-point DFT
>> G = fftshift(G);         % shift the DFT coefficients
>> G = t1*G;                % scale DFT such that
                            % DFT = CTFT
>> dw = 2*pi*f1/N;          % CTFT frequency resolution
>> w = -pi*f1:dw:pi*f1-dw;  % compute CTFT frequencies
>> stem(w,abs(G));          % plot CTFT magnitude spectrum
>> stem(w,angle(G));        % plot CTFT phase spectrum
```

Fig. 12.5. Spectral estimation of decaying exponential signal $g(t) = \exp(-0.5t)u(t)$ using the DFT in Example 12.6. (a) Estimated magnitude spectrum; (b) estimated phase spectrum.

The resulting plots are shown in Fig. 12.5, where we have limited the frequency axis to the range $-5\pi \leq \omega \leq 5\pi$. The magnitude and phase spectra plotted in Fig. 12.5 are fairly good estimates of the frequency characteristics of the decaying exponential signal listed in Table 5.3.

In Example 12.6, we used the CTFT $G(\omega)$ to determine the appropriate sampling rate. In most practical situations, however, the CTFTs are not known and one

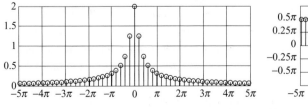

(a)

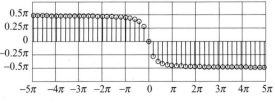

(b)

is forced to make an intelligent estimate of the bandwidth of the signal. If the frequency and time characteristics of the signal are not known, a high sampling rate and a large time window are arbitrarily chosen. In such cases, it is advised that a number of sampling rates and lengths be tried before finalizing the estimates.

Example 12.7
Using the DFT, estimate the frequency characteristics of the CT signal $h(t) = 2 \exp(j18\pi t) + \exp(-j8\pi t)$.

Solution
Following the procedure outlined in Section 12.1, the three steps involved in computing the CTFT are as follows.

Step 1: Impulse-train sampling The CT signal $h(t)$ consists of two complex exponentials with fundamental frequencies of 9 Hz and 4 Hz. The Nyquist sampling rate f_1 is therefore given by

$$f_1 \geq 2 \times 9 = 18 \text{ samples/s}.$$

We select a sampling rate of $f_1 = 32$ samples/s, or a sampling interval $T_1 = 1/32$ s. The DT approximation of $h(t)$ is given by

$$h[k] = h(kT_1) = 2e^{j18\pi k/32} + e^{-j8\pi k/32}.$$

Step 2: Time-limitation The DT sequence $h[k]$ is a periodic signal with fundamental period $K_0 = 32$. For periodic signals, it is sufficient to select the length of the rectangular window equal to the fundamental period. Therefore, N is set to 32.

Step 3: DFT computation The MATLAB code for computing the DFT of the truncated DT sequence is as follows.

```
>> f1 = 32;                    % set sampling rate
>> t1 = 1/f1;                  % set sampling interval
>> N = 32; k = 0:N-1;          % set length of DT sequence
>> h = 2*exp(j*18*pi*k/32) + exp(-j*8*pi*k/32);
                               % compute the DT sequence
>> H = fft(h);                 % determine the 32-point DFT
>> H = fftshift(H);            % shift the DFT coefficients
>> H = t1*H;                   % scale DFT such that
                               % DFT = CTFT
```

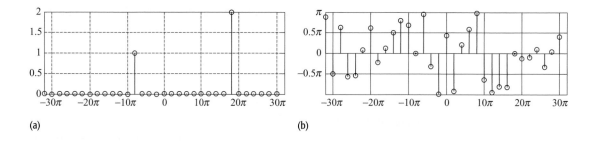

(a)   (b)

Fig. 12.6. Spectral estimation of decaying exponential signal $h(t)$ $= 2\exp(j18\pi t) + \exp(-j8\pi t)$ using the DFT in Example 12.7. (a) Estimated magnitude spectrum; (b) estimated phase spectrum.

```
>> dw = 2*pi*f1/N;          % CTFT frequency resolution
>> w = -pi*f1:dw:pi*f1-dw;  % compute CTFT frequencies
>> stem(w,abs(H));          % plot CTFT magnitude spectrum
>> stem(w,angle(H));        % plot CTFT phase spectrum
```

The resulting plots are shown in Fig. 12.6, and they have a frequency resolution of $\Delta\omega = 2\pi$. We know that the CTFT for $h(t)$ is given by

$$2e^{j18\pi t} + e^{-j8\pi t} \xleftrightarrow{\text{CTFT}} 2\delta(\omega - 18\pi) + \delta(\omega + 8\pi).$$

We observe that the two impulses at $\omega = -8\pi$ and 18π radians/s are accurately estimated in the magnitude spectrum plotted in Fig. 12.6(a). Also, the relative amplitude of the two impulses corresponds correctly to the area enclosed by these impulses in the CTFT for $h(t)$.

The phase spectrum plotted in Fig. 12.6(b) is unreliable except for the two frequencies $\omega = -8\pi$ and 18π radians/s. At all other frequencies, the magnitude $|H(\omega)|$ is zero, therefore the phase $<H(\omega)$ carries no information. This is because the phase is computed as the inverse tangent of the ratio between the imaginary and real components of $H(\omega)$. When $|H(\omega)|$ is close to zero, the argument of the inverse tangent is given by $\varepsilon_1/\varepsilon_2$, with ε_1 and ε_2 approaching zero. In such cases, incorrect results are obtained for the phase. The phase $<H(\omega)$ is therefore ignored when $|H(\omega)|$ is close to zero.

Example 12.8
Using the DFT, estimate the frequency characteristics of the CT signal $x(t) = 2\exp(j19\pi t)$.

Solution
The three steps involved in computing the CTFT are as follows.

Step 1: Impulse-train sampling The CT signal $x(t)$ constitutes a complex exponential with fundamental frequency 9.5 Hz. The Nyquist sampling rate f_1 is therefore given by

$$f_1 \geq 2 \times 9.5 = 19\,\text{samples/s}.$$

As in Example 12.7, we select a sampling rate of $f_1 = 32$ samples/s, or a sampling interval $T_1 = 1/32$ s. The DT approximation of $h(t)$ is given by

$$x[k] = x(kT_1) = 2e^{j19\pi k/32}.$$

Step 2: Time-limitation As in Example 12.7, we keep the length N of the rectangular window equal to 32.

Step 3: DFT computation The MATLAB code for computing the DFT of the truncated DT sequence is as follows:

```
>> f1 = 32;                    % set sampling rate
>> t1 = 1/f1;                  % set sampling interval
>> N = 32; k = 0:N-1;          % set length of DT sequence
                               % to N = 32
>> x = 2*exp(j*19*pi*k/32);    % compute the DT sequence
>> X = fft(x);                 % determine the 32-point DFT
>> X = fftshift(X);            % shift the DFT coefficients
>> X = t1*X;                   % scale DFT such that
                               % DFT = CTFT
>> dw = 2*pi*f1/N;             % CTFT frequency resolution
>> w = -pi*f1:dw:pi*f1-dw;     % compute CTFT frequencies
>> stem(w,abs(X));             % plot CTFT magnitude spectrum
```

The resulting magnitude spectrum is shown in Fig. 12.7(a), which has a frequency resolution of $\Delta\omega = 2\pi$ radians/s. Comparing with the CTFT for $x(t)$, which is given by

$$2e^{j19\pi t} \xleftarrow{\text{CTFT}} 2\delta(\omega - 19\pi),$$

we observe that Fig. 12.7(a) provides us with an erroneous result. This error is attributed to the poor resolution $\Delta\omega$ chosen to frequency-sample the CTFT. Since $\Delta\omega = 2\pi$, the frequency component of 19π present in $x(t)$ cannot be displayed accurately at the selected resolution. In such cases, the strength of the frequency component of 19π radians/s leaks into the adjacent frequencies, leading to non-zero values at these frequencies. This phenomenon is referred to as the *leakage* or *picket fence effect*.

Figure 12.7(b) plots the magnitude spectrum when the number N of samples in the discretized sequence is increased to 64. Since fft uses the same number M of samples to discretize the CTFT, the resolution $\Delta\omega = 2\pi T_1/M = \pi$ radians/s. The MATLAB code for estimating the CTFT is as follows:

```
>> f1 = 32; t1 = 1/f1; % set sampling rate and interval
>> N = 64; k = 0:N-1;  % set sequence length to N = 64
```

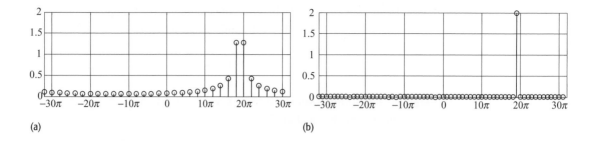

Fig. 12.7. Spectral estimation of complex exponential signal $x(t) = 2 \exp(j19\pi t)$ using the DFT in Example 12.8. (a) Estimated magnitude spectrum, with a 32-point DFT. (b) Same as part (a) except that a 64-point DFT is computed.

```
>> x = 2*exp(j*19*pi*k/32); % compute the DT sequence
>> X = fft(x);                % determine the 64-point DFT
>> X = fftshift(X);           % shift the DFT coefficients
>> X = 0.5*t1*X;              % scale DFT so DFT = CTFT
>> dw = 2*pi*f1/N;            % CTFT frequency resolution
>> w = -pi*f1:dw:pi*f1-dw;    % compute CTFT frequencies
>> stem(w,abs(X));            % plot CTFT magnitude spectrum
```

In the above code, we have highlighted the instructions that have been changed from the original version. In addition to setting the length N to 64 in the above code, we also note that the magnitude of the CTFT X is now being scaled by a factor of $0.5 \times T_1$. The additional factor of 0.5 is introduced because we are now computing the DFT over two consecutive periods of the periodic sequence $x[k]$. Doubling the time duration doubles the values of the DFT coefficients, and therefore a factor of 0.5 is introduced to compensate for the increase. Figure 12.7(b), obtained using a 64-point DFT, is a better estimate for the magnitude spectrum of $x(t)$ than Fig. 12.7(a), obtained using a 32-point DFT.

The DFT can also be used to estimate the DTFT of DT sequences. Examples 12.9 and 12.10 compute the DTFT of two aperiodic sequences.

Example 12.9
Using the DFT, calculate the DTFT of the DT decaying exponential sequence $x[k] = 0.6^k\, u[k]$.

Solution
Estimating the DTFT involves only Steps 2 and 3 outlined in Section 12.1.

Step 2: Time-limitation Applying a rectangular window of length $N = 10$, the truncated sequence is given by

$$x_{\mathrm{w}}[k] = \begin{cases} 0.6^k & 0 \le k \le 9 \\ 0 & \text{elsewhere.} \end{cases}$$

Table 12.1. Comparison between the DFT and DTFT coefficients in Example 12.9

DFT index, r	DTFT frequency, $\Omega_r = 2\pi r/N$	DFT coefficients, $X[r]$	DTFT coefficients, $X(\Omega)$
-5	$-\pi$	0.6212	0.6250
-4	-0.8π	$0.6334 + j0.1504$	$0.6373 + j0.1513$
-3	-0.6π	$0.6807 + j0.3277$	$0.6849 + j0.3297$
-2	-0.4π	$0.8185 + 0.5734$	$0.8235 + j0.5769$
-1	-0.2π	$1.3142 + j0.9007$	$1.3222 + j0.9062$
0	0	2.4848	2.5000
1	0.2π	$1.3142 - j0.9007$	$1.3222 - j0.9062$
2	0.4π	$0.8185 - j0.5734$	$0.8235 - j0.5769$
3	0.6π	$0.6807 - j0.3277$	$0.6849 - j0.3297$
4	0.8π	$0.6334 - j0.1504$	$0.6373 - j0.1513$

Step 3: DFT computation The MATLAB code for computing the DFT is as follows:

```
>> N = 10; k = 0:N-1;        % set sequence length
                             % to N = 10
>> x = 0.6.^k;               % compute the DT sequence
>> X = fft(x);               % calculate the 10-point DFT
>> X = fftshift(X);          % shift the DFT coefficients
>> w = -pi:2*pi/N:pi-2*pi/N; % compute DTFT frequencies
```

Table 12.1 compares the computed DFT coefficients with the corresponding DTFT coefficients obtained from the following DTFT pair:

$$0.6^k u[k] \xrightarrow{\text{DTFT}} \frac{1}{1 - 0.6e^{-j\Omega}}.$$

We observe that the values of the DFT coefficients are fairly close to the DTFT values.

Example 12.10
Calculate the DTFT of the aperiodic sequence $x[k] = [2, 1, 0, 1]$ for $0 \leq k \leq 3$.

Solution
Using Eq. (12.6), the DFT coefficients are given by

$$X[r] = [4, 2, 0, 2] \quad \text{for} \quad 0 \leq r \leq 3.$$

Mapping in the DTFT domain, the corresponding DTFT coefficients are given by

$$X(\Omega_r) = [4, 2, 0, 2] \quad \text{for} \quad \Omega_r = [0, 0.5\pi, \pi, 1.5\pi] \text{ radians/s.}$$

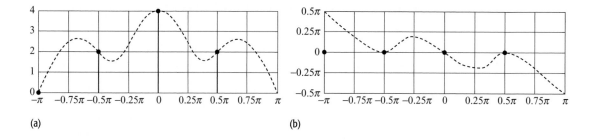

Fig. 12.8. Spectral estimation of DT sequences using the DFT in Example 12.10. (a) Estimated magnitude spectrum; (b) estimated phase spectrum. The dashed lines show the continuous spectrum obtained from the DTFT.

If instead the DTFT is to be plotted within the range $-\pi \le \Omega \le \pi$, then the DTFT coefficients can be rearranged as follows:

$$X(\Omega_r) = [4, 2, 0, 2] \quad \text{for} \quad \Omega_r = [-\pi, -0.5\pi, 0, 0.5\pi] \text{ radians/s.}$$

The magnitude and phase spectra obtained from the DTFT coefficients are sketched using stem plots in Figs. 12.8(a) and (b). For comparison, we use Eq. (11.28b) to derive the DTFT for $x[k]$. The DTFT is given by

$$X(\Omega) = \sum_{k=0}^{3} x[k]e^{-j\Omega k} = 2 + e^{-j\Omega} + e^{-j3\Omega}.$$

The actual magnitude and phase spectra based on the above DTFT expression are plotted in Figs. 12.8(a) and (b) respectively (see dashed lines). Although the DFT coefficients provide exact values of the DTFT at the discrete frequencies $\Omega_r = [0, 0.5\pi, \pi, 1.5\pi]$ radians/s, no information is available on the characteristics of the magnitude and phase spectra for the intermediate frequencies. This is a consequence of the low resolution used by the DFT to discretize the DTFT frequency Ω. Section 12.3.1 introduces the concept of zero padding, which allows us to improve the resolution used by the DFT.

12.3.1 Zero padding

To improve the resolution of the frequency axis Ω in the DFT domain, a commonly used approach is to append the DT sequences with additional zero-valued samples. This process is called *zero padding*, and for an aperiodic sequence $x[k]$ of length N is defined as follows:

$$x_{zp}[k] = \begin{cases} x[k] & 0 \le k \le N - 1 \\ 0 & N \le k \le M - 1. \end{cases}$$

The zero-padded sequence $x_{zp}[k]$ has an increased length of M. The frequency resolution $\Delta\Omega$ of the zero-padded sequence is improved from $2\pi/N$ to $2\pi/M$.

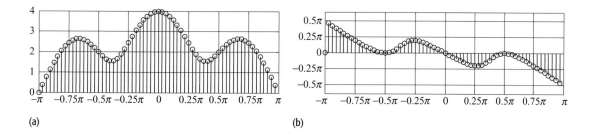

Fig. 12.9. Spectral estimation of zero-padded DT sequences using the DFT in Example 12.11. (a) Estimated magnitude spectrum; (b) estimated phase spectrum.

Example 12.11 illustrates the improvement in the DTFT achieved with the zero-padding approach.

Example 12.11

Compute the DTFT of the aperiodic sequence $x[k] = [2, 1, 0, 1]$ for $0 \le k \le 3$ by padding 60 zero-valued samples at the end of the sequence.

Solution

The MATLAB code for computing the DTFT of the zero-padded sequence is as follows:

```
>> N = 64; k = 0:N-1;           % set sequence length
                                % to N = 64
>> x = [2 1 0 1 zeros(1,60)];   % zero-padded sequence
>> X = fft(x);                  % determine the 64-point DFT
>> X = fftshift(X);             % shift the DFT coefficients
>> w = -pi:2*pi/N:pi-2*pi/N;    % compute DTFT frequencies
>> stem(w,abs(X));              % plot magnitude spectrum
>> stem(w,angle(X));            % plot the phase spectrum
```

The magnitude and phase spectra of the zero-padded sequence are plotted in Figs. 12.9(a) and (b), respectively. Compared with Fig. 12.8, we observe that the estimated spectra in Fig. 12.9 provide an improved resolution and better estimates for the frequency characteristics of the DT sequence.

12.4 Properties of the DFT

In this section, we present the properties of the M-point DFT. The length of the DT sequence is assumed to be $N \le M$. For $N < M$, the DT sequence is zero-padded with $M - N$ zero-valued samples. The DFT properties presented below are similar to the corresponding properties for the DTFT discussed in Chapter 11.

12.4.1 Periodicity

The M-point DFT of an aperiodic DT sequence with length N ($M \geq N$) is periodic with period M. In other words,

$$X[r] = X[r + M], \tag{12.22}$$

for $0 \leq r \leq M - 1$.

12.4.2 Orthogonality

The column vectors F_r of the DFT matrix F, defined in Section 12.2.2, form the basis vectors of the DFT. These vectors are orthogonal to each other and, for the M-point DFT, satisfy the following:

$$F_p^T F_q^* = \sum_{m=1}^{M} F_p(m, 1)[F_q(m, 1)]^* = \begin{cases} 1/M & \text{for } p = q \\ 0 & \text{for } p \neq q, \end{cases}$$

where the matrix F_p^T is the transpose of the matrix F_p and the matrix F_q^* is the complex conjugate of the matrix F_q.

12.4.3 Linearity

If $x_1[k]$ and $x_2[k]$ are two DT sequences with the following M-point DFT pairs:

$$x_1[k] \xrightarrow{\text{DFT}} X_1[r] \text{ and } x_2[k] \xrightarrow{\text{DFT}} X_2[r],$$

then the linearity property states that

$$a_1 x_1[k] + a_2 x_2[k] \xrightarrow{\text{DFT}} a_1 X_1[r] + a_2 X_2[r], \tag{12.23}$$

for any arbitrary constants a_1 and a_2, which may be complex-valued.

12.4.4 Hermitian symmetry

The M-point DFT $X[r]$ of a real-valued aperiodic sequence $x[k]$ is conjugate–symmetric about $r = M/2$. Mathematically, the Hermitian symmetry implies that

$$X[r] = X^*[M - r], \tag{12.24}$$

where $X^*[r]$ denotes the complex conjugate of $X[r]$.

In terms of the magnitude and phase spectra of the DFT $X[r]$, the Hermitian symmetry property can be expressed as follows:

$$|X[M - r]| = |X[r]| \quad \text{and} \quad <X[M - r] = -<X[r], \tag{12.25}$$

implying that the magnitude spectrum is even and that the phase spectrum is odd.

The validity of the Hermitian symmetry can be observed in the DFT plotted for various aperiodic sequences in Examples 12.2–12.11.

12.4.5 Time shifting

If $x[k] \xleftrightarrow{\text{DFT}} X[r]$, then

$$x[k - k_0] \xleftrightarrow{\text{DFT}} e^{-j2\pi k_0 r/M} X[r] \qquad (12.26)$$

for an M-point DFT and any arbitrary integer k_0.

12.4.6 Circular convolution

If $x_1[k]$ and $x_2[k]$ are two DT sequences with the following M-point DFT pairs:

$$x_1[k] \xleftrightarrow{\text{DFT}} X_1[r] \quad \text{and} \quad x_2[k] \xleftrightarrow{\text{DFT}} X_2[r],$$

then the circular convolution property states that

$$x_1[k] \otimes x_2[k] \xleftrightarrow{\text{DFT}} X_1[r]X_2[r] \qquad (12.27)$$

and

$$x_1[k]x_2[k] \xleftrightarrow{\text{DFT}} \frac{1}{M}[X_1[r] \otimes X_2[r]], \qquad (12.28)$$

where $\otimes$ denotes the circular convolution operation. Note that the two sequences must have the same length in order to compute the circular convolution.

Example 12.12
In Example 10.11, we calculated the circular convolution $y[k]$ of the aperiodic sequences $x[k] = [0, 1, 2, 3]$ and $h[k] = [5, 5, 0, 0]$ defined over $0 \leq k \leq 3$. Recalculate the result of the circular convolution using the DFT convolution property.

Solution
The four-point DFTs of the aperiodic sequences $x[k]$ and $h[k]$ are given by

$$X[r] = [6, -2 + j2, -2, -2 - j2]$$

and

$$H[r] = [10, 5 - j5, 0, 5 + j5]$$

for $0 \leq r \leq 3$. Using Eq. (12.27), the four-point DFT of the circular convolution between $x[k]$ and $h[k]$ is given by

$$x_1[k] \otimes x_2[k] \xleftrightarrow{\text{DFT}} [60, \ j20, \ 0 \ - j20].$$

Taking the inverse DFT, we obtain

$$x_1[k] \otimes x_2[k] = [15, \ 5, \ 15, \ 25],$$

which is identical to the answer obtained in Example 10.11.

12.4.7 Parseval's theorem

If $x[k] \xleftrightarrow{\text{DFT}} X[r]$, then the energy of the aperiodic sequence $x[k]$ of length N can be expressed in terms of its M-point DFT as follows:

$$E_x = \sum_{k=0}^{N-1} |x[k]|^2 = \frac{1}{M} \sum_{k=0}^{M-1} |X[r]|^2. \tag{12.29}$$

Parseval's theorem shows that the DFT preserves the energy of the signal within a scale factor of M.

12.5 Convolution using the DFT

In Section 10.6.1, we showed that the linear convolution $x_1[k] * x_2[k]$ between two time-limited DT sequences $x_1[k]$ and $x_2[k]$ of lengths K_1 and K_2, respectively, can be expressed in terms of the circular convolution $x_1[k] \otimes x_2[k]$. The procedure requires zero padding both $x_1[k]$ and $x_2[k]$ to have individual lengths of $K \geq (K_1 + K_2 - 1)$. It was shown that the result of the circular convolution of the zero-padded sequences is the same as that of the linear convolution.

Since computationally efficient algorithms are available for computing the DFT of a finite-duration sequence, the circular convolution property can be exploited to implement the linear convolution of the two sequences $x_1[k]$ and $x_2[k]$ using the following procedure.

(1) Compute the K-point DFTs $X_1[r]$ and $X_2[r]$ of the two time-limited sequences $x_1[k]$ and $x_2[k]$. The value of K is lower bounded by $(K_1 + K_2 - 1)$, i.e. $K \geq (K_1 + K_2 - 1)$.
(2) Compute the product $X_3[r] = X_1[r]X_2[r]$ for $0 \leq r \leq K - 1$.
(3) Compute the sequence $x_3[k]$ as the inverse DFT of $X_3[r]$. The resulting sequence $x_3[k]$ is the result of the linear convolution between $x_1[k]$ and $x_2[k]$.

The above approach is explained in Example 12.13.

Example 12.13
Example 10.13 computed the linear convolution of the following DT sequences:

$$x[k] = \begin{cases} 2 & k = 0 \\ -1 & |k| = 1 \\ 0 & \text{otherwise} \end{cases} \quad \text{and} \quad h[k] = \begin{cases} 2 & k = 0 \\ 3 & |k| = 1 \\ -1 & |k| = 2 \\ 0 & \text{otherwise,} \end{cases}$$

using the circular convolution method outlined in Algorithm 10.4 in Section 10.6.1. Repeat Example 10.13 using the DFT-based approach described above.

Table 12.2. Values of $X'[r]$, $H'[r]$ and $Y[r]$ for $0 \le r \le 6$ in Example 12.13

r	$X'[r]$	$H'[r]$	$Y[r]$
0	0	6	0
1	$0.470 - j0.589$	$-1.377 - j6.031$	$-4.199 - j2.024$
2	$-0.544 - j2.384$	$-2.223 + j1.070$	$3.760 + j4.178$
3	$-3.425 - j1.650$	$-2.901 - j3.638$	$3.933 + j17.247$
4	$-3.425 + j1.650$	$-2.901 + j3.638$	$3.933 - j17.247$
5	$-0.544 + j2.384$	$-2.223 - j1.070$	$3.760 - j4.178$
6	$0.470 + j0.589$	$-1.377 + j6.031$	$-4.199 + j2.024$

Solution

Step 1 Since the sequences $x[k]$ and $h[k]$ have lengths $K_x = 5$ and $K_y = 3$, the value of $K \ge (5 + 3 - 1) = 7$. We set $K = 7$ in this example:

padding $(K - K_x) = 4$ additional zeros to $x[k]$, we obtain
$$x'[k] = [-1, 2, -1, 0, 0, 0, 0];$$
padding $(K - K_h) = 2$ additional zeros to $h[k]$, we obtain
$$h'[k] = [-1, 3, 2, 3, -1, 0, 0].$$

The DFTs of $x'[k]$ are shown in the second column of Table 12.2, where the values for $X'[r]$ have been rounded off to three decimal places. Similarly, the DFTs of $h'[k]$ are shown in the third column of Table 12.2.

Step 2 The value of $Y[r] = X'[r]H'[r]$, for $0 \le r \le 6$, are shown in the fourth column of Table 12.2.

Step 3 Taking the inverse DFT of $Y[r]$ yields

$$y[k] = [0.998 \quad -5 \quad 5.001 \quad -1.999 \quad 5 \quad -5.002 \quad 1.001].$$

Except for approximation errors caused by the numerical precision of the computer, the above results are the same as those obtained from the direct computation of the linear convolution included in Example 10.13.

12.5.1 Computational complexity

We now compare the computational complexity of the time-domain and DFT-based implementations of the linear convolution between the time-limited sequences $x_1[k]$ and $x_2[k]$ with lengths K_1 and K_2, respectively. For simplicity, we assume that $x_1[k]$ and $x_2[k]$ are real-valued sequences with lengths K_1 and K_2, respectively.

Time-domain approach This is based on the direct computation of the convolution sum

$$y[k] = x_1[k] * x_1[k] = \sum_{m=-\infty}^{\infty} x_1[m]x_2[k-m],$$

which requires roughly $K_1 \times K_2$ multiplications and $K_1 \times K_2$ additions. The total number of floating point operations (flops) required with the time-domain approach is therefore given by $2K_1K_2$.

DFT-based approach Step 1 of the DFT-based approach computes two $K = (K_1 + K_2 - 1)$-point DFTs of the DT sequences $x_1[k]$ and $x_2[k]$. In Section 12.6, we show that the total number of flops required to implement a K-point DFT using fast Fourier transform (FFT) techniques is approximately $5K \log_2 K$. Therefore, Step 1 of the DFT-based approach requires a total of $10K \log_2 K$ flops.

Step 2 multiplies DFTs for $x_1[k]$ and $x_2[k]$. Each DFT has a length of $K = K_1 + K_2 - 1$ points; therefore, a total of K complex multiplications and $K - 1 \approx K$ complex additions are required. The total number of computations required in Step 2 is therefore given by $8K$ or $8(K_1 + K_2 - 1)$ flops.

Step 3 computes one inverse DFT based on the FFT implementation requiring $5K \log_2 K$ flops.

The total number of flops required with the DFT-based approach is therefore given by

$$15K \log_2 K + 6K \approx 15K \log_2 K \text{ flops},$$

where $K = K_1 + K_2 - 1$. Assuming $K_1 = K_2$, the DFT-based approach provides a computational saving of $O((\log_2 K)/K)$ in comparison with the direct computation of the convolution sum in the time domain. Table 12.3 compares the computational complexity of the two approaches for a few selected values of K_1 and K_2. The length K of the DFT should be equal to or greater than $(K_1 + K_2 - 1)$ depending on its value. Where $(K_1 + K_2 - 1)$ is not a power of 2, we have rounded $(K_1 + K_2 - 1)$ to the next higher integer that is a power of 2. In the second row, for example, $K_1 = 32$ and $K_2 = 5$, which implies that $(K_1 + K_2 - 1) = 36$. Since the radix-2 FFT algorithm, described in Section 12.6, can only be implemented for sequences with lengths that are powers of 2, K is set to 64. Based on the DFT-based approach, the number of flops required to compute the convolution of the two sequences is given by $(15 \times 64 \times \log_2 (64)) = 5760$. In Table 12.3, we observe that for sequences with lengths greater than 1000 samples, the DFT-based approach provides significant savings over the direct computation of the circular convolution in the time domain. If $x_1[k]$ and $x_2[k]$ are real-valued sequences, significant further savings (about 50%) can be achieved using the procedures mentioned in Problems 12.18 and 12.19.

Table 12.3. Comparison of the computational complexities of the time-domain versus the DFT-based approaches used to compute the linear convolution

Length K_1 of $x_1[k]$	Length K_2 of $x_2[k]$	Computational complexity, flops	
		Time domain ($2K_1 \times K_2$ flops)	DFT ($15K \log_2 K$ flops)
32	5	320	5760
32	32	2048	5760
1000	200	400 000	337 920
1000	1000	2 000 000	337 920
2000	2000	8 000 000	737 280

12.6 Fast Fourier transform

There are several well known techniques including the radix-2, radix-4, split radix, Winograd, and prime factor algorithms that are used for computing the DFT. These algorithms are referred to as the fast Fourier transform (FFT) algorithms. In this section, we explain the radix-2 decimation-in-time FFT algorithm.

To provide a general frame of reference, let us consider the computational complexity of the direct implementation of the K-point DFT for a time-limited complex-valued sequence $x[k]$ with length K. Based on its definition,

$$X[r] = \sum_{k=0}^{K-1} x[k]e^{-j(2\pi kr/K)}, \qquad (12.30)$$

K complex multiplications and $K - 1$ complex additions are required to compute a single DFT coefficient. Computation of all K DFT coefficients requires approximately K^2 complex additions and K^2 complex multiplications, where we have assumed K to be large such that $K - 1 \approx K$.

In terms of flops, each complex multiplication requires four scalar multiplications and two scalar additions, and each complex addition requires two scalar additions. Computation of a single DFT coefficient, therefore, requires $8K$ flops. The total number of scalar operations for computing the complete DFT is given by $8K^2$ flops.

We now proceed with the radix-2 FFT decimation-in-time algorithm. The radix-2 algorithm is based on the following principle.

Proposition 12.1 *For even values of K, the K-point DFT of a complex-valued sequence x[k] with length K can be computed from the DFT coefficients of two subsequences: (i) x[2k], containing the even-numbered samples of x[k], and (ii) x[2k + 1], containing the odd-numbered samples of x[k].*

Proof

Expressing Eq. (12.30) in terms of even- and odd-numbered samples of $x[k]$, we obtain

$$X[r] = \underbrace{\sum_{k=0,2,4,\ldots}^{K-1} x[k]e^{-j(2\pi kr/K)}}_{\text{Term I}} + \underbrace{\sum_{k=1,3,5,\ldots}^{K-1} x[k]e^{-j(2\pi kr/K)}}_{\text{Term II}}, \qquad (12.31)$$

for $0 \leq r \leq (M-1)$. Substituting $k = 2m$ in Term I and $k = 2m+1$ in Term II, Eq. (12.31) can be expressed as follows:

$$X[r] = \sum_{m=0,1,2,\ldots}^{K/2-1} x[2m]e^{-j(2\pi(2m)r/K)} + \sum_{m=0,1,2,\ldots}^{K/2-1} x[2m+1]e^{-j(2\pi(2m+1)r/K)}$$

or

$$X[r] = \sum_{m=0,1,2,\ldots}^{K/2-1} x[2m]e^{-j2\pi mr/(K/2)}$$

$$+ e^{-j(2\pi r/K)} \sum_{m=0,1,2,\ldots}^{K/2-1} x[2m+1]e^{-j2\pi mr/(K/2)}, \qquad (12.32)$$

where $\exp[-j2\pi(2m)r/K] = \exp[-j2\pi mr/(K/2)]$. By expressing $g[m] = x[2m]$ and $h[m] = x[2m+1]$, we can rewrite Eq. (12.32) in terms of the DFTs of $g[m]$ and $h[m]$:

$$X[r] = \underbrace{\sum_{m=0,1,2,\ldots}^{K/2-1} g[m]e^{-j2\pi mr/(K/2)}}_{=G[r]} + e^{-j2\pi r/K} \underbrace{\sum_{m=0,1,2,\ldots}^{K/2-1} h[m]e^{-j2\pi mr/(K/2)}}_{=H[r]}$$

$$(12.33)$$

or

$$X[r] = G[r] + W_K^r H[r], \qquad (12.34)$$

where W_K is defined as $\exp(-j2\pi/K)$. In FFT literature, W_K^r is generally referred to as the twiddle factor. Note that in Eqs. (12.33) and (12.34), $G[r]$ represents the ($K/2$)-point DFT of $g[k]$, the even-numbered samples of $x[k]$. Similarly, $H[r]$ represents the ($K/2$)-point DFT of $h[k]$, the odd-numbered samples of $x[k]$. Equation (12.34) thus proves Proposition 12.1.

Based on Eqs. (12.34), the procedure for determining the K-point DFT can be summarized by the following steps.

(1) Determine the ($K/2$)-point DFT $G[r]$ for $0 \leq r \leq (K/2 - 1)$ of the even-numbered samples of $x[k]$.

Fig. 12.10. Flow graph of a K-point DFT using two $(K/2)$-point DFTs for $K=8$.

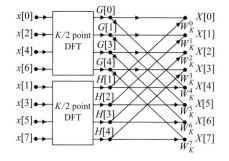

(2) Determine the $(K/2)$-point DFT $H[r]$ for $0 \le r \le (K/2-1)$ of the odd-numbered samples of $x[k]$.

(3) The K-point DFT coefficients $X[r]$ for $0 \le r \le (K-1)$ of $x[k]$ are obtained by combining the $K/2$ DFT coefficients $G[r]$ and $H[r]$ using Eq. (12.34a). Although the index r varies from zero to $K-1$, we only compute $G[r]$ and $H[r]$ over the range $0 \le r \le (K/2-1)$. Any outside value can be determined by exploiting the periodicity properties of $G[r]$ and $H[r]$, which state that

$$G[r] = G[r+K/2] \quad \text{and} \quad H[r] = H[r+K/2].$$

Figure 12.10 illustrates the flow graph for the above procedure for $K=8$-point DFT. In comparison with the direct computation of DFT using Eq. (12.30), Fig. 12.10 computes two $(K/2)$-point DFTs along with K complex additions and K complex multiplications. Consequently, $(K/2)^2 + K$ complex additions and $(K/2)^2 + K$ complex multiplications are required with the revised approach. For $K > 2$, it is easy to verify that $(K/2)^2 + K < K^2$; therefore, the revised approach provides considerable savings over the direct approach.

Assuming that K is a power of 2, Proposition 12.1 can be applied on Eq. (12.34) to compute the $(K/2)$-point DFTs $G[r]$ and $H[r]$ as follows:

$$G[r] = \underbrace{\sum_{\ell=0,1,2,\dots}^{K/4-1} g[2\ell]e^{-j(2\pi\ell r/(K/4))}}_{G'[r]} + W_{K/2}^r \underbrace{\sum_{\ell=0,1,2,\dots}^{K/4-1} g[2\ell+1]e^{-j(2\pi\ell r/(K/4))}}_{G''[r]}$$

$$(12.35)$$

and

$$H[r] = \underbrace{\sum_{\ell=0,1,2,\dots}^{K/4-1} h[2\ell]e^{-j(2\pi\ell r/(K/4))}}_{H'[r]} + W_{K/2}^r \underbrace{\sum_{\ell=0,1,2,\dots}^{K/4-1} h[2\ell+1]e^{-j(2\pi\ell r/(K/4))}}_{H''[r]}.$$

$$(12.36)$$

Fig. 12.11. Flow graphs of $(K/2)$-point DFTs using $(K/4)$-point DFTs. (a) $G[r]$; (b) $H[r]$.

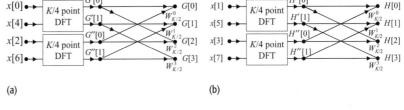

(a)   (b)

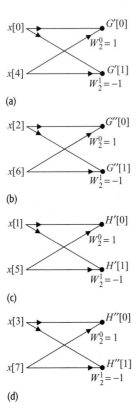

(a)

(b)

(c)

(d)

Fig. 12.12. Flow graphs of 2-point DFTs required for Fig. 12.11. (a) Top 2-point DFT $G'[0]$ and $G'[1]$ for Fig. 12.11(a). (b) Bottom 2-point DFT $G''[0]$ and $G''[1]$ for Fig 12.11(a). (c) Top 2-point DFT $H'[0]$ and $H'[1]$ for Fig 12.11(b). (d) Bottom 2-point DFT $H''[0]$ and $H''[1]$ for Fig 12.11(b).

Equation (12.35) expresses the $(K/2)$-point DFT $G[r]$ in terms of two $(K/4)$-point DFTs of the even- and odd-numbered samples of $g[k]$. Figure 12.11(a) illustrates the flow graph for obtaining $G[r]$ using Eq. (12.35). Similarly, Eq. (12.36) expresses the $(K/2)$-point DFT $H[r]$ in terms of two $(K/4)$-point DFTs of the even- and odd-numbered samples of $h[k]$, which can be implemented using the flow graph shown in Fig. 12.11(b). If K is a power of 2, then the above process can be continued until we are left with a 2-point DFT. For the aforementioned example with $K = 8$, the $(K/4)$-point DFTs in Fig. 12.11 can be implemented directly using 2-point DFTs. Using the definition of the DFT, the top left 2-point DFTs $G'[0]$ and $G'[1]$, for example, in Fig. 12.11(a) are expressed as follows:

$$G'[0] = x[0]\,\mathrm{e}^{-\mathrm{j}2\pi\ell r/2}\big|_{\ell=0,r=0} + x[4]\,\mathrm{e}^{-\mathrm{j}2\pi\ell r/2}\big|_{\ell=1,r=0} = x[0] + x[4]$$
(12.37)

and

$$G'[1] = x[0]\,\mathrm{e}^{-\mathrm{j}2\pi\ell r/2}\big|_{\ell=0,r=1} + x[4]\,\mathrm{e}^{-\mathrm{j}2\pi\ell r/2}\big|_{\ell=1,r=1} = x[0] - x[4].$$
(12.38)

The flow graphs for Eqs. (12.37) and (12.38) are shown in Fig. 12.12(a). By following this procedure, the flow diagrams for the remaining 2-point DFTs required in Fig. 12.11 are derived and are shown in Figs. 12.12(b)–(d). Because of their shape, the elementary flow graphs shown in Fig. 12.12 are generally referred to as the *butterfly* structures.

Combining the individual flow graphs shown in Figs. 12.10, 12.11, and 12.12, it is straightforward to derive the overall flow graph for the 8-point DFT, which is shown in Fig. 12.13; in this flow diagram, we have further reduced the number of operations for an 8-point DFT by noting that

$$W_{K/2}^r = \mathrm{e}^{-\mathrm{j}2\pi r/(K/2)} = \mathrm{e}^{-\mathrm{j}4\pi r/K} = W_K^{2r},$$

and by placing the common terms between the twiddle multipliers of the two branches, which are originating from the same node, before the source node.

12.6.1 Computational complexity

To derive the computational complexity of the decimation-in-time algorithm, we generalize the results obtained in Fig. 12.13, where K is set to 8. We observe that Fig. 12.13 consists of $\log_2 K = 3$ stages and that each stage

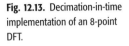

Fig. 12.13. Decimation-in-time implementation of an 8-point DFT.

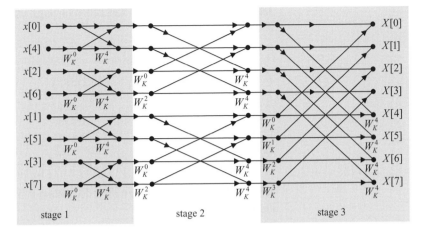

requires $K = 8$ complex multiplications and $K = 8$ complex additions. For example, stage 3 in Fig. 12.13 requires multiplications with twiddle factors W_K^0, W_K^1, W_K^2, W_K^3, and four W_K^4s. This is also obvious from Eq. (12.34), where in order to calculate the K-point DFT from two (K/2)-point DFTs, we need to perform K complex multiplications (with the twiddle factors) and approximately K complex additions. Therefore, the decimation-in-time FFT implementation for a K-point DFT requires a total of $K \log_2 K$ complex multiplications and $K \log_2 K$ complex additions.

Further reduction in the complexity of the decimation-in-time FFT implementation is obtained by observing that

$$W_K^{K/2} = \mathrm{e}^{-\mathrm{j}\pi} = -1. \qquad (12.39)$$

Note that multiplication by a factor of -1 can be performed by simply reversing the sign bit. It is observed from Fig. 12.13 that each stage contains four such multiplications (by a factor of W_K^4). In general, for a K-point FFT, $K/2$ such multiplications exist in each stage. Ignoring these trivial multiplications, the total number of complex multiplications for all K stages can be reduced to $0.5K \log_2 K$ complex multiplications. However, the number of complex additions stays the same at $K \log_2 K$. In other words, the complexity of a K-point FFT can be expressed as $0.5K \log_2 K$ butterfly operations where a *butterfly operation* includes one complex multiplication and two complex additions. Note that each complex multiplication requires a total of six flops (for four scalar multiplications and two scalar additions), and that each complex addition requires two flops (for two scalar additions). As each butterfly operation requires a total of ten flops, the overall complexity of the decimation-in-time FFT implementation is $5K \log_2 K$ flops.

Table 12.4 compares the number of computations for the direct implementation of Eq. (12.30) and the FFT implementation. As explained above, the number of scalar operations for the direct implementation is assumed to be $8K^2$ flops. whereas the number of scalar operations for the FFT implementation is

Table 12.4. Complexity of DFT calculation (in flops) with FFT and direct implementations

K	Number of flops		Increase in Speed
	FFT ($5K \log_2 K$)	direct ($8K^2$)	
32	800	8192	10.2
256	10 240	524 288	51.2
1024	51 200	8 388 608	163.8
8192	532 480	536 870 912	1 008.2

Table 12.5. Data reordering in radix-2 decimation-in-time FFT implementation

Original order, $x[k]$	Binary representation	Bit-reversed representation	
		binary	decimal
	$x[b_2 b_1 b_0]$	$x[b_0 b_1 b_2]$	$x_{re}[k]$
$x[0]$	$x[000]$	$x[000]$	$x[0]$
$x[1]$	$x[001]$	$x[100]$	$x[4]$
$x[2]$	$x[010]$	$x[010]$	$x[2]$
$x[3]$	$x[011]$	$x[110]$	$x[6]$
$x[4]$	$x[100]$	$x[001]$	$x[1]$
$x[5]$	$x[101]$	$x[101]$	$x[5]$
$x[6]$	$x[110]$	$x[011]$	$x[3]$
$x[7]$	$x[111]$	$x[111]$	$x[7]$

assumed to be $5K \log_2 K$ flops. For large values of K, say 8192, Table 12.4 illustrates a speed-up by up to a factor of 1000 with the FFT implementation. For real-valued sequences, the number of flops can be further reduced by exploiting the symmetry properties of the DFT. Further reduction in the complexity of the decimation-in-time FFT implementation can be obtained by ignoring multiplications by the twiddle factor W_K^0 as $W_K^0 = 1$.

12.6.2 Reordering of the input sequence

In Fig. 12.13, we observe that the input sequence $x[k]$ with length K has been arranged in an order that is considerably different from the natural order of occurrence. This arrangement is referred to as the bit-reversed order and is obtained by expressing the index k in terms of $\log_2 K$ bits and then reversing the order of bits such that the most significant bit becomes the least significant bit, and vice versa. For $K = 8$, the reordering of the input sequence is illustrated in Table 12.5.

The function myfft, available in the accompanying CD, implements the radix-2 decimation-in-time FFT algorithm. Direct computation of the DFT coefficients using Eq. (12.16) is also implemented and provided as a second function, `mydft`. The reader should confirm that the two functions compute the same result, with the exception that the implementation of `myfft` is computationally efficient

As mentioned earlier, MATLAB also provides a built-in function `fft` to compute the DFT of a sequence. Depending on the length of the sequence, the `fft` function chooses the most efficient algorithm to compute the DFT. For example, when the length of the sequence is a power of 2, it uses the radix-2 algorithm. On the other hand, if the length is such that a font method is not possible, it uses the direct method based on Eq. (12.15).

12.7 Summary

This chapter introduces the discrete Fourier transform (DFT) for time-limited sequences as an extension of the DTFT where the DTFT frequency Ω is discretized to a finite set of values $\Omega = 2\pi r/M$, for $0 \leq r \leq (M-1)$. The M-point DFT pair for a causal, aperiodic sequence $x[k]$ of length N is defined as follows:

DFT analysis equation $X[r] = \displaystyle\sum_{k=0}^{N-1} x[k] e^{-j(2\pi kr/M)}$ for $0 \leq r \leq M-1$;

DFT synthesis equation $x[k] = \dfrac{1}{M} \displaystyle\sum_{r=0}^{M-1} X[r] e^{j(2\pi kr/M)}$ for $0 \leq k \leq N-1$.

For $M = N$, Section 12.2 implements the synthesis and analysis equations of the DFT in the matrix-vector format as follows:

DFT synthesis equation $x = FX$;

DFT analysis equation $X = F^{-1}x$,

where F is defined as the DFT matrix given by

$$F = \begin{bmatrix} 1 & 1 & 1 & \cdots & 1 \\ 1 & e^{-j(2\pi/N)} & e^{-j(4\pi/N)} & \cdots & e^{-j(2(N-1)\pi/N)} \\ 1 & e^{-j(4\pi/N)} & e^{-j(8\pi/N)} & \cdots & e^{-j(4(N-1)\pi/N)} \\ \vdots & \vdots & \vdots & \ddots & \vdots \\ 1 & e^{-j(2(N-1)\pi/N)} & e^{-j(4(N-1)\pi/N)} & \cdots & e^{-j(2(N-1)(N-1)\pi/N)} \end{bmatrix}.$$

The columns (or equivalently the rows) of the DFT matrix define the basis functions for the DFT.

Section 12.3 used the M-point DFT $X[r]$ to estimate the CTFT spectrum $X(\omega)$ of an aperiodic signal $x(t)$ using the following relationship:

$$X(\omega_r) \approx \frac{MT_1}{N} X_2[r],$$

where T_1 is the sampling interval used to discretize $x(t)$, ω_r are the CTFT frequencies that are given by $2\pi r/(MT_1)$ for $-0.5(M-1) \leq r \leq 0.5(M-1)$, and N is the number of samples obtained from the CT signal. Similarly, the DFT $X[r]$ can be used to determine the DTFT $X(\Omega)$ of a time-limited sequence $x[k]$ of length N as

$$X(\Omega_r) = \frac{N}{M} X[r]$$

at discrete frequencies $\Omega_r = 2\pi r/M$, for $0 \leq r \leq M-1$.

Section 12.4 covered the following properties of the DFT.

(1) The periodicity property states that the M-point DFT of a sequence is periodic with period M.

(2) The orthogonality property states that the basis functions of the DFTs are orthogonal to each other.

(3) The linearity property states that the overall DFT of a linear combination of DT sequences is given by the same linear combination of the individual DFTs.

(4) The Hermitian symmetry property states that the DFT of a real-valued sequence is Hermitian. In other words, the real component of the DFT of a real-valued sequence is even, while the imaginary component is odd.

(5) The time-shifting property states that shifting a sequence in the time domain towards the right-hand side by an integer constant m is equivalent to multiplying the DFT of the original sequence by the complex exponential $\exp(-j2\pi m/M)$. Similarly, shifting towards the left-hand side by an integer m is equivalent to multiplying the DTFT of the original sequence by the complex exponential $\exp(j2\pi m/M)$.

(6) The time-convolution property states that the periodic convolution of two DT sequences is equivalent to the multiplication of the individual DFTs of the two sequences in the frequency domain.

(7) Parseval's theorem states that the energy of a DT sequence is preserved in the DFT domain.

Section 12.5 used the convolution property to derive alternative procedures for computing the convolution sum. Depending on the sequence lengths, these procedures may provide considerable savings over the direct implementation of the convolution sum.

Section 12.6 covers the decimation-in-time FFT implementation of the DFT. In deriving the FFT algorithm, we assume that the length N of the sequence equals the number M of samples in the DFT, i.e. $N = M = K$. We showed that if K is a power of 2, then the FFT implementations have a computational complexity of $O(K \log_2 K)$.

Problems

12.1 Calculate analytically the DFT of the following sequences, with length $0 \leq k \leq N - 1$:

(i) $x[k] = \begin{cases} 1 & k = 0, 3 \\ 0 & k = 1, 2 \end{cases}$ with length $N = 4$;

(ii) $x[k] = \begin{cases} 1 & k \text{ even} \\ -1 & k \text{ odd} \end{cases}$ with length $N = 8$;

(iii) $x[k] = 0.6^k$ with length $N = 8$;

(iv) $x[k] = u[k] - u[k - 8]$ with length $N = 8$;

(v) $x[k] = \cos(\omega_0 k)$ with $\omega_0 \neq 2\pi m/N$, $m \in Z$.

12.2 Calculate the DFT of the time-limited sequences specified in Examples 12.1(i)–(iv) using the matrix-vector approach.

12.3 Determine the time-limited sequence, with length $0 \leq k \leq N - 1$, corresponding to the following DFTs $X[r]$, which are defined for the DFT index $0 \leq r \leq N - 1$:

(i) $X[r] = [1 + j4, -2 - j3, -2 + j3, 1 - j4]$ with $N = 4$;

(ii) $X[r] = [1, 0, 0, 1]$ with $N = 4$;

(iii) $X[r] = \exp -j(2\pi k_0 r/N)$, where k_0 is a constant;

(iv) $X[r] = \begin{cases} 0.5N & r = k_0, N - k_0 \\ 0 & \text{elsewhere} \end{cases}$ where k_0 is a constant;

(v) $X[r] = e^{-j\pi r(m-1)/N} \dfrac{\sin(\pi r m/N)}{\sin(\pi r/N)}$ where $m \in Z$ and $0 < m < N$;

(vi) $X[r] = \left(\dfrac{r}{N}\right)$ for $0 \leq r \leq N - 1$.

12.4 In Problem 11.1, we determined the DTFT representation for each of the following DT periodic sequences using the DTFS. Using MATLAB, compute the DTFT representation based on the FFT algorithm. Plot the frequency characteristics and compare the computed results with the analytical results derived in Chapter 11.

(i) $x[k] = k$,   for $0 \leq k \leq 5$ and $x[k + 6] = x[k]$;

(ii) $x[k] = \begin{cases} 1 & 0 \leq k \leq 2 \\ 0.5 & 3 \leq k \leq 5 \\ 0 & 6 \leq k \leq 8 \end{cases}$ and $x[k + 9] = x[k]$;

(iii) $x[k] = 3 \sin\left(\dfrac{2\pi}{7}k + \dfrac{\pi}{4}\right)$;

(iv) $x[k] = 2 \exp\left(j\dfrac{5\pi}{3}k + \dfrac{\pi}{4}\right)$;

(v) $x[k] = \displaystyle\sum_{m=-\infty}^{\infty} \delta[k - 5m]$;

(vi) $x[k] = \cos(10\pi k/3)\cos(2\pi k/5)$;

(vii) $x[k] = |\cos(2\pi k/3)|$.

12.5 (a) Using the FFT algorithm in MATLAB, determine the DTFT representation for the following sequences. Plot the magnitude and phase spectra in each case.

(i) $x[k] = 2$;

(ii) $x[k] = \begin{cases} 3 - |k| & |k| < 3 \\ 0 & \text{otherwise}; \end{cases}$

(iii) $x[k] = k3^{-|k|}$;

(iv) $x[k] = \alpha^k \cos(\omega_0 k)u[k],\ |\alpha| < 1$;

(v) $x[k] = \alpha^k \sin(\omega_0 k + \phi)u[k],\ |\alpha| < 1$;

(vi) $x[k] = \dfrac{\sin(\pi k/5)\sin(\pi k/7)}{\pi^2 k^2}$;

(vii) $x[k] = \displaystyle\sum_{m=-\infty}^{\infty} \delta[k - 5m - 3]$;

(viii) $x[k] = \begin{cases} 3 - |k| & |k| < 3 \\ 0 & |k| = 3 \end{cases}$ and $x[k + 7] = x[k]$;

(ix) $x[k] = e^{j(0.2\pi k + 45°)}$;

(x) $x[k] = k3^{-k}u[k] + e^{j(0.2\pi k + 45°)}$.

(b) Compare the obtained results with the analytical results derived in Problem 11.4(a).

12.6 Using the FFT algorithm in MATLAB, determine the CTFT representation for each of the following CT functions. Plot the frequency characteristics and compare the results with the analytical results presented in Table 5.1.

(i) $x(t) = e^{-2t}u(t)$;

(ii) $x(t) = e^{-4|t|}$;

(iii) $x(t) = t^4 e^{-4t}u(t)$;

(iv) $x(t) = e^{-4t}\cos(10\pi t)u(t)$;

(v) $x(t) = e^{-t^2/2}$.

12.7 Prove the Hermitian property of the DFT.

12.8 Prove the time-shifting property of the DFT.

12.9 Prove the periodic-convolution property of the DFT.

12.10 Prove Parseval's theorem for the DFT.

12.11 Without explicitly determining the DFT $X[r]$ of the time-limited sequence

$$x[k] = [6 \quad 8 \quad -5 \quad 4 \quad 16 \quad 22 \quad 7 \quad 8 \quad 9 \quad 44 \quad 2],$$

compute the following functions of the DFT $X[r]$:

(i) $X[0]$;

(ii) $X[10]$;

(iii) $X[6]$;

(iv) $\displaystyle\sum_{r=0}^{10} X[r]$;

(v) $\displaystyle\sum_{r=0}^{10} |X[r]|^2$.

12.12 Without explicitly determining the the time-limited sequence $x[k]$ for the following DFT:

$$X[r] = [12, \quad 8+j4, \quad -5, \quad 4+j1, \quad 16, \quad 16, \quad 4-j1, \quad -5, \quad 8-j4],$$

compute the following functions of the DFT $X[r]$:

(i) $x[0]$;

(ii) $x[9]$;

(iii) $x[6]$;

(iv) $\displaystyle\sum_{r=0}^{9} x[k]$;

(v) $\displaystyle\sum_{r=0}^{9} |x[k]|^2$;

12.13 Given the DFT pair

$$x[k] \xleftrightarrow{\text{DFT}} X[r],$$

for a sequence of length N, express the DFT of the following sequences as a function of $X[r]$:

(i) $y[k] = x[2k]$;

(ii) $y[k] = \begin{cases} x[0.5k] & k \text{ even} \\ 0 & \text{elsewhere}; \end{cases}$

(iii) $y[k] = x[N-1-k]$ for $0 \le k \le N-1$;

(iv) $y[k] = \begin{cases} x[k] & 0 \le k \le N-1 \\ 0 & N \le k \le 2N-1; \end{cases}$

(v) $y[k] = (x[k] - x[k-2])e^{j(10\pi k/N)}$.

12.14 Compute the linear convolution of the following pair of time-limited sequences using the DFT-based approach. Be careful with the time indices of the result of the linear convolution.

(i) $x_1[k] = \begin{cases} k & 0 \le k \le 3 \\ 0 & \text{otherwise} \end{cases}$ and $x_2[k] = \begin{cases} 2 & -1 \le k \le 2 \\ 0 & \text{otherwise}; \end{cases}$

(ii) $x_1[k] = k$ for $0 \le k \le 3$ and $x_2[k] = \begin{cases} 5 & k = 0, 1 \\ 0 & \text{otherwise}; \end{cases}$

(iii) $x_1[k] = \begin{cases} 2 & 0 \le k \le 2 \\ 0 & \text{otherwise} \end{cases}$ and $x_2[k] = \begin{cases} k+1 & 0 \le k \le 4 \\ 0 & \text{otherwise}; \end{cases}$

(iv) $x_1[k] = \begin{cases} -1 & k = -1 \\ 1 & k = 0 \\ 2 & k = 1 \\ 0 & \text{otherwise} \end{cases}$ and $x_2[k] = \begin{cases} 3 & k = -1, 2 \\ 1 & k = 0 \\ -2 & k = 1, 3 \\ 0 & \text{otherwise}; \end{cases}$

(v) $x_1[k] = \begin{cases} |k| & |k| \le 2 \\ 0 & \text{otherwise} \end{cases}$ and $x_2[k] = \begin{cases} 2^{-k} & 0 \le k \le 3 \\ 0 & \text{otherwise}. \end{cases}$

12.15 Draw the flow graph for a 6-point DFT by subdividing into three 2-point DFTs that can be combined to compute $X[r]$. Repeat for the subdivision of two 3-point DFTs. Which flow graph provides more computational savings?

12.16 Draw a flow graph for a 10-point decimation-in-time FFT algorithm using two DFTs of size 5 in the first stage of the flow graph and five DFTs of size 2 in the second stage. Compare the computational complexity of the algorithm with the direct approach based on the definition.

12.17 Assume that $K = 3^3$. Draw the flow graph for a K-point decimation-in-time FFT algorithm consisting of three stages by using radix-3 as the basic building block. Compare the computational complexity of the algorithm with the direct approach based on the definition.

12.18 Consider two real-valued N-point sequences $x_1[k]$ and $x_2[k]$ with DFTs $X_1[r]$ and $X_2[r]$, respectively. Let $p[k]$ be an N-point complex-valued sequence such that $p[k] = x_1[k] + jx_2[k]$ and let the DFT of $p[k]$ be denoted by $P[r]$.
   (a) Show that the DFTs $X_1[r]$ and $X_2[r]$ can be obtained from the DFT $P[r]$.
   (b) Assume that $N = 2^m$ and that the decimation-in-time FFT algorithm discussed in Section 12.6 is used to calculate the DFT $P[r]$. Estimate the total number of flops required to calculate the DFTs $X_1[r]$ and $X_2[r]$ using the procedure in part (a).

12.19 Consider a real-valued N-point sequence $x[k]$, where N is a power of 2. Let $x_1[k]$ and $x_2[k]$ be two $N/2$-point real-valued sequences such that $x_1[k] = x[2k]$ and $x_2[k] = x[2k+1]$ for $0 \le k \le N/2 - 1$. Let the N-point DFT of $x[k]$ be denoted by $X[r]$ and let the $N/2$-point DFT of $x_1[k]$ and $x_2[k]$ be denoted by $X_1[r]$ and $X_2[r]$, respectively.
   (a) Determine $X[r]$ in terms of $X_1[r]$ and $X_2[r]$.
   (b) Estimate the total number of flops required to calculate $X[r]$ using the procedure discussed in Problem 12.18.

13 The z-transform

In Chapter 11, we introduced two frequency representations, namely the discrete-time Fourier series (DTFS) and the discrete-time Fourier transform (DTFT) for DT signals. These frequency representations are exploited to determine the output response of an LTID system. Unfortunately, the DTFT does not exist for all signals (e.g., periodic signals). In situations where the DTFT does not exist, an alternative transform, referred to as the *z-transform*, may be used for the analysis of LTID systems. Even for DT sequences for which the DTFT exists, the z-transforms are always real-valued, rational functions of the independent variable z provided that the DT sequences are real. In comparison, the DTFT is generally complex-valued. Therefore, using the z-transform simplifies the algebraic manipulations and leads to flow diagram representations of the DT systems, a pivotal step needed to fabricate the DT system in silicon. Finally, the DTFT can only be applied to a stable LTID system for which the impulse response is absolutely summable. Since the z-transform exists for both stable and unstable LTID systems, the z-transform can be used to analyze a broader range of LTID systems.

The difference between the DTFT and the z-transform lies in the choice of the independent variable used in the transformed domain. The DTFT $X(\Omega)$ of a DT sequence $x[k]$ uses the complex exponentials $e^{jk\Omega}$ as its basis function and maps $x[k]$ in terms of $e^{jk\Omega}$. The z-transform $X(z)$ expresses $x[k]$ in terms of z^k, where the independent variable z is given by $z = e^{(\sigma + j\Omega)}$. The z-transform is, therefore, a generalization of the DTFT, just as the Laplace transform is a generalization of the CTFT. In this chapter, we introduce the z-transform and illustrate its applications in the analysis of LTID systems.

This chapter is organized as follows. Section 13.1 defines the bilateral, also referred to as the two-sided, z-transform and illustrates the steps involved in its computation through a series of examples. For causal signals, the bilateral z-transform reduces to the one-sided, or unilateral, z-transform, which is covered in Section 13.2. Section 13.3 presents inverse methods of calculating the time-domain representation of the z-transform. The properties of the z-transform are derived in Section 13.4. Sections 13.5–13.9 cover various applications of the z-transform. Section 13.5 applies the z-transform to calculate the output of an LTID system from the input sequence and the impulse response of the LTID system. The relationship between the Laplace transform and the z-transform is discussed in Section 13.6. Stability analysis of the LTID system in the z-domain is presented in Section 13.7, while graphical techniques to derive the frequency response from the z-transform are discussed in Section 13.8. Section 13.9 compares the DTFT and z-transform in calculating the steady state and transient responses of an LTID system. Section 13.10 introduces important MATLAB library functions useful in computing the z-transform and in the analysis of LTID systems. Finally, the chapter is concluded in Section 13.11 with a summary of important concepts.

13.1 Analytical development

Section 11.1 defines the synthesis and analysis equations of the DTFT pair $x[k] \xleftrightarrow{\text{DTFT}} X(\Omega)$ as follows:

DTFT synthesis equation $\quad x[k] = \dfrac{1}{2\pi} \displaystyle\int\limits_{(2\pi)} X(\Omega) e^{j\Omega k} d\Omega;$ $\qquad\qquad$ (13.1)

DTFT analysis equation $\quad X(\Omega) = \displaystyle\sum_{k=-\infty}^{\infty} x[k] e^{-j\Omega k}.$ $\qquad\qquad$ (13.2)

To derive the expression for the bilateral z-transform, we calculate the DTFT of the modified version $x[k]e^{-\sigma k}$ of the DT signal. Based on Eq. (13.2), the DTFT of the modified signal is given by

$$\Im\left\{x[k]e^{-\sigma k}\right\} = \sum_{k=-\infty}^{\infty} x[k]e^{-\sigma k}e^{-j\Omega k} = \sum_{k=-\infty}^{\infty} x[k]e^{-(\sigma+j\Omega)k}. \qquad (13.3)$$

Substituting $e^{\sigma+j\Omega} = z$ in Eq. (13.3) leads to the following definition for the bilateral z-transform:

z-analysis equation $\quad X(z) = \Im\left\{x[k]e^{-\sigma k}\right\} = \displaystyle\sum_{k=-\infty}^{\infty} x[k]z^{-k}.$ $\qquad\qquad$ (13.4)

It may be noted that the summation in Eq. (13.4) is absolutely summable only for selected values of z. For other values of z, the infinite sum in Eq. (13.4) may not converge to a finite value, and hence $X(z)$ becomes infinite. The region in the complex z-plane, where summation (13.4) is finite, is referred to as the region of convergence (ROC) of the z-transform $X(z)$.

By following a similar derivation for the DTFT synthesis equation, Eq. (13.1), the expression for the inverse z-transform is given by

z-synthesis equation $\quad x[k] = \dfrac{1}{2\pi j} \displaystyle\oint_{C} X(z) z^{k-1} dz,$ $\qquad\qquad$ (13.5) where C is a closed contour traversed in the counterclockwise direction within the ROC. Solving Eq. (13.5) involves the application of contour integration techniques and is, therefore, seldom used directly. In Section 13.3, we will consider alternative approaches based on the look-up table, partial fraction expansion, and power series expansion to evaluate the inverse z-transform.

Collectively, Eqs. (13.4) and (13.5) form the bilateral z-transform pair, which is denoted by

$$x[k] \xleftrightarrow{z} X(z) \quad \text{or} \quad Z\{x[k]\} = X(z). \tag{13.6}$$

To illustrate the steps involved in computing the z-transform, we consider the following examples.

Example 13.1
Calculate the bilateral z-transform of the exponential sequence $x[k] = \alpha^k u[k]$.

Solution
Substituting $x[k] = \alpha^k u[k]$ in Eq. (13.4), we obtain

$$X(z) = \sum_{k=-\infty}^{\infty} \alpha^k u[k] z^{-k} = \sum_{k=0}^{\infty} (\alpha z^{-1})^k$$

$$= \begin{cases} \dfrac{1}{1 - \alpha z^{-1}} & |\alpha z^{-1}| < 1 \\ \text{undefined} & \text{elsewhere.} \end{cases}$$

In the above expression, if $|\alpha z^{-1}| \geq 1$ the bilateral z-transform has an infinite value. In such cases, we say that the z-transform is not defined. The set of values of z over which the bilateral z-transform is defined is referred to as the region of convergence (ROC) associated with the z-transform. In this example, the ROC for the z-transform pair

$$\alpha^k u[k] \xleftrightarrow{z} \frac{1}{1 - \alpha z^{-1}}$$

is given by

$$\text{ROC:} \left|\alpha z^{-1}\right| < 1 \quad \text{or} \quad |z| > |\alpha|.$$

Figure 13.1 highlights the ROC by shading the appropriate region in the complex z-plane.

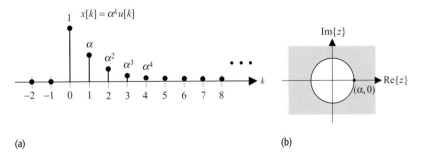

Fig. 13.1. (a) DT exponential sequence $x[k] = \alpha^k u[k]$; (b) the ROC, $|z| > |\alpha|$, associated with its bilateral z-transform. The ROC is shown as the shaded area and lies outside the circle of radius α.

Example 13.1 derives the bilateral z-transform of the exponential sequence $x[k] = \alpha^k u[k]$:

$$\alpha^k u[k] \overset{z}{\longleftrightarrow} \frac{1}{1 - \alpha z^{-1}}, \quad \text{with} \quad \text{ROC } |z| > |\alpha|.$$

Since no restriction is imposed on the magnitude of α, the bilateral z-transform of the exponential sequence exists for all values of α within the specified ROC. Recall that the DTFT of an exponential sequence exists only for $\alpha < 1$. For $\alpha \geq 1$, the exponential sequence is not summable and its DTFT does not exist. This is an important distinction between the DTFT and the bilateral z-transform. While the DTFT exists for a limited number of absolutely summable sequences, no such restrictions exist for the z-transform. By associating an ROC with the bilateral z-transform, we can evaluate the z-transform for a much larger set of sequences.

Example 13.2
Calculate the bilateral z-transform of the left-hand-sided exponential sequence $x[k] = -\alpha^k u[-k - 1]$.

Solution
For the DT sequence $x[k] = -\alpha^k u[-k - 1]$, Eq. (13.4) reduces to

$$X(z) = \sum_{k=-\infty}^{\infty} -\alpha^k u[-k - 1] z^{-k} = - \sum_{k=-\infty}^{-1} (\alpha z^{-1})^k.$$

To make the limits of summation positive, we substitute $m = -k$ in the above equation to obtain

$$X(z) = - \sum_{m=1}^{\infty} (\alpha^{-1} z)^m = \begin{cases} -\dfrac{\alpha^{-1} z}{1 - \alpha^{-1} z} & |\alpha^{-1} z| < 1 \\ \text{undefined} & \text{elsewhere,} \end{cases}$$

which simplifies to

$$X(z) = \begin{cases} \dfrac{1}{1 - \alpha z^{-1}} & |z| < |\alpha| \\ \text{undefined} & \text{elsewhere.} \end{cases}$$

The DT sequence $x[k] = -\alpha^k u[-k - 1]$ and the ROC associated with its z-transform are illustrated in Fig. 13.2.

In Examples 13.1 and 13.2, we have proved the following z-transform pairs:

$$\alpha^k u[k] \overset{z}{\longleftrightarrow} \frac{1}{1 - \alpha z^{-1}}, \quad \text{with} \quad \text{ROC } |z| > |\alpha|,$$

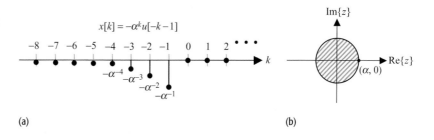

Fig. 13.2. (a) Non-causal function $x[k] = -\alpha^k u[-k-1]$; (b) its associated ROC, $|z| < |\alpha|$, shown as the shaded area excluding the circle, over which the bilateral z-transform exists.

and

$$-\alpha^k u[-k-1] \xleftrightarrow{z} \frac{1}{1 - \alpha z^{-1}}, \quad \text{with} \quad \text{ROC } |z| < |\alpha|.$$

Although the algebraic expressions for the bilateral z-transforms are the same for the two functions, the ROCs are different. This implies that a bilateral z-transform is completely specified only if both the algebraic expression and the associated ROC are included in its specification.

13.2 Unilateral z-transform

In Section 13.1, we introduced the bilateral z-transform, which may be used to analyze both causal and non-causal LTID systems. Since most physical systems in signal processing are causal, a simplified version of the bilateral z-transform exists in such cases. The simplified bilateral z-transform for causal signals and systems is referred to as the unilateral z-transform, and it is obtained by assuming $x[k] = 0$ for $k < 0$. The analysis equation, Eq. (13.4), simplifies as follows:

unilateral z-transform
$$X(z) = \sum_{k=0}^{\infty} x[k]z^{-k}. \tag{13.7}$$

Unless explicitly stated, we will, in subsequent discussion, assume the "unilateral" z-transform when referring to the z-transform. If the bilateral z-transform is being discussed, we will specifically state this. To clarify further the differences between the unilateral and bilateral z-transforms, we summarize the major points.

(1) The unilateral z-transform simplifies the analysis of causal LTID systems. Since most physical systems are naturally causal, we will mostly use unilateral z-transform in our computations. However, the unilateral z-transform cannot be used to analyze non-causal systems directly.
(2) For causal signals and systems, the unilateral and bilateral z-transforms are the same.

(3) The synthesis equation used for calculating the inverse of the unilateral z-transform is the same as Eq. (13.5) used for evaluating the inverse of the bilateral transform.

Example 13.3

Calculate the unilateral z-transform for the following sequences:

  (i) unit impulse sequence, $x_1[k] = \delta[k]$;
 (ii) unit step sequence, $x_2[k] = u[k]$;
(iii) exponential sequence, $x_3[k] = \alpha^k u[k]$;
 (iv) first-order, time-rising, exponential sequence, $x_4[k] = k\alpha^k u[k]$;
  (v) time-limited sequence, $x_5[k] = \begin{cases} 1 & k = 0, 1 \\ 2 & k = 2, 5 \\ 0 & \text{otherwise.} \end{cases}$

Solution

(i) By definition,

$$X_1(z) = \sum_{k=0}^{\infty} \delta[k]z^{-k} = \delta[0]z^0 = 1, \quad \text{ROC: entire z-plane.}$$

The z-transform pair for an impulse sequence is given by

$$\delta[k] \xleftrightarrow{z} 1, \quad \text{ROC: entire z-plane.}$$

(ii) By definition,

$$X_2(z) = \sum_{k=0}^{\infty} u[k]z^{-k} = \sum_{k=0}^{\infty} z^{-k} = \begin{cases} \dfrac{1}{1-z^{-1}} & \text{for } |z^{-1}| < 1 \\ \text{undefined} & \text{elsewhere.} \end{cases}$$

The z-transform pair for a unit step sequence is given by

$$u[k] \xleftrightarrow{z} \frac{1}{1-z^{-1}}, \quad \text{ROC: } |z| > 1.$$

In the above transform pair, note that the ROC $|z^{-1}| < 1$ is equivalent to $|z| > 1$ and consists of the region outside a circle of unit radius in the complex z-plane. This circle of unit radius, with the origin of the z-plane as the center, is referred to as the unit circle and plays an important role in the determination of the stability of an LTID system. We will discuss stability issues in Section 13.7.

(iii) By definition,

$$X_3(z) = \sum_{k=0}^{\infty} \alpha^k u[k]z^{-k} = \sum_{k=0}^{\infty} (\alpha z^{-1})^k = \begin{cases} \dfrac{1}{1-\alpha z^{-1}} & \text{for } |\alpha z^{-1}| < 1 \\ \text{undefined} & \text{elsewhere.} \end{cases}$$

The z-transform pair for an exponential sequence is therefore given by

$$\alpha^k u[k] \xleftrightarrow{z} \frac{1}{1-\alpha z^{-1}}, \quad \text{ROC: } |z| > |\alpha|.$$

In the above transform pair, the ROC $|\alpha z^{-1}| < 1$ is equivalent to $|z| > \alpha$ and consists of the region outside the circle of radius $|z| = \alpha$ in the complex z-plane. Example 13.1 derives the bilateral z-transform for the function $x_3[k] = \alpha^k u[k]$. Since the function is causal, the bilateral and unilateral z-transforms are identical.

(iv) By definition,

$$X_4(z) = \sum_{k=0}^{\infty} k\alpha^k u[k] z^{-k} = \sum_{k=0}^{\infty} k(\alpha z^{-1})^k.$$

Using the following result:

$$\sum_{k=0}^{\infty} k r^k = \frac{r}{(1-r)^2}, \quad \text{provided } |r| < 1,$$

the above summation reduces to

$$X_4(z) = \frac{\alpha z^{-1}}{(1-\alpha z^{-1})^2}, \quad \text{ROC: } |\alpha z^{-1}| < 1.$$

The z-transform pair for a time-rising, complex exponential is given by

$$k\alpha^k u[k] \xleftrightarrow{z} \frac{\alpha z^{-1}}{(1-\alpha z^{-1})^2} \text{ or } \frac{\alpha z}{(z-\alpha)^2}, \quad \text{ROC: } |z| > |\alpha|.$$

(v) Since the input sequence $x_5[k]$ is zero outside the range $0 \le k \le 5$, Eq. (13.4) reduces to

$$X_5(z) = \sum_{k=0}^{\infty} x[k] z^{-k} = x[0] + x[1]z^{-1} + x[2]z^{-2} + x[3]z^{-3} + x[4]z^{-4} + x[5]z^{-5}.$$

Substituting the values of $x_5[k]$ for the range $0 \le k \le 5$, we obtain

$$X_5(z) = 1 + z^{-1} + 2z^{-2} + 2z^{-5} \quad \text{ROC: entire z-plane, except } z = 0.$$

For finite-duration sequences, the ROC is always the entire z-plane except for the possible exclusion of $z = 0$ and $z = \infty$.

13.2.1 Relationship between the DTFT and the z-transform

Comparing Eq. (13.2) with Eq. (13.4), the DTFT can be expressed in terms of the bilateral z-transform as follows:

$$X(\Omega) = \sum_{k=-\infty}^{\infty} x[k] z^{-k} = X(z)|_{z=e^{j\Omega}}. \tag{13.8}$$

Since, for causal functions, the bilateral and unilateral z-transforms are the same, Eq. (13.8) is also valid for the unilateral z-transform for causal functions.

Equation (13.8) shows that the DTFT is a special case of the z-transform with $z = e^{j\Omega}$. The equality $z = e^{j\Omega}$ corresponds to the circle of unit radius ($|z| = 1$) in the complex z-plane. Equation (13.8) therefore implies that the

Table 13.1. Unilateral z-transform pairs for several causal DT sequences

DT sequence	z-transform with ROC
$x[k] = \dfrac{1}{2\pi\mathrm{j}} \oint_C X(z)z^{k-1}\mathrm{d}z$	$X(z) = \displaystyle\sum_{k=-\infty}^{\infty} x[k]z^{-k}$
(1) Unit impulse $\quad x[k] = \delta[k]$	1, ROC: entire z-plane
(2) Delayed unit impulse $\quad x[k] = \delta[k - k_0]$	z^{-k_0}, ROC: entire z-plane, except $z = 0$
(3) Unit step $\quad x[k] = u[k]$	$\dfrac{1}{1 - z^{-1}} = \dfrac{z}{z - 1}$, ROC: $|z| > 1$
(4) Exponential $\quad x[k] = \alpha^k u[k]$	$\dfrac{1}{1 - \alpha z^{-1}} = \dfrac{z}{z - \alpha}$, ROC: $|z| > |\alpha|$
(5) Delayed exponential $\quad x[k] = \alpha^{k-1} u[k - 1]$	$\dfrac{z^{-1}}{1 - \alpha z^{-1}} = \dfrac{1}{z - \alpha}$, ROC: $|z| > |\alpha|$
(6) Ramp $\quad x[k] = k u[k]$	$\dfrac{z^{-1}}{(1 - z^{-1})^2} = \dfrac{z}{(z - 1)^2}$, ROC: $|z| > 1$
(7) Time-rising exponential $\quad x[k] = k\alpha^k u[k]$	$\dfrac{\alpha z^{-1}}{(1 - \alpha z^{-1})^2} = \dfrac{\alpha z}{(z - \alpha)^2}$, ROC: $|z| > |\alpha|$
(8) Causal cosine $\quad x[k] = \cos(\Omega_0 k)u[k]$	$\dfrac{1 - z^{-1}\cos\Omega_0}{1 - 2z^{-1}\cos\Omega_0 + z^{-2}} = \dfrac{z[z - \cos\Omega_0]}{z^2 - 2z\cos\Omega_0 + 1}$, ROC: $|z| > 1$
(9) Causal sine $\quad x[k] = \sin(\Omega_0 k)u[k]$	$\dfrac{z^{-1}\sin\Omega_0}{1 - 2z^{-1}\cos\Omega_0 + z^{-2}} = \dfrac{z\sin\Omega_0}{z^2 - 2z\cos\Omega_0 + 1}$, ROC: $|z| > 1$
(10) Exponentially modulated cosine $\quad x[k] = \alpha^k \cos(\Omega_0 k)u[k]$	$\dfrac{1 - \alpha z^{-1}\cos\Omega_0}{1 - 2\alpha z^{-1}\cos\Omega_0 + \alpha^2 z^{-2}} = \dfrac{z[z - \alpha\cos\Omega_0]}{z^2 - 2\alpha z\cos\Omega_0 + \alpha^2}$, ROC: $|z| > |\alpha|$
(11) Exponentially modulated sine I $\quad x[k] = \alpha^k \sin(\Omega_0 k)u[k]$	$\dfrac{\alpha z^{-1}\sin\Omega_0}{1 - 2\alpha z^{-1}\cos\Omega_0 + \alpha^2 z^{-2}} = \dfrac{\alpha z\sin\Omega_0}{z^2 - 2\alpha z\cos\Omega_0 + \alpha^2}$, ROC: $|z| > \alpha$
(12) Exponentially modulated sine II $\quad x[k] = r\alpha^k \sin(\Omega_0 k + \theta)u[k]$, with $\alpha \in R$.	$\dfrac{A + Bz^{-1}}{1 + 2\gamma z^{-1} + \alpha^2 z^{-2}} = \dfrac{z(Az + B)}{z^2 + 2\gamma z + \gamma^2}$, ROC: $|z| > |\alpha|^{(a)}$

$^{(a)}$ Where $r = \sqrt{\dfrac{A^2\alpha^2 + B^2 - 2AB\gamma}{\alpha^2 - \gamma^2}}$, $\Omega_0 = \cos^{-1}\left(\dfrac{-\gamma}{\alpha}\right)$, and $\theta = \tan^{-1}\left(\dfrac{A\sqrt{\alpha^2 - \gamma^2}}{B - A\gamma}\right)$.

DTFT is obtained by computing the z-transform along the unit circle in the complex z-plane.

Table 13.1 lists the z-transforms for several commonly used sequences. Comparing Table 13.1 with Table 11.2, we observe that when the sequence is causal and its DTFT exists, the DTFT can be obtained from the z-transform by substituting $z = \mathrm{e}^{\mathrm{j}\Omega}$. Since the substitution $z = \mathrm{e}^{\mathrm{j}\Omega}$ can only be made if the ROC contains the unit circle, an alternative condition for the existence of the DTFT is the inclusion of the unit circle within the ROC of the z-transform. If the ROC of a z-transform does not include the unit circle, we cannot substitute $z = \mathrm{e}^{\mathrm{j}\Omega}$ and we say that its DTFT cannot be obtained from Eq. (13.8). For example, the ROC of the unit step function is given by $|z| > 1$, which does not contain the unit circle. Equation (13.8) is, therefore, not valid for the unit step function. This may also be verified from Table 11.2, where the DTFT of the unit step function is different from the value obtained by substituting $z = e^{j\Omega}$ in its z-transform. The DTFT of the unit step function in Table 11.2 contains the Dirac delta functions, which makes the amplitude of the DTFT infinite at certain frequencies. No Dirac delta functions exist in the z-transform of the unit step function. Likewise, the ROCs for the z-transforms of the sine and cosine waves do not contain the unit circle, and Eq. (13.8) is also not valid in these cases.

13.2.2 Region of convergence

As a side note to our discussion, we observe that the z-transform is guaranteed to exist at all points within the ROC. For example, consider the causal sinusoidal sequence $x[k] = \cos(0.2\pi k)u[k]$, whose z-transform is given in Table 13.1 as follows:

$$X(z) = \frac{1 - \cos(\Omega_0)z^{-1}}{1 - \cos(\Omega_0)z^{-1} + z^{-2}}, \quad \text{ROC: } |z| > 1,$$

with $\Omega_0 = 0.2\pi$. We are interested in calculating the values of its z-transform at two points $z_1 = 2 + j0.6$ and $z_2 = 0.8 + j0.6$. Since z_1 lies within the ROC, $|z| > 1$, the value of the z-transform at z_1 is given by

$$X(z) = \left. \frac{1 - \cos(0.2\pi)z^{-1}}{1 - \cos(0.2\pi)z^{-1} + z^{-2}} \right|_{z=2+j0.6} = 1.39 - j0.05.$$

However, the point $z_2 = 0.8 + j0.6$ lies outside the ROC, $|z| > 1$. Therefore, the z-transform of the causal sinusoidal sequence cannot be computed for z_2. In the following, we list the important properties of the ROC for the z-transform.

(1) The ROC consists of a 2D plane of concentric circles of the form $|z| > z_0$ or $|z| < z_0$. All entries in Table 13.1 have ROCs that are concentric circles.
(2) The ROC does not include any poles of the z-transform.

The poles of a z-transform are defined as the roots of its denominator polynomial. Since the value of the z-transform is infinite at the location of a pole, the ROC cannot include any pole. Property (2) can be verified for all entries in Table 13.1. Consider, for example, the unit step function, which has a single pole at $z = 1$. The ROC of the z-transform of the unit step function is given by $|z| > 1$ and does not include its pole ($z = 1$).

(3) The ROC of a right-hand-sided sequence ($x[k] = 0$ for $k < k_0$) is defined by the region outside a circle. In other words, the ROC of a right-hand-sided sequence has the form $|z| > z_0$.

Entries (3)–(12) in Table 13.1 are right-hand-sided sequences. Consequently, it is observed that the ROC for all these sequences is of the form $|z| > z_0$.

(4) The ROC of a left-hand-sided sequence ($x[k] = 0$ for $k > k_0$) is defined by the inside region of a circle. Mathematically, this implies that the ROC of a left-sided sequence has the form $|z| < z_0$.

In Example 13.2, we computed the ROC for the left-hand-sided exponential sequence $x[k] = -\alpha^k u[-k-1]$ as $|z| < \alpha$, which satisfies Property (4).

(5) The ROC of a double-sided (or bilateral) sequence, which extends to infinite values of k in both directions, is confined to a ring with a finite area and has the form $z_1 < |z| < z_2$.

An example of a double-sided sequence is $x[k] = \beta^k u[k] - \alpha^k u[-k-1]$. By applying the linearity property, which is formally derived in Section 13.4.1, it is observed that the z-transform is given by

$$\beta^k u[k] - \alpha^k u[-k-1] \xleftrightarrow{z} \frac{1}{1-\alpha z^{-1}} + \frac{1}{1-\beta z^{-1}}, \quad \text{ROC: } \beta < |z| < \alpha,$$

which satisfies Property (5).

(6) The ROC of a finite-length sequence ($x[k] = 0$ for $k < k_1, k > k_2$) is the entire z-plane except for the possible exclusion of the points $z = 0$ and $z = \infty$.

As an example of Property (6), we consider entries (1) and (2) of Table 13.1. Also, sequence $x_5[k]$ defined in Example 13.3 is a finite-length sequence. In such cases, we note that the ROC consists of the entire z-plane except for the possible exclusion of $z = 0$ and $z = \infty$.

13.3 Inverse z-transform

Evaluating the inverse of z-transform is an important step in the analysis of LTID systems. There are four commonly used methods to evaluate the inverse z-transform:

  (i) table look-up method;
 (ii) inversion formula method;
(iii) partial fraction expansion method;
(iv) power series method.

Evaluating the inverse z-transform using the inversion formula (method (ii)) involves contour integration, which is fairly complex and beyond the scope of the text. In this section, we cover the remaining three methods in more detail.

13.3.1 Table look-up method

In this method, the z-transform function $X(z)$ is matched with one of the entries in Table 13.1. As the transform pairs are unique, the inverse transform is readily obtained from the time-domain entry. For example, if the inverse z-transform of the function

$$X(z) = \frac{1}{1 - 0.3z^{-1}}, \quad \text{ROC:} |z| > 0.3$$

is required, we determine that the matching entry in Table 13.1 is given by the transform pair

$$\alpha^k u[k] \overset{z}{\longleftrightarrow} \frac{1}{1 - \alpha z^{-1}}, \quad \text{ROC:} |z| > \alpha.$$

Substituting $\alpha = 0.3$, the inverse z-transform of $X(z)$ is given by $x[k] = 0.3^k u[k]$. The scope of the table look-up method is limited to the list of z-transforms available in Table 13.1.

13.3.2 Inversion formula method

In this method, the inverse z-transform is calculated directly by solving the complex contour integral specified in the synthesis equation, Eq. (13.5). This approach involves contour integration, which is beyond the scope of the text.

13.3.3 Partial fraction method

In LTID signals and systems analysis, the z-transform of a function $x[k]$ generally takes the following rational form:

$$X(z) = \frac{N(z)}{D(z)} = \frac{b_m z^m + b_{m-1} z^{m-1} + \cdots + b_1 z + b_0}{z^n + a_{n-1} z^{n-1} + \cdots + a_1 z + a_0} \tag{13.9a}$$

or alternatively

$$X(z) = \frac{N'(z)}{D'(z)} = z^{m-n} \frac{b_m + b_{m-1} z^{-1} + \cdots + b_1 z^{-m+1} + b_0 z^{-m}}{1 + a_{n-1} z^{-1} + \cdots + a_1 z^{-n+1} + a_0 z^{-n}}. \tag{13.9b}$$

Note that the numerator $N(z)$ and denominator $D(z)$ in Eq. (13.9a) are polynomials of the complex function z. In this case, the inverse z-transform of $X(z)$ can be calculated using the partial fraction expansion method. The method consists of the following steps.

Step 1 Calculate the roots of the characteristic equation of the rational function Eq. (13.9a). The characteristic equation is obtained by equating the denominator $D(z)$ in Eq. (13.9a) to zero, i.e.

$$D(z) = z^n + a_{n-1} z^{n-1} + \cdots + a_1 z + a_0 = 0. \tag{13.10}$$

For an nth-order characteristic equation, there will be n first-order roots. Depending on the value of the coefficients $\{b_l\}$, $0 \le l \le n - 1$, roots $\{p_r\}$, $1 \le r \le n$, of the characteristic equation may be real-valued and/or complex-valued. By expressing $D(z)$ in the factorized form, the z-transform $X(z)$ is represented as follows:

$$\frac{X(z)}{z} \equiv \frac{N(z)}{z(z - p_1)(z - p_2) \cdots (z - p_{n-1})(z - p_n)}. \tag{13.11}$$

It may be noted that in Eq. (13.11) we represent $X(z)/z$, not $X(z)$, in terms of its poles. The reason for this will become clear after step 3.

Step 2 Using Heaviside's partial fraction expansion formula, explained in Appendix D, decompose $X(z)/z$ into a summation of the first- or second-order fractions. If no roots are repeated, $X(z)/z$ is decomposed:

$$\frac{X(z)}{z} = \frac{k_0}{z} + \frac{k_1}{z - p_1} + \frac{k_2}{z - p_2} + \cdots + \frac{k_{n-1}}{z - p_{n-1}} + \frac{k_n}{z - p_n}, \qquad (13.12)$$

where the coefficients $\{k_r\}, 0 \leq r \leq n$, are obtained from the following expression:

$$k_r = \left[(z - p_r)\frac{N(z)}{z D(z)}\right]_{z=p_r}. \qquad (13.13)$$

It may be noted that Eq. (13.13) appends roots $\{p_r\}, 1 \leq r \leq n$, of the characteristic equation, Eq. (13.10), with an additional root $p_0 = 0$ such that $n + 1$ partial fraction coefficients are obtained by solving Heaviside's expression.

If there are repeated roots, $X(z)$ takes a slightly different form (see Appendix D for more details). It is important to associate separate ROCs with each partial fraction term in Eq. (13.12). The ROC for each partial fraction term is determined such that the intersection of these individual ROCs results in the overall ROC specified for $X(z)$.

Multiplying both sides of Eq. (13.12) by z, we obtain

$$X(z) \equiv k_0 + k_1\frac{z}{z - p_1} + k_2\frac{z}{z - p_2} + \cdots + k_{n-1}\frac{z}{z - p_{n-1}} + k_n\frac{z}{z - p_n} \qquad (13.14a)$$

or

$$X(z) \equiv k_0 + k_1\frac{1}{1 - p_1 z^{-1}} + k_2\frac{1}{1 - p_2 z^{-1}} + \cdots + k_{n-1}\frac{1}{1 - p_{n-1}z^{-1}}$$
$$+ k_n\frac{1}{1 - p_n z^{-1}}. \qquad (13.14b)$$

Step 3 The inverse transform of $X(z)$ can now be calculated by calculating the inverse transform of each individual partial fraction in Eq. (13.14a) using the following transform pair (see Table 13.1):

$$\alpha^k u[k] \overset{z}{\longleftrightarrow} \frac{1}{1 - \alpha z^{-1}} = \frac{z}{z - \alpha}, \qquad \text{ROC: } |z| > \alpha,$$

and is given by

$$x[k] = k_0\delta[k] + k_1(p_1)^k u[k] + k_2(p_2)^k u[k] + \cdots + k_{n-1}(p_{n-1})^k u[k]$$
$$+ k_n(p_n)^k u[k]. \qquad (13.14c)$$

The reason for performing a partial fraction expansion of $X(z)/z$ and not of $X(z)$ should now be clear. It was done so that the transform pair in Eq. (13.14b) can readily be applied to calculate the inverse transform. Otherwise, we would be missing the factor of z in the numerator of Eq. (13.14a), and application of Eq. (13.14b) would have been more complicated.

To illustrate the aforementioned procedure (steps (1)–(3)) for evaluating the inverse z-transform using the partial fraction expansion, we consider the following example.

Example 13.4

The z-transform of three right-sided functions is given below. Calculate the inverse z-transform in each case.

(i) $X_1(z) = \dfrac{z}{z^2 - 3z + 2}$;

(ii) $X_2(z) = \dfrac{1}{(z - 0.1)(z - 0.5)(z + 0.2)}$;

(iii) $X_3(z) = \dfrac{2z(3z + 17)}{(z - 1)(z^2 - 6z + 25)}$.

Solution

(i) The characteristic equation of $X_1(z)$ is given by $z^2 - 3z + 2 = 0$, which has two roots, at $z = 1$ and 2. The z-transform $X_1(z)$ can therefore be expressed as follows:

$$\frac{X_1(z)}{z} = \frac{1}{z^2 - 3z + 2} \equiv \frac{k_1}{z - 1} + \frac{k_2}{z - 2}.$$

Using Heaviside's partial fraction expansion formula, the coefficients of the partial fractions k_1 and k_2 are given by

$$k_1 = \left[(z - 1) \frac{1}{(z - 1)(z - 2)} \right]_{z=1} = \left[\frac{1}{z - 2} \right]_{z=1} = -1$$

and

$$k_2 = \left[(z - 2) \frac{1}{(z - 1)(z - 2)} \right]_{z=2} = \left[\frac{1}{z - 1} \right]_{z=2} = 1.$$

The partial fraction expansion of $X_1(z)$ is therefore given by

$$X_1(z) = \underbrace{\frac{-z}{(z - 1)}}_{\text{ROC:}|z|>1} + \underbrace{\frac{z}{(z - 2)}}_{\text{ROC:}|z|>2} = \underbrace{\frac{-1}{(1 - z^{-1})}}_{\text{ROC:}|z|>1} + \underbrace{\frac{1}{(1 - 2z^{-1})}}_{\text{ROC:}|z|>2},$$

where the ROC is obtained by noting that each term in $X_1(z)$ corresponds to a right-hand-sided sequence. This follows directly from knowing that $x[n]$ is right-hand-sided; hence, each term in $X_1(z)$ should also correspond to a right-hand sequence. Calculating the inverse z-transform of $X_1(z)$, we obtain

$$x_1[k] = -u[k] + 2^k u[k] = (2^k - 1)u[k].$$

(ii) The characteristic equation of $X_2(z)$ has three roots, at $z = 0.1, 0.5$ and -0.2. Therefore, $X_2(z)/z$ can be expressed as follows:

$$\frac{X_2(z)}{z} = \frac{1}{z(z - 0.1)(z - 0.5)(z + 0.2)}$$

$$= \frac{k_0}{z} + \frac{k_1}{z - 0.1} + \frac{k_2}{z - 0.5} + \frac{k_3}{z + 0.2}.$$

The partial fraction coefficients k_0, k_1, k_2, and k_3 are given by

$$k_0 = \left[z \frac{1}{z(z - 0.1)(z - 0.5)(z + 0.2)} \right]_{z=0} = 100,$$

$$k_1 = \left[(z - 0.1) \frac{1}{z(z - 0.1)(z - 0.5)(z + 0.2)} \right]_{z=0.1} = -\frac{250}{3},$$

$$k_2 = \left[(z - 0.5) \frac{1}{z(z - 0.1)(z - 0.5)(z + 0.2)} \right]_{z=0.5} = \frac{50}{7},$$

and

$$k_3 = \left[(z + 0.2) \frac{1}{z(z - 0.1)(z - 0.5)(z + 0.2)} \right]_{z=-0.2} = -\frac{500}{21}.$$

The partial fraction expansion of $X_2(z)/z$ is therefore given by

$$\frac{X_2(z)}{z} = \frac{100}{z} - \frac{250}{3} \frac{1}{(z - 0.1)} + \frac{50}{7} \frac{1}{(z - 0.5)} - \frac{500}{21} \frac{1}{(z + 0.2)}$$

or

$$X_2(z) = 100 - \frac{250}{3} \frac{1}{(1 - 0.1z^{-1})} + \frac{50}{7} \frac{1}{(1 - 0.5z^{-1})} - \frac{500}{21} \frac{1}{(1 + 0.2z^{-1})}.$$

Using the pairs in Table 13.1 and assuming a right-hand-sided sequence, the inverse z-transform is given by

$$x_2[k] = 100\delta[k] + \left\{ -\frac{250}{3}(0.1)^k + \frac{50}{7}(0.5)^k - \frac{500}{21}(-0.2)^k \right\} u[k].$$

(iii) The characteristic equation of $X_3(z)$ has one real-valued root at $z = 1$ and two complex-conjugate roots at $z = 3 \pm j4$. Combining the complex roots in a quadratic term, $X_3(z)/z$ can be expressed as follows:

$$\frac{X_3(z)}{z} = \frac{2(3z + 17)}{(z - 1)(z^2 - 6z + 25)} \equiv \frac{k_1}{z - 1} + \frac{k_2 z + k_3}{z^2 - 6z + 25}.$$

Using Heaviside's partial fraction expansion formula, coefficient k_1 is given by

$$k_1 = \left[(z - 1) \frac{2(3z + 17)}{(z - 1)(z^2 - 6z + 25)} \right]_{z=1} = 2.$$

To determine the remaining partial fraction coefficients k_2 and k_3, we expand

$$\frac{2(3z + 17)}{(z - 1)(z^2 - 6z + 25)} \equiv \frac{2}{z - 1} + \frac{k_2 z + k_3}{z^2 - 6z + 25}$$

by cross-multiplying and equating the numerators, we obtain

$$2(3z + 17) \equiv 2(z^2 - 6z + 25) + (k_2 z + k_3)(z - 1).$$

Comparing coefficients of z^2 and z yields

coefficients of z^2 $0 \equiv 2 + k_2 \Rightarrow k_2 = -2$;

coefficients of z $6 \equiv -12 - k_2 + k_3 \Rightarrow k_3 = 16.$

The partial fraction expansion of $X_3(z)$ can therefore be expressed as follows:

$$\frac{X_3(z)}{z} = \frac{2}{z - 1} + \frac{-2z + 16}{z^2 - 6z + 25}$$

or

$$X_3(z) = 2\frac{z}{z - 1} - 2\frac{z(z - 5 \times 0.6)}{z^2 - 2 \times 5 \times z \times 0.6 + 5^2}$$
$$+ \frac{5}{2}\frac{z(5 \times 0.8)}{z^2 - 2 \times 5 \times z \times 0.6 + 5^2},$$

where the final rearrangement makes the three terms in the above expression consistent with entries (4), (10), and (11) of Table 13.1, with $\alpha = 5$, and $\cos(\Omega_0) = 0.6$ and $\sin(\Omega_0) = 0.8$. Assuming that the three terms represent right-hand-sided sequences, the inverse z-transform for each term is given by

term 1 $2\dfrac{z}{z - 1} \xleftarrow{z^{-1}} 2u[k]$;

term 2 $-2\left[\dfrac{z(z - 5 \times 0.6)}{z^2 - 2 \times 5 \times z \times 0.6 + 5^2}\right] \xleftarrow{z^{-1}} -2 \cdot \cos(\cos^{-1}(0.6)k) \cdot 5^k u[k]$;

term 3 $\dfrac{5}{2}\left[\dfrac{z(5 \times 0.8)}{z^2 - 2 \times 5 \times z \times 0.6 + 5^2}\right] \xleftarrow{z^{-1}} \dfrac{5}{2} \cdot \sin(\cos^{-1}(0.6)k) \cdot 5^k u[k].$

Substituting $\cos^{-1}(0.6) = 0.9273$, the three terms are combined as follows:

$$x_3[k] = 2u[k] - 2 \cdot 5^k \cos(0.9273k)u[k] + \frac{5}{2} \cdot 5^k \sin(0.9273k)\, u[k],$$

which can be simplified to

$$x_3[k] = \left\{2 + 3.2016 \times 5^k \cos(0.9273\, k - 2.25)\right\}u[k].$$

The DT sequences $x_1[k]$, $x_2[k]$, and $x_3[k]$ are plotted in Fig. 13.3 for duration $0 \leq k \leq 6$.

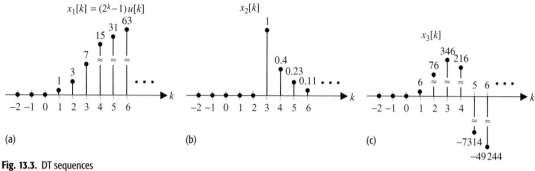

Fig. 13.3. DT sequences obtained in Example 13.4.

13.3.4 Power series method

When $X(z)$ is a rational function of the form in Eq. (13.9), the partial fraction expansion is a convenient method of calculating the inverse z-transform. At times, however, it may be difficult to expand $X(z)$ as partial fractions, especially when $X(z)$ is not a rational function. In such cases, we use the power series method. Alternatively, we may be interested in determining a few values of $x[k]$ for $k \geq 0$. The power series method is easy to apply since it does not require the evaluation of the complete inverse z-transform.

In the power series method, the transform $X(z)$ is expanded by long division as follows:

$$X(z) = \frac{N(z)}{D(z)} = a + bz^{-1} + cz^{-2} + dz^{-3} + \cdots . \tag{13.15a}$$

Taking the inverse z-transform of Eq. (13.15a), we obtain

$$x[k] = a\delta[k] + b\delta[k-1] + c\delta[k-2] + d\delta[k-3] + \cdots , \tag{13.15b}$$

which implies that $x[0] = a$, $x[1] = b$, $x[2] = c$, and $x[3] = d$. Additional samples of $x[k]$ can be obtained by determining additional terms in the quotient of Eq. (13.15a). We now illustrate the application of the power series method with an example.

Example 13.5

Calculate the first four non-zero values of the following right-sided sequences using the power series approach:

(i) $X_1(z) = \dfrac{z}{z^2 - 3z + 2}$;

(ii) $X_2(z) = \dfrac{1}{(z - 0.1)(z - 0.5)(z + 0.2)}$;

(iii) $X_3(z) = \dfrac{2z(3z + 17)}{(z - 1)(z^2 - 6z + 25)}$.

Solution

(i) Using long division, $X_1(z)$ can be expressed as follows:

$$
\begin{array}{r}
z^{-1} + 3z^{-2} + 7z^{-3} + 15z^{-4} \\
\hline
z^2 - 3z + 2 \quad | \quad z \\
z - 3 + 2z^{-1} \\
\hline
3 - 2z^{-1} \\
3 - 9z^{-1} + 6z^{-2} \\
\hline
7z^{-1} - 6z^{-2} \\
7z^{-1} - 21z^{-2} + 14z^{-3} \\
\hline
15z^{-2} - 14z^{-3} \\
15z^{-2} - 45z^{-3} + 30z^{-4}.
\end{array}
$$

In other words,

$$ X_1(z) = \frac{z}{z^2 - 3z + 2} = 0z^0 + z^{-1} + 3z^{-2} + 7z^{-3} + 15z^{-4} + \cdots . $$

Taking the inverse transform gives the following values for the first five samples of $x_1[k]$:

$$ x_1[0] = 0, \ x_1[1] = 1, \ x_1[2] = 3, \ x_1[3] = 7, \ x_1[4] = 15. $$

Note that the above values are consistent with Fig. 13.3(a) obtained in Example 13.4 (i).

(ii) Using long division, $X_2(z)$ can be expressed as follows:

$$
\begin{array}{r}
z^{-3} + 0.4z^{-4} + 0.23z^{-5} + 0.11z^{-6} \\
\hline
z^3 - 0.4z^2 - 0.07z + 0.01 \quad | \quad 1 \\
1 - 0.4z^{-1} - 0.07z^{-2} + 0.010z^{-3} \\
\hline
0.4z^{-1} + 0.07z^{-2} - 0.010z^{-3} \\
0.4z^{-1} - 0.16z^{-2} - 0.028z^{-3} + 0.0040z^{-4} \\
\hline
0.23z^{-2} + 0.018z^{-3} - 0.0040z^{-4} \\
0.23z^{-2} - 0.092z^{-3} - 0.0161z^{-4} + 0.0023z^{-5} \\
\hline
0.11z^{-3} + 0.0121z^{-4} - 0.0023z^{-5} \\
0.11z^{-3} + 0.0440z^{-4} - 0.0077z^{-5} + 0.0011z^{-5}.
\end{array}
$$

In other words,

$$ X_2(z) = \frac{1}{(z - 0.1)(z - 0.5)(z + 0.2)} $$
$$ = 0z^0 + 0z^{-1} + 0z^{-2} + z^{-3} + 0.4z^{-4} + 0.23z^{-5} + 0.11z^{-6} + \cdots . $$

Taking the inverse transform gives the following values for the first seven samples of $x_2[k]$:

$$ x_2[0] = 0, \ x_2[1] = 0, \ x_2[2] = 0, \ x_2[3] = 1, \ x_2[4] = 0.4, \ x_2[5] = 0.23, \ x_2[6] = 0.11. $$

This result is consistent with Fig. 13.3(b) obtained in Example 13.4(ii).

(iii) Using long division, $X_3(z)$ can be expressed as follows:

$$
\begin{array}{r}
6z^{-1} + 76z^{-2} + 346z^{-3} + 216z^{-4} \\
\hline
z^3 - 7z^2 + 31z - 25 \,\big|\, 6z^2 + 34z \\
6z^2 - 42z + 186 - 150z^{-1} \\
\hline
76z - 186 + 150z^{-1} \\
76z - 532 + 2356z^{-1} - 1900z^{-2} \\
\hline
346 - 2206z^{-1} + 1900z^{-2} \\
346 - 2422z^{-1} + 10726z^{-2} - 8650z^{-3} \\
\hline
216z^{-1} - 8826z^{-2} + 8650z^{-3} \\
216z^{-1} - 1512z^{-2} + 6696z^{-3} - 5400z^{-3}.
\end{array}
$$

In other words,

$$
X_3(z) = \frac{2z(3z+17)}{(z-1)(z^2-6z+25)} = 0z^0 + 6z^{-1} + 76z^{-2} + 346z^{-3} + 216z^{-4} + \cdots .
$$

Taking the inverse transform gives the following values for the first five samples of $x_3[k]$:

$$
x_3[0] = 0, \ x_3[1] = 6, \ x_3[2] = 76, \ x_3[3] = 346, \ x_3[4] = 216.
$$

The result is consistent with Fig. 13.3(c) obtained in Example 13.4(iii).

13.4 Properties of the z-transform

The unilateral and bilateral z-transforms have several interesting properties, which are used in the analysis of signals and systems. These properties are similar to the properties of the DTFT, which were covered in Section 11.4. In this section, we discuss several of these properties, including their proofs and applications, through a series of examples. A complete list of the properties is provided in Table 13.2. In most cases, we prove the properties for the unilateral z-transform. The proof for the bilateral z-transform follows along similar lines and is not included to avoid repetition.

13.4.1 Linearity

If $x_1[k]$ and $x_2[k]$ are two DT sequences with the following z-transform pairs:

$$
x_1[k] \overset{z}{\longleftrightarrow} X_1(z), \quad \text{ROC: } R_1
$$

and

$$
x_2[k] \overset{z}{\longleftrightarrow} X_2(z), \quad \text{ROC: } R_2,
$$

then

$$
a_1 x_1[k] + a_2 x_2[k] \overset{z}{\longleftrightarrow} a_1 X_1(z) + a_2 X_2(z), \quad \text{ROC: at least } R_1 \cap R_2. \quad (13.16)
$$

The linearity property is satisfied by both unilateral and bilateral z-transforms.

Table 13.2. Properties of the z-transform for transform pairs $x[k] \xleftrightarrow{z} X(z)$, ROC: R_x; $x[k]u[k] \xleftrightarrow{z} X^{(c)}(z)$, ROC: R_x; $x_1[k] \xleftrightarrow{z} X_1(z)$, ROC: R_1; $x_2[k] \xleftrightarrow{z} X_2(z)$, ROC: R_2

Properties	Time domain	z-domain	ROC
Linearity	$a_1 x_1[k] + a_2 x_2[k]$	$a_1 X_1(z) + a_2 X_2(z)$	at least $R_1 \cap R_2$
Time scaling	$x^{(m)}[k]$ for $m = 1, 2, 3, \ldots$	$X(z^m)$	$(R_x)^{1/m}$
Time shifting (non-causal)	$x[k-m]$	$z^m X(z)$	
Time shifting (causal)	$x[k-m]u[k-m]$	$z^m X^{(c)}(z)$	R_x, except for the possible deletion or addition of $z = 0$ or $z = \infty$
	$x[k+m]u[k]$	$z^m X^{(c)}(z) - z^m \sum_{k=0}^{m-1} x[k]z^{-k}$	
	$x[k-m]u[k]$	$z^{-m} X^{(c)}(z) + z^{-m} \sum_{k=1}^{m} x[-k]z^k$	
Frequency shifting	$e^{j\Omega_0 k} x[k]$	$X(e^{-j\Omega_0} z)$	R_x
Time differencing	$x[k] - x[k-1]$	$(1 - z^{-1})X(z)$	R_x, except for the possible deletion of the origin
Time accumulation	$y[k] = \sum_{m=0}^{k} x[m]$ [a]	$\dfrac{z}{z-1} X(z)$	$R_x \cap (|z| > 1)$
z-domain differentiation	$kx[k]$	$-z \dfrac{dX(z)}{dz}$	R_x
Time convolution	$x_1[k] * x_2[k]$	$X_1(z)X_2(z)$	at least $R_1 \cap R_2$
Initial-value theorem		$x[0] = \lim_{z \to \infty} X(z)$	provided $x[k] = 0$ for $k < 0$
Final-value theorem		$x[\infty] = \lim_{k \to \infty} x[k] = \lim_{z \to 1}(z - 1)X(z)$	provided $x[\infty]$ exists

[a] Provided that the sequence $y[k]$ has a finite value for all k.

Proof

Using Eq. (13.7), the z-transform of $a_1 x_1[k] + a_2 x_2[k]$ is calculated as follows:

$$Z\{a_1 x_1[k] + a_2 x_2[k]\} = \sum_{k=0}^{\infty} \{a_1 x_1[k] + a_2 x_2[k]\} z^{-k}$$

$$= a_1 \underbrace{\sum_{k=0}^{\infty} x_1[k]z^{-k}}_{X_1(z)} + a_2 \underbrace{\sum_{k=0}^{\infty} x_2[k]z^{-k}}_{X_2(z)},$$

which proves the algebraic expression, Eq. (13.16). To determine the ROC of the linear combination, we note that the z-transform $X_1(z)$ is finite within the specified ROC, R_1. Similarly, $X_2(z)$ is finite within its ROC, R_2. Therefore, the linear combination $a_1 X_1(z) + a_2 X_2(z)$ should be finite at least within region R

that represents the intersection of the two regions, i.e. $R = R_1 \cap R_2$. In certain cases, due to the interaction between $x_1[k]$ and $x_2[k]$, which may lead to cancelation of certain terms, the overall ROC R may be larger than the intersection of the two regions. On the other hand, if there is no common region between R_1 and R_2, the z-transform of $\{a_1x_1[k] + a_1x_2[k]\}$ does not exist.

13.4.2 Time scaling

As mentioned in Chapter 1, there are two types of scaling in the DT domain: decimation and interpolation.

13.4.2.1 Decimation

Because of the irreversible nature of the decimation operation, the z-transform of $x[k]$ and its decimated sequence $y[k] = x[mk]$ are not related to each other.

13.4.2.2 Interpolation

Section 1.3.2.2 defines the interpolation of $x[k]$ as follows:

$$x^{(m)}[k] = \begin{cases} x[k/m] & \text{if } k \text{ is a multiple of integer } m \\ 0 & \text{otherwise.} \end{cases}$$

The z-transform of an interpolated sequence is given by the following property. If $x[k] \overset{z}{\longleftrightarrow} X(z)$ with ROC R_x, then the z-transform $X^{(m)}(z)$ of $x^{(m)}[k]$ is given by

$$x^{(m)}[k] \overset{z}{\longleftrightarrow} X^{(m)}(z) = X(z^m), \quad \text{ROC: } (R_x)^{1/m} \tag{13.17}$$

for $2 \le m < \infty$. The interpolation property is satisfied by both unilateral and bilateral z-transforms.

Proof

$$Z\{x^{(m)}[k]\} = \sum_{k=0}^{\infty} x^{(m)}[k]z^{-k}$$

$$= x^{(m)}[0] + x^{(m)}[1]z^{-1} + \cdots + x^{(m)}[m]z^{-m} + x^{(m)}[m+1]z^{-(m+1)}$$
$$+ \cdots + x^{(m)}[2m]z^{-2m} + \cdots.$$

Based on Eq. (13.17), the interpolated sequence $x^{(m)}[k]$ is zero everywhere except when k is a multiple of m. This reduces the above transform as follows:

$$Z\{x^{(m)}[k]\} = x^{(m)}[0] + x^{(m)}[m]z^{-m} + x^{(m)}[2m]z^{-2m} + x^{(m)}[3m]z^{-3m} + \cdots.$$
$$= x[0] + x[1]z^{-m} + x[2]z^{-2m} + x[3]z^{-3m} + \cdots$$
$$= \sum_{k=0}^{\infty} x[k](z^m)^{-k} = X(z^m).$$

Because $X(z)$ is finite-valued within the region $z \in R_x$, $X(z^m)$ will have a finite value when $z^m \in R_x$ or $z \in (R_x)^{1/m}$.

13.4.3 Time shifting

The time-shifting property for a bilateral z-transform is as follows.

If $x[k] \xleftarrow{\text{bilateral } z} X(z)$ with ROC R_x, then

$$x[k-m] \xleftarrow{\text{bilateral } z} z^{-m} X(z), \qquad (13.18)$$

with ROC given by R_x except for the possible deletion or addition of $z = 0$ or $z = \infty$. The ROC is altered because of the inclusion of the z^m or z^{-m} term, which affects the roots of the denominator $D(z)$ in $X(z)$.

For causal sequences, the time-shifting property is more complicated. For any causal sequence $x[k]u[k]$ satisfying the DTFT pair

$$x[k]u[k] \xleftrightarrow{z} X(z)$$

and having the ROC R_x, the unilateral z-transform of the following time-shifted sequences are expressed as follows (for a positive integer m):

(a) $x[k-m]u[k-m] \xleftrightarrow{z} z^{-m}X(z);$ \hfill (13.19)

(b) $x[k+m]u[k] \xleftrightarrow{z} z^m X(z) - z^m \sum_{k=0}^{m-1} x[k]z^{-k};$ \hfill (13.20)

(c) $x[k-m]u[k] \xleftrightarrow{z} z^{-m}X(z) + z^{-m} \sum_{k=1}^{m} x[-k]z^k.$ \hfill (13.21)

In Eqs. (13.19)–(13.21), the ROC of the time-shifted sequences is given by R_x, except for the possible deletion or addition of $z = 0$ or $z = \infty$.

To illustrate the three time-shifting operations, consider a two-sided sequence $x[k] = \alpha^{|k|}$ with $|\alpha| < 1$, as illustrated in Fig. 13.4(a). Figures 13.4(b)–(d) illustrate the three time-shifting operations defined above in Eqs. (13.19)–(13.21) for $m = 2$.

Proof

We prove Eqs. (13.19)–(13.21) separately.

Equation (13.19)

$$Z\{x[k-m]u[k-m]\} = \sum_{k=0}^{\infty} x[k-m]u[k-m]z^{-k} = \sum_{k=m}^{\infty} x[k-m]z^{-k}.$$

Fig. 13.4. (a) Original DT
sequence $x[k] = \alpha^{|k|}$. Parts
(b)–(d) show sequences
obtained by time shifting the
sequence in part (a):
(b) $x[k - 2]u[k - 2]$;
(c) $x[k - 2]u[k]$;
(d) $x[k + 2]u[k]$.

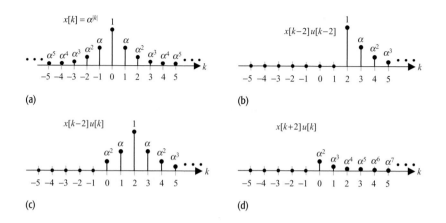

Substituting $p = k - m$, the above summation reduces to

$$Z\{x[k - m]u[k - m]\} = \sum_{p=0}^{\infty} x[p]z^{-(p+m)} = z^{-m}\sum_{p=0}^{\infty} x[p]z^{-p} = z^{-m}X(z).$$

Equation (13.20)

$$Z\{x[k + m]u[k]\} = \sum_{k=0}^{\infty} x[k + m]u[k]\,z^{-k} = \sum_{k=0}^{\infty} x[k + m]z^{-k}.$$

Substituting $p = k + m$, the above summation reduces to

$$Z\{x[k + m]u[k]\} = \sum_{p=m}^{\infty} x[p]z^{-(p-m)} = z^{m}\sum_{p=0}^{\infty} x[p]z^{-p} - z^{m}\sum_{p=0}^{m-1} x[p]z^{-p},$$

$$= z^{m}X(z) - z^{m}\sum_{k=0}^{m-1} x[k]z^{-k}.$$

Equation (13.21)

$$Z\{x[k - m]u[k]\} = \sum_{k=0}^{\infty} x[k - m]u[k]z^{-k} = \sum_{k=0}^{\infty} x[k - m]z^{-k}.$$

Substituting $p = k - m$, the above summation reduces to

$$Z\{x[k - m]u[k]\} = \sum_{p=-m}^{\infty} x[p]z^{-(p+m)}$$

$$= z^{-m}\sum_{p=0}^{\infty} x[p]z^{-p} + z^{-m}\sum_{p=-m}^{-1} x[p]z^{-p}.$$

$$= z^{-m}X(z) + z^{-m}\sum_{k=1}^{m} x[-k]z^{k}.$$

Example 13.6

Consider a non-causal DT sequence $x[k]$ with initial values $x[-1] = 11/6$ and $x[-2] = 37/36$. Express the z-transform of the function

$$g[k] = (x[k] - 5x[k-1] + 6x[k-2])u[k]$$

in terms of the z-transform $Z\{x[k]u[k]\} = X(z)$.

Solution

Applying the time-shifting property, Eq. (13.21), the z-transforms of $x[k-1]u[k]$ and $x[k-2]u[k]$ are given by

$$Z\{x[k-1]u[k]\} = z^{-1}X(z) + z^{-1}x[-1]z = z^{-1}X(z) + \frac{11}{6}$$

and

$$Z\{x[k-2]u[k]\} = z^{-2}X(z) + z^{-2}x[-1]z + z^{-2}x[-2]z^2$$
$$= z^{-2}X(z) + \frac{11}{6}z^{-1} + \frac{37}{36}.$$

Applying the linearity property, the z-transform of $g[k]$ is given by

$$G(z) = X(z) - 5\left[z^{-1}X(z) + \frac{11}{6}\right] + 6\left[z^{-2}X(z) + \frac{11}{6}z^{-1} + \frac{37}{36}\right]$$
$$= \left(1 - 5z^{-1} + 6z^{-2}\right)X(z) + 11z^{-1} - 3.$$

13.4.4 Time differencing

If $x[k] \xleftrightarrow{z} X(z)$ with ROC R_x, then the z-transform of the time-difference sequence $x[k] - x[k-1]$ is given by

$$x[k] - x[k-1] \xleftrightarrow{z} (1 - z^{-1})X(z), \qquad (13.22)$$

with the ROC given by R_x except for the possible deletion of $z = 0$. The time-differencing property can be proved easily by applying the linearity and time-shifting properties with $m = 1$. The time-differencing property is satisfied by both unilateral and bilateral z-transforms.

Example 13.7

Based on the z-transform pair

$$u[k] \xleftrightarrow{z} \frac{1}{1 - z^{-1}}, \qquad \text{ROC: } |z| > 1,$$

calculate the z-transform of the impulse function $x[k] = \delta[k]$ using the time-differencing property.

Solution

Using the time-differencing property, the z-transform of $u[k] - u[k-1]$ is given by

$$u[k] - u[k-1] \xleftrightarrow{z} (1 - z^{-1}) \cdot Z\{u[k]\}, \quad \text{ROC:} |z| > 1.$$

Substituting the value of $Z\{u[k]\} = 1/(1 - z^{-1})$ and noting that $u[k] - u[k-1] = \delta[k]$, we obtain

$$\delta[k] \xleftrightarrow{z} 1.$$

Since the z-transform of the unit impulse function is finite for all values of z, the ROC of the aforementioned z-transform pair is the entire z-plane.

13.4.5 z-domain differentiation

If $x[k] \xleftrightarrow{z} X(z)$ with ROC R_x, then

$$kx[k] \xleftrightarrow{z} -z\frac{\mathrm{d}X(z)}{\mathrm{d}z}, \quad \text{ROC:} R_x. \tag{13.23}$$

The z-domain differentiation property is satisfied by both unilateral and bilateral z-transforms.

Proof

By definition,

$$X(z) = \sum_{k=0}^{\infty} x[k]z^{-k}.$$

Differentiating both sides with respect to z yields

$$\frac{\mathrm{d}X(z)}{\mathrm{d}z} = \sum_{k=0}^{\infty} x[k]\,\frac{\mathrm{d}z^{-k}}{\mathrm{d}z} = \sum_{k=0}^{\infty} x[k](-k)z^{-k-1}.$$

Multiplying both sides by $-z$, we obtain

$$-z\frac{\mathrm{d}X(z)}{\mathrm{d}z} = \sum_{k=0}^{\infty} kx[k]z^{-k},$$

which proves Eq. (13.23).

Example 13.8

Given the z-transform pair

$$\alpha^k u[k] \xleftrightarrow{z} \frac{1}{1 - \alpha z^{-1}}, \quad \text{ROC:} |z| > |\alpha|,$$

calculate the z-transform of the function $k\alpha^k u[k]$.

Solution

We use the frequency-differentiation property,

$$ k\alpha^k x[k] \xleftrightarrow{\ z\ } -z \frac{d}{dz}\left[\frac{1}{1 - \alpha z^{-1}} \right], $$

which reduces to

$$ k\alpha^k x[k] \xleftrightarrow{\ z\ } \frac{\alpha z^{-1}}{(1 - \alpha z^{-1})^2} \quad \text{ROC: } |z| > |\alpha|. $$

13.4.6 Time convolution

If $x_1[k]$ and $x_2[k]$ are two arbitrary functions with the following z-transform pairs:

$$ x_1[k] \xleftrightarrow{\ z\ } X_1(z), \quad \text{ROC: } R_1 $$

and

$$ x_2[k] \xleftrightarrow{\ z\ } X_2(z), \quad \text{ROC: } R_2, $$

then the convolution property states that

$$ x_1[k] * x_2[k] \xleftrightarrow{\ z\ } X_1(z)X_2(z), \quad \text{ROC: at least } R_1 \cap R_2. \tag{13.24} $$

The convolution property is valid for both unilateral and bilateral z-transforms. The overall ROC of the convolved signals may be larger than the intersection of regions R_1 and R_2 because of the possible cancelation of some poles of the z-transforms of the convolved sequences.

Proof

By definition, the convolution of two sequences is given by

$$ x_1[k] * x_2[k] = \sum_{m=-\infty}^{\infty} x_1[m]x_2[k-m]. $$

Taking the z-transform of both sides yields

$$ x_1[k] * x_2[k] \xleftrightarrow{\ z\ } \sum_{k=-\infty}^{\infty} \sum_{m=-\infty}^{\infty} x_1[m]x_2[k-m]z^{-k}. $$

By interchanging the order of the two summations on the right-hand side of the transform pair, we obtain

$$ x_1[k] * x_2[k] \xleftrightarrow{\ z\ } \sum_{m=-\infty}^{\infty} x_1[m] \sum_{k=-\infty}^{\infty} x_2[k-m]z^{-k}. $$

Substituting $p = k - m$ in the inner summation leads to

$$ x_1[k] * x_2[k] \xleftrightarrow{\ z\ } \sum_{m=-\infty}^{\infty} x_1[m] \sum_{p=-\infty}^{\infty} x_2[p]z^{-(p+m)} $$

or

$$x_1[k] * x_2[k] \overset{z}{\longleftrightarrow} \sum_{m=-\infty}^{\infty} x_1[m] z^{-m} \sum_{p=-\infty}^{\infty} x_2[p] z^{-p},$$

which proves Eq. (13.24).

Like the DTFT convolution property discussed in Chapter 11, the time-convolution property of the z-transform provides us with an alternative approach to calculate the output $y[k]$ when a DT sequence $x[k]$ is applied at the input of an LTID system with the impulse response $h[k]$. The procedure for calculating the output $y[k]$ of an LTID system in the complex z-domain consists of the following four steps.

(1) Calculate the z-transform $X(z)$ of the input sequence $x[k]$. If the input sequence and the impulse response are both causal functions, then the unilateral z-transform is used. If either of the two functions is non-causal, the bilateral z-transform must be used.
(2) Calculate the z-transform $H(z)$ of the impulse response $h[k]$ of the LTID system. The z-transform $H(z)$ is referred to as the z-transfer function of the LTID system and provides a meaningful insight into the behavior of the system.
(3) Based on the convolution property, the z-transform $Y(z)$ of the resulting output $y[k]$ is given by the product of the z-transforms of the input signal and the impulse response of the LTID system. Mathematically, this implies that $Y(z) = X(z)H(z)$.
(4) Calculate the output response $y[k]$ in the time domain by taking the inverse z-transform of $Y(z)$ obtained in step (3).

Example 13.9

The exponential decaying sequence $x[k] = a^k u[k], 0 < a < 1$, is applied at the input of an LTID system with the impulse response $h[k] = b^k u[k], 0 < b < 1$. Using the z-transform approach, calculate the output of the system.

Solution

Based on Table 13.1, the z-transforms for the input sequence and the impulse response are given by

$$X(z) = \frac{1}{1 - az^{-1}} \quad \text{and} \quad H(z) = \frac{1}{1 - bz^{-1}}.$$

The z-transform of the output signal is, therefore, calculated as follows:

$$Y(z) = H(z)X(z) = \frac{1}{(1 - az^{-1})(1 - bz^{-1})}.$$

The inverse of $Y(z)$ takes two different forms depending on the values of a and b:

$$Y(z) = \begin{cases} \dfrac{1}{(1 - az^{-1})^2} & a = b \\[3mm] \dfrac{1}{(1 - az^{-1})(1 - bz^{-1})} & a \neq b. \end{cases}$$

We consider the two cases separately while calculating the inverse z-transform of $Y(z)$.

Case 1 ($a = b$) From Table 13.1, we know that

$$ka^k u[k] \stackrel{z}{\longleftrightarrow} \frac{az^{-1}}{(1 - az^{-1})^2}.$$

Applying the time-shifting property, we obtain

$$(k + 1)a^{k+1}u[k + 1] \stackrel{z}{\longleftrightarrow} \frac{a}{(1 - az^{-1})^2}.$$

The output response is therefore given by

$$y[k] = Z^{-1}\left\{\frac{1}{(1 - az^{-1})^2}\right\} = (k + 1)a^k u[k + 1],$$

which is the same as

$$y[k] = (k + 1)\,a^k u[k].$$

Case 2 ($a \neq b$) Using partial fraction expansion, the function $Y(z)$ is expressed as follows:

$$Y(z) = \frac{1}{(1 - az^{-1})(1 - bz^{-1})} \equiv \frac{A}{1 - az^{-1}} + \frac{B}{1 - bz^{-1}}, \quad (13.25)$$

where the partial fraction coefficients are given by

$$A = \left.\frac{1}{1 - bz^{-1}}\right|_{az^{-1}=1} = \frac{a}{a - b}$$

and

$$B = \left.\frac{1}{1 - az^{-1}}\right|_{bz^{-1}=1} = -\frac{b}{a - b}.$$

Substituting the values of A and B into Eq. (13.25) and taking the inverse DTFT yields

$$y[k] = \frac{a}{a - b} \times a^k u[k] - \frac{b}{a - b} \times b^k u[k] = \frac{1}{a - b}\left[a^{k+1} - b^{k+1}\right]u[k].$$

Combining case 1 with case 2, we obtain

$$y[k] = \begin{cases} (k + 1)a^k u[k] & a = b \\[2mm] \dfrac{1}{a - b}\left[a^{k+1} - b^{k+1}\right]u[k] & a \neq b, \end{cases} \quad (13.26)$$

which is identical to the result of Example 11.15.

13.4.7 Time accumulation

If $x[k] \xleftrightarrow{z} X(z)$ with ROC R_x, then

$$\sum_{m=0}^{k} x[m] \xleftrightarrow{z} \frac{z}{z-1} X(z), \quad \text{ROC: } R_x \cap (|z| > 1). \qquad (13.27)$$

Proof

To prove the time-accumulation property, we make use of the following convolution result:

$$\sum_{m=0}^{k} x[m] = x[k] * u[k].$$

Taking the z-transform of both sides and applying the time-convolution property yields

$$\sum_{m=0}^{k} x[m] \xleftrightarrow{z} X(z) Z\{u[k]\}.$$

In the above equation, we substitute the z-transform of the unit step function,

$$u[k] \xleftrightarrow{z} \frac{1}{1-z^{-1}}, \quad \text{ROC: } |z| > 1,$$

to obtain

$$\sum_{m=0}^{k} x[m] \xleftrightarrow{z} X(z) \frac{1}{1-z^{-1}},$$

which proves Eq. (13.27).

Example 13.10

Given the z-transform pair

$$u[k] \xleftrightarrow{z} \frac{1}{1-z^{-1}}, \quad \text{ROC: } |z| > 1,$$

calculate the z-transform of the function $ku[k]$ using the time-accumulation property.

Solution

Note that

$$ku[k] = \sum_{m=0}^{k} u[m] - u[k].$$

Calculating the z-transform of both sides and applying the time-accumulation property, we obtain

$$ku[k] \xleftrightarrow{z} \frac{z}{(z-1)(1-z^{-1})} - \frac{1}{1-z^{-1}},$$

which reduces to

$$ku[k] \xleftrightarrow{z} \frac{z^{-1}}{\left(1 - z^{-1}\right)^2},$$

which can be expressed in the following alternative form:

$$ku[k] \xleftrightarrow{z} \frac{z}{(z-1)^2}.$$

Note that the ROC for $ku[k]$ is the same as that for $u[k]$.

13.4.8 Initial- and final-value theorems

If $x[k] \xleftrightarrow{z} X(z)$ with ROC R_x, then

initial-value theorem $x[0] = \lim\limits_{z \to \infty} X(z),$ provided $x[k] = 0$ for $k < 0$;

$$(13.28)$$

final-value theorem $x[\infty] = \lim\limits_{k \to \infty} x[k] = \lim\limits_{z \to 1}(z-1)X(z),$

provided $x[\infty]$ exists. (13.29)

Note that the initial-value theorem is valid only for the unilateral z-transform as it requires the reference signal $x[k]$ to be zero for $k < 0$. The final-value theorem, however, may be used with either the unilateral or bilateral z-transform. It is possible to get a finite value from Eq. (13.29) even though $x[\infty]$ is undefined or equal to infinity. Readers are advised to check that $x[\infty]$ indeed converges to a finite value before using the final-value theorem. This generally happens if all poles of $(z-1)X(z)$ lie inside the unit circle.

Example 13.11
Given the following z-transforms of right-sided sequences, determine the initial and final values:

(i) $X_1(z) = \dfrac{z}{z^2 - 3z + 2}$;

(ii) $X_2(z) = \dfrac{1}{(z - 0.1)(z - 0.5)(z + 0.2)}$;

(iii) $X_3(z) = \dfrac{z^2(2z - 1.5)}{(z - 1)(z - 0.5)^2}$.

Solution
(i) Using the initial-value theorem,

$$x_1[0] = \lim\limits_{z \to \infty} X_1(z) = \lim\limits_{z \to \infty}\left[\frac{z}{z^2 - 3z + 2}\right] = \lim\limits_{z \to \infty}\left[\frac{1}{z - 3 + 2z^{-1}}\right] = 0.$$

Using the final-value theorem, we obtain

$$x_1[\infty] = \lim\limits_{z \to 1}(z-1)X_1(z) = \lim\limits_{z \to 1}\left[\frac{z(z-1)}{z^2 - 3z + 2}\right] = \lim\limits_{z \to 1}\left[\frac{z}{z - 2}\right] = -1.$$

From Example 13.4 part (i), where we determined $x_1[k]$, it can be verified that $x_1[0] = 0$. However, we obtain $x_1[\infty] = \infty$ from the result in Example 13.4, which is different from the result obtained above using the final-value theorem. Actually, in this case the final-value theorem cannot be applied as $x_1[\infty]$ is not finite. This can be guessed from the fact that $(z - 1)X(z)$ has a pole at $z = 2$, which is not inside the unit circle.

(ii) Using the initial-value theorem,

$$x_2[0] = \lim_{z \to \infty} X_2(z) = \lim_{z \to \infty} \left[\frac{1}{(z - 0.1)(z - 0.5)(z + 0.2)} \right] = 0.$$

Using the final-value theorem,

$$x_2[\infty] = \lim_{z \to 1}(z - 1)X_2(z) = \lim_{z \to 1} \left[\frac{(z - 1)}{(z - 0.1)(z - 0.5)(z + 0.2)} \right] = 0.$$

From the expression of $x_2[k]$ derived in Example 13.4 part (ii), it can be verified that the above values are indeed correct.

(iii) Using the initial-value theorem,

$$x_3[0] = \lim_{z \to \infty} X_3(z) = \lim_{z \to \infty} \left[\frac{z^2(2z - 1.5)}{(z - 1)(z - 0.5)^2} \right]$$

$$= \lim_{z \to \infty} \left[\frac{2 - 1.5z^{-1}}{(1 - z^{-1})(1 - 0.5z^{-1})^2} \right] = 2.$$

Using the final-value theorem,

$$x_3[\infty] = \lim_{z \to 1}(z - 1)X_3(z) = \lim_{z \to 1} \left[\frac{(z - 1)z^2(2z - 1.5)}{(z - 1)(z - 0.5)^2} \right]$$

$$= \lim_{z \to 1} \left[\frac{z^2(2z - 1.5)}{(z - 0.5)^2} \right] = 2.$$

By calculating the inverse z-transform of $X_3(z)$, we obtain

$$x_3[k] = (2 + k \times 2^{-k})u[k].$$

Based on the above expression, $x_3[0] = 2$ and $x_3[\infty] = 2$, which are indeed the values obtained using the initial- and final-value theorems.

13.5 Solution of difference equations

An important application of the z-transform is to solve linear, constant-coefficient difference equations. In Section 10.1, we used a time-domain approach to obtain the zero-input, zero-state, and overall solutions of difference equations. In this section, we discuss an alternative approach based on the z-transform. We illustrate the steps involved in the z-transform-based approach through Example 13.12.

Example 13.12

A causal system is represented by the following difference equation:

$$y[k+2] - 5y[k+1] + 6y[k] = 3x[k+1] + 5x[k]. \tag{13.30}$$

Calculate the output $y[k]$ for the input $x[k] = 2^{-k}u[k]$ and the initial conditions $y[-1] = 11/6$, $y[-2] = 37/36$.

Solution

Substituting $k - 2$ for k in Eq. (13.30), we obtain

$$y[k] - 5y[k-1] + 6y[k-2] = 3x[k-1] + 5x[k-2]. \tag{13.31}$$

Note that the input sequence $x[k] = 2^{-k}u[k]$ is causal, hence $x[-2] = x[-1] = 0$. Using the time-shifting property, Eq. (13.19), the z-transform of the right-hand side of Eq. (13.31) is given by

$$3x[k-1] + 5x[k-2] \overset{z}{\longleftrightarrow} 3z^{-1}X(z) + 5z^{-2}X(z).$$

Using the z-transform pair,

$$x[k] = 2^{-k}u[k] = 0.5^k u[k] \overset{z}{\longleftrightarrow} X(z) = \frac{1}{1 - 0.5z^{-1}},$$

the z-transform of the right-hand side of Eq. (13.31) is given by

$$3x[k-1] + 5x[k-2] \overset{z}{\longleftrightarrow} \frac{3z^{-1}}{1 - 0.5z^{-1}} + \frac{5z^{-2}}{1 - 0.5z^{-1}} = \frac{3z^{-1} + 5z^{-2}}{1 - 0.5z^{-1}}.$$

The output response is not causal as the initial conditions $y[-1]$ and $y[-2]$ are not zero. We are interested in determining the causal component $y[k]u[k]$ of the response $y[k]$. Let us denote the z-transform of $y[k]u[k]$ by $Y(z)$. Using the results in Example 13.6, the z-transform of the left-hand side of Eq. (13.31) is given by

$$y[k] - 5y[k-1] + 6y[k-2] \overset{z}{\longleftrightarrow} (1 - 5z^{-1} + 6z^{-2})Y(z) + (11z^{-1} - 3).$$

Equating the z-transforms of both sides of Eq. (13.31), we obtain

$$(1 - 5z^{-1} + 6z^{-2})Y(z) + (11z^{-1} - 3) = \frac{3z^{-1} + 5z^{-2}}{1 - 0.5z^{-1}},$$

which reduces to

$$\begin{aligned}
Y(z) &= \frac{3 - 11z^{-1}}{1 - 5z^{-1} + 6z^{-2}} + \frac{3z^{-1} + 5z^{-2}}{(1 - 0.5z^{-1})(1 - 5z^{-1} + 6z^{-2})} \\
&= \frac{(3 - 11z^{-1})(1 - 0.5z^{-1}) + 3z^{-1} + 5z^{-2}}{(1 - 0.5z^{-1})(1 - 5z^{-1} + 6z^{-2})} \\
&= \frac{3 - 9.5z^{-1} + 10.5^{-2}}{(1 - 0.5z^{-1})(1 - 2z^{-1})(1 - 3z^{-1})}.
\end{aligned}$$

Using partial fraction expansion, $Y(z)$ can be expressed as follows:

$$Y(z) = \frac{26}{15} \times \frac{1}{1 - 0.5z^{-1}} - \frac{7}{3} \times \frac{1}{1 - 2z^{-1}} + \frac{18}{5} \times \frac{1}{1 - 3z^{-1}}.$$

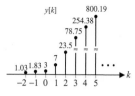

Fig. 13.5. Output response of the LTID system specified in Example 13.12.

Taking the inverse transform, we obtain

$$y[k] = \left[\frac{26}{15} \times 0.5^k - \frac{7}{3} \times 2^k + \frac{26}{15} \times 3^k\right] \text{ for } k > 0.$$

The output response is plotted in Fig. 13.5.

13.6 z-transfer function of LTID systems

In Chapters 10 and 11, we used the impulse response $h[k]$ and Fourier transfer function $H(\Omega)$ to represent an LTID system. An alternative representation for an LTID system is obtained by taking the z-transform of the impulse response:

$$h[k] \xleftrightarrow{z} H(z).$$

The $H(z)$ is referred to as the *z-transfer function* of the LTID system. In conjunction with the linear convolution property, Eq. (13.24), the z-transfer function $H(z)$ may be used to determine the output response $y[k]$ of an LTID system when an input sequence $x[k]$ is applied at its input. In the time domain, the output response $y[k]$ is given by

$$y[k] = x[k] * h[k]. \tag{13.32}$$

Taking the z-transform of both sides of Eq. (13.32), we obtain

$$Y(z) = X(z)H(z), \tag{13.33}$$

where $Y(z)$ and $X(z)$ are, respectively, the z-transforms of the output response $y[k]$ and the input sequence $x[k]$. Equation (13.33) provides us with an alternative definition for the transfer function as the ratio of the z-transform of the output response and the z-transform of the input signal. Mathematically, the transfer function $H(z)$ can be expressed as follows:

$$H(z) = \frac{Y(z)}{X(z)}. \tag{13.34}$$

The z-transfer function of an LTID system can be obtained from its difference equation representation, as described in the following.

Consider an LTID system whose input–output relationship is given by the following difference equation:

$$y[k + n] + a_{n-1}y[k + n - 1] + \cdots + a_0 y[k]$$
$$= b_m x[k + m] + b_{m-1}x[k + m - 1] + \cdots + b_0 x[k]. \tag{13.35}$$

By taking the z-transform of both sides of the above equation, we obtain

$$\{z^n + a_{n-1}z^{n-1} + \cdots + a_0\}Y(z) = \{b_m z^m + b_{m-1}z^{m-1} + \cdots + b_0\}X(z),$$

which reduces to the following transfer function:

$$H(z) = \frac{Y(z)}{X(z)} = \frac{b_m z^m + b_{m-1} z^{m-1} + \cdots + b_0}{z^n + a_{n-1} z^{n-1} + \cdots + a_0} \qquad (13.36a)$$

or alternatively as

$$H(z) = z^{m-n} \frac{b_m + b_{m-1} z^{-1} + \cdots + b_0 z^{-m}}{1 + a_{n-1} z^{-1} + \cdots + a_0 z^{-n}}. \qquad (13.36b)$$

13.6.1 Characteristic equation, poles, and zeros

The z-transfer function plays an important role in the analysis of LTID systems. In this section, we will define a few key concepts related to the z-transfer function.

Characteristic equation The characteristic equation for the transfer function, Eq. (13.36a), is defined as follows:

$$D(z) = a_n z^n + a_{n-1} z^{n-1} + \cdots + a_0 = 0. \qquad (13.37)$$

Zeros The zeros of the transfer function $H(z)$ of an LTID system are *finite* locations in the complex z-plane, where $|H(z)| = 0$. For the transfer function, Eq. (13.36a), the location of the zeros can be obtained by solving the following equation:

$$N(z) = b_m z^m + b_{m-1} z^{m-1} + \cdots + b_0 = 0. \qquad (13.38)$$

Since $N(z)$ is an mth-order polynomial, it will have m roots leading to m zeros.

Poles The poles of the transfer function $H(z)$ of an LTID system are defined as locations in the complex z-plane, where $|H(z)|$ has an infinite value. The poles corresponding to the transfer function, Eq. (13.36a), can be obtained by solving the characteristic equation, Eq. (13.37). Since $D(z)$ is an nth-order polynomial, it will have n roots leading to n poles.

Because $D(z)$ is an nth-order polynomial and $N(z)$ is an mth-order polynomial, the transfer function will have a total of n poles and m zeros. However, in some cases, the location of a pole may coincide with the location of a zero. In that case, the pole and zero will cancel each other, and the actual number of poles and zeros will be reduced. In order to calculate the zeros and poles, a transfer function is factorized and typically represented as follows:

$$H(z) = \frac{N(z)}{D(z)} = \frac{(z - z_1)(z - z_2) \cdots (z - z_m)}{(z - p_1)(z - p_2) \cdots (z - p_n)}, \qquad (13.39a)$$

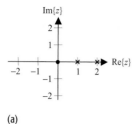

(a)

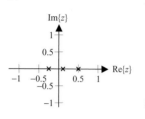

(b)

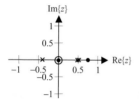

(c)

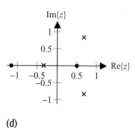

(d)

Fig. 13.6. Pole and zero plots for transfer functions in Example 13.13. Plot (a) corresponds to part (i) of Example 13.13; plot (b) corresponds to part (ii); plot (c) corresponds to part (iii); and plot (d) corresponds to part (iv). Also note that plot (c) includes double zeros at $z = 0$ and double poles at $z = 0.5$.

or alternatively as

$$H(z) = z^{m-n} \frac{(1 - z_1 z^{-1})(1 - z_2 z^{-1}) \cdots (1 - z_m z^{-1})}{(1 - p_1 z^{-1})(1 - p_2 z^{-1}) \cdots (1 - p_n z^{-1})}. \quad (13.39\text{b})$$

Example 13.13

Determine the poles and zeros of the following LTID systems:

(i) $H_1(z) = \dfrac{z}{z^2 - 3z + 2}$;

(ii) $H_2(z) = \dfrac{1}{(z - 0.1)(z - 0.5)(z + 0.2)}$;

(iii) $H_3(z) = \dfrac{z^2(2z - 1.5)}{(z + 0.4)(z - 0.5)^2}$;

(iv) $H_4(z) = \dfrac{z^2 + 0.7z + 1.6}{(z^2 - 1.2z + 1)(z + 0.3)}$.

Solution

(i) $H_1(z) = \dfrac{z}{z^2 - 3z + 2} = \dfrac{z}{(z - 1)(z - 2)}$.

There is one zero, at $z = 0$, and two poles, at $z = 1$ and 2.

(ii) $H_2(z) = \dfrac{1}{(z - 0.1)(z - 0.5)(z + 0.2)}$.

There is no zero, but there are three poles, at $z = 0.1, 0.5$, and -0.2.

(iii) $H_3(z) = \dfrac{z^2(2z - 1.5)}{(z + 0.4)(z - 0.5)^2}$.

There are three zeros, at $z = 0, 0$, and 0.75. There are three poles, at $z = -0.4$, 0.5, and 0.5.

(iv) $H_4(z) = \dfrac{(z - 0.5)(z + 1.2)}{((z - 0.6)^2 + 0.8^2)(z + 0.3)}$

$$= \dfrac{(z - 0.5)(z + 1.2)}{(z - 0.6 + \text{j}0.8)(z - 0.6 - \text{j}0.8)(z + 0.3)}.$$

There are two zeros, at $z = 0.5$ and -1.2. There are three poles, at $z = 0.6 - \text{j}0.8, 0.6 + \text{j}0.8$, and -0.3.

The poles and zeros of the above four systems are shown in Fig. 13.6. In the plot, $\times$ marks the position of a pole and $\bullet$ marks the position of a zero.

13.6.2 Determination of impulse response

The impulse response $h[k]$ of an LTID system can be obtained by calculating the inverse z-transform of the transfer function $H(z)$. Example 13.14 explains the steps involved in determining the impulse response.

Example 13.14

The input–output relationship of an LTID system is given by the following difference equation:

$$y[k+2] - \frac{3}{4}y[k+1] + \frac{1}{8}y[k] = 2x[k+2]. \qquad (13.40)$$

Determine the transfer function and the impulse response of the system.

Solution

Substituting $m = k + 2$, Eq. (13.40) can be written as follows:

$$y[m] - \frac{3}{4}y[m-1] + \frac{1}{8}y[m] = 2x[m].$$

Calculating the z-transform on both sides of the equation yields

$$Y(z) - \frac{3}{4}z^{-1}Y(z) + \frac{1}{8}z^{-2}Y(z) = 2X(z),$$

which results in the following transfer function:

$$H(z) = \frac{Y(z)}{X(z)} = \frac{2}{1 - \frac{3}{4}z^{-1} + \frac{1}{8}z^{-2}}.$$

To calculate the impulse response of the LTID system, consider the partial fraction expansion of $H(z)$ as

$$H(z) = \frac{2}{\left(1 - \frac{1}{2}z^{-1}\right)\left(1 - \frac{1}{4}z^{-1}\right)} \equiv \frac{4}{1 - \frac{1}{2}z^{-1}} - \frac{2}{1 - \frac{1}{4}z^{-1}}.$$

By calculating the inverse z-transform of both sides, the impulse response $h[k]$ is obtained:

$$h[k] = 4\left(\frac{1}{2}\right)^k u[k] - 2\left(\frac{1}{4}\right)^k u[k],$$

which is identical to the result obtained by Fourier technique in Example 11.18.

13.7 Relationship between Laplace and z-transforms

LTID signals and systems can be considered as special cases of LTIC signals and systems. Therefore, the Laplace transform can also be used to analyze such signals and systems. In this section, we derive the relationship between the Laplace and z-transforms.

If a DT sequence $x[k]$ is obtained by sampling a CT signal $x(t)$ with a sampling interval T, the CT sampled signal $x_s(t)$ may be expressed as follows:

$$x_s(t) = \sum_{k=-\infty}^{\infty} x(kT)\delta(t - kT),$$

Fig. 13.7. Using Laplace transform techniques to analyze LTID systems. (a) Reference LTID system; (b) equivalent LTIC system with CT input and output signals.

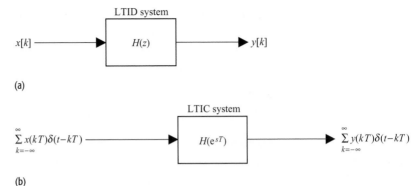

(a)

(b)

where $x(kT)$ are the sampled values of $x(t)$ which equals the DT sequence $x[k]$. Calculating the Laplace transform of $x_s(t)$, we obtain

$$X(s) = L\{x_s(t)\} = \sum_{k=-\infty}^{\infty} x(kT)L\{\delta(t-kT)\} = \sum_{k=-\infty}^{\infty} x(kT)e^{-kTs}.$$

Comparing $X(s)$ with the z-analysis equation,

$$X(z) = \sum_{k=-\infty}^{\infty} x[k]z^{-k},$$

it is clear that

$$X(s) = X(z)|_{z=e^{sT}} \tag{13.41a}$$

since $x[k] = x(kT)$. Equation (13.41a) illustrates the relationship between the Laplace transform $X(s)$ of a sampled function and the z-transform $X(z)$ of the DT sequence obtained from the samples. As illustrated in Fig. 13.7, an LTID system can be analyzed using an equivalent LTIC system. Figure 13.7(a) shows an LTID system with the z-transfer function $H(z)$ and sequence $x[k]$ applied at its input. The analysis of the LTID system can be completed in the s-domain with the LTIC system shown in Fig. 13.7(b). The transfer function of the LTIC system is given by

$$H(s) = H(z)|_{z=e^{sT}} \tag{13.41b}$$

and the DT input is transformed to an equivalent CT input of the form

$$x_s(t) = \sum_{k=-\infty}^{\infty} x(kT)\delta(t-kT).$$

The output in Fig. 13.7(b) can be calculated using CT analysis techniques. The resulting output $y(t)$ can then be transformed back into the DT domain using the relationship $y[k] = x(t)$ at $t = kT$.

Example 13.15
A DT system is represented by the following impulse response function:

$$h[k] = 0.5^5 u[k]. \tag{13.42}$$

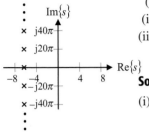

Fig. 13.8. Location of poles in the s-plane for the system in Example 13.15 with $T = 0.1$.

(i) Determine the z-transfer function of the system.

(ii) Determine the equivalent Laplace transfer function of the system.

(iii) Using the Laplace domain approach, determine if the system is stable.

Solution

(i) $H(z) = Z\{0.5^k u[k]\} = \dfrac{1}{1 - 0.5z^{-1}}$, or $\dfrac{z}{z - 0.5}$, ROC: $|z| > 0.5$.

(ii) Using Eq. (13.41b), the Laplace transfer function is given by

$$H(s) = H(z)|_{z=e^{sT}} = \frac{e^{sT}}{e^{sT} - 0.5}, \tag{13.43}$$

where T is the sampling interval.

(iii) A causal LTIC system is stable if all the poles corresponding to the Laplace transfer function lie in the left-hand half of the s-plane. Therefore, we will first calculate the pole locations in the s-plane, and then determine if the system is stable. The poles of the transfer function, Eq (13.43), are calculated from the characteristic equation as follows:

$$e^{sT} - 0.5 = 0 \Rightarrow e^{sT} = 0.5 \Rightarrow e^{(sT \pm j2\pi m)} = 0.5,$$

where $m = 0, 1, 2, \ldots$ Solving for the roots of this equation yields

$$s = \frac{1}{T}[\ln 0.5 \pm j2\pi m] \approx \frac{1}{T}[-0.693 \pm j2\pi m].$$

It is observed that an LTID system has an infinite number of poles in the s-domain. The locations of these poles for $T = 0.1$ are shown in Fig. 13.8. It is clear that these poles would lie in the left-half of the s-plane, irrespective of the value of the sampling interval T. The LTID system is, therefore, causal and stable.

Alternatively, the stability of the LTID system can be determined from its impulse response by noting that

$$\sum_{k=-\infty}^{\infty} |h[k]| = \sum_{k=-\infty}^{\infty} 0.5^k = 2 < \infty,$$

which satisfies the BIBO stability requirement derived in Chapter 10.

13.8 Stabilty analysis in the z-domain

In Example 13.15, the stability of an LTID system was determined by transforming its z-transfer function $H(z)$ to the Laplace transfer function $H(s)$ of an equivalent LTIC system and observing if the poles of $H(s)$ lie in the left-half s-plane. In this section, we derive a z-domain condition to check the stability of a system directly from its z-transfer function.

Consider a pole $z = p_z$ of an LTID system with the z-transfer function given by $H(z)$. Based on Eq. (13.41b), the location of the corresponding s-domain pole, $s = p_s$, of its equivalent LTIC system $H(e^{sT})$ is related to the location of the z-domain pole, $z = p_z$, of $H(z)$ by the following relationship:

$$p_z = e^{p_s T} \quad \text{or} \quad p_z = e^{\text{Re}\{p_s T\}} . e^{j \text{Im}\{p_s T\}}, \tag{13.44}$$

where $p_s T$ is decomposed into real and imaginary components as $\text{Re}\{p_s T\} + j \text{Im}\{p_s T\}$.

We consider two different cases. Case 1 refers to a stable system, which is not necessarily causal, while case 2 refers to a stable and causal system.

Case 1   Stable (not necessarily causal) LTID system   The LTIC stability condition for a stable system $H(s)$ is that the ROC of $H(s)$ must contain the vertical imaginary jω-axis in the complex s-plane. Since the ROC cannot contain any pole, in terms of the pole $s = p_s T$, this implies that $\text{Re}\{p_s T\} \neq 0$ such that no pole exists on the imaginary jω-axis. Substituting $\text{Re}\{p_s T\} \neq 0$ into Eq. (13.44) and calculating its magnitude yields

$$|p_z| = \underbrace{\left| e^{\text{Re}\{p_s T\}} \right|}_{\text{term I} \neq 1 \text{ if Re}\{p_s T\} \neq 0} \times \underbrace{\left| e^{j \text{Im}\{p_s T\}} \right|}_{\text{term II} = 1} \neq 1, \tag{13.45}$$

which implies that an LTID system $H(z)$ is stable if there is no pole on the unit circle of the z-plane. In terms of the ROC, it implies that the ROC must contain the unit circle for the system to be stable. The above condition does not assume the system to be causal, which is considered next.

Case 2   Stable and causal LTID system The LTIC stability condition for a stable and causal system $H(s)$ is that all poles of $H(s)$ must lie in the left-half of the complex s-plane. In terms of the pole $s = p_s T$, this implies that $\text{Re}\{p_s T\} < 0$. Substituting $\text{Re}\{p_s T\} < 0$ into Eq. (13.44) and taking the magnitude yields

$$|p_z| = \underbrace{\left| e^{\text{Re}\{p_s T\}} \right|}_{\text{term I} < 1 \text{ if Re }\{p_s T\} \, < \, 0} \times \underbrace{\left| e^{j \text{Im}\{p_s T\}} \right|}_{\text{term II} = 1} < 1. \tag{13.46}$$

Equation (13.46) states that *an LTID system $H(z)$ is stable if all poles lie within the unit circle*. Alternatively, the requirement for a causal and stable LTID system is stated as follows.

An LTID system will be absolutely BIBO stable and causal if and only if the ROC occupies the region outside and inclusive of the unit circle. In other words, the ROC for a stable and causal system is given by $|z| > z_0$, with $z_0 < 1$.

Example 13.16

Consider the LTID systems in Example 13.13. Considering various possibilities of the ROC, determine if the systems are absolutely BIBO state.

Solution

(i) Since

$$H_1(z) = \frac{z}{z^2 - 3z + 2} = \frac{z}{(z-1)(z-2)},$$

there are two poles of the LTID system, at $z = 1$ and 2. Since one pole lies on the unit circle, the ROC cannot contain the unit circle. The LTID system $H_1(z)$ is therefore not absolutely BIBO stable.

(ii) Since

$$H_2(z) = \frac{1}{(z-0.1)(z-0.5)(z+0.2)},$$

there are three poles, at $z = 0.1, 0.5$, and -0.2. There are four choices for the ROC, which are given by

ROC 1: $|z| < 0.1$. Such an implementation of the LTID system is not absolutely stable since the ROC does not contain the unit circle.

ROC 2: $0.1 < |z| < 0.2$. Such an implementation is not absolutely stable since the ROC does not contain the unit circle.

ROC 3: $0.2 < |z| < 0.5$. Such an implementation is not absolutely stable since the ROC does not contain the unit circle.

ROC 4: $|z| > 0.5$. Such an implementation is absolutely stable since the ROC contains the unit circle.

(iii) Since

$$H_3(z) = \frac{z^2(2z - 1.5)}{(z+0.4)(z-0.5)^2},$$

there are three poles, at $z = -0.4, 0.5$, and 0.5. There are three choices for the ROC, which are given by

ROC 1: $|z| < 0.4$. Such an implementation of the LTID system is not absolutely stable since the ROC does not contain the unit circle.

ROC 2: $0.4 < |z| < 0.5$. Such an implementation of the LTID system is not absolutely stable since the ROC does not contain the unit circle.

ROC 3: $|z| > 0.5$. Such an implementation of the LTID system is absolutely stable since the ROC contains the unit circle.

(iv) Since

$$H_4(z) = \frac{z^2 + 0.7z + 1.6}{(z^2 - 1.2z + 1)(z + 0.3)} = \frac{(z - 0.5)(z + 1.2)}{(z - 0.6 + j0.8)(z - 0.6 - j0.8)(z + 0.3)},$$

there are three poles, at $z = 0.6 - j0.8, 0.6 + j0.8$, and -0.3. The three choices of the ROC are given by

ROC 1: $|z| < 0.3$. Such an implementation of the LTID system is not absolutely stable since the ROC does not contain the unit circle.

ROC 2: $0.3 < |z| < |0.6 \pm \text{j}0.8|$ or $0.3 < |z| < 1$. Such an implementation of the LTID system is not absolutely stable since the ROC does not contain the unit circle.

ROC 3: $|z| > |0.6 \pm \text{j}0.8|$ or $|z| > 1$. Such an implementation of the LTID system is not absolutely stable since the ROC does not contain the unit circle.

13.8.1 Marginal stability

Equation (13.46) can be used to determine if a causal LTID system is absolutely stable. An absolutely stable and causal system has all poles inside the unit circle in the complex z-plane. On the contrary, if a causal system has one or more poles outside the unit circle then the system will not be absolutely stable. The impulse response of such a system includes a growing exponential function, making the system unstable. An intermediate case arises when a causal system has unrepeated poles on the unit circle and the remaining poles are inside the circle in the complex z-plane. Such a system is referred to as a marginally stable system. The condition for marginally stable and causal system is stated below.

A causal system with M unrepeated poles $p_m = a_m + \text{j}b_m$, $1 \le m \le M$, on the unit circle (such that $|p_m| = 1$) and all the remaining poles inside the unit circle in the z-plane is stable for all bounded input signals that do not include complex exponential terms of the form $\{\exp(\text{j}\Omega_m k)\}$, with $\Omega_m = \tan^{-1}(b_m/a_m)$, for $1 \le m \le M$. If any of the poles on the unit circle are repeated then the LTID system is unstable.

The following example demonstrates that a marginally stable system becomes unstable if the input signal includes a complex exponential $\exp(\text{j}\Omega_m)$ with frequency $\Omega_m = \tan^{-1}(b_m/a_m)$ corresponding to the location of the pole at $p_m = a_m + \text{j}b_m$ on the unit circle in the complex z-plane.

Example 13.17
A causal LTID system with transfer function given by

$$H(z) = \frac{1}{z^2 - z + 1} = \frac{1}{(z - 0.5 - \text{j}(\sqrt{3}/2))(z - 0.5 + \text{j}(\sqrt{3}/2))}$$

is a marginally stable system because of two unrepeated poles, at $z = 0.5 \pm \text{j}0.866$, on the unit circle. We will demonstrate the marginal stability by calculating the output for the following bounded input sequences:

(i) $x_1[k] = u[k]$;
(ii) $x_2[k] = \sin(\pi k/3)u[k]$.

Solution

(i) Taking the z-transform of the input sequence, we obtain

$$X_1(z) = \frac{z}{z-1}.$$

Applying the convolution property, the z-transform $Y_1(z)$ of the output response is given by

$$Y_1(z) = H(z)X_1(z) = \frac{z}{(z-1)(z^2-z+1)} = \frac{z^{-2}}{(1-z^{-1})(1-z^{-1}+z^{-2})}.$$

Taking the partial fraction expansion of $Y_1(z)$ yields

$$Y_1(z) = \frac{1}{1-z^{-1}} - \frac{1}{1-z^{-1}+z^{-2}}.$$

Using entries (3) and (12) of Table 13.1 (see Problem 13.5), we obtain

$$u[k] \overset{z}{\longleftrightarrow} \frac{1}{1-z^{-1}}$$

and

$$\frac{2}{\sqrt{3}} \sin\left(\frac{\pi k}{3} + \frac{\pi}{6}\right) u[k] \overset{z}{\longleftrightarrow} \frac{1}{1-z^{-1}+z^{-2}}.$$

Using the linearity property, the output $y_1[k]$ is given by

$$y_1[k] = \left[1 - \frac{2}{\sqrt{3}} \sin\left(\frac{\pi k}{3} + \frac{\pi}{6}\right)\right] u[k].$$

Note that the output response contains a unit step function and a sinusoidal term and is, therefore, bounded.

(ii) Taking the z-transform of the input sequence, we obtain

$$X_2(z) = \frac{(\sqrt{3}/2)z^{-1}}{1-z^{-1}+z^{-2}}.$$

Applying the convolution property, the z-transform $Y_2(z)$ of the output response is given by

$$Y_2(z) = H(z)X_2(z) = \frac{(\sqrt{3}/2)z^{-1}}{1-z^{-1}+z^{-2}} \cdot \frac{z^{-2}}{1-z^{-1}+z^{-2}} = \frac{(\sqrt{3}/2)z^{-3}}{(1-z^{-1}+z^{-2})^2}.$$

Using the frequency-differentiation property (see Problem 13.6), it can be shown that the following is a z-transform pair:

$$\left[\frac{2}{3} \sin\left(\frac{\pi}{3}k\right) - \frac{k}{\sqrt{3}} \sin\left(\frac{\pi}{3}k + \frac{\pi}{6}\right)\right] u[k] \overset{z}{\longleftrightarrow} \frac{(\sqrt{3}/2)z^{-3}}{(1-z^{-1}+z^{-2})^2}.$$

Therefore, the output response is given by

$$y_2[k] = \left[\frac{2}{3} \sin\left(\frac{\pi}{3}k\right) - \frac{k}{\sqrt{3}} \sin\left(\frac{\pi}{3}k + \frac{\pi}{6}\right)\right] u[k].$$

Note that the output is a growing sinusoid function because of the $k/\sqrt{3}$ scaling factor. Therefore, as k increases, the $|y_2[k]|$ increases without bound, leading to an unstable situation.

Table 13.3. Discrete frequencies corresponding to a few selected points along the unit circle in the z-domain

z-coordinates	$1 + j0$	$\frac{1}{\sqrt{2}} + j\frac{1}{\sqrt{2}}$	$0 + j1$	$-\frac{1}{\sqrt{2}} + j\frac{1}{\sqrt{2}}$	$-1 + j0$	$-\frac{1}{\sqrt{2}} - j\frac{1}{\sqrt{2}}$	$0 - j1$	$\frac{1}{\sqrt{2}} - j\frac{1}{\sqrt{2}}$
Frequency, Ω	0	$\pi/4$	$\pi/2$	$3\pi/4$	π	$5\pi/4$	$3\pi/2$	$7\pi/4$

In this example, we observe that the output response for the first input signal $x_1[k] = u[k]$ is bounded. On the other hand, the output produced by the second input, $x_2[k] = \sin(\pi k/3)u[k]$, is unbounded. Note that the second input is a sinusoidal sequence, which contains two complex exponentials:

$$\sin\left(\frac{\pi k}{3}\right) u[k] = \frac{1}{2j}\left[e^{j\pi k/3} - e^{-j\pi k/3}\right],$$

with discrete frequencies $\Omega_m = \pm\pi/3$. Since the frequencies of the complex exponentials are the same as the value of $\tan^{-1}(b_m/a_m) = \tan^{-1}(\pm\sqrt{3}/4) = \pm\pi/3$, determined from the poles, at $z = 0.5 \pm j\sqrt{3}/2$, on the unit circle, the output response is unbounded. This is consistent with the marginal stability condition mentioned above.

13.9 Frequency-response calculation in the z-domain

Based on Eq. (13.8), the DTFT transfer function is related to the z-transfer function by the following relationship:

$$H(\Omega) = \sum_{k=-\infty}^{\infty} h[k]z^{-k} = H(z)|_{z=e^{j\Omega}}, \qquad (13.47)$$

which may be used to derive the DTFT transfer function from the z-transfer function. Equation (13.47) has wider implications, as we discuss in the following.

(1) Taking the magnitude of both sides of the relationship $z = \exp(j\Omega)$ gives $|z| = 1$; therefore, Eq. (13.47) is only valid if the ROC of the z-transfer function contains the unit circle. Otherwise, the substitution $z = \exp(j\Omega)$ cannot be made and the DTFT transfer function does not exist.
(2) Equation (13.47) can also be used to compute the magnitude and phase spectra of the LTID system by evaluating the z-transfer function at different frequencies $(0 \le \Omega \le 2\pi)$ along the unit circle. The correspondence between the discrete frequency Ω and the z-coordinates is shown in Fig. 13.9. A selected subset of the discrete frequencies along the unit circle is shown in Table 13.3.

The computation of the magnitude and phase spectra from the z-transfer function is illustrated in the following example.

Fig. 13.9. Determination of the magnitude and phase spectra from the z-transfer function.

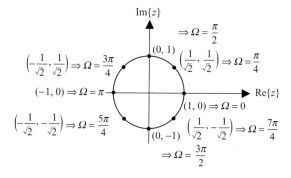

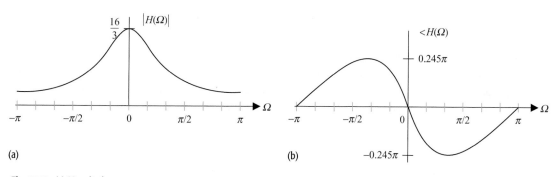

(a)   (b)

Fig. 13.10. (a) Magnitude spectrum and (b) phase spectrum of the LTID system considered in Example 13.18. The responses are shown in the frequency range $\Omega = [-\pi, \pi]$.

Example 13.18

Consider the system with z-transfer function given by

$$H(z) = \frac{2z^2}{z^2 - (3/4)z + (1/8)} = \frac{2}{1 - (3/4)z^{-1} + (1/8)z^{-2}}.$$

Calculate and plot the amplitude and phase spectra of the system.

Solution

The DTFT transfer function is given by

$$H(\Omega) = H(z)|_{z=e^{j\Omega}} = \frac{2}{1 - (3/4)e^{-j\Omega} + (1/8)e^{-j2\Omega}}.$$

The magnitude spectrum $|H(\Omega)|$ and the phase spectrum $<H(\Omega)$ are plotted in Fig. 13.10, which are identical to the spectra shown in Fig. 11.18.

13.10 DTFT and the z-transform

In Chapter 11 and in this chapter, we presented two different frequency-domain approaches to analyze DT signals and systems. The DTFT-based approach, introduced in Chapter 11, uses the real frequency Ω, whereas the z-transform-based approach uses the complex frequency $\sigma + j\Omega$. The output response of an LTID system can be computed using the convolution property of either the DTFT or the z-transform. In addition, the frequency-domain approach offers insight about the system characteristics, which is not readily available from the time-domain approach. However, an important issue is to determine which of the two transforms should be used to analyze the LTID system. Both approaches have their own advantages. Depending upon the application under consideration, the appropriate transform should be selected.

Example 13.19

Consider an LTID system represented by the unit impulse response $h[k] = 0.8^k u[k]$. Calculate the overall output and steady state output of the LTID system for the input sequence $x[k] = \cos(\pi k/3)u[k]$.

Solution

z-transform method Using Table 13.1, the z-transforms of the impulse response $h[k]$ and the input $x[k]$ are given by

$$H(z) = \frac{1}{1 - 0.8z^{-1}}$$

and

$$X(z) = \frac{1 - z^{-1}\cos(\pi/3)}{1 - 2z^{-1}\cos(\pi/3) + z^{-2}} = \frac{1 - 0.5z^{-1}}{1 - z^{-1} + z^{-2}}.$$

Using the convolution property, the z-transform of the output response is given by

$$Y(z) = H(z)X(z) = \frac{1 - 0.5z^{-1}}{(1 - 0.8z^{-1})(1 - z^{-1} + z^{-2})}.$$

By partial fraction expansion, the above expression becomes

$$Y(z) = \frac{2}{7} \times \frac{1}{1 - 0.8z^{-1}} + \frac{5}{7} \times \frac{1 + 0.5z^{-1}}{1 - z^{-1} + z^{-2}}$$

$$= \frac{2}{7} \times \frac{1}{1 - 0.8z^{-1}} + \frac{5}{7} \times \frac{1 - 0.5z^{-1}}{1 - z^{-1} + z^{-2}} + \frac{5}{7} \times \frac{z^{-1}}{1 - z^{-1} + z^{-2}}.$$

Taking the inverse z-transform, the output response is given by

$$y[k] = \frac{2}{7} \times 0.8^k u[k] + \frac{5}{7} \times \cos\left(\frac{\pi k}{3}\right) u[k] + \frac{10}{7\sqrt{3}} \times \sin\left(\frac{\pi k}{3}\right) u[k]$$

$$= \left[0.287(0.8)^k + 1.091\cos\left(\frac{\pi k}{3} - 0.857^r\right)\right] u[k]$$

where the superscript r indicates that the angle is expressed in radians.

The steady state output $y_{ss}[k]$ is computed by neglecting the transient term $(0.8)^k$, which decays to zero with time. The steady state output response is, therefore, given by

$$y_{ss}[k] = 1.091 \cos\left(\frac{\pi k}{3} - 0.857^r\right)u[k].$$

DTFT method   As in the CT case, the calculation of the actual output is difficult using the DTFT. However, the steady state value of the output can be easily calculated using DTFT. We have

$$H(\Omega) = \frac{1}{1 - 0.8e^{-j\Omega}}.$$

The value of the DTFT transfer function at $\Omega = \pi/3$, the fundamental frequency of the sinusoidal input, is given by

$$H(\Omega)|_{\Omega=\pi/3} = \frac{1}{1 - 0.8e^{-j(\pi/3)}} = 0.714 - j0.285 = 1.091e^{-j0.857},$$

implying that $|H(\Omega)| = 1.091$ and $<H(\Omega) = -0.857$ radians. Therefore, the steady state output response is given by

$$y_{ss}[k] = |H(\Omega)| \times \cos\left(\frac{\pi k}{3} + <H(\Omega)\right)u[k]$$

$$= 1.091 \cos\left(\frac{\pi k}{3} - 0.857^r\right)u[k].$$

Example 13.19 shows that the z-transform is a more convenient tool for transient analysis. For the steady state analysis, the z-transform does not offer much advantage over the DTFT. In signal processing applications, such as audio, image and video processing, the transients are generally ignored. In such applications, the DTFT is sufficient to analyze the steady state response. On the other hand, the transient analysis is important for applications such as control systems and process control. This is precisely the reason for the widespread use of the z-transform in digital control and system design, whereas the DTFT is preferred in signal processing applications.

13.11 Experiments with MATLAB

MATLAB provides several M-files for working with z-transforms. In this section, we explore five important functions, `residuez`, `residue`, `tf2zp`, `zp2tf`, and `zplane`. To illustrate the application of these M-files, we consider the following linear, constant-coefficient difference equation representation:

$$a_n y[k] + a_{n-1}y[k-1] + \cdots + a_0 y[k-n] = b_m x[k] + b_{m-1}x[k-1] + \cdots$$
$$+ b_0 x[k-m],$$

for modeling the relationship between the input sequence $x[k]$ and output response $y[k]$ of an LTID system. The above equation is a more general case of Eq. (13.35), where a_n was set to 1. Recall that Section 10.9 covered the

MATLAB file `filter` used to compute the output response $y[k]$ from specified sample values of the input sequence $x[k]$ and the ancillary conditions. In this section, we focus on the z-transfer function representation,

$$H(z) = \frac{Y(z)}{X(z)} = \frac{b_m + b_{m-1}z^{-1} + \cdots + b_0 z^{-m}}{a_n + a_{n-1}z^{-1} + \cdots + a_0 z^{-n}}, \qquad (13.48)$$

which can also be factorized as follows:

$$H(z) = \frac{Y(z)}{X(z)} = K\frac{(1 - z_0 z^{-1})(1 - z_1 z^{-1})\cdots(1 - z_M z^{-1})}{(1 - p_0 z^{-1})(1 - p_1 z^{-1})\cdots(1 - p_N z^{-1})}. \qquad (13.49)$$

Since MATLAB assumes that the numerator and denominator of the z-transfer function are expressed in increasing powers of z^{-1}, we prefer the aforementioned format for the z-transfer function.

13.11.1 Partial fraction expansion

To calculate the partial fraction expansion of a rational z-transfer function, MATLAB provides the `residuez` function, which has the following syntax:

```
>> [R,P,K] = residuez(B,A);
```

In terms of the transfer function in Eq. (13.48), the input variables B and A are defined as follows:

$$A = [a_n \ a_{n-1} \ldots a_0] \quad \text{and} \quad B = [b_m \ b_{m-1} \ldots b_0].$$

The output parameter R returns the values of the partial fraction coefficients, P returns the location of the poles, while K contains the direct term in the row vector.

Example 13.20

To illustrate the usage of the built-in function `residuez`, let us calculate the partial fraction expansion of the z-transfer function,

$$H(z) = \frac{2z(3z + 17)}{(z - 1)(z^2 - 6z + 25)},$$

considered in Example 13.4(iii).

Solution

Expressing the z-transfer function in increasing powers of z^{-1} yields

$$H(z) = \frac{6z^{-1} + 34z^{-2}}{1 - 7z^{-1} + 31z^{-2} - 25z^{-3}}.$$

The MATLAB code to determine the partial fraction expansion is given below. The explanation follows each instruction in the form of comments.

```
>> B = [0; 6; 34; 0];        % Coeff. of the numerator N(z)
>> A = [1; -7; 31; -25];     % Coeff. of the denominator D(z)
>> [R,P,K] = residuez(B,A)   % Calc. partial fraction expansion
```

The returned values are given by

```
     R = [-1.0000-1.2500j, -1.0000+1.2500j, 2.0000]
   P = [3.0000+4.0000j, 3.0000-4.0000j, 1.0000] and K=[].
```

The transfer function $H(z)$ can therefore be expressed as follows:

$$H(z) = \frac{-1 - j1.25}{1 - (3 + j4)z^{-1}} + \frac{-1 + j1.25}{1 - (3 - j4)z^{-1}} + \frac{2}{1 - z^{-1}}.$$

Alternative partial fraction expansion   Sometimes, it is desirable to perform the partial fraction in terms of the polynomials of z, instead of the polynomials of z^{-1}. In such cases, the MATLAB function residue is used. We solve Example 13.20 in terms of the alternative expression for the transfer function,

$$\frac{H(z)}{z} = \frac{6z + 34}{z^3 - 7z^2 + 31z - 25}.$$

The MATLAB code to determine the partial fraction expansion of the alternative expression is given below. As before, the explanation follows each instruction in the form of comments.

```
>> B = [0; 0; 6; 34];        % Coeff. of the numerator N(z)
>> A = [1; -7; 31; -25];     % Coeff. of the D(z)
>> [R,P,K] = residue(B,A)    % Calc. partial fraction expansion
```

The returned values are given by

```
     R = [-1.0000-1.2500j, -1.0000+1.2500j, 2.0000]
   P = [3.0000+4.0000j, 3.0000-4.0000j, 1.0000] and K = [].
```

The transfer function $H(z)$ can therefore be expressed as follows:

$$\frac{H(z)}{z} = \frac{-1 - j1.25}{z - (3 + j4)} + \frac{-1 + j1.25}{z - (3 - j4)} + \frac{2}{z - 1},$$

which is the same as result obtained in Example 13.4(iii).

13.11.2 Computing poles and zeros from the z-transfer function

MATLAB provides the built-in function tf2zp to calculate the location of the poles and zeros from the z-transfer function. Another function zplane can be used to plot the poles and zeros in the complex z-plane. In terms of Eq. (13.48), the syntaxes for these functions are given by

```
>> [Z,P,K] = tf2zp(B,A);     % Calculate poles and zeros
>> zplane(Z,P);              % plot poles and zeros,
```

where the input variables B and A are defined as follows:

$$A = [a_n \, a_{n-1} \ldots a_0] \text{ and } B = [b_m \, b_{m-1} \ldots b_0].$$

They are obtained from the transfer function given in Eq. (13.48). The vector Z contains the location of the zeros, vector P contains the location of the poles, while K returns a scalar providing the gain of the numerator.

Example 13.21

For the z-transfer function

$$H(z) = \frac{2z(3z + 17)}{(z - 1)(z^2 - 6z + 25)},$$

compute the poles and zeros and give a sketch of their locations in the complex z-plane.

Solution

The MATLAB code to determine the location of zeros and poles is listed below. The explanation follows each instruction in the form of comments.

```
>> B = [0, 6, 34, 0];     % Coefficients of the numerator N(z)
>> A = [1, -7, 31, -25];  % Coefficients of the denominator D(z)
>> [Z,P,K] = tf2zp(B,A)   % Calculate poles and zeros
>> zplane(Z,P)            % plot poles and zeros
```

The returned values are given by

```
Z = [0, -5.6667],
P = [3.0000+4.0000j 3.0000-4.0000j 1.0000] and K = 6.
```

The transfer function $H(z)$ can therefore be expressed as follows:

$$H(z) = 6\frac{z(z + 5.6667)}{(z - (3 + j4))(z - (3 - j4))(z - 1)}$$

$$= 6\frac{z^{-1}(1 + 5.6667z^{-1})}{(1 - (3 + j4)z^{-1})(1 - (3 - j4)z^{-1})(1 - z^{-1})}.$$

The pole–zero plot for $H(z)$ is shown in Fig. 13.11.

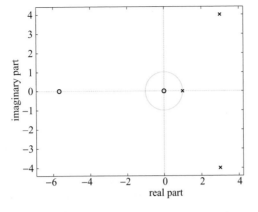

Fig. 13.11. Location of poles and zeros obtained in Example 13.21 using MATLAB

13.11.3 Computing the z-transfer function from poles and zeros

MATLAB provides the built-in function `zp2tf` to calculate the z-transfer function from poles and zeros. In terms of Eq. (13.49), the syntax for `zp2tf` is given by

```
>> [B,A] = zp2tf(Z,P,K);     % Calculate poles and zeros
```

where vector Z contains the location of the zeros, vector P contains the location of the poles, and K is a scalar providing the gain of the numerator. The numerator coefficients are returned in B and the denominator coefficients in A.

Example 13.22

Consider the poles and zeros calculated in Example 13.21. Using the values of the poles and, zeros and the gain factor, determine the transfer function $H(z)$.

Solution

The MATLAB code to determine the coefficients of the transfer function is listed below. The explanation follows each instruction in the form of comments.

```
>> Z = [0; -5.666667];     % Zeros in a column vector
>> P = [3+4*j; 3-4*j; 1];  % Poles in a column vector
>> K = 6;                  % Gain of the numerator
>> [B,A] = zp2tf(Z,P,K);   % Calculate poles and zeros
```

The returned values are given by

$$B = [0\ 6\ 34\ 0] \text{ and } A = [1\ -7\ 31\ -25],$$

which implies that the transfer function is given by

$$H(z) = \frac{6z^2 + 34z}{z^3 - 7z^2 + 31z - 25}.$$

The aforementioned expression is identical to the transfer function specified in Example 13.21.

13.12 Summary

In this chapter, we defined the bilateral z-transform for DT sequences as follows:

z-analysis equation $X(z) = \Im\{x[k]e^{-\sigma k}\} = \sum_{k=-\infty}^{\infty} x[k]z^{-k}.$

z-synthesis equation $x[k] = \dfrac{1}{2\pi j} \oint_C X(z)z^{k-1}dz.$

Unlike the DTFT, which requires the DT sequences to be absolutely summable for the DTFT to exist, the z-transform exists for a much larger set of DT sequences. Associated with the bilateral z-transform is a region of convergence (ROC) in the complex z-plane over which the z-transform is defined.

For causal sequences, the bilateral z-transform simplifies to the unilateral z-transform, defined in Section 13.2 as follows:

unilateral z-transform $X(z) = \sum_{k=0}^{\infty} x[k]z^{-k}.$

Section 13.3 introduced the look-up table, the partial fraction expansion, and the power-series-based approaches for determining the inverse z-transform of a rational function.

Section 13.4 presented the properties of the z-transform, which are summarized in the following.

(1) The linearity property states that the overall z-transform of a linear combination of DT sequences is given by the same linear combination of the individual z-transforms.
(2) The time-scaling property is only applicable for time-expanded (or interpolated) sequences. The time-scaling property states that interpolating a sequence in the time domain compresses its z-transform in the complex z-domain.
(3) The time-shifting property states that shifting a sequence in the time domain towards the right-hand side by an integer constant m is equivalent to multiplying the z-transform of the original sequence by a complex term z^{-m}. Similarly, shifting towards the left-hand side by integer m is equivalent to multiplying the z-transform of the original sequence with a complex term z^m.
(4) Time differencing is defined as the difference between an original sequence and its time-shifted version with a shift of one sample towards the right-hand side. The time-differencing property states that time-differencing a signal in the time domain is equivalent to multiplying its DTFT by a factor of $(1 - z^{-1})$.

(5) The z-domain-differentiation property states that differentiating the z-transform with respect to z and then multiplying with the variable $-z$ is equivalent to multiplying the original sequence by a factor of k.

(6) The time-convolution property states that the convolution of two DT sequences is equivalent to the multiplication of the z-transforms of the two sequences in the z-domain.

(7) The time-accumulation property is the converse of the time-differencing property. The accumulation property states that the z-transform of the running sum of a sequence is obtained by multiplying the z-transform of the original sequence by a factor of $z/(z - 1)$.

(8) The initial- and final-value theorems can be used to determine the initial value at $k = 0$ and final value at $k \to \infty$ directly from the z-transform of a DT sequence.

Section 13.5 covered the application of the z-transform in solving finite-difference equations and showed how the z-transfer function can be obtained from a difference equation of the following form:

$$H(z) = \frac{Y(z)}{X(z)} = \frac{b_m + b_{m-1}z^{-1} + \cdots + b_0 z^{-m}}{a_n + a_{n-1}z^{-1} + \cdots + a_0 z^{-n}}.$$

Section 13.6 defined the characteristic equation, poles, and zeros of an LTID system from the above rational expression of the z-transfer function. The characteristic equation for the transfer function is based on the denominator $D(z)$ of the z-transfer function $H(z)$ and is defined as follows:

$$D(z) = a_n z^n + a_{n-1} z^{n-1} + \cdots + a_0 = 0.$$

The roots of the characteristic equation define the poles of the LTID system as locations in the complex z-plane, where $|H(z)|$ has an infinite value. Similarly, the zeros of the transfer function $H(z)$ of an LTID system are *finite* locations in the complex z-plane where $|H(z)|$ approaches zero. If $N(z)$ is the numerator of $H(z)$, the zeros can be obtained by calculating the roots of the following equation:

$$N(z) = b_m z^m + b_{m-1} z^{m-1} + \cdots + b_0 = 0.$$

Sections 13.7 and 13.8 exploited the relationship between the z-transfer function $H(z)$ of an LTID system and the Laplace transfer function $H(s)$ of an equivalent LTIC system to derive the stability conditions for a causal and stable LTID system.

Section 13.9 showed how the magnitude and phase spectra can be obtained directly from the z-transform, while Section 13.10 compared the z-transfer-function-based analysis techniques with those based on the DTFT transfer function. We showed that the z-transform is a more convenient tool for transient analysis, while the DTFT is more appropriate for steady state analysis.

Finally, Section 13.11 illustrated some MATLAB library functions used to analyze the LTID systems in the complex z-domain.

Problems

13.1 Calculate the bilateral z-transform of the following non-causal functions:

    (i) $x_1[k] = 0.5^{k+1}u[k+5]$;

    (ii) $x_2[k] = (k+2)0.5^{|k|}$;

    (iii) $x_3[k] = |k+2| \times 0.5^{|k+2|}$;

    (iv) $x_4[k] = 3^{k+1}\cos\left(\dfrac{\pi}{3}k - \dfrac{\pi}{4}\right)u[-k+5]$.

13.2 Calculate the unilateral z-transform of the following causal functions:

    (i) $x_1[k] = \begin{cases} 1 & k = 10,\, 11 \\ 2 & k = 12,\, 15 \\ 0 & \text{otherwise}; \end{cases}$

    (ii) $x_2[k] = 3^{-k+2}u[k] + \displaystyle\sum_{m=1}^{4} m\delta[k-m]$;

    (iii) $x_3[k] = \sin\left(\dfrac{\pi k}{5} + \dfrac{\pi}{3}\right)u[k]$;

    (iv) $x_4[k] = 2^{-k}\sin\left(\dfrac{\pi k}{5} + \dfrac{\pi}{3}\right)u[k]$;

    (v) $x_5[k] = ku[k]$.

13.3 Using the partial fraction method, calculate the inverse z-transform of the following DT causal sequences:

    (i) $X_1(z) = \dfrac{z}{z^2 - 0.9z + 0.2}$;

    (ii) $X_2(z) = \dfrac{z}{z^2 - 2.1z + 0.2}$;

    (iii) $X_3(z) = \dfrac{z^2 + 2}{(z - 0.3)(z + 0.4)(z - 0.7)}$;

    (iv) $X_4(z) = \dfrac{z^2 + 2}{(z - 0.3)(z + 0.4)^2}$;

    (v) $X_5(z) = \dfrac{4z^{-1}}{1 - 5z^{-1} + 6z^{-2}}$;

    (vi) $X_6(z) = \dfrac{4z^{-2}}{10 - 6(z^1 + z^{-1})}$;

    (vii) $X_7(z) = \dfrac{2z^{-2}}{(1 - 4z^{-1})^2(1 - 2z^{-1})}$.

13.4 Using the power series expansion method, calculate the inverse z-transform of the DT causal sequences in Problem 13.3 for the first five non-zero values.

13.5 (a) Prove entry (12) of Table 13.1. (b) Using the proved result, derive the following z-transform pair used in Example 13.17(i):

$$\frac{2}{\sqrt{3}}\sin\left(\frac{\pi k}{3} + \frac{\pi}{6}\right)u[k] \overset{z}{\longleftrightarrow} \frac{1}{1 - z^{-1} + z^{-2}}.$$

13.6 (a) Using the z-domain-differentiation property and pairs (9) and (12) in Table 13.1, show that

(i) $k\sin(\Omega_0 k)u[k] \xleftrightarrow{z} \dfrac{z(z^2-1)\sin\Omega_0}{(z^2-2z\cos\Omega_0+1)^2}$,   ROC: $|z| > 1$;

(ii) $k\sin(\Omega_0 k + \theta)u[k] \xleftrightarrow{z} \dfrac{z[\sin(\Omega_0+\theta)z^2-2z\sin\theta-\sin(\Omega_0-\theta)]}{(z^2-2z\cos\Omega_0+1)^2}$,

ROC: $|z| > 1$.

(b) Using the above result, or otherwise, prove the following z-transform pair used in Example 13.17 (ii):

$$\left[\frac{2}{3}\sin\left(\frac{\pi}{3}k\right) - \frac{k}{\sqrt{3}}\sin\left(\frac{\pi}{3}k + \frac{\pi}{6}\right)\right]u[k] \xleftrightarrow{z} \frac{(\sqrt{3}/2)z}{(z^2-z+1)^2}$$

$$= \frac{(\sqrt{3}/2)z^{-3}}{(1-z^{-1}+z^{-2})^2}, \quad \text{ROC: } |z| > 1.$$

13.7 Using the time-shifting property and the results in Example 13.3(v), calculate the z-transform of the following function:

$$g[k] = \begin{cases} 1 & k = 10, 11 \\ 2 & k = 12, 15 \\ 0 & \text{otherwise.} \end{cases}$$

13.8 Prove the initial-value theorem stated in Section 13.4.8.

13.9 Prove the final-value theorem stated in Section 13.4.8.

13.10 Determine the z-transform of the following sequences using the specified property:

(i) $x[k] = (5/6)^k u[k-6]$, based on the z-transform pair (4) in Table 13.1 and the time-shifting property;

(ii) $x[k] = k(2/9)^k u[k]$, based on the z-transform pair (4) in Table 13.1 and the z-domain differentiation property;

(iii) $x[k] = ku(k)$, based on the z-transform pair (3) in Table 13.1 and the accumulation property;

(iv) $x[k] = e^k \sin(k)u[k]$, based on the z-transform pair (4) in Table 13.1 and the linearity property.

13.11 By selecting different ROCs, calculate four possible impulse responses of the transfer function

$$H(z) = \frac{1-z^{-1}}{(1-0.5z^{-1})(1-0.75z^{-1})(1-1.25z^{-1})}.$$

Determine the impulse response of the system that is stable. Is it causal? Why or why not?

13.12 You are given the unit impulse response of an LTID system,

$$h[k] = 5^{-k}u[k].$$

(i) Determine the impulse response $h_{inv}[k]$ of the inverse system that satisfies the property

$$h_{inv}[k] * h[k] = \delta[k].$$

(ii) Using any method, obtain the output $y[k]$ of the original system $h[k]$ for each of the following inputs: (a) $x_1[k] = u[k]$; (b) $x_2[k] = 5\delta[k-4] - 2\delta[k+4]$; and (c) $x_3[k] = e^{(k+2)}u[-k+2]$.

13.13 You are hired by a signal processing firm and you are hoping to impress them with the skills that you have acquired in this course. The firm asks you to design an LTID system that has the property that if the input is given by

$$x[k] = (1/3)^k u[k] - (1/4)^{k-1} u[k],$$

the output is given by

$$y[k] = (1/4)^k u[k].$$

(i) Determine the z-transfer function of the LTID system.
(ii) Determine the impulse response of the LTID system.
(iii) Determine the difference-equation representation of the LTID system.

13.14 The transfer function of a physically realizable system is as follows:

$$H(z) = \frac{1}{(1 - 0.3z^{-1})(1 - 0.5z^{-1})(1 - 0.7z^{-1})}.$$

(i) Determine the impulse response of the LTID system.
(ii) Determine the difference-equation representation of the LTID system.
(iii) Determine the unit step response of the LTID system by using the time-convolution property of the z-transform.
(iv) Determine the unit step response of the LTID system by convolving the unit step sequence with the impulse response obtained in part (i).

13.15 Given the difference equation

$$y[k] + y[k-1] + \frac{1}{4}y[k-2] = x[k] - x[k-2],$$

(i) determine the transfer function representing the LTID system;
(ii) determine the impulse response of the LTID system;
(iii) determine the output of the LTID system to the input $x[k] = (1/2)^k u[k]$ using the time-convolution property;
(iv) determine the output of the LTID system by convolving the input $x[k] = (1/2)^k u[k]$ with the impulse response obtained in part (ii).

13.16 Determine the output response of the following LTID systems with the specified inputs and impulse responses:

    (i) $x[k] = u[k+2] - u[-k-3]$ and $h[k] = u[k-5] - u[k-6]$;

    (ii) $x[k] = u[k] = u[k-9]$ and $h[k] = 3^{-k}u[k-4]$;

    (iii) $x[k] = 2^{-k}u[k]$ and $h[k] = k(u[k] - u[k-4])$;

    (iv) $x[k] = u[k]$ and $h[k] = 4^{-|k|}$;

    (v) $x[k] = 2^{-k}u[k]$ and $h[k] = 2^k u[-k-1]$.

13.17 When the DT sequence

$$x[k] = (1/4)^k \, u[k] + (1/3)^k u[k]$$

is applied at the input of a causal LTID system, the output response is given by

$$y[k] = 2\,(1/4)^k \, u[k] - 4\,(3/4)^k \, u[k].$$

    (i) Determine the z-transfer function $H(z)$ of the LTID system.

    (ii) Determine the impulse response $h[k]$ of the LTID system.

    (iii) Determine the difference-equation representation of the LTID system.

13.18 Consider an LTIC system with the following transfer function:

$$H(s) = \frac{e^{sT}}{e^{sT} - 0.3}.$$

Calculate the output response $y(t)$ of the LTIC system for the following input sequence:

$$f(t) = \sum_{k=0}^{\infty} (0.2)^{kT} \delta(t - kT).$$

13.19 Plot the poles and zeros of the following LTID systems. Assuming that the systems are causal, determine if the systems are BIBO stable.

    (i) $H(z) = \dfrac{z-2}{(z-0.6+j0.8)(z^2+0.25)}$;

    (ii) $H(z) = \dfrac{(z-2)(z-1)}{(z^2-2.5z+1)(z^2+0.25)}$;

    (iii) $H(z) = \dfrac{z-0.2}{(z+0.1)(z^2+4)}$;

    (iv) $H(z) = z^{-1} - 2z^{-2} + z^{-3}$;

    (v) $H(z) = \dfrac{(z^2+2.5z+0.9+j0.15)z}{z^3+(1.8+j0.3)z^2+(0.6+j0.6)z-0.2+j0.3}$;

    (vi) $H(z) = \dfrac{z^3-1.2z^2+2.5z+0.8}{z^6+0.3z^5+0.23z^4+0.209z^3+0.1066z^2-0.04162z-0.07134}$.

13.20 Consider an LTID system with the following transfer function:

$$H(z) = \frac{z}{z + 0.1}.$$

  (i) Using MATLAB, calculate the frequency response $H(e^{j\Omega})$ for $\Omega = [-\pi : \pi/20 : \pi]$. Plot the amplitude and phase spectra.
  (ii) If the DT signal $x[k] = 5\cos(\pi k/10)$ is passed through the system, what will be the steady state output of the system?

13.21 (a) Using MATLAB, determine the poles and zeros of the z-transfer functions specified in Problem 13.19. (b) Plot the location of poles and zeros in the complex z-plane using MATLAB.

13.22 (a) Using MATLAB, determine the partial fraction expansion of the z-transfer functions specified in Problem 13.19. (b) From the partial fraction expansion, calculate the impulse response function of the systems.

13.23 Assume that the functions in Problem 13.3 are z-transfer functions of some causal LTID systems. (a) Using MATLAB, determine the impulse responses of these systems. (b) Plot the impulse responses.

14 Digital filters

A digital filter is a system that transforms a sequence, applied at the input of the filter, by changing its frequency characteristics in a predefined manner. A convenient classification of digital filters is obtained by specifying the shape of their magnitude and phase spectra in the frequency domain. Based on the magnitude response, digital filters are classified into four important categories: lowpass, highpass, bandpass, and bandstop. A lowpass filter removes the higher-frequency components from an input sequence and is widely used to smooth out any sharp changes present in the sequence. An example of lowpass filtering is the elimination of the hissing noise present in magnetic audio cassettes. Since the background hissing noise contains higher frequency components than the music itself, a lowpass filter removes the hissing noise. A highpass filter eliminates the lower-frequency components and tends to emphasize sharp transitions in the input sequence. An application of highpass filtering is the detection of edges of different objects present in still images. While eliminating the smooth regions, represented by low frequencies, within each object, a highpass filter retains the boundaries between the objects. A bandpass filter allows a selected range of frequencies, referred to as the pass band, within the input sequence to be preserved at the output of the filter. All frequencies outside the pass band are eliminated from the input sequence. Bandpass filters are used, for example, in detecting the dual-tone multifrequency (DTMF) signals in digital telephone systems. As shown in Fig. 14.1, each DTMF key is represented by a pair of frequencies. At the receiver, a bank of bandpass filters, each tuned to one of the seven frequencies specified in Fig. 14.1, is used to determine the pressed key by isolating the pair of frequencies present in the transmitted signal. Bandstop filters are the converse of bandpass filters and allow all frequencies, except those in a specified stop band, to be retained at the output. An application of bandstop filters is to eliminate narrow-band noise, seen as bright and dark blotches in digital videos.

This chapter focuses on digital filters and introduces the basic filtering concepts and implementations useful in the design of digital filters. Section 14.1 describes four categories of frequency-selective filters, based on the

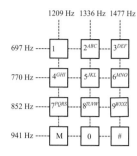

Fig. 14.1. Dual-tone
multifrequency (DTMF) signals
used in digital telephone
systems.

magnitude characteristics of the transfer function $H(\Omega)$. A second classification of digital filters is made on the basis of the length of the impulse response $h[k]$ and is covered in Section 14.2. Yet another classification of digital filters is made on the basis of the linearity of the phase $<H(\Omega)$, which is presented in Section 14.3. The impulse response of the ideal frequency-selective filters, considered in Section 14.1, is infinite, which makes them physically unrealizable. Section 14.4 describes realizable implementations of the ideal filters, which are causal. Sections 14.5–14.7 cover physical implementations of digital filters using special-purpose hardware confined to delays, adders, and scalar multipliers. During the actual implementation of digital filters in software or hardware, the filter coefficients can only be represented with finite precision. The impact of finite-precision arithmetic on the performance of digital filters is covered in Section 14.8. Important MATLAB library functions used in the analysis of digital filters are presented in Section 14.9. Finally, Section 14.10 concludes the chapter with a summary of the important concepts.

14.1 Filter classification

A digital filter is often classified on the basis of the magnitude and phase spectra derived from its transfer function. In this section, we consider a classification based on the shape of the magnitude spectrum of the filter. In the case of ideal filters, the shape of the magnitude spectrum is rectangular with a sharp transition between the range of frequencies passed and the range of frequencies blocked by the filter. The range of frequencies passed by the filter is referred to as the pass band of the filter, while the range of blocked frequencies is referred to as the stop band.

14.1.1 Ideal lowpass filter

The transfer function $H_{ilp}(\Omega)$ of an ideal lowpass filter, with a cut-off frequency of Ω_c, is given by

$$H_{ilp}(\Omega) = \begin{cases} 1 & |\Omega| \leq \Omega_c \\ 0 & \Omega_c < |\Omega| \leq \pi, \end{cases} \qquad (14.1a)$$

which has a pass band of $|\Omega| \leq \Omega_c$ and a stop band of $\Omega_c \leq |\Omega| \leq \pi$. Since the frequency $\Omega = \pi$ is the highest frequency present in the DTFT, the lowpass filter removes the higher frequencies in the range of $\Omega_c < |\Omega| \leq \pi$. The magnitude response of the lowpass filter is shown in Fig. 14.2(a). It is observed that the lowpass filter has a unity gain in the pass band and zero gain in the stop band. Sometimes, a lowpass filter has a pass band gain different from unity. If the gain is greater than one, the pass band signal is amplified, if the gain is less than one, the pass band signal is attenuated.

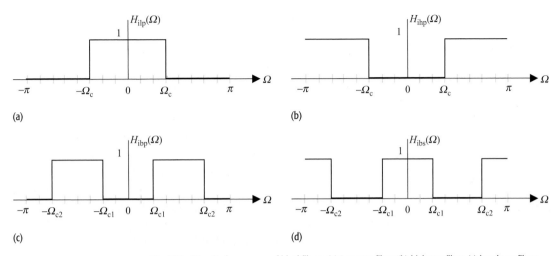

(a)

(b)

(c)

(d)

Fig. 14.2. Magnitude response of ideal filters. (a) Lowpass filter; (b) highpass filter; (c) bandpass filter; (d) bandstop filter.

The impulse response $h_{ilp}[k]$ of the ideal lowpass filter is obtained by calculating the inverse DTFT of Eq. (14.1a) and is given by

$$h_{ilp}[k] = \frac{\sin(k\Omega_c)}{k\pi} = \frac{\Omega_c}{\pi} \text{ sinc}\left(\frac{k\Omega_c}{\pi}\right). \tag{14.1b}$$

14.1.2 Ideal highpass filter

The transfer function $H_{ihp}(\Omega)$ of an ideal highpass filter, with a cut-off frequency of Ω_c, is given by

$$H_{ihp}(\Omega) = \begin{cases} 0 & |\Omega| < \Omega_c \\ 1 & \Omega_c \le |\Omega| \le \pi, \end{cases} \tag{14.2a}$$

which has a pass band of $\Omega_c \le |\Omega| \le \pi$ and a stop band of $|\Omega| < \Omega_c$. From the magnitude response of the highpass filter shown in Fig. 14.2(b), it is clear that the highpass filter blocks the lower frequencies $|\Omega| < \Omega_c$, while the higher frequencies $\Omega_c \le |\Omega| \le \pi$ are passed with a unity gain.

The transfer function $H_{ihp}(\Omega)$ of an ideal highpass filter is related to the transfer function $H_{ilp}(\Omega)$ of an ideal lowpass filter as follows:

$$H_{ihp}(\Omega) = 1 - H_{ilp}(\Omega), \tag{14.2b}$$

provided that the cut-off frequencies Ω_c of both filters are the same. Calculating the inverse DTFT of Eq. (14.2b), the impulse response $h_{ihp}[k]$ of the ideal highpass filter is obtained:

$$h_{ihp}[k] = \delta[k] - h_{ilp}[k]. \tag{14.3a}$$

Substituting the expression for $h_{\text{ilp}}[k]$ given in Eq. (14.1b) into the above equation, the impulse response $h_{\text{ihp}}[k]$ can be expressed as follows:

$$h_{\text{ihp}}[k] = \delta[k] - h_{\text{ilp}}[k] = \delta[k] - \frac{\Omega_c}{\pi} \, \text{sinc}\left(\frac{k\Omega_c}{\pi}\right). \tag{14.3b}$$

14.1.3 Ideal bandpass filter

The transfer function $H_{\text{ibp}}(\Omega)$ of an ideal bandpass filter, with cut-off frequencies of Ω_{c1} and Ω_{c2}, is given by

$$H_{\text{ibp}}(\Omega) = \begin{cases} 1 & \Omega_{c1} \leq |\Omega| \leq \Omega_{c2} \\ 0 & \Omega_{c1} < |\Omega| \quad \text{and} \quad \Omega_{c2} < |\Omega| \leq \pi, \end{cases} \tag{14.4a}$$

which has a pass band of $\Omega_{c1} \leq |\Omega| \leq \Omega_{c2}$ and a stop band of $|\Omega| \leq \Omega_{c1}$ and $\Omega_{c2} \leq |\Omega| \leq \pi$. The magnitude response of the ideal bandpass filter is shown in Fig. 14.2(c).

The transfer function $H_{\text{ibp}}(\Omega)$ is expressed in terms of the transfer functions of two ideal lowpass filters:

$$H_{\text{ibp}}(\Omega) = H_{\text{ilp1}}(\Omega)\Big|_{\text{cut-off freq}=\Omega_{c2}} - H_{\text{ilp2}}(\Omega)\Big|_{\text{cut-off freq}=\Omega_{c1}}. \tag{14.4b}$$

Calculating the inverse DTFT of Eq. (14.4a), the impulse response $h_{\text{ibp}}[k]$ of the ideal bandpass filter can be expressed as follows:

$$h_{\text{ibp}}[k] = h_{\text{ilp1}}[k]\Big|_{\Omega_c=\Omega_{c2}} - h_{\text{ilp2}}[k]\Big|_{\Omega_c=\Omega_{c1}}. \tag{14.4c}$$

Substituting the expression for $h_{\text{ilp}}[k]$ given in Eq. (14.1b) into the above equation, the impulse response $h_{\text{ibp}}[k]$ of the ideal bandpass filter can be expressed as follows:

$$h_{\text{ibp}}[k] = \frac{\Omega_{c2}}{\pi} \, \text{sinc}\left(\frac{k\Omega_{c2}}{\pi}\right) - \frac{\Omega_{c1}}{\pi} \, \text{sinc}\left(\frac{k\Omega_{c1}}{\pi}\right). \tag{14.4d}$$

Equation (14.4b) shows that a bandpass filter can be formed by a parallel configuration of two lowpass filters. The first lowpass filter in the parallel configuration should have a cut-off frequency of Ω_{c2}, while the second lowpass filter has a cut-off frequency of Ω_{c1}. Other configurations of bandpass filters are also possible, such as a series combination of a lowpass and a highpass filter.

14.1.4 Ideal bandstop filter

The transfer function $H_{\text{ibs}}(\Omega)$ of an ideal bandstop filter, with cut-off frequencies Ω_{c1} and Ω_{c2}, is given by

$$H_{\text{ibs}}(\Omega) = \begin{cases} 0 & \Omega_{c1} \leq |\Omega| \leq \Omega_{c2} \\ 1 & |\Omega| < \Omega_{c1} \quad \text{and} \quad \Omega_{c2} < |\Omega| \leq \pi, \end{cases} \tag{14.5a}$$

which has a pass band of $|\Omega| < \Omega_{c1}$ and $\Omega_{c2} < |\Omega| \leq \pi$ and a stop band of $\Omega_{c1} \leq |\Omega| \leq \Omega_{c2}$. The magnitude response of the ideal bandstop filter is shown in Fig. 14.2(d).

Table 14.1. Impulse response of ideal lowpass, highpass, bandpass, and bandstop filters in terms of normalized cut-off frequencies, $\Omega_n = \Omega_c/\pi$
The pass-band gain is assumed to be unity. For bandpass and bandstop filters, there are two cut-off frequencies, and $\Omega_{n2} > \Omega_{n1}$

Filter Type	Normalized cut-off frequency	Ideal filter impulse response
Lowpass	Ω_n	$h_{\text{ilp}}[k] = \Omega_n \, \text{sinc}[k\Omega_n]$
Highpass	Ω_n	$h_{\text{ilp}}[k] = \delta[k] - \Omega_n \, \text{sinc}[k\Omega_n]$
Bandpass	Ω_{n1}, Ω_{n2}	$h_{\text{ibp}}[k] = \Omega_{n2} \, \text{sinc}[k\Omega_{n2}] - \Omega_{n1} \, \text{sinc}[k\Omega_{n1}]$
Bandstop	Ω_{n1}, Ω_{n2}	$h_{\text{ibs}}[k] = \delta[k] - \Omega_{n2} \, \text{sinc}[k\Omega_{n2}] + \Omega_{n1} \, \text{sinc}[k\Omega_{n1}]$

The transfer function $H_{\text{ibs}}(\Omega)$ of an ideal bandstop filter is related to the transfer function $H_{\text{ibp}}(\Omega)$ of an ideal bandpass filter by

$$H_{\text{ibs}}(\Omega) = 1 - H_{\text{ibp}}(\Omega), \qquad (14.5b)$$

provided that the the cut-off frequencies Ω_{c1} and Ω_{c2} of both filters are the same. Calculating the inverse DTFT of Eq. (14.5b), the impulse response $h_{\text{ibs}}[k]$ of the ideal bandstop filter is obtained:

$$\begin{aligned} h_{\text{ibs}}[k] &= \delta[k] - h_{\text{ibp}}[k]\big|_{\Omega_c=\Omega_{c2},\Omega_{c1}} \\ &= \delta[k] - h_{\text{ilp1}}[k]\big|_{\Omega_c=\Omega_{c2}} + h_{\text{ilp2}}[k]\big|_{\Omega_c=\Omega_{c1}} \qquad (14.6) \\ &= \delta[k] - \frac{\Omega_{c2}}{\pi} \, \text{sinc}\left(\frac{k\Omega_{c2}}{\pi}\right) + \frac{\Omega_{c1}}{\pi} \, \text{sinc}\left(\frac{k\Omega_{c1}}{\pi}\right). \end{aligned}$$

Equation (14.6) shows that a bandstop filter can be formed by a parallel configuration of two lowpass filters having cut-off frequencies Ω_{c2} and Ω_{c1}.

The impulse responses of the four types of frequency-selective ideal filters discussed above are summarized in Table 14.1 in terms of the normalized cut-off frequencies. It is observed that the impulse responses primarily include one or two sinc functions and that all four types of ideal filters are non-causal.

14.2 FIR and IIR filters

A second classification of digital filters is made on the length of their impulse response $h[k]$. The length (or width) of a digital filter is the number N of samples k beyond which the impulse response $h[k]$ is zero in both directions along the k-axis. A filter of length N is also referred to as an N-tap filter.

A finite impulse response (FIR) filter is defined as a filter whose length N is finite. On the other hand, if the length N of the filter is infinite, the filter is called an infinite impulse response (IIR) filter. Below, we provide examples of FIR and IIR filters with length N specified in the parentheses.

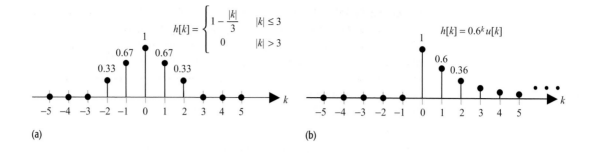

(a)   (b)

Fig. 14.3. (a) FIR filter; (b) IIR filter.

FIR filters

Triangular sequence $h[k] = \begin{cases} 1 - \dfrac{|k|}{3} & |k| \le 3 \\ 0 & \text{elsewhere} \end{cases}$ $(N = 5)$;

shifted impulse sequence $h[k] = 0.1\delta[k-2] + \delta[k] + 0.2\delta[k-2]$

$(N = 5)$;

exponentially decaying triangular sequence $h[k] = \displaystyle\sum_{m=-5}^{5} 0.4^{|k|}\delta[k-m]$

$(N = 11)$;

decaying impulses $h[k] = \displaystyle\sum_{m=0}^{10\,000} \dfrac{1}{m+1}\delta[k-m]$ $(N = 10\,001)$.

IIR filters

Causal decaying exponential $h[k] = 0.6^k u[k]$ $(N = \infty)$;

causal decaying sinusoidal $h[k] = 0.5^k \sin(0.2\pi k)u[k]$ $(N = \infty)$.

Other examples of IIR filters include non-causal ideal filters as shown in Table 14.1.

Figure 14.3(a) plots the triangular sequence with length $N = 5$ as an example of the FIR filter. Likewise, Fig. 14.3(b) plots the causal decaying exponential sequence with infinite length as an example of the IIR filter. An important consequence of a finite-length impulse response $h[k]$ is observed during the determination of the output response of an FIR filter resulting from a finite-length input sequence. Since the output response is obtained by the convolution of the impulse response and the input sequence, the output of an FIR filter is finite in length if the input sequence itself is finite in length. On the other hand, an IIR filter produces an output response that is always infinite in length.

A second consequence of the finite length of the FIR filters is observed in the stability characteristics of such filters. Recall that an LTID system with impulse response function $h[k]$ is BIBO stable if

$$\sum_{k=-\infty}^{\infty} |h[k]| < \infty.$$

Since the FIR filter is non-zero for only a limited number of samples k, the stability criterion is always satisfied by an FIR filter. As IIR filters contain an infinite number of impulse functions, even if the amplitudes of the constituent impulse functions are finite, the summation $\sum h[k]$ in an IIR filter may not be finite. In other words, it is not guaranteed that an IIR filter will always be stable. Therefore, care should be taken when designing IIR filters so that the filter is stable.

The implementation cost, typically measured by the number of delay elements used, is another important criterion in the design of filters. IIR filters are implemented using a feedback loop, in which the number of delay elements is determined by the order of the IIR filter. The number of delay elements used in FIR filters depends on their length, and so the implementation cost of such filters increases with the number of filter taps. An FIR filter with a large number of taps may therefore be computationally infeasible.

14.3 Phase of a digital filter

In Section 14.1, we introduced ideal frequency-selective filters as having rectangular magnitude response with sharp transitions between the pass band and stop band. The phase of ideal filters is assumed to be zero at all frequencies. An ideal filter is physically unrealizable because of the sharp transitions between the pass bands and stop bands and also because of the zero phase. In this section, we illustrate the effect of the phase on the performance of digital filters. In particular, we show that distortionless transmission within the pass band can be achieved by using a filter having a linear phase within the pass band.

Consider the following sinusoidal sequence:

$$x[k] = A_1 \cos(\Omega_1 k) + A_2 \cos(\Omega_2 k) + A_3 \cos(\Omega_3 k),$$

consisting of three tone frequencies $\Omega_1 < \Omega_2 < \Omega_3$ applied at the input of a physically realizable lowpass filter with the frequency response $H(\Omega)$ illustrated in Fig. 14.4. The magnitude spectrum $|H(\Omega)|$ of the filter is shown by a solid line, while the phase spectrum $<H(\Omega)$ is shown by a dashed line. The filter has a cut-off frequency Ω_c, such that $\Omega_2 < \Omega_c < \Omega_3$, and the cut-off frequency lies within the transition band. Based on the frequency response $H(\Omega)$ shown

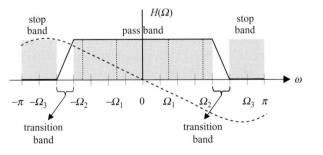

Fig. 14.4. Physically realizable lowpass filter with transition bands and non-zero phase.

in Fig. 14.4, the magnitudes and phases of the transfer function at the tone frequencies are given by

frequency $\Omega = \pm\Omega_1$ $|H(\Omega)| = 1, <H(\Omega) = \mp m_1\Omega_1;$

frequency $\Omega = \pm\Omega_2$ $|H(\Omega)| = 1, <H(\Omega) = \mp m_2\Omega_2;$

frequency $\Omega = \pm\Omega_3$ $|H(\Omega)| = 0, <H(\Omega) = \mp m_3\Omega_3;$

where m_1, m_2, and m_3 are the slopes of the phase response at $\Omega = \Omega_1$, Ω_2 and Ω_3, respectively.

Using the convolution property, the DTFT of the output of the filter is given by

$$Y(\Omega) = A_1\pi[\delta(\Omega - \Omega_1)e^{jm_1\Omega} + \delta(\Omega - \Omega_1)e^{-jm_1\Omega}]$$
$$+ A_2\pi[\delta(\Omega + \Omega_2)e^{jm_2\Omega} + \delta(\Omega - \Omega_2)e^{-jm_2\Omega}]$$
$$+ A_3\pi[\delta(\Omega + \Omega_3) + \delta(\Omega - \Omega_3)] \cdot 0.$$

Taking the inverse DTFT of the above equation, we obtain

$$y[k] = A_1\cos(\Omega_1(k - m_1)) + A_2\cos(\Omega_2(k - m_2)).$$

For $m_1 \neq m_2$, the input tones $A_1\cos(\Omega_1 k)$ and $A_2\cos(\Omega_2 k)$ are delayed unequally and the output sequence $y[k]$ is a distorted version of the sinusoidal components present within the pass band of the filter. To retain the shape of the pass-band components, each sinusoidal term $A_1\cos(\Omega_1 k)$ and $A_2\cos(\Omega_2 k)$ in $y[k]$ should be delayed equally, i.e. $m_1 = m_2$. In signal processing, the following two types of delays are defined:

phase delay $d_p = -\phi(\Omega)/\Omega;$

group delay $d_g = -\dfrac{d\phi(\Omega)}{d\Omega};$

where $\phi(\Omega)$ is the phase of the filter transfer function, i.e. $\phi(\omega) = \angle H(\omega)$. In other words, the *phase delay* (d_p) is defined as the phase divided by the frequency, whereas the *group delay* (d_g) is defined as the derivative of the phase with respect to frequency. From the above definitions, it is observed that the delay of a filter will be constant if the phase $\phi(\Omega)$ of the filter is a linear function of frequency. A filter is said to have a linear phase response if it satisfies the following relationships.

$$\phi(\Omega) = -\alpha\Omega, \quad \text{or} \quad \phi(\Omega) = -\alpha\Omega + \beta.$$

The first condition ensures that the filter has constant phase and group delay, whereas the second condition ensures only constant group delay. Although it is desirable to have both constant group and phase delays, a constant group delay is generally sufficient in many applications.

Based on the above discussion, the conditions for distortionless filtering, where the pass-band components are retained precisely at the filter output, are enlisted as follows.

(1) The pass-band gain of the filter should be the same for all frequency components present in the input signal that lie within the pass-band of the filter.
(2) The phase $<H(\Omega)$ of the filter should be linear for all input frequency components that lie within the pass band of the filter.
(3) The stop-band gain of the filter should be zero within the stop band of the filter.

Conditions (1)–(3) are valid for distortionless transmission within the pass bands of both FIR and IIR filters and are checked by plotting the magnitude and phase spectra of the filters. For FIR filters, the linear phase condition (condition (2)) can also be checked directly from the impulse response $h[k]$ as explained next.

14.3.1 Linear-phase FIR filters

Consider an FIR filter with impulse response $h[k]$, which is non-zero within the range $0 \leq k \leq N - 1$. The z-transform of the FIR filter is expressed as follows:

$$H(z) = \sum_{k=0}^{N-1} h[k]z^{-k} = h[0] + h[1]\,z^{-1} + h[2]\,z^{-2} + \cdots + h[N-1]\,z^{-(N-1)}.$$

$$(14.7)$$

The following proposition provides sufficient conditions for the phase linearity of an FIR filter.

Proposition 14.1 *If the impulse response function of an N-tap filter, with z-transfer function given by Eq. (14.7), satisfies either of the following relationships:*

symmetrical impulse response $h[k] = h[N-1-k]$; $(14.8a)$

antisymmetrical impulse response $h[k] = -h[N-1-k]$, $(14.8b)$

then the frequency response function can be represented as follows:

$$H(\Omega) = G(\Omega)e^{j(-\alpha\Omega+\beta)},$$

$$(14.9)$$

where $G(\Omega)$ is a real-valued function of Ω, $\alpha = (N-1)/2$, and β is a constant that can be either zero or $\pi/2$. Depending on the symmetry/anti-symmetry and even/odd length of $h[k]$, the FIR filters can be divided into four types: type 1, type 2, type 3 and type 4. Table 14.2 defines these four types of filters and the corresponding $G(\Omega)$ and β values. It is observed that type 1 and type 2 filters have constant phase and group delays, whereas type 3 and type 4 filters only have constant group delay.

Table 14.2. Linear-phase FIR filter types and the corresponding $G(\Omega)$ and β values

The coefficients $a[k]$ and $b[k]$ in column 4 are defined as follows: $a[0] = h[(N-1)/2]$, $a[k] = 2h[(N-1)/2 - k]$, $b[k] = 2h[N/2 - k]$

Type of FIR filter	Length, N	Symmetry	$G(\Omega)$	β
Type 1	odd	$h[k] = h[N-1-k]$	$\displaystyle\sum_{k=0}^{(N-1)/2} a[k]\cos(\Omega k)$	0
Type 2	even	$h[k] = h[N-1-k]$	$\displaystyle\sum_{k=1}^{N/2} b[k]\cos[\Omega(k-0.5)]$	0
Type 3	odd	$h[k] = -h[N-1-k]$	$\displaystyle\sum_{k=1}^{(N-1)/2} a[k]\sin(\Omega k)$	$\pi/2$
Type 4	even	$h[k] = -h[N-1-k]$	$\displaystyle\sum_{k=1}^{N/2} b[k]\sin[\Omega(k-0.5)]$	$\pi/2$

Proof

We prove Proposition 14.1 for type 1 filters. The proof for type 2, type 3, and type 4 filters follows along the same lines.

By substituting $z = \exp(j\Omega)$ in Eq. (14.7), we get

$$H(\Omega) = h[0] + h[1]\,e^{-j\Omega} + \cdots + h[N-2]\,e^{-j(N-2)\Omega} + h[N-1]\,e^{-j(N-1)\Omega}.$$

Taking $\exp(j(N-1)\Omega/2)$ common from the left-hand side of the above equation yields

$$H(\Omega) = e^{-j(N-1)\Omega/2}\big[h[0]\,e^{j(N-1)\Omega/2} + h[1]\,e^{j(N-3)\Omega/2}$$
$$+ \cdots + h[N-2]\,e^{-j(N-3)\Omega/2} + h[N-1]\,e^{-j(N-1)\Omega/2}\big]. \quad (14.10)$$

We now pair the first term with the last term, the second term with the second last term, and so on for the remaining terms. Note that for a type 1 filter, N has an odd value and $h[k] = h[N-1-k]$. By pairing terms in Eq. (14.10), we obtain

$$H(\Omega) = e^{-j(N-1)\Omega/2}\Big[\big(h[0]\,e^{j(N-1)\Omega/2} + h[N-1]\,e^{-j(N-1)\Omega/2}\big)$$
$$+ \big(h[1]\,e^{j(N-3)\Omega/2} + h[N-2]e^{-j(N-3)\Omega/2}\big) + \cdots$$
$$+ \Big(h\Big[\frac{N-1}{2}-1\Big]e^{j\Omega} + h\Big[\frac{N-1}{2}+1\Big]e^{-j\Omega}\Big) + h\Big[\frac{N-1}{2}\Big]\Big].$$

Because $h[k] = h[N-1-k]$, the above equation reduces as follows:

$$H(\Omega) = e^{-j(N-1)\Omega/2}\Big[2h[0]\cos\Big(\frac{(N-1)\Omega}{2}\Big) + 2h[1]\cos\Big(\frac{(N-3)\Omega}{2}\Big)$$
$$+ \cdots + 2h\Big[\frac{N-1}{2}-1\Big]\cos(\Omega) + h\Big[\frac{N-1}{2}\Big]\Big]$$
$$= e^{-j(N-1)\Omega/2}\Big\{h\Big[\frac{N-1}{2}\Big] + \sum_{k=0}^{(N-3)/2} 2h[k]\cos\Big[\Omega\Big(\frac{N-1}{2}-k\Big)\Big]\Big\}.$$

Table 14.3. Examples of FIR filters with linear and non-linear phase

Number of taps, N	z-transfer function, $H(z)$	Phase (linear or non-linear)	Phase value
4	$1 - 2z^{-1} + 2z^{-2} - z^{-3}$	type 4, linear	$-1.5\Omega + \pi/2$
3	$1 - z^{-2}$	type 3, linear	$-\Omega + \pi/2$
3	$1 + 2z^{-1} + 2z^{-2}$	non-linear	
4	$1 + 2z^{-1} + 2z^{-2} + z^{-3}$	type 2, linear	-1.5Ω
4	$1 + 2z^{-1} - 2z^{-2} + z^{-3}$	non-linear	
5	$1 + 2z^{-1} + 3z^{-2} + 2z^{-3} + z^{-4}$	type 1, linear	-2Ω
5	$1 + 2z^{-1} + 3z^{-2} + 2z^{-3} - z^{-4}$	non-linear	

Substituting $m = \frac{(N-1)}{2} - k$ in the above equation, we obtain

$$H(\Omega) = e^{-j(N-1)\Omega/2} \left\{ h\left[\frac{N-1}{2}\right] + \sum_{m=1}^{(N-1)/2} 2h\left[\frac{N-1}{2} - m\right] \cos(m\Omega) \right\}$$

$$= e^{-j(N-1)\Omega/2} \left\{ h\left[\frac{N-1}{2}\right] + \sum_{k=1}^{(N-1)/2} 2h\left[\frac{N-1}{2} - k\right] \cos(k\Omega) \right\}$$

$$= e^{-j(N-1)\Omega/2} \left\{ \sum_{k=0}^{(N-1)/2} a[k] \cos(k\Omega) \right\},$$

where $a[0] = h[(N-1)/2]$ and $a[k] = 2h[(N-1)/2 - k]$. It is observed that the derived $H(\Omega)$ matches with Eq. (14.9), with $\alpha = (N-1)/2$ and $G(\Omega)$ given in Table 14.2.

Example 14.1
Determine if the FIR filters specified in column 2 of Table 14.3 have linear phase or not. Also determine the value of the phase.

Solution
The phase linearity can be determined using the conditions given in Eq. (14.8). The third column of Table 14.3 shows whether a filter has linear phase and the type of linear-phase. The phase function, i.e. $(-\alpha\Omega + \beta)$ in Eq. (14.9), is shown in the fourth column.

To confirm the results of the last two entries of Table 14.3, Fig. 14.5 plots the magnitude and phase spectra of the FIR filter specified in the second to last row of Table 14.3. The phase plot in Fig. 14.5(b) confirms that the FIR filter has a linear phase. Since a phase of π is the same as that of $-\pi$, the sharp transitions at $\Omega = \pm 0.5\pi$ are not discontinuities but correspond to the same value. The magnitude spectrum illustrates non-uniform gains within the pass band and stop bands, implying that the FIR filter is not an ideal lowpass filter despite having a linear phase.

Fig. 14.5. Example of an FIR filter $H(z) = 1 + 2z^{-1} + 3z^{-2} + 2z^{-3} + z^{-4}$ with linear phase. (a) Magnitude spectrum; (b) phase spectrum.

Fig. 14.6. Example of an FIR filter $H(z) = 1 + 2z^{-1} + 3z^{-2} + 2z^{-3} - z^{-4}$ with non-linear phase. (a) Magnitude spectrum; (b) phase spectrum.

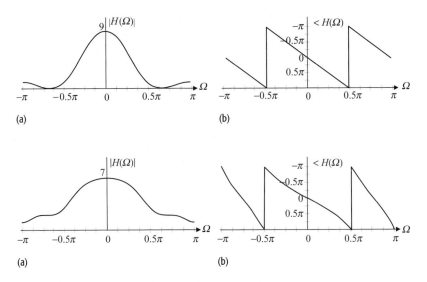

Likewise, Fig. 14.6 plots the magnitude and phase spectra of the FIR filter specified in the last row of Table 14.3. The phase plot shown in Fig. 14.6(b) confirms that the FIR filter has a non-linear phase.

14.4 Ideal versus non-ideal filters

Table 14.1 shows the impulse response of four types of frequency-selective ideal filters. It is observed that the ideal impulse responses are non-zero for $k < 0$. Therefore, these ideal filters are non-causal and hence physically non-realizable. It is, however, possible to realize a non-causal filter by applying an appropriate delay. To elaborate, let us consider the transfer function of an ideal lowpass filter shown in Eq. (14.1a) in a slightly different form as follows:

$$H_{\text{ilp}}(\Omega) = \begin{cases} e^{-jm\Omega} & |\Omega| \leq \Omega_c \\ 0 & \Omega_c < |\Omega| \leq \pi, \end{cases} \tag{14.11}$$

where a linear-phase component of $\exp(-jm\Omega)$, is included within the pass band. The variable m is a constant that corresponds to the delay of the filter. The impulse response $h_{\text{ilp}}[k]$ of the ideal lowpass filter is obtained by taking the inverse DTFT of Eq. (14.11), and is given by

$$h_{\text{ilp}}[k] = \frac{\sin((k-m)\Omega_c)}{(k-m)\pi} = \frac{\Omega_c}{\pi}\text{sinc}\left(\frac{(k-m)\Omega_c}{\pi}\right). \tag{14.12}$$

Figure 14.7 plots the impulse response $h_{\text{lp}}[k]$. As illustrated in Fig. 14.7, the impulse response $h_{\text{lp}}[k]$ of the ideal lowpass filter has an infinite length and is still non-causal. The ideal lowpass filter is therefore not physically realizable, irrespective of the value of delay m. Since the magnitude of the impulse response decays in both directions from its origin, $k = m$, a simple method to derive a

Fig. 14.7. Impulse response of an ideal lowpass filter with a cut-off frequency of Ω_c and delay m.

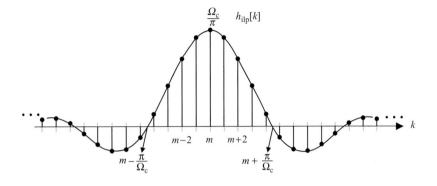

causal implementation of the ideal lowpass filter is to truncate its impulse response on either side of its origin. We consider two such implementations:

FIR implementation I

$$h_1[k] = \begin{cases} \dfrac{\Omega_c}{\pi}\operatorname{sinc}\left(\dfrac{(k-m)\Omega_c}{\pi}\right) & m - 70 \le k \le m + 70 \\ 0 & \text{elsewhere;} \end{cases} \tag{14.13a}$$

FIR implementation II

$$h_2[k] = \begin{cases} \dfrac{\Omega_c}{\pi}\operatorname{sinc}\left(\dfrac{(k-m)\Omega_c}{\pi}\right) & m - 10 \le k \le m + 10 \\ 0 & \text{elsewhere.} \end{cases} \tag{14.13b}$$

The length of the truncated FIR approximation is 141 in Eq. (14.13a) and 21 in Eq. (14.13b). The magnitude spectra for the two implementations are shown in Fig. 14.8. Compared with the ideal lowpass filter, we observe three significant changes in the causal implementations.

(1) The gain within the pass band of the causal implementations is no longer constant but includes several oscillating ripples, referred to as the pass-band ripples. The distortion caused by the pass-band ripples is significantly higher when the truncated length is small. Compared with Eq. (14.13a) with a truncated length of 141, Eq. (14.13b) has a length of 21 and results in a higher ripple distortion.

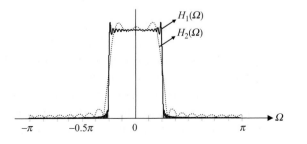

Fig. 14.8. Magnitude spectrum of FIR implementations $h_1[k]$ and $h_2[k]$ obtained by truncating the impulse response $h_{ilp}[k]$ of an ideal lowpass filter.

Fig. 14.9. Specifications of a practical lowpass filter with three modifications from an ideal lowpass filter. First, a pass-band ripple of δ_p is included about the unity pass-band gain. Then a stop-band ripple of δ_s is included. Finally, a transition band of $\Omega_s - \Omega_p$ allows for smooth transition between the stop and pass bands.

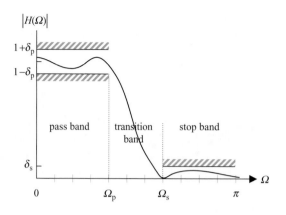

(2) Unlike the ideal lowpass filter, the FIR implementations have a significant transition band between the pass band and the stop band. The width of the transition band depends upon the length of the FIR implementations. The smaller the truncated length, the larger the width of the transition region.

(3) The gain within the stop band of the causal implementations is no longer zero but contains ripples, referred to as the stop-band ripples. As in the pass band, the distortion produced by the stop-band ripples is higher when the truncated length is smaller.

Since ideal filters are not physically realizable, a practical implementation of these filters is obtained by allowing acceptable variations in the magnitude response within the pass band and stop band. In addition, a transition band is included between the pass band and the stop band so that the magnitude response of the filter can drop off smoothly. Figure 14.9 specifies the magnitude response of a practical lowpass filter with the following characteristics:

pass band $\quad (1 - \delta_p) \leq |H(\Omega)| \leq (1 + \delta_p) \quad$ for $|\Omega| \leq \Omega_p$;

transition band $\quad 0 \leq |H(\Omega)| \leq 1 \quad$ for $\Omega_p < |\Omega| \leq \Omega_s$;

stop band $\quad 0 \leq |H(\Omega)| \leq \delta_s \quad$ for $\Omega_s \leq |\Omega| \leq \pi$.

The objective of a good design is to obtain a filter with limited ripples within the pass band and stop band, narrow transition bandwidth, and a linear phase at a reasonable implementation cost. Such an objective is self-contradictory. For example, a smaller transition band requires a relatively longer FIR filter or, alternatively, a higher-order IIR filter. In the case of FIR filters, the complexity of the filter is directly proportional to its length. Keeping the transition band small therefore results in a higher cost. Likewise, for IIR filters, the complexity depends upon the order of the filter. Increasing the order of the IIR filter to reduce the transition bandwidth increases the implementation cost. The design

Table 14.4. Impulse response of a 21-tap FIR filter

k	0, 20	1, 19	2, 18	3, 17	4, 16	5, 15	6, 14	7, 13	8, 12	9, 11	10
h[k]	−0.0014	0.0015	0.0066	0.0081	−0.0059	−0.0330	−0.0411	0.121	0.1320	0.2619	0.3183

process generally involves some trade-offs between the desired characteristics of the specified digital filter. We will revisit this issue in Chapters 15 and 16, where we introduce several design techniques for the FIR and IIR filters.

Examples 14.2 and 14.3 consider the FIR and IIR filters.

Example 14.2

Calculate the transfer function of a causal DT FIR filter whose impulse response $h[k]$ is specified in Table 14.4. Determine and plot the magnitude spectrum of the FIR filter. What are the values of the stop-band ripple δ_s and the transition bandwidth?

Solution

The impulse response of the FIR filter is plotted in Fig. 14.10(a). To determine the frequency characteristics of the filter, we determine the z-transfer function of the FIR filter:

$$H(z) = \sum_{k=0}^{20} h[k]z^{-k}.$$

Fig. 14.10. FIR filter in Example 14.2. (a) Impulse response $h[k]$; (b) magnitude spectrum $|H(\Omega)|$; (c) phase spectrum $<H(\Omega)$; (d) magnitude spectrum $|H(\Omega)|$ in decibels.

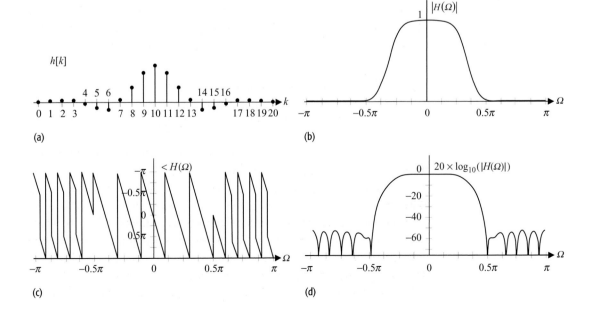

Substituting $z = \exp(j\Omega)$, the Fourier transfer function of the FIR filter is given by

$$H(\Omega) = \sum_{k=0}^{20} h[k] e^{-jk\Omega},$$

which is used to plot the magnitude and phase spectra of the FIR filter in Figs. 14.10(b) and (c). It is observed that the gain of the filter is close to unity at low frequencies ($\Omega \approx 0$), while the gain is zero at high frequencies ($\Omega \approx \pi$). Therefore, the impulse response $h[k]$ represents a lowpass filter. Also, Fig. 14.10(c) illustrates that the phase of the FIR is piecewise linear.

Without knowing the exact values of the pass and stop bands, it is difficult to determine the exact values of the stop-band ripple δ_s and the transition bandwidth. An intelligent guess can be made by looking at the Bode plot of the FIR filter. Recall that the Bode plot is the same as the magnitude spectrum except that the magnitude $|H(\Omega)|$ of the filter is expressed in decibels (dB) as follows:

$$\text{gain in dB} = 20 \log_{10}(|H(\Omega)|).$$

From the Bode plot shown in Fig. 14.10(d), we observe that the maximum value of $|H(\Omega)|$ within the stop band is approximately -52 dB. Expressed on a linear scale, the stop-band ripple δ_s is given by

$$\delta_s \text{ (dB)} = 20 \log_{10}(\delta_s) = -52 \quad \Rightarrow \quad \delta_s = 10^{-2.6} = 0.0025.$$

Figure 14.10(d) also provides approximate estimations of the pass band and stop band as follows:

$$\text{pass band } (0 \leq |\Omega_p| \leq 0.5) \quad \text{and} \quad \text{stop band } (1.5 \leq |\Omega_s| \leq \pi).$$

The transition band is therefore given by $0.5 < |\Omega| < 1.5$.

Example 14.3
The transfer function of a DT IIR filter is given by

$$H(z) = \frac{0.12z}{z^2 - 1.2z + 0.32}.$$

Determine and sketch the impulse response $h[k]$ of the filter. Determine and plot the magnitude response of the IIR filter.

Solution
The characteristic equation of $H(z)$ is given by $z^2 - 1.2z + 0.32 = 0$, which has two roots, at $z = 0.8$ and 0.4. The z-transfer function $H(z)$ can therefore be expressed as follows:

$$\frac{H(z)}{z} = \frac{0.12}{z^2 - 1.2z + 0.32} \equiv \frac{k_1}{z - 0.8} + \frac{k_2}{z - 0.4}.$$

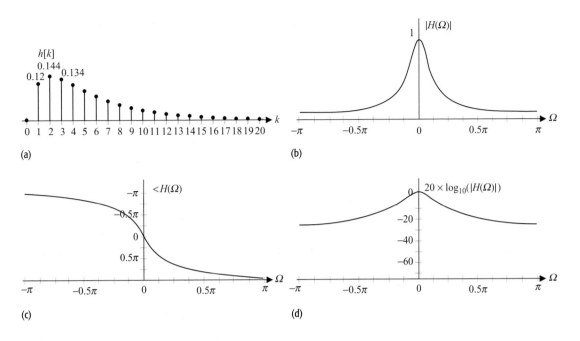

Fig. 14.11. IIR filter in Example 14.3. (a) Impulse response $h[k]$; (b) magnitude spectrum $|H(\Omega)|$; (c) phase spectrum $<H(\Omega)$; (d) magnitude spectrum $|H(\Omega)|$ in decibels.

Using Heaviside's partial fraction formula the coefficients of the partial fractions k_1 and k_2 are given by

$$k_1 = \left[(z - 0.8) \frac{0.12}{(z - 0.8)(z - 0.4)} \right]_{z=0.8} = \left[\frac{0.12}{z - 0.4} \right]_{z=0.8} = 0.3$$

and

$$k_2 = \left[(z - 0.4) \frac{0.12}{(z - 0.8)(z - 0.4)} \right]_{z=0.4} = \left[\frac{0.12}{z - 0.8} \right]_{z=0.4} = -0.3.$$

The partial fraction expansion of $H(z)$ is therefore given by

$$H(z) = \frac{0.3z}{z - 0.8} + \frac{-0.3z}{z - 0.4}$$

Taking the inverse z-transform of $H(z)$ yields

$$h[k] = 0.3[(0.8)^k - (0.4)^k]u[k].$$

which is plotted in Fig. 14.11(a). Note that the IIR filter has infinite length, as expected.

The Fourier transfer function of the IIR filter is obtained by substituting $z = \exp(j\Omega)$:

$$H(\Omega) = \frac{0.12e^{-j\Omega}}{1 - 1.2e^{-j\Omega} + 0.32e^{-j2\Omega}}.$$

The magnitude spectrum of the IIR filter is plotted in Figs. 14.11(b) and (d). Since the gain of the filter is unity at low frequencies (around $\Omega \approx 0$) and close to zero at high frequencies (around $\Omega \approx \pi$), the impulse response $h[k]$

represents a lowpass filter. Figure 14.11(c) illustrates that the phase of the IIR filter is non-linear; therefore, the IIR filter introduces distortion within the pass band.

14.5 Filter realization

In the preceding chapters, we presented several different techniques to calculate the output of a DT system. In the time domain, the output response $y[k]$ can be determined from its input $x[k]$ either by solving a linear, constant-coefficient, difference equation of the following form:

$$a_0 y[k] + a_1 y[k-1] + \cdots + a_N y[k-N]$$
$$= b_0 x[k] + b_1 x[k-1] + \cdots + b_M x[k-M]$$

or, alternatively, by calculating the convolution sum between the input $x[k]$ and the impulse response $h[k]$. The convolution sum is given by

$$y[k] = x[k] * h[k] = \sum_{m=-\infty}^{\infty} x[m]h[k-m].$$

In the frequency domain, the convolution property is used to express the convolution sum in terms of the transfer function $H(\Omega)$ and the CTFT $X(\Omega)$ of the input as follows:

$$Y(\Omega) = X(\Omega)H(\Omega),$$

from which the output $y[k]$ can be determined by calculating the inverse CTFT of $Y(\Omega)$. On digital computers and specialized DSP boards, the output of a digital filter is generally obtained by iteratively evaluating the recurrence formula,

$$y[k] = -\frac{1}{a_0}(+a_1 y[k-1] + \cdots + a_N y[k-N])$$
$$+ \frac{1}{a_0}(b_0 x[k] + b_1 x[k-1] + \cdots + b_M x[k-M]),$$

derived from the difference equation. Implementing the recurrence formula requires delaying the samples of the input and output sequences, multiplying the sample values with constant coefficients, and adding the resulting products. In other words, we require three mathematical operations, shift or delay, multiplication, and addition, to solve a difference equation iteratively. In the following, we introduce the schematic representation of these three fundamental operations.

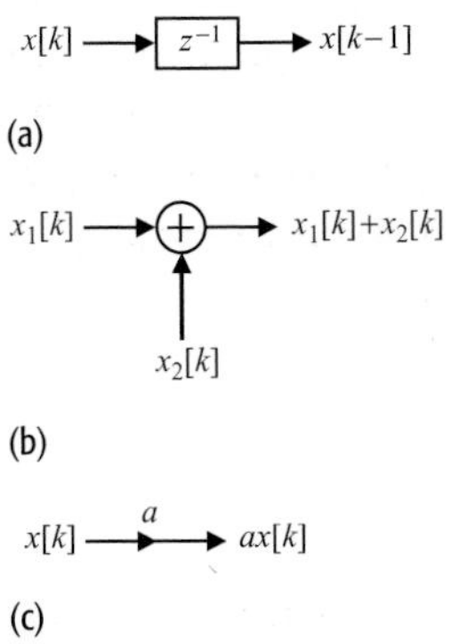

Fig. 14.12. Fundamental elements for building digital implementations for FIR and IIR filters. (a) Unit delay element; (b) adder; (c) constant-coefficient multiplier.

14.5.1 Shift or delay operator

On digital computers and specialized DSP boards, the shift operation is implemented using a cascaded combination of delay elements. The schematic representation of a unit delay element is illustrated in Fig. 14.12(a), where the transfer function of the block is given by $H(z) = z^{-1}$. The impulse response $h[k]$ of the unit delay element is given by $h[k] = \delta[k-1]$. The output is therefore given by $x[k] * \delta[k-1] = x[k-1]$. If a delay of more than one sample is required, several unit delay elements may be cascaded together in a series configuration.

14.5.2 Adder

On digital devices, adders are typically implemented using combinational or sequential circuits consisting of registers and logic gates. The schematic representation of an adder is illustrated in Fig. 14.12(b), where the input sequences $x_1[k]$ and $x_2[k]$ produce an output $x_1[k] + x_2[k]$.

14.5.3 Multiplication by a constant

On digital devices, multipliers are typically implemented using sequential circuits consisting of registers, shift delays, and logic gates. The schematic representation of a constant multiplier is shown in Fig. 14.12(c), where the input sequence $x[k]$ is multiplied with a constant a, producing an output $ax[k]$.

In the following sections, we sketch signal flow graphs for efficient implementations of both FIR and IIR digital filters using the aforementioned elements, referred to as the fundamental elements. By manipulating the signal flow graphs, we present several different but equivalent structures for the same transfer function. We also demonstrate the effect of finite-precision arithmetic on the gain–frequency characteristics of digital filters, and provide several design tips to alleviate the problems arising from finite-precision arithmetic.

14.6 FIR filters

A causal FIR filter, of finite length N and having non-zero values in the range $0 \le k \le (N-1)$, is represented by the following transfer function:

$$H(z) = \sum_{k=0}^{N-1} h[k]z^{-k} = h[0] + h[1]z^{-1} + h[2]z^{-2} + \cdots + h[N-1]z^{-(N-1)}$$

$$(14.14)$$

or, alternatively by a difference equation obtained by solving the convolution sum:

$$y[k] = \sum_{m=0}^{N-1} h[k]x[k-m]$$

$$= h[0]x[k] + h[1]x[k-1] + \cdots + h[N-1]x[k-(N-1)]. \quad (14.15)$$

Fig. 14.13. Direct form for causal FIR filters of length N.

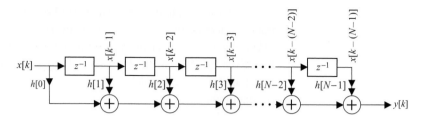

There are several flow graph representations of the FIR filter. In the following, we discuss some of them.

14.6.1 Direct form

The flow graph for direct form is achieved by implementing Eq. (14.15) directly. In direct form, the constant multipliers are the same as the coefficients of the difference equation, Eq. (14.15). The direct form of the flow graph for a causal FIR filter is shown in Fig. 14.13. Since the cost of implementation of a filter is directly proportional to the number of fundamental elements used, we include a count of these elements for each flow graph. The number of the fundamental elements used in Fig. 14.13 is shown in the second row of Table 14.5.

The flow graph for the direct form resembles a tapped delay line used frequently in communication systems for channel equalization. The filter shown in Fig. 14.13 is therefore referred to as a tapped delay line filter or sometimes as a transversal filter.

14.6.2 Cascaded form

The flow graph for the cascaded form is achieved by expressing Eq. (14.14) in terms of a product of quadratic terms:

$$H(z) = h[0] \prod_{n=1}^{\left\lceil \frac{N-1}{2} \right\rceil} (1 + b_{1n}z^{-1} + b_{2n}z^{-2}). \qquad (14.16)$$

Factorizing $H(z)$ in terms of quadratic terms ensures coefficients b_{1n} and b_{2n} to be real-valued provided that the impulse response $h[k]$ is also real-valued. Had linear factors been considered in Eq. (14.16) there would be no guarantee for the coefficients of the linear factors to be real-valued, even with real-valued $h[k]$. The upper limit $\lceil (N-1)/2 \rceil$ in the summation in Eq. (14.16) represents a ceiling operation, which equals $(N-1)/2$ if N is odd. If N is even, the upper limit equals $N/2$ with $b_{2n} = 0$ for the last product term.

The flow graph of the cascaded form is achieved by considering $\lceil (N-1)/2 \rceil$ substructures and cascading the substructures together in a series configuration. The resulting flow graph is shown in Fig. 14.14. The number of fundamental elements used in Fig. 14.14 is shown in the third row of Table 14.5.

Table 14.5. Number of elements required to implement different types of FIR filter structures

Structure	two-input address	Unit delays	Constant multipliers
Direct form	$N-1$	$N-1$	N
Cascaded form	$N-1$	$N-1$	N
Linear phase filters	$N-1$	$N-1$	$N/2$ (N even) $(N+1)/2$(N odd)

Fig. 14.14. Cascaded form for causal FIR filters of length N.

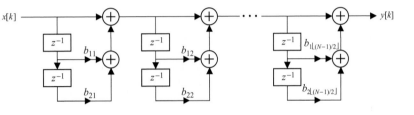

14.6.3 Linear-phase FIR filters

As proved in Proposition 14.1, an N-tap linear phase FIR filter satisfies the following symmetry condition:

$$h[k] = h[N-1-k] \quad \text{or} \quad h[k] = -h[N-1-k].$$

For the symmetry condition $h[k] = h[N-1-k]$, we show that the condition can be used to reduce the number of constant multipliers. The derivation for the antisymmetry condition, $h[k] = -h[N-1-k]$, follows along similar lines.

If the length N of the filter is even, Eq. (14.14) is rearranged as follows:

$$H(z) = h[0]\left(1 + z^{-(N-1)}\right) + h[1]\left(z^{-1} + z^{-(N-2)}\right)$$
$$+ \cdots + h\left[\frac{N}{2} - 1\right]\left(z^{-(N/2-1)} + z^{-(N/2)}\right).$$

On the other hand, if the length N of the filter is odd, Eq. (14.14) is rearranged as follows:

$$H(z) = h[0]\left(1 + z^{-(N-1)}\right) + h[1]\left(z^{-1} + z^{-(N-2)}\right)$$
$$+ \cdots + h\left[\frac{N-1}{2} - 1\right]\left(z^{-((N-1)/2-1)} + z^{-((N-1)/2+1)}\right)$$
$$+ h\left[\frac{N-1}{2}\right]z^{-(N-1)/2}.$$

Using the above equations, the flow graphs of the linear-phase FIR filter satisfying the symmetry condition is shown in Fig. 14.15. Both even and odd values of length N are considered. The numbers of fundamental elements required are shown in the fourth row of Table 14.5. It is observed that the number of constant

Fig. 14.15. Flow graphs for linear-phase FIR filters. (a) Length N is odd; (b) length N is even.

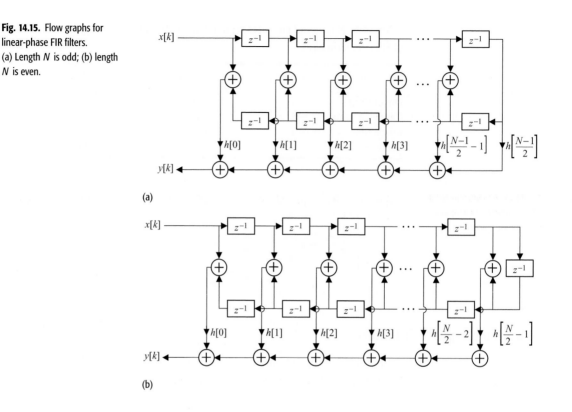

(a)

(b)

multipliers is roughly half that required in direct form or cascaded form. The number of unit delay and addition elements, however, stays the same.

14.6.4 Transposed forms

Alternative flow graphs for implementations in Sections 14.6.1–14.6.3 can be realized by applying the transpose operation. Transposition of a flow graph is achieved by (i) interchanging the role of the input and output; (ii) reversing the directions of all branches within a flow graph; and (iii) replacing the source nodes by adders, and vice versa. Note that the number of fundamental elements required to implement a filter does not change if the transposed form is used for implementation. We explain the principle of transposition with an example.

Example 14.4

Implement direct form and cascaded configurations of the flow graph for the FIR filter with transfer function given by

$$H(z) = -0.3 - 0.4z^{-1} + 1.4z^{-2} - 0.4z^{-3} - 0.8z^{-4}.$$

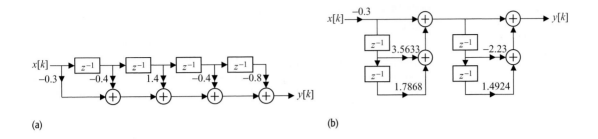

(a)

(b)

Fig. 14.16. (a) Direct form I and (b) cascaded configurations for the FIR filter in Example 14.4.

Using transposition, derive an alternative configuration from the cascaded implementation.

Solution

The flow graph for the direct form is shown in Fig. 14.16(a). For the cascaded configuration, we factorize $H(z)$ as follows:

$$H(z) = -0.3\,(1 + 2.9595z^{-1})(1 + 0.6038z^{-1})(1 - (1.1150 - j0.4992)z^{-1})$$
$$\times(1 - (1.1150 + j0.0.4992)z^{-1}).$$

Expressing $H(z)$ as a product of quadratic terms, we obtain

$$H(z) = -0.3(1 + 3.5633z^{-1} + 1.7868z^{-2})(1 - 2.23z^{-1} + 1.4924z^{-2}),$$

which has the flow graph illustrated in Fig. 14.16(b).

The alternative configuration for the cascaded form, obtained by applying the transposition principle, is shown in Fig. 14.17 using two steps. Step 1 interchanges the role of the input and output, reverses the directions of all branches, and replaces the source nodes with adders. Similarly, the adders are replaced by source nodes. The resulting configuration is shown in Fig. 14.17(a), where the input is on the right-hand side of the flow graph and the output is on the left-hand side. Figure 14.17(b) is a reordered version of Fig. 14.17(a) with the input and output, rearranged to the standard right-hand and left-hand sides, respectively.

Fig. 14.17. Transpose configurations of flow graph in Fig. 14.16(b).

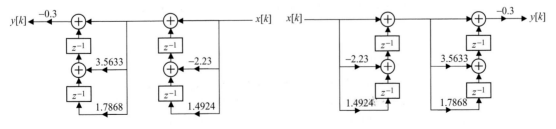

(a)   (b)

Table 14.6. Number of elements required to implement different types of IIR filter structures

M and N are, respectively, the degree of the numerator and denominator polynomials in $H(z)$, as shown in Eq. (14.17)

Structure	Unit delays	Two-input adders	Constant multipliers
Direct form I	$M + N$	$M + N$	$M + N + 1$
Direct form II	$\max(M, N)$	$M + N$	$M + N + 1$
Cascaded form	$\max(M, N)$	$M + N$	$M + N + 1$
Parallel form	$\max(M, N)$	$M + N$ $(M \geq N)$	$M + N + 1$ $(M \geq N)$
		$2N$ $(M < N)$	$2N + 1$ $(M < N)$

Finally, it should be noted that $H(z)$ does not represent a linear-phase FIR filter. As such, the linear-phase configuration cannot be derived for this filter.

The direct form and cascaded implementations of the FIR filters can be extended to the IIR filters, which are discussed in Section 14.7.

14.7 IIR filters

The transfer function of an IIR filter is given by

$$H(z) = \frac{b_0 + b_1 z^{-1} + \cdots + b_M z^{-M}}{1 + a_1 z^{-1} + \cdots + a_N z^{-N}}, \qquad (14.17)$$

where the coefficient a_0 of the constant term in the denominator is normalized to one. Based on Eq. (14.17), an IIR filter can alternatively be modeled by the linear, constant-coefficient difference equation given by

$$y[k] + a_1 y[k - 1] + \cdots + a_N y[k - N]$$
$$= b_0 x[k] + b_1 x[k - 1] + \cdots + b_M x[k - M]. \qquad (14.18)$$

There are four major architectures to implement the IIR filters, which are considered in the following.

14.7.1 Direct form I

To derive the IIR realization of the transfer function, Eq. (14.17), we implement the numerator and denominator functions, defined as follows:

numerator $N(z) = b_0 + b_1 z^{-1} + \cdots + b_M z^{-M};$

denominator $D(z) = 1 + a_1 z^{-1} + \cdots + a_N z^{-N},$

separately. The resulting flow graph is shown in Fig. 14.18, where the first structure represents $N(z)$ and the second structure represents $D(z)$. The numbers of fundamental elements required in direct form I are shown in the second row of Table 14.6.

Fig. 14.18. Direct form I for IIR filters where numerator polynomial $N(z)$ and denominator polynomial $D(z)$ are implemented as cascaded systems. The degree M of the numerator is assumed to the same as the degree N of the denominator.

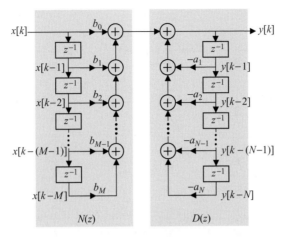

Fig. 14.19. Direct form I realization for the IIR filter in Example 14.5.

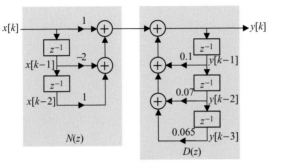

Example 14.5

Implement the direct form I realization of an IIR filter with the following transfer function:

$$H(z) = \frac{z^3 - 2z^2 + z}{z^3 - 0.1z^2 - 0.07z - 0.065}. \qquad (14.19)$$

Solution

The transfer function $H(z)$ can be represented as follows:

$$H(z) = \frac{1 - 2z^{-1} + z^{-2}}{1 - 0.1z^{-1} - 0.07z^{-2} - 0.065z^{-3}},$$

with the difference equation given by

$$\begin{aligned}
y[k] &= x[k] - 2x[k-1] + x[k-2] - \{-0.1y[k-1] \\
&\quad - 0.07y[k-2] - 0.065y[k-3]\} \\
&= x[k] - 2x[k-1] + x[k-2] + 0.1y[k-1] \\
&\quad + 0.07y[k-2] + 0.065y[k-3].
\end{aligned}$$

The flow graph using direct form I is illustrated in Fig. 14.19.

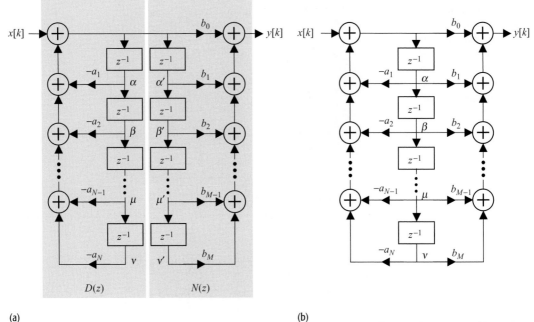

Fig. 14.20. Direct form II for IIR filters where degrees of the numerator (M) and denominator (N) are assumed to be the same.

14.7.2 Direct form II

Direct form II is realized by noting that the order of structures $N(z)$ and $D(z)$ can be interchanged as for any two systems in a series combination. The resulting flow graph is shown in Fig. 14.20(a). Since nodes α and α' have the same polarity, these nodes can be merged by replacing the top two delay elements by one delay element. Similarly, nodes β and β' can be merged, and so on for the rest of the adjacent nodes below the delays in structures $D(z)$ and $N(z)$. The resulting flow diagram is referred to as direct form II and is illustrated in Fig. 14.20(b). The number of fundamental elements required in direct form II is shown in the third row of Table 14.6.

A flow graph that requires the minimum number of delay elements, multipliers, and adders to implement a filter is referred to as a *canonical structure*. It can be shown that the implementation complexity of an arbitrary IIR filter with a numerator of degree M and a denominator of degree N cannot be less than the complexity of the flow graph for direct form II shown in Fig. 14.20(b). Therefore, direct form II with the flow graph shown in Fig. 14.20(b), is a canonical architecture. On the other hand, direct form II with the flow graph shown in Fig. 14.20(a) is a non-canonical architecture.

Fig. 14.21. Direct form II
architecture for the IIR filter in
Example 14.6.

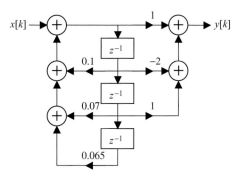

Example 14.6
Implement the filter in Example 14.5 using the direct form II realization.

Solution
The flow graph for direct form II realization is shown in Fig. 14.21.

14.7.3 Cascaded form

The flow graph for the cascaded form is achieved by expressing the numerator
and denominator polynomials in Eq. (14.17) in terms of a product of quadratic
terms:

$$H(z) = b_0 \frac{\displaystyle\prod_{m=1}^{\lceil \frac{M}{2} \rceil} (1 + b_{1m}z^{-1} + b_{2m}z^{-2})}{\displaystyle\prod_{n=1}^{\lceil \frac{N}{2} \rceil} (1 + a_{1n}z^{-1} + a_{2n}z^{-2})}. \tag{14.20}$$

Fig. 14.22. Cascaded form
architecture for IIR filters.

Factorizing $H(z)$ in terms of quadratic terms ensures coefficients b_{1n} and b_{2n}
to be real-valued provided that the impulse response $h[k]$ is also real-valued.

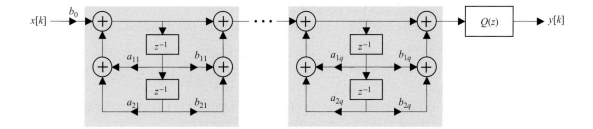

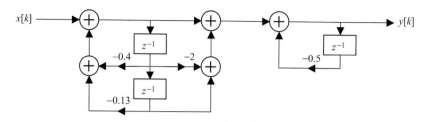

$x[k]$ ⟶ ⊕ ⟶ ⊕ ⟶ ⊕ ⟶ $y[k]$

z^{-1} -0.4 -2 -0.5 z^{-1}

z^{-1} -0.13

Fig. 14.23. Cascaded form architecture for the IIR filter in Example 14.7.

In general, the quadratic terms may be coupled together in the following form:

$$H(z) = b_0 \frac{(1 + b_{11}z^{-1} + b_{21}z^{-2})}{(1 + a_{11}z^{-1} + a_{21}z^{-2})} \times \frac{(1 + b_{12}z^{-1} + b_{22}z^{-2})}{(1 + a_{12}z^{-1} + a_{22}z^{-2})}$$

$$\times \cdots \times \frac{(1 + b_{1q}z^{-1} + b_{2q}z^{-2})}{(1 + a_{1q}z^{-1} + a_{2q}z^{-2})} \times Q(z), \qquad (14.21)$$

where $q = \min(\lceil N/2 \rceil, \lceil (M/2 \rceil)$, and $Q(z)$ represents the uncoupled terms arising from unequal values of degree N and M. The first q quadratic terms in Eq. (14.21) are implemented using a cascaded configuration of the direct form II realization, while $Q(z)$ may be implemented in either direct form I or direct form II realization. The flow graph for Eq. (14.21) is shown in Fig. 14.22. The numbers of fundamental elements required in cascaded form are shown in the fourth row of Table 14.6.

Example 14.7

Implement the filter in Example 14.5 using the cascaded form.

Solution

The transfer function $H(z)$ is expressed as follows:

$$H(z) = \frac{1 - 2z^{-1} + z^{-2}}{1 - 0.1z^{-1} - 0.07z^{-2} - 0.065z^{-3}}$$

$$= \frac{(1 - z^{-1})(1 - z^{-1})}{(1 - 0.5z^{-1})[1 - (-0.2 + j0.3)z^{-1}][1 - (-0.2 - j0.3)z^{-1}]}.$$

Note that if the filter is implemented using only first-order filters, the filter coefficients will be complex. In order to avoid complex values for the filter coefficients, the complex roots are combined into a quadratic term as follows:

$$H(z) = \frac{1 - 2z^{-1} + z^{-2}}{1 + 0.4z^{-1} + 0.13z^{-2}} \times \frac{1}{1 + 0.5z^{-1}}. \qquad (14.22)$$

The flow diagram for Eq. (14.22) is shown in Fig. 14.23, where we have omitted scalar multiplications where the multiplier is unity.

Table 14.7. Comparison of the number of fundamental elements in flow graphs obtained from different forms for the IIR filter implemented in Examples 14.4–14.7

Form	Number of		
	unit delays	scalar multipliers	dual-input adders
Direct form I	5	4	5
Direct form II	3	4	5
Cascaded form	3	4	5
Parallel form	3	6	5

14.7.4 Parallel form

In this form, IIR filters are implemented as a parallel combination of first- and/or second-order filters. To derive the parallel realization, the transfer function $H(z)$ is expressed in terms of its partial fractions:

$$H(z) \equiv Q(z) + \frac{k_1}{1 - p_1 z^{-1}} + \frac{k_2}{1 - p_2 z^{-1}} + \cdots + \frac{k_N}{1 - p_N z^{-1}}, \quad (14.23)$$

where $k_1, k_2, \ldots, k_N$ are partial fraction coefficients, obtained from Heaviside's formula, and $p_1, p_2, \ldots, p_N$ are the poles of $H(z)$. To prevent complex-valued coefficients, Eq. (14.23) is expressed in terms of quadratic terms as follows:

$$H(z) = Q(z) + \sum_{n=1}^{\left\lceil \frac{N}{2} \right\rceil} \frac{b_{1n} + b_{2n} z^{-1}}{1 + a_{1n} z^{-1} + a_{2n} z^{-2}}. \quad (14.24)$$

If the degree N of the denominator in $H(z)$ is odd, $a_{2n} = b_{2n} = 0$ (for $n = \lceil N/2 \rceil$). The parallel form of the IIR filter is illustrated in Fig. 14.24. The number of fundamental elements required in the parallel form are shown in the fifth row of Table 14.6. Note that the parallel architecture has the same complexity as the direct form II and cascade architectures when $N = M$. If the numerator and the denominator are not of the same degree, a larger number of scalar multipliers and two-input adders are required.

Example 14.8
Implement the IIR filter in Example 14.5 using the parallel form.

Solution
Using partial fraction expansion, the transfer function $H(z)$ is expressed as follows:

$$H(z) \equiv \frac{k_1}{1 - 0.5z^{-1}} + \frac{k_{21} + k_{22}z^{-1}}{1 + 0.4z^{-1} + 0.13z^{-2}},$$

where the partial fraction coefficients are determined as $k_1 = 0.431$, $k_{21} = 0.569$, and $k_{22} = -1.8879$. Figure 14.25 shows the parallel form of the IIR filter.

Fig. 14.24. Parallel form architecture for IIR filters.

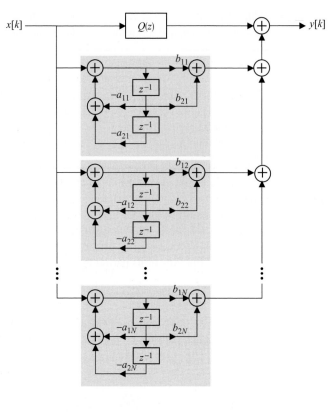

Fig. 14.25. Parallel form architecture for the IIR filter in Example 14.8.

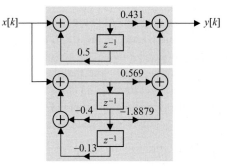

In Table 14.7, we compare the different realizations of the IIR filter specified in Example 14.5. Trivial scalar multiplications, where the scalar multiplier is unity, are ignored. The cascaded form yields the minimum number of fundamental elements used. This, however, is valid only for Example 14.5 and is not true in general.

14.7.5 Transposed forms

As was the case for FIR filters, alternative flow graphs for the implementations in Sections 14.7.1–14.7.4 can be realized by applying the transpose operation.

14.7.6 Choice of structures

The direct form II, cascaded and parallel forms are referred to as canonical structures and have roughly the same implementation complexity. The actual complexity of each form of realization depends on the transfer function under consideration. Table 14.7 compares the four structures in terms of the number of unit delays, scalar multipliers, and dual-input adders for the filter considered in Examples 14.4–14.7. It is observed that the direct form I requires the largest number of delay elements. The direct form II, cascaded, and parallel structures require an identical number of delay elements and adders. However, the cascaded form needs to implement the lowest number of multipliers. This is because there are two multipliers that perform multiplication by a factor of one. These unity multipliers need not be implemented. The parallel structure requires the largest number of multipliers.

Irrespective of the arithmetic complexity, all of these realizations should provide identical outputs for the same input. As we shall see in the following section, the filter coefficients are implemented using finite precision. The impact of finite-precision arithmetic on the performance of digital filters is the focus of our discussion in Section 14.8. The following are some empirical observations that should be kept in mind when choosing a particular realization.

(i) When the poles of the transfer function lie close to each other or close to the unit circle in the complex z-plane, direct form realizations, with filter coefficients represented using finite precision, produce large deviations from the output of an exact filter.

(ii) The order in which the first- and second-order systems are implemented in cascaded forms affects the output of the filter in finite-precision implementations. Changing the order may reduce the deviation from the output of an exact filter.

(iii) Pairing of complex poles and zeros is important for all cascaded and parallel realizations.

(iv) In cascaded realizations, scalar multipliers between different systems may be required to prevent the partial fraction coefficients from becoming too large or too small.

14.8 Finite precision effect

Figure 14.26 illustrates the processing of analog signals with digital systems. The analog signal $y(t)$ produced by such a system contains distortions from several sources, including

(i) analog-to-digital conversion (ADC) noise;

(ii) finite-precision approximation of filter coefficients;

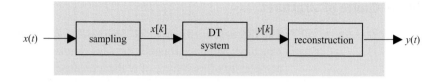

Fig. 14.26. Processing of analog
signals with digital filters

(iii) round-off errors;
(iv) register overflow.

These effects are considered in the following.

14.8.1 Analog-to-digital conversion (ADC) noise

The process of encoding the analog signal $x(t)$ into a DT signal $x[k]$, quantized
to a fixed number of bits, involves discarding the higher resolution information
of the analog signal. The resulting distortion is referred to as the analog-to-
digital (ADC) noise. The amount of ADC noise is inversely proportional to the
number of bits used in the quantization process. For example, assume that the
true value of a sample is given by 0.875 364 573 894 562 234 5. If the sample
is quantized by a 3-bit uniform quantizer with a peak-to-peak range of ± 1 V,
the sample value would be quantized to 0.9375, leading to an ADC noise of
$-0.062\,135\,426\,105\,44$. If instead an 8-bit uniform quantizer is used, the sample
value would be approximated to 0.878 906 25 with an ADC noise of -0.003
541 676 105 44. The ADC noise can, therefore, be reduced by using a higher-
resolution quantizer with a larger number of reconstruction levels, but it can
never be eliminated. The ADC noise causes the analog signal $y(t)$, recovered
from the processed digital sequence $y[k]$, to deviate from the output signal
produced by a completely analog system, which is equivalent to the schematic
representation of Fig. 14.26.

A second error introduced by the quantizer is referred to as the saturation
noise, which occurs when the input signal $x(t)$ exceeds the peak-to-peak operating range for which the quantizer is designed. Since the range of the saturation
noise is unlimited, the saturation noise is more objectionable than the ADC
noise.

14.8.2 Finite-precision approximation of filter coefficients

The filter coefficients designed from a given specification are analog and have
infinite precision. When the filter coefficients are represented using a finite
number of bits, quantization noise is introduced. As a result, the characteristics
of the digital filter may change considerably from the design specifications.

A common standard used for representing floating point numbers on a digital
computer is the IEEE 754 floating point standard, which uses 32 bits in the
single-precision mode. The representation for the 32-bit IEEE standard is shown

Table 14.8. Representation used in the 32-bit IEEE 754 floating point single-precision standard

31	30	29	28	27	26	25	24	23	22	21	20	19	18	17	16	15	14	13	12	11	10	9	8	7	6	5	4	3	2	1	0
s (1bit)	exponent (8 bits)								significand (23 bits)																						

Table 14.9. IEEE 754 floating point representation for the decimal number for -0.75_{ten}

31	30	29	28	27	26	25	24	23	22	21	20	19	18	17	16	15	14	13	12	11	10	9	8	7	6	5	4	3	2	1	0
1	0	1	1	1	1	1	1	0	1	0	0	0	0	0	0	0	0	0	0	0	0	0	0	0	0	0	0	0	0	0	0
s	exponent								significand																						

in Table 14.8, where a single-precision, floating point number in IEEE 754 standard is represented in scientific notation as follows:

$$(-1)^{s} \times (1 + 0.\text{significand}) \times 2^{(\text{exponent}-127)}. \qquad (14.25)$$

Note that the s-bit represents the sign of the floating point. The s-bit is set to unity for negative numbers and to zero for positive numbers. The significand specifies the decimal fraction, while the exponent represents the power in terms of 2. As an example, consider the IEEE 754 binary representation of the decimal number -0.75, represented by -0.75_{ten}. The binary representation of -0.75_{ten} is given by

$$-0.75_{\text{ten}} = -0.11_{\text{two}},$$

which in scientific notation is represented by

$$-0.75_{\text{ten}} = -1.1_{\text{two}} \times 2^{-1}.$$

Comparing with Eq. (14.25), the values of the exponent and significand are given by

$$0.\text{significand} = 0.1_{\text{two}} \quad \text{and} \quad \text{exponent} = 126 \quad \text{or} \quad 011\,111\,10_{\text{two}}.$$

The single-precision representation for -0.75_{ten} is specified in Table 14.9.

To derive the resolution of the 32-bit single-precision arithmetic, we calculate the two smallest numbers that can be represented by Eq. (14.25). The smallest number is given by

$$(-1)^{1} \times (1 + 0.111\,111\,11_{\text{two}}) \times 2^{-127}$$
$$= -1 \times 1.996\,093\,75 \times 2^{-127} = -1.173\,198\,463\,418\,338 \times 10^{-38}.$$

The next smallest number represented by the 32-bit single-precision arithmetic is

$$(-1)^{1} \times (1 + 0.111\,111\,10_{\text{two}}) \times 2^{-127}$$
$$= -1 \times 1.992\,187\,50 \times 2^{-127} = -1.170\,902\,576\,014\,388 \times 10^{-38}.$$

The resolution of the 32-bit single-precision arithmetic is therefore the difference of these numbers:

$$-1.173\,198\,463\,418\,338 \times 10^{-38} - (-1.170\,902\,576\,014\,388 \times 10^{-38})$$
$$= -2.295\,887\,403\,950\,041 \times 10^{-41}.$$

If the hardware allows for IEEE 754 single precision, then the quantization error is proportional to $-2.295\,887\,403\,950\,041 \times 10^{-41}$. Generally, specialized DSP boards are restricted to a smaller number of bits than the IEEE 32-bit single-precision representation.

The addition of quantization noise into the filter is a non-linear process. The detailed analysis of the effect of the quantization noise on the performance of the filter is beyond the scope of this text. In the following, Example 14.9 illustrates the effect of finite-precision arithmetic on the magnitude response of a 21-tap FIR filter.

14.8.3 Round-off errors

Because of the limited resolution of DSP boards, the output response of a filter cannot be accurately represented. In the 32-bit signed IEEE 754 floating point standard, the resolution of each sample of the output response is restricted to $2.295\,887\,403\,950\,041 \times 10^{-41}$. The distortion due to rounding off sample values to the resolution allowed by the DSP board is less damaging than the finite-precision representation of the filter coefficients. In the latter case, the distortion is substantially magnified. Still, the round-off errors in the sample values should be considered in the analysis of a filter performance.

14.8.4 Arithmetic overflow

Arithmetic overflow occurs during multiplication, division, or addition, when the final answer falls outside the range of the DSP board. For example, the dynamic range of the 32-bit signed IEEE 754 floating point standard is restricted to a maximum value of $2.0_{ten} \times 10^{38}$ and a minimum value of $-2.0_{ten} \times 10^{38}$. If the result of any mathematical operation between the two floating point numbers falls outside this range, then an overflow occurs.

Example 14.9
Consider the 21-tap FIR filter with impulse response as shown in Table 14.10, where each coefficient is represented by 14 decimal digits. The FIR filter is implemented on a DSP board, which uses finite-precision arithmetic given by

$$(-1)^s \times (0 + 0.\text{significand}),$$

where the significand represents the decimal fraction of the number and is limited to a fixed number of bits. There are no bits allocated for the exponent.

Table 14.10. Finite impulse response
$h[k]$ of the 21-tap FIR filter specified in
Example 14.9

k	$h[k]$
10	0.318 348 783 765 15
9,11	0.261 850 185 125 51
8,12	0.132 021 415 468 16
7,13	0.012 135 562 150 39
6,14	−0.041 086 983 052 48
5,15	−0.032 969 416 668 68
4,16	−0.005 898 263 640 95
3,17	0.008 055 858 168 72
2,18	0.006 608 361 295 03
1,19	0.001 494 396 943 68
0, 20	−0.001 385 507 671 95

Table 14.11. Impulse response of the FIR filter in Example 14.9 with 4-bit and 8-bit finite precisions

		$h[k]$		
k	Exact	8-bit binary representation	4-bit precision	8-bit precision
10	0.318 348 783 765 15	0.010 100 01	0.3125	0.316 406 25
9, 11	0.261 850 185 125 51	0.010 000 11	0.25	0.261 718 75
8, 12	0.132 021 415 468 16	0.001 000 01	0.125	0.128 906 25
7, 13	0.012 135 562 150 39	0.000 000 11	0	0.011 718 75
6, 14	−0.041 086 983 052 48	−0.000 010 10	0	−0.039 062 5
5, 15	−0.032 969 416 668 68	−0.000 010 00	0	−0.031 25
4, 16	−0.005 898 263 640 95	−0.000 000 01	0	−0.003 906 25
3, 17	0.008 055 858 168 72	0.000 000 10	0	0.007 812 5
2, 18	0.006 608 361 295 03	0.000 000 01	0	0.003 906 25
1, 19	0.001 494 396 943 68	0.000 000 00	0	0
0, 20	−0.001 385 507 671 95	0.000 000 00	0	0

Calculate the filter coefficients with the significand restricted to a total of 7 bits and where 1 bit is allocated for the sign. Plot the magnitude response of the filter. Repeat for a 3-bit significand with 1 bit allocated for the sign.

Solution

The filter coefficients with the 4-bit and 8-bit finite-precision arithmetic are shown in Table 14.11. We illustrate how we derived the result for the filter coefficient $h[10] = 0.318\,348\,783\,765\,15$. The remaining entries can be derived by following the procedure specified for $h[10]$.

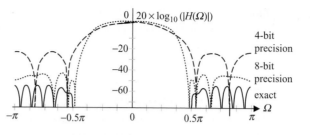

The binary representation for $h[10] = 0.318\,348\,783\,765\,15$ is given by

$$0.31834878376515_{\text{ten}} = 0.010\,100\,010\,111\,111\,1\ldots_{\text{two}}.$$

For 4-bit precision, the finite-precision representation of $h[10]$ is given by

$$(-1)^0 \times (0 + 0.0101)_{\text{two}} = 2^{-2} + 2^{-4} = 0.3125.$$

For 8-bit precision, the finite-precision representation of $h[10]$ is given by

$$(-1)^0 \times (0 + 0.010\,100\,01)_{\text{two}} = 2^{-2} + 2^{-4} + 2^{-8} = 0.316\,406\,25.$$

In deriving the above values, the finite-precision representations are truncated to
the available number of bits. Alternatively, the numerical values can be rounded
off to the nearest available level in each representation. The latter reduces the
quantization noise.

In Table 14.7, we observe that several filter coefficients are reduced to zero.
With 8-bit precision, the values of $h[0], h[1], h[19]$, and $h[20]$ are all represented
by zero. With 4-bit precision, a total of 16 values within the ranges $0 \le k \le 7$
and $13 \le k \le 20$ are reduced to zero. In other words, the FIR filter becomes a
17-tap filter with 8-bit precision and a 5-tap filter with 4-bit precision.

A comparison of the frequency characteristics for the three filters, with coefficients listed in Table 14.11, is shown in Fig. 14.27. Noticeable differences in
the magnitude spectrum are observed in the three implementations. The width
of the transition band increases substantially for the FIR filter represented with
4-bit precision. The stop-band ripple also increases with the finite-precision
filters. The original filter has a minimum attenuation of 50 dB in the stop band.
The minimum attenuation is decreased to 40 dB with 8-bit finite precision and
to 20 dB with 4-bit precision. In fact, it is difficult to describe the 4-bit finite-
precision filter as a lowpass filter since the higher-frequency components pass
through the system with comparatively little attenuation.

Increasing the number of bits used in the finite-precision representation generally improves the approximation of the original filter characteristics. However,
the increase in precision also increases the implementation cost.

14.9 MATLAB examples

In Chapter 13, we introduced a MATLAB M-file `residuez` for the partial fraction expansion of a given rational function. Similarly, the M-file `tf2zp` was introduced to calculate the location of poles and zeros for a given transfer function. These M-files can also be used to derive the cascaded and parallel forms of the transfer function. We illustrate the application of these M-files by deriving the cascaded and parallel forms for the transfer function,

$$H(z) = \frac{z^3 - 2z^2 + z}{z^3 - 0.1z^2 - 0.07z - 0.065} = \frac{1 - 2z^{-1} + z^{-2}}{1 - 0.1z^{-1} - 0.07z^{-2} - 0.065z^{-3}},$$

considered in Example 14.5.

14.9.1 Parallel form

The MATLAB code to determine the partial fraction expansion is given below. The explanation follows each instruction in the form of comments.

```
>> B = [1 −2 1 0];               % Coefficients of the
                                 %  numerator of H(z)
>> A = [1 −0.1 −0.07 −0.065];    % Coefficients of the
                                 %  denominator of H(z)
>> [R, P, K] = residuez(B, A);   % Calculate partial
                                 %  fraction expansion
```

The returned values are given by

```
        R = [0.4310 0.2845+3.3362j 0.2845−3.3362j]
  P = [0.5000 −0.2000+0.3000j −0.2000−0.3000j] and K = 0.
```

The transfer function $H(z)$ can therefore be expressed as follows:

$$H(z) = \frac{0.4310}{1 - 0.5z^{-1}} + \frac{0.2845 + j3.3362}{1 - (-0.2 + j0.3)z^{-1}} + \frac{0.2845 - j3.3362}{1 - (-0.2 - j0.3)z^{-1}}.$$

To eliminate complex-valued coefficients, we combine the complex poles as follows:

$$H(z) = \frac{0.4310}{1 - 0.5z^{-1}} + \frac{0.5690 - 1.8879z^{-1}}{1 + 0.4z^{-1} + 0.13z^{-2}}.$$

The partial fraction expansion is then implemented using the parallel form as shown in Fig. 14.25.

14.9.2 Series form

The MATLAB code to determine the poles and zeros of $H(z)$ is given by

```
>> B = [0 1 -2 1];              % The numerator of H(z)
>> A = [1 -0.1 -0.07 -0.065];   % The denominator of H(z)
>> [Z, P, K] = tf2zp(B, A);     % Calculate poles and
                                % zeros
```

The locations of the poles and zeros are given by

$$Z = [0\ 1\ 1]$$
$$P = [0.5000\ \ -0.2000+0.3000j\ \ -0.2000-0.3000j]\ \text{and}\ K = 1.$$

The transfer function $H(z)$ can therefore be expressed as follows:

$$H(z) = 1\frac{(1 - 0z^{-1})(1 - 1z^{-1})(1 - 1z^{-1})}{(1 - 0.5z^{-1})(1 - (-0.2 + j0.3)z^{-1})(1 - (-0.2 - j0.3)z^{-1})}$$

$$= \frac{(1 - z^{-1})^2}{(1 - 0.5z^{-1})(1 - (-0.2 + j0.3)z^{-1})(1 - (-0.2 - j0.3)z^{-1})}.$$

Combining the complex roots in the denominator, the cascaded configuration is given by

$$H(z) = \frac{1 - z^{-1}}{1 - 0.5z^{-1}} \times \frac{1 - z^{-1}}{1 + 0.4z^{-1} + 0.13z^{-2}}.$$

The cascaded configuration is then implemented using the series form as shown in Fig. 14.23.

14.10 Summary

Chapter 14 defined digital filters as systems used to transform the frequency characteristics of the DT sequences, applied at the input of the filter, in a predefined manner. Based on the magnitude spectrum $|H(\Omega)|$, Section 14.1 classifies filters in four different categories. A lowpass filter removes the higher-frequency components above a cut-off frequency Ω_c from an input sequence, while retaining the lower-frequency components $\Omega \le \Omega_c$. A highpass filter is the converse of the lowpass filter and removes the lower-frequency components below a cut-off frequency Ω_c from an input sequence, while retaining the higher-frequency components $\Omega \ge \Omega_c$. A bandpass filter retains a selected range of frequency components between the lower cut-off frequency Ω_{c1} and the upper cut-off frequency Ω_{c2} of the filter. A bandstop filter is the converse of the bandpass filter, which rejects the frequency components between the lower cut-off frequency Ω_{c1} and the upper cut-off frequency Ω_{c2} of the filter. All other frequency components are retained at the output of the bandstop filter.

Section 14.2 introduces a second classification of digital filters based on the length of the impulse response $h[k]$ of the digital filter. Finite impulse response

(FIR) filters have a finite length impulse response, while the length of infinite impulse response (IIR) filters is infinite. The ideal frequency-selective filters, introduced in Section 14.1, are not practically realizable because of constant gains within the pass band and stop band, the sharp transitions between the pass band and the stop band, and because of the zero phase. Sections 14.3 and 14.4 explore practical realizations of the ideal filter obtained by allowing some variations in the pass-band and stop-band gains, introducing a linear-phase within the pass band, and by leaving some transitional bandwidth between the pass band and the stop band. The transition bandwidth allows the filter characteristics to change gradually. Section 14.3 also proved the following sufficient condition for ensuring a linear phase for FIR filters. If the impulse response function of an N-tap filter, with z-transfer function given by Eq. (14.7), satisfies either of the following relationships:

symmetrical impulse response $h[k] = h[N - 1 - k]$;

antisymmetrical impulse response $h[k] = -h[N - 1 - k]$,

then the phase $<H(\Omega)$ of the filter is linearly proportional to the frequency.

When implementing digital filters on digital computers or specialized DSP boards, the output of a digital filter is obtained by solving the following recursive formula:

$$y[k] = -\frac{1}{a_0}(a_1 y[k - 1] + \cdots + a_N y[k - N])$$
$$+ \frac{1}{a_0}(b_0 x[k] + b_1 x[k - 1] + \cdots + b_M x[k - M]).$$

Based on the aforementioned formula, Sections 14.5–14.7 derived physical realizations of digital filters using three fundamental elements: a two-input adder, a scalar multiplier, and a unit delay. For FIR filters, direct form, series form, and parallel forms are derived in Section 14.6. The series form is obtained by factorizing the transfer function in terms of a product of quadratic polynomials and then cascading the transfer function for the individual quadratic polynomials. The parallel form is obtained by partial fraction expansion of the transfer function. Section 14.7 derived similar realizations for IIR filters. For both FIR and IIR filters, alternative flow diagrams are obtained by applying the transpose operation. Transposition of a flow graph is achieved by (i) interchanging the role of the input and output; (ii) reversing the directions of all branches within a flow graph; and (iii) replacing the source nodes with adders and the adders with source nodes.

Direct form II, the series form, and the cascaded form are defined as canonical representations since there forms, in general, use the minimal number of fundamental elements, whereas direct form I is referred to as a non-canonical representation. Irrespective of the arithmetic complexity, all of these realizations provide identical outputs for the same input sequence.

During the actual realization of digital filters in software or hardware, the filter coefficients are implemented with finite precision. Section 14.8 discussed the impact of finite-precision arithmetic on the performance of digital filters. It is observed that the effect of finite-precision arithmetic varies from one realization of the filter to another. We list in the following some empirical observations that should be kept in mind while choosing a particular realization.

(i) When the poles of the transfer function lie close to each other or close to the unit circle in the complex z-plane, direct form realizations, with filter coefficients represented using finite precision, produce large deviations from the output of an exact filter.

(ii) The order in which the first- and second-order systems are implemented in cascaded forms affects the output of the filter in finite-precision implementations. Changing the order may reduce the deviation from the output of an exact filter.

(iii) Pairing of complex poles and zeros is important for all cascaded and parallel realizations.

(iv) In cascaded realizations, scalar multipliers between different systems may be required to prevent the partial fraction coefficients from becoming too large or too small.

Section 14.9 presented two MATLAB functions, residuez and tf2zp, for deriving the physical realizations of digital filters.

Problems

14.1 Determine if the filters represented by the following transfer functions are (a) FIR or IIR, and (b) causal or non-causal. If a filter is FIR, determine if its phase is linear.

   (i) $H(z) = 0.7 + 0.2z^{-1} + 0.8z^{-2}$;

   (ii) $H(z) = \dfrac{1}{3}(z + 1 + z^{-1})$

   (iii) $H(z) = \dfrac{0.7 + 0.2z^{-1} + 0.8z^{-2}}{1 + 0.5z^{-1} - 0.24z^{-2}}$;

   (iv) $H(z) = \dfrac{1 - 0.1z^{-1} - 0.06z^{-2}}{1 + 0.2z^{-1}}$.

14.2 Consider two filters with transfer functions given by

$$H_1(z) = 1 + 2z^{-1} + 3z^{-2} + 2z^{-3} + z^{-4} \quad \text{and}$$
$$H_2(z) = 1 + 2z^{-1} + 3z^{-2} + 2z^{-3} - z^{-4}.$$

Fig. P14.5. The FIR system for
Problem 14.5.

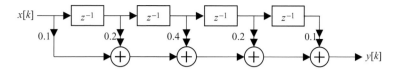

(i) Determine and plot the frequency characteristics of the filters.

(ii) If the sequence $x[k] = \cos(0.5k) + \cos(k)$ is applied at the input of filter $H_1(z)$, determine the output of the filter from the frequency characteristics obtained in (i).

(iii) Repeat (ii) for filter $H_2(z)$. What advantage do you see with the linear-phase filter?

14.3 Consider a digital filter with impulse response given by

$$h[k] = \begin{cases} 1/3 & -1 \leq k \leq 1 \\ 0 & \text{otherwise.} \end{cases}$$

(i) Calculate the transfer function of the filter.

(ii) Sketch the amplitude and phase responses of the filter with respect to frequency.

(iii) How will you classify this filter – lowpass, bandpass, bandstop, or highpass?

(iv) Does it have a linear phase?

14.4 Consider a digital filter with transfer function given by

$$H(z) = \frac{0.7 + 0.2z^{-1} + 0.8z^{-2}}{1 + 0.5z^{-1} - 0.24z^{-2}}.$$

(i) Plot the impulse response and the frequency characteristics of the filter.

(ii) From the frequency characteristics, determine the maximum magnitude of the pass-band and stop-band ripples and the transition bandwidth.

14.5 Given the flow graph in Fig. P14.5, calculate the transfer function and the impulse response of the LTI system of the realization. From the transfer function, calculate the magnitude and phase spectra for the filter.

14.6 The flow graph of Fig. P14.5 can be implemented by using only three scalar multipliers. Sketch the flow graph which uses three scalar multipliers without increasing the number of delay elements or two-input adders.

14.7 Repeat Problem 14.5 for the flow graph shown in Fig. P14.7.

14.8 Draw the flow graphs for (i) the direct form and (ii) the cascaded form for an FIR filter with a transfer function given by

$$H(z) = 0.4 - 0.8z^{-1} + 0.4z^{-2}.$$

Fig. P14.7. The IIR system for
Problem 14.7.

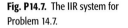

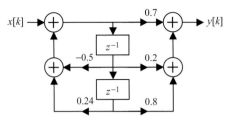

14.9 Using the principle of transposition for the flow graphs, derive two alternative representations for the FIR filter specified in Problem 14.8.

14.10 Draw the linear-phase flow graph for the FIR filter specified in Problem 14.8.

14.11 The transfer function of an IIR filter is given by

$$H(z) = (1 - 0.25z^{-1})^8.$$

Draw the flow graphs based on the following forms: (i) cascade of eight first-order FIR systems; (ii) cascade of four second-order FIR systems; (iii) cascade of two third-order FIR systems and one second-order FIR system; (iv) cascade of two fourth-order FIR systems; (v) cascade of one sixth-order FIR system and one second-order FIR system. Compare the computational complexity of each realization.

14.12 The transfer function of an IIR filter, with impulse response given by

$$h[k] = 0.5^k \sin\left(\frac{\pi}{4}k\right)u[k],$$

is given by the following expression:

$$H(z) = \frac{0.5z\sin\left(\dfrac{\pi}{4}\right)}{z^2 - 2 \times 0.5\cos\left(\dfrac{\pi}{4}\right)z + 0.25} \approx \frac{0.3536z}{z^2 - 0.7071z + 0.25}.$$

Draw the flow graphs for (i) direct form I, (ii) direct form II, (iii) the cascaded form, and (iv) the parallel form realizations of the IIR filter.

14.13 Using the principle of transposition for the flow graphs, derive four alternative flow graph representations for the IIR filter specified in Problem 14.12.

14.14 The transfer function of a digital system is given by

$$H(z) = \frac{1 - 0.8z^{-1} + 0.15z^{-2}}{1 - 0.7z^{-1} - 0.18z^{-2}}.$$

Draw the flow graphs for (i) direct form I, (ii) direct form II, (iii) the cascaded form, and (iv) the parallel form realizations of the IIR filter.

14.15 Using the principle of transposition for the flow graphs, derive four alternative flow graph representations for the IIR filter specified in Problem 14.14.

14.16 An allpass filter has a constant gain for all frequencies, i.e. $|H(\Omega)| = 1$.
    (i) Show that the transfer functions

$$H_1(z) = \frac{\alpha_1 + z^{-1}}{1 + \alpha_1 z^{-1}} \quad \text{and} \quad H_2(z) = \frac{\alpha_2 + \alpha_1 z^{-1} + z^{-2}}{1 + \alpha_1 z^{-1} + \alpha_2 z^{-2}}$$

    represent allpass filters.
    (ii) Sketch the flow graph for the first-order allpass filter $H_1(z)$, which uses a single scalar multiplier.
    (iii) Sketch the flow graph for the second-order allpass filter $H_2(z)$ with only two scalar multipliers. There is no restriction on the number of unit delay elements or two-input adders in each case.

14.17 The impulse response of an LTID system is given by

$$h[k] = \begin{cases} \alpha^k & 0 \le k \le 9 \\ 0 & \text{elsewhere.} \end{cases}$$

    (i) Draw the flow graph for the above LTID system with no feedback paths.
    (ii) The z-transfer function for the above impulse response is given by

$$H(z) = \frac{1 - \alpha^{10} z^{-10}}{1 - \alpha z^{-1}}.$$

    Draw the flow graph of the IIR system specified by this transfer function.
    (iii) Compare the two implementations with respect to the number of delays, scalar multipliers and two-input adders.

14.18 Implement the filter with transfer function given by

$$H(z) = 0.4 - 0.8z^{-1} + 0.4z^{-2}$$

with finite-precision arithmetic given by

$$(-1)^s \times (0 + 0.\text{significand}),$$

where the significand represents the decimal fraction of the coefficients and is limited to 3 bits with 1 bit allocated for the sign. Compare the magnitude response of the original filter with the magnitude response of the filter implemented with finite-precision representation.

14.19 Repeat Problem 14.14 for the following transfer function:
$$H(z) = \frac{1 - 0.8z^{-1} + 0.15z^{-2}}{1 - 0.7z^{-1} - 0.18z^{-2}}.$$

14.20 Repeat Problem 14.14 for the following transfer function:
$$H(z) = \frac{0.5z \sin\left(\dfrac{\pi}{4}\right)}{z^2 - \cos\left(\dfrac{\pi}{4}\right)z + 0.25}.$$

15 FIR filter design

In Chapter 14, we defined frequency-selective filters as systems that modify the frequency components of the input signals in a predefined manner. Further classification of frequency-selective filters is based on the length N of their impulse responses $h[k]$. If the length N of the impulse response of a frequency-selective filter is finite, the filter is referred to as a finite impulse response (FIR) filter. If the length N is infinite, the frequency-selective filter is referred to as an infinite impulse response (IIR) filter. In this chapter, we consider the design of frequency-selective FIR filters.

The design of digital filters involves three distinct stages. Stage 1 describes the desired specifications of the frequency characteristics of the filter. Based on the specified frequency characteristics, stage 2 derives the transfer function $H(z)$, or the impulse response $h[k]$, of the filter. Finally, stage 3 develops the canonical realization of the filter using one of the several forms presented in Chapter 14. While deriving the impulse response $h[k]$ of an FIR filter in stage 2, the following two conditions must also be satisfied.

(1) Causality condition. This implies that the impulse response $h[k]$ of an FIR filter is zero for $k < 0$. This will ensure a causal, and hence a physically realizable, filter.

(2) Linear-phase condition. This implies that the impulse response $h[k]$ of an FIR filter of length N is symmetrical or anti-symmetrical, i.e. $h[k] = \pm h[N - 1 - k]$. The linear-phase condition ensures that no distortion is introduced in the input frequency components lying within the pass band of the FIR filter.

Generally, FIR filters are designed directly from the impulse response of an ideal lowpass filter. Section 15.1 describes the windowing approach, where an appropriate window function $w[k]$ is used to truncate the impulse response of an ideal lowpass filter to a finite length N. The specifications of the FIR filter, along with the characteristics of the window function, are used to calculate the length N of the FIR filter. Sections 15.2–15.4 extend the windowing approach to the design of highpass, bandpass, and bandstop FIR filters. The FIR filter design techniques, based on the windowing function, can result in several alternative designs, all of which satisfy the given specifications. Section 15.5 presents the Parks–McClellan method, which recursively computes the optimal filter for a given length N. Section 15.6 presents several library functions available in MATLAB to design FIR filters. Finally, the chapter is concluded in Section 15.7.

15.1 Lowpass filter design using windowing method

In Section 14.1, it was shown that the impulse response of an ideal lowpass filter is a sinc function, and therefore that an ideal lowpass filter is non-causal and IIR. In Section 14.4, it was shown that a causal lowpass FIR filter can be obtained by delaying the ideal impulse response by m time units (see Fig. 15.1a) and truncating the impulse response. To generate an N-tap FIR filter, the truncation of the ideal impulse response is performed as follows:

$$h_{lp}[k] = \begin{cases} h_{ilp}[k] = \dfrac{\Omega_c}{\pi} \, \text{sinc}\left(\dfrac{(k-m)\Omega_c}{\pi} \right) & 0 \le k \le N-1 \\ 0 & \text{elsewhere;} \end{cases} \tag{15.1}$$

where the value of m in Eq. (15.1) is selected to be $(N-1)/2$. This approach of designing an FIR filter is referred to as the *windowing method*, and is shown in Fig. 15.1. Note that the impulse response $h_{lp}[k]$ of the resulting FIR filter is non-zero only within the range $0 \le k \le N-1$. In addition, the impulse response $h_{lp}[k]$ is symmetrical about $k = (N-1)/2$, i.e.

$$h[k] = h[N-1-k], \tag{15.2}$$

and satisfies the linear-phase condition given in Eq. (14.8a). If N is an odd-valued integer, the resulting FIR filter is a type 1 linear-phase filter with an integer delay m. On the other hand, if N is an even-valued integer, the resulting FIR filter is a type 2 linear-phase filter with a fractional delay m.

Truncating the impulse response of an ideal lowpass filter affects the frequency characteristics of the ideal lowpass filter. In addition to introducing ripples within the pass and stop bands, the truncation leads to a transition band between the pass band and the stop band. In the following subsection, we analyze the effect of truncating the impulse response $h_{lp}[k]$ of the ideal lowpass filter with a rectangular window of length N.

15.1.1 Rectangular window

The rectangular window of length N, centered at $k = (N-1)/2$, is defined as follows:

$$w_{\text{rect}}[k] = \begin{cases} 1 & 0 \le k \le N-1 \\ 0 & \text{otherwise,} \end{cases} \tag{15.3}$$

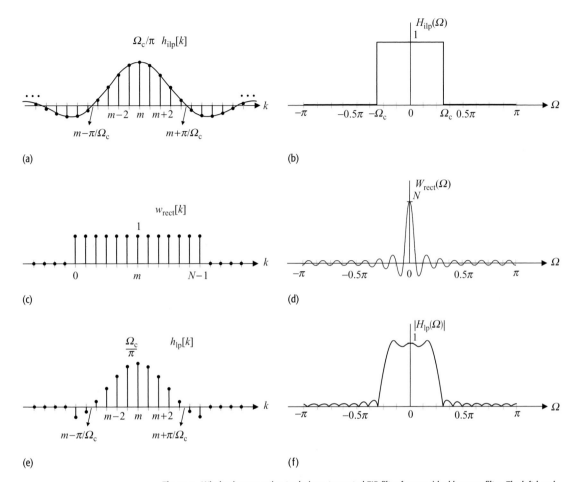

Fig. 15.1. Windowing operation to derive a truncated FIR filter from an ideal lowpass filter. The left-hand column (plots (a), (c), and (e)) represents the windowing operation in the time domain, and the right-hand column (plots (b), (d), and (f)) represents the windowing operation in the frequency domain. (a) Impulse response $h_{ilp}[k]$ of an ideal lowpass filter. (b) Magnitude spectrum $|H_{ilp}(\Omega)|$ of an ideal lowpass filter. (c) Rectangular window $w_{rect}[k]$. (d) DTFT $W_{rect}(\Omega)$ of the rectangular window. (e) Impulse response $h_{lp}[k] = h_{ilp}[k]w_{rect}[k]$ of the truncated lowpass filter. (f) Magnitude spectrum $|H_{lp}(\Omega)| = |(1/2\pi)H_{ilp}(\Omega) * W_{rect}(\Omega)|$ of the truncated lowpass filter.

where we have assumed that the length N of the windowing function is odd. Taking the DTFT of Eq. (15.3) results in the following frequency characteristics for the rectangular window:

$$W_{rect}(\Omega) = e^{-j(N-1)\Omega/2} \times \frac{\sin(N\Omega/2)}{\sin(\Omega/2)}. \tag{15.4}$$

The rectangular window $w_{rect}[k]$ and its magnitude spectrum $|W_{rect}(\Omega)|$ are illustrated in Figs. 15.1(c) and (d), respectively. The narrow lobe, centered at $\Omega = 0$, in $W_{rect}(\Omega)$ is referred to as the main lobe, while the lobes on each side of the main lobe are referred to as the side lobes of the rectangular window.

Truncating the impulse response $h_{ilp}[k]$ of the ideal lowpass filter to length N is the same as multiplying the impulse response $h_{ilp}[k]$ by the rectangular window in the time domain. The truncation operation is, therefore, modeled as follows:

$$h_{lp}[k] = h_{ilp}[k]w_{rect}[k], \tag{15.5}$$

with $m = (N - 1)/2$. The result of the truncation step is illustrated in Fig. 15.1(e). Since multiplication in the time domain is equivalent to convolution in the frequency domain, the transfer function of the truncated FIR filter is given by

$$H_{lp}(\Omega) = \frac{1}{2\pi}[W_{rect}(\Omega) * H_{ilp}(\Omega)] = \frac{1}{2\pi} \int\limits_{(2\pi)} H_{ilp}(\theta)W_{rect}(\theta - \Omega)\,d\theta, \tag{15.6}$$

which results in the magnitude spectrum shown in Fig. 15.1(f). Comparing the magnitude spectrum $|H_{ilp}(\Omega)|$ of the ideal lowpass filter with the magnitude spectrum $|H_{lp}(\Omega)|$ of the truncated lowpass filter, we note three major differences. First, there are significant ripples within the pass band of the truncated lowpass filter. Secondly, the magnitude spectrum of the truncated lowpass filter does not change abruptly in between the pass band and stop band. In fact, a transition band of finite width appears. Thirdly, there are additional ripples within the stop band of the truncated lowpass filter. The appearance of ripples in the pass band and stop band is referred to as the *Gibbs phenomenon*. In order to reduce the ripples and eliminate the transition band, the DTFT $W_{rect}(\Omega)$ should be a narrow impulse function. This would imply that the length N of the windowing function is very large, increasing the implementation complexity of the truncated lowpass filter.

In Fig. 15.1(c), we observe that the rectangular window has abrupt truncations outside the range $0 \le k \le N - 1$. The pass-band and stop-band ripples, as well as the transition band, can be decreased by selecting alternative windows that taper smoothly to zero from the peak value of 1 at $k = (N - 1)/2$. Section 15.1.2 discusses several alternatives to the rectangular window.

15.1.2 Commonly used windows

There are a number of alternatives to the rectangular window. A few popular choices are defined in the following.

Bartlett (triangular) window

$$w_{bart}[k] = \begin{cases} \dfrac{2k}{N - 1} & 0 \le k \le (N - 1)/2 \\ 2 - \dfrac{2k}{N - 1} & (N - 1)/2 < k \le N - 1 \\ 0 & \text{otherwise.} \end{cases} \tag{15.7}$$

Fig. 15.2. Commonly used windows of length N.

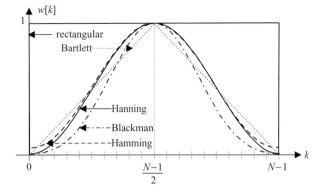

Generalized Hamming window   For $0 < \alpha < 1$,

$$w_{\text{gene}}[k] = \begin{cases} \alpha - (1 - \alpha) \cos\left[\dfrac{2\pi k}{N - 1}\right] & 0 \le k \le N - 1 \\ 0 & \text{otherwise.} \end{cases} \tag{15.8}$$

Hamming window

$$w_{\text{hamm}}[k] = \begin{cases} 0.54 - 0.46 \cos\left[\dfrac{2\pi k}{N - 1}\right] & 0 \le k \le N - 1 \\ 0 & \text{otherwise.} \end{cases} \tag{15.9}$$

Hanning window

$$w_{\text{hann}}[k] = \begin{cases} 0.5 - 0.5 \cos\left[\dfrac{2\pi k}{N - 1}\right] & 0 \le k \le (N - 1) \\ 0 & \text{otherwise.} \end{cases} \tag{15.10}$$

Blackman window

$$w_{\text{blac}}[k] = \begin{cases} 0.42 - 0.5 \cos\left[\dfrac{2\pi k}{N - 1}\right] + 0.08 \cos\left[\dfrac{4\pi k}{N - 1}\right] & 0 \le k \le N - 1 \\ 0 & \text{otherwise.} \end{cases} \tag{15.11}$$

The shapes of the windows are shown in Fig. 15.2, where, for convenience of illustration, continuous plots are used. In reality, the windows are a function of the DT variable k. It may be noted that the Hamming and Hanning windows are special cases of the generalized Hamming window. For the Hamming window, variable α in Eq. (15.8) of the generalized Hamming window equals 0.54. Similarly, for the Hanning window, variable α in Eq. (15.8) equals 0.5.

    The DTFTs of the aforementioned windows are shown in Fig. 15.3, where the vertical axis represents the magnitude of the DTFTs based on the decibel (dB) scale. The two important parameters used in the FIR filter design are (i) the width of the main lobes of the DTFT of the windows; (ii) the relative strength of the highest value side lobe with respect to the main lobe. The width of the main lobe is defined as the distance between the nearest zero crossings of the main lobe, while the relative side lobe strength is defined as the difference in dB between the magnitudes of the highest value side lobe and the main lobe.

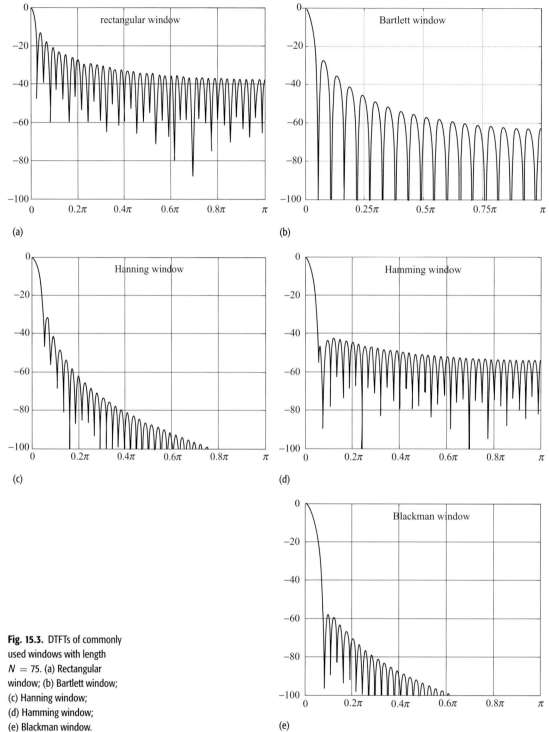

Fig. 15.3. DTFTs of commonly
used windows with length
$N = 75$. (a) Rectangular
window; (b) Bartlett window;
(c) Hanning window;
(d) Hamming window;
(e) Blackman window.

Table 15.1. Comparison of the properties of the commonly used windows

Window	Width of main lobe	Peak side lobe amplitude[a] (dB)	Max. stop/pass-band error $20\log_{10}(\delta)$	Kaiser window[b] β	transition width
Rectangular	$4\pi/N$	-13.3	-21	0	$1.81\pi/(N-1)$
Bartlett	$8\pi/(N-1)$	-26.5	-25	1.33	$2.37\pi/(N-1)$
Hanning	$8\pi/(N-1)$	-31.4	-44	3.86	$5.01\pi/(N-1)$
Hamming	$8\pi/(N-1)$	-42.6	-53	4.86	$6.27\pi/(N-1)$
Blackman	$12\pi/(N-1)$	-58.0	-74	7.04	$9.19\pi/(N-1)$

[a]The peak side lobe magnitude in column 3 is relative to the magnitude of the main lobe.
[b]The last two columns for the Kaiser window are explained in Section 15.1.5.

The second and third columns of Table 15.1 compare these two parameters for the commonly used windows as a function of the length N of the window. The fourth column of Table 15.1 quantifies the maximum difference between the magnitude spectra within the pass and stop bands of the ideal lowpass filter and the causal FIR filter obtained from the windowing method. In other words, it provides an upper bound on the values of the ripples in the pass and stop bands of the causal FIR filter. For example, the maximum pass- and stop-band error of –21dB for the rectangular window implies that the pass- and stop-band ripples are confined to –21dB in the FIR filter obtained with the rectangular window.

In filter design, we prefer to minimize the transition band and reduce the strength of the ripples. These are conflicting requirements, as we see next. To minimize the transition band in the FIR filter, the main lobe width of the windows should be as small as possible. To reduce the pass-band and stop-band ripples in the FIR filter, the area enclosed by the side lobes (in other words, the relative strength of the side lobes) of the windows should be small. Table 15.1 illustrates that these two requirements are contradictory. The rectangular window has the smallest width main lobe, but the relative strength of its highest side lobe with respect to the main lobe is the largest. As a result, for the rectangular window, the transition bandwidth is small, but the ripple magnitude is large. On the contrary, the relative strength of the side lobe for the Blackman window is the smallest, but the width of its main lobe is the largest. In other words, for the Blackman window, the transition bandwidth is large, but the ripple magnitude is small.

In the following example, we illustrate the effect of the rectangular and Hamming windows on the frequency characteristics of an ideal lowpass filter truncated with these windows.

Example 15.1

Calculate the impulse response of an ideal DT lowpass filter with radian cut-off frequency $\Omega_c = 1$. From the ideal filter, design two 21-tap FIR filters with $\Omega_c = 1$ using the rectangular and Hamming windows.

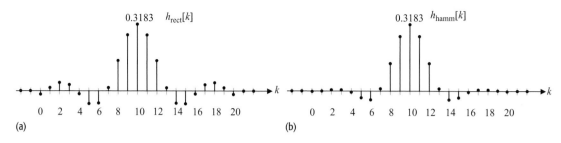

Fig. 15.4. Impulse response of FIR filters obtained by truncating the impulse response of the ideal lowpass filter with (a) a rectangular window and (b) a Hamming window.

Solution

Substituting $\Omega_c = 1$ in Eq. (14.12), the impulse response of an ideal lowpass filter is given by

$$h_{\text{ilp}}[k] = \frac{\sin(k - m)}{(k - m)\pi} = \frac{1}{\pi} \text{sinc}\left(\frac{k - m}{\pi}\right), \tag{15.12}$$

where $m = (N - 1)/2 = 10$. The expressions for the rectangular and Hamming windows with 21 taps are as follows:

$$\text{rectangular window} \quad w_{\text{rect}}[k] = \begin{cases} 1 & 0 \le k \le 20 \\ 0 & \text{otherwise}; \end{cases} \tag{15.13}$$

$$\text{Hamming window} \quad w_{\text{hamm}}[k] = \begin{cases} 0.54 - 0.46\cos\left(\dfrac{2\pi k}{20}\right) & 0 \le k \le 20 \\ 0 & \text{otherwise}. \end{cases} \tag{15.14}$$

The FIR filters are obtained by multiplying the impulse response of the ideal lowpass filter by the expressions for the rectangular and Hamming windows. The resulting impulse responses are as follows:

$$\text{rectangular window} \quad h_{\text{rect}}[k] = \begin{cases} \dfrac{1}{\pi} \text{sinc}\left(\dfrac{k - 10}{\pi}\right) & 0 \le k \le 20 \\ 0 & \text{otherwise}; \end{cases} \tag{15.15}$$

$$\text{Hamming window} \quad h_{\text{hamm}}[k]$$

$$= \begin{cases} \dfrac{1}{\pi} \text{sinc}\left(\dfrac{(k - 10)}{\pi}\right)\left(0.54 - 0.46\cos\left[\dfrac{2\pi k}{20}\right]\right) & 0 \le k \le 20 \\ 0 & \text{otherwise}. \end{cases} \tag{15.16}$$

The impulse responses for FIR filters obtained by truncating the ideal lowpass filter impulse response with the rectangular and Hamming windows are shown in Fig. 15.4. Although the two impulse responses have the same value at $k = 10$, the impulse response $h_{\text{hamm}}[k]$, shown in Fig. 15.4(b), decays more rapidly as we move away from the central point ($k = 10$) and is different from the impulse response $h_{\text{rect}}[k]$, shown in Fig. 15.4(a). Typically, the pass-band gain of the truncated FIR filters, obtained from the ideal lowpass filters using the windowing method, is not unity, as desired. To prove this, we calculate the value

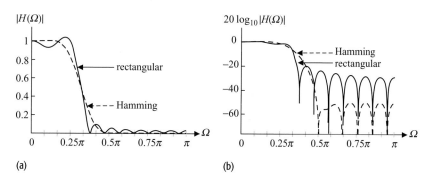

Fig. 15.5. Magnitude spectra of the FIR filters obtained by truncating the impulse response of the ideal lowpass filter with the rectangular and Hamming windows. The magnitude spectrum of the FIR filter obtained from the rectangular window is plotted as a solid line, and the magnitude spectrum of the FIR filter obtained from the Hamming window is plotted as a dashed line. (a) Plotted using a linear scale for the gain. (b) Plotted using a dB scale for the gain.

of $H_{lp}(0)$ by substituting $\Omega = 0$ in the DTFT $H_{lp}(\Omega)$:

$$H_{lp}(0) = \sum_{k=-\infty}^{\infty} h_{lp}[k]e^{-j\Omega k}\bigg|_{\Omega=0} = \sum_{k=-\infty}^{\infty} h_{lp}[k]. \qquad (15.17)$$

Equation (15.17) can therefore be used to calculate the pass-band gain at $\Omega = 0$. Using the values of the samples plotted in Figs. 15.4(a) and (b), the values of the gain of the two truncated filters at $\Omega = 0$ are given by

rectangular window $\qquad H_{\text{rect}}(0) = \sum_{k=-\infty}^{\infty} h_{\text{rect}}[k] = 0.9754;$

Hamming window $\qquad H_{\text{hamm}}(0) = \sum_{k=-\infty}^{\infty} h_{\text{hamm}}[k] = 0.9982.$

To ensure a unity gain in the pass band, the impulse response corresponding to the rectangular window in Eq. (15.15) is normalized by a factor of 0.9754. Similarly, the impulse response corresponding to the Hamming window is normalized by a factor of 0.9982. The resulting magnitude spectra of the two normalized FIR filters are shown in Fig. 15.5, where the gains of the filter are plotted on a linear scale in Fig. 15.5(a) and on a logarithmic scale in Fig. 15.5(b). It is observed that the dc gain, defined as the gain of the filter at $\Omega = 0$, for both filters is unity. The rectangular window results in higher pass-band and stop-band ripples. However, the rectangular window provides a smaller transition band than the Hamming window.

From Fig. 15.5(b), we quantify the gain in the stop band for the FIR filter obtained using the Hamming window and compare its value with the stop-band gain for the FIR filter obtained using the rectangular window. The maximum gain in $|H_{\text{hamm}}(\Omega)|$ is less than -50 dB in the stop band ($\Omega > 0.49\pi$). Equivalently, we can also say that the minimum attenuation in the stop band of the Hamming window is greater than 50 dB. The maximum gain in $|H_{\text{hamm}}(\Omega)|$,

Fig. 15.6. Desired specifications of a lowpass filter.

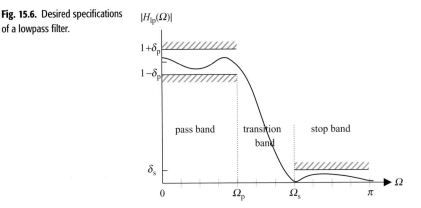

obtained from the rectangular window, is about -22 dB for $\Omega > 0.37\pi$. In other words, the Hamming window attenuates the higher-frequency components of the input signals more strongly than the rectangular window. As discussed earlier, this improvement in the stop-band attenuation is at the expense of a higher transitional bandwidth in the truncated FIR filter obtained from the Hamming window.

15.1.3 Design of FIR lowpass filters

We now list the main steps involved in the design of FIR filters using the windowing method. The design specifications for a lowpass filter are illustrated in Fig. 15.6 and are given by

pass band $(0 \leq \Omega \leq \Omega_{\mathrm{p}})$ $\quad (1 - \delta_{\mathrm{p}}) \leq |H_{\mathrm{lp}}(\Omega)| \leq (1 - \delta_{\mathrm{p}})$;

stop band $(\Omega_{\mathrm{s}} < \Omega \leq \pi)$ $\qquad\qquad 0 \leq |H_{\mathrm{lp}}(\Omega)| \leq \delta_{\mathrm{s}}.$

Expressed in decibels (dB), $20\log_{10}(\delta_{\mathrm{p}})$ is referred to as the pass-band ripple or the peak approximation error within the pass band. Similarly, $20\log_{10}(\delta_{\mathrm{s}})$ is referred to as the stop-band ripple or the peak approximation error in the stop band. The stop-band ripple can also be expressed in terms of the stop-band attenuation as $-20\log_{10}(\delta_{\mathrm{s}})$ dB.

For digital filters, the pass and stop bands are generally specified in the DT frequency Ω domain, which is limited to the range $0 \leq \Omega \leq 2\pi$. A DT system may also be used to process a CT signal. The schematic representation of such a system was shown in Fig. 9.1. In such cases, it is possible that the pass and stop bands of the overall system are specified in the CT frequency ω domain and we are required to compute the transfer function of the DT system shown as the central block in Fig. 9.1. We assume that the sampling frequency f_0 used in the analog to digital (A/D) converter is known. The following nine steps design an FIR filter using the windowing method.

Step 1 Calculate the normalized cut-off frequency of the filter based on the following expressions:

DT frequency Ω specifications

cut-off frequency, $\Omega_c = 0.5(\Omega_p + \Omega_s)$;

normalized cut-off frequency, $\Omega_n = \Omega_c/\pi$;

CT frequency ω (or f) specifications

cut-off frequency, $\omega_c = 0.5(\omega_p + \omega_s)$ or $f_c = 0.5(f_p + f_s)$;

normalized cut-off frequency, $\Omega_n = \frac{\omega_c}{0.5\omega_0} \overset{\Delta}{=} \frac{f_c}{0.5 f_0}$.

Note that the for CT specifications, ω_p and ω_s denote the pass-band and stop-band edge frequencies in radians/s, and f_p and f_s denote the pass-band and stop-band edge frequencies in Hz, respectively. The above frequency normalization scales the DT frequency range $[0, \pi]$ to $[0, 1]$. For CT, the frequency range $[0, 0.5\omega_0]$ (in radians/s) or $[0, 0.5 f_0]$ (in Hz) is scaled to $[0, 1]$. The normalized cut-off frequency Ω_n can have a value in the range $[0, 1]$.

Step 2 The impulse response of an ideal lowpass filter is given by

$$h_{\text{ilp}}[k] = \frac{\sin((k - m)\Omega_c)}{(k - m)\pi} = \frac{\Omega_c}{\pi} \operatorname{sinc}\left(\frac{(k - m)\Omega_c}{\pi}\right) = \Omega_n \operatorname{sinc}((k - m)\Omega_n),$$

where $\Omega_c = \pi \Omega_n$ and $m = (N - 1)/2$, where N is the filter length to be calculated in step 6. Note that the DT filter impulse response $h_{\text{ilp}}[k]$ primarily depends on the normalized frequency Ω_n. If the DT filter is used to process DT signals obtained using different sampling rates, the CT cut-off frequency will change depending on the sampling frequency, but the Ω_n will remain same.

Step 3 Calculate the minimum attenuation A using $A = \min(\delta_p, \delta_s)$ and then convert it to the dB scale.

Because of the nature of the windowing method and the inherent symmetry in the window functions, the resulting FIR filter has identical attenuations of A dB in both the pass and stop bands. If $\delta_p > \delta_s$, the designed filter will satisfy the pass-band attenuation requirement and exceed the stop-band attenuation requirement. Conversely, if $\delta_s > \delta_p$, then the filter will satisfy the stop-band attenuation requirement and exceed the pass-band attenuation requirement.

Step 4 Use the first three columns of Table 15.2 to choose the window type for the specified attenuation A.

In Table 15.2, the attenuation A, specified in the first two columns, is relative to the pass-band gain. For a given value of A, more than one choice of the window type is possible. With a minimum attenuation requirement of 20 dB, for example, any of the four windows may be selected. Although the higher

Table 15.2. Selection of the type of window based on the attenuation values obtained from step 3

Minimum attenuation (A)		Type of window	Transition bandwidth, $\Delta\Omega_n$
dB	linear scale		
≤ 20	≤ 0.1	rectangular	$1.8/N$
≤ 40	≤ 0.01	Hanning	$6.2/N$
≤ 50	≤ 0.003	Hamming	$6.6/N$
≤ 70	$\leq 0.000\,03$	Blackman	$11/N$

attenuation windows (Hanning, Hamming, or Blackman) reduce the pass- and stop-band ripples, the transition bands of the resulting FIR filters obtained with these windows are larger than the transition band of the FIR filter obtained with the rectangular window.

The first two columns of Table 15.2 are approximated directly from the fourth column of Table 15.1, which lists the stop-band attenuation. The last column of Table 15.2 is based on empirical observations.

Step 5 Calculate the normalized transition bandwidth for the FIR filter using the following expressions:

DT frequency specifications

transition BW, $\Delta\Omega_c = \Omega_s - \Omega_p$;

normalized transition BW, $\Delta\Omega_n = \Delta\Omega_c/\pi$;

CT frequency specifications

transition BW, $\Delta\omega_c = \omega_s - \omega_p$ or $\Delta f_c = f_s - f_p$;

normalized transition BW, $\Delta\Omega_n = \dfrac{\Delta\omega_c}{0.5\omega_0} = \dfrac{\Delta f_c}{0.5 f_0}$.

Step 6 Using the last column of Table 15.2, determine the minimum length N of the filter for the computed transitional bandwidth $\Delta\Omega_n$ obtained in step 5 and the window function selected in step 4.

Step 7 Determine the expression $w[k]$ for the window function using the window type selected in step 4 for length N obtained in step 6. The expression for the rectangular window is given in Eq. (15.3), while the expressions for the remaining window functions are specified in Eqs. (15.7)–(15.11).

Step 8 Derive the impulse response $h_{lp}[k]$ of the FIR filter:

$$h_{lp}[k] = h_{ilp}[k]w[k].$$

If the pass-band gain $|H_{lp}(0)|$ at $\Omega = 0$, given by $\sum h_{lp}[k]$, is not equal to one, we normalize $h_{lp}[k]$ with $\sum h_{lp}[k]$, where $\sum$ denotes the summation operation.

Step 9 Confirm that the impulse response $h_{lp}[k]$ satisfies the initial specifications by plotting the magnitude spectrum $|h_{lp}(k)|$ of the FIR filter obtained in step 8.

We illustrate the working of the aforementioned FIR filter design algorithm in Example 15.2.

Example 15.2

Figure 9.1 is used to process a CT signal with a digital filter. The overall characteristics of the CT system, modeled with Fig. 9.1, are specified below:

(i)   pass-band edge frequency $(\omega_p) = 3\pi$ kradians/s (or 1500 Hz);
(ii) stop-band edge frequency $(\omega_s) = 4\pi$ kradians/s (or 2000 Hz);
(iii) minimum stop-band attenuation, $-20\log_{10}(\delta_s) = 50$ dB;
(iv) sampling frequency $(f_0) = 8$ ksamples/s.

Design the DT system in Fig. 9.1 based on the aforementioned CT specifications.

Solution

Step 1 suggests that the cut-off frequency of the filter is given by

$$\omega_c = 0.5(\omega_p + \omega_s) = 3.5\pi \text{ kradians/s}.$$

Using $\omega_0 = 2\pi f_0 = 16\pi \times 10^3$, the normalized cut-off frequency is given by

$$\Omega_n = \omega_c/(0.5\omega_0) = \left(3.5\pi \times 10^3\right) / \left(0.5 \times 2\pi \times 8 \times 10^3\right) = 0.4375.$$

Based on step 2, the impulse response of the ideal lowpass filter with the normalized cut-off frequency $\Omega_n = 0.4375$ is given by

$$h_{ilp}[k] = 0.4375 \text{ sinc}(0.4375(k - m))$$

with m set to $(N - 1)/2$. The value of N is determined in step 6.

Step 3 determines the minimum attenuation A to be 50 dB.

Step 4 determines the type of window. For the minimum stop-band attenuation of 50 dB, Table 15.2 limits our choice to either the Hamming or Blackman window. We select the Hamming window because its length N will be lower than that of the Blackman window.

Step 5 computes the normalized transition bandwidth:

$$\Delta\Omega_n = \Delta\omega_c/(0.5\omega_0) = (4\pi - 3\pi) \times 10^3/(0.5 \times 2\pi \times 8 \times 10^3) = 0.1250.$$

Step 6 evaluates the length N of the Hamming window:

$$6.6/N = 0.1250 \Rightarrow N = 8 \times 6.6 = 52.8.$$

Ceiling off the length of the window to the nearest larger odd integer, we obtain $N = 53$.

Fig. 15.7. Magnitude spectrum of the FIR filter designed in Example 15.2.

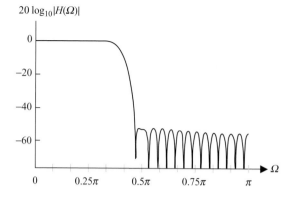

Step 7 derives the expression for the Hamming window of length $N = 53$:

$$w_{\text{hamm}}[k] = \begin{cases} 0.54 - 0.46\cos\left[\dfrac{2\pi k}{52}\right] & 0 \le k \le 52 \\ 0 & \text{otherwise.} \end{cases}$$

Step 8 gives the impulse response of the FIR filter:

$$h_{\text{lp}}[k] = \begin{cases} 0.4375\,\text{sinc}(0.4375(k - 26))\left\{0.54 - 0.46\cos\left[\dfrac{2\pi k}{52}\right]\right\} & 0 \le k \le 52 \\ 0 & \text{otherwise.} \end{cases}$$

Since $\sum h_{\text{lp}}[k] = 0.9998 \approx 1$, the impulse response $h_{\text{lp}}[k]$ of the FIR filter is not normalized with $\sum h_{\text{lp}}[k]$. The magnitude spectrum of the FIR filter is plotted in Fig. 15.7 using a dB scale. We observe that the pass-band frequency components below $\Omega = 1.5$ kHz are passed without any attenuation. The minimum attenuation in the stop band is also observed to be less than 50 dB.

15.1.4 Kaiser window

As shown in Table 15.1, the minimum stop-band attenuation δ in the FIR filter obtained using either the rectangular, Bartlett, Hamming, Hanning, or Blackman window is fixed. In most cases, the selected window surpasses the required specifications for the attenuation δ. Consider, for example, the design of an FIR filter with the minimum attenuation specified at 60 dB. Table 15.2 determines that only the Blackman window can be used, and it exceeds the minimum attenuation requirement by 10 dB. There is no alternative choice available, and the selection of the Blackman window is an overkill achieved at the cost of a wider transition band. Several advanced windows, such as Lanczos, Tukey, Dolph–Chebyshev and Kaiser windows have been proposed, which provide control over the stop-band ripple δ by means of an additional parameter characterizing the window. In this section, we introduce the Kaiser window and outline the steps for designing FIR filters with the Kaiser window.

Fig. 15.8. Kaiser windows of length $N = 51$ for different values of the shape control parameter β.

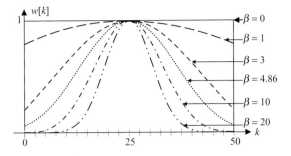

The Kaiser window is based on the zeroth-order Bessel function of the first kind and is defined as follows:

$$w_{\text{kaiser}}[k] = \begin{cases} \dfrac{I_0\left[\beta\left(\sqrt{1 - [(k - m)/m]^2}\right)\right]}{I_0[\beta]} & 0 \leq k \leq N - 1 \\ 0 & \text{otherwise,} \end{cases} \quad (15.18)$$

where $m = (N - 1)/2$, N is the length of the filter, and $I_0[\cdot]$ represents the zeroth-order Bessel function of the first kind, which can be approximated by

$$I_0[\beta] \approx 1 + \sum_{r=1}^{\infty}\left[\frac{(\beta/2)}{r!}\right]^2. \quad (15.19)$$

The parameter β is referred to as the *shape control parameter*. By varying β with respect to the window's length N, the shape of the Kaiser window can be adjusted to trade the amplitude of the side lobe for the width of the main lobe of the DTFT of the Kaiser window. Figure 15.8 illustrates the variations in the shape of the Kaiser window as β varies from 0 to 20. The length N of the window is kept constant at 51. From Fig. 15.8, we observe that the Kaiser window can be used to approximate any of the rectangular, Bartlett, Hamming, Hanning, or Blackman windows by appropriately selecting the value of β. When $\beta = 0$, for example, the shape of the Kaiser window is identical to the rectangular window. Similarly, when $\beta = 4.86$, the shape of the Kaiser window is almost identical to the Hamming window. Since the shape of the window also determines the maximum ripples within the pass and stop bands, parameter β is also referred to as the *ripple control parameter*.

We now explain the last two columns included in Table 15.1 As explained earlier, the Kaiser window can be used to approximate the five basic windows covered in Section 15.1.2. The second to last column in Table 15.1 specifies the value of the shape control parameter β for which the Kaiser window approaches the basic windows. Setting $\beta = 4.86$, for example, will cause the shape of the Kaiser window to be similar to that of the Hamming window. The last column lists the width of the transition band of the FIR filter obtained by using the Kaiser window. For $\beta = 4.86$, the Kaiser window would approach the Hamming window. The transition band of the resulting FIR filter obtained by truncating the ideal lowpass filter to length N with the Kaiser window is given by $6.27\pi/(N-1)$. We can explain the remaining entries in the last two columns of Table 15.1 in a similar fashion.

15.1.5 Lowpass filter design steps using the Kaiser window

The steps involved in designing a lowpass FIR filter using the Kaiser window are similar to those in the filter design outlined in Section 15.1.3, except for steps 4, 6, and 7. Below, we only include a brief description of steps 1–3, which are common to the two algorithms. The steps that are different are explained in more detail.

Step 1 Calculate the normalized cut-off frequency Ω_n of the filter. See step 1 of Section 15.1.3 for details.

Step 2 Determine the expression for the impulse response of an ideal lowpass filter:

$$h_{\text{ilp}}[k] = \Omega_n \,\text{sinc}(\Omega_n(k-m)),$$

where $m = (N-1)/2$ and N is the length of the FIR filter, which is calculated in step 6.

Step 3 Calculate the minimum attenuation A on a dB scale using $A = \min(\delta_p, \delta_s)$.

Step 4 Based on the value of A obtained in step 3, calculate the shape parameter β from the following:

$$\beta = \begin{cases} 0 & A \le 21\ \text{dB} \\ 0.5842(A-21)^{0.4} + 0.0789(A-21) & 21\ \text{dB} < A < 50\ \text{dB} \\ 0.1102(A-8.7) & A \ge 50\ \text{dB}. \end{cases}$$
$$(15.20)$$

The above expression was derived empirically by J. F. Kaiser, who came up with the specifications of the Kaiser window.

Step 5 Calculate the normalized transitional bandwidth $\Delta\Omega_n$ for the FIR filter. See step 5 of Section 15.1.3 for details.

Step 6 The length N of the Kaiser window is calculated from the following expression:

$$N \ge \frac{A - 7.95}{2.285\pi \times \Delta\Omega_n}.$$
$$(15.21)$$

Equation (15.21) was also derived by Kaiser from empirical observations. Select an appropriate value of N and then calculate $m = (N-1)/2$.

Step 7 Determine the Kaiser window by substituting the values of β (obtained in step 4) and m (obtained in step 6) into Eq. (15.18). Let the determined Kaiser window be denoted by $w_{\text{kaiser}}[k]$.

Step 8 The impulse response of the FIR filter is given by:

$$h_{\text{lp}}[k] = h_{\text{ilp}}[k]w_{\text{kaiser}}[k]. \qquad (15.22)$$

If the pass-band gain $|H_{\text{lp}}(0)|$ at $\Omega = 0$, given by $\sum h_{\text{lp}}[k]$, is not equal to one, we normalize $h_{\text{lp}}[k]$ with $\sum h_{\text{lp}}[k]$.

Step 9 Confirm that the impulse response $h_{\text{lp}}[k]$ satisfies the initial specifications by plotting the magnitude spectrum $|H_{\text{lp}}(\Omega)|$ of the FIR filter obtained in step 8.

Example 15.3 uses the above algorithm to design an FIR filter using the Kaiser window.

Example 15.3
Using the Kaiser window, design the FIR filter specified in Example 15.2.

Solution
Following steps 1–3 of Example 15.2, we determine the following values for the normalized cut-off frequency, impulse response of the ideal lowpass filter, and minimum attenuation A:

$$\Omega_n = 0.4375; h_{\text{lp}}[k] = 0.4375 \, \text{sinc}(0.4375(k - m)); A = 50 \, \text{dB}.$$

The value of m in the impulse response is set to $(N - 1)/2$.
Step 4 of Section 15.1.5 determines the value of β:

$$\beta = 0.1102(A - 8.7) = 4.5513.$$

Step 5 computes the normalized transition bandwidth:

$$\Delta\Omega_n = \Delta\omega_c/(0.5\omega_0) = (4\pi - 3\pi) \times 10^3/(0.5 \times 2\pi \times 8 \times 10^3) = 0.1250.$$

Using Step 6, the length of the Kaiser window is given by

$$N \geq \frac{A - 7.95}{2.285\pi \times \Delta\Omega_n} = \frac{50 - 7.95}{2.285\pi \times 0.125} = 46.8619,$$

which is rounded off to the closest higher odd number as 47.

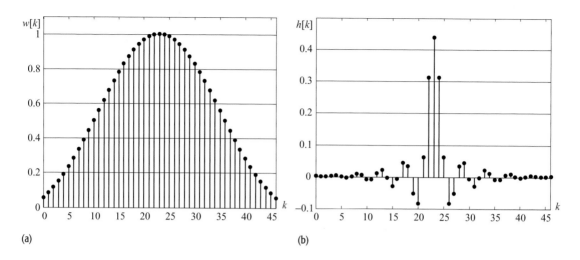

(a)

(b)

Fig. 15.9. (a) Kaiser window $w[k]$ of length $N = 47$ and $\beta = 4.5513$. (b) Impulse response $h[k]$ of the FIR filter obtained by multiplying the ideal lowpass filter impulse response by the Kaiser window in Example 15.3.

Substituting $\beta = 4.5513$ and $N = 47$ in Eq. (15.18), the expression for the Kaiser window is given by

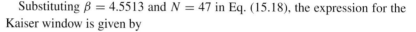

$$w_{\text{kaiser}}[k] = \begin{cases} \dfrac{I_0\left[4.5513\left(\sqrt{1 - \left[(k-23)/23\right]^2}\right)\right]}{I_0[4.5513]} & 0 \leq k \leq 46 \\ 0 & \text{otherwise.} \end{cases}$$

The impulse response of the FIR filter is then given by $h[k] = h_{\text{ilp}}[k]\, w_{\text{kaiser}}[k]$.

Figure 15.9(a) plots the time-domain representation of the Kaiser window of length $N = 47$ and shape control parameter $\beta = 4.5513$, and Fig. 15.9(b) plots the impulse response of the FIR filter.

The frequency characteristics of the FIR filter are shown in Fig. 15.10. Since $\sum h[k] = 0.9992 \approx 1$, the impulse response $h[k]$ of the FIR filter is not normalized by $\sum h[k]$. It is observed that the FIR filter meets the design specification. The stop-band attenuation is lower than 50 dB.

By comparing the results of Example 15.3 with those of Example 15.2, we observe that the FIR filter obtained from the Kaiser window in Example 15.3 has a smaller length $N = 47$ than the FIR filter obtained from the Hamming window

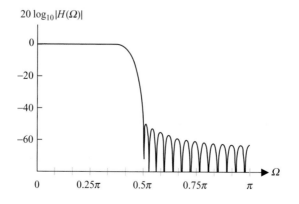

Fig. 15.10. Magnitude response of the lowpass FIR filter designed in Example 15.3.

in Example 15.2, which has a length of $N = 53$. Therefore, the Kaiser window provides an FIR filter with a lower implementational cost. This reduction in cost is attributed to the flexibility in the Kaiser window of closely meeting the stop-band attenuation of 50 dB. The stop-band attenuation in the Hamming window is fixed to 60 dB and cannot be varied.

Example 15.4

Design a lowpass FIR filter based on the following specifications:

(i)   cut-off frequency $\Omega_c = 0.3636\pi$ radians/s;
(ii) transition width $\Delta\Omega_c = 0.0727\pi$ radians/s;
(iii) pass-band ripple $20\log_{10}(1 + \delta_p) \leq 0.07$ dB;
(iv) stop-band attenuation $-20\log_{10}(\delta_s) \geq 40$ dB.

Solution

The specifications for the digital filter are specified in the DT frequency Ω domain.

Step 1 suggests that the normalized cut-off frequency is given by

$$\Omega_n = (0.3636\pi)/\pi = 0.3636.$$

Step 2 determines the impulse response of the ideal lowpass filter with the normalized cut-off frequency $\Omega_n = 0.3636$:

$$h_{ilp}[k] = 0.3636\,\text{sinc}(0.3636(k - m)),$$

with m set to $(N - 1)/2$.

Step 3 determines the value of the minimum attenuation A. The pass-band ripple $20\log_{10}(1 + \delta_p)$ is limited to 0.07 dB. Expressed on a linear scale, we obtain $\delta_p \leq 0.0081$. Similarly, the stop-band ripple $20\log_{10}(\delta_s)$ is limited to -40 dB, which implies $\delta_s \leq 0.01$. The value of the minimum attenuation is therefore given by $A = \min(0.0081, 0.01) = 0.0081$. Expressed in decibels, the value of the minimum attenuation is $-20\log_{10}(A) = 41.83$ dB.

Step 4 determines the value of the shape control parameter β from Eq. (15.20):

$$\beta = 0.5842(A - 21)^{0.4} + 0.0789(A - 21) = 3.6115.$$

Step 5 computes the normalized transition bandwidth:

$$\Delta\Omega_n = \Delta\Omega_c/\pi = 0.0727.$$

Step 6 determines the length of the Kaiser window:

$$N \geq \frac{A - 7.95}{2.285\pi \times \Delta\Omega_n} = \frac{41.83 - 7.95}{2.285\pi \times 0.0727} = 64.92,$$

which is rounded off to the closest higher odd number as 65.

Fig. 15.11. Magnitude response of the lowpass FIR filter designed in Example 15.4.

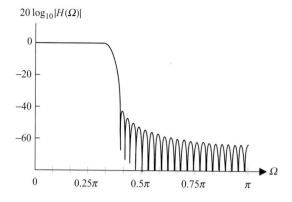

Substituting $\beta = 3.6115$ and $N = 65$ in Eq. (15.18), the expression for the Kaiser window is given by

$$
w_{\text{kaiser}}[k] = \begin{cases} \dfrac{I_0\left[3.6115\left(\sqrt{1-[(k-32)/32]^2}\right)\right]}{I_0\,[3.6115]} & 0 \le k \le 64 \\[2mm] 0 & \text{otherwise.} \end{cases}
$$

The impulse response of the FIR filter is then given by $h[k] = h_{\text{ilp}}[k]w_{\text{kaiser}}[k]$. The magnitude response of the FIR filter is plotted in Fig. 15.11 using a dB scale.

15.2 Design of highpass filters using windowing

The windowing method is not restricted to design of lowpass FIR filters; it can be generalized to design other types of FIR filters. Section 15.2 considers highpass FIR filters, and Sections 15.3 and 15.4 extend the windowing method to bandpass and bandstop FIR filters, respectively.

The transfer function of an ideal highpass filter was defined in Section 14.1.2, and is reproduced here for convenience:

$$
H_{\text{ihp}}(\Omega) = \begin{cases} 0 & |\Omega| < \Omega_{\text{c}} \\ 1 & \Omega_{\text{c}} \le |\Omega| \le \pi. \end{cases} \tag{15.23}
$$

It was shown in Section 14.1.2 that the impulse response of a highpass filter can be related to the impulse response of a lowpass filter with the same cut-off frequency, and is given by Eqs. (14.3a) and (14.3b). As shown in Table 14.1, the impulse response of the ideal highpass filter with a normalized cut-off frequency Ω_{n} is given by

$$
h_{\text{ihp}}[k] = \delta[k] - \Omega_{\text{n}}\,\text{sinc}[\Omega_{\text{n}}k]. \tag{15.24}
$$

Fig. 15.12. Desired
specifications of a highpass filter.

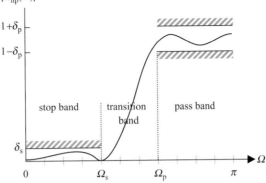

Fig. 15.12. Desired
specifications of a highpass filter.

The filter with this impulse response is non-causal and hence non-realizable. By applying a delay m, the impulse response of an ideal highpass filter is obtained:

$$h_{\text{ihp}}[k] = \delta[k - m] - \Omega_n \, \text{sinc}[\Omega_n(k - m)]. \qquad (15.25)$$

Given the impulse response of an ideal highpass filter, we can use the windowing method to design a highpass FIR filter. The specifications for the highpass FIR filter are illustrated in Fig. 15.12 and are given by

stop band $(0 \leq \Omega \leq \Omega_s)$ $\qquad\qquad\qquad 0 \leq |H_{\text{hp}}(\Omega)| \leq \delta_s;$

pass band $(\Omega_p < \Omega \leq \pi)$ $\qquad\qquad (1 - \delta_p) \leq |H_{\text{hp}}(\Omega)| \leq (1 + \delta_p).$

The steps involved in the design of a highpass FIR filter are given in the following.

Step 1   Calculate the normalized cut-off frequency Ω_n of the filter:

cut-off frequency $\qquad\qquad\qquad\qquad \Omega_c = 0.5 \left(\Omega_p + \Omega_s \right);$

normalized cut-off frequency $\qquad\qquad\quad \Omega_n = \Omega_c / \pi.$

Step 2   Determine the expression for the impulse response of an ideal highpass filter:

$$h_{\text{ihp}}[k] = \delta[k - m] - \Omega_n \sin[\Omega_n(k - m)], \qquad (15.26)$$

where $\Omega_c = \pi \Omega_n$ and $m = (N - 1)/2$, where N is the length of the FIR filter.

Step 3   Calculate the minimum attenuation A on a dB scale using $A = \min(\delta_p, \delta_s)$.

Step 4   Calculate the normalized transition band $\Delta\Omega_n$ for the FIR filter:

transition BW $\qquad\qquad\qquad\qquad\qquad \Delta\Omega_c = (\Omega_p - \Omega_s);$

normalized transition BW $\qquad\qquad\qquad \Delta\Omega_n = \Delta\Omega_c / \pi.$

Step 5 Design an appropriate window with parameters A and $\Delta\Omega_n$ using the procedures mentioned in Section 15.1.3 or Section 15.1.5. Denote this window by $w[k]$.

Step 6 Derive the impulse response of the FIR filter:

$$h_{hp}[k] = h_{ihp}[k]w[k]. \tag{15.27}$$

We now derive the condition for the gain $|H(\pi)|$ to be equal to one. Substituting $\Omega = \pi$ in the DTFT $H(\Omega)$, we obtain

$$H(\pi) = \left. \sum_{k=0}^{N-1} h_{hp}[k]e^{-jk\Omega} \right|_{\Omega=\pi} = \sum_{k=0,2,\dots}^{N-1} h_{hp}[k] - \sum_{k=1,3,\dots}^{N-1} h_{hp}[k]. \tag{15.28}$$

In other words, the difference between the sum of the even-numbered samples of $h[k]$ and the sum of the odd-numbered samples of $h[k]$ should equal one. If not, we calculate the normalized impulse response $h'_{hp}[k] = h_{hp}[k]/H(\pi)$.

Step 7 Confirm that the impulse response $h_{hp}[k]$ satisfies the initial specifications by plotting the magnitude spectrum $|H_{hp}(\Omega)|$ of the FIR filter obtained in step 6.

We observe that the above algorithm is similar to the design of a lowpass filter, except that the impulse response of the ideal lowpass filter is replaced by the impulse response of the ideal highpass filter. Example 15.5 uses the above algorithm to design a highpass FIR filter using the Kaiser window.

Example 15.5

Design a highpass FIR filter, using the Kaiser window, with the following specifications:

(i) pass-band edge frequency $\Omega_p = 0.5\pi$ radians/s;
(ii) stop-band edge frequency $\Omega_s = 0.125\pi$ radians/s;
(iii) pass-band ripple $20\log_{10}(1 + \delta_p) \le 0.01$ dB;
(iv) stop-band attenuation $-20\log_{10}(\delta_s) \ge 60$ dB.

Plot the frequency characteristics of the designed filter.

Solution

The cut-off frequency Ω_c of the filter is given by $\Omega_c = 0.5(0.125\pi + 0.5\pi) = 0.3125\pi$. The normalized cut-off frequency Ω_n of the filter is $\Omega_c/\pi = 0.3125$. The impulse response of the ideal high pass filter with a cut-off frequency of 0.3125 is given by

$$h_{ihp}[k] = \delta[k - m] - 0.3125 \, \text{sinc}(0.3125(k - m)). \tag{15.29}$$

To determine the minimum attenuation A, we calculate δ_p and δ_s. Since $20\log_{10}(1 + \delta_p) <= 0.01$ dB, the pass-band ripple δ_p should be less than

Fig. 15.13. Magnitude response
of the highpass FIR filter
designed in Example 15.5.

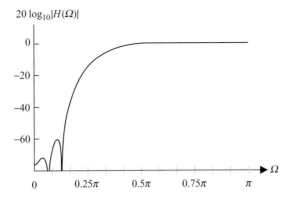

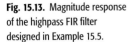

Fig. 15.13. Magnitude response of the highpass FIR filter designed in Example 15.5.

$10^{0.01/20} - 1 = 0.0012$, while δ_s should be less than $10^{-60/20} - 1 = 0.001$. The minimum attenuation A is therefore given by $A = \min(0.0012, 0.001) = 0.001$, or 60 dB.

The shape parameter is evaluated from Eq. (15.20) as follows:

$$\beta = 0.1102(A - 8.7) = 5.6533.$$

The transition band $\Delta\Omega_c$ for the FIR filter is $\Omega_p - \Omega_s = 0.375\pi$. The normalized transition band $\Delta\Omega_n$ is therefore given by $\Delta\Omega_c/\pi = 0.375$. Using $\Delta\Omega_n = 0.375$, the length N of the Kaiser window is given by

$$N \geq \frac{60 - 7.95}{2.285\pi \times 0.375} = 19.3354.$$

Rounding off to the higher closest odd number, we obtain $N = 21$.

The expression for the Kaiser window is given by

$$w_{\text{kaiser}}[k] = \begin{cases} \dfrac{I_0\left[5.6533\left(\sqrt{1 - [(k - 10)/10]^2}\right)\right]}{I_0[5.6533]} & 0 \leq k \leq 20 \\ 0 & \text{otherwise.} \end{cases} \quad (15.30)$$

The impulse response of the highpass FIR filter is given by

$$h_{\text{hp}}[k] = h_{\text{ihp}}[k]w_{\text{kaiser}}[k],$$

where $h_{\text{ihp}}[k]$ is specified in Eq. (15.29) with $m = 10$ and $w_{\text{kaiser}}[k]$ is given in Eq. (15.30). The filter gain at $\Omega = \pi$ is given by

$$H_{\text{hp}}(\pi) = \sum_{k=0,2,\dots}^{N-1} h_{\text{hp}}[k] - \sum_{k=1,3,\dots}^{N-1} h_{\text{hp}}[k] = 1.0002.$$

As $H(\pi) \approx 1$, the coefficients of $h[k]$ need not be normalized.

The magnitude response of the highpass FIR filter is plotted in Fig. 15.13, which verifies that the initial specifications of the filter are satisfied.

In Example 15.5 we designed a highpass FIR filter directly from the given specifications. An alternative procedure to design a highpass FIR filter is to exploit

Fig. 15.14. Desired specifications of a bandpass filter.

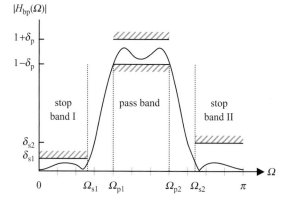

Eq. (14.2b) and implement $H_{lp}(\Omega)$ instead. Based on the frequency characteristics of the highpass FIR filter illustrated in Fig. 15.12, the specifications of the $H_{lp}(\Omega)$ in Eq. (14.2b) are given by

pass band $(0 \leq \Omega \leq \Omega_s)$ $(1 - \delta_s) \leq |H_{lp}(\Omega)| \leq (1 + \delta_s)$;

stop band $(\Omega_p < \Omega \leq \pi)$ $0 \leq |H_{lp}(\Omega)| \leq \delta_p$.

The impulse response of the lowpass FIR filter $h_{lp}[k]$ is then transformed to the impulse response $h_{hp}[k]$ of the highpass FIR filter using the following equation:

$$h_{hp}[k] = \delta[k - m] - h_{lp}[k].$$

15.3 Design of bandpass filters using windowing

The design specifications for bandpass filters are specified in Fig. 15.14 and are given by

stop band I $(0 \leq \Omega \leq \Omega_{s1})$ $0 \leq |H(\Omega)| \leq \delta_{s1}$;

stop band II $(\Omega_{s2} \leq \Omega \leq \pi)$ $0 \leq |H(\Omega)| \leq \delta_{s2}$;

pass band $(\Omega_{p1} < \Omega \leq \Omega_{p2})$ $(1 - \delta_p) \leq |H(\Omega)| \leq (1 + \delta_p)$,

where we assume that the values of ripples δ_{s1} and δ_{s2} allowed in the two stop bands are different. The algorithm used to design a bandpass FIR filter using windowing is similar to the design for the highpass filter described in Section 15.2, except that the impulse response of an ideal bandpass filter is used in step 2.

    The transfer function of an ideal bandpass filter was defined in Section 14.1.3, and is reproduced here for convenience:

$$H_{ibp}(\Omega) = \begin{cases} 1 & \Omega_{c1} \leq |\Omega| \leq \Omega_{c2} \\ 0 & |\Omega| < \Omega_{c1} \quad \text{and} \quad \Omega_{c2} \leq |\Omega| \leq \pi. \end{cases} \qquad (15.31)$$

As shown in Table 14.1, the impulse response of the ideal bandpass filter with normalized cut-off frequencies of Ω_{n1}, Ω_{n2} ($\Omega_{n2} > \Omega_{n1}$) is given by

$$h_{ibp}[k] = \Omega_{n2}\,\text{sinc}[\Omega_{n2}k] - \Omega_{n1}\,\text{sinc}[\Omega_{n1}k]. \tag{15.32}$$

As the filter with this impulse response is non-causal, we apply a delay of m, and the modified impulse response is obtained:

$$h_{ibp}[k] = \Omega_{n2}\,\text{sinc}[\Omega_{n2}(k - m)] - \Omega_{n1}\,\text{sinc}[\Omega_{n1}(k - m)]. \tag{15.33}$$

The steps for designing a bandpass filter using windowing are as follows.

Step 1 Calculate the two normalized cut-off frequencies Ω_{n1} and Ω_{n2} of the bandpass filter:

cut-off frequencies $\Omega_{c1} = 0.5(\Omega_{p1} + \Omega_{s1})$ and $\Omega_{c2} = 0.5(\Omega_{p2} + \Omega_{s2})$;

normalized cut-off frequencies $\Omega_{n1} = \Omega_{c1}/\pi$ and $\Omega_{n2} = \Omega_{c2}/\pi$.

Step 2 Determine the impulse response of the ideal bandpass filter by substituting the values of Ω_{n1} and Ω_{n2} in Eq. (15.33).

Step 3 Calculate the minimum attenuation A on a dB scale using $A = \min(\delta_p, \delta_{s1}, \delta_{s2})$.

Step 4 Calculate the normalized transition bandwidth $\Delta\Omega_n$ for the FIR filter:

transitional BW $\Delta\Omega_{c1} = (\Omega_{p1} - \Omega_{s1})$ and $\Delta\Omega_{c2} = (\Omega_{s2} - \Omega_{p2})$;

normalized transition BW $\Delta\Omega_n = \min(\Delta\Omega_{c2}, \Delta\Omega_{c1})/\pi$.

Step 5 Design an appropriate window with parameters A and $\Delta\Omega_n$ using the procedures mentioned in Section 15.1.3 or Section 15.1.5. Denote this window by $w[k]$.

Step 6 Derive the impulse response of the FIR filter:

$$h_{bp}[k] = h_{ibp}[k]w[k]. \tag{15.34}$$

Step 7 Confirm that the impulse response $h_{bp}[k]$ satisfies the initial specifications by plotting the magnitude spectrum $|H_{bp}(\Omega)|$ of the FIR filter obtained in step 6.

Example 15.6 illustrates the working of the above algorithm by designing a bandpass FIR filter using the Kaiser window.

Example 15.6

Design a bandpass FIR filter, using Kaiser window, with the following specifications:

(i) pass-band edge frequencies, $\Omega_{p1} = 0.375\pi$ and $\Omega_{p2} = 0.5\pi$ radians/s;
(ii) stop-band edge frequencies, $\Omega_{s1} = 0.25\pi$ and $\Omega_{s2} = 0.625\pi$ radians/s;
(iii) stop-band attenuations, $\delta_{s1} > 50$ dB and $\delta_{s2} > 50$ dB.

Plot the gain–frequency characteristics of the designed bandpass filter.

Solution

The cut-off frequencies of the bandpass filter are given by

$$\Omega_{c1} = 0.5\,(0.25\pi + 0.375\pi) = 0.3125\pi$$

and

$$\Omega_{c2} = 0.5\,(0.5\pi + 0.625\pi) = 0.5625\pi.$$

The normalized cut-off frequencies are given by $\Omega_{n1} = \Omega_{c1}/\pi = 0.3125$ and $\Omega_{n2} = \Omega_{c2}/\pi = 0.5625$. The impulse response of an ideal bandpass filter is given by

$$h_{\text{ibp}}[k] = 0.5625\,\text{sinc}[0.5625(k - m)] - 0.3125\,\text{sinc}[0.3125(k - m)].$$

$$(15.35)$$

Since only the stop-band attenuations are specified, and these are both equal to 50 dB, the minimum attenuation $A = 50$ dB.

The shape parameter β of the Kaiser window is computed to be

$$\beta = 0.1102(50 - 8.7) = 4.5513.$$

The transition bands $\Delta\Omega_{c1}$ and $\Delta\Omega_{c2}$ for the bandpass FIR filter are given by

$$\Delta\Omega_{c1} = 0.375\pi - 0.25\pi = 0.125\pi$$

and

$$\Delta\Omega_{c2} = 0.625\pi - 0.5\pi = 0.125\pi,$$

which lead to the normalized transition BW of $\Delta\Omega_n = 0.125$.
The length N of the Kaiser window is given by

$$N \geq \frac{50 - 7.95}{2.285\pi \times 0.125} = 46.8619.$$

Rounded to the closest higher odd number, $N = 47$, and the value of m in Eq. (15.35) is 23. The expression for the Kaiser window is as follows:

$$w_{\text{kaiser}}[k] = \begin{cases} \dfrac{I_0\left[4.5513\left(\sqrt{1 - [(k - 23)/23]^2}\right)\right]}{I_0[4.5513]} & 0 \leq k \leq 46 \\ 0 & \text{otherwise.} \end{cases} \quad (15.36)$$

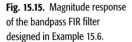

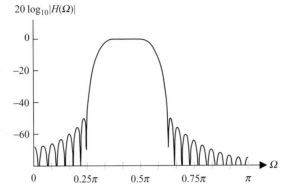

The impulse response of the bandpass FIR filter is given by

$$h_{bp}[k] = h_{ibp}[k]w_{kaiser}[k].$$

where $h_{ibp}[k]$ is specified in Eq. (15.35) with $m = 23$ and $w_{kaiser}[k]$ is specified in Eq. (15.36).

The magnitude spectrum of the bandpass FIR filter is plotted in Fig. 15.15. It is observed that the bandpass filter satisfies the design specifications.

In Example 15.6, we designed a bandpass FIR filter directly. As for the highpass FIR filter, an alternative procedure to design a bandpass FIR filter is to exploit Eq. (14.4e) and implement two lowpass FIR filters with impulse responses $H_{lp1}(k)$ and $H_{lp2}(k)$. The specifications for the two lowpass filters should be carefully derived such that the pass- and stop-band ripples of the combined system are limited to values allowed in the original bandpass filter's specifications.

15.4 Design of a bandstop filter using windowing

As illustrated in Fig. 15.16, the design specifications for a bandstop filter are given by

pass band I $(0 \leq \Omega \leq \Omega_{p1})$ $(1 - \delta_{p1}) \leq |H_{bs}(\Omega)| \leq (1 + \delta_{p1})$;

pass band II $(\Omega_{p2} \leq \Omega \leq \pi)$ $(1 - \delta_{p2}) \leq |H_{bs}(\Omega)| \leq (1 + \delta_{p2})$;

stop band $(\Omega_{s1} < \Omega \leq \Omega_{s2})$ $0 \leq |H_{bs}(\Omega)| \leq \delta_s$.

The steps involved in the design of a bandpass FIR filter using windowing are similar to those specified for the bandpass filter in Section 15.3.

The transfer function of an ideal bandstop filter was defined in Section 14.1.4, and is reproduced here for convenience:

$$H_{ibs}(\Omega) = \begin{cases} 0 & \Omega_{c1} \leq |\Omega| \leq \Omega_{c2} \\ 1 & |\Omega| < \Omega_{c1} \text{ and } \Omega_{c2} < |\Omega| \leq \pi, \end{cases} \quad (15.37)$$

Fig. 15.16. Desired
specifications of a bandstop
filter.

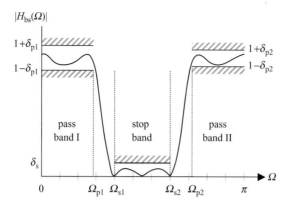

As shown in Table 14.1, the impulse response of the ideal bandstop filter with normalized cut-off frequencies of Ω_{n1}, Ω_{n2} $(\Omega_{n2} > \Omega_n)$ is given by

$$h_{ibs}[k] = \delta[k] - \Omega_{n2}\, \text{sinc}[\Omega_{n2}k] + \Omega_{n1}\, \text{sinc}[\Omega_{n1}k]. \qquad (15.38)$$

By applying a delay m, the modified impulse response of an ideal bandpass filter is obtained:

$$h_{ibs}[k] = \delta[k - m] - \Omega_{n2}\, \text{sinc}[\Omega_{n2}(k - m)] + \Omega_{n1}\, \text{sinc}[\Omega_{n1}(k - m)].$$
$$(15.39)$$

In the following example, we illustrate the steps involved in designing a practical bandstop filter using the windowing method.

Example 15.7
Design a bandstop FIR filter, using a Kaiser window, with the following specifications:

(i) pass-band edge frequencies, $\Omega_{p1} = 0.25\pi$ and $\Omega_{p2} = 0.625\pi$ radians/s;
(ii) stop-band edge frequencies, $\Omega_{s1} = 0.375\pi$ and $\Omega_{s2} = 0.5\pi$ radians/s;
(iii) stop-band attenuations, $\delta_{s1} > 50$ dB and $\delta_{s2} > 50$ dB.

Solution
The cut-off frequencies of the bandpass filter are given by

$$\Omega_{c1} = 0.5\,(0.25\pi + 0.375\pi) = 0.3125\pi$$

and

$$\Omega_{c2} = 0.5\,(0.5\pi + 0.625\pi) = 0.5625\pi.$$

The normalized cut-off frequencies are given by $\Omega_{n1} = 0.3125$ and $\Omega_{n2} = 0.5625$. The impulse response of an ideal bandpass filter is given by

$$h_{ibs}[k] = \delta[k - m] - 0.5625\, \text{sinc}[0.5625(k - m)]$$
$$+ 0.3125\, \text{sinc}[0.3125(k - m)]. \qquad (15.40)$$

The minimum attenuation $A = 50$ dB. Therefore, The shape parameter β of the Kaiser window is computed as

$$\beta = 0.1102(50 - 8.7) = 4.5513.$$

The transition bands $\Delta\Omega_{c1}$ and $\Delta\Omega_{c2}$ for the bandpass FIR filter are given by

$$\Delta\Omega_{c1} = (0.375\pi - 0.25\pi) = 0.125\pi \text{ and}$$
$$\Delta\Omega_{c2} = (0.625\pi - 0.5\pi) = 0.125\pi,$$

which leads to the normalized transition BW of $\Delta\Omega_n = 0.125$.
  The length N of the Kaiser window is given by

$$N \geq \frac{50 - 7.95}{2.285\pi \times 0.125} = 46.8619.$$

Rounded to the closest higher odd number, $N = 47$, and the value of m in Eq. (15.40) is 23.
  The expression for the Kaiser window is as follows:

$$w_{kaiser}[k] = \begin{cases} \dfrac{I_0\left[4.5513\left(\sqrt{1 - [(k - 23)/23]^2}\right)\right]}{I_0[4.5513]} & 0 \leq k \leq 46 \\ 0 & \text{otherwise.} \end{cases} \quad (15.41)$$

The impulse response of the bandstop FIR filter is given by

$$h_{bs}[k] = h_{ibs}[k]w_{kaiser}[k],$$

where $h_{ibs}[k]$ is specified in Eq. (15.40) with $m = 23$ and $w_{kaiser}[k]$ is as shown in Eq. (15.41).
  The magnitude response of the bandstop FIR filter is plotted in Fig. 15.17. It is observed that the bandstop filter satisfies the design specifications.

In the above example, a bandstop FIR filter was designed directly. As for the highpass and bandpass FIR filters, an alternative design procedure (see Eq. 14.6) is to express the transfer function of a bandstop FIR filter in terms of the transfer functions of two lowpass filters as follows:

$$h_{ibs}[k] = \delta[k - m] - h_{ilp1}[k]|_{\Omega_c=\Omega_{c2}} + h_{ilp2}[k]|_{\Omega_c=\Omega_{c1}}. \quad (15.42)$$

The specifications for these two lowpass filters are derived from the given design specifications of the bandpass filter. As for bandpass FIR filters, the specifications of the lowpass filters should be carefully assigned such that the pass- and stop-band ripples of the combined system satisfy the original bandstop filter's specifications.

15.5 Optimal FIR filters

Designing an FIR filter using the windowing approach is simple but suffers from one severe limitation. The minimum attenuation obtained in the stop

Fig. 15.17. Magnitude response of the bandstop FIR filter designed in Example 15.7.

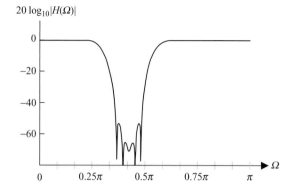

band of the FIR filter is fixed for the elementary window types covered in Section 15.1.2. The Kaiser window, introduced in Section 15.1.4, provides some flexibility in controlling the stop-band attenuation by introducing an additional design parameter β. However, there is no guarantee that the FIR filter, designed with either the elementary windows or the Kaiser window, is optimal. In this section, we introduce a computational optimization procedure for the design of FIR filters. The procedure is commonly referred to as the Parks–McClellan algorithm, which iteratively minimizes the absolute value of the error:

$$\varepsilon(\Omega) = W(\Omega)\,[H_{\mathrm{d}}(\Omega) - H(\Omega)], \tag{15.42}$$

where $H_{\mathrm{d}}(\Omega)$ is the transfer function of the desired or ideal filter, whose frequency characteristics are to be approximated, $H(\Omega)$ is the transfer function of the approximated FIR filter, and $W(\Omega)$ is a weighting function introduced to emphasize the relative importance of various frequency components of the filter. For a lowpass filter, for example, a logical way to select the values of the weighting function is to set

$$\text{lowpass filter} \qquad W(\Omega) = \begin{cases} 1/\delta_{\mathrm{p}} & 0 \le \Omega \le \Omega_{\mathrm{p}} \\ 0 & \Omega_{\mathrm{p}} < \Omega < \Omega_{\mathrm{s}} \\ 1/\delta_{\mathrm{s}} & \Omega_{\mathrm{s}} \le \Omega \le \pi. \end{cases} \tag{15.43}$$

Equation (15.43) implies that if the condition for the pass-band ripple is more stringent (i.e. smaller) than the condition for the stop-band ripple, the weighting function allocates a higher weight to the pass band than to the stop band, and vice versa. Zero weight is associated with the transition band, which means that the weighting function does not care about the characteristics of the FIR filter in the transition band as long as the filter's gain changes steadily between the pass and stop bands. Scaling Eq. (15.43) with δ_{s}, the normalized weighting function is given by

$$\text{lowpass filter} \qquad W(\Omega) = \begin{cases} \delta_{\mathrm{s}}/\delta_{\mathrm{p}} & 0 \le \Omega \le \Omega_{\mathrm{p}} \\ 0 & \Omega_{\mathrm{p}} < \Omega < \Omega_{\mathrm{s}} \\ 1 & \Omega_{\mathrm{s}} \le \Omega \le \pi. \end{cases} \tag{15.44}$$

The weighting function for highpass, bandpass, and bandstop filters can be derived in a similar fashion. Given a weighting function, the Parks–McClellan algorithm seeks to solve the following optimization problem:

$$\min_{\{h[k]\}} \left[\max_{\Omega \in S} \ |\varepsilon(\Omega)| \right], \tag{15.45}$$

where S is defined as a set of discrete frequencies chosen within the pass and stop bands. For a lowpass filter, the set of frequencies that can be included in S should lie in the following range:

lowpass filter $\qquad S = \left[0 \le \Omega \le \Omega_p \right] \cup [\Omega_s \le \Omega \le \pi] \tag{15.46}$

Similarly, the sets S of discrete frequencies are carefully selected over the pass and stop bands for other types of filters.

Because Eq. (15.45) minimizes a cost function, which is the maximum of the error $\varepsilon(\Omega)$, Eq. (15.45) is also referred to as the minimax optimization problem. The goal in solving Eq. (15.45) is to determine the set of coefficients for the impulse response $h[k]$ of the optimal FIR filter of length N.

It was shown in Proposition 14.1 (see Section 14.3.1) that if the filter coefficients of an FIR filter are symmetric or anti-symmetric, the phase response of the filter is a linear function of frequency, and the transfer function can be expressed as follows:

$$H(\Omega) = G(\Omega) e^{j(\beta - \alpha \Omega)}, \tag{15.47}$$

where $\alpha = (N-1)/2$, β is a constant, and $G(\Omega)$ is a real-valued function. Table 14.2 shows the values of β and $G(\Omega)$ for four types of linear-phase FIR filters.

The Parks–McClellan algorithm exploits Proposition 14.1 to solve the minimax optimization problem, as explained in the following. For various types of linear-phase FIR filter, $G(\Omega)$ is a summation of a finite number of sinusoidal terms of the form $\cos(\Omega k)$ or $\sin(\Omega k)$, which themselves can be expressed as polynomials of $\cos(\Omega)$. For example, $\cos(\Omega k)$ can be expressed as a kth-order polynomial of $\cos(\Omega)$, which, for $k = 2$ and 3, can be expressed as follows:

$$\cos(2\Omega) = 2\cos^2(\Omega - 1);$$
$$\cos(3\Omega) = 4\cos^3(\Omega) - \cos(\Omega).$$

It is observed from Table 14.2 that the function $G(\Omega)$ can be expressed as a sum of several higher-order terms $\cos(\Omega k)$ or $\sin(\Omega k)$. Therefore, $G(\Omega)$ can also be expressed as a polynomial of $\cos(\Omega)$. It can be shown that the error function $\varepsilon(\Omega)$ in Eq. (15.42), corresponding to linear-phase FIR filters, can also be expressed as a polynomial of $\cos(\Omega)$. Parks and McClellan applied the alternation theorem from the theory of polynomial approximation to solve the minimax optimization problem. For convenience, we first express the alternation theorem in the context of polynomial approximation, and later we show its adaptation to the minimax optimization problem.

15.5.1 Alternation theorem

Let S be a compact subset on the real axis x and let $D(x)$ be a desired function of x which is continuous on S. Let $D(x)$ be approximated by $P(x)$, an Lth-order polynomial of x, which is given by

$$P(x) = \sum_{m=0}^{L} c_m x^m. \tag{15.48a}$$

Define the approximation error $\varepsilon(x)$ and the amplitude of the maximum error value $\varepsilon_{\max}$ on S as follows:

$$\varepsilon(x) = W(x)[D(x) - P(x)]: \tag{15.48b}$$

$$\varepsilon_{\max} = \arg \max_{x \in S} |\varepsilon(x)|. \tag{15.48c}$$

A necessary and sufficient condition for $P(x)$ to be the unique Lth-order polynomial minimizing $\varepsilon_{\max}$ is that $\varepsilon(x)$ exhibits at least $L + 2$ alternations. In other words, there must exist $L + 2$ values of x, $\{x_1 < x_2 < \cdots x_{L+2}\} \in S$ such that $\varepsilon(x_m) = -\varepsilon(x_{m+1}) = \pm\varepsilon_{\max}$.

Note that the minimax optimization problem for optimal filter design fits very well in the framework of the alternation theorem. In the filter design problem, S is the subset of DT frequencies, $D(x)$ is the desired filter response, $P(x)$ is the approximated filter response, and $\varepsilon_{\max}$ is the maximum deviation between the desired and approximated filter response. Therefore, the FIR filter obtained using minimax optimization is also expected to exhibit alternations in its frequency response. However, note that $G(\Omega)$ is a polynomial of $\cos(\Omega)$ and not of Ω. This issue can be addressed by using the mapping function $x = \cos(\Omega)$. In this case, the frequency space $\Omega = [0, \pi]$ can be mapped to $x = [-1, 1]$, and the optimization problem can be reformulated around x to calculate the optimal filter coefficients. It can be shown that the alternation in the frequency response of the optimal filters is still applicable.

Based on the above discussion, the alternation theorem can be restated for the minimax optimization problem as follows. Consider the following minimax optimization problem:

$$\{h[k], 0 \le \overset{\min}{k} \le (N-1)\} \left[\max_{\Omega \in S} |\varepsilon(\Omega)| \right]; \tag{15.49a}$$

$$\varepsilon(\Omega) = W(\Omega) \left[H_{\mathrm{d}}(\Omega) - \underbrace{G(\Omega) e^{-j\Omega(N-1/2)} e^{j\phi}}_{H(\Omega)} \right]. \tag{15.49b}$$

where S is a set of discrete extremal frequencies chosen within the pass and stop bands, $W(\Omega)$ is a positive weighting function, $H_{\mathrm{d}}(\Omega)$ is the transfer function of the ideal filter with a unity gain within the pass band and a zero gain within the stop band, and $G(\Omega)$ is a polynomial of $\cos(\Omega)$ with degree L, which is uniquely specified by the impulse response $h[k]$. Let $\varepsilon_{\max}$ denote the maximum value of the error $|\varepsilon(\Omega)|$. The polynomial $G(\Omega)$, which best approximates $H_d(\Omega)$ (i.e. minimizes $\varepsilon_{\max}$), produces the error function $\varepsilon(\Omega)$ that must satisfy the following property. There should be at least $L + 2$ discrete frequencies $\{\Omega_1 < \Omega_2 < \cdots < \Omega_{L+2}\} \in S$ at which the maximum and minimum peak values of the error alternate, i.e. $\varepsilon(\Omega_{m+1}) = -\varepsilon(\Omega_m) = \varepsilon_{\max}$.

Before presenting some examples of the application of the alternation theorem, we briefly comment on the degree L of the error function $\varepsilon(\Omega)$ in the FIR filters. The value of L is determined by evaluating the highest power of $\cos(\Omega)$ in the $G(\Omega)$ function of the filters. For the four types of FIR filters with length N, the value of L is specified as follows:

type I FIR filters $\qquad\qquad L = \dfrac{N-1}{2}$;

type II FIR filters $\qquad\qquad L = \dfrac{N-2}{2}$;

type III FIR filters $\qquad\qquad L = \dfrac{N-3}{2}$;

type IV FIR filters $\qquad\qquad L = \dfrac{N-2}{2}$.

The alternation theorem states that the minimum number of alternations for the optimal FIR filter should be at least $L + 2$. The actual number of alternations in an optimal FIR may, however, exceed the minimum number specified by the alternation theorem. An optimally designed lowpass or highpass filter can have up to $L + 3$ alternations, while an optimal bandpass or bandstop filter can have up to $L + 5$ alternations.

Example 15.8

The magnitude spectra of two lowpass FIR filters with lengths $N = 13$ and 20 are, respectively, shown in Figs. 15.18(a) and (b), where the filter gain within the pass and stop bands is enclosed within a frame box. Determine if the two filters satisfy the alternation theorem.

Solution

Figure 15.18(a) shows the frequency response of a type I FIR filter with length $N = 13$. The degree L of $\cos(\Omega)$ in the polynomial $\varepsilon(\Omega)$ is given by $L = (13 - 1)/2 = 6$. Based on the alternation theorem, there should be at least $L + 2 = 8$ alternations in polynomial $\varepsilon(\Omega)$. Note that the absolute value of error $|\varepsilon(\Omega)|$ is the difference $|H(\Omega) - H_d(\Omega)|$, where $H_d(\Omega)$ has a unity gain within the pass band and zero gain within the stop band. Therefore, counting the number of alternations in $\varepsilon(\Omega)$ is the same as counting the number of alternations in $H(\Omega)$ with respect to the pass- and stop-band ripples. From Fig. 15.18 we observe that there are indeed eight alternations (shown by $\times$ symbols) in $H(\Omega)$. One of these alternations occurs at the pass-band edge frequency Ω_p,

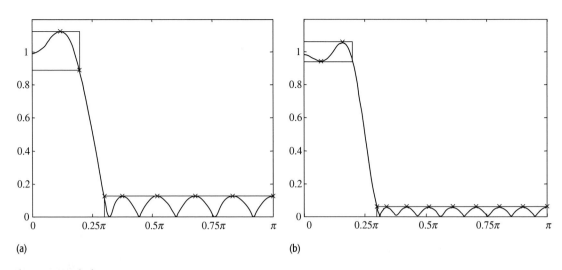

(a)   (b)

Fig. 15.18. Magnitude spectrum of lowpass FIR filters. (a) Type I FIR filter of length $N = 13$. (b) Type II FIR filter of length $N = 20$.

and two of these alternations occur at the stop-band edge frequencies Ω_s and π. In other words, Fig. 15.18(a) satisfies the alternation theorem.

Figure 15.18(b) shows the frequency response of a type II FIR filter with length $N = 20$. The degree L of $\cos(\Omega)$ in polynomial $\varepsilon(\Omega)$ is given by $L = (20 - 2)/2 = 9$. Based on the alternation theorem, there should be at least $L + 2 = 11$ alternations in polynomial $\varepsilon(\Omega)$. In Fig. 15.18(b), we observe 12 alternations in $H(\Omega)$, which exceed the minimum required number of alternations. Therefore, Fig. 15.18(b) satisfies the alternation theorem.

15.5.2 Parks–McClellan algorithm

In this section we present steps of the Parks–McClellan algorithm for designing optimal filters. In this discussion, we will consider only type I filters. Algorithms for other types of filters can be obtained in the same manner. To derive the Parks–McClellan algorithm, the approximated error function in Eq. (15.49b) is expressed as follows:

$$G(\Omega) + \frac{\varepsilon(\Omega)}{W(\Omega)} \approx H_{\mathrm{d}}(\Omega). \qquad (15.50)$$

For type I filters, we obtain $G(\Omega)$ from Table 14.2 as follows:

$$G(\Omega) = h\left[\frac{N-1}{2}\right] + 2 \sum_{k=1}^{(N-1)/2} h\left[\frac{N-1}{2} - k\right] \cos(\Omega k).$$

Since we are interested in calculating $(N - 1)/2 + 1$, or $L + 1$, coefficients of $h[k]$ in $G(\Omega)$ and the value of the maximum error $\varepsilon_{\max}$, we pick $L + 2$

discrete frequencies $\{\Omega_1 < \Omega_2 < \cdots < \Omega_{L+2}\} \in S$, and solve Eq. (15.50) at the selected frequencies. Assuming that the selected frequencies are the extremal frequencies at which the maximum error changes between its peak value of $\pm\varepsilon_{max}$, Eq. (15.50) reduces to

$$G(\Omega_m) + \frac{1}{W(\Omega_m)}(-1)^m \varepsilon_{max} = H_d(\Omega_m). \tag{15.51}$$

for $1 \le m \le (L+2)$. The resulting set of $(L+2)$ simultaneous equations is as follows:

$$\underbrace{\begin{bmatrix} 1 & \cos(\Omega_1) & \cdots & \cos(L\Omega_1) & -1/W(\Omega_1) \\ 1 & \cos(\Omega_2) & \cdots & \cos(L\Omega_2) & -1/W(\Omega_2) \\ \vdots & \vdots & \ddots & \vdots & \vdots \\ 1 & \cos(\Omega_{L+1}) & \cdots & \cos(L\Omega_{L+1}) & (-1)^{L+1}/W(\Omega_{L+1}) \\ 1 & \cos(\Omega_{L+2}) & \cdots & \cos(L\Omega_{L+2}) & (-1)^{L+2}/W(\Omega_{L+2}) \end{bmatrix}}_{\Delta(\cos(k\Omega))} \begin{bmatrix} h\left[\frac{N-1}{2}\right] \\ 2h\left[\frac{N-1}{2}-1\right] \\ \vdots \\ 2h[0] \\ \varepsilon_{max} \end{bmatrix} = \begin{bmatrix} H_d(\Omega_1) \\ H_d(\Omega_2) \\ \vdots \\ H_d(\Omega_{L+1}) \\ H_d(\Omega_{L+2}) \end{bmatrix}.$$

$$\tag{15.52}$$

Once the extremal frequencies $\{\Omega_1 < \Omega_2 < \cdots < \Omega_{L+2}\}$ are known, Eq. (15.52) can be used to solve for the coefficients of the FIR filter. The extremal frequencies are computed using the Remez algorithm, which is based on Eq. (15.52) (though it does not solve the simultaneous equations explicitly) and consists of the following steps.

> Initialization: pick $\{\Omega_1 < \Omega_2 < \cdots < \Omega_{L+2}\} \in S$ evenly over the pass and stop bands.
> Given: transfer function $H_d(\Omega)$ of the ideal filter and the weighting function $W(\Omega)$.

Step 1 Solve Eq. (15.52) to calculate ε_{max}. To compute ε_{max}, we do not need to solve the complete set of simultaneous equations given in Eq. (15.52). Instead the following expression, obtained from Eq. (15.52) is solved:

$$\varepsilon_{max} = \frac{(-1)^{L+3}}{|\Delta(\cos(k\Omega))|} \begin{vmatrix} 1 & \cos(\Omega_1) & \cdots & \cos(L\Omega_1) & H_d(\Omega_1) \\ 1 & \cos(\Omega_2) & \cdots & \cos(L\Omega_2) & H_d(\Omega_2) \\ \vdots & \vdots & \ddots & \vdots & \vdots \\ 1 & \cos(\Omega_{L+1}) & \cdots & \cos(L\Omega_{L+1}) & H_d(\Omega_{L+1}) \\ 1 & \cos(\Omega_{L+2}) & \cdots & \cos(L\Omega_{L+2}) & H_d(\Omega_{L+2}) \end{vmatrix},$$

$$\tag{15.53}$$

where $|\Delta(\cdot)|$ denotes the determinant of the matrix $\Delta(\cdot)$.

Step 2 Substituting the value of ε_{max} determined in step 1, compute the values of $G(\Omega_m)$ at discrete frequencies $\{\Omega_1 < \Omega_2 < \cdots < \Omega_{L+2}\}$ using Eq. (15.51).

Step 3 Using the values of $G(\Omega_m)$ computed in step 2, sketch a line plot of $G(\Omega)$ as a function of Ω by interpolating intermediate values of $G(\Omega)$. Generally, $G(\Omega)$ is interpolated over a large grid of discrete frequencies within S.

Step 4 Using the line plot of $G(\Omega)$ obtained in step 3, sketch the line plot of $\varepsilon(\Omega)$ as a function of Ω using the following expression:

$$\varepsilon(\Omega) = W(\Omega)[H_{\mathrm{d}}(\Omega) - G(\Omega)],$$

derived from Eq. (15.50).

Step 5 Update the $L + 2$ extremal frequencies $\{\Omega_1 < \Omega_2 < \cdots < \Omega_{L+2}\} \in S$ by determining the $L + 2$ maxima and minima in $\varepsilon(\Omega)$ plotted in step 4.

Step 6 Check if the $L + 2$ maxima and minima observed in step 5 have the same value. If they do, then the alternation theorem is satisfied and the updated frequencies $\{\Omega_1 < \Omega_2 < \cdots < \Omega_{L+2}\}$ can be used to solve Eq. (15.52) for the filter coefficients. If not, then go back to step 1 and repeat steps 1–6.

The Parks–McClellan algorithm, highlighted in the aforementioned discussion, designs a lowpass FIR filter. Extension to other types of FIR filters is straightforward, provided that the required filter can be expressed in terms of a lowpass filter. Equation (15.24) illustrates how the design of a highpass filter can be transformed to the design of a lowpass filter. Similarly, Eqs. (15.32) and (15.38), respectively, provide transformations for bandpass and bandstop filters. Once the specifications of the required filter are expressed in terms of a lowpass filter, the impulse response of the optimal lowpass FIR filter is computed using the Parks–McClellan algorithm. The impulse response of the required FIR filter is then calculated from the impulse response of the optimal lowpass FIR filter.

15.6 MATLAB examples

The design algorithms covered in this chapter are incorporated as library functions in most signal processing software packages. In this section, we introduce several M-files available in MATLAB for the design of FIR filters. In particular, we cover `rectwin`, `bartlett`, `hann`, `hamming`, and `blackman` functions, which are used to implement the elementary windows covered in Section 15.1. In addition, we consider the `fir1` function to derive the impulse response of the FIR filter. The `kaiser` function used to design FIR filters using the Kaiser window and the `firpm` function used to implement the Parks–McClellan algorithm are also presented in this section. In each case, we write the MATLAB code for the design of the FIR filter specified in Example 15.2.

For convenience, the specifications of the lowpass filter in Example 15.2 are repeated below:

pass-band edge frequency $(\omega_p) = 3\pi$ kradians/s,

stop-band edge frequency $(\omega_s) = 4\pi$ kradians/s,

maximum allowable pass-band ripple $- 20 \log_{10}(\delta_p) = 25$ dB,

    i.e. $\delta_p = 0.0562$,

minimum stop-band attenuation $-20 \log_{10}(\delta_s) = 50$ dB,

    i.e. $\delta_s = 0.0032$,

sampling frequency $(f_0) = 8$ ksamples/s.

Example 15.9

Design the lowpass FIR filter considered in Example 15.2 using the rectangular, Bartlett, Hanning, Hamming, and Blackman windows. Sketch and compare the magnitude response of the resulting FIR filters.

Solution

As shown in Example 15.2, the values of the normalized cut-off frequency and the normalized transition bandwidth for the lowpass filter are given by $\Omega_n = 0.4375$ and $\Delta\Omega_n = 0.125$, respectively.

Since the minimum stop-band attenuation is 50 dB, only the Hamming and Blackman windows may be used for the filter design. The value of length N of the FIR filters for the two windows is given by

Hamming window $6.6/N = 0.1250 \Rightarrow N = 6.6/0.125 = 52.8$;

Blackman window $11/N = 0.1250 \Rightarrow N = 11/0.125 = 88$.

MATLAB provides the `fir1` function to derive the impulse response of the FIR filter. The syntax for the `fir1` function is given by

```
fir_coeff = fir1(order, norm_cut_off, type,window);
```

where the input argument `order` denotes the order of the FIR filter. For an FIR filter of length N, the order is given by $N - 1$. The input argument `norm_cut_off` specifies the normalized cut-off frequency of the FIR filter. Its value should lie between zero and one. The input argument `type` specifies the type of the FIR filter. Two possible choices for `type` are `'low'` for the lowpass FIR filter and `'high'` for the highpass FIR filter. Finally, the input argument `window` accepts coefficients $w[k]$ of the window type being used in the FIR filter design. Any of the elementary windows covered in Section

15.1 can be used by naming the appropriate window function. The syntaxes for various types of length-N window functions are as follows:

```
>> win_coeff = rectwin(N);   % rectangular window
>> win_coeff = bartlett(N);  % bartlett window
>> win_coeff = hann(N);      % hanning window
>> win_coeff = hamming(N);   % hamming window
>> win_coeff = blackman(N);  % blackman window
```

For Example 15.2, the MATLAB code for the design of the FIR filter using the Hamming window is given by

```
% lowpass filter design using Hamming window
>> wn = 0.4375;                    % Normalized cut-off
                                   % frequency
>> N = 53;                         % Hamming Window
>> h_hamm = fir1(N-1,wn,'low',hamming(N));
                                   % Impulse response of
                                   % the LPF
>> w = 0:0.001*pi:pi;              % discrete frequencies
                                   % for response
>> H_hamm = freqz(h_hamm,1,w);     % transfer function
>> plot(w,20*log10(abs(H_hamm)));  % magnitude response
>> axis([0 pi -120 20]);           % set axis
>> title('FIR filter using Hamming window')
>> grid on
```

The magnitude response of the FIR filter obtained with the Hamming window is shown in Fig. 15.19(a). Note that the magnitude response satisfies the filter specifications.

The MATLAB code for the design of the FIR filter using the Blackman window is similar, except for a few minor changes, which are shown below.

```
% lowpass filter design using Blackman window
>> wn = 0.4375;            % Normalized cut-off
                           % frequency
>> N = 88;                 % Blackman Window size
>> h_black = fir1(N-1,wn,'low',blackman(N));
                           % Impulse response of
                           % the LPF
>> w = 0:0.001*pi:pi;      % discrete frequencies
                           % for response
```

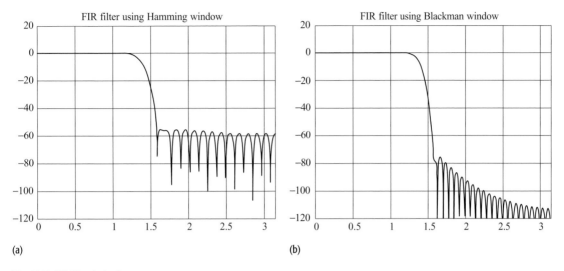

Fig. 15.19. FIR filter design for
Example 15.9 using MATLAB.
(a) Hamming window
(b) Blackman window.

```
>> H_black = freqz(h_black,1,w);    % transfer function
>> plot(w,20*log10(abs(H_black)));  % magnitude response
>> axis([0 pi -120 20]);            % set axis
>> title('FIR filter using Blackman window');
>> grid on
```

The magnitude response of the FIR filter obtained with the Blackman window is
shown in Fig. 15.19(b). On comparing with Fig. 15.19(a), we note that the stop-
band attenuation in Fig. 15.19(b) is higher. The improvement in the stop-band
attenuation is the result of the shape of the Blackman window.

Although the above example uses only the Hamming and Blackman windows,
any of the elementary windows covered in Section 15.1 can be used by specifying the appropriate window coefficients in the `fir1` function.

Example 15.10
Design the lowpass FIR filter considered in Example 15.3 using the Kaiser
window. Sketch and compare the magnitude response of the resulting FIR filter
with those of the FIR filters obtained in Example 15.3.

Solution
As shown in Example 15.3, the normalized cut-off frequency $\Omega_n = 0.4375$ and
the normalized transition bandwidth $\Delta\Omega_n = 0.1250$. The design parameters for
the Kaiser window were calculated as $\beta = 4.5513$ and $N = 47$.
    The MATLAB code for the design of the FIR filter using the Kaiser window
is similar to the MATLAB code in Example 15.9. The major difference is in the
`fir1` instruction, where the window argument is now replaced by the `kaiser`

Fig. 15.20. FIR filter design for
Example 15.10 with the Kaiser
window using MATLAB.

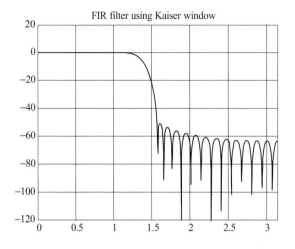

function. A Kaiser window of length N and shape parameter β can be generated
by the following instruction:

```
Win = kaiser(N, beta)
```

The MATLAB code is given below

```
% lowpass filter design using Kaiser window
>> wn = 0.4375;                    % Normalized Cutoff
                                   % frequency
>> N = 47;                         % Kaiser Window length
>> beta = 4.5513;                  % Kaiser Shape control
                                   % parameter
>> h_kaiser = fir1(N-1,wn,'low',kaiser(N,beta));
                                   % Impulse response of
                                   % the LPF
>> w = 0:0.001*pi:pi;              % discrete frequencies
                                   % for response
>> H_kaiser = freqz(h_kaiser,1,w); % transfer function
>> plot(w,20*log10(abs(H_kaiser))); % magnitude response
>> axis([0 pi -120 20]);           % set axis
>> title('FIR filter using Kaiser window');
>> grid on
```

The magnitude response of the FIR filter obtained with the Kaiser window is
shown in Fig. 15.20. Compared with Figs. 15.19(a) and (b), we note that the
minimum stop-band attenuation in Fig. 15.20 is exactly 50 dB. Being able to
provide the exact specified attenuation, the Kaiser window is able to reduce
the length of the lowpass FIR filter to 47. Among the three filters, the FIR

filter obtained from the Kaiser window is therefore the least expensive from the implementation perspective.

For the design of optimal filters, MATLAB has incorporated the firpm function, which has the following syntax:

```
fir_coefficients = firpm(order,range_norm_cut_off,
  f_response,wmatrix);
```

where the input argument order denotes the order of the FIR filter. The second input argument rang_norm_cut_off is a vector that specifies the edges of the normalized cut-off frequency of the FIR filter. All elements of this vector should have a value between zero and one. For a lowpass filter, the elements of the rang_norm_cut_off vector are given by

```
rang_norm_cut_off = [0, pass_band_cut_off, stop_band_cut_off,
  1];
```

The third input argument f_response specifies the four gains of the FIR filter at the four frequencies specified in the rang_norm_cut_off vector. For a lowpass filter, the value of the f_response vector is given by

```
f_response =[1, 1, 0, 0];
```

Finally, the fourth input argument wmatrix specifies the weight matrix. Since wmatrix has one entry per band, it is half the length of rang_norm_cut_off and f_response vectors.

Example 15.11 illustrates the design of an optimal FIR filter using the firpm function.

Example 15.11
Examples 15.9 and 15.10 designed an FIR filter using rectangular. Hamming, and Kaiser windows with a given set of design specifications. It was shown in Example 15.10 that an FIR filter of length 47, designed using a Kaiser window, satisfies the design specifications. Design the optimal FIR filter of length 47 using the Parks–McClellan algorithm and compare the magnitude frequency response with that of the FIR filter obtained using the Kaiser window.

Solution
The values of the normalized pass- and stop-band edge frequencies are given by

normalized pass-band edge frequency $\quad \Omega_p = (3\pi \times 10^3)/(0.5 \times 2\pi \times 8 \times 10^3)$
$$= 0.375;$$
normalized stop-band edge frequency $\quad \Omega_s = (4\pi \times 10^3)/(0.5 \times 2\pi \times 8 \times 10^3)$
$$= 0.5.$$

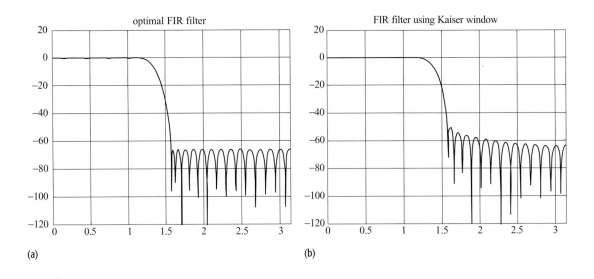

(a)   (b)

Fig. 15.21. Optimal FIR filter designed in Example 15.11 using MATLAB. (a) Optimal FIR filter of length $N = 47$. (b) FIR filter of length $N = 47$ using the Kaiser window.

The MATLAB code for the design of the optimal FIR filter is similar to the MATLAB code in Example 15.8, except for the use of the `firpm` function, which replaces the `fir1` function:

```
% optimal lowpass filter design using Parks-McClellan
% algorithm
>> sz = 47;                          % Length of FIR filter
>> range_norm_cut_off = [0, 0.375, 0.5, 1];
                                     % normalized cut-off
                                     % frequencies
>> f_response = [1, 1, 0, 0];        % gains at the cut-off
                                     % frequencies
>> wmatrix = [0.0032/0.0562, 1];     % weight matrix
>> h_optimal = firpm(sz-1, range_norm_cut_off, f_response,
   wmatrix);                         % Impulse response of
                                     % the optimal LPF
                                     % FIR filter
>> w = 0:0.001*pi:pi;                % discrete frequencies
>> H_optimal = freqz(h_optimal,1,w); % transfer function
>> plot(w,20*log10(abs(H_optimal))); % magnitude response
>> axis([0 pi -120 20]);             % set axis
>> title('optimal FIR filter');
>> grid on
```

The magnitude response of the optimal FIR filter obtained from the above code is shown in Fig. 15.21(a). Comparing Fig. 15.21(a) with the magnitude response of the FIR filter obtained from the Kaiser window shown in Fig. 15.21(b), the following differences are noted.

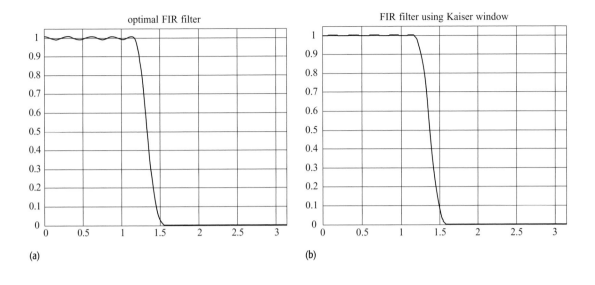

optimal FIR filter   FIR filter using Kaiser window

(a)   (b)

Fig. 15.22. Same as Fig. 15.21 except the frequency responses are plotted on a linear scale.

(1) The stop-band ripples in Fig. 15.21(a) have a uniform peak value of roughly 70 dB, which is about 20 dB less than the maximum stop-band ripple value in Fig. 15.21(b). The stop-band attenuation of the optimal FIR filter is therefore higher than that for the filter obtained from the Kaiser window.

(2) As illustrated in Fig. 15.22(a), where the magnitude response of the optimal FIR filter is plotted on a linear scale, there are noticeable pass-band ripples in the magnitude response of the optimal FIR filter. Figure 15.22(b) plots the magnitude response of the FIR filter obtained from the Kaiser window, where the pass-band ripples are negligible. The improvement in the stop-band attenuation of the optimal FIR filter can therefore be attributed to the pass-band ripples that the optimal filter has incorporated. The optimal FIR filter distributes the distortion between the pass and stop bands. The FIR filter obtained from the Kaiser window has most distortion concentrated in the stop band, which leads to higher ripples (or lesser attenuation) within its stop band.

(3) Finally, we observe that the transition bands in the two FIR filters are roughly of the same width.

15.7 Summary

This chapter presented techniques for designing causal FIR filters. The ideal frequency-selective filters presented in Chapter 14 are physically unrealizable because of strict constraints on the pass- and stop-band gains of the filter and also because of a sharp transition between the pass and stop bands. Practical implementations of the ideal filters are obtained by allowing acceptable variations (or ripples) within the pass and stop bands. In addition, a transition band is included between the pass and stop bands so that the filter gain can drop off smoothly.

Section 15.1 introduced the windowing approach used to design FIR filters from the ideal frequency-selective filters. The windowing approach truncates the impulse response $h[k]$ of an ideal filter, with a linear-phase component of $\exp(-jm\Omega)$, to a finite length N within the range $0 \leq k \leq (N-1)$. The value of m in the phase component is selected to be $(N-1)/2$ such that the filter coefficients in the causal FIR filter are symmetrical with respect to m. Common elementary windows used to design FIR filters are the rectangular, Bartlett, Hamming, Hanning, and Blackman windows. The selection of type of window depends upon the maximum value of the pass- and stop-band ripples. The length N of the window is determined from the allowable width of the transition band.

The minimum stop-band attenuation in the FIR filter obtained from the elementary windows is fixed. In most cases, the selected window surpasses the given specification on the stop-band attenuation and the resulting FIR filter is therefore of higher computational complexity than required. Section 15.2 introduced the Kaiser window, which provides control over the stop-band attenuation by including an additional design parameter, referred to as the shape control parameter β. The order of the FIR filter designed by the Kaiser window is significantly smaller than those of the FIR filters obtained using the elementary window functions.

The FIR design techniques covered in Sections 15.1 and 15.2 are applicable to all types of frequency-selective filters such as the lowpass, highpass, bandpass, and bandstop filters. Common convention, however, is to express the transfer functions of the highpass, bandpass, and bandstop filters in terms of the transfer function of the lowpass filter. Using the resulting relationships, the design of any type of filter can be reduced to the design of one or more lowpass filters. Section 15.3 covered design techniques for highpass FIR filters. We covered design algorithms using the original highpass filter specifications as well as techniques that transform the problem of designing a highpass FIR filter to designing a lowpass FIR filter. Similarly, Section 15.4 presented design techniques for bandpass FIR filters, while Section 15.5 designed bandstop FIR filters.

The windowing approaches produce a suboptimal design. Section 15.5 introduced a computational procedure based on the Parks–McClellan algorithm that exploits the inherent structure, expressed in Proposition 14.1 for the linear-phase FIR filters. The Parks–McClellan algorithm computes the best FIR filter of length N that minimizes the maximum absolute difference between the transfer function $H_{\mathrm{d}}(\Omega)$ of the ideal filter and the transfer function $H(\Omega)$ of the corresponding FIR filter. Mathematically, the Parks–McClellan algorithm solves the minimax optimization problem, which finds the set of filter coefficients that minimizes the maximum error between the desired frequency response and the actual frequency response. According to Proposition 14.1, the frequency response of a linear-phase filter can be expressed as a polynomial of $\cos(\Omega)$. It can also be shown that error $\varepsilon(\Omega)$ between the desired and actual frequency response is also a polynomial of $\cos(\Omega)$. The Parks–McClellan algorithm uses the alternation theorem, which provides the following condition for the optimal design of $H(\Omega)$.

The transfer function $H(\Omega)$, which best approximates $H_d(\Omega)$ in the minimax sense, produces an error function $\varepsilon(\Omega)$ with at least $L + 2$ discrete extremal frequencies $\{\Omega_1 < \Omega_2 < \cdots < \Omega_{L+2}\} \in S$ in $\varepsilon(\Omega)$ that alternate between the maximum and minimum peak values of the error, i.e. $\varepsilon(\Omega_{m+1}) = -\varepsilon(\Omega_m) = \varepsilon_{max}$, where ε_{max} is the maximum value of error $|\varepsilon(\Omega)|$.

The Parks–McClellan algorithm is available as a library function in most signal processing packages. Section 15.7 covered the `firpm` function used to design optimal FIR filters in MATLAB using the Parks–McClellan algorithm. In addition, we introduced other library functions including `rectwin`, `bartlett`, `hann`, `hamming`, `blackman`, and `kaiser` functions used to implement the elementary windows covered in Sections 15.1 and 15.2. The `fir1` function used to derive the impulse response of an FIR filter is also covered.

Problems

15.1 The ideal DT differentiator is commonly used to differentiate a CT signal directly from its samples. The transfer function of a DT differentiator is given by

$$H_{diff}(\Omega) = j\Omega e^{-jm\Omega} \quad 0 \le |\Omega| \le \pi.$$

Determine the impulse response $h_{diff}[k]$ of the ideal differentiator.

15.2 A system with the block schematic shown in Fig. 9.1 is used to process a CT signal with a digital filter. The A/D converter has a sampling rate of 8000 samples/s. Design the ideal digital filter if the overall transfer function of Fig. 9.1 represents an ideal lowpass filter with a cut-off frequency of 2 kHz. Repeat for the sampling rates of 16 000 samples/s and 44 100 samples/s.

15.3 Calculate the amplitude of the 5-tap ($N = 5$) rectangular, Hanning, Hamming, and Blackman windows. Sketch the window functions.

15.4 The specifications for a lowpass filter are given as follows:

$$\text{pass-band edge frequency} = 0.25\pi;$$
$$\text{stop-band edge frequency} = 0.55\pi;$$
$$\text{minimum stop-band attenuation} = 35 \text{ dB}.$$

Determine which of the elementary windows mentioned in Table 15.2 would satisfy these specifications. For the permissible choices, determine the lengths N of the windows that meet the width requirement for the transition band.

15.5 Repeat Problem 15.4 for the Kaiser window.

15.6 Determine the impulse response of an ideal discrete-time lowpass filter with a cut-off frequency of $\Omega_c = 1$ radian/s. Using a rectangular window, truncate the length N of the ideal filter to 51. Plot the impulse response and amplitude frequency characteristics of the FIR filter.

15.7 Repeat Problem 15.6 for the Hamming window and compare the resulting FIR filter with the FIR filter obtained from the rectangular window in that problem.

15.8 Design the digital FIR filter, shown as the central block and labeled as the DT system in Fig. 9.1, if the specifications of the overall system are given as follows (the overall CT system is a lowpass filter):

$$\text{pass-band edge frequency} = 10.025\,\text{kHz};$$
$$\text{width of the transition band} = 1\,\text{kHz};$$
$$\text{minimum stop-band attenuation} = 45\,\text{dB};$$
$$\text{sampling rate} = 44.1\,\text{ksamples/s}.$$

(a) Determine the possible types of windows that may be used.
(b) Assuming that the Hamming window is used to design the FIR filter, plot the impulse response $h[k]$ of the resulting FIR filter.
(c) Plot the amplitude–frequency characteristics of the FIR filter on both absolute and logarithmic scales.

15.9 Repeat Problem 15.8 for a Kaiser window.

15.10 Using the Kaiser window, design a highpass FIR filter based on the following specifications:

$$\text{pass-band edge frequency} = 0.64\pi;$$
$$\text{width of the transition band} = 0.3\pi;$$
$$\text{maximum pass-band ripple} < 0.002;$$
$$\text{maximum stop-band ripple} < 0.005.$$

Use MATLAB to confirm that the designed FIR filter satisfies the given specifications:

15.11 Using the Kaiser window, design a bandpass FIR filter based on the following specifications:

$$\text{pass-band edge frequencies} = 0.4\pi \text{ and } 0.6\pi;$$

stop-band edge frequencies $= 0.2\pi$ and 0.8π;

maximum pass-band ripple <0.02;

maximum stop-band ripple <0.009.

Use MATLAB to confirm that the designed bandpass FIR filter satisfies the given specifications.

15.12 Using the Kaiser window, design a bandstop FIR filter based on the following specifications:

stop-band edge frequencies $= 0.3\pi$ and 0.7π;

pass-band edge frequencies $= 0.4\pi$ and 0.6π;

maximum pass-band ripple <0.05;

maximum stop-band ripple <0.05.

Use MATLAB to confirm that the designed bandstop FIR filter satisfies the given specifications.

15.13 Equation (15.44) defines the expression for the normalized weighting function used in the design of a lowpass filter using the Parks–McClellan algorithm. Derive the expressions for the normalized weighting functions for highpass, bandpass, and bandstop filters.

15.14 For a type I FIR filter of length N, show that the degree L of the error function $\varepsilon(\Omega)$ defined in Eq. (15.42) is given by $(N-1)/2$.

15.15 Repeat Problem 15.14 for a type II FIR filter of length N by showing that the degree L of the error function $\varepsilon(\Omega) = \Gamma(\cos(\Omega))$ defined in Eq. (15.42) is given by $(N-2)/2$.

15.16 Repeat Problem 15.14 for a type III FIR filter of length N by showing that the degree L of the error function $\varepsilon(\Omega)$ defined in Eq. (15.42) is given by $(N-3)/2$.

15.17 Repeat Problem 15.14 for a type IV FIR filter of length N by showing that the degree L of the error function $\varepsilon(\Omega)$ defined in Eq. (15.42) is given by $(N-2)/2$.

15.18 Truncate the impulse response of an ideal bandstop FIR filter with edge frequencies of 0.25π and 0.75π with a 20-tap rectangular window filter. Plot the magnitude response of the resulting FIR filter and compare the frequency characteristics with a 40-tap FIR filter.

15.19 Using MATLAB, determine the impulse response of the FIR filters designed in Problems 15.4 and 15.5. Sketch the magnitude response and ensure that the FIR filters satisfy the given specifications. Comment on the complexity and frequency characteristics of the designed filters.

15.20 Using MATLAB, determine the impulse response of the optimal FIR filter for the specifications provided in Problem 15.4. You may use the Kaiser window to determine the length of the optimal FIR filter. Sketch the magnitude response of the optimal FIR filter and compare its frequency characteristics with those of the FIR filters plotted in Problem 15.18.

15.21 Show that the alternation theorem is satisfied for the magnitude response of the optimal FIR filter designed in Problem 15.20.

15.22 Using the `fir1` function in MATLAB, design a 41-tap lowpass filter with a normalized cut-off frequency of $\Omega_n = 0.55$ using (i) rectangular; (ii) Hamming; (iii) Blackman; and (iv) Kaiser (with $\beta = 4$) windows. Plot the amplitude–frequency characteristics for the four filters. For each plot, determine (i) the maximum pass-band ripple; (ii) the peak side lobe gain; and (iii) the transition bandwidth. Assume that the transition band is a band where the filter gain drops from –2 dB to –20 dB.

15.23 Using the `fir1` function in MATLAB, design a 45-tap linear-phase bandpass FIR filter with pass-band edge frequencies of 0.45π and 0.65π, stop-band edge frequencies of 0.15π and 0.9π, maximum pass-band attenuation of 0.1 dB, and minimum stop-band attenuation of 40 dB. Use the Kaiser window for your design and sketch the frequency characteristics of the resulting filter.

15.24 The `fir2` function in MATLAB is used to design FIR filters with arbitrary frequency characteristics. Using `fir2`, design a 95-tap FIR filter with the following frequency characteristics:

$$|H(\Omega)| = \begin{cases} 0.85 & 0 \le |\Omega_n| \le 0.15 \\ 0.55 & 0.20 \le |\Omega_n| \le 0.45 \\ 1 & 0.55 \le |\Omega_n| \le 0.75 \\ 0.5 & 0.78 \le |\Omega_n| \le 1, \end{cases}$$

where Ω_n is the normalized DT frequency. Use MATLAB to confirm that the designed FIR filter satisfies the given specifications.

16   IIR filter design

Based on the length of the impulse response $h[k]$, Chapter 14 classified digital (or "discrete-time") filters into two categories: finite impulse response (FIR) filters and infinite impulse response (IIR) filters. The design techniques for the FIR filter, with an impulse response $h[k]$ of finite length, were covered in Chapter 15. In this chapter, we present design methodologies for the IIR filters. A common technique used to design IIR filters is based on mapping the DT frequency specifications $H(\Omega)$ of the IIR filters in the Ω domain to the CT frequency specifications $H(\omega)$ specified in the ω domain. Based on the transformed specifications, a CT filter is designed, which is then transformed back to the original DT frequency Ω domain to obtain the transfer function of the required IIR filter. In this chapter, we present two different DT to CT frequency transformations. The first method is referred to as the impulse invariance transformation, which provides a linear transformation between the DT and CT frequency domains. At times, the impulse invariance transformation suffers from aliasing, which may lead to deviations from the original DT specifications. An alternative to the impulse invariance transformation is the bilinear transformation, which is a non-linear mapping between the CT and DT frequency domains. The bilinear transformation eliminates aliasing to a large extent.

A classical problem in the design of digital filters is the selection between FIR and IIR filters. While both types of filters can be used to satisfy a given set of specifications, the order N of IIR filters is in general much lower than that of FIR filters. As a consequence of the lower order N, the IIR filters have reduced implementation complexity and less propagation delay when compared with FIR filters designed for the same specifications. However, IIR filters are implemented using feedback loops, resulting in transfer functions with a significant number of poles. IIR filters are, therefore, susceptible to instability issues when realized on finite-precision DSP boards. In addition, IIR filters have a non-linear phase, whereas FIR filters can be designed with a linear phase. An appropriate digital filter type is selected based on the requirement of a given application.

The organization of this chapter is as follows. IIR filter design principles are introduced in Section 16.1. Sections 16.2 and 16.3 present design principles of lowpass IIR filters based on the frequency transformation methods. In Section 16.2, we introduce the impulse invariance transformation, and in Section 16.3 we present the bilinear transformation. The analytical design procedure is illustrated through a series of examples. We also provide the MATLAB code, which can also be used in the design of IIR filters. Section 16.4 covers the design techniques for bandpass, bandstop, and highpass filters. Finally, Section 16.5 compares the frequency characteristics of IIR filters with those of FIR filters designed for the same specifications. Section 16.6 presents a summary of the important concepts covered in the chapter.

16.1 IIR filter design principles

As specified in Chapter 14, the transfer function of an IIR filter is given by

$$H(z) = \frac{b_0 + b_1 z^{-1} + \cdots + b_M z^{-M}}{1 + a_1 z^{-1} + \cdots + a_N z^{-N}}, \tag{16.1}$$

where b_r, for $0 \leq r \leq M$, and a_r, for $0 \leq r \leq N$, are known as the filter coefficients. In Eq. (16.1), we have also normalized the coefficient a_0 (corresponding to $r = 0$) in the denominator to unity. Based on Eq. (16.1), the IIR filter can alternatively be modeled by the following linear, constant-coefficient difference equation:

$$y[k] + a_1 y[k - 1] + \cdots + a_N y[k - N] = b_0 x[k] + b_1 x[k - 1]$$
$$+ \cdots + b_M x[k - M]. \tag{16.2}$$

The objective of the IIR filter design is to calculate a set of filter coefficients b_r, for $0 \leq r \leq M$, and a_r, for $1 \leq r \leq N$, such that the frequency characteristics of the IIR filter match the design specifications. IIR filter design can, therefore, be viewed as a mathematical optimization problem.

A popular method used to design an IIR filter is based on converting its desired frequency specifications $H(\Omega)$ into the CT frequency domain. Using the CT design techniques for the Butterworth, Chebyshev, or elliptic filters covered in Chapter 6, the transfer function $H(s)$ of the analog filter is determined. The z-transfer function $H(z)$ of the desired IIR filter is then obtained by transforming $H(s)$ back into the DT domain. Such transformation approaches yield closed-form transfer functions for the IIR filters.

A number of transformations have been proposed to convert the transfer function $H(s)$ of the CT (or analog) filter into the z-transfer function $H(z)$ of the IIR filter such that the frequency characteristics of the CT filter in the s-plane are preserved for the IIR filter in the z-plane. These transformations include the following methods:

(a) finite-difference discretization of differential equations;
(b) mapping poles and zeros from the s-plane to the z-plane;
(c) impulse invariance method;
(d) bilinear transformation.

The finite-difference discretization of differential equations is a straightforward method to derive difference equation representations for digital filters. First, the s-transfer functions, obtained by using the CT filter design techniques, are used to calculate the input–output relationship of the equivalent CT filter. These relationships are generally in the form of linear, constant-coefficient differential equations, and are discretized to obtain difference equations that represent the input–output relationships of the designed DT filters.

In the second method, referred to as the matched z-transform technique, the s-plane poles and zeros of a designed CT filter are mapped to the z-plane. The s-plane poles and zeros are then used to derive the transfer function $H(z)$ of the digital IIR filter.

The impulse invariance method samples the impulse response $h(t)$ of an LTIC system to derive the impulse response $h[k]$ of the corresponding LTID system. Finally, the bilinear transformation provides a one-to-one, non-linear mapping from the s-plane to the z-plane. The impulse invariance and bilinear transformations are the focus of this chapter. In Section 16.2, we cover the impulse invariance method followed by the bilinear transformation, in Section 16.3.

16.2 Impulse invariance

To derive the impulse invariance transformation, we approximate the impulse response $h(t)$ of a CT filter with its sampled representation,

$$h(t) \approx \sum_{n=-\infty}^{\infty} h(t)\delta(t - nT) = \sum_{n=-\infty}^{\infty} h(nT)\delta(t - nT), \qquad (16.3)$$

obtained by sampling $h(t)$ with an impulse train $\sum \delta(t - nT)$. Clearly, the approximation in Eq. (16.3) improves as the sampling interval $T \to 0$. The DT impulse response $h[k]$ of the equivalent IIR filter is obtained from the samples $h(kT)$ and is given by

$$h[k] = h(kT) = \sum_{n=-\infty}^{\infty} h(nT)\delta(k - n). \qquad (16.4)$$

Comparing the expressions for the Laplace transform of Eq. (16.3) given by

Laplace transform $\qquad H(s) = \sum_{n=-\infty}^{\infty} h(nT)\mathrm{e}^{-nTs} \qquad (16.5)$

Fig. 16.1. Impulse invariance
transformation from the s-plane
(a) to the z-plane (b).

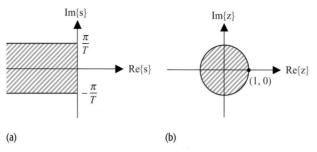

(a)   (b)

and the z-transform of Eq. (16.4) given by

z-transform $$H(z) = \sum_{k=-\infty}^{\infty} h(nT)z^{-n},$$ (16.6)

we note that the two expressions are equal provided

$$z = e^{Ts}.$$ (16.7)

In terms of real and imaginary components of $s = \sigma + j\omega$, Eq. (16.7) can be expressed as follows:

$$z = e^{\sigma T} e^{j\omega T}.$$ (16.8)

Equation (16.7) provides a mapping between the DT variable z and the CT variable s. The mapping, commonly referred to as the impulse invariance transformation, is illustrated in Fig. 16.1, where we observe that the s-plane region

$$\text{Re}\{s\} = \sigma < 0 \quad \text{and} \quad |\text{Im}\{s\}| = |\omega| < \pi/T,$$

shown as the shaded region in Fig. 16.1(a) maps into the interior of the unit circle $|z| < 1$ shown in Fig 16.1(b). Equations (16.7) and (16.8) can also be used to derive the following observations.

Right-half s-plane Re$\{s\} > 0$ Taking the absolute value of Eq. (16.8) yields

$$|z| = |e^{\sigma T}| \cdot |e^{j\omega T}| = |e^{\sigma T}|.$$ (16.9)

In the right-half s-plane, $\text{Re}\{s\} = \sigma > 0$, resulting in $|z| > 1$. Therefore, the right-half s-plane is mapped to the exterior of the unit circle.

Origin $s = 0$ Substituting $s = 0$ into Eq. (16.7) yields $z = 1$. The origin $s = 0$ in the s-plane is therefore mapped to the coordinate $(1, 0)$ in the z-plane.

Imaginary axis Re$\{s\} = 0$ Taking the absolute value of Eq. (16.8) and substituting $\text{Re}\{s\} = \sigma = 0$ yields $|z| = 1$. The imaginary axis $\text{Re}\{s\} = 0$ is therefore mapped on to the unit circle $|z| = 1$.

Left-half s-plane Re{s} < 0 Substituting $\text{Re}\{s\} = \sigma < 0$ in Eq. (16.9) yields $|z| < 1$. Therefore, we observe that the left-half s-plane is mapped to the interior of the unit circle. We now show that the mapping $z = e^{sT}$ is not a unique one-to-one mapping and that different strips of width $2\pi/T$ are mapped into the same region within the unit circle $|z| < 1$.

Consider the set of points $s = \sigma_0 + j2k\pi/T$, with $k = 0, \pm 1, \pm 2, \ldots$, in the s-plane. Substituting $s = \sigma_0 + j2k\pi/T$ in Eq. (16.7) yields

$$z = e^{T(\sigma_0 + j2k\pi/T)} = e^{\sigma_0 T} e^{j2k\pi} = e^{\sigma_0 T}. \qquad (16.10)$$

In other words, the set of points $s = \sigma_0 + j2k\pi/T$ are all mapped to the same point $z = \exp(\sigma_0 T)$ in the z-plane. Equation (16.8) is, therefore, not a unique one-to-one mapping, and different strips of width $2\pi/T$ in the left-half s-plane are mapped to the same region within the interior of the unit circle.

We now illustrate the procedure used to obtain an equivalent $H(z)$ from an impulse response $h(t)$ through Examples 16.1 and 16.2.

Example 16.1

Use the impulse invariance method to convert the s-transfer function

$$H(s) = \frac{1}{s + \alpha}$$

into the z-transfer function of an equivalent LTID system.

Solution

Calculating the inverse Laplace transform of $H(s)$ yields

$$h(t) = e^{-\alpha t} u(t).$$

Using impulse train sampling with a sampling interval of T, the impulse response of the LTID system is given by

$$h(kT) = e^{-\alpha kT} u(kT) \quad \text{or} \quad h[k] = e^{-\alpha kT} u[k].$$

The z-transform function of the equivalent LTID system is given by

$$H(z) = z\{h[k]\} = \frac{1}{1 - e^{-\alpha T} z^{-1}}, \quad \text{ROC} : |z| > e^{-\alpha T}.$$

Figure 16.2 compares the impulse response $h(t)$ and transfer function $H(s)$ of the LTIC system with the impulse response $h[k]$ and transfer function $H(z)$ of the equivalent LTID system obtained using the impulse invariance method. A sampling period of $T = 0.1$ s and $\alpha = 0.5$ are used. Comparing the CT impulse response $h(t)$, plotted in Fig. 16.2(a), with the DT impulse response $h[k]$, plotted in Fig. 16.2(c), we observe that $h[k]$ is a sampled version of $h(t)$, and the shapes of the impulse responses are fairly similar to each other.

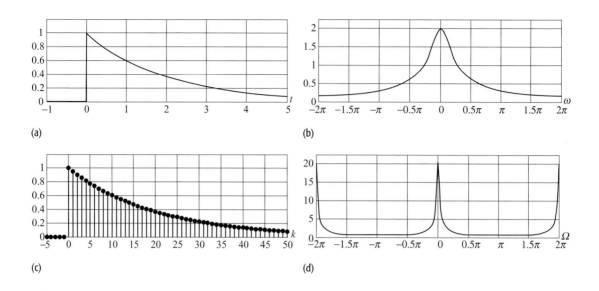

(a)

(b)

(c)

(d)

Fig. 16.2. Impulse invariance method used for transforming analog filters to digital filters in Example 16.1. (a) Impulse response $h(t)$ and (b) magnitude spectrum $H(\omega)$ of the analog filter. (c) Impulse response $h[k]$ and (d) magnitude spectrum $H(\Omega)$ of the transformed digital filter.

Comparing the magnitude spectrum $|H(\omega)|$ of the LTIC system with the magnitude spectrum $|H(\Omega)|$ of the LTID system plotted in Figs. 16.2(b) and (d), respectively, we observe two major differences. First, the magnitude spectrum $|H(\Omega)|$ is periodic with a period of 2π. Secondly, the magnitude spectrum $|H(\Omega)|$ is scaled by a factor of $1/T$ in comparison with $|H(\omega)|$. In order to obtain a DT filter with a DC amplitude gain of the same value as that of the CT filter, we multiply the sampled impulse response $h[k]$ by a factor of T:

$$h[k] = Th(kT) = Te^{-\alpha kT}u(k). \tag{16.11}$$

Alternatively, the following transform pair can be used for the impulse invariance transformation:

$$\frac{1}{s+\alpha} \xrightarrow{\text{impulse invariance}} \frac{T}{1-e^{-\alpha T}z^{-1}} \quad \text{or} \quad \frac{zT}{z-e^{-\alpha T}}. \tag{16.12}$$

Example 16.2 illustrates the application of Eq. (16.12) in transforming a Butterworth lowpass filter into a digital lowpass filter.

Example 16.2

Consider the following Butterworth filter:

$$H(s) = \frac{81.6475}{s^2 + 12.7786s + 81.6475}.$$

Use the impulse invariance transformation to derive the transfer function of the equivalent digital filter.

Solution

Expressing the transfer function of the CT filter as follows:

$$H(s) = 12.7786 \frac{6.3894}{(s + 6.3893)^2 + 6.3894^2}, \qquad (16.13)$$

and calculating the inverse Laplace transform, the impulse response of the CT filter is given by

$$h(t) = 12.7786 e^{-6.3893t} \sin(6.3894t) u(t). \qquad (16.14)$$

Using Eq. (16.11) to derive the impulse response of the DT filter, we obtain

$$h[k] = Th(kT) = 12.7786T \, e^{-6.3893kT} \sin(6.3894kT) u(kT) \qquad (16.15)$$

or

$$h[k] = 12.7786T \, e^{-6.3893\,kT} \sin(6.3894kT) u[k]. \qquad (16.16)$$

Calculating the z-transform of Eq. (16.16), the transfer function of the DT filter is given by (see Problem 16.2)

$$H(z) = \frac{12.7786T \, e^{-6.3893\,T} \sin(6.3894T) z}{z^2 - 2z \, e^{-6.3893T} \cos(6.3894T) z + e^{-2 \times 6.3893T}}. \qquad (16.17)$$

Alternative solution Equation (16.17) can also be derived by using the impulse invariance transformation specified in Eq. (16.12). Using partial fraction expansion, $H(s)$ is expressed as follows:

$$H(s) = \frac{12.7786}{2j} \left[\frac{1}{s + 6.3893 - j6.3894} - \frac{1}{s + 6.3893 + j6.3894} \right], \qquad (16.18)$$

which is then transformed to the z-domain using Eq. (16.12)

$$H(z) = \frac{12.7786}{2j} \left[\frac{zT}{z - e^{-(6.3893 - j6.3894)T}} - \frac{zT}{z - e^{-(6.3893 + j6.3894)T}} \right].$$

It is straightforward to show that the above expression reduces to Eq. (16.17).

Selection of the sampling interval To choose an appropriate sampling interval T, we need to analyze the magnitude spectrum of $h(t)$. Substituting $s = j\omega$ in Eq. (16.13), we obtain

$$H(\omega) = 12.7786 \frac{6.3894}{(j\omega + 6.3893)^2 + 6.3894^2} = \frac{81.6489}{(81.6489 - \omega^2) + j12.7788\omega}, \qquad (16.19)$$

which leads to the following magnitude spectrum:

$$|H(\omega)| = \frac{81.6489}{\sqrt{(81.6489 - \omega^2)^2 + 163.2977\omega^2}}. \qquad (16.20)$$

Table 16.1. DT filters obtained in Example 16.2 for different values of the sampling interval T

The magnitude spectra of these transfer functions are plotted in Figs. 16.3(b)–(e)

T	$H(z)$
0.1	$H(z) = \dfrac{0.4023z}{z^2 - 0.8475z + 0.2786}$
0.0348	$H(z) = \dfrac{0.0785z}{z^2 - 1.5619z + 0.6410}$
0.01	$H(z) = \dfrac{0.0077z}{z^2 - 1.8724z + 0.8800}$
0.001	$H(z) = \dfrac{8.113 \times 10^{-5}z}{z^2 - 1.9872z + 0.9873}$

The peak value of the magnitude spectrum $|H(\omega)|$ occurs at $\omega = 0$, with a value $|H(0)| = 1$. Also, the magnitude spectrum $|H(\omega)|$ is a monotonically decreasing function with respect to ω. Assuming that the maximum frequency present in the function $h(t)$ is approximated as ω_{max} such that $|H(\omega)| \leq 0.01$ for $|\omega| \geq \omega_{max}$, it can be shown that $\omega_{max} = 90.4$ radians/s. Using the Nyquist sampling rate, the sampling interval is therefore given by

$$T \leq \frac{1}{2f_{max}} = \frac{2\pi}{2\omega_{max}} = 0.348 \text{ s.}$$

Table 16.1 compares the transfer function of the transformed DT filters obtained by substituting different values of the sampling intervals T into Eq. (16.17). The transfer functions of the DT filters for different values of $T = 0.1, 0.0348, 0.01$, and 0.001 are given in Table 16.1. A comparison of the magnitude spectra of the four transfer functions is illustrated in Fig. 16.3. We make the following observations.

(1) Although the shapes of the magnitude spectra (Figs. 16.3(b)–(d)) of the digital filters appear to be different, they are all valid representations of the magnitude spectrum of the analog filter (Fig. 16.3(a)). Substituting $s = j\omega$ and $z = e^{j\Omega}$ into Eq. (16.7) yields

$$\Omega = \omega T.$$

The 3 dB frequency Ω_0 of the digital implementations therefore depends upon the sampling interval T. Based on the 3 dB frequency $\omega_0 = 9.03$ radians/s, the values of Ω_0 are given by 0.2874π radians/s for $T = 0.1$ s, by 0.1π radians/s for $T = 0.0348$ s, by 0.0287π radians/s for $T = 0.01$ s, and by 0.0029π radians/s for $T = 0.001$ s.

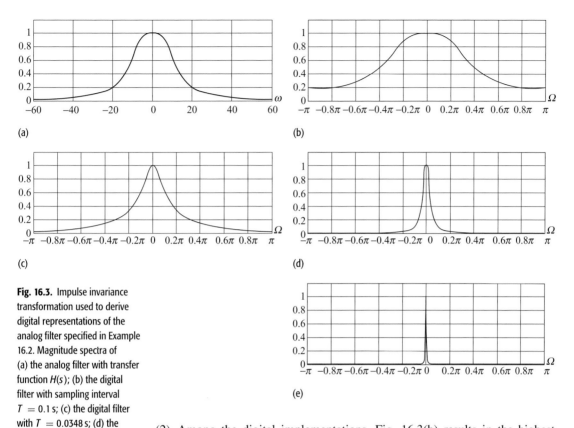

Fig. 16.3. Impulse invariance transformation used to derive digital representations of the analog filter specified in Example 16.2. Magnitude spectra of (a) the analog filter with transfer function $H(s)$; (b) the digital filter with sampling interval $T = 0.1$ s; (c) the digital filter with $T = 0.0348$ s; (d) the digital filter with $T = 0.01$ s; (e) the digital filter with $T = 0.001$ s.

(2) Among the digital implementations, Fig. 16.3(b) results in the highest gain (i.e. lowest attenuation) at the stop-band frequency $\Omega = \pm\pi$ radians/s. Since the sampling interval ($T = 0.1$ s) is greater than the Nyquist bound ($T = 0.0348$ s), Fig. 16.3(b) suffers from aliasing, which increases the gain within the stop band. In using impulse invariance transformation, it is critical that the effects of the aliasing be considered within the stop band.

16.2.1 Impulse invariance transformation using MATLAB

MATLAB provides a library function impinvar to transform CT transfer functions into the DT domain using the impulse variance method. We illustrate the application of impinvar for Example 16.2 with the sampling interval T set to 0.1 s. The MATLAB code for the transformation is as follows:

```
>> num = [0 0 81.6475];        % numerator of CT filter
>> den = [1 12.7786 81.6475]; % denominator of CT filter
>> T = 0.1;
>> Fs = 1/T;                   % sampling rate
>> [numz,denz] = impinvar (num,den,Fs);
                               % numerator & denominator
                               % of DT filter
```

The above MATLAB code results in the following values for the coefficients of $H(z)$:

```
numz = [0 0.4023 0] and denumz = [1 -0.8475 0.2786],
```

which correspond to the following transfer function:

$$H(z) = \frac{0.4023z}{z^2 - 0.8475z + 0.2786}.$$

The above expression is the same as the one obtained analytically, and it is included in row 1 of Table 16.1.

16.2.2 Look-up table

Examples 16.1 and 16.2 present direct methods to compute the impulse response $h[k]$, or correspondingly the transfer function $H(z)$, of the DT filter by sampling the impulse response $h(t)$ of an analog filter. The process can be simplified further in cases where the transfer function $H(s)$ of the analog filter is a rational function. In such cases, the transfer function $H(s)$ can be expressed in terms of partial fractions as follows:

$$H(s) = \sum_{r=1}^{N} \frac{k_r}{s + \alpha_r}, \tag{16.21}$$

where k_r is the coefficient of the rth partial fraction. Applying the impulse invariance transformation, Eq. (16.12), the transfer function $H(z)$ of the digital filter is given by

$$H(z) = \sum_{r=1}^{N} \frac{k_r z}{z - e^{-\alpha_r T}}. \tag{16.22}$$

Table 16.2 lists a number of commonly occurring s-domain terms and the equivalent representation in the z-domain. We now list the steps involved in the design of digital IIR filters using the impulse invariance transformation.

16.2.3 IIR filter design using impulse invariance transformation

The steps involved in designing IIR filters using the impulse invariance transformation are as follows.

Step 1 Using $\Omega = \omega T$, transform the specifications of the digital filter from the DT frequency Ω domain to the CT frequency ω domain. For convenience, we choose $T = 1$.

Step 2 Using the analog filter techniques (see Chapter 7), design an analog filter $H(s)$ based on the transformed specifications obtained in step 1.

Step 3 Using the impulse invariance transformation specified in Eq. (16.12),

$$\frac{1}{s + \alpha} \xleftarrow{\text{impulse invariance}} \frac{T}{1 - e^{-\alpha T} z^{-1}} \quad \text{or} \quad \frac{zT}{z - e^{-\alpha T}}$$

Table 16.2. Analog to-digital transformation using impulse invariance method

CT domain		DT domain	
$H(s)$	$h(t)$	$h[k]$	$H(z)$
1	$\delta(t)$	$T\delta[k]$	T
$\dfrac{1}{s}$	$u(t)$	$Tu[k]$	$\dfrac{T}{1-z^{-1}} = \dfrac{Tz}{z-1}$
$\dfrac{1}{s^2}$	$tu(t)$	$kT^2u[k]$	$\dfrac{T^2z^{-1}}{(1-z^{-1})^2} = \dfrac{T^2z}{(z-1)^2}$
$\dfrac{1}{s+\alpha}$	$e^{-\alpha t}u(t)$	$Te^{-\alpha kT}u[k]$	$\dfrac{T}{(1-e^{-\alpha T}z^{-1})} = \dfrac{Tz}{(z-e^{-\alpha T})}$
$\dfrac{1}{(s+\alpha)^2}$	$te^{-\alpha t}u(t)$	$kT^2e^{-\alpha kT}u[k]$	$\dfrac{T^2e^{-\alpha T}z^{-1}}{(1-e^{-\alpha T}z^{-1})^2} = \dfrac{T^2e^{-\alpha T}z}{(z-e^{-\alpha T})^2}$
$\dfrac{s+\alpha}{(s+\alpha)^2+\beta^2}$	$e^{-\alpha t}\cos(\beta t)u(t)$	$Te^{-\alpha kT}\cos(\beta kT)u[k]$	$\dfrac{Tz[z-e^{-\alpha T}\cos(\beta T)]}{z^2-2e^{-\alpha T}\cos(\beta T)z+e^{-2\alpha T}}$
$\dfrac{\beta}{(s+\alpha)^2+\beta^2}$	$e^{-\alpha t}\sin(\beta t)u(t)$	$Te^{-\alpha kT}\sin(\beta kT)u[k]$	$\dfrac{Tze^{-\alpha T}\sin(\beta T)}{z^2-2e^{-\alpha T}\cos(\beta T)z+e^{-2\alpha T}}$

or the look-up table approach, derive the z-transfer function $H(z)$ from the s-transfer function $H(s)$.

Step 4 Confirm that the z-transfer function $H(z)$ obtained in step 3 satisfies the design specifications by plotting the magnitude spectrum $|H(\Omega)|$. If the design specifications are not satisfied, increase the order N of the analog filter designed in step 2 and repeat steps 2–4.

We now illustrate the application of the above algorithm in Example 16.3.

Example 16.3
Design a lowpass IIR filter with the following specifications:

pass band $(0 \leq |\Omega| \leq 0.25\pi$ radians/s) $\qquad 0.8 \leq |H(\Omega)| \leq 1;$
stop band $(0.75\pi \leq |\Omega| \leq \pi$ radians/s) $\qquad |H(\Omega)| \leq 0.20.$

Solution
Choosing the sampling interval $T = 1$, step 1 transforms the given specifications of the DT filter into the corresponding specifications for the CT filter:

pass band $(0 \leq |\omega| \leq 0.25\pi$ radians/s) $\qquad 0.8 \leq |H(\omega)| \leq 1;$
stop band $(|\omega| > 0.75\pi$ radians/s) $\qquad |H(\omega)| \leq 0.20.$

Step 2 designs the analog filter based on the transformed specifications. We use the Butterworth filter, whose design procedure is outlined in Section 7.3.1.1.

Design of the analog Butterworth filter To determine the order N of the filter, we calculate the gain terms:

$$G_p = \frac{1}{(1 - \delta_p)^2} - 1 = 0.5625$$

and

$$G_s = \frac{1}{(\delta_s)^2} - 1 = 24.$$

The order N of the filter is therefore given by

$$N = \frac{1}{2} \times \frac{\ln(G_p/G_s)}{\ln(\omega_p/\omega_s)} = \frac{1}{2} \times \frac{\ln(0.5625/24)}{\ln(0.25\pi/0.75\pi)} = 1.7083.$$

Using Table 7.2, the transfer function for the normalized Butterworth filter of order $N = 2$ is given by

$$H(S) = \frac{1}{S^2 + 1.414S + 1}.$$

Equation (7.32) determines the cut-off frequency ω_c of the Butterworth filter from the stop-band constraint as follows:

$$\omega_c = \frac{\omega_s}{(G_s)^{0.5/N}} = \frac{0.75\pi}{24^{0.25}} = 0.3389\pi \text{ radians/s.}$$

The transfer function $H(s)$ of the required analog lowpass filter is given by

$$H(s) = H(S)|_{S=s/\omega_c} = \frac{1}{S^2 + 1.414S + 1}\bigg|_{S=s/0.3389\pi}$$

$$= \frac{1.1332}{s^2 + 1.5055s + 1.1332},$$

which can be expressed as follows:

$$H(s) = 1.5053 \frac{0.7528}{(s + 0.7528)^2 + 0.7528^2}.$$

Using Table 16.2, step 3 derives the z-transfer function as follows:

$$H(z) = 1.5053 \frac{ze^{-0.7528}\sin(0.7528)}{z^2 - 2e^{-0.7528}\cos(0.7528)z + e^{-2(0.7528)}},$$

which simplifies to

$$H(z) = 1.5053 \frac{0.3220z}{z^2 - 0.6875z + 0.2219}.$$

Step 4 computes the magnitude spectrum by substituting $z = \exp(j\Omega)$. The resulting plot is shown in Fig. 16.4(a), where we observe that the magnitude spectrum satisfies the pass-band requirements, though the dc gain of the filter is not equal to unity. The stop band requirement is not satisfied, however, as the

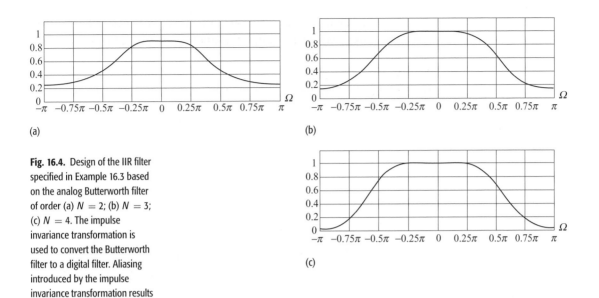

(a)

(b)

Fig. 16.4. Design of the IIR filter specified in Example 16.3 based on the analog Butterworth filter of order (a) $N = 2$; (b) $N = 3$; (c) $N = 4$. The impulse invariance transformation is used to convert the Butterworth filter to a digital filter. Aliasing introduced by the impulse invariance transformation results in a considerably higher order ($N = 4$) Butterworth filter to meet the design specifications.

(c)

gain $|H(\Omega)|$ of the filter is greater than 0.20 at the stop-band corner frequency of 0.75π radians/s. The above procedure is repeated for a Butterworth filter of order $N = 3$.

Iteration 2 for Butterworth filter of order $N = 3$ The transfer function for the normalized Butterworth filter of order $N = 3$ is obtained from Table 7.2 as follows:

$$H(S) = \frac{1}{(S + 1)(S^2 + S + 1)}.$$

The cut-off frequency ω_c of the Butterworth filter is obtained from the stop-band constraint:

$$\omega_c = \frac{\omega_s}{(G_s)^{0.5/N}} = \frac{0.75\pi}{24^{1/6}} = 0.4416\pi \text{ radians/s.}$$

The transfer function $H(s)$ of the required analog lowpass filter is given by

$$H(s) = H(S)|_{S=s/\omega_c} = \frac{1}{(S + 1)(S^2 + S + 1)}\bigg|_{S=s/0.4416\pi}$$

$$= \frac{2.6702}{s^3 + 2.7747s^2 + 3.8494s + 2.6702}.$$

Expanding $H(s)$ in terms of partial fractions and using Table 16.2, we can derive the z-transfer function of the equivalent digital filter as follows:

$$H(z) = \frac{0.4695z^2 + 0.1907z}{z^3 - 0.6106z^2 + 0.3398z - 0.0624}.$$

The above derivation is left as an exercise for the reader in Problem 16.3(a).

Figure 16.4(b) plots the magnitude spectrum $|H(\Omega)|$ of the third-order filter. We observe that the attenuation is increased at the stop-band corner frequency of

0.75π radians/s, but that it is still greater than the specified value. We therefore repeat the above procedure for a Butterworth filter of order $N = 4$.

Iteration 3 for Butterworth filter of order $N = 4$ The transfer function for the normalized Butterworth filter of order $N = 4$ is obtained from Table 7.2 as follows:

$$H(S) = \frac{1}{(s^2 + 0.7654s + 1)(s^2 + 1.8478s + 1)}.$$

The cut-off frequency ω_c of the Butterworth filter is obtained from the stop-band constraint:

$$\omega_c = \frac{\omega_s}{(G_s)^{0.5/N}} = \frac{0.75\pi}{24^{1/8}} = 0.5041\pi \text{ radians/s.}$$

The transfer function $H(s)$ of the required analog lowpass filter is given by

$$H(s) = H(S)|_{S=s/\omega_c} = \frac{1}{(s^2 + 0.7654s + 1)(s^2 + 1.8478s + 1)}\bigg|_{S=s/0.5041\pi},$$

which reduces to

$$H(s) = \frac{6.2902}{s^4 + 4.1383s^3 + 8.5630s^2 + 10.3791s + 6.2902}.$$

Problem 16.3(b) derives the z-transfer function of the equivalent digital filter as follows:

$$H(z) = \frac{0.3298z^3 + 0.4274z^2 + 0.0427z}{z^4 - 0.4978z^3 + 0.3958z^2 - 0.1197z + 0.0159}.$$

Figure 16.4(c) plots the magnitude spectrum $|H(\Omega)|$ of the fourth-order filter. We observe that both pass-band and stop-band requirements are satisfied by the Butterworth filter of order $N = 4$.

Impulse invariance transformation using MATLAB Starting with the analog Butterworth filter, the IIR filters in Example 16.3 can also be designed using the MATLAB function impinvar. The syntax to call the function is given by

```
[numz,denumz] = impinvar(nums,denums,fs)
```

where nums and denums specify the coefficients of the numerator and denominator of the analog filter and fs is the sampling rate in samples/s. For Example 16.3, the MATLAB code is given by

```
>> fs = 1;                        % fs = 1/T = 1
>> nums = [1.1332];               % numerator of CT filter
>> denums = [1 1.5055 1.1332];    % denominator of CT filter
>> [numz,denumz] = impinvar (nums,denums,fs);
                                  % coefficients of the DT
                                    % filter
```

which returns the following values:

```
numz = 0.4848 and denumz = [1.0000 -0.6876 0.2219].
```

The transfer function of the second-order IIR filter is given by

$$H(z) = \frac{0.4848z}{z^2 - 0.6875z + 0.2219},$$

which yields the same expression as the one derived in Example 16.3.

For the third-order Butterworth filter, the MATLAB code for the impulse invariance transformation is given by

```
>> fs = 1;                        % fs = 1/T = 1
>> nums = [2.6702];               % numerator of the CT filter
>> denums = [1   2.7747   3.8494   2.6702];
                                  % denominator of the CT filter
>> [numz,denumz] = impinvar (nums,denums,fs);
                                  % coeffs of the DT filter
```

which returns the following values:

```
numz = [0 0.4695 0.1907] and denumz = [1.0000 -0.6106
        0.3398 -0.0624].
```

The transfer function of the third-order IIR filter is given by

$$H(z) = \frac{0.4695z^2 + 0.1907z}{z^3 - 0.6106z^2 + 0.3398z - 0.0624}.$$

Similarly, the MATLAB code for transforming the fourth-order Butterworth filter is given by

```
>> fs = 1;              % fs = 1/T = 1
>> nums = [6.2902];     % numerator of the CT filter
>> denums = [1    4.1383    8.5603    10.3791    6.2902];
                        % denominator of CT filter
>> [numz,denumz] = impinvar (nums,denums,fs);
                        % coefficients of the DT filter
```

which returns the following values:

$$numz = [0 \quad 0.3298 \quad 0.4276 \quad 0.0428]$$

$$denumz = [1 \ -0.4977 \quad 0.3961 \quad -0.1197 \quad 0.0159].$$

The transfer function of the fourth-order IIR filter is given by

$$H(z) = \frac{0.3298z^3 + 0.4276z^2 + 0.0428z}{z^4 - 0.4977z^3 + 0.3958z^2 - 0.1197z + 0.0159}.$$

The above expression is similar to the one obtained in Example 16.3 for the fourth-order Butterworth filter.

16.2.4 Limitations of impulse invariance method

As illustrated in Example 16.3, the impulse invariance method introduces aliasing while transforming an analog filter to a digital filter. Since the analog filter is not band-limited, the impulse invariance transformation would always introduce aliasing in the digital domain. Therefore, a higher-order DT filter is generally required to satisfy the design constraints. Section 16.3 introduces a second transformation, known as the bilinear transformation, to eliminate the effect of aliasing.

16.3 Bilinear transformation

The bilinear transformation provides a one-to-one mapping from the s-plane to the z-plane. The mapping equation is given by

$$s = k\frac{z-1}{z+1}, \tag{16.23}$$

where k is the normalization constant given by $2/T$, where T is the sampling interval. To derive the frequency characteristics of the bilinear transformation, we substitute $z = \exp(j\Omega)$ and $s = j\omega$ in Eq. (16.23). The resulting expression

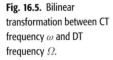

Fig. 16.5. Bilinear
transformation between CT
frequency ω and DT
frequency Ω.

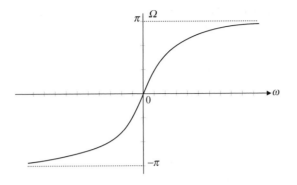

is given by

$$\omega = k \tan \frac{\Omega}{2} \quad \text{or} \quad \Omega = 2 \tan^{-1} \frac{\omega}{k}, \tag{16.24}$$

which is plotted in Fig. 16.5. We observe that the transformation is highly
non-linear since the positive CT frequencies within the range $\omega = [0, \infty]$ are
mapped to the DT frequencies $\Omega = [0, \pi]$. Similarly, the negative CT frequencies $\omega = [-\infty, 0]$ are mapped to the DT frequencies $\Omega = [-\pi, 0]$. This non-
linear mapping is known as frequency warping, and is illustrated in Fig. 16.6,
where an analog lowpass filter is transformed into a digital lowpass filter using
Eq. (16.24) with $k = 1$. Since the CT frequency range $[-\infty, \infty]$ in Fig. 16.5
is mapped on to the DT frequency range $[-\pi, \pi]$, there is no overlap between
adjacent replicas constituting the magnitude response of the digital filter. Frequency warping, therefore, eliminates the undesirable effects of aliasing from
the transformed digital filter. We now show how different regions of the s-plane
are mapped onto the z-plane.

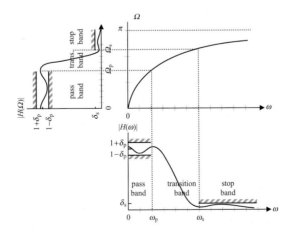

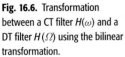

Fig. 16.6. Transformation
between a CT filter $H(\omega)$ and a
DT filter $H(\Omega)$ using the bilinear
transformation.

16.3.1 Mapping between the s-plane and the z-plane

For $k = 1$, Eq. (16.23) can be represented in the following form:

$$z = \frac{1 + s}{1 - s}.$$ (16.25)

Substituting $s = \sigma + j\omega$ into Eq. (16.25), we obtain

$$z = \frac{1 + \sigma + j\omega}{1 - \sigma - j\omega},$$ (16.26)

with an absolute value given by

$$|z| = \sqrt{\frac{(1 + \sigma)^2 + \omega^2}{(1 - \sigma)^2 + \omega^2}}.$$ (16.27)

By substituting different values of $s = \sigma + j\omega$ corresponding to the right-half, left-half, and imaginary axes of the s-plane in Eq. (16.27), we derive the following observations.

Left-half s-plane $(\sigma < 0)$ For $\sigma < 0$, we observe that the value of the denominator $(1 - \sigma)^2 + \omega^2$ in Eq. (16.27) exceeds the value of the numerator $(1 + \sigma)^2 + \omega^2$, resulting in $|z| < 1$. In other words, the bilinear transformation maps the left-half of the s-plane to the interior of the unit circle within the z-plane.

Right-half s-plane $(\Omega > 0)$ For $\sigma > 0$, the value of the numerator $(1 + \sigma)^2 + \omega^2$ in Eq. (16.27) exceeds the value of the denominator $(1 - \sigma)^2 + \omega^2$, resulting in $|z| > 1$. Consequently, the bilinear transformation maps the right-half of the s-plane to the exterior of the unit circle within the z-plane.

Imaginary axis $(\sigma = 0)$ For $\sigma = 0$, the denominator and numerator in Eq. (16.27) are equal, resulting in $|z| = 1$. The bilinear transformation maps the imaginary axis of the s-plane onto the unit circle within the z-plane.

Note that the mapping in Eq. (16.25) is a one-to-one mapping, which means that no two points in the s-plane will map to the same point in the z-plane, and vice versa.

16.3.2 IIR filter design using bilinear transformation

The steps involved in designing IIR filters using the bilinear transformation are as follows.

Step 1 Using Eq. (16.24), $\omega = k \tan(\Omega/2)$, transform the specifications of the digital filter from the DT frequency (Ω) domain to the CT frequency (ω) domain. For convenience, we choose $k = 1$.

Step 2 Using the analog filter design techniques, design an analog filter $H(s)$ based on the transformed specifications obtained in step 1.

Step 3 Using the bilinear transformation $s = (z - 1)/(z + 1)$ (obtained by rearranging Eq. (16.25) to express z in terms of s), derive the z-transfer function $H(z)$ from the s-transfer function $H(s)$.

Step 4 Confirm that the z-transfer function $H(z)$ obtained in step 3 satisfies the design specifications by plotting the magnitude spectrum $|H(\Omega)|$. If the design specifications are not satisfied, increase the order N of the analog filter designed in step 2 and repeat from step 2.

We now illustrate the application of the above algorithm in Example 16.4.

Example 16.4
Repeat Example 16.3 using the bilinear transformation.

Solution
Choosing $k = 1$ (sampling interval $T = 2$), step 1 transforms the pass-band and stop-band corner frequencies into the CT frequency domain:

pass-band corner frequency $\quad \omega_p = \tan(0.5\Omega_p) = \tan(0.5 \times 0.25\pi)$
$$= 0.4142 \text{ radians/s};$$

stop-band corner frequency $\quad \omega_s = \tan(0.5\Omega_s) = \tan(0.5 \times 0.75\pi)$
$$= 2.4142 \text{ radians/s}.$$

The transformed specifications of the CT filter are given by

pass-band ($0 \leq |\omega| \leq 0.4142$ radians/s) $\qquad 0.8 \leq |H(\omega)| \leq 1$;
stop-band ($|\omega| > 2.4142$ radians/s) $\qquad |H(\omega)| \leq 0.20$.

Step 2 designs the analog filter based on the transformed specifications. As in Example 16.3, we use the Butterworth filter. The gain terms for the filter stay the same as in Example 16.3:

$$G_p = \frac{1}{(1 - \delta_p)^2} - 1 = 0.5625$$

and

$$G_s = \frac{1}{(\delta_s)^2} - 1 = 24$$

The order N of the filter is given by

$$N = \frac{1}{2} \times \frac{\ln(G_p/G_s)}{\ln(\omega_p/\omega_s)} = \frac{1}{2} \times \frac{\ln(0.5625/24)}{\ln(0.4142/2.4142)} = 1.0646,$$

Fig. 16.7. Magnitude response $|H(\Omega)|$ of the lowpass filter designed in Example 16.4 using the bilinear transformation.

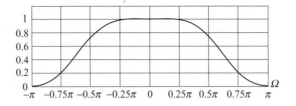

which is rounded up to $N = 2$. Using Table 7.2, the transfer function for the normalized Butterworth filter of order $N = 2$ is given by

$$H(S) = \frac{1}{S^2 + 1.414S + 1}.$$

Using Eq. (7.31) to determine the cut-off frequency ω_c of the Butterworth filter, we obtain

$$\omega_c = \frac{\omega_s}{(G_s)^{0.5/N}} = \frac{2.4142}{24^{0.25}} = 1.0907 \text{ radians/s}.$$

The transfer function $H(s)$ of the required analog lowpass filter is given by

$$H(s) = H(S)|_{S=s/\omega_c} = \frac{1}{S^2 + 1.414S + 1}\bigg|_{S=s/1.0907}$$

$$= \frac{1.1897}{s^2 + 1.5421s + 1.1897}.$$

Step 3 derives the z-transfer function of the digital filter using the bilinear transformation:

$$H(z) = H(s)|_{s=(z-1)/(z+1)}$$

$$= \frac{1.1897(z + 1)^2}{(z - 1)^2 + 1.5421(z - 1)(z + 1) + 1.1897(z + 1)^2},$$

which simplifies to

$$H(z) = \frac{0.3188z^2 + 0.6375z + 0.3188}{z^2 + 0.1017z + 0.1734}.$$

Step 4 computes the magnitude spectrum by substituting $z = \exp(j\Omega)$. The resulting plot is shown in Fig. 16.7, where we observe that the magnitude spectrum satisfies the specified pass-band and stop-band requirements.

Bilinear transformation using MATLAB The `bilinear` function is provided in MATLAB to transform a CT filter to a DT filter using the bilinear transformation. The syntax for calling the `bilinear` function is similar to that of the `impinvar` function and is given by

```
[numz,denumz] = bilinear(nums,denums,fs)
```

where `nums` and `denums` specify the coefficients of the numerator and denominator of the analog filter and `fs` is the sampling rate in samples/s. For Example 16.4, the MATLAB code is given by

```
>> fs = 0.5;                    % fs = 1/T = k/2 = 0.5
>> nums = [1.1897];             % numerator of the CT filter
>> denums = [1 1.5421 1.1897]; % denominator of CT filter
>> [numz,denumz] = bilinear (nums,denums,fs);
                                % coefficients of DT filter
```

which returns the values

```
numz = [0.3188   0.6376   0.3188];
denumz = [1.0000   0.1017   0.1735],
```

which are the same as the coefficients obtained in Example 16.4.

Filter design using MATLAB Several additional functions are provided in MATLAB for directly determining the transfer function of the digital filters. The `buttord` and `butter` functions, introduced in Chapter 7, can also be used to compute IIR filters in the digital domain. The `buttord` function computes the order N and cut-off frequency `wn` of the Butterworth filter, and the `butter` function computes the coefficients of the numerator and denominator of the z-transfer function of the Butterworth filter. For lowpass filters, the syntaxes for calling the `buttord` and `butter` functions are given by

```
buttord function:     [N, wn] = buttord(wp, ws, rp, rs);
butter function:      [numz, denumz] = butter(N, wn),
```

where N is the order of the lowest-order digital Butterworth filter that loses no more than `rp` dB in the pass band and has at least `rs` dB of attenuation in the stop band. The frequencies `wp` and `ws` are the pass-band and stop-band edge frequencies, normalized between zero and unity, where unity corresponds to π radians/s. Similarly, `wn` is the normalized cut-off frequency for the Butterworth filter. The matrix `numz` contains the coefficients of the numerator, while matrix `denumz` contains the coefficients of the denominator of the transfer function of the Butterworth filter.

For Example 16.4, the MATLAB code is given by

```
>> [N,wn] = buttord(0.25,0.75,20*log10(0.8),20*log10
   (0.20));
>> [numz,denumz] = butter(N,wn);
```

which results in the following coefficients:

```
numz = [0.3188   0.6376   0.3188];
denumz = [1.0000 0.1017   0.1735],
```

which are identical to those obtained analytically in Example 16.4.

16.4 Designing highpass, bandpass, and bandstop IIR filters

In the following examples, we design the highpass, bandpass, and bandstop IIR filters.

Example 16.5

Example 15.5 designed a highpass FIR filter for the following specifications:

(i) pass-band edge frequency $\Omega_p = 0.5\pi$ radians/s;
(ii) stop-band edge frequency $\Omega_s = 0.125\pi$ radians/s;
(iii) pass-band ripple ≤ 0.01 dB;
(iv) stop-band attenuation ≥ 60 dB.

Design an IIR filter with the same specifications.

Solution

Choosing $k = 1$ (sampling interval $T = 2$), step 1 transforms the pass-band and stop-band corner frequencies into the CT frequency domain:

pass-band corner frequency $\quad \omega_p = \tan(0.5\Omega_p) = \tan(0.25\pi) = 1$ radian/s;
stop-band corner frequency $\quad \omega_s = \tan(0.5\Omega_s) = \tan(0.0625\pi) = 0.1989$ radians/s.

Step 2 designs the analog filter based on the transformed specifications. In Chapter 7, we presented the design methodology for deriving the transfer function of the analog highpass filter analytically. Here, we use MATLAB to calculate the analog elliptic filter based on the above specifications:

```
>> wp = 1; ws = 0.1989;  Rp = 0.01; Rs = 60 ;
>> [N,wn] = ellipord (wp,ws,Rp,Rs, 's');
                        % Order and cut off frequency
                        % of the analog elliptic filter
>> [nums,denums]=ellip (N,Rp,Rs,wn,'high','s');
                        % Tx function of the analog
                        % elliptic filter
```

which yields the following transfer function for the analog filter:

$$H(s) = \frac{0.9988s^4 + 0.0542s^2 + 0.000373}{s^4 + 1.872s^3 + 1.824s^2 + 1.04s + 0.3732}.$$

Step 3 derives the z-transfer function of the digital filter using the bilinear transformation. This is achieved by using the bilinear function in MATLAB.

```
>> [numz,denumz] = bilinear (nums,denums,0.5) % DT Filter
```

The resulting filter is given by

$$H(z) = \frac{0.1725z^4 - 0.6539z^3 + 0.9638z^2 - 0.6539z + 0.1725}{z^4 - 0.6829z^3 + 0.7518z^2 - 0.138z + 0.0468}.$$

Fig. 16.8. Magnitude response
of the DT highpass filter
designed in Example 16.5.

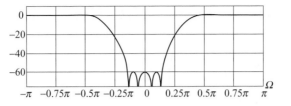

Figure 16.8 shows the magnitude response of the designed filter. We observe
that the pass-band and stop-band specifications are both satisfied.

Example 16.6

Example 15.6 designed a bandpass FIR filter with the following specifications:

(i) pass-band edge frequencies, $\Omega_{p1} = 0.375\pi$ and $\Omega_{p2} = 0.5\pi$ radians/s;
(ii) stop-band edge frequencies, $\Omega_{s1} = 0.25\pi$ and $\Omega_{s2} = 0.625\pi$ radians/s;
(iii) stop-band attenuations, $\delta_{s1} > 50$ dB and $\delta_{s2} > 50$ dB.

Design an IIR filter with the same specifications.

Solution

Choosing $k = 1$ (sampling interval $T = 2$), step 1 transforms the pass-band and
stop-band corner frequencies into the CT frequency domain:

pass-band corner frequency I $\omega_{p1} = \tan(0.5\Omega_{p1}) = \tan(0.1875\pi) = 0.6682$ radians/s;
pass-band corner frequency II $\omega_{p2} = \tan(0.5\Omega_{p2}) = \tan(0.25\pi) = 1$ radian/s;
stop-band corner frequency I $\omega_{s1} = \tan(0.5\Omega_{s1}) = \tan(0.125\pi) = 0.4142$ radians/s;
stop-band corner frequency II $\omega_{s2} = \tan(0.5\Omega_{s2}) = \tan(0.3125\pi) = 1.4966$ radians/s.

Step 2 designs an analog filter for the aforementioned specifications. We can
either use the analytical techniques developed in Chapter 7 or use a MATLAB
program. In the following, we calculate an analog elliptic filter for the given
specifications using MATLAB. Since the pass-band ripple is not specified, we
assume that it is given by 0.03 dB. The MATLAB code is given by

```
>> wp = [0.6682 1]; ws = [0.4142   1.4966];
>> Rp = 0.03; Rs = 50;
>> [N, wn] = ellipord(wp,ws,Rp,Rs,'s');
>> [nums,denums] = ellip(N,Rp,Rs,wn,'s');
```

which results in an eighth-order elliptic filter with the following transfer
function:

$$H(s) = \frac{0.001(3.164s^8 + 30.27s^6 + 57.02s^4 + 13.51s^2 + 0.6308)}{s^8 + 0.7555s^7 + 3.07s^6 + 1.634s^5 + 3.229s^4 + 1.092s^3 + 1.371s^2 + 0.2254s + 0.1994}.$$

Step 3 derives the z-transfer function of the digital filter using the bilinear transformation. This is achieved by using the `bilinear` function in MATLAB.

Fig. 16.9. Magnitude response
of the DT bandpass filter
designed in Example 16.6.

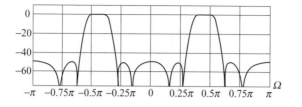

```
>> [numz,denumz]=bilinear(nums,denums,0.5) % DT Filter
```

The resulting filter is given by

$$H(z) = \frac{0.001\left(8.317z^8 - 6.94z^7 + 4.236z^6 - 5.952z^5 + 13.52z^4 - 5.952z^3 + 4.236z^2 - 6.94z + 8.317\right)}{z^8 - 1.389z^7 + 3.714z^6 - 3.356z^5 + 4.685z^4 - 2.693z^3 + 2.397z^2 - 0.7107z + 0.4106}.$$

Figure 16.9 shows the magnitude response of the designed filter, which illustrates that the pass-band and stop-band specifications are both satisfied.

Example 16.7

Example 15.7 designed a bandstop FIR filter with the following specifications:

  (i) pass-band edge frequencies, $\Omega_{p1} = 0.25\pi$ and $\Omega_{p2} = 0.625\pi$ radians/s;
 (ii) stop-band edge frequencies, $\Omega_{s1} = 0.375\pi$ and $\Omega_{s2} = 0.5\pi$ radians/s;
(iii) stop-band attenuations, $\delta_{s1} > 50$ db and $\delta_{s2} > 50$ dB.

Design an IIR filter with the same specifications.

Solution

Choosing $k = 1$ (sampling interval $T = 2$), step 1 transforms the pass-band and stop-band corner frequencies into the CT frequency domain:

pass-band corner frequency I $\omega_{p1} = \tan(0.5\Omega_{p1}) = \tan(0.125\pi) = 0.4142$ radians/s;
pass-band corner frequency II $\omega_{p2} = \tan(0.5\Omega_{p2}) = \tan(0.3125\pi) = 1.4966$ radians/s;
stop-band corner frequency I $\omega_{s1} = \tan(0.375\Omega_{s1}) = \tan(0.1875\pi) = 0.6682$ radians/s;
stop-band corner frequency $\omega_{s2} = \tan(0.5\Omega_{s2}) = \tan(0.25\pi) = 1$ radian/s.

Step 2 designs an analog filter for the aforementioned specifications. In the following, we use MATLAB to derive the analog elliptic filter for the transformed specifications and an assumed pass-band ripple of 0.03 dB:

```
>> wp = [0.4142 1.4966]; ws = [0.6682   1];
>> Rp = 0.03; Rs = 50;
>> [N,wn] = ellipord(wp,ws,Rp,Rs,'s');
>> [nums,denums] = ellip(N,Rp,Rs,wn,'stop','s');
```

The resulting elliptic filter is of the eighth order and has the following transfer function:

$$H(s) = \frac{0.9966s^8 + 2.8s^6 + 2.854s^4 + 1.25s^2 + 0.1987}{s^8 + 2.137s^7 + 5.15s^6 + 5.926s^5 + 6.747s^4 + 3.96s^3 + 2.3s^2 + 0.6377s + 0.1994}.$$

Step 3 derives the z-transfer function of the digital filter using the bilinear function.

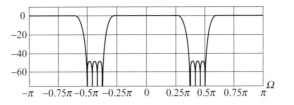

```
>> [numz,denumz]=bilinear(nums,denums,0.5); % DT Filter
```

The resulting DT filter is given by

$$H(z) = \frac{0.2887z^8 - 0.4484z^7 + 1.363z^6 - 1.372z^5 + 2.149z^4 - 1.372z^3 + 1.363z^2 - 0.4484z + 0.2887}{z^8 - 1.096z^7 + 1.977z^6 - 1.519z^5 + 1.78z^4 - 0.8638z^3 + 0.6172z^2 - 0.1739z + 0.09751}.$$

Figure 16.10 shows the magnitude response of the designed bandstop filter. We observe that both the pass-band and stop-band specifications are satisfied by the bandstop filter.

16.5 IIR and FIR filters

A classical problem in the design of digital filters is the selection between FIR and IIR filters since both types of filters can be used to satisfy a given set of specifications. In this section, we compare IIR and FIR filters with respect to three criteria: stability, implementation complexity, and delay.

16.5.1 Stability

Stability is a major concern in the design of filters. When designing digital filters, care must be taken to ensure that the designed filters are absolutely BIBO stable to prevent infinite outputs. Recall that an LTID system is stable if its poles lie inside the unit circle in the z-plane. Since the only poles in FIR filters lie at the origin ($z = 0$), FIR filters are always BIBO stable. On the other hand, IIR filters have non-trivial poles because of feedback loops and therefore may run into stability issues.

Use of finite-precision DSP boards places a severe limitation on the type of IIR filters that can be used. Even if the designed IIR filter is stable, quantization of the filter coefficients can adversely affect its stability. To illustrate the effect of quantization on the stability of the filter, consider the following four filters.

(1) Lowpass filter (arbitrary):

$$H(z) = \frac{0.001(3.5747z^7 - 13.649z^6 + 20.9446z^5 - 10.7188z^4 - 10.7188z^3 + 20.9446z^2 - 13.649z + 3.5747)}{z^7 - 5.9664z^6 + 15.5383z^5 - 22.8594z^4 + 20.49z^3 - 11.1881z^2 + 3.4416z - 0.46}.$$

(2) Highpass filter (Example 16.5):

$$H(z) = \frac{0.1725z^4 - 0.6539z^3 + 0.9638z^2 - 0.6539z + 0.1725}{z^4 - 0.6829z^3 + 0.7518z^2 - 0.138z + 0.0468}.$$

Table 16.4. Implementation complexity of FIR and IIR filters
Note that N corresponds to the order of a DT filter

		Number of two-input adders	Number of scalar multipliers	Unit delay elements
Highpass filter	FIR ($N = 20$)	21	10	20
(Examples 15.5/16.5)	IIR ($N = 4$)	8	9	4
Bandpass filter	FIR ($N = 46$)	47	24	46
(Examples 15.6/16.6)	IIR ($N = 8$)	16	17	8
Bandstop filter	FIR ($N = 46$)	47	24	46
(Examples 15.7/16.7)	IIR ($N = 8$)	16	17	8

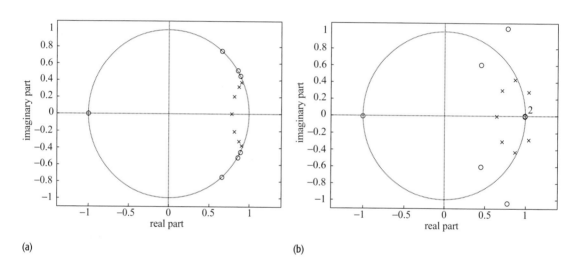

(a)   (b)

Fig. 16.12. Locations of the poles and zeros of the lowpass filter specified as item 1 in section 16.5.1 (a) Before quantization; (b) after quantization of coefficients.

the direct form II realizations, while the IIR filters are implemented using the linear implementation (see Section 14.6.3).

It is observed in Table 16.4 that the complexity of IIR filters is significantly lower than that for the corresponding FIR filters. For example, the highpass FIR filter requires 21 additions, 10 scalar multiplications, and 20 unit delays. On the other hand, the highpass IIR filter requires only 8 additions, 9 multiplications, and 4 unit delays. The difference is more conspicuous for the bandpass and bandstop filters, where the orders of the FIR filters are much larger than the corresponding orders of the IIR filters.

In summary, for applications such as image and video processing, where a smaller-order FIR filter can satisfy the design specifications, FIR filters are generally chosen. In other applications, such as acoustics, a filter with a long impulse response in the range of 2000 samples is required. In such cases, the FIR filter provides a large implementation complexity compared with that for an IIR filter designed with the same specifications. Between these two extremes, there are a large number of applications where an appropriate filter (FIR or IIR) is chosen based on implementation cost and robustness.

16.5.3 Delay

The propagation delay between the time an input signal is applied and the time when the output appears is another important factor in filter selection. Because of the larger number of implementation elements, the FIR filters generally have a larger delay than the IIR filters.

16.6 Summary

This chapter presented transformation techniques, namely the impulse invariance and bilinear transformations, used to design IIR filters. These transformation techniques are based on converting the frequency specifications $H(\Omega)$ of IIR filters from the DT frequency Ω domain into the CT frequency specifications $H(\omega)$. Based on the CT frequency specifications, a CT filter with transfer function $H(s)$ is designed, which is then transformed back into the original DT frequency Ω domain to obtain the transfer function $H(z)$ of the required IIR filter. Section 16.2 introduced the impulse invariance transformation used to design lowpass filters. The impulse invariance method uses a linear expression,

$$\Omega = \omega T,$$

where T is the sampling interval, to convert DT specifications to the CT domain. Because of the sampling process, the impulse invariance method suffers from aliasing when transforming the analog filter $H(s)$ to the digital filter $H(z)$. A consequence of aliasing is that the order N of the designed filter $H(z)$ is much higher than the optimal design. To prevent aliasing, Section 16.3 presented the bilinear transformation, which transforms the DT specifications to the CT frequency domain using the following expression:

$$\omega = k\tan(\Omega/2) \quad \text{or} \quad \Omega = 2\tan^{-1}(\omega/k).$$

The transfer function $H(s)$ of the CT filter is then transformed into the z-domain using the following transformation:

$$s = \frac{1}{k}\frac{z-1}{z+1},$$

in which k is generally set to unity. Section 16.4 extended the design techniques to highpass, bandpass, and bandstop filters.

A comparison of IIR and FIR filters was presented in Section 16.5. We demonstrated that the order of the FIR filter is generally higher than that for IIR filters for the same design specifications. Therefore, the implementation cost of IIR filters is generally lower than for FIR filters. In addition, IIR filters generally have a lower delay. However, a major limitation in the use of IIR filters is the stability. Because IIR filters are implemented using feedback loops, they have non-zero poles. Care should be taken in designing IIR filters by ensuring that the poles are well inside the unit circle; this achieves good relative stability. FIR filters have trivial poles (at $z = 0$) and are always stable.

Another approach taken to design IIR filters is referred to as the direct design method, which derives the filter recursively using a least-squares method. Unlike the analog prototyping method, the direct design method is not constrained to the standard lowpass, highpass, bandpass or bandstop configurations. Filters with an arbitrary, perhaps multiband, frequency response are also possible. In MATLAB the `yulewalk` function designs IIR digital filters by performing a least-squares fit in the time domain. For more details on FIR filter design using direct design method, refer to refs. [1] and [2].[†]

Problems

16.1 Using the impulse invariance transformation and a sampling interval of $T = 0.1\,\text{s}$, convert the following analog transfer functions to their equivalent digital transfer functions:

(a) $H(s) = \dfrac{s + 2}{(s + 4)(s^2 + 4s + 3)}$;

(b) $H(s) = \dfrac{s^2 + 9s + 20}{(s + 2)(s^2 + 4s + 3)}$;

(c) $H(s) = \dfrac{s^3 + s^2 + 6s + 14}{(s^2 + s + 1)(s^2 + 2s + 5)}$.

16.2 Derive the following z-transform pair used in Example 16.2:

$$12.7786T e^{-6.3893\,kT} \sin(6.3894kT)u[k]$$

$$\overset{z}{\longleftrightarrow} \frac{12.7786T e^{-6.3893\,T} \sin(6.3894T)z}{z^2 - 2z e^{-6.3893\,T} \cos(6.3894T)z + e^{-2\times 6.3893\,T}}.$$

16.3 (a) Use the impulse invariance method to show that the analog transfer function given by

$$H(s) = \frac{2.6702}{s^3 + 2.7747s^2 + 3.8494s + 2.6702}$$

[†] [1] B. Friedlander and B. Porat, the modified Yule–Walker method of ARMA spectral estimation, *IEEE Transactions on Aerospace Electronic Systems* (1984), **AES-20**(2), 158–173.
[2] L. B. Jackson, *Digital Filters and Signal Processing*, 3rd edn. Kluwer Academic Publishers (1996), Chap. 10, pp. 345–355.

results in the following z-transfer function:

$$H(z) = \frac{0.4695z^2 + 0.1907z}{z^3 - 0.6106z^2 + 0.3398z - 0.0624}$$

as stated in Example 16.3 for the third-order Butterworth filter.

(b) Use the impulse invariance method to show that the analog transfer function given by

$$H(s) = \frac{6.2902}{s^4 + 4.1383s^3 + 8.5630s^2 + 10.3791s + 6.2902}$$

results in the following z-transfer function:

$$H(z) = \frac{0.3298z^3 + 0.4274z^2 + 0.0427z}{z^4 - 0.4978z^3 + 0.3958z^2 - 0.1197z + 0.0159}$$

as stated in Example 16.3 for the fourth-order Butterworth filter.

16.4 Using the impulse invariance transformation, design a lowpass IIR Butterworth filter based on the following specifications:

$$\text{pass-band edge frequency} = 0.64\pi;$$
$$\text{width of transition band} = 0.3\pi;$$
$$\text{maximum pass-band ripple} < 0.002;$$
$$\text{maximum stop-band ripple} < 0.005.$$

16.5 Repeat Problem 16.4 for a highpass IIR Butterworth filter.

16.6 Figure 9.1 shows a schematic for processing CT signals using DT systems. The overall system should have the CT frequency characteristics as follows:

$$\text{overall CT system is a lowpass filter;}$$
$$\text{pass-band edge frequency} = 3\pi \text{ kradians/s;}$$
$$\text{width of the transition band} = 4\pi \text{ kradians/s;}$$
$$\text{minimum stop-band attenuation} > 50 \text{ dB}$$
$$\text{maximum pass-band attenuation} < 0.03 \text{ dB}$$
$$\text{sampling rate} = 8 \text{ ksamples/s,}$$

Design a digital IIR filter that will provide the above characteristics using the following steps.

(a) Derive the DT specifications from the CT specifications using the impulse invariance transformation with $T = 1/8 \times 10^{-3}$ s.

(b) Design the digital IIR filter using a CT elliptic filter and the bilinear transformation.

16.7 Repeat Problem 16.1 for the bilinear transformation.

16.8 Design a lowpass IIR Butterworth filter specified in Problem 16.4 using the bilinear transformation.

16.9 Design a highpass IIR Butterworth filter specified in Problem 16.5 using the bilinear transformation.

16.10 Using the bilinear transformation, design a highpass IIR filter based on the following specifications:

$$\text{pass-band edge frequency} = 0.64\pi;$$
$$\text{width of transition band} = 0.3\pi;$$
$$\text{maximum pass-band ripple} < 0.002;$$
$$\text{maximum stop-band ripple} < 0.005.$$

16.11 Using the bilinear transformation, design a bandpass IIR filter based on the following specifications.

$$\text{pass-band edge frequencies} = 0.4\pi \text{ and } 0.6\pi;$$
$$\text{stop-band edge frequencies} = 0.2\pi \text{ and } 0.8\pi;$$
$$\text{maximum pass-band ripple} < 0.02;$$
$$\text{maximum stop-band ripple} < 0.009.$$

16.12 Using the bilinear transformation, design a bandstop IIR filter based on the following specifications:

$$\text{pass-band edge frequencies} = 0.3\pi \text{ and } 0.7\pi;$$
$$\text{stop-band edge frequencies} = 0.4\pi \text{ and } 0.6\pi;$$
$$\text{maximum pass-band ripple} < 0.05;$$
$$\text{maximum stop-band ripple} < 0.05.$$

16.13 Consider the lowpass filter design, using the bilinear transformation and analog Butterworth filter in Example 16.4. Repeat the IIR filter design using (i) Chebyshev Type 1 and (ii) Chebyshev Type 2 CT filters. Plot the frequency characteristics of the designed DT filter.

16.14 Consider the highpass filter design using the bilinear transformation and analog elliptical filter in Example 16.5. Repeat the IIR filter design using (i) Chebyshev Type 1 and (ii) Chebyshev Type 2 CT filters. Plot the frequency characteristics of the designed DT filter.

16.15 Consider the bandpass filter design using the bilinear transformation and analog elliptical filter in Example 16.6. Repeat the IIR filter design using (i) Chebyshev Type 1 and (ii) Chebyshev Type 2 CT filters. Plot the frequency characteristics of the designed DT filter.

16.16 Consider the bandstop filter design using the bilinear transformation and analog elliptical filter in Example 16.7. Repeat the IIR filter design using (i) Butterworth and (ii) Chebyshev Type 2 CT filters. Plot the frequency characteristics of the designed DT filter.

16.17 Quantize the coefficients of the bandpass filters obtained in Problem 16.15 with a resolution of three decimal points. Are the filters with quantized coefficients stable?

16.18 Quantize the coefficients of the bandstop filters obtained in Problem 16.16 with a resolution of three decimal points. Are the filter with quantized coefficients stable?

16.19 Repeat Problem 16.18 with a resolution of one decimal point.

16.20 By plotting the poles of the highpass filter obtained in Problem 16.10, determine if the filter is absolutely stable. Quantize the coefficients of the filter with a resolution of three decimal points. Are the filter with quantized coefficients stable?

16.21 By plotting the poles of the bandpass filter obtained in Problem 16.11, determine if the filter is absolutely stable. Quantize the coefficients of the filter with three decimal points accuracy. Is the filter with quantized coefficients stable?

16.22 By plotting the poles of the bandstop filter obtained in Problem 16.12, determine if the filter is absolutely stable. Quantize the coefficients of the filter with three decimal points accuracy. Is the filter with quantized coefficients stable?

16.23 Compare the implementation complexity of the highpass FIR filter designed in Example 15.5 and the IIR filters designed in Problem 16.14.

16.24 Compare the implementation complexity of the bandpass FIR filter designed in Example 15.6 and the IIR filters designed in Problem 16.15.

16.25 Compare the implementation complexity of the bandstop FIR filter designed in Example 15.7 and the IIR filters designed in Problem 16.16.

16.26 Using the MATLAB filter design functions, confirm the transfer functions derived in Problems 16.10–16.16.

17 Applications of digital signal processing

With the increasing availability of digital computers and specialized digital hardware, digital signal processing offers a cost-effective alternative to many traditional analog signal processing applications. The digital approach is particularly attractive due to its adaptability and immunity to variations in the operating conditions. Since the operation of digital systems does not depend upon the exact value of the input signals or the constituent digital components, digital signal processing allows precise replication where the same operation can be repeated a large number of times, if required. In contrast, analog signal processing suffers from deviations caused by degradation in the performance of the analog components and changes in the operating conditions. Digital implementations are also adaptable to changes in the specifications of the system. By modifying the software, different specifications can be implemented by the same digital hardware. An analog system, on the other hand, has to be redesigned every time the specifications of the system change.

This chapter reviews elementary applications of digital signal processing in the field of spectral estimation, audio and musical signal processing, and image processing. Our aim is to motivate readers to explore the use of digital signal processing in applications of interest to them. Section 17.1 introduces spectral estimation, in which the spectral content of a non-stationary signal is estimated from a limited number of signal realizations. Sections 17.2, 17.3, and 17.4 consider audio signal processing, including spectral estimation, filtering, and compression of audio signals. As an example of multidimensional signal processing, we consider digital image processing in Sections 17.5, 17.6, and 17.7. Finally, Section 17.8 concludes the chapter with a summary of important concepts.

17.1 Spectral estimation

Estimating the frequency content of a signal, commonly referred to as spectral analysis or spectral estimation, is an important step in signal processing

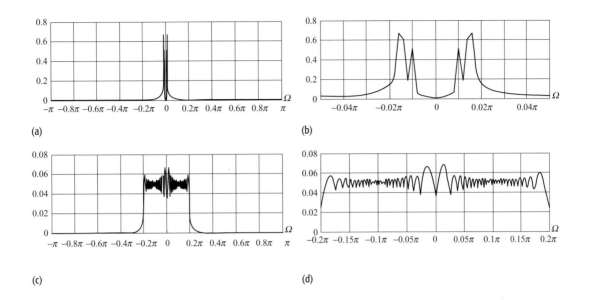

(a)

(b)

(c)

(d)

Fig. 17.1. DFT used to estimate the frequency content of stationary and non-stationary signals in Example 17.1.
(a) Magnitude sepctrum of $x_1[k]$. (b) Enlarged version of part (a) in the frequency range $-0.05\pi \leq \Omega \leq 0.05\pi$.
(c) Magnitude spectrum of $x_2[k]$. (d) Enlarged version of part (c) in the frequency range $-0.2\pi \leq \Omega \leq 0.2\pi$.

applications. For most signals of interest, the discrete Fourier transform (DFT) provides a convenient approach for spectral estimation. Example 17.1 highlights the DFT-based approach for two test signals.

Example 17.1
Using the DTFT, estimate the spectral content of the following DT signals:

(a) $x_1[k] = \cos(0.01\pi k) + 2\cos(0.015\pi k)$;
(b) $x_2[k] = \cos(0.0001\pi k^2)$,

from observations made over the interval $0 \leq k \leq 1000$.

Solution
(a) The magnitude spectrum of $x_1[k]$ based on the DFT is plotted over the frequency range $-\pi \leq \Omega \leq \pi$ in Fig. 17.1(a) with the magnified version shown in Fig. 17.1(b), where the frequency range $-0.05\pi \leq \Omega \leq 0.05\pi$ is enhanced. By looking at the peak values in Fig. 17.1(b), it is clear that the frequencies $\Omega_1 = 0.01\pi$ and $\Omega_2 = 0.015\pi$ radians/s are the dominant frequencies in the signal. On a relative scale, the frequency component $\Omega_2 = 0.015\pi$ has a higher strength compared with the frequency component $\Omega_1 = 0.01\pi$.

(b) The magnitude spectrum of $x_2[k]$ based on the DFT over the frequency range $-\pi \leq \Omega \leq \pi$ is plotted in Fig. 17.1(c), with the magnified version shown in Fig. 17.1(d), where the frequency range $-0.2\pi \leq \Omega \leq 0.2\pi$ is enhanced. From the subplots, it seems that all frequencies within the range $-0.2\pi \leq \Omega \leq 0.2\pi$ are fairly significant in $x_2[k]$. To confirm the validity of our estimation, let us calculate the instantaneous frequency of the signal.

Note that the phase of $x_2[k]$ is given by $\theta_0 = 0.0001\pi k^2$. By differentiating the phase θ_0 with respect to k, the instantaneous frequency is obtained as $\omega_0 = 0.0002\pi k$. The instantaneous frequency ω_0 is a function of time k, and increases proportionately as k increases. However, this time-varying nature of the frequency is not obvious from the magnitude spectrum shown in Fig. 17.1(c). Since the DFT averages the frequency components over all time k, the DFT provides a misleading result in this case.

Example 17.1 shows that the DFT magnitude spectrum based approach is convenient for estimating the spectral content of a stationary signal comprising sinusoidal components with fixed frequencies. However, it may provide misleading results for non-stationary signals, where the instantaneous frequency changes with time. In other words, it is difficult to visualize the time evolution of frequency in the DFT magnitude spectrum. The short-time Fourier transform is defined in Section 17.1.1 to address this limitation of DFT.

17.1.1 Short-time Fourier transform

In order to estimate the time evolution of the frequency components present in a signal, the short-time Fourier transform (STFT) parses the signal into smaller segments. The DFT of each segment is calculated separately and plotted as a function of time k. The STFT is therefore a function of both frequency Ω and time k. Mathematically, the STFT of a DT signal $x[k]$ is defined as follows:

$$X_s(\Omega, b) = \sum_{k=-\infty}^{\infty} x[k]g^*[k - b]e^{-j\Omega k}, \qquad (17.1)$$

where the subscript s in $X_s(\Omega, b)$ denotes the STFT and b indicates the amount of shift in the time-localized window $g[k]$ along the time axis. Typical windows used to calculate the STFT are rectangular, Hanning, Hamming, Blackman, and Kaiser windows. Compared to the rectangular window, the tapered windows, such as Hanning and Blackman, reduce the amount of ripple and are generally preferred.

In most cases, the time shift b is selected such that successive STFTs are taken over adjacent samples of $x[k]$ and there is some overlap of samples between successive STFTs. As discussed earlier, the STFT is a function of two variables: the frequency Ω and the central location of the window. It is typically plotted as an image plot, known as a spectrogram, with frequency Ω varying along the y-axis and the time (i.e. the center of the window function) varying along the x-axis. The intensity values of the image plot show the relative strength of various frequency components in the original signal.

Example 17.2
Plot the spectrogram of the signal $x_2[k] = \cos(0.0001\pi k^2)$ for duration $k = [0, 39\,999]$.

Solution

In order to calculate the STFT, let us choose the Hanning window function of length $N_w = 901$ samples to parse the data sequence of length $N_s = 40\,000$. Further assume that the overlap N_o between two consecutive windows to be $N_o = 600$ samples. The total number of complete windows is given by

$$M = \left\lfloor \frac{N_s - N_o}{N_w - N_o} \right\rfloor = 130. \tag{17.2}$$

The $p = 0$ window is centered at sample $k = 450$; the $p = 1$ window is centered at $450 + (901 - 600) = 751$; the $p = 2$ window is centered at $750 + (901 - 600) = 1052$. In general, a window p is centered at

$$k = \frac{N_w - 1}{2} + p(N_w - N_o) = 450 + 301p \tag{17.3}$$

for $0 \leq p \leq 129$. To obtain improved resolution in the frequency domain and to use the FFT algorithm efficiently, we zero-pad each time-windowed signal by 123 zero samples to make the total length of each segment equal 1024, which is a power of 2.

Note that the DFT of each zero-padded time-windowed signal will have a total of 1024 coefficients in the frequency domain. As the signal is real, the DFT coefficients will satisfy the Hermitian symmetry property. In other words, the amplitude spectrum is even-symmetric and we can ignore the second half of the spectrum which corresponds to the negative frequencies. So, we choose the first 513 coefficients out of a total of 1024 DFT coefficients corresponding to each windowed signal. The spectrogram is therefore a 2D matrix of size 513 $\times$ 130 samples. Each of the 130 columns will represent the amplitude spectrum of the signal at the time instant given by Eq. (17.3). Each row contains the amplitude of the 513 DFT coefficients. Note that the first coefficient ($r = 0$) represents frequency $\Omega = 0$ and the last ($r = 512$) coefficient represents frequency $\Omega = \pi$, with the intermediate frequencies given by

$$\Omega_r = \frac{r}{512} \times \pi \tag{17.4}$$

for $0 \leq r \leq 512$. The resulting spectrogram is shown in Fig. 17.2, where the black intensity points represent lower magnitudes and the light intensity points represent higher magnitudes. Note that the spectrogram is wrapped around the frequency range $[0, \pi]$. Figure 17.2 illustrates that the frequency of the chirp signal increases linearly with time.

In Example 17.2, we selected values for the window length and the overlap period on an *ad hoc* basis. The choice of the window size is important as it provides a trade-off between the resolution obtained in the frequency domain and the localization in the time domain. A larger window allows us to observe a signal for a longer period of time before we calculate the DFT. As a result, it provides a higher frequency resolution in the spectrogram. On the other hand, a shorter time window provides a better localization in time but a poor frequency

Fig. 17.2. Spectrogram of the chirp signal $x_2[k] = \cos(0.0001\pi k^2)$ from Example 17.2.

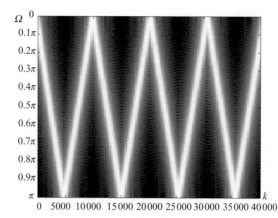

Fig. 17.2. Spectrogram of the chirp signal $x_2[k] = \cos(0.0001\pi k^2)$ from Example 17.2.

resolution. A longer window, therefore, generates a narrow-band spectrogram while a shorter window generates a wide-band spectrogram. Similarly, the overlap chosen between two consecutive windows provides continuity and reduces sharp transitions in the spectrogram.

17.1.2 Spectrogram computation using MATLAB

In MATLAB, the signal processing toolbox includes the function `specgram` for calculating the spectrogram of a signal. The spectrogram in Example 17.2 is computed using the following code:

```
>> k = [0:39999];
>> x2= cos(0.0001*pi*k.*k)   ;
>> Fs = 1;
>> Nwind = 901; Nfft = 1024; Noverlap = 600;
>> [spgram, F, T] = specgram(x2, Nfft, Fs, hanning(Nwind),
    Noverlap);
>> imagesc([0 length(x2)/Fs], 2*pi*F,
    20*log10(abs(spgram) + eps));
>> colormap(gray)
```

The MATLAB function `imagesc` displays the spectrogram using a color map. We can set the color map to gray using the last command in the code.

17.1.3 Random signals

The signals that we have studied so far are referred to as deterministic signals. Such signals can be specified by unique mathematical expressions, allowing us to calculate them precisely for all time. A second category consists of signals that cannot be predicted precisely in advance, which are collectively referred to as random or stochastic signals. Individual values of stochastic signals carry little information, and therefore statistical averages such as mean, autocorrelation, and power spectral density are commonly used to specify stochastic signals. We start by defining the statistical mean and autocorrelation commonly used to define a stochastic signal. If $x[k]$, $x[k_1]$, $x[k_2]$ are discrete random variables taking on values from the set $\{x_m, -\infty \leq m \leq \infty\}$ at times k, k_1, and k_2, respectively, the mean and autocorrelation functions are defined as follows:

mean
$$E\{x[k]\} = \sum_{m=-\infty}^{\infty} x_m P[x[k] = x_m]; \tag{17.5}$$

autocorrelation $\quad R_{xx}[k_1, k_2] = E\{x[k_1]x[k_2]\}$
$$= \sum_{m=-\infty}^{\infty} \sum_{n=-\infty}^{\infty} x_m x_n P[x[k_1] = x_m; \ x[k_2] = x_n]. \tag{17.6}$$

In Eqs. (17.5) and (17.6), the operator E denotes the expectation and $P[x[k] = x_m]$ is the probability that $x[k]$ takes on the value x_m. Likewise, $P[x[k_1] = x_m; \ x[k_2] = x_n]$ refers to the joint probability for random signals $x[k_1]$ and $x[k_2]$ observed at time instants k_1 and k_2. Estimating the mean and autocorrelation of a stochastic signal is difficult in general. In many applications, random signals satisfy the following two properties.

(1) The mean $E\{x[k]\}$ is constant and independent of time.
(2) The autocorrelation $E\{x[k_1]x[k_2]\}$ depends upon the duration between the observation instants k_1 and k_2. In other words, the autocorrelation is independent of the observation instants and is only determined by the duration between the two observations.

Such signals are referred to as wide-sense stationary (WSS) random signals. Sometimes, these are referred to as weak-sense stationary or second-order stationary random signals. Mathematically, the aforementioned two properties of the WSS signals can be expressed as follows:

mean
$$E\{x[k]\} = \mu_x; \tag{17.7}$$

autocorrelation $\quad R_{xx}[k_1, k_2] = R_{xx}[k_1 - k_2] = R_{xx}[m]. \tag{17.8}$

The DTFT of the autocorrelation $R_{xx}[m]$ of a WSS signal is referred to as the power spectral density, which is defined as follows:

power spectral density
$$S_{xx}(\Omega) = \sum_{m=-\infty}^{\infty} R_{xx}[m]e^{-j\Omega m}. \tag{17.9}$$

Equations (17.8) and (17.9) are widely used to estimate the spectral content of WSS signals, and the equations require the probability density functions to estimate the spectral content, which is generally not known in most signal processing applications. In the following, we present a method, based on the periodogram, to estimate the spectral content of stochastic signals from a finite number of observations.

17.1.4 Periodogram

The periodogram method is similar to the spectrogram method and exploits the STFT for spectrum estimation using a window function $g[k]$ of length N_w and centered at $k = b$. The time-windowed sequence $u_b[k]$, centered at $k = b$, is given by

$$u_b[k] = x\left[k + b - \left\lfloor \frac{N_w}{2} \right\rfloor\right] g[k], \; 0 \le k \le (N_w - 1). \tag{17.10}$$

The DFT of $u_b[k]$ is given by

$$U_b(\Omega) = \sum_{k=0}^{N_w - 1} u_b[k] e^{-j\Omega k}. \tag{17.11}$$

The periodogram method estimates the power spectrum $P_{xx}(\Omega)$ using the following equation:

$$\hat{P}_{xx}(\Omega) = \frac{1}{\mu^2} |U_b(\Omega)|^2, \tag{17.12}$$

where μ is referred to as the norm of the window function $g[k]$ and is calculated as follows:

$$\mu = \sqrt{\sum_k g^2[k]}. \tag{17.13}$$

While computing the STFT, different window functions attenuate the original samples of the signal $x[k]$ by different amounts. Inclusion of a scaling factor of $1/\mu^2$ in Eq. (17.12) reduces the bias introduced by a particular window function.

If $g[k]$ is a rectangular window, the estimate of the power spectrum $P_{xx}(\Omega)$ computed with Eq. (17.12) is called the *periodogram*. For all other windows, the estimate is referred to as the *modified periodogram*.

In its current form, Eq. (17.11) calculates the N_w-point DFT that produces DTFT values for a set of equally spaced N_w frequency points within the range $\Omega = [0, 2\pi]$. As for the spectrogram, we can zero-pad the time-windowed sequence and increase the DFT length to obtain a denser plot in the frequency domain.

17.1.5 Average periodogram

To estimate the power spectrum, Eq. (17.12) uses a single window with duration of $0 \le k \le (N_w-1)$ within the input signal $x[k]$. Improved results are obtained if several estimates from different locations of the signal are obtained and the resulting values are averaged. Starting from duration $0 \le k \le (N_w-1)$, the first iteration computes the periodogram from $x[k]$ within the specified duration. In the second iteration, the window is moved forward by $(N_w - N_o)$ samples such that there is an overlap of N_o between successive windows. The new location of the window is given by $(N_\text{w} - N_\text{o} - 1) \leq k \leq (2N_\text{w} - N_\text{o} - 2)$ for the second iteration, which is used to compute the periodogram for the second duration. The process is repeated until the entire signal is parsed and the average value of the periodogram is selected as the estimate of the power spectrum. This method, based on averaging the values of the power spectrum obtained from different periodograms, is referred to as the Welch estimate of the periodogram.

In the signal processing toolbox of MATLAB, the built-in function `psd` estimates the power spectrum of a signal using the periodogram approach. The following example illustrates the use of the `psd` function.

Example 17.3
Estimate the power spectral density of the following signal:

$$x[k] = 3\cos(0.2\pi k) + 2\cos(0.3\pi k) + r[k], \qquad (17.14)$$

where $r[k]$ is a white noise with Gaussian distribution with a variance of 4.

Solution
Note that the signal $x[k]$ includes a deterministic component consisting of the two sinusoids and a random component. The following code generates a realization of $x[k]$ and estimates the power spectrum:

```
>> k = [0:6000];
>> x = 3*cos(0.2*pi*k) + 2*cos(0.4*pi*k) +
  2*randn(size(k));
>> Fs = 2 ; nwind = length(x);
>> nfft = length(x); noverlap = 0 ;
>> [PxxNoAvg, F] = psd(x, nfft, Fs, rectwin(nwind),
>> noverlap); Fs = 2; nwind=301;
>> nfft = 512; noverlap = floor(4*nwind/5) ;
>> [PxxWelch, F] = psd(x, nfft, Fs,
  hanning(nwind),noverlap);
```

The random component $r[k]$ is generated using the MATLAB function `randn`. As the variance of the random component is 4, we multiply `randn` by the standard deviation, which equals 2. Figure 17.3 shows the first 201 samples of an example of $x[k]$. Over different simulations, the signal $x[k]$ may have slight variations due to the presence of the random component.

The MATLAB code computes the power spectrum in two ways. The first estimate `PxxNoAvg` represents the power spectrum obtained by calculating the DFT of the entire signal. Note that there is no averaging in this case. The second estimate, `PxxWelch`, represents the power spectrum obtained by the Welch method, where the signal is parsed into shorter sequences with a Hanning

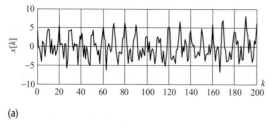

(a)

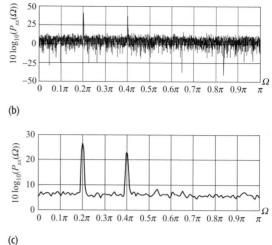

(b)

Fig. 17.3. Estimating the power spectrum of a random signal using the periodogram approach. (a) Original random signal. (b) Power spectrum obtained from periodogram with no averaging. (c) Power spectrum obtained from periodogram with overlap and averaging based on the Welch method.

(c)

window of size 301. Two consecutive windows have an overlap of 240 samples, resulting in a total of 94 data windows. Each of these sequences is zero-padded with 211 zero-valued samples and the DFT is calculated. The averaged power spectrum is then obtained by averaging all 94 power spectra.

The resulting power spectra are shown in Figs. 17.3(b) and (c). Although both spectra exhibit peaks at $\Omega = 0.2\pi$ and 0.4π the estimate PxxNoAvg contains a substantial amount of noise. Since the estimate PxxWelch averages the power spectrum, most of the noise is canceled out. However, averaging also reduces the magnitudes of peaks at $\Omega = 0.2\pi$ and 0.4π in PxxWelch. In the latter case, the peaks are not as pronounced as the peaks in PxxNoAvg.

17.2 Digital audio

Since the 1980s, digital audio has become a very popular multimedia format for several applications, including the audio CD, teleconferencing, and digital movies. With the enormous growth of the World Wide Web (WWW), audio processing techniques such as filtering, equalization, noise suppression, compression, and synthesis are being used increasingly. In this section, we focus on three aspects of audio processing: spectrum estimation, audio filtering, and audio compression. We start by discussing how audio is stored in files and played back in MATLAB.

17.2.1 Digital audio fundamentals

Sound is a physical phenomenon induced by vibrations of physical matter, such as the excitation of a violin string, clapping of hands, and movement of our vocal tract. The vibrations in the matter are transferred to the surrounding air resulting

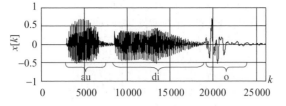

Fig. 17.4. Waveform of a digital audio signal stored in the testaudio1.wav file.

in the propagation of pressure waves. The human auditory system processes the air waves and uses the information contained in the pressure variations to extract audio information from the wave. It is possible to process sound waves directly, as in a microphone, which converts sound to electrical signals that are amplified and played back using a loudspeaker. The term audio refers to electronically recorded or reproduced sound, while digital audio is obtained by the sampling and quantization of an analog audio signal. The waveform of an audio signal is shown in Fig. 17.4.

An audio signal is described using two properties. The first property is pitch, which describes the shrillness of sound. Pitch is directly related to the frequency of the audio signal and the two terms are used interchangeably. The second property is the loudness, which measures the amplitude or intensity of the audio signal using the decibel (dB) scale. Generally, the audible intensity of an audio signal varies between 0 and 140 dB, where 0 dB represents the lower threshold of hearing, below which a human auditory system is incapable of hearing any sound. Typical office environments have an ambient audio level of about 70 dB. Audio above 120 dB is very loud and is injurious to humans.

Sound generated from physical phenomena contains frequency in the range 0–10 GHz. Since the human auditory system is only intelligible to sound frequencies between 20 Hz and 20 kHz, most audio signals record sound within this audible range and neglect any higher-frequency components. For example, the digital audio stored on an audio compact disc is obtained by filtering the CT audio by a lowpass filter with a cut-off frequency of 20 kHz, and the filtered signal is sampled using a sampling rate of 44.1 ksamples/s. The number of quantization levels used to produce digital audio depends upon the application and may vary from 4096 levels obtained with a 12-bit quantizer, to 65 536 levels with a 16-bit quantizer, to 4 million levels with a 24-bit quantizer. Higher numbers of quantization levels result in lower distortion and more precise reproduction of the original sound.

17.2.2 Formats for storing digital audio

Digital audio is available in a wide variety of formats, such as the au, wav, and mp3 formats. Both au and wav formats store audio in the uncompressed form, while mp3 compresses audio using Layer 3 of the MPEG-1 audio compression standard. In this section, we will focus on the au and wav formats. Typically, a digital audio file stored in the au format has an .au extension, while digital audio stored in the wav format has a .wav extension.

MATLAB provides a number of library functions to read and write audio files stored in the au and wav formats. For the au format, MATLAB provides the `auread` and `auwrite` functions to read and write an audio file, respectively. Likewise, the `wavread` and `wavwrite` functions are available to read and write an audio file in the wav format. The following code reads the audio file "testaudio1.wav" using the `wavread` function. There are three output arguments to the `wavread` function. The first argument x is an array where the audio signal is restored. For mono (single-channel) audio signals, x is a 1D vector. For stereo (dual-channel) signals, x is a 2D array corresponding to the number of signals played by the two speakers. The second argument Fs represents the sampling rate, while `nbit` represents the number of bits per sample.

```
>> %Reading the input audio file
>> infile = 'f:\MATLAB\signal\       % audio file
>> testaudio1.wav';
>> [x, Fs, nbit] = wavread(infile);  % x = signal
                                     % Fs = sampling rate
                                     % nbit = number of
                                     % bits per sample
```

The above MATLAB program will produce a 1D array x with dimension $26\,079 \times 1$. In other words, the audio signal is a mono signal and contains $26\,079$ samples. The sampling rate is 22.05 ksamples/s and the signal is quantized using an 8-bit quantizer. The waveform of the audio signal stored in the testaudio1.wav file is shown in Fig. 17.4. To play the audio signal stored in x, we use the `sound` or `soundsc` function available in MATLAB as follows:

```
>> sound(x,Fs);
```

The `soundsc` function normalizes the entries of vector x so that the sound is played as loud as possible without clipping. The mean value is also removed. After playing the vector x obtained from testaudio1.wav, you should recognize that the file contains the spoken word "audio." Relating the word "audio" to Fig. 17.4, we observe that the waveform has three distinct segments. The first segment represents the syllable "au," the second segment represents the syllable "di," and the last segment represents "o." Some silent intervals, represented by near-zero-amplitude waveforms, are also observed in the plot.

17.2.3 Spectral analysis of speech signals

In Section 17.1, we presented techniques for estimating the spectral content of a nonstationary signal. Audio signals such as speech, music, and ambient

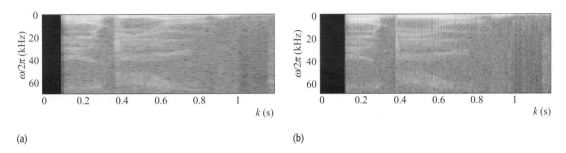

(a)   (b)

Fig. 17.5. Spectrograms of the
speech signal recorded in
testaudio1.wav.
(a) Narrow-band spectrogram;
(b) wide-band spectrogram.

sound are examples of non-stationary signals. Therefore, the techniques presented in Section 17.1 can also be used to estimate the spectral content of audio
signals.

To calculate the spectrogram of the audio signal stored in testaudio1.wav, we
use the following MATLAB code:

```
>> %Reading the input audio file
>> infile = 'testaudio1.wav';        % audio file
>> [x, Fs, nbit] = wavread(infile);  % x = signal
                                     % Fs = sampling rate
                                     % nbit = number of
                                     % bits per sample
>> nfft = 1024; nwind = 1024; noverlap = 768;
>> [spgram,F,T] = specgram(x, nfft,Fs,hanning(nwind),
       noverlap);
>> spgramdB = 20*log10 (abs (spgram) + eps);
>> imagesc([0 length (x)/Fs], 2*pi*F, spgrandB);
>> colormap(gray)
```

The above code calculates the spectrogram using a window size of 1024, shown
in Fig. 17.5(a). As the window size is a power of 2, we choose to calculate the
DFT without any zero padding. For the audio signal testaudio1.wav, the
sampling rate of the signal is given by 22 050 samples/s. A window size of
1024 samples therefore corresponds to a duration of $1024/22\,050 = 0.0461$ s.
Hence, the time resolution of the spectrogram is limited to 46 ms.

The frequency resolution in the spectrogram plotted in Fig. 17.5(a) is obtained
by dividing the sampling frequency by the total number of samples in the frequency domain, which gives $22\,050/1024 = 21.53$ Hz. During the computation
of the spectrogram, it is possible to trade-off time resolution for the frequency
resolution, and vice versa. To improve the time resolution of the spectrogram
in Fig. 17.5(a), we decrease the window size to 256 with an overlap of 128
samples between two successive windows:

```
>> %Reading the input audio file
>> infile = 'testaudio1.wav';        % audio file
>> [x, Fs, nbit] = wavread(infile);  % x = signal
                                     % Fs = sampling rate
                                     % nbit = number of
                                     % bits per sample
>> nfft = 256; nwind = 256; noverlap = 128;
>> [spgram,F,T] = specgram(x,nfft, Fs,hanning(nwind),
     noverlap);
>> spgramdB = 20*log10 (abs (spgram) + eps);
>> imagesc([0 length (x)/Fs], 2*pi*F, spgrandB);
>> colormap(gray)
```

The resulting spectrogram is shown in Fig. 17.5(b). Choosing a window size of
256 samples improves the time resolution to 11.6 ms. However, the frequency
resolution is reduced to $22\,050/256 = 86.13$ Hz. Comparing the two histograms
in Fig. 17.5, we observe that the time resolution of Fig. 17.5(b) is better than that
of Fig. 17.5(a). However, the improvement in the time resolution is obtained at
the cost of the frequency resolution. Clearly, Fig. 17.5(b) has a relatively lower
frequency resolution compared with that of Fig. 17.5(a). Therefore, Fig. 17.5(a),
with a better frequency resolution, is considered a narrow-band spectrogram,
whereas Fig. 17.5(b), with a lower frequency resolution, is considered a wide-
band spectrogram.

17.2.4 Power spectrum

Using the techniques discussed in Section 17.1.5, the power spectrum of the
speech signal stored in vector x obtained from the testaudio1.wav file can
be computed using the psd function available in MATLAB as follows:

```
>> nwind=512; nfft = 512; noverlap = floor(3*nwind/4) ;
>> [Pxx, F] = psd(x, nfft, Fs, hanning(nwind),noverlap);
>> plot(F,10*log10(Pxx));
```

The resulting power spectrum is shown in Fig. 17.6, where we observe that
most of the energy of the signal is concentrated in the frequency band 0–2 kHz.

17.2.5 Spectral analysis of music signals

In this section, we analyze the spectral content of the music signal stored in
testaudio2.wav using the spectrogram and periodogram methods. The

Fig. 17.6. Power spectrum of the speech signal stored in the `testaudio1.wav` file.

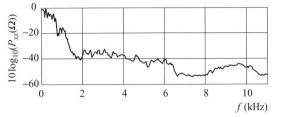

music signal is read using the following MATLAB code and the resulting time-varying waveform of the music signal is plotted in Fig. 17.7(a):

```
>> %Reading the input audio file
>> infile = 'testaudio2.wav';       % audio file
>> [x, Fs, nbit] = wavread(infile); % Fs = sampling rate,
                                     % nbit = # bits/sample
>> plot(1/Fs*[0:length(x)-1],x);
>> nfft=1024; nwind=1024; noverlap=512;
>> [spgram, F, T] = specgram(x,nfft,Fs,hanning(nwind)
      noverlap);
>> imagesc([0 length(x)/Fs], F/1000, 20*log10
      (abs(spgram) + eps));
>> colormap(gray)
>> [Pxx, F] = psd(x,nfft,Fs, hanning(nwind),noverlap);
>> plot(F,10*log10(Pxx));
```

The resulting spectrogram is shown in Fig. 17.7(b), where the horizontal axis represents time and the vertical axis represents frequency. As the speech signal is real-valued, the spectrum is plotted for the positive frequencies only. Since the bright intensity regions represent higher energy, it can be seen that the signal has most energy at the lower frequencies.

The average periodogram of the music signal is plotted in Fig. 17.7(c). It is observed that the peak power of about 6.5 dB occurs at 100 Hz and that the power decreases as the frequency is increased.

17.3 Audio filtering

Frequency-selective filtering emphasizes certain frequency components by attenuating the remaining frequency components present in a signal. Four types of digital filters, namely lowpass, highpass, bandpass, and bandstop filters, were covered in Chapters 14–16. In this section, we process audio signals using these digital filters.

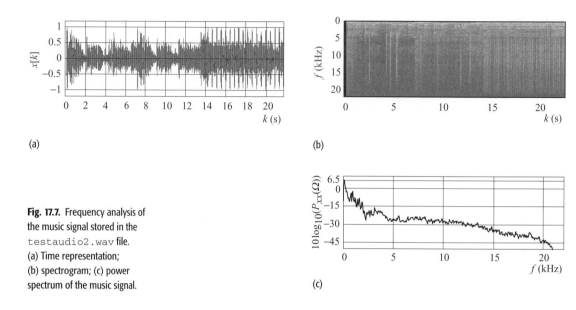

Fig. 17.7. Frequency analysis of the music signal stored in the `testaudio2.wav` file. (a) Time representation; (b) spectrogram; (c) power spectrum of the music signal.

Example 17.4

Consider the audio signal stored in the `bell.wav` file, which was sampled at a sampling rate of 22 050 samples/s and quantized using an 8-bit quantizer. The power spectral density, shown in Fig. 17.8(b), illustrates that the signal has frequency components across the entire 0–11 025 Hz frequency range. We now process the audio signal with the lowpass, highpass, and bandpass filters.

Lowpass filtering A lowpass FIR filter with a cut-off frequency of 3 kHz and order 64 is designed using the `fir1` MATLAB library function. The following MATLAB code designs the lowpass filter:

```
>> filtLow = fir1(64,3000/        % Filter: Order = 64
   (Fs/2));                        % cutoff = 3kHz
>> w = 0:0.001*pi:pi;             % discrete frequencies for
                                   % spectrum
>> HLpf = freqz(filtLow,1,w);     % transfer function
>> plot(w*Fs/(2*pi),20*log10      % magnitude spectrum
   (abs(HLpf) + eps));
```

By default, the `fir1` function uses the Hamming window. Since the `fir1` function accepts normalized frequencies, the cut-off frequency is normalized with half the sampling frequency. The magnitude spectrum of the resulting lowpass filter is shown in Fig. 17.9(a).

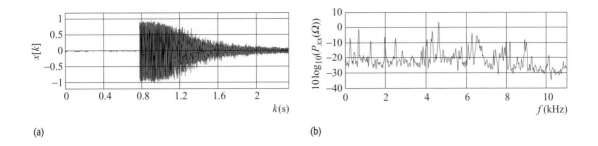

Fig. 17.8. Audio signal stored in the `bell.wav` file. (a) Time representation; (b) power spectrum.

To derive the output of the lowpass filter when the audio signal stored in `bell.wav` is applied at the input of the filter, the following MATLAB code is used:

```
>> xLpf = filter(filtLow,1,x);    % Lowpass filtered audio
                                  % signal
```

To hear the resulting audio signal and plot its power spectrum, we use the following MATLAB code:

```
>> sound(xLpf,Fs);               % Play filtered sound
>> nfft=1024; nwind=1024; noverlap=512;
>> [Pxx, F] = psd(xLpf,nfft,Fs, hanning(nwind),noverlap);
>> plot(F,10*log10(Pxx));
```

Fig. 17.9. Lowpass filtering of the audio signal stored in the `bell.wav` file. (a) Frequency characteristics of a 64-tap FIR lowpass filter designed using a Hamming window with a cut-off frequency of 3000 Hz. (b) Power spectrum of the filtered signal.

Listening to the lowpass filtered sound, we observe that the sound is less shrill with a lower pitch. This is also apparent from the power spectrum shown in Fig. 17.9(b), where we observe that the frequency components above 3 kHz have a much lower magnitude than the corresponding frequency components of the original bell sound.

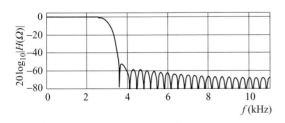

(a)

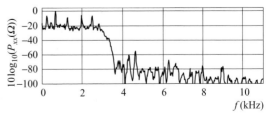

(b)

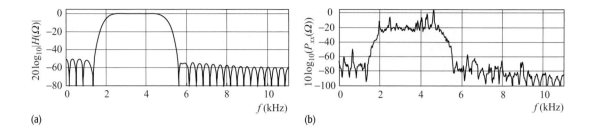

(a)　　　　　　　　　　　　　　　　　　(b)

Fig. 17.10. Bandpass filtering of the audio signal stored in the `bell.wav` file. (a) Frequency characteristics of a 64-tap FIR bandpass filter designed using a Hamming window with cut-off frequencies of 2000 and 5000 Hz. (b) Power spectrum of the filtered signal.

Bandpass filtering As was the case for the lowpass filter, we design the bandpass filter using the `fir1` command. The MATLAB code is given below.

```
>> fBp = fir1(64,[2000        %Filter: order = 64
   5000]/(Fs/2));             % cutoff = [2 5]kHz
>> w = 0:0.001*pi:pi;         % discrete frequencies for
                              % spectrum
>> HBpf = freqz(fBp,1,w);     % transfer function
>> plot(w*Fs/(2*pi),20*log
   10(abs(HBpf) + eps));      % magnitude spectrum
```

The magnitude spectrum of the bandpass filter is plotted in Fig. 17.10(a), which filters the bell sound using the following MATLAB code:

```
>> xBpf = filter(fBp,1,x);    % Bandpass filtered audio
                              % signal
>> sound(xBpf,Fs);            % Play filtered sound
>> nfft=1024; nwind=1024; noverlap=512;
>> [Pxx, F] = psd(xBpf,nfft, Fs,hanning(nwind),noverlap);
>> plot(F,10*log10(Pxx + eps));
```

The power spectrum of the resulting bandpass signal is plotted in Fig. 17.10(b). We see that the frequency components within the pass band of [2000 5000] Hz are retained in the filtered signal. The remaining frequency components are attenuated by the bandpass filter.

Highpass filtering The highpass filter with a cut-off frequency of 4 kHz is designed using the following MATLAB code:

```
>> fHp = fir1(64,4000/(Fs/2),'high');
                              % Filter: order = 64
                              % cutoff = 4kHz
>> w = 0:0.001*pi:pi;         % discrete frequencies for
                              % spectrum
```

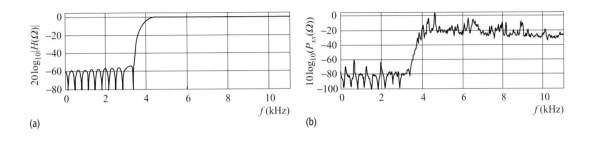

Fig. 17.11. Highpass filtering of the audio signal stored in the `bell.wav` file. (a) Frequency characteristics of a 64-tap FIR highpass filter, with cut-off frequency of 4000 Hz, designed using a Hamming window. (b) Power spectrum of the filtered signal.

```
>> HHpf = freqz(fHp,1,w);        % transfer function
>> plot(w*Fs/(2*pi),20*log10    % magnitude spectrum
    (abs(HHpf) + eps));
```

The magnitude spectrum of the highpass filter is plotted in Fig. 17.11(a), which filters the bell sound using the following code:

```
>> xHpf = filter(fHp,1,x);       % Highpass filtered audio
                                  % signal
>> sound(xHpf,Fs)                % play the sound
>> nfft=1024; nwind=1024; noverlap=512;
>> [Pxx, F] = psd(xHpf,nfft, Fs,hanning(nwind),noverlap);
>> plot(F,10*log10(Pxx + eps));
```

The power spectrum of the highpass filtered signal is shown in Fig. 17.11(b), where we observe that the frequency components below 4 kHz are strongly attenuated. The higher frequency components are left unattenuated. The observation is confirmed on playing the filtered sound, which sounds shriller, with a higher pitch than the original bell sound.

Example 17.4 demonstrates the effects of lowpass, bandpass, and highpass filtering on an audio signal. The following example uses a bandstop filter to eliminate noise from a noisy signal.

Example 17.5

Consider the audio signal stored in the `testaudio3.wav` file with the time-domain representation shown in Fig. 17.12(a). The audio signal is sampled at a sampling rate of 22 050 samples/s. Using the average periodogram method discussed in Section 17.1.5, the power spectral density of the audio signal is estimated and plotted in Fig. 17.12(b). From the power spectral density plot, we observe that there is a sharp peak at 8 kHz, which is identified as noise corrupting the audio signal. The noise can be heard if we play the audio signal.

To suppress the noise, we use a bandstop filter of order 128 with a stop band that ranges from 7800–8200 Hz. The order of the bandstop filter is chosen arbitrarily in this example. In more sophisticated applications, the order is

Part III Discrete-time signals and systems

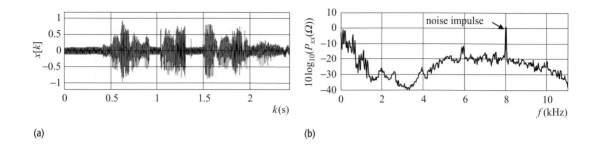

Fig. 17.12. Noise-corrupted signal stored in the `testaudio3.wav` file. (a) Time representation; (b) power spectrum.

computed from the amount of attenuation required within the stop band. Using MATLAB, the transfer function of the bandpass filter is computed as follows:

```
>> wc =[7800 8200]/11025;       % Normalized cutoff
                                % frequency
>> fBs = fir1(128,wc,'stop');   % order-128 filter, 129 tap
>> w = 0:0.001*pi:pi;           % discrete frequencies
                                % for spectrum
>> HBs = freqz(fBs,1,w);        % transfer function
>> plot(w*Fs/(2*pi),20*log10 (abs(HBs)));
                                % magnitude spectrum
```

The magnitude spectrum of the resulting bandstop filter is plotted in Fig. 17.13(a), which shows strong attenuation at 8 kHz. The gain at the remaining frequencies is close to unity. The noisy signal is filtered with the bandstop filter and the power spectral density of the filtered signal is calculated using the following MATLAB code:

Fig. 17.13. Bandstop filtering to eliminate noise from the noise corrupted signal shown in Fig. 17.12. (a) Frequency characteristics of a 129-tap FIR bandstop filter, with cut-off frequencies of [7800 8200] Hz, designed using a Hamming window. (b) Power spectrum of the filtered signal.

```
>> xBsf = filter(fBs,1,x);      % Bandstop filtered audio
                                % signal
>> nfft=1024; nwind=1024; noverlap=512;
>> [Pxx, F] = psd (xBsf,nfft,Fs,hanning (nwind),noverlap);
>> plot(F,10*log10(Pxx));
```

The power spectral density of the filtered output is shown in Fig. 17.13(b), which shows a strong attenuation in the noise impulse present at 8 kHz. On playing the filtered signal, we observe that the effects of the noise have been

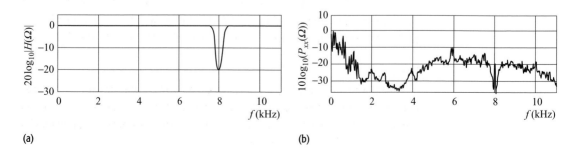

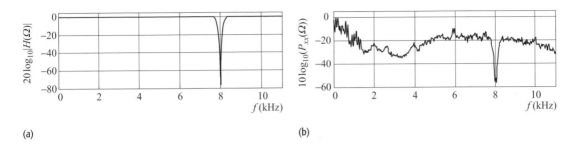

(a)   (b)

Fig. 17.14. Bandstop filtering to eliminate noise from the noise-corrupted signal shown in Fig. 17.12. (a) Frequency characteristics of a 201-tap FIR bandstop filter, with cut-off frequencies of [7800 8200] Hz, designed using a Hamming window. (b) Power spectrum of the filtered signal.

reduced, but not completely eliminated. Therefore, we increase the order of the bandstop FIR filter to 200. Using the above code with the order set to 200, we compute the impulse response of the 201-tap bandstop FIR filter. The magnitude spectrum of the filter is plotted in Fig. 17.14(a). The power spectral density of the filtered signal obtained from the 201-tap bandstop filter is shown in Fig. 17.14(b). On playing the filtered signal, we observe that the noise component has been successfully suppressed. However, the suppression of noise is at the cost of eliminating certain frequency components which neighbor the frequency of the impulse noise.

17.4 Digital audio compression

Audio data in the raw format requires a large number of bits for representation. For example, the CD-quality stereo audio requires a data rate of 176.4 kbytes/s for transmission or storage. This data rate is not supported by many networks, including the internet, hence real-time audio applications cannot be supported if the audio data are transmitted in the raw format. Similarly, storing at a data rate of 176.4 kbytes/s requires a large storage capacity, even to save a five-minute session. Compressing audio is therefore imperative for real-time audio transmission or for storing an audio session of meaningful length. Audio compression is defined as the process through which digital audio can be represented by a lower number of bits. Most compression techniques can be classified into two categories, namely lossy compression and lossless compression. While lossless techniques are ideal as they allow perfect reconstruction of audio, they limit the amount of compression that can be achieved. Lossy techniques exploit the psychoacoustic characteristics of the human auditory system and achieve higher compression by eliminating audio components that are not audible to humans. In this section, we present the basic principles of audio compression. Example 17.6 emphasizes the need for audio compression.

Example 17.6

(a) A stereo (dual-channel) audio signal is to be transmitted through a 56 kbps network in real time. If the sampling rate of the digital audio signal is 22.05 ksamples/s, what is the maximum average number of bits that can be used to represent an audio sample?

(b) If the quantizer uses 8 bits/sample for each channel, what is the maximum allowable sampling rate such that the audio signal can be transmitted over a 56 kbps network?

(c) Calculate the compression ratio required to transmit the stereo audio signal through a 56 kbps channel if the sampling rate is given by 22.05 ksamples/s and the quantizer uses 8 bits/sample.

Solution

(a) Assuming that the quantizer uses n bits to represent each sample,

$$\text{number of bits produced per second} = n \text{ bits/sample} \times 22\,050 \text{ samples/s}$$
$$\times 2 \text{ channels} = 44\,100n \text{ bps}$$

Equating this with the transmission rate of 56 kbps, we obtain

$$n = 56\,000/44\,100 = 1.27 \text{ bits/sample.}$$

(b) Assuming that the sampling rate is given by f_s samples/s,

$$\text{number of bits produced per second} = 8 \text{ bits/sample} \times f_s \text{ samples/s}$$
$$\times 2 \text{ channels} = 16 f_s \text{ bits/s.}$$

Equating this with the transmission rate of 56 kbps, we obtain

$$f_s = 56\,000/16 = 3500 \text{ samples/s.}$$

(c) To determine the compression ratio, we first calculate the number of bits produced per second:

$$\text{number of bits produced per second} = 8 \text{ bits/sample} \times 22\,050 \text{ samples/s}$$
$$\times 2 \text{ channels} = 352\,800 \text{ samples/s.}$$

The compression ratio is therefore given by

$$\text{compression ratio} = \frac{\text{number of bits per second in the raw data}}{\text{number of bits per second in the compressed data}}$$
$$= \frac{352\,800}{5600} = 6.3.$$

Example 17.6 demonstrates that digital audio can be transmitted over a low-capacity transmission channel in real time using three different approaches. The first approach reduces the number of bits used to represent each sample. This approach is not useful as it reduces the number of quantization levels such that considerable distortion is introduced into the transmitted audio. The second approach uses a low sampling rate, which is not practical as the sampling rate is dependent on the maximum frequency present in the audio signal. The maximum frequency of the audio signal can be reduced by lowpass filtering, but this will again introduce distortion. The third approach compresses the raw audio data. Compression of digital audio is achieved by eliminating redundancy present in a signal. There are primarily three types of redundancies present in an audio signal that may be exploited.

Statistical redundancy In most audio signals, samples with lower magnitudes have a higher probability of occurrence than samples with higher magnitude. In such cases, an entropy coding scheme, such as the Huffman code, can be used to allocate fewer bits to frequently occurring values and a higher number of bits to the other values. This reduces the bit rate for representing audio signals when compared with a coding scheme with an equal number of bits allocated per sample.

Temporal redundancy Neighboring audio samples typically have a strong correlation between themselves such that the value of a sample can be predicted with fairly high accuracy from the last few sample values. Predictive coding schemes exploit this temporal redundancy by subtracting the predicted value from the actual sample value. The resulting difference signal is then compressed using an entropy based coding scheme, such as the dictionary or Huffman codes.

Psychoacoustics redundancy There are many idiosyncrasies in the human auditory system. For example, the sensitivity of the human auditory system is maximum for frequencies within the 2000–4000 Hz band and the sensitivity decreases above or below this band. In addition, a strong frequency component masks the neighboring weaker frequency components. The unequal frequency sensitivity and masking properties are exploited to compress the audio.

In the following section, we present a simplified audio compression technique, known as the differential pulse-code modulation (DPCM) technique. To achieve compression, the DPCM reduces the temporal redundancy present in an audio signal.

17.4.1 Differential pulse-code modulation

Most audio signals encoded with pulse-code modulation (PCM) exhibit a strong correlation between neighboring samples. This is especially true if the signal is sampled above the Nyquist sampling rate. Figure 17.15 plots 30 samples of an audio signal stored in the chord.wav file. We observe that the neighboring samples are correlated such that their values are fairly close to each other. In DPCM, an audio sample $s[k]$ is predicted from the past samples. An M-order predictor calculates the predicted value of an audio sample at time instant k using the following equation:

$$\hat{s}[k] = \sum_{m=1}^{M} \alpha_m s[k-m], \qquad (17.15)$$

where $s[k-m]$ is the value of the audio sample at time instant $k-m$ and α_m are the predictor coefficients. The DPCM encoder quantizes the prediction

Fig. 17.15. Selected samples (sample 700 to 730) of the audio signal stored in the `chord.wav` file. The neighboring samples exhibit a strong correlation between themselves.

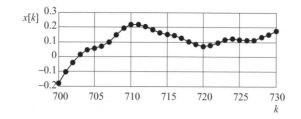

error as follows:

$$e[k] = s[k] - \sum_{m=1}^{M} \alpha_m s[k-m], \qquad (17.16)$$

which is followed by a lossless entropy coding scheme. The DPCM decoder takes the inverse of the above steps in the reverse order. Since the actual sample values $s[k-m]$ are not accessible at the decoder, the decoder uses the reconstructed values. In order to use the same prediction model at the encoder and decoder, Eq. (17.16) is modified as follows:

$$e[k] = s[k] - \sum_{m=1}^{M} \alpha_m s'[k-m], \qquad (17.17)$$

Fig. 17.16. Schematic of differential pulse-code modulator used for lossy compression. (a) DPCM encoder used to compress a signal; (b) DPCM decoder used to reconstruct a signal. The difference $e[k]$ between the original input signal $s[k]$ and its predicted value $\hat{s}[k]$ is quantized and transmitted to the receiver.

where $s'[k-m]$ is the reconstructed value of the audio sample $s[k-m]$. The values of the predictor coefficients α_m are usually estimated based on a maximum likelihood (ML) estimator. Alternatively, a universal prediction model may be used where the predictor coefficients are kept constant for different audio signals. Examples of the universal prediction models include the following:

first-order prediction model $\hat{s}[k] = 0.97s'[k-1];$ (17.18)

second-order prediction model $\hat{s}[k] = 1.8s'[k-1] - 0.84s'[k-2];$ (17.19)

third-order prediction model $\hat{s}[k] = 1.2s'[k-1] + 0.5s'[k-2]$

$$\qquad\qquad\qquad\qquad\qquad - 0.78s'[k-3]. \qquad (17.20)$$

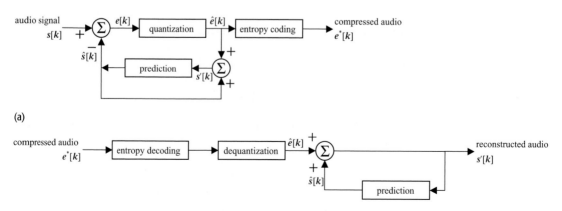

(a)

(b)

The block diagrams of DPCM encoding and decoding systems are shown in Fig. 17.16. Example 17.7 illustrates various steps of the DPCM coding.

Example 17.7

Assume that the first four samples of a digital audio sequence are given by [70, 75, 80, 82]. The audio samples are encoded using DPCM with the first-order predictor defined in Eq. (17.18). The error samples obtained by subtracting the predicted sample values from the actual audio sample values are divided by a quantization factor of 2 and then rounded to the nearest integer. Determine the values of the reconstructed signal.

Solution

In DPCM, the first sample value is encoded independent of other samples in the sequence. In this example, we assume that the first audio sample, at $k = 0$, with a value of 70 is encoded without any quantization error. In other words, $e[0] = \hat{e}[0] = 0$ and the reconstructed sample value $s'[0] = 70$.

At $k = 1$, the predicted sample, the associated error, and the quantized error are given by

predicted value $\qquad$ $\hat{s}[1] = 0.97 \times 70 = 67.9$;

error $\qquad$ $e[1] = 75 - 67.9 = 7.1$;

quantized error $\qquad$ $\hat{e}[1] = \text{round}(7.1/2) = 4$.

The reconstructed value of the sample at $k = 1$ is therefore given by

$$s'[1] = 0.97 \times 70 + 4 \times 2 = 75.9.$$

At $k = 2$, the predicted sample, the associated error, and the quantized error are given by

predicted value $\qquad$ $\hat{s}[2] = 0.97 \times 75.9 = 73.623$;

error $\qquad$ $e[2] = 80 - 73.623 = 6.377$;

quantized error $\qquad$ $\hat{e}[2] = \text{round}(6.377/2) = 3$.

The reconstructed value of the sample at $k = 2$ is therefore given by

$$s'[2] = 0.97 \times 75.9 + 3 \times 2 = 79.623.$$

At $k = 3$, the predicted sample, the associated error, and the quantized error are given by

predicted value $\qquad$ $\hat{s}[3] = 0.97 \times 79.623 = 77.2343$;

error $\qquad$ $e[3] = 82 - 77.2343 = 4.7657$;

quantized error $\qquad$ $\hat{e}[3] = \text{round}(4.7657/2) = 2$.

Table 17.1. Various steps of DPCM coding for Example 17.7

	Time index, k			
	0	1	2	3
Original signal, $s[k]$	70	75	80	82
Error signal, $e[k]$	0	$75 - 67.9 = 7.1$	$80 - 3.6 = 6.4$	$82 - 7.2 = 4.8$
Quantized error signal, $\hat{e}[k]$	0	$7.1/2 = 4$	$6.4/2 = 3$	$4.8/2 = 2$
Reconstructed error	0	$4 \times 2 = 8$	$3 \times 2 = 6$	$2 \times 2 = 4$
Reconstructed signal, $s'[k]$	70	$67.9 + 8 = 75.9$	$73.6 + 6 = 79.6$	$77.2 + 4 = 81.2$
Reconstruction error	0	-0.9	0.4	0.8
Predicted signal for next sample	$70 \times 0.97 = 67.9$	$75.9 \times 0.97 = 73.6$	$79.6 \times 0.97 = 77.2$	$81.2 \times 0.97 = 78.8$

The reconstructed value of the sample at $k = 2$ is therefore given by

$$s'[3] = 77.2343 + 2 \times 2 = 81.2343.$$

The values of the audio samples reconstructed from DPCM are given by

$$[70, 75.9, 79.623, 81.2343],$$

which implies that the following distortion is introduced by DPCM:

$$[0, -0.9, 0.377, 0.7657].$$

The above steps are summarized in Table 17.1. The third row contains the quantized values of the error signal, which is compressed with a lossless scheme and transmitted to the receiver.

17.4.2 Audio compression standards

The DPCM compression scheme, as described in Section 17.4.1, is a primitive audio compression method that provides a low compression ratio. Several more efficient compression techniques have been developed since the 1980s. In order to achieve compatibility between the compressed bit streams, several audio compression standards have been developed by the International Organization for Standardization (ISO) and the International Telecommunication Union (ITU). These audio compression standards can be broadly classified into two categories: the low-bit-rate audio coders for telephony, such as G.711, G.722, and G.729 developed by the ITU, and the general-purpose high-fidelity audio coders, such as the moving pictures expert group (MPEG) audio standards, developed by the ISO and included in MPEG-1, MPEG-2, and MPEG-4.

The ISO standards are generic audio compression standards designed for general-purpose audio. These standards provide a trade-off between compression ratio and quality. For example, the MPEG-1 audio algorithm has three layers. Layer 1 is the simplest algorithm and provides moderate compression. Layer 2 has moderate complexity and provides a higher compression than Layer 1.

Layer 3 has the highest complexity and provides the best performance. Note that the MPEG-1 Layer 3 standard is also referred to as the MP3 standard.

In addition to the ITU G.7xx and ISO MPEG standards, a few other standards have been developed. For example, Dolby Laboratories have developed multi-channel high-fidelity audio coding standards such as AC-2 and AC-3 coders. The AC-3 standard has been adopted in the standard and high-definition digital television standard in North America. Readers are referred to more advanced texts for details on audio compression standards.

17.5 Digital images

Digital images have become a part of our daily lives. In this section, we present a brief overview of digital images and the techniques used to represent them.

17.5.1 Image fundamentals

A still monochrome image is defined in terms of its intensity or brightness i as a function of the spatial coordinates (x, y). A still image is, therefore, a 2D function $i(x, y)$. For analog images, coordinates (x, y) have a continuous value.

A discrete image $i[m, n]$ is obtained by sampling the intensity $i(x, y)$ along a rectangular grid $M = [m \Delta x, n \Delta y]$ with resolutions of Δx along the horizontal axis and Δy along the vertical axis. Each discrete point $[m \Delta x, n \Delta y]$ along the rectangular grid is referred to as a picture element, or pixel. A digital image $i[m, n]$ is an extension of the discrete image, where the intensity i is quantized by a uniform quantizer. The number of quantization levels varies from one application to another and depends upon the precision required. Most digital images are quantized using an 8-bit quantizer, leading to 128 quantization levels. Medical images require higher precision and are quantized using a 12- or 16-bit quantizer. Color images are further extensions of discrete images, where the intensities of the three primary colors are measured at each pixel. Color images are therefore represented in terms of three components $r[m, n]$, $g[m, n]$, and $b[m, n]$, where intensities are denoted by $r[m, n]$ for red, $g[m, n]$ for green, and $b[m, n]$ for blue.

As an example of still images, the back cover of this book illustrates a 450×366 pixel test image, referred to as "train," using three different quantization levels. The first figure shows the train image in the black and white (BW) format, where a single bit is used to represent each pixel. Bit 0 represents the lowest intensity (black), while bit 1 represents the highest intensity (white). The total number of bits used to represent the BW image is given by 1 bit/pixel $\times$ (450×366) pixels $= 164\,700$ bits. To provide more details, the second figure uses 8-bit quantization for each pixel, leading to a total number of 8 bit/pixel $\times$ (450×366) pixels $= 1\,317\,600$ bits. The third figure shows the train image in the color format, where each pixel is represented in terms of the intensities of the three primary colors. The color representation of the train image requires 8 bit/color $\times$ 3 color/pixel $\times$ (450 $\times$ 366) pixels $= 3\,952\,800$ bits.

A final extension of discrete images is obtained by measuring the color intensities $r[m, n]$, $g[m, n]$, and $b[m, n]$ at discrete time k. Exploiting the persistence of vision and showing continuously recorded images at a uniform rate provides the impression of a video. A digital video is therefore defined in terms of the three color components $r[m, n, k]$, $g[m, n, k]$, and $b[m, n, k]$. In this section, we limit ourselves to 8-bit, monochrome, still images $i[m, n]$. However, the techniques are generalizable to color images and videos.

17.5.2 Sampling of coordinates

Chapter 9 defined the Nyquist rate as the minimum sampling rate that can be used to sample a time-varying CT signal without introducing any distortion during reconstruction. For a baseband signal, the Nyquist rate is twice the maximum frequency present in the signal. For analog images, the principle can be extended to the spatial coordinates (x, y) in two dimensions to obtain a discrete image. The minimum sampling rates are given by the Nyquist rates and are computed from the maximum frequencies in the two directions.

17.5.3 Image formats

Like digital audio, images are available in a wide variety of formats, including pgm, ppm, gif, jpg, and tiff. In each format, a digital image is stored as a 1D stream of numbers. The difference in the format lies in the manner in which the image data is compressed before storage. The portable graymap (PGM) format is used for storage of gray-level images, where raw data is stored without compression in the ASCII or binary representations. A few bytes of header information included before the image data describe the format of the file, the representation (ASCII or binary) used and the number of rows and columns in the image. The portable pixmap (PPM) format is an extension of the PGM format for color images, where the intensities of the three primary colors are stored for each pixel.

The graphical interface (GIF) format uses a compression algorithm to reduce the size of the data file. It is limited to 8-bit (256) color images and hence is suitable for images with a few distinctive colors. It supports interlacing and simple animation, and it can also support grayscale images using a gray palette.

The joint photograph expert group (JPEG) format uses transform-based compression and provides the user with the capability of setting the desired compression ratio.

The tagged image file (TIFF) format supports different types of images, including monochrome, grayscale, 8-bit and 24-bit RGB images that are tagged. The images can be stored with or without compression.

MATLAB provides two library functions `imread` and `imwrite`, respectively, to read and write images. These functions can read and write the image files in several different formats. The following code shows the syntax for calling these functions:

```
>> x = imread('rini.jpg'); % x is a 2-D "uint" type array
>> size(x);                % displays the size of the image
>> imshow(x);              % displays the image
>> xd = double(x);         % xd is the image array
                             % with double precision
>> xmax = max(max(xd))
>> x_bright=uint8(xd*2);   % increases brightness of image
>> imwrite(x_bright,'rini_bright. jpg','jpg',
      'Quality',80) ;
```

The above code loads the Rini test image from the `rini.jpg` file and displays the image in Fig. 17.17(a) using the `imshow` function. The `imread` function used in the code returns an array stored as unsigned integers with 8-bit precision. To carry out any arithmetic operation on the image, we need to convert the data to other data types. The instruction `double` changes the data type from unsigned integer to double. The instruction `max` determines the maximum gray level present in the image, and for the Rini image the value of `xmax` is given by 124. As `xmax` has a low value, the image has low brightness, as observed in Fig. 17.17(a). A possible way to improve the brightness of the image is to increase the intensity level of the whole image linearly. In an 8-bit image, the maximum gray level is 255. Therefore, we scale up the gray values by a factor of 2, which is achieved by multiplying the intensity by a factor of 2. This is followed by the conversion of the gray values to the `uint8` type. The brightened image represented by the matrix `x_bright` is shown in Fig. 17.17(b). The last line of the MATLAB code stores the brightened image in the JPEG format with filename `rini_bright.jpg` using the `imwrite` function. Note that the JPEG format compresses the gray image based on the specified quality factor, which is a number between 0 and 100. A high value for the quality factor implies higher quality with low compression, while a low value of the quality factor implies lower quality with high compression. Using a quality factor of 80, the processed rini image is compressed to a file size of 27 kbytes. Compared with the original `rini.jpg` file, which has a size of 180 kbytes, this implies a compression ratio of 6.66.

17.5.4 Spectral analysis of images

Like real audio signals, natural images are non-stationary signals. The frequency content of the images is estimated by extending the 1D spectral analysis techniques, presented in Section 17.1, to two dimensions. Here, we discuss the average periodogram approach to calculate the power spectrum of a 2D image.

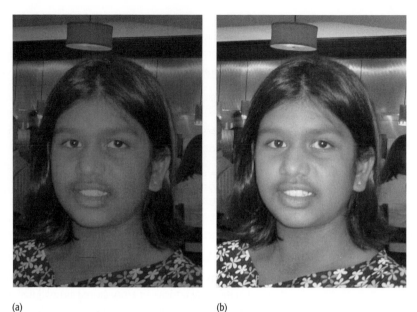

(a)   (b)

Step 1 Parse the input image into smaller 2D segments by applying a 2D window $g[m, n]$. Depending upon the application, the parsed segments may or may not have overlapping pixels.

Step 2 Compute the 2D DFT $I(\Omega_m, \Omega_n)$ of each image segment $i(m, n)$, which is used to estimate the power spectrum based on the following expression:

$$\hat{P}_I(\Omega) = \frac{1}{\mu^2} |I(\Omega_m, \Omega_n)|^2, \tag{17.21}$$

where μ is the norm of the 2D window function defined as follows:

$$\mu = \sqrt{\sum_m \sum_n g^2[m, n]}. \tag{17.22}$$

Step 3 The average power spectrum is obtained by averaging the waveforms obtained in step 2.

We illustrate the steps involved in computing the power spectrum with the following example.

Example 17.8
Consider the synthetic image, referred to as the sinusoidal grating, defined by the following equation:

$$i(x, y) = 127 \cos[2\pi(4x + 2y)]. \tag{17.23}$$

Discretizing the analog image with a sampling rate of 20 samples/spatial unit in each direction, the DT image is given by

$$i[m, n] = 127 \cos[2\pi(4m + 2n)/20] \qquad (17.24)$$

for $0 \leq m, n \leq 255$. Compute the power spectrum of the DT image using the average periodogram approach.

Solution

We plot the DT image modeled in Eq. (17.24) using the following MATLAB code:

```
>> m = [0:1:255];              % x-coordinates
>> n = [0:1:255];              % y-coordinates
>> [mgrid, ngrid] =
     meshgrid(m,n);            % determine the 2D meshgrid
>> I = 127*cos(2*pi*(4*mgrid + 2*ngrid)/20);
                               % pixel intensities
>> imagesc(I);                 % sketch image
>> axis image;
>> colormap (gray);
```

The resulting image is shown in Fig. 17.18(a). The power spectrum is calculated using the 2D Bartlett window of size (64×64) pixels, an overlap of (48×48) pixels between adjacent windows, and a (256×256)-point DFT for each parsed subimage. The MATLAB code used to compute the power spectrum is given by

```
>> m = [0:1:255];           % x-coordinates
>> n = [0:1:255];           % y-coordinates
>> [mgrid, ngrid]           % determine the
     = meshgrid(m,n);         % 2D meshgrid
>> I = 127*cos(2*pi*(4*mgrid + 2*ngrid)/20);
                            % pixel intensities
% 2D Bartlett window
>> x = bartlett(64);
>> for i = 1:64
       zx(i,:) = x' ;
       zy(:,i) = x ;
>> end
>> bartlett2D = zx .* zy;
>> mesh(bartlett2D)         % displaying 2D
                            % Bartlett window
```

Fig. 17.18. (a) Synthetic sinusoidal grating. (b) Power spectrum of the synthetic sinusoidal grating.

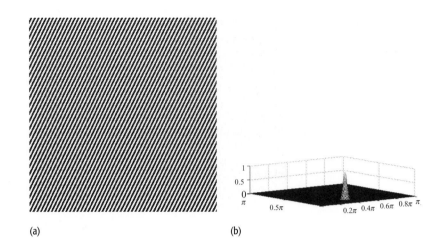

(a)   (b)

```
% calculate power spectrum
>> P = zeros(256,256);
>> for (i = 1:64:255)
        for (j = 1:64:255)
                Isub = I(i:i+63,j:j+63).*bartlett2D;
                P = P + fft2(Isub,256,256);
        end
end
% mesh plot with x and y-axis scaled by pi
>> mesh([1:128]*2/256,[1:128]*2/256,
    abs(P(1:128,1:128)/max(max(P))).^2);
```

Figure 17.18(b) illustrates a sharp peak at the horizontal frequency $\Omega_x = 0.4\pi$ and at the vertical frequency $\Omega_y = 0.2\pi$. This observation is consistent with the mathematical model, Eq. (17.23), used to construct the synthetic image. Unlike the earlier power spectrum plots, we use a linear scale along the z-axis in Fig. 17.18(b).

The above MATLAB code is modified to construct the power spectrum of a real test image, referred to as the Lena image. The test image has dimensions of 512×512 pixels and is illustrated in Fig. 17.19(a) along with its power spectrum in Fig. 17.19(b). In computing the power spectrum, a 2D Bartlett window of dimension 128×128 with an overlap of 96×96 pixels, and a (256×256)-point DFT is used. The dB scale is used along the z-axis to plot the power spectrum. Real images typically include most frequencies and hence the power spectrum in Fig. 17.19(b) exhibits an almost uniform distribution over all frequencies in the horizontal and vertical directions.

Fig. 17.19. (a) Original
(512 × 512) pixel Lena image.
(b) Power spectrum of the Lena
image.

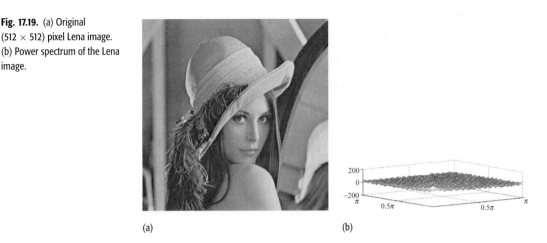

(a)   (b)

17.6 Image filtering

Real images consist of a combination of smooth regions and active regions
with edges. In smooth regions, the intensity values of the pixels do not change
significantly. Therefore, the smooth regions represent lower-frequency components in the 2D frequency space. On the other hand, the intensity values in
the active regions change significantly over edges. The active regions represent
higher-frequency components. Extracting the low- and high-frequency components from a real image has important applications in image processing. In this
section, we introduce frequency-selective filtering in two dimensions.

The mathematical model for filtering a 2D image $g[m, n]$ by a filter with an
impulse response $h[m, n]$ is given by

$$y[m, n] = g[m, n] * h[m, n] = \sum_{q=-\infty}^{\infty} \sum_{r=-\infty}^{\infty} g[m - q, n - r]h[q, r], \quad (17.25)$$

where $y[m, n]$ is the output response of the filter and $*$ denotes the convolution
operation. Alternatively, the filtering can be performed in the frequency domain
using the following equation:

$$Y(\Omega_x, \Omega_y) = G(\Omega_x, \Omega_y)H(\Omega_x, \Omega_y), \quad (17.26)$$

where $G(\Omega_x, \Omega_y)$ is the Fourier transform of the input image, $H(\Omega_x, \Omega_y)$ is the
2D transfer function of the filter, and $Y(\Omega_x, \Omega_y)$ is the Fourier transform of the
resulting output. Like 1D filters, 2D filters can be broadly classified into four
categories: lowpass, bandpass, highpass, and bandstop filters. Some examples
of these filters are given in the following.

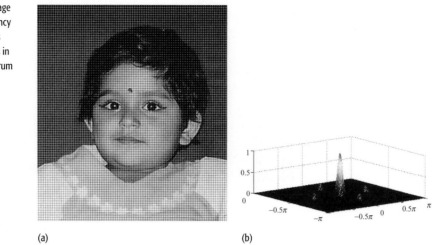

(a)   (b)

17.6.1 Lowpass filtering

Lowpass filtering is widely used in many image processing applications. Some
applications include reducing high-frequency noise that is corrupting an image,
band-limiting the frequency component of an image prior to decimation, and
smoothing the rough edges of an image. In Example 17.9 we provide an example
of lowpass filtering in the spatial domain.

Example 17.9

Figure 17.20(a) shows a noise-corrupted image, referred to as Ayantika. Show
that:

(a) the image has high-frequency noise by plotting the power spectrum;
(b) the lowpass filter with the following impulse response:

$$h[m, n] = \frac{1}{64} \begin{bmatrix} 1 & 2 & 3 & 2 & 1 \\ 2 & 3 & 4 & 3 & 2 \\ 3 & 4 & 5 & 4 & 3 \\ 2 & 3 & 4 & 3 & 2 \\ 1 & 2 & 3 & 2 & 1 \end{bmatrix} \qquad (17.27)$$

eliminates the high-frequency noise from the image.

Solution

The MATLAB code used to plot the power spectrum is given by

```
>> I = imread('ayantika.tif');
>> I = double(I);
>> I = I - mean(mean(I));
```

```
% 2D Bartlett window
>> x = bartlett(32);
>> for i = 1:32
        zx(i,:) = x';
        zy(:,i) = x;
>> end
>> bartlett2D = zx .* zy;
>> n = 0;
% calculate power spectrum
>> P = zeros(256,256);
>> for (i = 1:16:320)
        for (j = 1:16:288)
            Isub = I(i:i+31,j:j+31).*bartlett2D;
            P = P + fftshift(fft2(Isub,256,256));
            n = n + 1;
        end
>> end
>> Pabs = (abs(P)/n).^2;
>> mesh([-128:127]*2/256,[-128:127]*2/256,Pabs/
        max(max(Pabs))));
```

The resulting power spectrum is shown in Fig. 17.20(b), where we see peaks at frequencies $[\Omega_x, \Omega_y]$ given by [0, 0], [0, ±0.5π], and [±0.5π, 0]. The peak at [0, 0] corresponds to the dc gain, whereas the remaining peaks are because of the additive noise that has corrupted the image. We now attempt to eliminate the noise with a lowpass filter.

Figure 17.21(a) shows the magnitude spectrum of the filter with the impulse response specified in Eq. (17.27). We use the following MATLAB code to plot the magnitude spectrum:

```
>> h = 1/64*[1 2 3 2 1; 2 3 4 3 2; 3 4 5 4 3; 2 3 4 3 2; 1
        2 3 2 1];
>> H = fftshift(fft2(h,256,256));
% magnitude spectrum with 256-pt fft
% 2D mesh plot with frequency axis normalized to pi
>> mesh([-128: 127]*2*1/256, [-128:127]*2*1/256, abs(H));
```

Since the filter provides a higher gain at the lower frequencies and lower gain at higher frequencies, it is clear that Fig. 17.21(a) corresponds to a lowpass filter. Note that the gain at frequencies [0, ±0.5π] and [±0.5π, 0] is zero, therefore the lowpass filter would eliminate the additive noise. The filter2

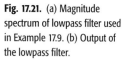

Fig. 17.21. (a) Magnitude spectrum of lowpass filter used in Example 17.9. (b) Output of the lowpass filter.

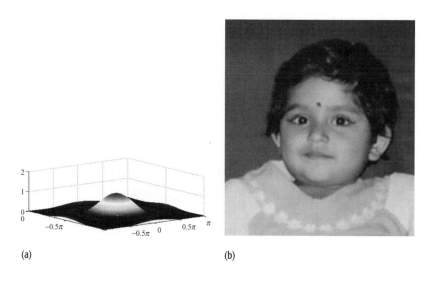

(a)   (b)

function is used to compute the output of the lowpass filter using the following code:

```
>> Y = filter2(h,I);
>> imagesc(Y);
>> axis image; colormap (gray);
```

The resulting output is plotted in Fig. 17.21(b). It is observed that the horizontal and vertical strips have been suppressed by the lowpass filter. However, the lowpass filter also suppresses some high-frequency components other than noise. Therefore, the quality of the filtered image degrades marginally, as observed at the edges. The image in Fig. 17.20(a) has crisp edges, whereas the edges in Fig. 17.21(b) are somewhat blurred.

17.6.2 Highpass filtering

Highpass filtering is used to detect the edges or suppress the low-frequency noise in an image. At times, highpass filters are also used to sharpen the edges of an image. Example 17.10 illustrates one application of highpass filtering.

Example 17.10
Consider the pepper image shown in Fig. 17.22(a). Show that the filter with the impulse response

$$h[m, n] = \frac{1}{9} \begin{bmatrix} -1 & -1 & -1 \\ -1 & 8 & -1 \\ -1 & -1 & -1 \end{bmatrix} \tag{17.28}$$

extracts the edges of the image.

Fig. 17.22. (a) Original 512 ×
512 pixels peppers image.
(b) Magnitude response of the
highpass filter with impulse
response shown in Eq. (17.28).
(c) Output of the highpass filter.

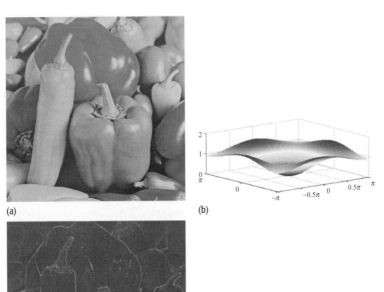

(a)   (b)

(c)

Solution

The following MATLAB code is used to plot the magnitude spectrum:

```
>> h = 1/9*[-1 -1 -1; -1 8 -1; -1 -1 -1];
% magnitude spectrum with 256-point fft
>> H = fftshift(fft2(h,256,256));
% 2D mesh plot with frequency axis normalized to pi
>> mesh([-128:127]*2*1/256, [-128:127]*2*1/256, abs(H));
```

The magnitude frequency response of the filter is shown in Fig. 17.22(b). Since
the gain of the filter is almost zero at low frequencies and unity at higher
frequencies, Eq. (17.28) models a highpass filter. The output of the highpass
filter is obtained using the following code:

```
>> I = imread('peppers.tif');
>> Y = filter2(h,I);
>> imagesc(Y);
>> axis image
```

Figure 17.22(c) shows the output of the highpass filter. From the image plot, it is
clear that the highpass filter extracts the edges, eliminating the smooth regions
(low-frequency components) of the image.

17.7 Image compression

Raw data from digital images requires large disk space for storage. Image compression reduces the amount of data needed to represent an image. As in audio compression, image compression techniques are grouped into lossless and lossy categories. With lossless compression, exact reconstruction of the original image is possible. However, the amount of compression that can be achieved with lossless compression is limited. Lossy compression introduces controlled distortion to increase the compression ratio. The redundancies exploited during image compression are classified into the following categories.

Statistical redundancy The values of pixels in natural images have a non-uniform probability distribution of occurrences such that some values occur more frequently than others. Some compression can be achieved by allocating fewer bits to represent pixels that occur more frequently and more bits to represent pixels that occur less frequently.

Spatial redundancy In real images, the value of a pixel is highly correlated to its neighboring pixels. Image compression exploits such spatial redundancy to compress the image.

Psychovisual redundancy The human visual system (HVS) is less sensitive to certain features within an image. For example, slight variations in the pixel intensities within a uniform region cannot be perceived by the HVS. Image compression exploits such psychovisual redundancy to remove features from the image whose presence or absence is inconceivable to the HVS.

17.7.1 Predictive coding

Predictive coding exploits spatial redundancy to compress an image. Instead of encoding the original pixels, predictive-coding schemes calculate the difference between the actual pixel values and the estimated pixel values predicted from the neighboring pixels. The resulting difference or error image is instead encoded. Since the difference image has lower correlation than the original image, more compression is achieved by encoding the difference image. Predictive coding may use a universal model or a localized model derived from the reference image. Examples of universal predictive models are listed below:

first-order prediction $\quad \hat{i}[m, n] = i[m, n - 1];$ $\qquad$ (17.29)

$\qquad\qquad\qquad\quad \hat{i}[m, n] = i[m - 1, n];$ $\qquad$ (17.30)

second-order prediction $\quad \hat{i}[m, n] = 0.48i[m, n - 1] + 0.48i[m - 1, n]$

$\qquad\qquad\qquad\qquad\qquad\qquad\qquad\qquad\qquad\qquad$ (17.31)

third-order prediction $\quad \hat{i}[m, n] = 0.33i[m, n - 1] + 0.33i[m - 1, n]$

$\qquad\qquad\qquad\qquad\quad + 0.33i[m - 1, n - 1].$ $\qquad$ (17.32)

Predictive compression techniques can be considered as an extension of DPCM in two dimensions. Example 17.11 illustrates the use of the third-order predictive model in compressing still images.

Example 17.11

Consider the Sanjukta image shown in Fig. 17.23(a). The first 4×4 pixels of the image are given by

$$
i[m, n] = \begin{bmatrix}
156 & 157 & 154 & 149 \\
156 & 159 & 159 & 155 \\
153 & 158 & 160 & 159 \\
149 & 154 & 157 & 156
\end{bmatrix}. \tag{17.33}
$$

Using the predictors in Eqs. (17.30) and (17.31) for the first row and column, respectively, and the predictor in Eq. (17.32) to predict the remaining values, calculate the error in the reconstructed image. In your calculations, assume that the quantizer divides the difference image by a quantization factor $Q = 3$ and rounds to the nearest integer before quantization.

Solution

Using zero boundary conditions, the predicted sample value, the prediction error, the quantized error, and the reconstructed sample value at $m = 0, n = 0$ are given by

predicted value $\qquad\qquad\qquad \hat{i}[0, 0] = 0;$

error $\qquad\qquad\qquad e[0, 0] = i[0, 0] - \hat{s}[0, 0] = 156;$

quantized error $\qquad\qquad\qquad \hat{e}[0, 0] = \text{round}(156/3) = 52.$

reconstructed value $\quad i'[0, 0] = \hat{i}[0, 0] + 3 \times \hat{e}[0, 0] = 0 + 3 \times 52 = 156.$

For spatial location $m = 0, n = 1$, the predicted sample value, the prediction error, the quantized error, and the reconstructed sample value are given by

predicted value $\qquad\qquad\qquad \hat{i}[0, 1] = i'[0, 0] = 156;$

error $\qquad\qquad\qquad e[0, 1] = i[0, 1] - \hat{i}[0, 1] = 157 - 156 = 1;$

quantized error $\qquad\qquad\qquad \hat{e}[0, 1] = \text{round}(1/3) = 0;$

reconstructed value $\qquad i'[0, 1] = \hat{i}[0, 1] + 3 \times \hat{e}[0, 1] = 156.$

For spatial location $m = 0, n = 2$, the predicted sample, the prediction error, the quantized error, and the reconstructed sample value are given by

predicted value $\qquad\qquad\qquad \hat{i}[0, 2] = i'[0, 1] = 156;$

error $\qquad\qquad\qquad e[0, 2] = i[0, 2] - \hat{i}[0, 2] = 154 - 156 = -2;$

quantized error $\qquad\qquad\qquad \hat{e}[0, 2] = \text{round}(-2/3) = -1;$

reconstructed value $\quad i'[0, 2] = \hat{i}[0, 2] + 3 \times \hat{e}[0, 2] = 156 - 3 = 153.$

For spatial location $m = 0, n = 3$, the predicted sample value, the prediction error, the quantized error, and the reconstructed sample value are given by

predicted value
$$\hat{i}[0, 3] = i'[0, 2] = 153;$$

error
$$e[0, 3] = i[0, 3] - \hat{i}[0, 3] = 149 - 153 = -4;$$

quantized error
$$\hat{e}[0, 3] = \text{round}(-4/3) = -1;$$

reconstructed value
$$i'[0, 3] = \hat{i}[0, 3] + 3 \times \hat{e}[0, 3] = 153 - 3 = 150.$$

For spatial location $m = 1, n = 0$, the predicted sample value, the prediction error, the quantized error, and the reconstructed sample value are given by

predicted value
$$\hat{i}[1, 0] = i'[0, 0] = 156;$$

error
$$e[1, 0] = i[1, 0] - \hat{i}[1, 0] = 156 - 156 = 0;$$

quantized error
$$\hat{e}[1, 0] = \text{round}(0/3) = 0;$$

reconstructed value
$$i'[1, 0] = \hat{i}[1, 0] + 3 \times \hat{e}[1, 0] = 156 + 0 = 156.$$

For spatial location $m = 1, n = 1$, the predicted sample value, the prediction error, the quantized error, and the reconstructed sample value are given by

predicted value
$$\hat{i}[1, 1] = 0.33(i'[1, 0] + i'[0, 1] + i'[0, 0])$$
$$= 0.33 \times 468 = 154.44;$$

error
$$e[1, 1] = i[1, 1] - \hat{i}[1, 1] = 159 - 154.44 = 4.56;$$

quantized error
$$\hat{e}[1, 1] = \text{round}(4.56/3) = 2;$$

reconstructed value
$$i'[1, 1] = \hat{i}[1, 1] + 3 \times \hat{e}[1, 1] = 154.44 + 6 = 160.44.$$

For spatial location $m = 1, n = 2$, the predicted sample value, the prediction error, the quantized error, and the reconstructed sample value are given by

predicted value
$$\hat{i}[1, 2] = 0.33(i'[1, 1] + i'[0, 2] + i'[0, 1])$$
$$= 0.33 \times 469.44 = 154.92;$$

error
$$e[1, 2] = i[1, 2] - \hat{i}[1, 2] = 159 - 154.92 = 4.08;$$

quantized error
$$\hat{e}[1, 2] = \text{round}(4.08/3) = 1;$$

reconstructed value
$$i'[1, 2] = \hat{i}[1, 2] + 3 \times \hat{e}[1, 2] = 154.92 + 3 = 157.92.$$

For spatial location $m = 1, n = 3$, the predicted sample value, the prediction error, the quantized error, and the reconstructed sample value are given by

predicted value
$$\hat{i}[1, 3] = 0.33(i'[1, 2] + i'[0, 3] + i'[0, 2])$$
$$= 0.33 \times 460.92 = 152.10;$$

error
$$e[1, 3] = i[1, 3] - \hat{i}[1, 3] = 155 - 152.10 = 2.90;$$

quantized error
$$\hat{e}[1, 3] = \text{round}(2.90/3) = 1;$$

reconstructed value
$$i'[1, 3] = \hat{i}[1, 3] + 3 \times \hat{e}[1, 3] = 152.10 + 3 = 155.10.$$

Similarly, the pixel values at other locations can be calculated using the above procedure. The computed values are as follows:

$$
i'[m, n] = \begin{bmatrix} 156 & 156 & 153 & 150 \\ 156 & 160.4 & 157.9 & 155.1 \\ 153 & 157.9 & 160.2 & 159.2 \\ 150 & 155.1 & 156.2 & 156.9 \end{bmatrix}.
$$

Subtracting the aforementioned values from the original values given in Eq. (17.33) gives the following values for the error image:

$$
\hat{e}[m, n] = \begin{bmatrix} 0 & 1 & 1 & -1 \\ 0 & -1.4 & 1.1 & -0.1 \\ 0 & 0.1 & -0.2 & -0.2 \\ -1 & -1.1 & 0.8 & -0.9 \end{bmatrix}.
$$

In image compression, the mean square error (MSE) is typically used to measure the quantitative quality of a compressed image $i'[m, n]$. The MSE is defined as follows

$$
\text{MSE} = \frac{1}{MN} \sum_{m=0}^{M-1} \sum_{m=0}^{N-1} [i[m, n] - i'[m, n]],
$$

where $i[m, n]$ is the pixel intensity of the original image having $(M \times N)$ dimensions. For Example 17.11, the MSE is given by 0.6206.

In DPCM, the first pixel is referred to as the reference pixel and is typically encoded directly with $e[0, 0] = 0$. The remaining pixels are encoded using the error image, which is typically divided by a quantization factor Q before encoding. To achieve quantization, the entire dynamic range of the error image is divided into 2^B intervals and each interval is represented by B bits. Typically, B is kept small to achieve a large compression ratio. Figure 17.23 shows two reconstructed Sanjukta test images processed at two different compression ratios. Figure 17.23(b) is compressed with a quantization factor $Q = 5$ and $B = 4$. Similarly, Fig. 17.23(c) is compressed with a quantization factor $Q = 16$ and $B = 2$. Higher compression introduces more distortion in Fig. 17.23(c), which is illustrated by the lower subjective quality of Fig. 17.23(c) when compared with that of Fig. 17.23(b). The superior quality of Fig. 17.23(b) can also be quantified by computing the MSE. Figure 17.23(b) has a reconstruction MSE of 6, while Fig. 17.23(c) has a MSE of 44. If required, the quantized error values can be further encoded using a variable-length code or an entropy code to achieve more compression.

17.7.2 Image compression standards

DPCM provides moderate compression. Several techniques, such as transform coding, arithmetic coding, and object-based techniques, have been developed to achieve performances superior to DPCM. In addition, several image compression standards have been developed by the International Organization for

(a)

(b)

(c)

(d)

(e)

Fig. 17.23. Subjective quality of two DPCM encoded images. (a) Sanjukta image. (b) Reconstructed image after DPCM compression with a quantization factor Q of 5 and a 4-bit quantizer. (c) Same as (b) except the quality factor Q is set to 16 and a 2-bit quantizer is used. (d) Difference between the original image and reconstructed image shown in (b). (e) Difference between the original image and the reconstructed image shown in (c). The MSE associated with image (b) is 6, while the MSE associated with image (c) is 44.

Standardization (ISO) and the International Telecommunication Union (ITU) to ensure compatibility between different compressed bit streams. A popular ISO image compression standard is referred to as the JPEG standard, selected as an acronym for the Joint Photographic Experts Group, the ISO subcommittee responsible for developing the standard.

The JPEG standard algorithm encodes both gray level and color images. In this standard, an image is decorrelated using the discrete cosine transform (DCT). The DCT coefficients are quantized and the quantized coefficients are encoded using a combination of run length and Huffman coding. The size of the compressed bit stream is varied by changing the quality factor Q, which has a value between 1 and 100. The highest quality representation is obtained using a quality factor of 100, and the lowest quality representation is obtained using quality factor of 1. A high quality factor ensures superior perceived quality, but the compression is limited. Conversely, a low quality factor increases compression, but at the expense of quality.

The image processing toolbox in MATLAB includes a simplified version of the JPEG encoder and decoder, which allows images to be encoded at different quality factors Q. If x is a 2D array containing the gray values of a test image, the following command:

```
>> imwrite(x,'test_70.jpg','jpg','Quality', 70);
```

creates the JPEG compressed image "test_70.jpg" with a quality factor of 70. The following example illustrates the compression performance of the JPEG encoder and decoder.

Example 17.12
Consider the 8-bit Sanjukta image shown in Fig. 17.23(a). Using the `imwrite` command, generate different compressed JPEG images with quality factors 100, 50, 25, 10, and 5. Determine the compression ratio in each case and plot the reconstructed images.

Solution
The following MATLAB code creates compressed images with different quality factors:

```
>> x = imread('sanjukta_gray.tif');
>> imwrite(x,'sanjukta_100.jpg','jpg','Quality', 100) ;
>> imwrite(x,'sanjukta_50.jpg','jpg','Quality', 50) ;
>> imwrite(x,'sanjukta_25.jpg','jpg','Quality', 25) ;
>> imwrite(x,'sanjukta_10.jpg','jpg','Quality', 10) ;
>> imwrite(x,'sanjukta_5.jpg','jpg','Quality', 5) ;
```

The raw image has 126 672 pixels, with each pixel represented using 8 bits. Therefore, the uncompressed image size is 126 672 bytes or 126.7 kbytes. The sizes of the compressed files determined from the compressed files and their respective compression ratio are provided in Table 17.2. Table 17.2 illustrates

(a)

(b)

(c)

(d)

(e)

(f)

Fig. 17.24. Subjective quality of JPEG compressed images using different quality factors. (a) Original image; (b) quality factor 100; (c) quality factor 50; (d) quality factor 25; (e) quality factor 10; (f) quality factor 5.

Table 17.2. Comparison of JPEG compression performance for `sanjukta_gray` image

Quality factor, Q	Name of the compressed file	Size of the file (kbytes)	Compression ratio	MSE
100	sanjukta_100	41.7	3	0.05
50	sanjukta_50	11.1	11	12.34
25	sanjukta_25	4.8	26	18.98
10	sanjukta_10	2.9	43	36.69
5	sanjukta_5	2.3	54	76.05

that decreasing the quality factor increases the compression ratio at the cost of the reconstruction quality, apparent from the increase in MSE.

To provide a subjective comparison, the reconstructed images are shown in Fig. 17.24. We observe that the perceived quality of the reconstructed images degrades with the decrease in the quality factor. In other words, there is a trade-off between quality and size of the compressed file.

17.8 Summary

This chapter presented applications of digital signal processing in audio and image processing. Digital signals, including audio, images, and video, are random in nature. Section 17.2 presented an overview of spectral analysis methods for random signals based on the short-time Fourier transform, spectrogram, and periodogram. Section 17.3 covered fundamentals of audio signals, their storage format, and spectral analysis of audio signals. Filtering of audio signals was covered in Section 17.3, and principles of audio compression were presented in Section 17.4.

Section 17.5 extended digital signal processing to 2D signals. In particular, we introduced digital images, their storage format, and the spectral analysis of image signals. Section 17.6 covers 2D filtering, including the application of lowpass filters to eliminate high-frequency noise and highpass filters for edge detection. In each case, we presented examples of image filtering through MATLAB. Section 17.7 introduced principles of image compression including the 2D differential pulse-code modulation (DPCM) and the Joint Photographic Expert Group (JPEG) standard. Using MATLAB, we compared the performance of JPEG at different compression ratios.

Problems

17.1 Consider the following deterministic signal:

$$x[k] = 2\sin(0.2\pi k) + 3\cos(0.5\pi k).$$

Using a DFT magnitude spectrum, estimate the spectral content of $x[k]$ for the following cases: (a) a 20-point DFT and a sample size of $0 \le k \le 19$; (b) a 32-point DFT and a sample size of $0 \le k \le 31$; (c) a 64-point DFT and a sample size of $0 \le k \le 31$; (d) a 128-point DFT and a sample size of $0 \le k \le 31$; and (e) a 128-point DFT and a sample size of $0 \le k \le 63$. Comment on the leakage effect in each case.

17.2 Calculate and plot the amplitude spectra of the following DT signals:
  (i) $x_1[k] = \cos(0.25\pi k)$, $0 \le k \le 2000$;
  (ii) $x_2[k] = \cos(2.5 \times 10^{-4}\pi k^2$, $0 \le k \le 2000$;
  (iii) $x_3[k] = \cos(2.5 \times 10^{-7}\pi k^3)$, $0 \le k \le 11000$.
  Comment on the spectral content of the signals.

17.3 Calculate and plot the spectrograms of the three signals considered in Problem 17.2. Compare the results with those obtained in Problem 17.2.

17.4 Using MATLAB, estimate the power spectral density of the following signal:

$$x[k] = 2\cos(0.4\pi k + \theta_1) + 4\cos(0.8\pi k + \theta_2),$$

where θ_1 and θ_2 are independent random variables with uniform distribution between $[0, \pi]$. Use a sample realization of $x[k]$ with 10 000 samples, the Bartlett window with length 1024, an overlap of 600 samples, and the average Welch approach.

17.5 Determine the frequency content of the audio signal "chord.wav", provided in the accompanying CD using (i) a spectrogram and (ii) an average periodogram.

17.6 Consider the "testaudio4.wav" file provided in the accompanying CD. Load the audio signal using the wavread function available in MATLAB.
  (a) What is the sampling rate used to discretize the signal? What is the total number of samples stored in the file?
  (b) How many bits are used to represent each sample?
  (c) Is the audio signal stored in the mono or stereo format?
  (d) Estimate the power spectrum of the signal

17.7 Repeat Problem 17.6 for "testaudio3.wav" provided in the accompanying CD.

17.8 Repeat Problem 17.6 for "bell.wav" provided in the accompanying CD.

17.9 Repeat Problem 17.6 for "test44k.wav" provided in the accompanying CD.

17.10 Repeat Example 17.7 for the following audio samples:

$$x_1[k] = [66, 67, 68, 69] \text{ and } x_2[k] = [66, 72, 61, 56].$$

Show that the reconstruction error is greater for the second case, where the neighboring audio samples are less correlated.

17.11 Consider the "`girl.jpg`" file provided in the accompanying CD. Read the image using the `imread` function available in MATLAB.
   (a) What are the dimensions of the image stored in the "`girl.jpg`" file?
   (b) What are the maximum and minimum values of the intensity of the pixels stored in the file?
   (c) Sketch the image using the `imagesc` function available in MATLAB.
   (d) Calculate and plot the 2D power spectrum of the image to illustrate the dominant spatial frequency components of the image.

17.12 Consider the 2D filter defined by the following impulse response:

$$h[m, n] = \frac{1}{16} \begin{bmatrix} 1 & 1 & 1 & 1 \\ 1 & 1 & 1 & 1 \\ 1 & 1 & 1 & 1 \\ 1 & 1 & 1 & 1 \end{bmatrix}.$$

   (a) Show that $h[m, n]$ is a lowpass filter by sketching its magnitude spectrum using the mesh plot.
   (b) Assume that the image stored in "`girl.jpg`" is applied at the input of the filter $h[m, n]$. Determine and sketch the output image.
   (c) Calculate the 2D power spectrum of the filtered image. Comparing this with the result of Problem 17.11 (d), highlight how the high-frequency components have been attenuated in the filtered image.

17.13 Repeat Problem 17.12 for the 2D filter with the following impulse response:

$$h[m, n] = \frac{1}{3.2764} \begin{bmatrix} 0 & 0 & 0.0221 & 0 & 0 \\ 0 & 0.1563 & 0.3907 & 0.1563 & 0 \\ 0.0221 & 0.3907 & 1 & 0.3907 & 0.0221 \\ 0 & 0.1563 & 0.3907 & 0.1563 & 0 \\ 0 & 0 & 0.0221 & 0 & 0 \end{bmatrix}.$$

17.14 Consider the 2D filter defined by the following impulse response:

$$h[m, n] = \frac{1}{9} \begin{bmatrix} -1 & -1 & -1 \\ -1 & 8 & -1 \\ -1 & -1 & -1 \end{bmatrix}.$$

   (a) Show that $h[m, n]$ is a highpass filter by sketching its magnitude spectrum using the mesh plot.
   (b) Assume that the image stored in "`girl.jpg`" is applied at the input of the filter $h[m, n]$. Determine and sketch the output image.

Show that the highpass filtering leads to the detection of edges in the image.

(c) Calculate the 2D power spectrum of the filtered image. Comparing this with the result of Problem 17.11 (d), highlight how the low-frequency components have been attenuated in the filtered image.

17.15 Repeat Problem 17.14 for the 2D filter with the following impulse response:

$$h[m, n] = \frac{1}{6.21} \begin{bmatrix} 0 & 0 & -0.0442 & 0 & 0 \\ 0 & -0.3126 & -0.7815 & -0.3126 & 0 \\ -0.0442 & -0.7815 & 4.5532 & -0.7815 & -0.0442 \\ 0 & -0.3126 & -0.7815 & -0.3126 & 0 \\ 0 & 0 & -0.0442 & 0 & 0 \end{bmatrix}.$$

17.16 Repeat Example 17.11 for the following selections of (4×4) pixels:

$$i_1[m, n] = \begin{bmatrix} 156 & 157 & 158 & 159 \\ 150 & 151 & 151 & 150 \\ 153 & 155 & 154 & 156 \\ 155 & 154 & 157 & 156 \end{bmatrix} \quad \text{and}$$

$$i_2[m, n] = \begin{bmatrix} 156 & 177 & 148 & 189 \\ 160 & 171 & 181 & 150 \\ 123 & 125 & 174 & 196 \\ 175 & 164 & 147 & 156 \end{bmatrix}.$$

Show that the reconstruction error is greater for the second case, where the neighboring pixels are less correlated.

17.17 Compress the image stored in the file "lena.tif" in the accompanying CD using the JPEG standard with quality factors set to 80, 60, 40, 20, and 10. Determine the compression ratio for different quality factors and show that the subjective quality deteriorates as the quality factor is decreased. Compute the mean square error for the compressed images.

Appendix A **Mathematical preliminaries**

A.1 Trigonometric identities

$$e^{\pm jt} = \cos t \pm j \sin t$$

$$\cos t = \frac{1}{2}[e^{jt} + e^{-jt}]$$

$$\sin t = \frac{1}{2j}[e^{jt} - e^{-jt}]$$

$$\cos\left(t \pm \frac{\pi}{2}\right) = \mp \sin t$$

$$\sin\left(t \pm \frac{\pi}{2}\right) = \pm \cos t$$

$$\sin 2t = 2 \sin t \cos t$$

$$\cos^2 t + \sin^2 t = 1$$

$$\cos^2 t - \sin^2 t = \cos 2t$$

$$\cos^2 t = \frac{1}{2}(1 + \cos 2t)$$

$$\sin^2 t = \frac{1}{2}(1 - \cos 2t)$$

$$\cos^3 t = \frac{1}{4}(3 \cos t + \cos 3t)$$

$$\sin^3 t = \frac{1}{4}(3 \sin t - \sin 3t)$$

$$\cos(t \pm \theta) = \cos t \cos \theta \mp \sin t \sin \theta$$

$$\sin(t \pm \theta) = \sin t \cos \theta \pm \cos t \sin \theta$$

$$\tan(t \pm \theta) = \frac{\tan t \pm \tan \theta}{1 \mp \tan t \tan \theta}$$

$$\sin t \sin \theta = \frac{1}{2}[\cos(t - \theta) - \cos(t + \theta)]$$

$$\cos t \cos \theta = \frac{1}{2}[\cos(t + \theta) + \cos(t - \theta)]$$

$$\sin t \cos\theta = \frac{1}{2}[\sin(t+\theta) + \sin(t-\theta)]$$

$$a\cos t + b\sin t = C\cos(t+\theta),\ C = \sqrt{a^2+b^2},\ \theta = \tan^{-1}(-b/a)$$

$$a\cos(mt) + b\sin(mt) = \sqrt{a^2+b^2}\cos(mt-\theta),\ \theta = \tan^{-1}\frac{b}{a}$$

$$a\cos(mt) + b\sin(mt) = \sqrt{a^2+b^2}\sin(mt+\phi),\ \phi = \tan^{-1}\frac{a}{b}$$

A.2 Power series

$$\ln(1+t) = t - \frac{t^2}{2} + \frac{t^3}{3} - \frac{t^4}{4} + \cdots$$

$$e^t = 1 + t + \frac{t^2}{2!} + \frac{t^3}{3!} + \frac{t^4}{4!} + \cdots$$

$$\sin t = t - \frac{t^3}{3!} + \frac{t^5}{5!} - \frac{t^7}{7!} + \cdots$$

$$\cos t = 1 - \frac{t^2}{2!} + \frac{t^4}{4!} - \frac{t^6}{6!} + \cdots$$

$$\tan t = t + \frac{t^3}{3} + \frac{2t^5}{15} + \frac{17t^7}{315} + \cdots$$

$$\sin^{-1} t = t + \frac{1}{2}\frac{t^3}{3} + \frac{1.3}{2.4}\frac{t^5}{5} + \cdots$$

A.3 Series summation

Arithmetic series

$$\sum_{n=1}^{n}[a + (n-1)d] = \frac{N}{2}[2a + (N-1)d]$$

$$\sum_{n=1}^{n} n = 1 + 2 + \cdots + N = \frac{N(N+1)}{2}$$

Geometric series

$$\sum_{n=0}^{N} ar^n = \frac{a(1-r^{N+1})}{1-r}$$

$$\sum_{n=0}^{N-1} \exp\left[j\frac{2\pi kn}{N}\right] = \begin{cases} 0 & 1 \le k \le (N-1) \\ N & k = 0, \end{cases}$$

$$\sum_{n=0}^{\infty} r^n = \frac{1}{1-r},\ |r| < 1$$

$$\sum_{n=0}^{\infty} nr^n = \frac{r}{(1-r)^2},\ |r| < 1$$

The geometric progression (GP) series sum of the form

$$S = \sum_{n=0}^{N} ar^n = a + ar + ar^2 + \cdots + ar^N$$

is used frequently in this text while dealing with the discrete-time signals. Note that the factor r can be real, imaginary, or complex.

A.4 Limits and differential calculus

$$\lim_{t \to \infty} t^{-\alpha} \ln t = 0, \quad \operatorname{Re}(\alpha) > 0$$
$$\lim_{t \to 0} t^{\alpha} \ln t = 0, \quad \operatorname{Re}(\alpha) > 0$$

L'Hôpital's Rule:

If $\lim_{t \to a} x(t) = \lim_{t \to a} y(t) = 0$ or $\lim_{t \to a} x(t) = \lim_{t \to a} y(t) = \infty$, and $\lim_{t \to a} \dfrac{x'(t)}{y'(t)}$ has a finite value, then $\lim_{t \to a} \dfrac{x(t)}{y(t)} = \lim_{t \to a} \dfrac{x'(t)}{y'(t)}$

$$\frac{d}{dt}\left\{\frac{1}{g(t)}\right\} = -\frac{1}{g^2(t)}\frac{dg(t)}{dt}$$
$$\frac{d}{dt}\left\{\frac{h(t)}{g(t)}\right\} = \frac{1}{g^2(t)}\left[g(t)\frac{dh(t)}{dt} - h(t)\frac{dg(t)}{dt}\right]$$

A.5 Indefinite integrals

$$\int u \, dv = uv - \int v \, du$$
$$\int f(t)g(t) \, dt = f(t)\int g(t)dt - \int\left[\frac{df}{dt}\int g(t)dt\right]dt$$
$$\int \cos at \, dt = \frac{1}{a}\sin at + C, a \neq 0$$
$$\int \sin at \, dt = -\frac{1}{a}\cos at + C, a \neq 0$$
$$\int \cos^2 at \, dt = \frac{t}{2} + \frac{\sin 2at}{4a} + C, a \neq 0$$
$$\int \sin^2 at \, dt = \frac{t}{2} - \frac{\sin 2at}{4a} + C, a \neq 0$$
$$\int t \cos at \, dt = \frac{1}{a^2}(\cos at + at \sin at) + C, a \neq 0$$
$$\int t \sin at \, dt = \frac{1}{a^2}(\sin at - at \cos at) + C, a \neq 0$$

$$\int t^2 \cos at \, dt = \frac{1}{a^3}(2at \cos at - 2\sin at + a^2t^2 \sin at) + C, a \neq 0$$

$$\int t^2 \sin at \, dt = \frac{1}{a^3}(2at \sin at - 2\cos at - a^2t^2 \cos at) + C, a \neq 0$$

$$\int \cos at \cos bt \, dt = \frac{\sin(a-b)t}{2(a-b)} + \frac{\sin(a+b)t}{2(a+b)} + C, a^2 \neq b^2$$

$$\int \sin at \sin bt \, dt = \frac{\sin(a-b)t}{2(a-b)} - \frac{\sin(a+b)t}{2(a+b)} + C, a^2 \neq b^2$$

$$\int \sin at \cos bt \, dt = -\left[\frac{\cos(a-b)t}{2(a-b)} + \frac{\cos(a+b)t}{2(a+b)}\right] + C, a^2 \neq b^2$$

$$\int \sin^{-1} at \, dt = t \sin^{-1} at + \frac{1}{a}\sqrt{1 - a^2t^2} + C, a \neq 0$$

$$\int \cos^{-1} at \, dt = t \cos^{-1} at - \frac{1}{a}\sqrt{1 - a^2t^2} + C, a \neq 0$$

$$\int e^{at} \, dt = \frac{1}{a}e^{at} + C, a \neq 0$$

$$\int b^{at} \, dt = \frac{b^{at}}{a \ln b} + C, a \neq 0, b > 0, b \neq 1$$

$$\int te^{at} \, dt = \frac{e^{at}}{a^2}(at - 1) + C, a \neq 0$$

$$\int t^2 e^{at} \, dt = \frac{e^{at}}{a^3}(a^2t^2 - 2at + 2) + C, a \neq 0$$

$$\int t^n e^{at} \, dt = \frac{1}{a}t^n e^{at} - \frac{n}{a}\int t^{n-1} e^{at} \, dt, a \neq 0$$

$$\int t^n b^{at} \, dt = \frac{t^n b^{at}}{a \ln b} - \frac{n}{a \ln b}\int t^{n-1} b^{at} \, dt, a \neq 0, \ b > 0, b \neq 1$$

$$\int e^{at} \sin bt \, dt = \frac{e^{at}}{a^2 + b^2}(a \sin bt - b \cos bt) + C$$

$$\int e^{at} \cos bt \, dt = \frac{e^{at}}{a^2 + b^2}(a \cos bt + b \sin bt) + C$$

$$\int t^n \ln at \, dt = \frac{t^{n+1}}{n + 1} \ln at - \frac{t^{n+1}}{(n + 1)^2} + C, n \neq -1$$

$$\int \frac{1}{t^2 + a^2} \, dt = \frac{1}{a} \tan^{-1}\left(\frac{t}{a}\right) + C, a \neq 0$$

$$\int \frac{t}{t^2 + a^2} \, dt = \frac{1}{2} \ln(t^2 + a^2) + C$$

Appendix B Introduction to the complex-number system

In this appendix, we introduce some elementary concepts that define complex numbers. In presenting the material, it is anticipated that most readers have some prior exposure to complex numbers, so the information presented here serves primarily as a review. The appendix is organized as follows. In Section B.1, we review the definition of real numbers and then survey their arithmetic properties, including some basic operations like addition, subtraction, multiplication, and division. Section B.2 extends the arithmetic operations to complex numbers, and Section B.3 introduces its geometric structure using the 2D Cartesian representation. Section B.4 presents an alternative representation, referred to as the polar representation for complex numbers. Section B.5 concludes the appendix.

B.1 Real-number system

A real-number system $\Re$ is a set of all real numbers, which is defined in terms of two basic operations: addition and multiplication. For two arbitrarily selected real numbers $a, b \in \Re$, these basic operations are given by

addition $\qquad\qquad\qquad\qquad\qquad s_1 = a + b;$ $\qquad\qquad$ (B.1)

multiplication $\qquad\qquad\qquad\qquad m_1 = a \times b,$ $\qquad\qquad$ (B.2)

such that $s_1, m_1 \in \Re$. The remaining arithmetic operations, for example, subtraction and division, are expressed in terms of Eqs (B.1) and (B.2) as follows:

subtraction $\qquad\qquad\qquad s_2 = a - b = a + (-b);$ $\qquad\qquad$ (B.3)

division $\qquad\qquad\qquad m_2 = a/b = a \times (1/b),$ $\qquad\qquad$ (B.4)

such that $s_2, m_2 \in \Re$. The real number $-b$ is referred to as the additive inverse of b since $b + (-b) = 0$. Likewise, the real number $1/b$ is referred to as the multiplicative inverse of b since $b \times (1/b) = 1$. For $\Re$ to represent a complete set of real numbers, it must satisfy the following properties.

Fig. B.1. Representation of a real-number system using a 1D straight line.

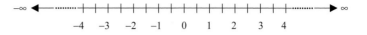

$$-\infty \xleftarrow{\hspace{1cm}} \underset{-4 \quad -3 \quad -2 \quad -1 \quad 0 \quad 1 \quad 2 \quad 3 \quad 4}{\vert\vert\vert\vert\vert\vert\vert\vert\vert\vert\vert\vert\vert\vert\vert\vert\vert} \xrightarrow{\hspace{1cm}} \infty$$

(i) The addition of two real numbers $a, b \in \Re$ produces a unique real number $s_1 \in \Re$.

(ii) Subtracting a real number $a \in \Re$ from another real number $b \in \Re$ produces a unique real number $s_2 \in \Re$.

(iii) Multiplication of two real numbers $a, b \in \Re$ produces a unique real number $m_1 \in \Re$.

(iv) Dividing a real number $a \in \Re$ by another real number $b \in \Re$, $b \neq 0$, produces a unique real number $m_2 \in \Re$.

Frequently, a real-number system is modeled graphically using a 1D straight line, as illustrated in Fig. B.1. Each point on the line represents a real number. The 1D line is packed with real numbers such that an uncountable number of real numbers exists between two arbitrarily selected points on the line.

B.2 Complex-number system

Let j be the root of the equation $x^2 + 1 = 0$, such that $j = \sqrt{-1}$. In terms of j, a complex number x is defined as

$$x = a + jb, \quad \text{such that } x \in C, \tag{B.5}$$

where a and b represent two real numbers, $a, b \in \Re$, and C denote a set of all possible complex numbers. Equation (B.5) is referred to as the rectangular or Cartesian representation of the complex number x. From Eq. (B.5), it is straightforward to deduce the following:

(i) The real component of the complex number x is a. This is denoted by $\Re(x) = a$.

(ii) The imaginary component of the complex number x is b. This is denoted by $\Im(x) = b$.

In the following, we define the basic arithmetic operations between two complex numbers. In our definitions, we use the following two operands: $x_1 = a_1 + jb_1$ and $x_2 = a_2 + jb_2$, with $x_1, x_2 \in C$ and $a_1, a_2, b_1, b_2 \in \Re$.

B.2.1 Addition

Addition of two complex numbers is defined as follows:

$$\begin{aligned} x_1 + x_2 &= (a_1 + jb_1) + (a_2 + jb_2) \\ &= (a_1 + a_2) + j(b_1 + b_2). \end{aligned} \tag{B.6}$$

In other words, when adding two complex numbers the real and imaginary components are added separately.

B.2.2 Subtraction

The definition of subtraction follows the same lines as that for addition. Subtracting a complex number x_2 from x_1 is defined as follows:

$$
\begin{aligned}
x_1 - x_2 &= (a_1 + jb_1) - (a_2 + jb_2) \\
&= (a_1 - a_2) + j(b_1 - b_2).
\end{aligned}
$$
(B.7)

As for addition, the real and imaginary components are subtracted separately.

B.2.3 Multiplication

Multiplication of two complex numbers x_1 and x_2 is defined as follows:

$$
\begin{aligned}
x_1 x_2 &= (a_1 + jb_1)(a_2 + jb_2) \\
&= a_1 a_2 + jb_1 a_2 + ja_1 b_2 + j^2 b_1 b_2 \\
&= (a_1 a_2 - b_1 b_2) + j(b_1 a_2 + a_1 b_2),
\end{aligned}
$$
(B.8)

where the final expression is obtained by noting that $j^2 = -1$.

B.2.4 Complex conjugation

From Eq. (B.8), it is easy to deduce that

$$
(a_1 + jb_1)(a_1 - jb_1) = (a_1)^2 + (b_1)^2.
$$
(B.9)

In other words, the imaginary component is eliminated. The complex number $x_1^* = a_1 - jb_1$ is referred to as the complex conjugate of $x_1 = a_1 + jb_1$, and vice versa. Equation (B.9) leads to the definition of the modulus or magnitude of a complex number, which is discussed next.

B.2.5 Modulus

The modulus (or magnitude) of a complex number $x_1 = a_1 + jb_1$ is defined as follows:

$$
|x_1| = \sqrt{x_1 x_1^*} = \sqrt{(a_1)^2 + (b_1)^2}.
$$
(B.10)

B.2.6 Division

Dividing two complex numbers is more complicated. To divide x_1 by x_2, we multiply both the numerator and denominator by the complex conjugate of x_2 and expand the numerator and denominator separately using the definition of multiplication from Section B.2.3; i.e.

$$\frac{x_1}{x_2} = \frac{a_1 + jb_1}{a_2 + jb_2} = \frac{(a_1 + jb_1)}{(a_2 + jb_2)} \cdot \frac{(a_2 - jb_2)}{(a_2 - jb_2)}$$
$$= \frac{a_1 a_2 + b_1 b_2}{a_2^2 + b_2^2} + j\frac{a_2 b_1 - a_1 b_2}{a_2^2 + b_2^2}, \quad (B.11)$$

where the final expression is obtained by noting that $j^2 = -1$. We illustrate these concepts with an example.

Example B.1

Two complex numbers are given by $x = 5 + j7$ and $y = 2 - j4$. Calculate (i) $\Re(x)$, $\Im(x)$, $\Re(y)$, $\Im(y)$; (ii) $x + y$; (iii) $x - y$; (iv) xy; (v) x^*, y^*; (vi) $|x|$, $|y|$; and (vii) x/y.

Solution

(1) The real and imaginary components of the complex number x are $\Re(x) = 5$ and $\Im(x) = 7$. Likewise, the real and imaginary components of y are $\Re(y) = 2$ and $\Im(y) = -4$.

(2) Adding x and y yields

$$x + y = (5 + j7) + (2 - j4) = (5 + 2) + j(7 - 4) = 7 + j3.$$

Since addition is commutative, the order of the operands does not matter, i.e. $x + y = y + x$.

(3) Subtracting y from x yields

$$x - y = (5 + j7) - (2 - j4) = (5 - 2) + j(7 - (-4)) = 3 + j11.$$

Subtraction is not commutative. In fact, $x - y = -(y - x)$.

(4) Multiplication of x and y is performed as follows:

$$xy = (5 + j7)(2 - j4) = 10 + j14 - j20 - j^2 28$$
$$= (10 + 28) + j(14 - 20) = 38 - j6.$$

Multiplication is commutative, therefore $xy = yx$.

(5) The complex conjugate of the complex number $x = 5 + j7$ is $x^* = 5 - j7$. Likewise, the complex conjugate of $y = 2 - j4$ is $y^* = 2 + j4$.

(6) The modulus of $x = 5 + j7$ is given by $|x| = \sqrt{5^2 + 7^2} = \sqrt{74}$. Likewise, the modulus of $y = 2 - j4$ is $|y| = \sqrt{2^2 + (-4)^2} = \sqrt{20}$.

(7) Dividing x by y yields

$$\frac{x}{y} = \frac{5 + j7}{2 - j4} = \frac{(5 + j7)}{(2 - j4)} \cdot \frac{(2 + j4)}{(2 + j4)}$$
$$= \frac{(5)(2) - (7)(4)}{2^2 + 4^2} + j\frac{(7)(2) + (5)(4)}{2^2 + 4^2} = -\frac{18}{20} + j\frac{34}{20}.$$

B.3 Graphical interpertation of complex numbers

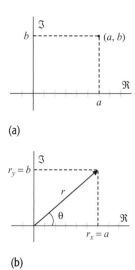

Fig. B.2. Graphical representations for a complex number $x = a + jb$. (a) Cartesian representation; (b) polar representation.

Any complex number $x = a + jb$ can be associated with an ordered pair of real numbers (a, b), i.e.

$$x = (a + jb) \longleftrightarrow (a, b). \tag{B.12}$$

The ordered pair of numbers (a, b) is represented by a point in the Cartesian coordinate system as shown in Fig. B.2(a), in which the horizontal axis represents the real component $\Re$ of the complex number and the vertical axis represents the imaginary component $\Im$ of the complex number. Alternatively, the complex number x can be associated with a vector $\vec{r}$ originating from the coordinate $(0, 0)$ and extending to the point (a, b). The rules for vector addition and subtraction can be used to add and subtract complex numbers, and vice versa. Since the two representations are equivalent, it is common to map a complex number to a vector.

Similar to the rectangular and polar representations of a vector, there are two alternative and equivalent representations for complex numbers. The rectangular representation was introduced in Section B.2. The polar representation is derived in Section B.4 by using Fig. B.2(b) and applying the geometric properties associated with vectors. Here, we define the notation used in the derivation of the polar representation. The length or magnitude of the vector $\vec{r}$, shown in Fig. B.2(b), is denoted by $|\vec{r}|$, or simply r. The angle that the vector $\vec{r}$ makes with the positive horizontal axis is denoted by θ. The projection of the vector $\vec{r}$ onto the horizontal axis is denoted by r_x, while the projection on the vertical axis is denoted by r_y. In terms of r and θ, the two projections are given by

$$r_x = r \cos \theta \quad \text{and} \quad r_y = r \sin \theta. \tag{B.13}$$

Using Pythagoras's theorem, it is straightforward to prove that the length or magnitude r of vector $\vec{r}$ is given by

$$r = \sqrt{r_x^2 + r_y^2}, \tag{B.14}$$

and the angle θ that the vector makes with the horizontal axis is given by

$$\theta = \tan^{-1}(r_y/r_x). \tag{B.15}$$

B.4 Polar representation of complex numbers

To derive the polar representation of a complex number, we base our discussion on Euler's formula:[†]

$$e^{j\theta} = \cos \theta + j \sin \theta. \tag{B.16}$$

The polar representation of a complex number $x = a + jb$ is then defined as

$$x = re^{j\theta}, \tag{B.17}$$

[†] Euler's formula is named after Leonhard Euler (1707–1783), a prolific eighteenth century Swiss mathematician and physicist.

where r represents the magnitude or length of the vector $\vec{r}$ obtained by mapping the complex number x onto the Cartesian plane. The length r and angle θ associated with vector $\vec{r}$ are obtained from Eqs (B.14) and (B.15) with $r_x = a$ and $r_y = b$. We demonstrate the conversion of a complex number from one representation to another with a series of examples.

Example B.2

Converting rectangular format into polar format   Consider a complex number $x = 2 + j4$. Clearly, x is represented in the rectangular format. To derive its equivalent polar format, we map the complex number into the Cartesian plane and calculate the parameters r and θ. Using Eqs (B.14) and (B.15), we obtain

$$r = \sqrt{2^2 + 4^2} = \sqrt{20}$$

and

$$\theta = \tan^{-1}(4/2) = 0.35\pi \text{ radians.}$$

The polar representation of $x = 2 + j4$ is $x = \sqrt{20}e^{j0.35\pi}$.

Example B.3

Converting polar format into rectangular format   Consider a complex number in the polar format $x = 4e^{j\pi/3}$. The rectangular representation of x is derived using Eq. (B.13) as

$$a = r_x = 4\cos\left(\frac{\pi}{3}\right) = 2$$

and

$$b = r_y = 4\sin\left(\frac{\pi}{3}\right) = 2\sqrt{3}.$$

The rectangular representation of $x = 4e^{j\pi/3}$ is $x = 2 + j2\sqrt{3}$.

In terms of polar representations, the basic arithmetic operations between two complex numbers $x_1 = r_1 e^{j\theta_1}$ and $x_2 = r_2 e^{j\theta_2}$ are defined as follows.

B.4.1 Addition

Addition of two complex numbers in polar format:

$$
\begin{aligned}
x_1 + x_2 &= r_1 e^{j\theta_1} + r_2 e^{j\theta_2} = (r_1\cos\theta_1 + jr_1\sin\theta_1) + (r_2\cos\theta_2 + jr_2\sin\theta_2) \\
&= (r_1\cos\theta_1 + r_2\cos\theta_2) + j(r_1\sin\theta_1 + r_2\sin\theta_2) \\
&= \sqrt{(r_1\cos\theta_1 + r_2\cos\theta_2)^2 + (r_1\sin\theta_1 + r_2\sin\theta_2)^2} \\
&\quad \times \exp\left[j\tan^{-1}\left(\frac{r_1\sin\theta_1 + r_2\sin\theta_2}{r_1\cos\theta_1 + r_2\cos\theta_2}\right)\right] \\
&= \sqrt{r_1^2 + r_2^2 + 2r_1 r_2\cos(\theta_1 - \theta_2)} \\
&\quad \times \exp\left[j\tan^{-1}\left(\frac{r_1\sin\theta_1 + r_2\sin\theta_2}{r_1\cos\theta_1 + r_2\cos\theta_2}\right)\right].
\end{aligned} \tag{B.18}
$$

B.4.2 Subtraction

Subtraction of two complex numbers in polar format:

$$
\begin{aligned}
x_1 - x_2 &= r_1 e^{j\theta_1} - r_2 e^{j\theta_2} \\
&= (r_1 \cos\theta_1 - r_2 \cos\theta_2) + j(r_1 \sin\theta_1 - r_2 \sin\theta_2) \\
&= \sqrt{(r_1 \cos\theta_1 - r_2 \cos\theta_2)^2 + (r_1 \sin\theta_1 - r_2 \sin\theta_2)^2} \\
&\quad \times \exp\left[j\tan^{-1}\left(\frac{r_1 \sin\theta_1 - r_2 \sin\theta_2}{r_1 \cos\theta_1 - r_2 \cos\theta_2} \right) \right] \\
&= \sqrt{r_1^2 + r_2^2 - 2r_1 r_2 \cos(\theta_1 - \theta_2)} \\
&\quad \times \exp\left[j\tan^{-1}\left(\frac{r_1 \sin\theta_1 - r_2 \sin\theta_2}{r_1 \cos\theta_1 - r_2 \cos\theta_2} \right) \right].
\end{aligned}
\tag{B.19}
$$

B.4.3 Multiplication

Multiplication of two complex numbers x_1 and x_2 in polar format:

$$
\begin{aligned}
x_1 x_2 &= r_1 e^{j\theta_1} \cdot r_2 e^{j\theta_2} \\
&= r_1 r_2 e^{j(\theta_1 + \theta_2)}.
\end{aligned}
\tag{B.20}
$$

B.4.4 Complex conjugation

The complex conjugate of complex number x_1 is given by

$$
x_1^* = r_1 e^{-j\theta_1}.
\tag{B.21}
$$

B.4.5 Modulus

The modulus (or magnitude) of a complex number $x_1 = r_1 e^{j\theta_1}$ is $|x_1| = r_1$.

B.4.6 Division

Dividing two complex numbers in polar format:

$$
\frac{x_1}{x_2} = \frac{r_1 e^{j\theta_1}}{r_2 e^{j\theta_2}} = \frac{r_1}{r_2} e^{j(\theta_1 - \theta_2)}.
\tag{B.22}
$$

Before we end this section, we note that both rectangular and polar formats have their advantages. It is easier to add or subtract complex numbers in the rectangular format. Multiplication and division are, however, simpler in the polar representation. We illustrate the concepts discussed in Section B.4 with the following example.

Example B.4

Consider the two complex numbers

$$
x = 5 + j7 = \sqrt{74} e^{j0.3026\pi}
$$

and

$$y = 2 - j4 = \sqrt{20}e^{-j0.3524\pi}.$$

Repeat Example B.1 but by selecting one of the two formats (rectangular or polar) for which the arithmetic operation is computationally simpler.

Solution

(1) The real and imaginary components of the complex number x are obtained from the rectangular format, i.e. $\Re(x) = 5$ and $\Im(x) = 7$. Likewise, for y the components are $\Re(y) = 2$ and $\Im(y) = -4$.

(2) Addition of x and y is performed in the rectangular format as follows:

$$\begin{aligned} x + y &= (5 + j7) + (2 - j4) \\ &= (5 + 2) + j(7 - 4) \\ &= 7 - j3. \end{aligned}$$

If polar format is required, we can express the above answer for $(x + y)$ in the polar format as $x + y = \sqrt{58}e^{j\tan^{-1}(-3/7)} = 7.62e^{-j0.13\pi}$.

(3) Subtraction is also performed in the rectangular format as follows:

$$\begin{aligned} x - y &= (5 + j7) - (2 - j4) \\ &= (5 - 2) + j(7 - (-4)) \\ &= 3 - j11. \end{aligned}$$

Converting the above answer into polar form, we obtain $x - y = \sqrt{130}e^{j\tan^{-1}(-11/3)} = 11.40e^{-j0.415\pi}$.

(4) Multiplication of x and y is performed in the polar format as follows:

$$\begin{aligned} xy &= \sqrt{74}e^{j0.3026\pi} \cdot \sqrt{20}e^{-j0.3524\pi} \\ &= \sqrt{1480}e^{-j0.0498\pi}. \end{aligned}$$

The rectangular format is $xy = \sqrt{1480}(\cos(0.0498\pi) + j\sin(-0.0498\pi)) = 38 - j6$.

(5) In rectangular format, the complex conjugate of the complex number $x = 5 + j7$ is $x^* = 5 - j7$. Likewise, the complex conjugate of $y = 2 - j4$ is $y^* = 2 + j4$ in rectangular format. The complex conjugates in polar format are $x^* = \sqrt{74}e^{-j0.3026\pi}$ and $y^* = \sqrt{20}e^{j0.3524\pi}$.

(6) The moduli of x and y are obtained directly from the polar format as $|x| = \sqrt{74}$ and $|y| = \sqrt{20}$.

(7) Dividing x by y is performed in polar format, yielding

$$\frac{x}{y} = \frac{\sqrt{74}e^{j0.3026\pi}}{\sqrt{20}e^{-j0.3524\pi}} = \sqrt{3.7}e^{j0.655\pi},$$

which, in rectangular format, is $\sqrt{3.7}(\cos(0.655\pi) + j\sin(0.655\pi)) = -0.9 + j1.7$.

B.5 Summary

Complex numbers in rectangular and polar formats were reviewed. Basic arithmetic operations such as addition, subtraction, multiplication, division, and complex conjugation were illustrated in both rectangular and polar domains.

Problems

B.1 Calculate the polar representations for (a) 1; (b) j; (c) − 1; (d) −j; (e) 3 + j4; (f) 8 − j6; and (g) 12 + j4.

B.2 Calculate the rectangular representations for (a) $11\exp(j2\pi)$; (b) $125\exp(j\pi/2)$; (c) $72\exp(-j\pi)$; (d) $125\exp(j\pi/8)$; (e) 25.47 $\exp(-j3\pi/4)$; and (f) $0.85\exp(-j\pi/4)$.

B.3 Consider the complex function

$$g(t) = \frac{2 + j3t}{1 + j2t}.$$

Plot the magnitude and phase of the function $g(t)$ each as a function of the independent variable t.

B.4 Determine and sketch the roots of the equation $e^x + 10 = 0$ in the Cartesian plane. [Hint: The polar representation for $-10 = e^{\ln(10)+j(2m+1)\pi}$.]

B.5 Prove the following identities:

(i) $\cos\theta = \dfrac{e^{j\theta} + e^{-j\theta}}{2}$;

(ii) $\sin\theta = \dfrac{e^{j\theta} - e^{-j\theta}}{2j}$;

(iii) $e^{jm\pi} = (-1)^m$ and $e^{j(2m\pi+\theta)} = e^{j\theta}$;

(iv) $\cos\theta = 1 - \dfrac{\theta^2}{2!} + \dfrac{\theta^4}{4!} - \dfrac{\theta^6}{6!} + \cdots$;

(v) $\sin\theta = \theta - \dfrac{\theta^3}{3!} + \dfrac{\theta^5}{5!} - \dfrac{\theta^7}{7!} + \cdots$.

Appendix C Linear constant-coefficient differential equations

It was shown in Chapters 2 and 3 that linear constant-coefficient differential equations play an important role in LTIC systems analysis. In this appendix, we review a direct method for solving differential equations of the form

$$\sum_{k=0}^{n} a_k \frac{d^k y(t)}{dt^k} = \sum_{k=0}^{m} b_k \frac{d^k x(t)}{dt^k}, \tag{C.1}$$

where the a_ks and b_ks are constants, and the derivatives

$$y(t), \frac{dy(t)}{dt}, \frac{d^2 y(t)}{dt^2}, \dots, \frac{d^{n-1} y(t)}{dt^{n-1}} \tag{C.2}$$

of the output signal $y(t)$ are known at a given time instant, say $t = t_0$. We will use the compact notation $\dot{y}(t)$ to denote the first derivative of $y(t)$ with respect to t. Therefore, $\dot{y}(t) = dy/dt$, $\ddot{y}(t) = d^2 y/dt^2$, and similarly for the higher-order derivatives. In the context of LTIC systems, the differential equation, Eq. (C.1), provides a linear relationship between the input signal $x(t)$ and the output $y(t)$. The values of the derivatives of $y(t)$, Eq. (C.2), for such LTIC systems are typically specified at $t_0 = 0$ and are referred to as the initial conditions. The highest derivative in Eq. (C.1) denotes the order of the differential equation. Equation (C.1) is therefore either of order n or m.

The method discussed in this appendix is direct, in the sense that it solves Eq. (C.1) in the time domain and does not require calculation of any transforms. The direct approach expresses the output $y(t)$ described by a differential equation as the sum of two components:

(i) zero-input response $y_{zi}(t)$ associated with the initial conditions;
(ii) zero-state response $y_{zs}(t)$ associated with the applied input $x(t)$.

The zero-input response $y_{zi}(t)$ is the component of the output $y(t)$ of the system when the input is set to zero. The zero-input response describes the manner in which the system dissipates any energy or memory of the past as specified by the initial conditions. The zero-state response $y_{zs}(t)$ is the component of the output $y(t)$ of the system with initial conditions set to zero. It describes the behavior of the system forced by the input. In the following, we outline the procedure to evaluate the zero-input and zero-state responses.

C.1 Zero-input response

The zero-input response $y_{zi}(t)$ is the output of the system when the input is zero. Hence, $y_{zi}(t)$ is the solution to the following homogeneous differential equation:

$$\sum_{k=0}^{n} a_k \frac{d^k y(t)}{dt^k} = 0, \tag{C.3}$$

with known initial conditions

$$y(t), \frac{dy(t)}{dt}, \frac{d^2 y(t)}{dt^2}, \ldots, \frac{d^n y(t)}{dt^n} \quad \text{at } t = 0. \tag{C.4}$$

To determine the zero-input response $y_{zi}(t)$, assume that the zero-input response is given by $y_{zi}(t) = Ae^{st}$, substitute $y_{zi}(t)$ in the homogeneous differential equation, Eq. (C.3), and solve the resulting equation. We illustrate the procedure for calculating the homogeneous solution by considering an example.

Example C.1
Consider a CT system modeled by the following differential equation:

$$\frac{d^2 y}{dt^2} + 5\frac{dy}{dt} + 4y(t) = 3x(t). \tag{C.5}$$

Compute the zero-input response of the system for initial conditions $y(0) = 2$ and $\dot{y}(0) = -5$.

Solution
Substituting $y_{zi}(t) = Ae^{st}$ in the homogeneous equation

$$\frac{d^2 y}{dt^2} + 5\frac{dy}{dt} + 4y(t) = 0, \tag{C.6}$$

obtained by setting input $x(t) = 0$, yields

$$Ae^{st}(s^2 + 5s + 4) = 0. \tag{C.7}$$

Ignoring the trivial solution, i.e. assuming $Ae^{st} \neq 0$, Eq. (C.7) reduces to the following quadratic equation, referred to as the characteristic equation, in s:

$$s^2 + 5s + 4 = 0, \tag{C.8}$$

which has two roots at $s = -1, -4$. The zero-input solution is given by

$$y_{zi}(t) = A_0 e^{-t} + A_1 e^{-4t}, \tag{C.9}$$

where A_0 and A_1 are constants to be determined from the given initial conditions. Substituting the initial conditions in Eq. (C.9) yields

$$A_0 + A_1 = 2,$$
$$-A_0 - 4A_1 = -5, \qquad (C.10)$$

which has solution $A_0 = 1$ and $A_1 = 1$. The zero-input response for Eq. (C.5) is therefore given by

$$y_{zi}(t) = e^{-t} + e^{-4t}. \qquad (C.11)$$

C.1.1 Repeated roots

The form of the zero-input response changes slightly when the characteristic equation has repeated roots. If a root $s = a$ is repeated J times, then we include J distinct terms in the zero-input response associated with a by using the following J functions:

$$e^{at}, te^{at}, t^2e^{at}, \ldots, t^{J-1}e^{at}. \qquad (C.12)$$

The zero-input response of an LTIC system is then given by

$$y_{zi}(t) = A_0e^{at} + A_1te^{at} + A_2t^2e^{at} + \cdots + A_{J-1}t^{J-1}e^{at}. \qquad (C.13)$$

The procedure for calculating the homogeneous solution for differential equations with repeated roots is illustrated in Example C.2.

Example C.2
Consider a CT system modeled by the following differential equation:

$$\frac{d^3y}{dt^2} + 4\frac{d^2y}{dt^2} + 5\frac{dy}{dt} + 2y(t) = x(t). \qquad (C.14)$$

Compute the zero-input response of the system for initial conditions $y(0) = 4$, $\dot{y}(0) = -5$, $\ddot{y}(0) = 9$.

Solution
By substituting $y_{zi}(t) = Ae^{st}$ in the homogeneous representation for Eq. (C.14), we obtain the following characteristic equation:

$$s^3 + 4s^2 + 5s + 2 = 0, \qquad (C.15)$$

which has three roots at $s = -1, -2, -2$. The zero-input solution is therefore given by

$$y_{zi}(t) = A_0e^{-t} + A_1e^{-2t} + A_2te^{-2t}, \qquad (C.16)$$

where A_0, A_1, and A_2 are constants determined from the given initial conditions. Substituting the initial conditions into Eq. (C.16) yields

$$
\begin{aligned}
A_0 + A_1 &= 4, \\
-A_0 + A_1 - 2A_2 &= -5, \\
A_0 - 2A_1 + 4A_2 &= 9,
\end{aligned}
\tag{C.17}
$$

which has solution $A_0 = 1$, $A_1 = 2$, and $A_2 = 3$. The zero-input response for Eq. (C.14) is therefore given by

$$
y_{zi}(t) = e^{-t} + 2e^{-2t} + 3te^{-2t}.
\tag{C.18}
$$

C.1.2 Complex roots

Solving a characteristic equation may give rise to complex roots of the form $s = a + jb$. Typically, a homogeneous differential equation, Eq. (C.3), with real coefficients, has complex roots in conjugate pairs. In other words, if $s = a + jb$ is a root of the characteristic equation obtained from Eq. (C.3) then $s = a - jb$ must also be a root of the characteristic equation. For such complex roots, the zero-input response can be modified to the following form:

$$
y_{zi}(t) = A_0 e^{at} \cos(bt) + A_1 e^{at} \sin(bt).
\tag{C.19}
$$

Example C.3
Compute the zero-input response of a system represented by the following differential equation:

$$
\frac{d^4 y}{dt^4} + 2\frac{d^2 y}{dt^2} + 1 = x(t),
\tag{C.20}
$$

with initial conditions $y(0) = 2$, $\dot{y}(0) = 2$, $\ddot{y}(0) = 0$, $\dddot{y}(0) = -4$.

Solution
Substituting $y_{zi}(t) = Ae^{st}$ in the homogeneous representation for Eq. (C.20) results in the following characteristic equation:

$$
s^4 + 2s^2 + 1 = 0.
\tag{C.21}
$$

The roots of the characteristic equation are given by $s = j, j, -j$, and $-j$. Note that the roots are not only complex but also repeated. The zero-input solution is given by

$$
y_{zi}(t) = A_0 \cos(t) + A_1 t \cos(t) + A_2 \sin(t) + A_3 t \sin(t),
\tag{C.22}
$$

where A_0, A_1, A_2, and A_3 are constants. To calculate these constants, we substitute the following initial conditions:

$$
\begin{aligned}
A_0 & & & = 2, \\
& A_1 + A_2 & & = 2, \\
-A_0 & & + 2A_3 & = 0, \\
& -3A_1 - A_2 & & = -4,
\end{aligned}
\qquad \text{(C.23)}
$$

which has solution $A_0 = 2$, $A_1 = 1$, $A_2 = 1$, and $A_3 = 1$. The zero-input response for the system in Eq. (C.20) is therefore given by

$$
y_{zi}(t) = 2\cos(t) + t\cos(t) + \sin(t) + t\sin(t). \qquad \text{(C.24)}
$$

C.2 Zero-state response

The zero-state response $y_{zs}(t)$ depends upon the input signal $x(t)$ subject to zero initial conditions. The zero-state response consists of two components: (i) the homogeneous component $y_{zs}^{(h)}(t)$ and (ii) the particular component $y_{zs}^{(p)}(t)$. The homogeneous component is obtained by following the procedure used to solve for the zero-input response but with zero initial conditions. The particular component of the zero-state response is obtained from a look-up table such as Table C.1. For example, if the input signal is $x(t) = Ke^{-at}$, then the particular component of the zero-state response is assumed to be $y_{zs}^{(p)}(t) = Ce^{-at}$. The constant C is determined such that $y_{zi}(t)$ satisfies the system's differential equation. The procedure for computing the zero-state response is illustrated in Example C.4.

Example C.4
Consider the system specified by the differential equation given in Example C.1:

$$
\frac{d^2y}{dt^2} + 5\frac{dy}{dt} + 4y(t) = 3x(t). \qquad \text{(C.25)}
$$

Compute the zero-state response of the system for the input signal $x(t) = \cos tu(t)$.

Solution
The homogeneous and particular components of the zero-state response $y_{zi}(t)$ are solved separately in three steps as follows.

Step 1 Compute the homogeneous component $y_{zs}^{(h)}(t)$ The solution for the homogeneous component is similar to the zero-input response of the system. Using the result of Eq. (C.9), the homogeneous component of the zero-input

Table C.1. Zero-state response corresponding to common input signals

Input	Particular component of the zero-state response
Impulse function, $K\delta(t)$	$C\delta(t)$
Unit step function, $Ku(t)$	$Cu(t)$
Exponential, Ke^{-at}	Ce^{-at}
Sinusoidal, $A\cos(\omega_0 t + \phi)$	$C_0\cos(\omega_0 t) + C_1\sin(\omega_0 t)$

response is given by

$$y_{zs}^{(h)}(t) = B_0 e^{-t} + B_1 e^{-4t}, \tag{C.26}$$

where B_0 and B_1 are constants.

Step 2 Determine the particular component $y_{zs}^{(p)}(t)$ The particular component is obtained by consulting Table C.1. For the input signal $x(t) = \cos t\, u(t)$, the particular component of the zero-state response is of the form $y_{zs}^{(p)}(t) = C_0 \cos t + C_1 \sin t$ for $t > 0$. Substituting the particular component in Eq. (C.25) yields

$$(-5C_0 + 3C_1)\sin t + (3C_0 + 5C_1)\cos t = 3\cos t. \tag{C.27}$$

Equating the cosine and sine terms on the left- and right-hand sides of the equation, we obtain the following simultaneous equations:

$$-5C_0 + 3C_1 = 0,$$
$$3C_0 + 5C_1 = 3, \tag{C.28}$$

with solution $C_0 = 9/34$ and $C_1 = 15/34$. The particular component $y_{zs}^{(p)}(t)$ of the zero-state response is given by

$$y_{zs}^{(p)}(t) = \frac{9}{34}\cos t + \frac{15}{34}\sin t \quad \text{for} \quad t > 0. \tag{C.29}$$

Step 3 Determine the zero-state response from $y_{zs}(t) = y_{zs}^{(h)}(t) + y_{zs}^{(p)}(t)$. The zero-state response is the sum of the homogeneous and particular components, and is given by

$$y_{zs}(t) = (B_0 e^{-t} + B_1 e^{-4t}) + \frac{9}{34}\cos t + \frac{15}{34}\sin t, \tag{C.30}$$

where B_0 and B_1 are obtained by inserting zero initial conditions, $y(0) = 0$ and $\dot{y}(0) = 0$. This leads to the following simultaneous equations:

$$B_0 + B_1 = -\frac{9}{34},$$
$$B_0 + 4B_1 = \frac{15}{34}, \tag{C.31}$$

with solution $B_0 = -1/2$ and $B_1 = 4/17$. The zero-state response of Eq. (C.25) is

$$y_{zs}(t) = -\frac{1}{2}e^{-t} + \frac{4}{17}e^{-4t} + \frac{9}{34}\cos t + \frac{15}{34}\sin t. \qquad (C.32)$$

This approach for finding the particular component of the zero-state response is modified when the input is of the same form as one of the terms in the homogeneous component of the zero-state response. We illustrate with an example where we outline the modified procedure for calculating the particular component of the zero-state response.

Example C.5

Repeat Example C.4 for the input signal $x(t) = 2e^{-t}$.

Solution

The homogeneous component for the zero-state response is given by Eq. (C.26):

$$y_{zs}^{(h)}(t) = B_0 e^{-t} + B_1 e^{-4t},$$

where B_0 and B_1 are constants. The input signal $x(t) = 2e^{-t}$. Based on Table C.1, the particular component is of the form $y_{zs}^{(p)}(t) = Ce^{-t}$, which is similar to the first term in the homogeneous component. In such a scenario, we assume a particular component that is different from the first term of the homogeneous component. To achieve this, we multiply the particular component by the lowest power of t that will make the particular component different from the first term of the homogeneous component. The particular component, in this example, is therefore given by $y_{zs}^{(p)}(t) = Cte^{-t}$. In order to evaluate the value of constant C, we substitute the particular component in the system's differential equation and solve for C; it is found that $C = 3$. The overall zero-state response is therefore given by

$$y_{zs}(t) = B_0 e^{-t} + B_1 e^{-4t} + 3te^{-t}, \qquad (C.33)$$

where the values of B_0 and B_1 are computed using zero initial conditions. The resulting simultaneous equations are given by

$$\begin{aligned} B_0 + B_1 &= 0, \\ B_0 + 4B_1 &= -3, \end{aligned} \qquad (C.34)$$

which has solution $B_0 = 1$ and $B_1 = -1$. The overall zero-state response is given by

$$y_{zs}(t) = e^{-t} - e^{-4t} + 3te^{-t}.$$

C.3 Complete response

The complete response of an LTIC system is the sum of the homogeneous and particular components. The procedure for calculating the complete response consists of the following steps.

(1) Compute the zero-input response $y_{zi}(t)$ of the system using the given initial conditions.
(2) Compute the zero-state response $y_{zs}(t)$ of the system using zero initial conditions and the input signal. The zero-state response is obtained by determining its homogeneous and particular components.
(3) Add the zero-input and zero-state responses of the systems to get the complete response.

Example C.6
Calculate the output of an LTIC system represented by the following differential equation:

$$\frac{d^2 y}{dt^2} + 5\frac{dy}{dt} + 4y(t) = 3x(t), \tag{C.35}$$

for the input signal $x(t) = \cos t\, u(t)$ and the initial conditions $y(0) = 2$ and $\dot{y}(0) = -5$.

Solution
The zero-input response was calculated in Example C.1 and is given by Eq. (C.11), repeated below:

$$y_{zi}(t) = e^{-t} + e^{-4t}. \tag{C.36}$$

The zero-state response was calculated in Example C.4 and is given by Eq. (C.32), repeated below:

$$y_{zs}(t) = -\frac{1}{2}e^{-t} + \frac{4}{17}e^{-4t} + \frac{9}{34}\cos t + \frac{15}{34}\sin t. \tag{C.37}$$

The complete response is the sum of Eqs (C.36) and (C.37) and is given by

$$y(t) = \frac{1}{2}e^{-t} + \frac{21}{17}e^{-4t} + \frac{9}{34}\cos t + \frac{15}{34}\sin t \tag{C.38}$$

for $t \geq 0$.

Appendix D **Partial fraction expansion**

An alternative approach to convolution, used in calculating the output response of a linear time-invariant (LTI) system, is to calculate the product of appropriately selected transforms of the convolving signals and then evaluate the inverse transform of the product. In most cases, the transform-based approach is more convenient as it leads to a closed-form solution. It is therefore important to develop methods to compute the inverse of a specified transform to determine the output response of the LTI system in the time domain. For transforms that can be expressed as a rational function of two polynomials, the partial fraction expansion simplifies the evaluation of the inverse transform by expressing the rational function as a summation of simpler terms whose inverse is obtained from a look-up table. This appendix focuses on the partial fraction expansion of a rational function. The partial fraction expansion techniques for the four transforms, namely the Laplace transform, the continuous-time Fourier transform (CTFT), the z-transform, and the discrete-time Fourier transform (DTFT), covered in the text are presented separately in Sections D.1–D.4.

D.1 Laplace transform

Consider a function $X(s)$ of the form

$$X(s) = \frac{N(s)}{D(s)} = \frac{b_m s^m + b_{m-1} s^{m-1} + \cdots + b_1 s + b_0}{a_n s^n + a_{n-1} s^{n-1} + \cdots + a_1 s + a_0}, \qquad (\text{D.1})$$

where the numerator $N(s)$ is a polynomial of degree m and the denominator $D(s)$ is a polynomial of degree n. If $m \geq n$, we can divide $N(s)$ by $D(s)$ and express $X(s)$ in an alternative form as follows:

$$X(s) = \sum_{\ell=0}^{m-n} \alpha_\ell s^\ell + \frac{N_1(s)}{D(s)}. \qquad (\text{D.2})$$

If $m < n$, there is no summation term in Eq. (D.2) and $N_1(s) = N(s)$. The partial fraction expansion represents the rational fraction $N_1(s)/D(s)$ as a summation of simpler terms.

The first step in obtaining the partial fraction expansion is to factorize the denominator polynomial and express the function $X(s)$ as follows:

$$\frac{N_1(s)}{D(s)} = \frac{N_1(s)}{(s - p_1)(s - p_2) \cdots (s - p_n)}, \tag{D.3}$$

where $p_1, p_2, \ldots, p_n$ are the n roots of the characteristic equation,

$$D(s) = a_n s^n + a_{n-1} s^{n-1} + \cdots + a_1 s + a_0 = 0. \tag{D.4}$$

If $X(s)$ represent the transfer function of an LTIC system, then the roots p_1, $p_2, \ldots, p_n$ of the characteristic equation are the poles of the system. The partial fraction expansion expresses Eq. (D.3) as the following summation:

$$\frac{N_1(s)}{D(s)} = \frac{k_1}{s - p_1} + \frac{k_2}{s - p_2} + \cdots + \frac{k_n}{s - p_n}, \tag{D.5}$$

where k_r, for $1 \leq r \leq n$, is referred to as the coefficient (also known as the residue) of the rth partial fraction. Depending on the nature of the poles, different procedures are used to compute the partial fraction coefficients k_r. We consider two cases in the following sections.

D.1.1 First-order poles The poles $p_1, p_2, \ldots, p_n$ are of the first order if they are not repeated. In such cases, the value of the rth partial fraction coefficients k_r can be calculated from the Heaviside formula:[†]

$$k_r = \left[(s - p_r) \frac{N_1(s)}{D(s)} \right]_{s = p_r}. \tag{D.6}$$

We illustrate the application of the formula with four examples.

Example D.1

For the function

$$X(s) = \frac{4s^2 + 20s - 2}{s^3 + 3s^2 - 6s - 8}, \tag{D.7}$$

(i) calculate the partial fraction expansion;

(ii) based on your answer to (i), calculate the inverse Laplace transform of $X(s)$.

[†] This formula is named after Oliver Heaviside (1850–1925), an English electrical engineer, mathematician, and physicist, who developed techniques for applying the Laplace transforms to the solution of differential equations.

Solution

(i) The characteristic equation of $X(s)$ is given by

$$s^3 + 3s^2 - 6s - 8 = 0,$$

which has roots at $s = -1$, 2, and -4. The partial fraction expansion of $X(s)$ is therefore given by

$$X(s) = \frac{4s^2 + 20s - 2}{s^3 + 3s^2 - 6s - 8} \equiv \frac{k_1}{s+1} + \frac{k_2}{s-2} + \frac{k_3}{s+4}.$$

Using the Heaviside formula, the residues k_r are given by

$$k_1 = \frac{4s^2 + 20s - 2}{(s-2)(s+4)}\bigg|_{s=-1} = \frac{4 - 20 - 2}{-9} = 2,$$

$$k_2 = \frac{4s^2 + 20s - 2}{(s+1)(s+4)}\bigg|_{s=2} = \frac{16 + 40 - 2}{3 \times 6} = 3,$$

and

$$k_3 = \frac{4s^2 + 20s - 2}{(s+1)(s-2)}\bigg|_{s=-4} = \frac{64 - 80 - 2}{(-3) \times (-6)} = \frac{-18}{18} = -1.$$

Substituting the values of the partial fraction coefficients k_1, k_2, and k_3, we obtain

$$X(s) = \frac{2}{s+1} + \frac{3}{s-2} - \frac{1}{s+4}. \tag{D.8}$$

(ii) Assuming the function $x(t)$ to be causal or right-sided, we use Table 6.1 to determine the inverse Laplace transform $x(t)$ of the $X(s)$ as follows:

$$x(t) = \left(2e^{-t} + 3e^{2t} - e^{-4t}\right)u(t). \tag{D.9}$$

Example D.2

For the function

$$X(s) = \frac{6s^2 + 11s + 26}{s^3 + 4s^2 + 13s}, \tag{D.10}$$

(i) calculate the partial fraction expansion;

(ii) based on your answer to (i), calculate the inverse Laplace transform of $X(s)$.

Solution

(i) The characteristic equation of $X(s)$ is given by

$$s^3 + 4s^2 + 13s = 0,$$

which has roots at $s = 0$, $-2 + j3$, and $-2 - j3$. The partial fraction expansion of $X(s)$ is therefore given by

$$X(s) = \frac{6s^2 + 11s + 26}{s^3 + 4s^2 + 13s} \equiv \frac{k_1}{s} + \frac{k_2}{s + 2 + j3} + \frac{k_3}{s + 2 - j3}. \qquad (\text{D.11})$$

Note that in this case, there are two complex-conjugate poles at $s = -2 \pm j3$. Using the Heaviside formula, the residues k_r are given by

$$k_1 = \left[s \frac{6s^2 + 11s + 26}{s(s + 2 + j3)(s + 2 - j3)} \right]_{s=0} = 2,$$

$$k_2 = \left[(s + 2 + j3) \frac{6s^2 + 11s + 26}{s(s + 2 + j3)(s + 2 - j3)} \right]_{s=-2-j3} = 2 - j\frac{5}{6},$$

and

$$k_3 = \left[(s + 2 + j3) \frac{6s^2 + 11s + 26}{s(s + 2 + j3)(s + 2 - j3)} \right]_{s=-2+j3} = 2 + j\frac{5}{6}.$$

Substituting the values of the partial fraction coefficients k_1, k_2, and k_3, we obtain

$$X(s) = \frac{2}{s} + \frac{2 - j\frac{5}{6}}{s + 2 + j3} + \frac{2 + j\frac{5}{6}}{s + 2 - j3}. \qquad (\text{D.12})$$

(ii) Assuming the function $x(t)$ to be causal or right-sided, we use Table 6.1 to determine the inverse Laplace transform $x(t)$ of the $X(s)$ as follows:

$$x(t) = \left[2 + \left(2 - j\frac{5}{6} \right) e^{-(2+j3)t} + \left(2 + j\frac{5}{6} \right) e^{-(2-j3)t} \right] u(t)$$

$$= \left[2 + e^{-2t} \left\{ \left(2 - j\frac{5}{6} \right) e^{-j3t} + \left(2 + j\frac{5}{6} \right) e^{j3t} \right\} \right] u(t)$$

$$= \left[2 + e^{-2t} \left\{ 2 \left(e^{j3t} + e^{-j3t} \right) + \frac{j5}{6} \left(e^{j3t} - e^{-j3t} \right) \right\} \right] u(t)$$

$$= \left[2 + e^{-2t} \left\{ 4\cos(3t) - \frac{5}{3} \sin(3t) \right\} \right] u(t)$$

$$= \left[2 + 4e^{-2t} \cos(3t) - \frac{5}{3} e^{-2t} \sin(3t) \right] u(t). \qquad (\text{D.13})$$

In Example D.2, the complex-valued poles of the Laplace transform $X(s)$ occur in conjugate pairs. This is true, in general, for any polynomial with real-valued coefficients. Although the Heaviside formula may be used to determine the values of the partial fraction residues corresponding to the complex poles, the procedure is often complicated due to complex algebra. Below, we present another procedure, which expresses such complex-valued and conjugate poles in terms of a quadratic term in the partial fraction expansion.

Example D.3

Repeat Example D.2 by expressing the complex-valued poles as a quadratic term.

Solution

(i) Combining the complex-valued terms in Eq. (D.11),

$$X(s) = \frac{6s^2 + 11s + 26}{s^3 + 4s^2 + 13s} = \frac{k_1}{s} + \frac{k_2}{s + 2 + j3} + \frac{k_3}{s + 2 - j3},$$

$$= \frac{k_1}{s} + \frac{(k_2 + k_3)s + 2(k_2 + k_3)}{(s + 2)^2 - (j3)^2}.$$

Since k_2 and k_3 are constants, their linear combinations can be replaced with other constants. Substituting $k_2 + k_3 = A_1$ and $k_2 + k_3 = A_2$, we obtain

$$X(s) = \frac{6s^2 + 11s + 26}{s^3 + 4s^2 + 13s} \equiv \frac{k_1}{s} + \frac{A_1 s + A_2}{s^2 + 4s + 13}. \qquad (D.14)$$

It may be noted that the above expression could have been obtained directly by factorizing the denominator,

$$s^3 + 4s^2 + 13s = s(s^2 + 4s + 13),$$

and writing the partial fraction expansion of $X(s)$ in terms of two terms, one with a linear polynomial s in the denominator and the other with a quadratic polynomial $(s^2 + 4s + 13)$.

The partial fraction coefficient k_1 of the linear polynomial denominator is obtained using the Heaviside formula as follows:

$$k_1 = \left[s \frac{6s^2 + 11s + 26}{s(s^2 + 4s + 13)} \right]_{s=0} = 2.$$

In order to calculate the remaining coefficients A_1 and A_2, we substitute $k_1 = 2$ in Eq. (D.14). Cross-multiplying and equating the numerators in Eq. (D.5), we obtain

$$6s^2 + 11s + 26 = 2(s^2 + 4s + 13) + (A_1 s + A_2)s$$

or

$$(A_1 + 2)s^2 + (A_2 + 8)s + 26 = 6s^2 + 11s + 26.$$

Equating the coefficients of the polynomials of the same degree on both sides of the above equation, we obtain:

coefficient of s^2 $(A_1 + 2) = 6, \quad A_1 = 4;$

coefficient of s $(A_2 + 8) = 11, A_2 = 3.$

Substituting the values of the partial fraction coefficients k_1, A_1, and A_2 in Eq. (D.14) yields

$$X(s) = \frac{2}{s} + \frac{4s+3}{(s+2)^2+9}. \tag{D.15}$$

(ii) The Laplace transform $X(s)$ is rearranged:

$$X(s) = \frac{2}{s} + \frac{4(s+2)}{(s+2)^2+9} - \frac{5}{3}\frac{3}{(s+2)^2+9},$$

such that the second and third terms are in the same form as entries (13) and (14) in Table 6.1. Taking the inverse transform gives the following transform pairs:

$$x(t) = \left[2 + 4e^{-2t}\cos(3t) - \frac{5}{3}e^{-2t}\sin(3t)\right]u(t). \tag{D.16}$$

Note that the inverse Laplace transform $x(t)$ obtained in Eq. (D.13) is identical to the answer obtained in Example D.2. The procedure followed in Example D.3 avoids complex numbers and is preferable. In cases where the roots of the characteristic equations are complex-valued, we will express the Laplace transform directly in terms of partial fraction terms with quadratic denominators.

Example D.4
For the function

$$H(s) = \frac{2s^3 + 10s^2 + 8s - 18}{s^3 + 3s^2 - 6s - 8}, \tag{D.17}$$

(i) calculate the partial fraction expansion;
(ii) based on your answer to (i), calculate the inverse Laplace transform of $X(s)$.

Solution
(i) Since the degree of both the numerator and denominator polynomials is 3, we divide the numerator polynomial by the denominator polynomial and express $H(s)$ as follows:

$$H(s) = 2 + \underbrace{\frac{4s^2 + 20s - 2}{s^3 + 3s^2 - 6s - 8}}_{X(s)}.$$

The second term in $H(s)$ is the same as the rational fraction $X(s)$ specified in Example D.1. Using the results of Example D.1, the partial fraction expansion of $H(s)$ is given by

$$H(s) = 2 + \frac{2}{s+1} + \frac{3}{s-2} - \frac{1}{s+4}. \tag{D.18}$$

(ii) Assuming that the inverse Laplace transform $x(t)$ is right-sided, we use Table 6.1 to determine the inverse Laplace transform $x(t)$ of the $X(s)$:

$$h(t) = 2\delta(t) + (2e^{-t} + 3e^{2t} - e^{-4t})u(t). \tag{D.19}$$

D.1.2 Higher-order poles The residues in a partial fraction can be calculated using the Heaviside formula in Eq. (D.4) when the poles are not repeated. However, when there are multiple poles at the same location, Eq. (D.4) cannot be directly used to calculate the coefficients corresponding to the fractions at multiple pole locations. To illustrate the partial fraction expansion for repeated poles, consider a Laplace transform $X_1(s)$ with $r - 1$ unrepeated poles at $s = p_1, p_2, \ldots, p_{r-1}$ and q repeated poles at $s = p_r$. To be consistent with the rational fraction expression in Eq. (D.1), $r - 1 + q = n$. The Laplace transform $X_1(s)$ can be expressed as follows:

$$\frac{N_1(s)}{D(s)} = \frac{N_1(s)}{(s - p_1)(s - p_2) \cdots (s - p_{r-1})(s - p_r)^q}. \tag{D.20}$$

The partial fraction expansion of the above rational function is given by

$$\frac{N_1(s)}{D(s)} = \frac{k_1}{s - p_1} + \frac{k_2}{s - p_2} + \cdots + \frac{k_{r-1}}{s - p_{r-1}} + \frac{k_{r,1}}{s - p_r} + \frac{k_{r,2}}{(s - p_r)^2}$$
$$+ \cdots + \frac{k_{r,q}}{(s - p_r)^q}. \tag{D.21}$$

The coefficients $k_1, k_2, k_3, \ldots$ and k_{r-1} corresponding to the unrepeated roots can be calculated using the Heaviside formula, Eq. (D.6). The last coefficient $k_{r,q}$ can also be calculated using Eq. (D.6) as follows:

$$k_{r,q} = \left[(s - p_r)^q \frac{N_1(s)}{D(s)} \right]_{s=p_r}. \tag{D.22}$$

However, the coefficients $k_{r,m}$ for $1 \leq m \leq (q-1)$, corresponding to the repeated poles, cannot be calculated using Eq. (D.6). Instead, these coefficients are calculated using the following formula

$$k_{r,m} = \frac{1}{(q - m)!} \left[\frac{d^{q-m}}{ds^{q-m}} (s - p_r)^q \frac{N_1(s)}{D(s)} \right]_{s=p_r} \quad \text{for } 1 \leq m \leq (q - 1). \tag{D.23}$$

Example D.5
For the function

$$X(s) = \frac{s^3 + 10s^2 + 27s + 20}{(s + 1)(s + 2)^3}, \tag{D.24}$$

(i) calculate the partial fraction expansion;
(ii) based on your answer to (i), calculate the inverse Laplace transform of $X(s)$.

Solution

(i) The partial fraction expansion of Eq. (D.24) is given by

$$X(s) = \frac{s^3 + 10s^2 + 27s + 20}{(s+1)(s+2)^3} \equiv \frac{k_1}{s+1} + \frac{k_{2,1}}{s+2} + \frac{k_{2,2}}{(s+2)^2} + \frac{k_{2,3}}{(s+2)^3}.$$

The partial fraction coefficient k_1 is calculated using the Heaviside formula, Eq. (D.6), as follows:

$$k_1 = \left. \frac{s^3 + 10s^2 + 27s + 20}{(s+2)^3} \right|_{s=-1} = \frac{2}{1} = 2.$$

The partial fraction coefficient $k_{r,3}$ is calculated using Eq. (D.22) as follows:

$$k_{2,3} = \left. \frac{s^3 + 10s^2 + 27s + 20}{s+1} \right|_{s=-2} = \frac{-8 + 40 - 54 + 20}{-1} = 2.$$

The remaining partial fraction coefficients are calculated using Eq. (D.22) as follows:

$$
\begin{aligned}
k_{2,2} &= \left\{ \frac{1}{(3-2)!} \frac{d}{ds} \left[\frac{s^3 + 10s^2 + 27s + 20}{s+1} \right] \right\}_{s=-2} \\
&= \left\{ \frac{1}{(s+1)^2} \left[(s+1)\frac{d}{ds}(s^3 + 10s^2 + 27s + 20) \right. \right. \\
&\qquad \left. \left. -(s^3 + 10s^2 + 27s + 20)\frac{d}{ds}(s+1) \right] \right\}_{s=-2} \\
&= \left\{ \frac{1}{(s+1)^2}[(s+1)(3s^2 + 20s + 27) - (s^3 + 10s^2 + 27s + 20)] \right\}_{s=-2} \\
&= \left\{ \frac{1}{(s+1)^2}[2s^3 + 13s^2 + 20s + 7] \right\}_{s=-2} = 3
\end{aligned}
$$

and

$$
\begin{aligned}
k_{2,1} &= \left\{ \frac{1}{(3-1)!} \frac{d^2}{ds^2} \left[\frac{s^3 + 10s^2 + 27s + 20}{s+1} \right] \right\}_{s=-2} \\
&= \left\{ \frac{1}{2} \frac{d}{ds} \left[\frac{2s^3 + 13s^2 + 20s + 7}{(s+1)^2} \right] \right\}_{s=-2} \\
&= \frac{1}{2} \left\{ \frac{1}{(s+1)^4} \left[(s+1)^2 \frac{d}{ds}(2s^3 + 13s^2 + 20s + 7) \right. \right. \\
&\qquad \left. \left. -(2s^3 + 13s^2 + 20s + 7)\frac{d}{ds}(s+1)^2 \right] \right\}_{s=-2} \\
&= \frac{1}{2} \left\{ \underbrace{\frac{1}{(s+1)^4}}_{=1} \left[\underbrace{(s+1)^2}_{=1} \underbrace{(6s^2 + 26s + 20)}_{=-8} \right. \right. \\
&\qquad \left. \left. - \underbrace{(2s^3 + 13s^2 + 20s + 7)}_{=3} \underbrace{(2s+2)}_{=-2} \right] \right\}_{s=-2} \\
&= \frac{1}{2}\{-8 + 6\} = -1.
\end{aligned}
$$

Therefore, the partial fraction expansion for $X(s)$ is given by

$$X(s) = \frac{2}{s+1} - \frac{1}{s+2} + \frac{3}{(s+2)^2} + \frac{2}{(s+2)^3}. \qquad \text{(D.25)}$$

(ii) Assuming that the inverse Laplace transform $x(t)$ is right-sided, we use Table 6.1 to determine the inverse Laplace transform $x(t)$ of the $X(s)$:

$$\begin{aligned} x(t) &= (2e^{-t} - e^{-2t} + 3te^{-2t} + t^2e^{-2t})u(t) \\ &= [2e^{-t} + (t^2 + 3t - 1)e^{-2t}]u(t). \end{aligned} \qquad \text{(D.26)}$$

D.2 Continuous-time Fourier transform

The partial fraction expansion method, described above, may also be applied to decompose the CTFT functions to a summation of simpler terms. Consider the following rational function for CTFT:

$$X(\omega) = \frac{N(\omega)}{D(\omega)} = \frac{b_m(j\omega)^m + b_{m-1}(j\omega)^{m-1} + \cdots + b_1(j\omega) + b_0}{a_n(j\omega)^n + a_{n-1}(j\omega)^{n-1} + \cdots + a_1(j\omega) + a_0}, \qquad \text{(D.27)}$$

where the numerator $N(\omega)$ is a polynomial of degree m and the denominator $D(\omega)$ is a polynomial of degree n. If $m \geq n$, we can divide $N(\omega)$ by $D(\omega)$ and express $X(\omega)$ as follows:

$$X(\omega) = \sum_{\ell=0}^{m-n} \alpha_\ell(j\omega)^{-\ell} + \underbrace{\frac{N_1(\omega)}{D(\omega)}}_{X_1(\omega)}. \qquad \text{(D.28)}$$

The procedure for decomposing $X_1(\omega)$ in simpler terms remains the same as that discussed for the Laplace transform, except that the expansion is now made with respect to $(j\omega)$. For example, if the denominator polynomial $D(\omega)$ has n first-order, non-repeated roots, $p_1, p_2, \ldots, p_n$, such that

$$X_1(\omega) = \frac{N_1(\omega)}{D(\omega)} = \frac{N_1(\omega)}{(j\omega - p_1)(j\omega - p_2) \cdots (j\omega - p_n)}, \qquad \text{(D.29)}$$

the function $X_1(\omega)$ may be decomposed as follows:

$$\frac{N_1(\omega)}{D(\omega)} = \frac{k_1}{j\omega - p_1} + \frac{k_2}{j\omega - p_2} + \cdots + \frac{k_n}{j\omega - p_n}, \qquad \text{(D.30)}$$

where the partial fraction coefficients k_r are calculated using the Heaviside formula:

$$k_r = \left[(j\omega - p_r) \frac{N_1(\omega)}{D(\omega)} \right]_{j\omega = p_r}. \qquad \text{(D.31)}$$

Using the CTFT pair

$$e^{-at}u(t) \overset{\text{CTFT}}{\longleftrightarrow} \frac{1}{a + j\omega},$$

the inverse CTFT of Eq. (D.30) is given by

$$x_1(t) = (k_1 e^{p_1 t} + k_2 e^{p_2 t} + \cdots + k_n e^{p_n t})u(t). \tag{D.32}$$

Similarly, the complex roots and repeated roots may be expanded in partial fractions by following the procedure outlined for the Laplace transform.

Example D.6
Using the partial fraction method, calculate the inverse CTFT of the following function:

$$X(\omega) = \frac{2(j\omega) + 7}{(j\omega)^3 + 10(j\omega)^2 + 31(j\omega) + 30}. \tag{D.33}$$

Solution
The characteristic equation of $X(\omega)$ is given by

$$(j\omega)^3 + 10(j\omega)^2 + 31(j\omega) + 30 = 0,$$

which has roots at $j\omega = -2$, -3, and -5. The partial fraction expansion of $X(\omega)$ is therefore given by

$$X(\omega) = \frac{2(j\omega) + 7}{(j\omega + 2)(j\omega + 3)(j\omega + 5)} \equiv \frac{k_1}{j\omega + 2} + \frac{k_2}{j\omega + 3} + \frac{k_3}{j\omega + 5}.$$

The partial fraction coefficients are calculated using the Heaviside formula:

$$k_1 = \left[(j\omega + 2) \frac{2(j\omega) + 7}{(j\omega + 2)(j\omega + 3)(j\omega + 5)}\right]_{j\omega = -2} = 1,$$

$$k_2 = \left[(j\omega + 3) \frac{2(j\omega) + 7}{(j\omega + 2)(j\omega + 3)(j\omega + 5)}\right]_{j\omega = -3} = -\frac{1}{2},$$

and

$$k_3 = \left[(j\omega + 5) \frac{2(j\omega) + 7}{(j\omega + 2)(j\omega + 3)(j\omega + 5)}\right]_{j\omega = -5} = -\frac{1}{2}.$$

Therefore, the partial fraction expansion of $X(\omega)$ is given by

$$X(\omega) = \frac{1}{j\omega + 2} - \frac{1}{2} \frac{1}{(j\omega + 3)} - \frac{1}{2} \frac{1}{(j\omega + 5)}. \tag{D.34}$$

Using Table 5.2, the inverse DTFT $x(t)$ of $X(\omega)$ is given by

$$x(t) = \left[e^{-2t} - \frac{1}{2}e^{-3t} - \frac{1}{2}e^{-5t}\right]u(t). \tag{D.35}$$

Example D.7

Using the partial fraction method, calculate the inverse CTFT of the following function:

$$X(\omega) = \frac{4(j\omega)^2 + 20(j\omega) + 19}{(j\omega)^3 + 5(j\omega)^2 + 8(j\omega) + 4}. \tag{D.36}$$

Solution

The characteristic equation of $X(\omega)$ is given by

$$(j\omega)^3 + 5(j\omega)^2 + 8(j\omega) + 4 = 0,$$

which has roots at $j\omega = -1$, -2, and -2. The partial fraction expansion of $X(\omega)$ is therefore given by

$$X(\omega) = \frac{4(j\omega)^2 + 20(j\omega) + 19}{(j\omega)^3 + 5(j\omega)^2 + 8(j\omega) + 4} \equiv \frac{k_1}{(j\omega + 1)} + \frac{k_{2,1}}{(j\omega + 2)} + \frac{k_{2,2}}{(j\omega + 2)^2}.$$

The partial fraction coefficients k_1 and $k_{2,2}$ are calculated using the Heaviside formula:

$$k_1 = \left[(j\omega + 1) \frac{4(j\omega)^2 + 20(j\omega) + 19}{(j\omega + 1)(j\omega + 2)^2} \right]_{j\omega = -1} = 3$$

and

$$k_{2,2} = \left[(j\omega + 2)^2 \frac{4(j\omega)^2 + 20(j\omega) + 19}{(j\omega + 1)(j\omega + 2)^2} \right]_{j\omega = -2} = 5.$$

The remaining partial fraction coefficient is calculated using Eq. (D.23):

$$k_{2,1} = \frac{1}{(2-1)!} \left[\frac{d}{d(j\omega)} \frac{4(j\omega)^2 + 20(j\omega) + 19}{(j\omega + 1)} \right]_{j\omega = -2}, \tag{D.37}$$

where the differentiation is with respect to $j\omega$. To simplify the notation for differentiation, we substitute $s = j\omega$ in Eq. (D.37) to obtain:

$$k_{2,1} = \frac{1}{(2-1)!} \left[\frac{d}{ds} \frac{4s^2 + 20s + 19}{(s+1)} \right]_{s=-2}$$

$$= \left[\frac{(s+1)(8s+20) - (4s^2 + 20s + 19)}{(s+1)^2} \right]_{s=-2} = 1.$$

The partial fraction expansion of $X(\omega)$ is therefore given by

$$X(\omega) = \frac{4(j\omega)^2 + 20(j\omega) + 19}{(j\omega)^3 + 5(j\omega)^2 + 8(j\omega) + 4} = \frac{3}{(j\omega + 1)} + \frac{1}{(j\omega + 2)} + \frac{5}{(j\omega + 2)^2}. \tag{D.38}$$

Using Table 5.2, the inverse CTFT $x(t)$ of $X(\omega)$ is given by

$$x(t) = [3e^{-t} + e^{-2t} + 5te^{-2t}]u(t). \tag{D.39}$$

D.3 Discrete-time Fourier transform

To illustrate the partial fraction expansion of the DTFT, consider the following rational function:

$$X(\Omega) = \frac{N(\Omega)}{D(\Omega)} = \frac{b_m e^{jm\Omega} + b_{m-1} e^{j(m-1)\Omega} + \cdots + b_1 e^{j\Omega} + b_0}{a_n e^{jn\Omega} + a_{n-1} e^{j(n-1)\Omega} + \cdots + a_1 e^{j\Omega} + a_0}, \qquad \text{(D.40)}$$

where the numerator $N(\Omega)$ is a polynomial of degree m and the denominator $D(\Omega)$ is a polynomial of degree n. An alternative representation for Eq. (D.40) is obtained by dividing both the numerator and the denominator by $e^{jn\Omega}$ as follows:

$$X(\Omega) = \frac{N(\Omega)}{D(\Omega)} = e^{j(m-n)\Omega} \cdot \underbrace{\frac{b_m + b_{m-1} e^{-j\Omega} + \cdots + b_1 e^{-j(m-1)\Omega} + b_0 e^{-jm\Omega}}{a_n + a_{n-1} e^{-j\Omega} + \cdots + a_1 e^{-j(n-1)\Omega} + a_0 e^{-jn\Omega}}}_{X'(\omega)} \cdot$$

$$\text{(D.41)}$$

We need to express Eq. (D.41) in simpler terms using the partial fraction expansion with respect to $e^{-j\Omega}$. To simplify the factorization process, we substitute $z = e^{j\Omega}$:

$$X(z) = z^{(m-n)} \cdot \frac{b_m + b_{m-1} z^{-1} + \cdots + b_1 z^{-(m-1)} + b_0 z^{-m}}{a_n + a_{n-1} z^{-1} + \cdots + a_1 z^{-(n-1)} + a_0 z^{-n}}. \qquad \text{(D.42)}$$

The process for the partial fraction expansion of Eq. (D.41) is the same as for the CTFT and Laplace transform, except that the expansion is performed with respect to z^{-1}. Below we illustrate the process with an example.

Example D.8

Using the partial fraction method, calculate the inverse CTFT of the following function:

$$X(\Omega) = \frac{N(\Omega)}{D(\Omega)} = \frac{2e^{j2\Omega} - 5e^{j\Omega}}{e^{j2\Omega} - (4/9)e^{j\Omega} + (1/27)}. \qquad \text{(D.43)}$$

Solution

Dividing both the numerator and the denominator of Eq. (D.43) by $e^{j2\Omega}$ yields

$$X(\Omega) = \frac{2 - 5e^{-j\Omega}}{1 - (4/9)e^{-j\Omega} + (1/27)e^{-2j\Omega}}.$$

Substitute $z = e^{j\Omega}$ in the above equation to obtain

$$X(z) = \frac{2 - 5z^{-1}}{1 - (4/9)z^{-1} + (1/27)z^{-2}},$$

with the characteristic equation

$$1 - \frac{4}{9}z^{-1} + \frac{1}{27}z^{-2} = 0 \quad \text{or} \quad z^2 - \frac{4}{9}z + \frac{1}{27} = 0,$$

which has two poles at $z = 1/3$ and $1/9$. The partial fraction expansion of $X(z)$ is therefore given by

$$X(z) = \frac{2 - 5z^{-1}}{(1 - (1/3)z^{-1})(1 - (1/9)z^{-1})} \equiv \frac{k_1}{1 - (1/3)z^{-1}} + \frac{k_2}{1 - (1/9)z^{-1}}.$$

Using the Heaviside formula, the partial fraction coefficients are given by

$$k_1 = \left[(1 - (1/3)z^{-1}) \frac{2 - 5z^{-1}}{(1 - (1/3)z^{-1})(1 - (1/9)z^{-1})} \right]_{z^{-1}=3} = -\frac{39}{2}$$

and

$$k_2 = \left[(1 - (1/9)z^{-1}) \frac{2 - 5z^{-1}}{(1 - (1/3)z^{-1})(1 - (1/9)z^{-1})} \right]_{z^{-1}=9} = \frac{43}{2}.$$

The partial fraction expansion of Eq. (D.43) is given by

$$X(z) = -\frac{39}{2} \frac{1}{1 - (1/3)z^{-1}} + \frac{43}{2} \frac{1}{1 - (1/9)z^{-1}}.$$

We substitute $z = e^{j\Omega} = z$ to express the above equation in terms of the discrete frequency Ω as follows:

$$X(\Omega) = -\frac{39}{2} \frac{1}{1 - (1/3)e^{-j\Omega}} + \frac{43}{2} \frac{1}{1 - (1/9)e^{-j\Omega}}.$$

Using Table 11.2, the inverse DTFT $x[k]$ of $X(e^{j\Omega})$ is given by

$$x(t) = \left[-\frac{39}{2} \left(\frac{1}{3} \right)^k + \frac{43}{2} \left(\frac{1}{9} \right)^k \right] u[k]. \tag{D.44}$$

D.4 The z-transform

The partial fraction expansion method can also be applied to evaluate the inverse transform of the z functions. Consider a z function of the following form:

$$X(z) = \frac{N(z)}{D(z)} = \frac{b_m z^m + b_{m-1} z^{m-1} + \cdots + b_1 z + b_0}{a_n z^n + a_{n-1} z^{n-1} + \cdots + a_1 z + a_0} \tag{D.45}$$

or

$$X(z) = \frac{N(z)}{D(z)} = z^{m-n} \frac{b_m + b_{m-1} z^{-1} + \cdots + b_1 z^{-(m-1)} + b_0 z^{-m}}{a_n + a_{n-1} z^{-1} + \cdots + a_1 z^{-(n-1)} + a_0 z^{-n}}. \tag{D.46}$$

Either of the two forms, Eq. (D.45) or Eq. (D.46), may be used to calculate the partial fraction expansion and eventually the inverse z-transform. If we use the format specified in Eq. (D.45), the partial fraction of the function $X(z)/z$ is performed with respect to z. As illustrated in Example D.9, the partial fraction of $X(z)/z$ leads to expansion terms for which the inverse z-transform is readily available in Table 13.1. If instead Eq. (D.46) is used, the partial fraction of the function $X(z)$ is performed with respect to z^{-1}. We illustrate the procedure for both formats in Examples D.9 and D.10.

Example D.9

Using Eq. (D.45) for the partial fraction expansion, calculate the inverse z-transform of the following function:

$$X(z) = \frac{z^2 - 3z}{z^3 - z^2 + 0.17z + 0.028}. \tag{D.47}$$

Solution

The transform $X(z)$ is expressed in the following form:

$$\frac{X(z)}{z} = \frac{z - 3}{z^3 - z^2 + 0.17z + 0.028}, \tag{D.48}$$

which has poles at $z = -0.1, 0.4,$ and 0.7. The partial fraction expansion of Eq. (D.48) is given by

$$\frac{X(z)}{z} = \frac{z - 3}{z^3 - z^2 + 0.17z + 0.028} \equiv \frac{k_1}{z + 0.1} + \frac{k_2}{z - 0.4} + \frac{k_3}{z - 0.7}.$$

The partial fraction coefficients are calculated using the Heaviside formula:

$$k_1 = \left[(z + 0.1)\frac{z - 3}{(z + 0.1)(z - 0.4)(z - 0.7)}\right]_{z=-0.1} = -\frac{31}{4},$$

$$k_2 = \left[(z - 0.4)\frac{z - 3}{(z + 0.1)(z - 0.4)(z - 0.7)}\right]_{z=0.4} = \frac{52}{3},$$

and

$$k_3 = \left[(z - 0.7)\frac{z - 3}{(z + 0.1)(z - 0.4)(z - 0.7)}\right]_{z=0.7} = -\frac{115}{12}.$$

The partial fraction expansion is given by

$$\frac{X(z)}{z} = -\frac{31}{4}\frac{1}{(z + 0.1)} + \frac{52}{3}\frac{1}{(z - 0.4)} - \frac{115}{12}\frac{1}{(z - 0.7)}$$

or

$$X(z) = -\frac{31}{4}\frac{z}{(z + 0.1)} + \frac{52}{3}\frac{z}{(z - 0.4)} - \frac{115}{12}\frac{z}{(z - 0.7)}.$$

Assuming a right-sided sequence, the inverse z-transform $x[k]$ of the $X(z)$ is given by

$$x[k] = \left[-\frac{31}{4}(-0.1)^k + \frac{52}{3}(0.4)^k - \frac{115}{12}(0.7)^k\right]u\,[k].$$

Example D.10

Using Eq. (D.46) for the partial fraction expansion, calculate the inverse z-transform of the following function:

$$X(z) = \frac{z^2 - 3z}{z^3 - z^2 + 0.17z + 0.028}.$$

Solution

The transform $X(z)$ is expressed in the following form:

$$X(z) = \frac{z^{-1} - 3z^{-2}}{1 - z^{-1} + 0.17z^{-2} + 0.028z^{-3}}, \qquad (D.49)$$

which has poles at $z = -0.1$, 0.4, and 0.7. The partial fraction expansion of Eq. (D.49) is given by

$$X(z) = \frac{z^{-1} - 3z^{-2}}{1 - z^{-1} + 0.17z^{-2} + 0.028z^{-3}}$$

$$\equiv \frac{k_1}{1 + 0.1z^{-1}} + \frac{k_2}{1 - 0.4z^{-1}} + \frac{k_3}{1 - 0.7z^{-1}}.$$

The partial fraction coefficients are calculated using the Heaviside formula:

$$k_1 = \left[(1 + 0.1z^{-1})\frac{z^{-1} - 3z^{-2}}{(1 + 0.1z^{-1})(1 - 0.4z^{-1})(1 - 0.7z^{-1})}\right]_{z^{-1}=-10} = -\frac{31}{4},$$

$$k_2 = \left[(1 - 0.4z^{-1})\frac{z^{-1} - 3z^{-2}}{(1 + 0.1z^{-1})(1 - 0.4z^{-1})(1 - 0.7z^{-1})}\right]_{z^{-1}=10/4} = \frac{52}{3},$$

and

$$k_3 = \left[(1 - 0.7z^{-1})\frac{z^{-1} - 3z^{-2}}{(1 + 0.1z^{-1})(1 - 0.4z^{-1})(1 - 0.7z^{-1})}\right]_{z^{-1}=10/7} = -\frac{115}{12}.$$

The partial fraction expansion is given by

$$X(z) = -\frac{31}{4}\frac{k_1}{(1 + 0.1z^{-1})} + \frac{52}{3}\frac{k_2}{(1 - 0.4z^{-1})} - \frac{115}{12}\frac{k_3}{(1 - 0.7z^{-1})}.$$

Assuming a right-sided sequence, the inverse z-transform $x[k]$ of the $X(z)$ is given by

$$x[k] = \left[-\frac{31}{4}(-0.1)^k + \frac{52}{3}(0.4)^k - \frac{115}{12}(0.7)^k\right]u[k].$$

Appendix E Introduction to MATLAB

E.1 Introduction

MATLAB, an abbreviation for the term "MATrix LABoratory," is a powerful computing environment for numerical calculations and multidimensional visualization. It has become a *de facto* industry standard for developing engineering applications for several reasons. First, MATLAB reduces programming to data processing abstraction. Instead of becoming bogged down with the intrinsic details of programming, as required with other high-level languages, it allows the user to focus on the theoretical concepts. Developing code in MATLAB takes a fraction of the time necessary with other programming languages. Secondly, it provides a rich collection of library functions, referred to as toolboxes, in virtually every field of engineering. The user can access the library functions to build the required application. Thirdly, it supports multidimensional visualization that allows experimental data to be rendered graphically in a comprehensible format.

In this appendix we provide a brief introduction to MATLAB. Our intention is to introduce the basic capabilities of MATLAB so that the reader can start working on the problems contained in this text. In the following discussion, MATLAB commands and results are shown in "Courier" font with the commands preceded by the >> prompt. Results returned by MATLAB in response to the typed commands are also shown in the "Courier" font but are *not* preceded by the >> prompt.

Starting a MATLAB session

MATLAB is available on a variety of computing platforms. On an IBM compatible PC, a MATLAB session can be initiated by selecting the MATLAB program or double clicking on its icon. In an X-window system, MATLAB is invoked by typing the complete path to the executable file of MATLAB at the shell prompt. Before using MATLAB, it is recommended that you create a subdirectory named ⟨matlab⟩ (all lower case letters for case-sensitive operating systems) in your home directory. Any file placed in this subdirectory can be accessed from within the MATLAB environment without specifying the complete path of the file.

MATLAB includes a comprehensive combination of demos to illustrate the offered features and capabilities to its users. In order to explore the demo, just type demo at the command line of the MATLAB environment indicated by the >> prompt:

```
>> demo
```

This will open the MATLAB demo window. Follow the interactive options by clicking on the features that interest you. In most cases, the MATLAB code used to generate the demo is also included for illustration.

Help in MATLAB

MATLAB provides a useful built-in help facility. You can access help either from the command line or by clicking on the graphical "Help" menu. On the command line, the format for obtaining help on a particular MATLAB function is to type help followed by the name of the function. For example, to learn more about the plot function, type the following instruction in the MATLAB command window:

```
>> help plot
```

If the name of the function is not known beforehand, you can use the lookfor command followed by a keyword that identifies the function being searched, to enlist the available MATLAB functions with the specified keyword. For example, all MATLAB functions with the keyword "Fourier" can be listed by typing the following command:

```
>> lookfor Fourier
```

On execution of the above command, MATLAB returns the following list, specifying the names of the functions and a brief comment on their capabilities:

```
FFT Discrete Fourier transform.
FFT2 Two-dimensional discrete Fourier Transform.
FFTN N-dimensional discrete Fourier Transform.
IFFT Inverse discrete Fourier transform.
IFFT2 Two-dimensional inverse discrete Fourier trans-
form.
IFFTN N-dimensional inverse discrete Fourier transform.
XFOURIER Graphics demo of Fourier series expansion.
DFTMTX Discrete Fourier transform matrix.
```

E.2 Entering data into MATLAB

Data can be entered in the MATLAB as a scalar quantity, a row or column vector, and a multidimensional array. In each case, both real and complex numbers can be entered. As required in other high-level languages, there is no need to declare the type of a variable before assigning data to it. For example, variable a can be assigned the value $(6 + j8)$ by typing the following command:

```
>> a = 6 + j*8
```

On the execution of the above command, MATLAB returns the following answer:

```
a = 6.0000 + 8.0000i
```

In the above command, we did not allocate any value to j, yet MATLAB recognized it as a complex operator with value $j^2 = 1$. There is a whole range of special words that are used by MATLAB either as the name of functions or variables. These include pi, i, j, Inf, NaN, sin, cos, tan, exp, and rem. Type help elfun to list the names that are used by MATLAB to specify the built-in functions and variables. The value of any of these special words can be changed by assigning a new value to it. For example,

```
>> sin = 1
```

allocates the value of 1 to the variable sin. The MATLAB definition of the trigonometric sine is overwritten by our command. To check the current status of the runtime environment of MATLAB, type whos at the prompt:

```
>> whos
```

MATLAB returns the following answer:

```
Name      Size     Bytes     Class
a         1x1      16        double array (complex)
sin       1x1      8         double array
Grand total is 2 elements using 24 bytes
```

Alternatively, the command who can also be used to list the name of defined variables in the MATLAB runtime environment. The command who does not provide additional details such as the size and class of each variable. In the preceding discussions, we overwrote the sin function and allocated a value of 1 to it. Consequently, we cannot access the MATLAB built-in function sin to evaluate the sine of an angle. To clear our definition of sin, we can use the following command:

```
>> clear sin
```

The original definition of sin is restored in the MATLAB environment. Typing

```
>> sin(pi/6)
```

calls up the built-in sin function with $\pi/6$ as the input argument. Recall that the variable pi is a built-in variable that has been assigned the value of 3.141 596 25. MATLAB returns

```
ans =
    0.5000
```

after execution of the sin command. For additional information on the sin function, type help sin. To allocate the returned value of sin(pi/6) to variable x, for example, type

```
>> x = sin(pi/6)
```

which returns

```
x =
    0.5000
```

In the above examples, MATLAB displays the result of each instruction. The display can be suppressed by inserting a semicolon at the end of each instruction. For example, the command

```
>> x = sin(pi/6);
```

initializes x = 0.5000 without displaying the end result.

Most common arithmetic operations are available in MATLAB. These include + (add), − (subtract), * (multiply), / (divide), ^ (power), .* (array multiplication), and ./ (array division). For complex numbers, in addition to the aforementioned operators, MATLAB provides a collection of library functions that can be used to perform more complex operations. These are illustrated through the following example, where a brief explanation of each instruction is included as a comment. In MATLAB, the segment of line after the % sign on the same line are treated as comments and ignored during execution. The returned value is enclosed in parentheses and is also included with the explanation.

```
>> x = 2.3 - 4.7*i;      % Initializes x as a complex
                         % variable.
>> x_magn = abs(x);      % magnitude of x, (5.2326)
>> x_phas = angle(x);    % phase of x in radians/s, (1.1157)
>> x_real = real(x);     % Real component of x, (2.3)
>> x_imag = imag(x);     % Imaginary component of x, (-4.7)
>> x_conj = conj(x);     % Complex conjugate of x,
                         % (2.3 + 4.7i)
```

MATLAB also provides a set of functions for decimal numbers. If applied to integers, these functions do not make any changes. On the other hand, if these functions are applied to complex numbers, each operation is performed individually on the real and imaginary component. Below we provide a selected list.

```
>> x = 2.3 - 4.7*i;        % Initializes x as a complex
                           % variable
>> x_round = round(x);     % rounds to nearest
                           % integer, (2 - 5i)
>> x_fix = fix(x)          % rounds to nearest integer
                           % towards zero, (2 - 4i)
>> x_floor = floor(x)      % rounds down (towards negative
                           % infinity), (2 - 5i)
>> x_ceil = ceil(x)        % rounds up (towards positive
                           % infinity) (3 - 4i)
```

We now consider initialization of multidimensional arrays through a series of examples.

Example E.1

Consider the two row vectors

$$f = \begin{bmatrix} 1, & 4, & -2, & (3 - 2i) \end{bmatrix}$$

and

$$g = \begin{bmatrix} -3, & (5 + 7i), & 6, & 2 \end{bmatrix}.$$

Perform the following mathematical operations in MATLAB on vectors f and g:

(i) addition, $r1 = f + g$;

(ii) dot product, $r2 = f \cdot g$;

(iii) mean, $r3 = \dfrac{1}{4} \sum\limits_{k=1}^{4} f(k)$;

(iv) average energy, $r4 = \dfrac{1}{4} \sum\limits_{k=1}^{4} |f(k)|^2$;

(v) variance, $r5 = \dfrac{1}{4} \sum\limits_{k=1}^{4} |f(k) - r3|^2$, where $r3$ is defined in (iii).

Solution

The MATLAB code to solve part (i) is given below with comments following the % sign:

```
>> f = [1 4 -2 3-2*i];      % initialize f
>> g = [-3 5+7*i 6 2];      % initialize g
>> r1 = f + g               % Calculate the sum of f and g
                            % The result is displayed due
                            % to the absence of a
                            % semicolon at the end of the
                            % instruction
```

results in the following value for r1:

```
r1 =
-2.0000 9.0000+7.0000i 4.0000 5.0000-2.0000i
```

which can be confirmed by direct addition of vectors f and g.

    (ii) To compute part (ii), we use the MATLAB function dot as follows:

```
>> r2 = dot(f,g)       % dot returns dot product btw f and g
```

which returns

```
r2 =
11.0000+32.0000i
```

An alternative approach to compute the dot product is to multiply the row vector f by the conjugate transpose of g. The transpose is needed to make the two vectors conformable for multiplication. You may verify that the instruction

```
>> r2 = g*f';          % alternative expression for calculating
                       % the dot product. Operator ' denotes
                       % complex-conjugate transpose
```

returns the same value as above.

    (iii) The instruction for part (iii) is as follows:

```
>> r3 = sum(f)/length(f)    % sum(f) adds all row entries
                            % of vector f. length(f)
                            % returns no. of entries in f
```

which returns

```
r3 =
1.5000 − 0.5000i
```

    (iv) The instruction for part (iv) is as follows:

```
>> r4 = sum(f.*conj(f))/length(f)
                       % Operation f.*g does an element by
                       % element multiplication of vectors f
                       % and g. Operation conj(f) takes complex
                       % conjugate of each entry in f
```

which returns

```
r4 =
8.5000
```

(v) To compute part (v), we can modify the code in part (iv) by preceding it with the following instruction:

```
>> f_zero_mean = f - mean(f);% mean(f) computes the
                             % average value of f
>> r5 = sum(f_zero_mean.*conj(f_zero_mean))/length
   (f_zero_mean)
```

which returns

```
r5 =
6
```

As a final note to our introduction on vectors, the second element of vector f can be accessed by the instruction

```
>> f(2)
```

which returns

```
ans =
4
```

A range of elements within a vector can be accessed by specifying the integer index numbers of the elements. To access elements 1 and 2 of row vector f, for example, we can type the instruction

```
>> f(1:1:2);
```

Similarly, the odd number elements in f can be accessed by the instruction

```
>> x = f(1:2:length(f));
```

where we have assigned the returned value to a new variable x. Code `1:2:length(f)` is referred to as a range-generating statement that generates a row vector. The first element of the row vector is specified by the left-most number (1 in our example). The next element in the row vector is obtained by adding the middle element (2 in our example) to the first element and proceeding all the way till the limit (`length(f)`) is reached. The middle element (2 in our example) specifies the increment, while the third element (`length(f)`) is the ending index. If the increment is missing, MATLAB assigns a default value of 1 to it. As another example, the range-generating statement `1:11` produces the row vector `[1 2 3 4 5 6 7 8 9 10 11]`. Further, the starting index, increment, or ending index can also be real-valued numbers. The range-generating statement `[0.1:0.1:0.9]` produces the row vector `[0.1 0.2 0.3 0.4 0.5 0.6 0.7 0.8 0.9]`.

Example E.2

Initialize the following matrix:

$$A = \begin{bmatrix} 2 & 4 & -1 & 0 \\ 5 & 2 & 3 & 9 \end{bmatrix}$$

and take the pseudo-inverse of A, defined as $A^+ = (A^T A)^{-1} A^T$ with T denoting the conjugate transpose operation.

Solution

The following MATLAB code initializes matrix A:

```
>> A = [2 4 -1 0;5 2 3 9];   % The semicolon inside square
                             % parenthesis separates
                             % adjacent rows of a matrix
```

An alternative but longer set of instructions for the initialization of A is as follows:

```
>> A(1,1)=2; A(1,2)=4; A(1,3)=-1; A(1,4)=0;
>> A(2,1)=5; A(2,2)=2; A(2,3)=-3; A(2,4)=9;
```

To calculate the pseudo-inverse of A, the following instruction may be used:

```
>> Ainverse = inv(A'*A)*A';   % Function inv calculates
                              % inverse of a matrix
                              % while ' denotes conjugate
                              % transpose
```

which returns a warning that the matrix is singular. From linear algebra, we know that the inverse of a matrix only exits if it is non-singular, hence the pseudo-inverse does not exist for the above choice of A.

Example E.3

Initialize the following discrete-time function:

$$f[k] = 2^* \cos\left(\frac{\pi}{15}{}^* k\right) \quad \text{for} \quad 0 \le k \le 30.$$

Solution

As in other high-level languages, we can use a `for` statement to initialize the function f. The code is given by

```
for k = 0:1:30,
    f(k+1) = 2*cos(pi/15*k);      % In MATLAB, the index of a
                                   % vector or a matrix must
end                                % not be zero.
```

In MATLAB, the index of a vector or matrix cannot be zero. Therefore, we use two row vectors k and f to store the DT function. The row vector k specifies the time indices at which function f is evaluated, while f contains the value of the DT function at the corresponding time index stored in k. The above initialization can also be performed in MATLAB more quickly and in a much more compact way.

```
clear             % user-defined variables are cleared
k = 0:30;         % k is a row vector of dimensions 1x30
f = 2*cos(5*k)    % f has the same dimensions as k
```

returns the following answer:

```
f =
  Columns 1 through 7
    2.0000 0.5673 -1.6781 -1.5194 0.8162 1.9824 0.3085
  Columns 8 through 14
   -1.8074 -1.3339 1.0506 1.9299 0.0443 -1.9048 -1.1249
  Columns 15 through 21
    1.2666 1.8435 -0.2208 -1.9688 -0.8961 1.4603 1.7246
  Columns 22 through 28
   -0.4819 -1.9980 -0.6516 1.6284 1.5754 -0.7346 -1.9922
  Columns 29 through 31
   -0.3956 1.7677 1.3985
```

In terms of execution time, implementation 2 is more efficient than the first implementation. Since MATLAB is an interpretive language, loops take a long time to be executed. An efficient MATLAB code avoids loops and, if possible, replaces them with matrix or vector multiplications.

Example E.4
Initialize the following DT function:

$$g[k] = f[k] \quad \text{for} \quad 0 \le k \le 6.$$

Solution
In the above example, it has been assumed that the matrix f has been initialized as per Example E.3. The following MATLAB code will initialize row vector g:

```
>> g = f(1:7);
```

If missing, a default value of 1 is assumed as the increment in the range-generating statement $(1:7)$. Therefore, $g(1:7)$ is equivalent to $g(1:1:7)$.

E.3 Control statements

MATLAB supports several other loop statements (while, switch, etc.) as well as the if-else statement. In functionality, these statements are similar to their counterparts in C but the syntax is slightly different. In the following, we provide examples for some of the loop and conditional statements by providing analogy with the C code. Readers who are unfamiliar with C can skip the C instructions and study the explanatory comments that follow.

Example E.5

Consider the following set of instructions in C:

```
int X[2][2] ={ {2, 5},{4,6} };     /* initialize matrix X */
int Y[2][2] ={ {1, 5},{6,-2} };    /* initialize matrix Y */
int Z[2][2];                            /* declare Z */
for (m = 1; m <= 2; m++) {
    Z[m][n] = X[m][n] + Y[m][n]; /* Z = X + Y */
}
```

Write down the equivalent MATLAB code for the above instructions. Can the MATLAB code be simplified?

Solution

Implementation 1 Following a step-by-step conversion of the C code into MATLAB yields

```
>> X = [2 5; 4 6]                    % X is initialized
>> Y = [1 5; 6 -2]                   % Y is initialized
>> for m = 1:2,
      for n = 1:2,
        Z(m,n)  = X(m,n)+Y(m,n);
      end
   end
```

Implementation 2 The for loops in MATLAB can be replaced by the while statement as follows:

```
>> X = [2 5; 4 6]       % X is initialized
>> Y = [1 5; 6 -2]      % Y is initialized
>> m = 1;
>> while (m < 3),
```

```
    n = 1;
    while (n < 3),
       Z(m,n) = X(m,n)+Y(m,n);
       n = n + 1;
    end
    m = m + 1;
end
```

Implementation 3   We can avoid the two `for` or `while` loops by performing a direct sum of matrices X and Y as follows:

```
>> X = [2 5; 4 6]      % X is initialized
>> Y = [1 5; 6 -2]     % Y is initialized
>> Z = X + Y;
```

Compared with the first two implementations, the third implementation is cleaner and faster.

Example E.6

Consider the following set of instructions in C:

```
int a = 15;  /* initialize scalar a */
int x;       /* declare x */
if (a > 0)
  x = 5;     /* initialize x to 5 if a > 0*/
else
  x = 100;   /* initialize x to 5 if a <= 0 */
```

Write down the equivalent MATLAB code.

Solution

Following a step-by-step conversion, we obtain the following equivalent set of instructions in MATLAB:

```
>> a = 15;
>> if a > 0,
      x = 5;
   else,
      x = 100
   end
```

While using the conditional statements, relational operators such as equal to, not equal to, or less than are generally required in the code. MATLAB provides six basic relational operators which are defined in Table E.1.

Table E.1. Relational operations available in
MATLAB

Relational operator	Definition
<	less than
>	greater than
==	equal to
~=	not equal to
<=	less than or equal to
>=	greater than or equal to

E.4 Elementary matrix operations

MATLAB provides several built-in functions to manipulate matrices. In the
following, we provide a brief description of some of the important matrix operations. Consider the instruction

```
>> f = exp(0.05*[1:30]); % Initialize row vector f
```

which initializes the row vector f according to the following definition:

$$f[k] = e^{0.05k} \quad \text{for} \quad 1 \leq k \leq 30.$$

The following MATLAB instructions provide examples of basic arithmetic
operations performed on a row or column vector. Comments against each
instruction provide a brief description of the instruction, with the value returned
by MATLAB enclosed in parenthesis:

```
>> f_max = max(f);        % Maximum value in f (4.4817)
>> f_min = min(f);        % Minimum value in f (1.0513)
>> f_sum = sum(f);        % Sum of all entries in f (71.3891)
>> f_prod = prod(f);      % Product of entries in
                          f (1.2513e+10)
>> f_mean = mean(f);      % Mean of entries in f (2.3796)
>> f_var = var(f);        % Variance of entries in f (1.0578)
>> f_size = size(f);      % Dimensions of f ([1 30])
>> f_length = length(f);  % Length of f (30)
>> fprintf('\nThe min value of all matrix elements =
   %f\n', f_min);         % Prints the variable f_min
```

The fprintf instruction at the end of the code is used to print the value of
the variable f_min onto the screen. It returns

```
   The min value of all matrix elements = 1.051300
```

The aforementioned instructions can alternatively be used for matrices and
higher dimensional arrays. The syntax stays the same, but the result may be

different. For matrices, for example, the specified operation is performed on each column of the matrix and a row vector is returned as the answer. For example, consider the matrix F initialized by the following instruction:

```
>> F = magic(5);      % magic(N) returns an (N x N) matrix
                      % with entries between 1 through
                      % N^2 having equal row, column, and
                      % diagonal sums
```

For matrix F, the values indicated in the comments are returned:

```
>> F_max = max(F);    % Maximum value along each column
                      % [23 24 25 21 22]
>> F_min = min(F);    % Minimum value along each column
                      % [ 4 5 1 2 3]
>> F_sum = sum(F);    % Sum of entries along each column
                      % [65 65 65 65 65]
>> F_prod = prod(F);  % Product of entries along each
                        % column
                      % [172040 155520 43225 94080 142560]
>> F_mean = mean(F);  % Mean of entries along each column
                      % [13 13 13 13 13]
>> F_var = var(F);    % Variance of entries along each
                        % column
                      % [52.5 65.0 90.0 65.0 52.5]
>> F_size = size(F);  % Dimensions of F; [5 5]
>> F_length
   = length(F);       % Returns number of rows in F (5)
```

For completeness, we also include a list of some basic matrix operations, some of which were introduced in Section E.1:

```
>> X = [2 5; 4 6];    % Initailize (2 x 2) matrix X
>> Y = [1 5; 6 -2];   % Initailize (2 x 2) matrix Y
>> Zsum = X + Y;      % Adds matrices of equal dimensions.
                      % Returns [3 10; 10 4]
>> Zdif = X - Y;      % Subtracts matrices of equal
                        % dimensions;
                      % Returns [1 0; -2 8]
>> Zprod = X*Y;       % Multiplies matrices conformable for
                      % multiplication; Returns
                        % [32 0; 40 8].
```

```
>> Ztran = X';          % Calculates transpose of X
                        % Returns [2 4; 5 6]
>> Zinv = inv(X);       % Inverts X
                        % Returns [-0.75 0.62; 0.50 -0.25]
>> Zarraymul = X.*Y;    % Element by element multiplication
                        % Returns [2 25; 24 -12]
>> Zarraydiv = X./Y;    % Element by element division
                        % Returns [2 1; 0.6667 -3]
>> Zpower1 = X.^2;      % Each element is raised to power
                          % by 2
                        % Returns [4 25; 26 36]
>> Zpower2 = X.^Y;      % Each element in X is raised to
                          % power by its corresponding
                          % element in Y
                        % Returns [2 3125; 4096 0.028]
```

E.5 Plotting functions

MATLAB supports multidimensional visualization that allows experimental data to be rendered graphically in a comprehensible format. In this section, we will focus on 2D plots for continuous-time and discrete-time variables. Readers should check the demo for more advanced graphics including 3D plots.

Example E.7
Plot the following function:

$$f[k] = 2\cos(0.5k)$$

as a function of k for the range $-20 \le k \le 20$.

Solution
The following set of MATLAB instructions will generate and plot the function:

```
>> k = -20:20;          % Initializes k as a (1 x 41)
                          % row vector
>> f = 2*cos(0.5*k);    % Initializes f as cos(0.5k)
>> figure(1);           % selects figure 1 where plot
                          % is drawn
>> plot(k,f); grid on;  % CT plot of f (ordinate)
                          % versus k (abscissa)
                        % Grid is turned on
>> xlabel('k');         % Sets label of X-axis to k
>> ylabel('f[k]');      % Sets label of Y-axis to f[k]
>> axis([-25 25 -3 3])  % Plot is viewed in the range
                          % given by
```

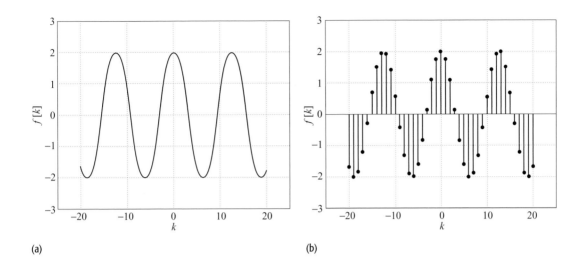

Fig. E.1. Plots of
$f[k] = 2\cos(0.5k)$ versus k in
the range $-20 \le k \le 20$.
(a) CT plot; (b) stem DT plot.

```
>> print -dtiff plot.tiff
```
```
% [x-min x-max y-min y-max]
% Saves figure in the file
%  "plot.tiff" in
% the TIFF format
```

These instructions produce a continuous plot cosine wave, as shown in Fig. E.1. It is also possible to construct a discrete-time plot using the stem function:

```
>> figure(2)
>> stem(k,f,'filled');      % DT plot; option 'filled'
                            % fills the circles at the
                            % top of vertical bars
>> xlabel('k');             % Sets label of X-axis to k
>> ylabel('f[k]');          % Sets label of Y-axis to f[k]
>> axis([-25 25 -3 3])
>> print -dtiff plot2.tiff
```

Both plot and stem functions have a variety of options available, which may be selected to change the appearance of the figures. The reader is encouraged to explore these options by seeking help on these functions in MATLAB. In addition, there are several other 2D graphical functions in MATLAB. These include semilogx, semilogy, loglog, bar, hist, polar, stairs, rose, errorbar, compass, and pie.

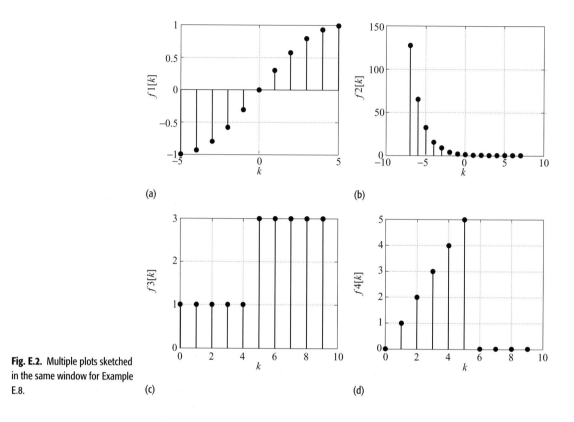

Fig. E.2. Multiple plots sketched in the same window for Example E.8.

Plotting multiple graphs in one figure

MATLAB provides the function subplot to sketch multiple graphs in one figure. We demonstrate the application of the subplot function through an example.

Example E.8

Plot the following functions over the specified range in one figure:

(a) $f_1[k] = \sin(0.1\pi k)$ for $-5 \leq k \leq 5$;

(b) $f_2[k] = 2^{-k}$ for $-7 \leq k \leq 7$;

(c) $f_3[k] = \begin{cases} 1 & (0 \leq k \leq 4) \\ 3 & (5 \leq k \leq 9) \end{cases}$;

(d) $f_4[k] = \begin{cases} k & (0 \leq k \leq 5) \\ 0 & (6 \leq k \leq 9) \end{cases}$.

Solution

The following set of MATLAB instructions plots the four functions illustrated in Fig. E.2.

```
>> % Part (a)
>> figure(5)                    % Select figure 5 for plots
>> clf                          % Clear figure 5
>> k = [-5:5];                  % k = [-5 -4 ...0 ...4 5]
>> f1 = sin(0.1*pi*k);          % Calculate function f1
>> subplot(2,2,1);             % Divides fig 5 into (m = 2)
                                % vertical and (n = 2)
                                % horizontal sub-figures.
                                % The last argument (p = 1)
                                % accesses sub-figures
                                % (1 <= p <= m*n).

>> stem(k,f1,'filled');
   grid on;                     % DT plot of f1 versus k
>> xlabel('k') ;               % Label of X-axis
>> ylabel('f1[k]')              % Label of Y-axis
>> % Part (b)
>> k = [-7:7];                  % k overwritten to
                                % [-7 -6 ...0 ...6 7]

>> f2 = 2. ^  (-k) ;            % Calculate function f2
>> subplot(2,2,2);             % Select p = 2 sub-figure
>> stem(k,f2,'filled');
   grid on;                     % DT plot of f2 versus k
>> xlabel('k');                % Label of X-axis
>> ylabel('f2[k]');            % Label of Y-axis
>> % Part (c)
>> k = [0:9];                   % k overwritten to
                                %  [0 1 ...8 9]
>> f3 = [1 1 1 1 1 3 3 3 3 3];  % Calculate function f3
>> subplot(2,2,3);             % Select p = 3 sub-figure
>> stem(k,f3,'filled');
    grid on;                    % DT plot of f3 versus k
>> xlabel('k');                % Label of X-axis
>> ylabel('f3[k]');            % Label of Y-axis
>> % Part (d)
>> k = [0:9];
>> f4 = [0 1 2 3 4 5 0 0 0 0];  % Calculate function f4
>> subplot(2,2,4);             % Select p = 4 sub-figure
>> stem(k,f4, 'filled');
    grid;                       % DT plot of f2 versus k
>> xlabel('k');                % Label of X-axis
>> ylabel('f4[k]');            % Label of Y-axis
>> print -dtiff plot.tiff;      % Save the figure as a
                                % TIFF file
```

E.6 Creating MATLAB functions

In the preceding examples, we have used MATLAB in an interactive mode with each instruction individually typed at the command prompt. MATLAB allows for the creation of M-files, where instructions can be stored in a file. An M-file can be of two types, called *scripts* and *functions*. A script is a list of MATLAB instructions that are saved in file with a . m extension. The script file can access the variables defined in the MATLAB workspace. Likewise, all variables declared in the script are accessible to the workspace. For example, the instructions to solve part (a) of Example E.8 can be stored in a file myfirstplot.m as follows:

```
% Content of script myfirstplot.m
% Part (a)
figure(5)              % Select figure 5 for plots
clf                    % Clear figure 5
k = [-5:5];            % k = [-5 -4 ...0 ...4 5]
f1 = sin(0.1*pi*k);    % Calculate function f1
subplot(2,2,1);        % Divides fig 5 into (m = 2) vertical
                       % and (n = 2) horizontal sub-figures
                       % The last argument (p = 1) accesses
                       % sub-figures (1 <= p <= m*n)
stem(k,f1,'filled');
   grid on;            % DT plot of f1 versus k
xlabel('k');           % Label of X-axis
ylabel('f1[k]')        % Label of Y-axis
```

To execute myfirstplot.m, simply type the name of the M-file (myfirstplot in this case) at the command prompt. By executing the function whos, you can determine that all variables defined in myfirstplot.m are part of the MATLAB workspace.

A function in MATLAB is a special type of script file that can accept input arguments and return output arguments. Variables declared within a function are local to the function. Likewise, none of the variables defined in the MATLAB working environment are accessible by the function unless these variables are explicitly passed as an input argument to the function. A function file must follow a specific format. The first line defines the function by specifying a name for the function and indicates the number of input and output arguments. Immediately following the definition, lines that begin with a comment symbol (%) are printed when help is requested on the function. As an example, we modify script myfirstplot.m into a function in the following.

```
function [f1] = myfirstplot(k)
% USAGE: [f1] = myfirstplot(k)
% Plots f1 = sin(0.1*pi*k) as a function of k in subplot
% (2,2,1) where k = row vector containing the indices
% where f1 is to be defined f1 is the output row vector

figure(5)              % Select figure 5 for plots
clf                    % Clear figure 5
f1 = sin(0.1*pi*k);    % Calculate function f1
subplot(2,2,1);        % Divides fig 5 into (m = 2) vertical
                       % and (n = 2) horizontal sub-figures
                       % The last argument (p = 1) accesses
                       % sub-figures. (1 <= p <= m*n)
stem(k,f1,'filled');
    grid on;           % DT plot of f1 versus k
xlabel('k') ;          % Label of X-axis
ylabel('f1[k]')        % Label of Y-axis
end
```

Once a function has been created, it must be saved in a file whose name is same as the defined name of the function. In our example, the aforementioned function must be saved in a file myfirstplot.m. The calling format for a function is the same as one would use to access a MATLAB built-in function. To access myfirstplot, the following instructions must be typed at the MATLAB prompt:

```
>> m = [-5:5];         % Define the input argument
>> [y] = myfirstplot(m);  % Output value is returned to y
                       % with subplot plotted in
                         % figure 5
```

E.7 Summary

In this appendix, a working introduction to MATLAB is provided. The intent is to introduce the basic capabilities of MATLAB to the reader. MATLAB supports hundreds of built-in functions from linear algebra, numerical analysis, polynomial algebra, and numerical optimization. These built-in functions are supported in both the student and full version of MATLAB, and do not require any toolboxes. A list of built-in functions is available on the Mathworks website (www.mathworks.com). Readers are encouraged to visit the website and explore MATLAB in more detail.

Appendix F **About the CD**

This book is accompanied by a CD that includes material for supplementary reading, MATLAB code used in the text, and data used in different simulations. The organization of the CD is shown in Table F.1.

In Table F.1, we have assumed that the CD drive is mapped to the shortcut "CD." Check the appropriate shortcut to the CD drive on your computer. For example, if the CD drive is mapped to the shortcut "F," replace "CD" in the aforementioned paths to the folders with "F" such that the path to the interactive programs is specified by F:\InteractEnv. The other two folders can be found in a similar way. In the following we provide additional information on each folder.

F.1 Interactive environment

The "InteractEnv" folder contains three interactive learning objects used to explain the operations of convolution integral, convolution sum, and digital filtering. While the first two learning objects developed to explain convolution integral and sum are based on Macromedia Flash, the third learning object uses a graphical interface environment based on MATLAB.

F.1.1 Convolution

Convolution is an important signal processing operation, which is extensively used to compute the output of linear time-invariant systems. The graphical approach to solve the convolution integral in the CT domain was presented in Section 3.5. Likewise, the steps involved in computing the convolution sum in the DT domain were explained in Section 10.5. To help understand the two convolution operations, the CD includes two Shockwave Flash animations, one each for the convolution integral and convolution sum.

The learning object for the convolution integral convolves the following CT signal:

$$x(t) = u(t + 0.5) - u(t - 1) \text{ with } h(t) = u(t + 0.5) - u(t + 1)$$

Table F.1. Organization of the CD

Folder	Comments
CD:\ InteractEnv	contains interactive programs explaining important concepts such as the convolution integral, the convolution sum, and digital filtering
CD:\Data	contains selected audio clips and images used in MATLAB simulations
CD:\ MATLAB Codes	contains MATLAB functions used in the text

Table F.2. Values of the sequence $y[k]$

k	−5	−4	−3	−2	−1	0	1	2
$y[k]$	8	12	14	15	15	7	3	1

and describes the graphical approach to derive the output of the LTIC system. By analytical computation, it is straightforward to derive the following expression for the output:

$$y(t) = \begin{cases} t+1 & -1 \le t < 0.5 \\ -t+2 & 0.5 \le t < 2 \\ 0 & \text{otherwise.} \end{cases}$$

The learning object for the convolution sum uses the following DT sequences:

$$x[k] = u[k+2] - u[k-3] \text{ with } h[k] = 2^{-k}(u[k+3] - u[k-1])$$

and describes the graphical approach to derive the output of the LTID system. All non-zero values of the output sequence $y[k]$ are specified in Table F.2. In order to run the two animations, you should open a web browser, such as Netscape or Internet Explorer (IE), with the Flash Player incorporated within the browser. If the Flash Player is not incorporated, it can be downloaded and installed from http:/www.macro.media.com, which is the official website of Macromedia. In the following, we highlight the procedure for the convolution integral through a series of steps.

Step 1 Open the internet browser (Netscape or IE) by selecting the program from the task bar. Within the browser, select the "File" option from the extreme top left menu and click on the "Open" option. This opens a dialog box, where you can provide the complete path to the convolution integral animation and choose a file. Browse to the convolution animation and select it. In our case, the path to the animation for the convolution integral is given by

CD:\InteractEnv\convolution\ConvolutionIntegral.swf

where CD specifies the drive name to the CD-ROM. After the execution of step 1, a frame similar to that in Fig. F.1 would be displayed on the computer screen.

Fig. F.1. Initial Flash window for convolution integral.

Fig. F.1. Initial Flash window for convolution integral.

Step 2 The frame displayed in step 1 has three subwindows. The top subwindow on the left-hand side plots the figures graphically, while the top subwindow on the right displays different steps involved in computing the convolution integral. The step being executed is highlighted, with the explanation included in the bottom subwindow. To interact with the animation, three options are available. Clicking on the "previous step" option moves the animation back by one frame, showing the result of the previous step. Clicking on the "next step" moves the animation forward by one frame, while clicking on the "reset" option initializes the animation to the start.

Step 3 Play the animation according to your speed and try to understand all operations performed to compute the result of the convolution integral. Once the animation has been completely played, a frame similar to that in Fig. F.2 would appear on the computer screen.

The procedure for running the convolution sum animation is identical to that of the convolution integral. Once this animation has been completely played, a frame similar to that in Fig. F.3 would appear on the computer screen.

F.1.2 Digital audio filtering

To explain digital filtering, the CD provides a set of MATLAB programs used to create a digital audio filtering interactive environment (DAFIE). The programs

Fig. F.2. Final frame for the
learning object explaining the
convolution integral operation.

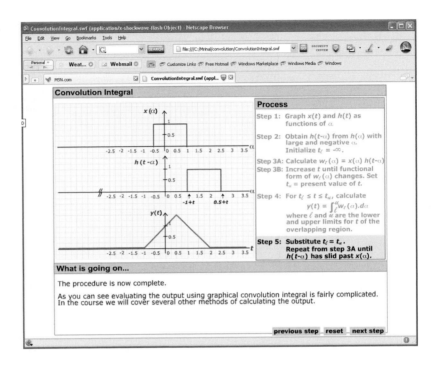

are available in the following folder:

CD:\InteractEnv\filter

where CD specifies the drive name to the CD-ROM. DAFIE is a graphical user
interface (GUI), which may be used to select an audio file, read the signal,
and manipulate the signal in different ways. The following four functions are
primarily used to create the interactive environment:

```
dafie.m          % main program for generating DAFIE
localbutton.m    % function that selects the operation
                 % using local buttons
designfilter.m   % designs filters based on the specs
                 % provided by the user
openfile.m       % opens a dialog box to select an input
                 % audio file
```

The main program `dafie` uses the built-in MATLAB function `uicontrol`
to create the user interface. When the main program `dafie` is run, an interactive window is created. A snapshot of the window is shown in Fig. F.4. The
interactive window consists of three subwindows: Command, Comments, and
Graphics. The Command subwindow controls the environment through a series
of buttons. A brief description on the functionality of each button is as follows.

Fig. F.3. Final frame for the learning object explaining the convolution sum operation.

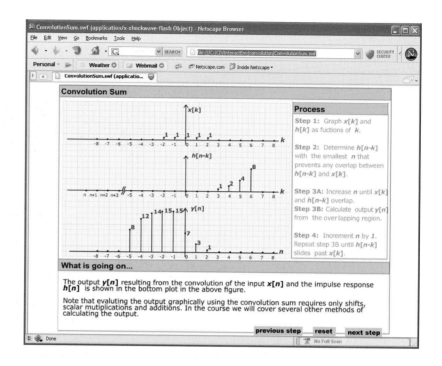

Read File: Loads the input signal from a sound file stored in the wav format.

Plot Signal: Plots the loaded signal in the Graphics window.

Play Signal: Plays the audio signal. The user must have a sound card and speakers to hear the audio.

Signal Spectrum: Computes the power spectrum of the audio signal and displays it in the Graphics subwindow. The power spectrum is calculated by parsing the audio signal in segments. Each segment has a length of 1024 samples with an overlap of 512 in between the neighboring segments. Section 17.2.3 explains the steps involved in computing the power spectrum of a signal.

Design Filter: Designs a DT filter and displays the coefficients of the filter. If the selected filter is of the FIR type, then the impulse response $h[k]$ of the filter is plotted in the Graphics window. If the selected filter is of the IIR type, then the coefficients of the numerator and denominator of the transfer function of the filter are displayed using the stem plot. DAFIE provides the option of selecting one of the Bartlett, Hamming, Hanning, Blackman, or Kaiser windows in designing the FIR filter with the number of taps limited to 201. For IIR filters, the choices are limited to the Butterworth or Chebyshev type II filter with a stop-band attenuation of at least 50 dB and pass-band ripples limited to a maximum level of 2 dB. The number of taps is ignored for the IIR filters.

Fig. F.4. The DAFIE environment for digital audio filtering.

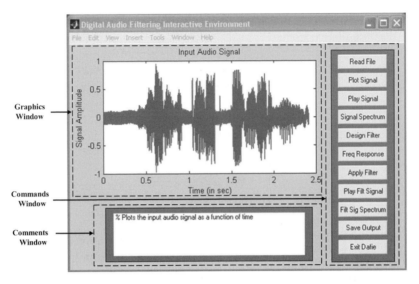

Freq Response: Calculates and plots the magnitude spectrum of the designed filter. The magnitude spectrum is displayed in the Graphics window.

Apply Filter: Filters the input signal and plots the resulting output signal as a function of time.

Play Filt Signal: Plays the output (filtered) signal as audio.

Filt Sig Spectrum: Computes the power spectrum of the output signal and displays it in the Graphics window.

Save Output: Stores the output signal as audio in the file ⟨output.wav⟩ in the working directory. If you choose this command, ensure that you have write permission to the current working directory.

Exit Dafie: Exits the DAFIE, ending the program.

F.2 Data

The Data folder in the CD contains two subfolders. These subfolders contain different audio clips and images used in the text. The audio clips are stored in the wav format with the .wav extension. The images are stored in the TIFF (also referred to as the TIF) format, where the image data is stored without any distortions. A list of the audio clips and images included in the CD is provided in the following.

Audio clips (CD:\Data\audio)

```
bell.wav      % Audio sampled at 22.05 kHz and
                 % quantized to 8-bits
test44k.wav   % Audio sampled at 44.1 kHz and
                 % quantized to 8-bits
```

```
noisy_audio1.wav % Audio signal corrupted with
                 % narrowband noise
noisy_audio2.wav % Audio signal corrupted with
                 % wideband noise
```

Gray images (CD:\Data\image)

```
{ayantika.tif, lena.tif,    % Images used in this book
  rini.jpg, sanjukta.tif,
  train.jpg}
{castle.jpg, eiffel.jpg,    % Other images given for
  girl.jpg, sounio.jpg}     % solving problems
```

Note that images with tif/tiff extension include no distortion. On the other hand, images with jpg extension are compressed using JPEG codec.

Color images (CD:\Data\image\color)

```
{castle, eiffel, gardern, girl,   % Selected color images
  lena, sanjukta, sounio,         % in JPG/TIFF format
  stadium, train}
```

F.3 MATLAB codes

The CD includes the MATLAB codes used in various examples in the text. In the following, we provide a listing of the names of the functions arranged in terms of their inclusion in different chapters.

Chapter 1

```
Example_01_23.m   % plots several CT functions using
                  % subplot and plot
Example_01_24.m   % plots several DT sequences using
                  % subplot and stem
```

Chapter 3

```
Example_03_12.m   % solves first order differential
                  % equation
Example_03_13.m   % solves second order differential
                  % equation
myfunc1.m         % defines a first order differential
                  % equation
myfunc2.m         % computes vector of derivatives
```

Chapter 5

```
bodeplot.m          % plots BODE plot of a transfer
                    % function in section 5.10.2
myctft.m            % calculates CTFT of a function in
                    % section 5.10.1
myinvctft.m         % calculates inverse CTFT of a function
                    % in section 5.10.1
section_5_10_1.m    % calculates CTFT of a function in
                    % section 5.10.1
```

Chapter 7

```
Example_07_5.m      % calculates and plots frequency
                    % response of Butterworth filter
Example_03_7.m      % calculates and plots frequency
                    % response of Chebyshev I filter
Example_07_8.m      % calculates and plots frequency
                    % response of Chebyshev II filter
Example_03_9.m      % calculates and plots frequency
                    % response of elliptic filter
Example_03_10.m     % designs highpass filter and plots
                    % frequency response
Example_07_11.m     % designs bandpass filter and plots
                    % frequency response
Example_03_12.m     % designs bandstop filter and plots
                    % frequency response
```

Chapter 8

```
ImmuneSystem1.mdl   % Simulink model for stable immune
                    % system
ImmuneSystem2.mdl   % Simulink model for unstable immune
                    % system
```

Chapter 10

```
Example_10_17.m     % calculates system output using direct
                    % method in Example 10.17
Example_10_18.m     % calculates system output using direct
                    % method in Example 10.18
Example_10_19.m     % calculates system output using conv
                    % function in Example 10.19
```

Chapter 12

Example_12_6.m	% calculates freq. charac. of decaying % exponential function using dft
Example_12_7.m	% calculates freq. charac. of two % complex exponential functions % using dft
Example_12_8.m	% calculates frequency characteristics % using N=32
Example_12_8_N64.m	% calculates frequency characteristics % using N=64
Example_12_9.m	% calculates dft of a decaying % exponential function
Example_12_11.m	% calculates DTFT of an aperiodic % sequence
mydft.m	% calculates dft using direct % calculation
myfft.m	% calculates dft using radix-2 fft % method
tfft.m	% test program to compare mydft and % myfft functions

Chapter 13

Example_13_20.m	% calculates partial fraction % coeffs of H(z)=B(z)/A(z)
Example_13_21.m	% calculates poles and zeros and plots % them in the z-plane
Example_13_22.m	% calculates transfer function of a % system from its poles and zeros

Chapter 14

section14_9_1.m	% calculates the partial fraction % coefficients in section 14.9.1
section14_9_2.m	% calculates the zeros and poles of a % transfer function in section 14.9.2

Chapter 15

Example_15_9.m	% designs lowpass FIR filter using % Hamming/Blackman windows
Example_15_10.m	% designs lowpass FIR filter using % Kaiser window
Example_15_11.m	% designs lowpass FIR filter using % Parks-McClellan algorithm

Chapter 16

```
Example_16_2.m      % converts a CT filter to DT using
                    %   impulse invariance method
Example_16_3.m      % converts a CT filter to DT using
                    %   impulse invariance method
Example_16_4.m      % converts a CT filter to DT using
                    %   bilinear transformation
Example_16_5.m      % designs highpass IIR filter using
                    %   CT elliptic filter and bilinear
                    %   transform
Example_16_6.m      % designs bandpass IIR filter using
                    %   CT elliptic filter and bilinear
                    %   transform
Example_16_7.m      % designs bandstop IIR filter using
                    %   CT elliptic filter and bilinear
                    %   transform
```

Chapter 17

```
Example_17_2.m      % calculates spectrogram of a DT signal
Example_17_3.m      % calculates power spectral density
                    %   using Welch method

Example_17_4.m      % filters (lowpass, bandpass,highpass)
                    %   an audio signal
Example_17_5.m      % bandstop filters an audio signal
Example_17_8.m      % calculates 2-D spectrum of a grating
                    %   image
Example_17_9.m      % spectral analysis and lowpass
                    %   filtering of an image
Example_17_10.m     % highpass filters an image
Example_17_11.m     % predictive coding of an image
Example_17_12.m     % JPEG compression of an image with
                    %   different quality factors
Section_17_2_3.m    % calculates power spectral density of
                    %   an audio signal
Section_17_2_5.m    % calculates power spectral density of
                    %   a music signal
Section_17_5_3.m    % reads and manipulates an image
```

Appendix E

```
Example_E_7.m       % plots a CT and a DT function
Example_E_8.m       % plots several functions in one figure
```

Bibliography

In the following, we have included selected textbooks and reference books on subjects related to signals and systems.

Signals and systems

S. R. Devasahayam, *Signals and Systems in Biomedical Engineering: Signal Processing and Physiological Systems Modeling*. Kluwer Academic/Plenum Publishers (2000).

S. Haykin and B. V. Veen, *Signals and Systems*. 2nd edn. Wiley (2002).

H. Hsu, *Schaum's Outline of Signals and Systems*. McGraw-Hill (1995).

B. P. Lathi, *Signal Processing and Linear Systems*. Oxford University Press (2000).

A. V. Oppenheim, A. S. Willsky, and S. Hamid, *Signals and Systems*, 2nd edn. Prentice Hall (1996).

R. E. Ziemer, *Signals and Systems: Continuous and Discrete*, 4th edn. Prentice Hall (1998).

Digital signal processing and filtering

A. Antoniou, *Digital Filters: Analysis, Design and Applications*, 2nd edn. McGraw-Hill (2001).

L. B. Jackson, *Digital Filters and Signal Processing*, 3rd edn. Kluwer Academic Publishers (1996).

S. K. Mitra, *Digital Signal Processing: A Computer-Based Approach*, 2nd edn. McGraw-Hill (2001).

A. V. Oppenheim, R. W. Schafer, and J. R. Buck, *Discrete-Time Signal Processing*, 2nd edn. Prentice Hall (1999).

T. W. Parks and C. S. Burrus, *Digital Filter Design*. Wiley-Interscience (1987).

J. G. Proakis and D. K. Manolakis, *Digital Signal Processing: Principles, Algorithms and Applications*, 3rd edn. Prentice Hall (1995).

Electrical circuits

R. L. Boylestad, *Introductory Circuit Analysis*, 10th edn. Prentice Hall (2002).

A. M. Davis, *Linear Circuit Analysis*. Thomson Engineering (1998).

J. O. Malley, *Schaum's Outline of Basic Circuit Analysis*, 2nd edn. McGraw-Hill (1992).

W. D. Stanley, *Network Analysis with Applications*, 4th edn. Prentice Hall (2002).

Communications

A. B. Carlson, P. B. Crilly, and J. Rutledge, *Communication Systems*, 4th edn. McGraw-Hill (2001).

S. Haykin, *Communications Systems*, 4th edn. Wiley (2000).

M. Schwartz, *Information Transmission, Modulation and Noise*. McGraw Hill (1980).

Multimedia

B. Furht, S. W. Smoliar, and H. Zhang, *Video and Image Processing in Multimedia Systems*. Kluwer Academic Publishers (1995).

R. C. Gonzalez and R. E. Woods, *Digital Image Processing*, 2nd edn. Prentice Hall (2002).

M. K. Mandal, *Multimedia Signals and Systems*. Kluwer Academic Publishers (2003).

John Watkinson, *The MPEG Handbook*. Focal Press (2001).

U. Zölzer, *Digital Audio Signal Processing*. John Wiley & Sons (1997).

Systems and control

D. Basmadjian, *Mathematical Modeling of Physical Systems: An Introduction*. Oxford University Press (2002).

R. C. Dorf and R. H. Bishop, *Modern Control Systems*, 10th edn. Prentice Hall (2004).

B. C. Kuo and F. Golnaraghi, *Automatic Control Systems*, 8th edn. Wiley (2002).

N. S. Nise, *Control Systems Engineering*, 4th edn. Wiley (2003).

Mathematics

E. O. Brigham, *The Fast Fourier Transform and its Applications*. Prentice Hall (1988).

G. A. Korn and T. M. Korn, *Mathematical Handbook for Scientists and Engineers: Definitions, Theorems, and Formulas for Reference and Review*, 2nd edn. Dover Publications (2000).

E. Kreyszig, *Advanced Engineering Mathematics*, 8th edn. Wiley (1998).

K. A. Stroud and D. J. Booth, *Engineering Mathematics*, 5th edn. Industrial Press (2001).

Index